Khanna's

Outlines of
CHEMICAL & PETROLEUM ENGINEERING

DR. ALAPATI SURYANARAYANA
(Retired Prof. & HOD Chem. Engg. Dept. N.I.T. Rourkela)
Presently Prof. & HOD Chem. Engg. Dept., RRSCET,
Ex. Principal TRRECW
Muthangi, Medak DT., Telangana – 502300

DR. VIKAS MAHTO
Associate Professor
Department of Petroleum Engineering
Indian School of Mines, Dhanbad-826004 (India)

KHANNA BOOK PUBLISHING CO. (P) LTD.
Publisher, of Science, Technology and Engineering Books
4C/4344, Ansari Road, Darya Ganj, New Delhi-110002
Phone: 011-23244447-48 **Mobile:** +91-99109 09320
E-mail: contact@khannabooks.com
Website: www.khannabooks.com

Price : ₹ 499.00

KHANNA'S OUTLINES OF
CHEMICAL & PETROLEUM ENGINEERING
Alapati Suryanarayana, Vikas Mahto

ISBN: 978-93-86173-19-5

Edition: 2018

Published by:

KHANNA BOOK PUBLISHING CO. (P) LTD.
4C/4344, Ansari Road, Darya Ganj, New Delhi-110 002
Phone: 011-23244447-48 **Mobile:** +91-9910909320
E-mail: contact@khannabooks.com

Printed in India by:
S.P.S. Printers & Binders, Delhi

Preface

This book provides broad outlines of chemical engineering and petroleum engineering in which key concepts are taken from different standard textbooks, industry manuals and lectures of distinguished persons. It covers each and every aspect of chemical and allied industries with special emphasis on petroleum industries.

This book is divided into 21 chapters which contain summaries of various topics and necessary technical information required for the successful completion of undergraduate and postgraduate programmes of chemical and petroleum engineering. Different types of objective questions with their answers given in all the chapters. The requisite data and necessary equations are also included in this book for solving the conceptual and difficult numerical problems. These relevant questions may be used for the self assessment of knowledge acquired by the candidate and also for promoting his fundamental understanding of chemical and petroleum engineering.

The content of this book will be very useful to fresh young graduates for setting their goal to become a perfect chemical engineer or petroleum engineer and it will also make their journey toward that goal easier. It will especially help those graduate students who have chemical engineering background and willing to make their career in petroleum industries, It will also serve as a necessary tool or instrument for chemical and petroleum engineers working in the different industries.

–Authors

Acknowledgement

The authors have to thank the contributors of this book. Namely, Prof. J. S. N. Murthy, Prof. G. K. Roy, Prof. P. Rath and Prof. S. C. Naik.

Sincere thanks to our Chairman, Dr. T. Rammohan Reddy, Director Dr. M. Prabhakar and management of RRS College of Engineering & Technology in successfully bringing out this book.

Authors also express their heartfelt thanks to Prof. D.C. Panigrahi, Prof. T. Kumar, Prof. S. Laik, Prof. V.P. Sharma and Prof. A.K. Pathak for their inspirational guidance and encouragement for writing this book.

—Authors

Contents

CHAPTER 1

Stoichiometry

1. Simultaneous equations can be solved graphically also. This is by giving values and plotting the lines or curves. The intersection points are the solutions.
2. Graphical integration is used when analytical integration can not be done. The area under the curve between the limits is the integral value.
3. Log-log coordinates or log-log graph paper is used to find the constants in equations of the form, $y = ax^b$ where a, b are constants ($\log y = \log a + b \log x$).
4. Semilog coordinates/ semilog graph paper are used for equations of the form. $y = ab^{cx}$, $K = Ae^{-E/RT}$; ($\log y = (c \log b)x + \log a$)
5. Triangular diagrams or triangular coordinates are used to represent three components, e.g. liquid-liquid extraction.
6. Conversion of units and equations :
7. Units : Cp—heat capacity—1 BTU/lb.°F (FPS) = 1 cal/g.°C (CGS)

   D—internal diameter—1 foot (FPS) = 30.48 cm (CGS)

   G—Mass velocity—1 lb/sec.sq.ft (FPS) = 0.01356 g/sec.sq. cm (CGS)

   h—heat transfer coefficient—1 BTU/sec.sq.ft °F (FPS) = 0.4883 g.cal/sec.cm^2°C (CGS)

   K—thermal conductivity—1 BTU/sec.ft°F (FPS) = 14.882 g.cal/sec.cm^2°C (CGS)

   μ—viscosity—1 lb/ft.sec (FPS) = 14.882 g/sec.cm (CGS)
8. Material balance is based on law of conservation of mass.
9. Stoichiometry covers the mass and volumetric relations and compositions taking place in Chemical Reactions.
10. A system refers to a substance or a group of substances under consideration in stoichiometry and an unit or units in other subjects, e.g. hydrogen, oxygen, water.
11. A process refers to the changes taking place in a system, e.g. hydrogen combining with oxygen to form water.
12. Extensive properties depend on mass, e.g. volume, mass.
13. Intensive properties do not depend on mass, e.g. temperature, density.
14. In the chemical reaction : $3\,Fe + 4\,H_2O \rightarrow Fe_3\,O_4 + 4H_2$: 3 × 55.84 = 167.52 kg of iron reacts with 4 × 18 = 72 kg of steam (water) to give products of 231.5 kg of iron oxide and 4 × 2 = 8 kg of hydrogen.
15. For ideal gases volume percent = mole percent.
16. Kg moles of a substance = kg/Molecular weight.

17. 1.0 g mole of a gas at 0°C and 760 mm Hg occupies 22.414 liters of volume. 1 lb.mole of a gas at 32°F and 760 mm Hg (1 atm) pressure occupies 359 cubic feet.

18. Weight percent of A in a mixture = w_A / w *100; where w_A is weight of A in kg/lb/g etc. and w is the total weight of the mixture in kg/lb/g etc.

19. Volume percent of A = 100* volume of $A(V_A)/V$ (total volume) in same units.

20. Mole percent of A = [moles of A/Total moles in mixture] *100.

21. Degrees Baume = 145 – 145/G where G is greater than one.

22. Degrees Baume = 100/G – 130 where G is specific gravity of liquid less than 1.

23. Degree API = 141.5/G –131.5

24. Thermodynamics deals with the transformation of energy from one form to another, e.g. heat to work.

25. Degrees Twaddel : 200(G – 1.0) : G is greater than zero.

26. First law of thermodynamics is law of conservation of energy.

27. The translational energy of a gas molecule is dependent on temperature only.

28. Ideal gas : $PV = n\ RT$; $P_1\ V_1/T_1 = P_2\ V_2/T_2$.

29. Dalton's law : P (total pressure) = $pA + pB +$ .....where pA, pB, pC are partial pressures.

30. Amagat's aw : V(total volume) = $VA + VB + VC +$ .....

31. Ideal gas law is valid at low pressure and high temperature.

32A. The temperature at which a liquid has its vapour pressure is equal to total pressure (ambient pressure) is boiling point.

32B. Clausius-Clapeyron equation is $\ln P/P_0 = \lambda/R\ (1/T_0 - 1/T)$ where P is vapour pressure T is temperature.

33. Reduced conditions : $Tr = T/Tc$; $Pr = P/Pc$; $Vr = V/Vc$ where c is critical value.

34. The boiling point of a non-homogeneous mixture must be lower than that of any of its separate components.

35. Raoult's law : $Pt = PA + PB + PC$ —; $ya = pA/Pt$; $pA = PA * xA$; $pB = P_B * x_B$ where p is partial pressure and P is vapour pressure.

36. Relative saturation % = pA/PA * 100 where pA is partial pressure and PA is vapour pressure of A.

37. Saturation % = Y/Ys * 100 where Y is humidity and Ys is saturation humidity.

38. Distribution coefficient = $K = CE/CR$ where CE is concentration of solute in extract phase and CR is the concentration of solute in raffinate phase.

39. In material balance when chemical reactions are present then it is desirable to use molar units.

40. In material balance when no chemical reactions are present then it is preferable to use mass units.

41. Recycling increases yields, enrich a product to conserve heat or to improve operation.

42. Bypassing of a fluid stream is done for accurate control in concentration or flow rate.

43. $Cp - Cv = R$; $Cp = a + bT + cT* T$; C = specific heat, R = gas constant, T = temperature.

44. Trouton's rule : $k = \lambda_b/Tb$ where Tb is boiling point temperature and λ_b is latent heat of vaporization and b is boiling point.

45. Kistiakowsky equation : $\lambda_b/Tb = 8.75 + 4.571 \log Tb$

46. ΔH reaction = $\Sigma\Delta H$ reactants – $\Sigma\Delta H$ products.

$Qp - Qv = -\Delta ng\ RT$; ΔH is enthalpy change; Δng = gas moles change : R = gas constant; T = temperature; $\Delta ng = n$ products $- n$ reactants.

47. Freundlich adsorption equation : $w = k\ p^n$

Langmuir equation : $V = V_m\ bp/\ (1 + bp)$; w = amount of solute abdorbed per unit weight of adsorbent, p = partial pressure of gas molecules, V = volume of gas adsorbed per unit mass of adsorbent; V_m = volume of gas adsorbed per unit mass of adsorbent to cover surface with a layer of one molecule thick; n, k, b are constants.

QUESTIONS

SECTION—A

1. At eutectic point the number of phases present is :
(a) One (b) Two
(c) Three (d) Many

2. The heat capacities of substances :
(a) Decrease with temperature increase
(b) Increase with temperature increase
(c) Temperature has no effect.

3. Kopp's rule deals with :
(a) Heat capacities (b) Thermal conductivity
(c) Viscosities

4. For ideal gas $C_P - C_V$ is equal to :
(a) Zero (b) R
(c) $2R$ (d) $3/2R$

5. For monoatonic gas C_V is equal to :
(a) $3/2\ R$ (b) R
(c) $2R$ (d) $3R$

6. One Newton is equal to :
(a) 10^3 dynes (b) 10 dynes
(c) 10^2 dynes (d) 10^5 dynes

7. N.T.P. means :
(a) 1 atm. absolute pressure and 15°C (b) 1 atm. gauge pressure and 0°C
(c) 1 atm. absolute pressure and 0°C.

8. S.T.P. means :
(a) 1 atm. absolute pressure and 15.5°C (b) 1 atm. gauge pressure and 15.5°C
(c) 1 atm. absolute pressure and 0°C (zero).

9. For gases :
(a) Volume percent is equal to weight % (b) Volume percent is equal to mole %
(c) Volume percent is not equal to mole %

10. Inputs = outputs, is valid :
(a) For system without chemical reaction
(b) For system at steady state with chemical reaction
(c) For system at steady state without chemical reaction and losses.

11. Total energy at a point consists of :
(a) Internal energy (b) Potential energy
(c) Kinetic energy (d) Pressure energy
(e) *a*, *b*, *c*, and *d*

12. Maximum excess air is required by :
(a) Solid fuels (b) Liquid fuels
(c) Gaseous fuel

13. 1 kg/cm^2 is equl to :
(a) 5 m water (b) 1 m water
(c) 760 mm water (d) 10 m water

14. Air has 21% and 79% of oxygen and nitrogen by volume respectively. What is the average molecular weight?
(a) 29 (b) 28.84
(c) 29.3

15. 1° Brix is equivalent to a sugar solution containing :
(a) 10% sugar (b) 1% sugar
(c) 0.1% sugar (d) 0.01% sugar

16. When a solute is dissolved in a solvent, the vapour pressure of the resultant solution is :
(a) Equal to that of the pure solvent
(b) Less than that of the pure solvent
(c) More than that of the pure solvent

17. Boiling point of a solution when compared to that of the corresponding solvent is :
(a) Almost the same (b) Less than that of the solvent
(c) More than that of the solvent

18. An ideal solution is the one which obeys :
(a) Raoult's law (b) Amagat's law
(c) Charles's law (d) Dalton's law

19. A bypass stream for a process is useful because :
(a) It improves conversion
(b) It increases the product yield
(c) It ensures better control over the process.

20. A recycle ratio is defined as :
(a) Recycle stream/ gross feed stream
(b) Recycle stream/ fresh feed stream
(c) Gross feed stream/recycle stream.

SECTION—B

1. For ideal gas internal energy is independent of .....and.....
2. Kistiakowsky equation deals with the relation of.....and.....
3. Trouton's rule deals with the relation of.....and.....
4. The heat of reactions at constant pressure and volume are related by the equation.....
5. Internal energy of a substance consists of..... energy, ... energy and ..... energy.
6. The fundamental quantities of dimensions are ....., ..., ... and .....
7. The two additional fundamental quantities introduced in S.I. units are..... and .....
8. One bar is ..... kg force/cm^2.
9. A process flow sheet shows ..... and .....
10. Antoine equation for calculating vapour pressure is .....
11. Calorific value of a fuel is the..... of fuel.
12. Molarity is defined as the .....
13. Normality is defined as the.....
14. Molality is defined as the .....
15. The equilvanet weight of sodium carbonate (Na_2CO_3) having molecular weight 106 is .....
16. Heat of reaction depends on ..... and .....
17. Recycle stream is purged for .....
18. A graph relating to the Antoine equation is called .....
19. With the help of Kopp's rule, the heat capacity of a solid compound is calculated from .....
20. The effect of temperature on heat of reaction is given by ..... equation.

SECTION—C

1. What are the various specific gravity scales?
2. What are Dalton and Amagat's law of gases?
3. What are the uses of recycle?
4. Define standard heat of reaction.
5. What is meant by heat of formation?
6. What is the law of Hess?
7. Define heat of combustion.
8. What is meant by theoretical flame temperature?
9. Write four equations of state.
10. What is a limiting reactant?
11. What is an excess reactant?
12. Give the expression for Clausius-Clapeyron equation.
13. The available nitrogen content in a urea sample is 45% by weight. Find the actual urea content of the sample.
14. 200 g of KCl is mixed with 600 g of NaCl. Find the mole percentage of each in the mixture.

15. Calculate the average molecular weight of a flue gas having the following composition by volume.

$$CO_2 - 13.1\%, O_2 - 7.7\% \text{ and } N_2 - 79.2\%$$

16. Calculate the density of a gas mixture of hydrogen and oxygen at 725 mm Hg and 30°C containing 11.1% Hydrogen by weight.

17. An alcohol whose molecular weight is 74 analyzes as C – 64.87% W, H – 13.51% W and O – 21.62% W, what is the formula?

18. The gas acetylene is produced according to the following reaction by treating calcium carbide with water

$$CaC_2 + 2H_2O \rightarrow C_2H_2 + Ca(OH)_2$$

Calculate the number of hours of service that can be derived from 1 kg of carbide in an acetylene lamp burning 30 litres of gas per hour at a temperature of 20°C and a pressure of 740 mm Hg.

19. In the Deacon process for the manufacture of chlorine, a dry mixture of hydrochloric acid gas is passed over a heated catalyst which promotes oxidations of the acid. Air is used in 30% excess that theoretically required. Calculate the weight of air supplied per kg. of acid and the composition (by weight) of the inlet gas mixture.

20. By electrolyzing a mixed brine, a mixture of gases is obtained at the cathode having the following percentage composition by weight :

$$Cl_2 - 66, Br_2 - 29, O_2 - 5$$

Calculate the specific gravity of the mixture.

21. 100 m³ of a boiler flue gas has the following composition by volume at N.T.P.

$$CO_2 - 14.5, O_2 - 3.5, N_2 - 82.0$$

Calculate the moles and weight percentage of the gases in the mixture.

22. Sulphur dioxide reacts with pure oxygen to form SO_3. If the reaction is carried out with twice as much oxygen necessary for complete conversions, but under low temperature and pressure the reaction goes only to 60% conversion. Calculate the average molecular weight of the product mixture.

23. Calculate the total pressure and composition of vapours in contact with a solution at 100°C containing 35% benzene, 40% toluene and 25% ortho-xylene by weight. Vapour pressure at 100°C are : Benzene = 1340 mm Hg, Toluene = 560 mm Hg, ortho-xylene = 210 mm Hg.

24. Calculate the standard heat of reaction, ΔH_{25} of the following reaction :

$$CaC_2(s) + 2H_2O(l) = Ca(OH)_2(s) + C_2H_2(g)$$

$$\text{for } CaC_2(s), \Delta H_f = -15000 \text{ cal}$$

$$H_2O(l), \Delta H_f = -68{,}317 \text{ cal}$$

$$Ca(OH)_2(s), \Delta H_f = -235{,}800 \text{ cal}$$

$$C_2H_2(f), \Delta H_f = +54194 \text{ cal.}$$

25. Chlorinated diphenyl is heated from 100 to 200°C at the rate of 5000 kg/hr in a heater. The specific heat of the fluid is $C = 0.275 + 0.00035\, t$ kcal/ (kg.°C) where t is °C. Calculate the heat to be supplied to the fluid.

26. A drier uses 8.0 kgf/cm² steam at the rate of 2000 kg/hr. The condensate at its saturation temp. is discharged to the atmosphere through a steam trap. Assuming no heat loss, calculate the quantity of flash steam produced at atmospheric pressure due to let down of the pressure. Enthalpy of steam at 8.0 kgf/cm^2 is 171.4 kcal/kg and at 1 atm pr is 100.04 kcal/kg (condensate). Enthalpy of flash steam at 1 atm pr. is 639.11 kcal/kg.

27. Calculate heat of formation of ethylene gas at 25°C using the heat of combustion data.

Component	Phase	Heat of combustion, kacl/kg mole at 25°C
C	Solid	–94.05
H_2	Gas	–54.796
Ethylene	Gas	–316.190

28. It is required to make 2000 kg mixed acid containing 60% H_2SO_4, 32% HNO_3 and 8% water by blending (*a*) the spent acid containing 11.3% HNO_3, 44.4% H_2SO_4 and 44.3% H_2O (*b*) HNO_3 and (*c*) aqueous 98% H_2SO_4. All percentages are by weight. Calculate the amounts of *a*, *b* and *c* to be mixed.

29. Tallow is saponified with caustic soda giving rise to sodium stearate and glycerine. The molecular weight of tallow, sodium hydroxide, sodium stearate and glycerine are 890, 40, 309 and 92 respectively. Calculate the theoretical amount of caustic soda required and glycerine produced for 200 kg of tallow?

30. What is the yield of glauber salt (Na_2SO_4, 10 H_2O) if pure 32% solution is cooled to 20°C without evaporation loss? Solubility of Na_2SO_4 in water at 20°C is 19.4 g/100 g water. Mol. wt of Na_2SO_4 is 142.

SECTION—D

1. Pure O_2 is mixed with air to produce an enriched air containing 50% volume O_2. The ratio of moles of air to O_2 used is :

(a) 1.72 (b) 0.58

(c) 0.50 (d) 0.2

2. Methanol is produced by the reaction of CO with hydrogen:

$$CO + H_2 \rightarrow CH_3OH$$

Only 15% of carbon monoxide entering the reactor is converted to methanol. The methanol formed is condensed and recovered completely. The unreacted CO and H_2 are recycled back to the reactor. The feed will contain H_2 and CO in the ratio of 2 : 1. For 3200 kg/hour of methanol produced, calculate :

(a) Kg. mole/hr of fresh feed (b) Kg. mole/hr of recycle gas

(Mol. wt of CH_3OH is 32)

3. Ideal gas law is applicable at :

(a) Low T, low P (b) High T, high P

(c) Low T, high P (d) High, T, low P

4. Find ΔH_{298} for the reaction.

$$C_3H_8\,(g) + 5O_2(g) \rightarrow 3CO_2\,(g) + 4H_2O\,(l)$$

If the standard enthalpies of formation ΔH_f, of C_3H_8 (*g*), CO_2(*g*) and H_2O (*l*) are – 125.5, –393.3, and – 284.5 kJ/ mole respectively at one atmosphere, then :

What is the value of ΔE^0_{298} for the above reaction?

5. At the triple point of a pure substance, the number of degrees of freedom is :

(a) 0 (b) 1

(c) 2 (d) 3

6. A vessel of volume 1.0 m^3 contains a mixture of liquid water and steam in equilibrium at 1.0 bar. Given that 90% of the volume is occupied by the steam, find the dryness fraction of the mixture. Assume at 1.0 bar, $v_f = 0.001$ m^3/kg. and $v_g = 1.7$ m^3/kg.

7. In the vicinity of triple point, the vapor pressures of liquid and solid ammonia are respectively given by.

$\ln P = 15.16 - 3063/T$ and $\ln P = 18.70 - 3754/T$

where P is in atmospheres and T is in kelvin. What is the temperature at the triple point?

8. 1000 kg of a solution containing 50% by weight of a salt dissolved in it is cooled. 400 kg of anhydrous salt is separated out. The solubility of the salt at the lower temperature in kg/100 kg of water is :

(a) 80 (b) 50

(c) 40 (d) 20

9. Methane is completely burnt with air. The maximum possible volume percent of carbon dioxide. (on dry basis) in the flue gases is :

(a) 11.7 (b) 21.0

(c) 44.0 (d) 28.0

10. Iron pyrites (FeS_2) is burned with air in 100% excess of that required to oxidize all iron to Fe_2O_3 and all sulphur dioxide. Calculate the composition of exit gases, if 80% of sulphur is oxidised to sulphur dioxide and the rest to sulphur dioxide. All iron is oxidised to Fe_2O_3.

11. The kinetic energy of gas molecule is zero at :

(a) 0°C (b) 273°C

(c) 100°C (d) –273°C

12. An evaporator while concentrating an aqueous solution from 10 to 40% solids evaporates 3000 kg of water. The amount of solids handled by the system in kg is :

(a) 400 (b) 900

(c) 4600 (d) 300

13. 1000 kg of wet solids are to be dried from 60% to 20% moisture (by weight). The mass of moisture removed in kg is :

(a) 520 (b) 200

(c) 400 (d) 500

14. Pure Co is mixed with 100% excess air and burnt. Only 80% Co burns. The reactants are at 100°C and the products are at 300°C. Calculate the amount of heat added or removed per kg mole of Co fed to the reactor. Data : Mean Molar specific heats between 25°C and T°C (given below) in kJ/ (kg. mole. °K) are :

Gas	$T = 100°C$	$T = 300°C$
CO	29.22	30.61
CO_2	—	43.77

O_2	29.64	30.99
N_2	29.17	29.66

Standard heat of formation at 25°C is in kJ/kg. mole are : CO → – 110524; CO_2 → – 393514

15. In the system shown in figure below each stream contains three components.

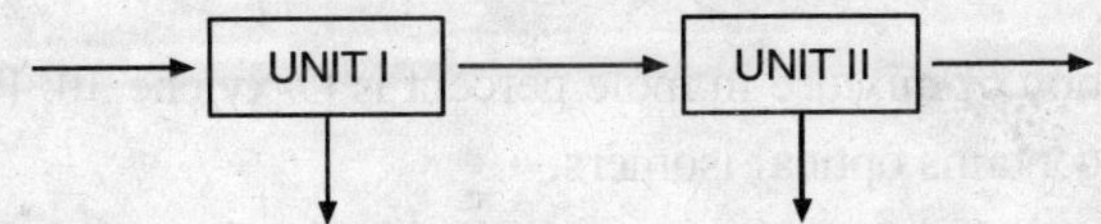

The maximum number of independent material balances is :

(a) 3 (b) 4

(c) 6 (d) 9

16. When an unsaturated air-water mixture is heated at constant pressure then,

(a) Partial pressure of water vapor increases

(b) The specific humidity decreases

(c) The relative humidity increases

(d) The relative humidity decreases

17. A flow sheet is given below.

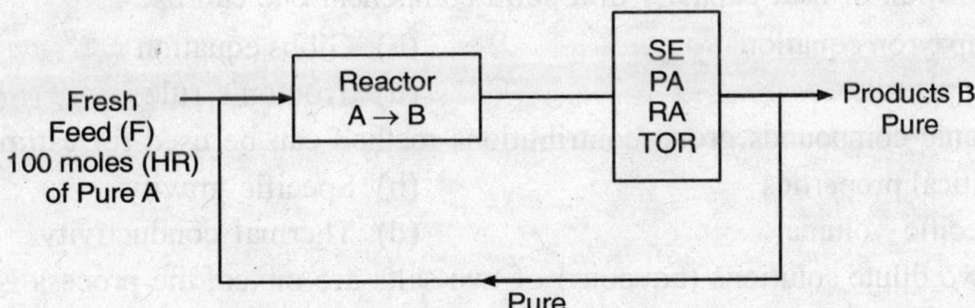

If the single pass (once through) conversion of A to B is 20%, then the Rate of Recycle (R) in moles/HR is :

(a) 300 (b) 400

(c) 500 (d) 600

18. Sea water is desalinated by reverse osmosis as shown

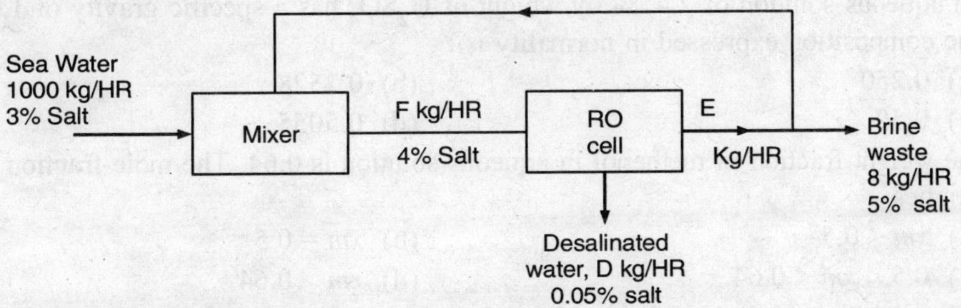

Calculate R/F. All compositions are on mass-basis.

19. A liquid mixture contains 30% O-xylene, 60% P-xylene and 10% m-xylene (all percentages w/w) Which of the following statements would be true in respect of this mixture :

(a) The mixture exhibits an azeotrope at 101.3 kpw.

(b) The composition of the mixture, in percent by volume is : O-xylene 30, P-xylene 60, and m-xylene 10.

(c) The composition of mixture in mole percent is O-xylene 30; P-xylene 60, and m-xylene 10.

(d) The mixture contains optical isomers.

20 Pure ethanol vapour is fed to a reactor packed with alumina catalyst, at the rate of 100 k mol. per hour. The reactor products comprise : Ethylene; 95 K. mole/h, water vapor : 97.5 K. mol/h and diethyle ether : 2.5 K mol/h. The reactions occuring can be represented by :

$$C_2H_5\,OH \rightarrow C_2H_4 + H_2O$$

$$2C_2H_5OH \rightarrow C_2H_5 - O - C_2H_5 + H_2O$$

The percent conversion of ethanol in reactor is :

(a) 100.0 (b) 97.5

(c) 95.0

SECTION–E

1. For estimation of heat capacity of a solid component one can use :

(a) Clapeyron equation (b) Gibbs equation

(c) Kopp's rule (d) Trouton's rule

2. For organic compounds group contributions method can be used for estimation of :

(a) Critical properties (b) Specific gravity

(c) Specific volume (d) Thermal conductivity.

3. When two dilute solutions (aqueous) of two salts are mixed, the process is associated with :

(a) Decrease in temperature (b) Increase in temperature

(c) No change in temperature (d) Change in temperature which is a function of composition.

4. A lubricant 100 times more viscous than water would have a viscosity (in Pa.s).

(a) 0.01 (b) 0.10

(c) 1 (d) 10

5. An aqueous solution of 2.45% by weight of H_2SO_4 has a specific gravity of 1.011 g/cubic cm. The composition expressed in normality is :

(a) 0.250 (b) 0.2528

(c) 0.50 (d) 0.5055

6. The weight fraction of methanol in aqueous solution is 0.64. The mole fraction of methanol xm satisfies :

(a) $xm < 0.5$ (b) $xm = 0.5$

(c) $0.5 < xm < 0.64$ (d) $xm > 0.64$

7. Viscosity of water at 40°C lies in the range of :

(a) 0.001 to 0.002 kg/m.s. (b) 0.0005 to 0.001 kg/m.s.

(c) 1 to 2 kg/m.s. (d) 0.5 to 1 kg/s.m.

ANSWERS

SECTION–A

1. (c)	2. (b)	3. (a)	4. (b)	5. (a)
6. (d)	7. (c)	8. (a)	9. (b)	10. (c)
11. (e)	12. (a)	13. (d)	14. (b)	15. (b)
16. (b)	17. (c)	18. (a)	19. (c)	20. (b)

SECTION–B

1. Pressure and volume.
2. Latent heat and boiling point.
3. Latent heat and boiling point.
4. $Q_p = Q_v - \Delta ng, RT$.
5. Translation, vibrational, rotational.
6. Length, mass, time and temperature.
7. Power (in watts) and force (in Newtons),
8. 1.01972.
9. The flow of various materials and utilities.
10. Log10 $P = A + B/(t + c)$ where A, B and C are constants, P vapour pressure, in mm Hg and t is temperature in °C.
11. Heat of combustion.
12. Number of g moles of solute dissolved in one litre of solution.
13. Number of g equivalents dissolved in one litre of solution.
14. Number of g moles of solute dissolved in one kilogram of solvent.
15. 53.
16. Pressure and temperature.
17. Limiting the inerts.
18. Cox chart.
19. The atomic heat capacities of its constituent elements.
20. Kirchhoff's.

SECTION—C

1. (a) Baume gravity scale : Degree Baume; $(140/G) - 130$ (For liquids lighter than water.) G = specific gravity at 60°F. Degree Baume = $145 - (145/G)$ (for liquids heavier than water.)
   (b) API scale : Degree API = $(141.5/G) - 131.5$
   (c) Twaddell scale : Degree Twaddell = $200\ (G - 1.0)$,
   (d) Brix = $(400/G) - 400$.

2. Law of Dalton : The total pressure exerted by a gaseous mixture is equal to the sum of partial pressures of the components.

$$P = P_A + P_B + P_C + \ldots\ldots \text{ where } P \text{ is}$$

total pressure and P_A, P_B ..... etc. are partial pressures.
3. To increase yields, to enrich a product, to conserve heat or to improve operation.
4. Standard heat of reaction is defined as the change in enthalpy resulting from the procedure of the reaction under pressure of 1 atm, starting and ending with all materials at a constant temperature of 25°C.
5. The heat of formation of a chemical compound is a special case of the standard heat of a chemical reaction where in the reactants are the necessary elements and the compound in question is the only product formed.
6. The total change in enthalpy of a system is dependent on the temperature, pressure, state of aggregation and state of combination in the beginning and at the end of reaction and is independent of the number of intermediate chemical reactions involved.
7. The heat of combustion of a substance is the heat of reaction resulting from the oxidation of the substance with molecular oxygen.
8. The temperature attained when a fuel is burned in air or oxygen without gain or loss of heat.
9. (a) Ideal gas law, $PV = nRT$.
   (b) Van der Waals equation $(p + a/V^2)(V - b) = RT$
   (c) Beattie-Bridgeman equation,

$$PV = RT + \beta/V + \gamma/V^2 + \ldots\ldots$$

   (d) Berthelot equation,

$$P = \frac{RT}{V-b} - \frac{a}{TV^2}$$

10. A limiting reactant is one which decides the conversion in the reaction.
11. An excess reactant is the one which is in excess amount over the stoichiometric requirement of the reactant.
12. $\ln\frac{P}{P_o} = \frac{\lambda}{R}\left(\frac{1}{T_0} - \frac{1}{T}\right)$

   P = vapour pressure, T = absolute temperature
   λ = latent heat.
13. Molecular formula of urea = $CO(NH_2)_2$
   Molecular weight = 60.
   28 parts of N_2 = 60 parts of urea

   45 parts = $\frac{60}{23} \times 45$ = 96.43 parts of urea

   Urea content of the sample = 96.43%

14. Molecular weight NaCl = 58.5
    Molecular weight KCl = 74.5
    Moles of NaCl = 600/58.5 = 10.26
    Moles of KCl = 200/74.5 = 2.68
    Total no. of mol. = 12.94
    Mole % NaCl = 79.29, Mole % KCl = 20.71.
15. Basis : 1 gram mole of the flue gas.

Constituents	*g. mole*	*weight g*
CO_2	0.131	5.76
O_2	0.077	2.46
N_2	0.792	22.18
	1.00	30.4

Average molecular weight of the flue gas = 30.4.

16. Basis : 1 g of the mixture.
    H_2 = 0.11 g = 0.0555 g. mole
    O_2 = 0.889 g = 0.0278 g. mole
    Total = 0.0833 g. mole
    Volume at NTP = 0.0833 × 22.4 = 1.866 lit.
    Volume at 30°C and 735 mm Hg.

$$= \frac{760}{735} \times \frac{303}{273} \times 1.866 = 2.141 \text{ lit.}$$

17. 1 g. mole of the compound (i.e. 74 g.)

Constituents	Amount g.	g. atom
C	48.0	4.0
H	10.0	10.0
O	16.0	1.0

The formula is C_4H_9OH.

18. $CaC_2 \rightarrow 2H_2O \rightarrow C_2H_2 + Ca(OH)_2$
    64 kg of CaC_2 yields 26 kg of acetylene
    1 kg of CaC_2 will yield 26/64 = 0.406 kg acetylene
    = 0.406/26 = 0.0156 kg. mol.
    Assuming ideal gas behaviour for C_2H_2
    Volume at NTP = 0.0156 × 22400 = 349.44 lit,
    Volume at 20°C and 740 mm Hg.
    = 349.4 × 293/273 × 760/740 = 385.1 lit,
    No. of hourse of service by the acetylene lamp
    = 385.1/30 = 12.84

19. Basis : 1 kg of dry HCl gas

$4HCl + O_2 \rightarrow 2Cl_2 + 2H_2O$

4 × 36.5 kg of HCl require 32 kg of O_2

1 kg of HCl will require 32/(4 × 36.5) = 0.219 kg.

Theoretical air required = 0.219 × (100/23) = 0.953 kg.

Actual air supplied = 0.953 × 1.3 = 1.239 kg.

Composition of inlet gas mixture—

Constituent,	Weight kg	Weight %
HCl	1.000	44.66
Air	1.239	55.34
	2.239	100.00

20. Basis : 1 kg of the gas mixture at NTP—

Constitutions	Wt., kg.	kg. mol.
Cl_2	0.66	0.0093
Br_2	0.29	0.0018
O_2	0.05	0.0016
	Total = 1.00	0.0127

Volume of gas mixture of NTP = 0.0127 × 22400

= 284.48 c.c.

Density at NTP = 1000/284.48 = 3.515 g/lit

Mol. weight of air = 28.84

Density of air at NTP (assuming ideal behaviour)

= 28.84/22.4 = 1.288 g/lit.

21. Basis : 100 m³ of boiler flue gas at NTP.

Constituents	Vol., m³	Amount, kg. mol.	Amount, kg	Weight %
CO_2	14.5	0.647	28.47	20.94
O_2	3.5	0.156	4.9	3.67
N_2	82.0	3.661	102.51	75.39
Total =	100.0	4.64	135.97	100.00

Ideal gas behaviour is assumed for the individual component.

22. Bais : 1 kg mole of SO_2

$$SO_2 + \frac{1}{2} \rightarrow SO_3$$

$$O_2 \text{ supplied} = 2 \times \frac{1}{2} = 1.0 \text{ kg mole}$$

SO_2 converted to SO_3 = 0.6 mole

SO_2 unconverted = 1.0 – 0.6 = 0.4 kg mole

O_2 used = $0.6 \times \frac{1}{2}$ = 0.3 kg mole

O_2 unused = 1.0 – 0.3 = 0.7 kg mole

Product analysis :

Constituents	kg. mole	mole %
SO_2	0.4	23.53
SO_3	0.6	36.29
O_2	0.7	41.18
Total =	1.7	100.00

Average molecular weight of the product mixture

= 64 × 0.2353 + 80 × 0.3629 + 32 × 0.4118

= 15.06 + 28.23 + 13.18 = 56.47

23. Basis : 100 kg solution

Benzene = 35/78 = 0.449 kg mole.

Toluene = 40/92 = 0.435 kg mole.

O-xylene = 216/106 = 0.236 kg mole.

Total = 1.120 kg mole.

Benzene partial pressure = 340 × 0.449/1.120 = 536 mm Hg

Toluene partial pressure = 560 × 0.435/1.120 = 217 mm Hg

O-xylene partical pressure = 210 × 0.236/1.120 = 44 mm Hg

797 mm Hg

Mole % composition in liquid	vapor %
Benzene = (0.449/1.120) × 100 = 40.1	536/797 = 67.3
Toluene = (0.435/1.120) × 100 = 38.8	217/9 = 27.2
O-xylene = (0.236/1.12) × 100 = 21.1	44/797 = 5.5
100.0	100.0

24. ΔH_{25} = – 235,800 + 54,194 – (– 15,000) – 2 (–6831.7)

= – 299,720 cal.

25. Basis : 1 kg of chlorinated diphenyl

Sensible heat change = Q = $\int C dt$

$$= \int_{100}^{200} (0.275 + 0.00035t)dt.$$

$$= 0.275\ (200 - 100) + -\frac{0.00035}{2}(200^2 - 100^2)$$

= 27.5 + 5.25 = 32.75 kcal/kg

For mass flow rate of 5000 kg/hr.

$Q = 5000 \times 32.75 = 163760$ kcal/hour.

26. Basis : 2000 kg/hour condensate at 8 kfg/cm^2 and at condensation temperature.

The additional heat of condensate is converted into latent heat of vaporization and flash takes place.

If x is the quantity of flash steam in kg. produced, condensate at 1 atm.

$= (2000 - x)$kg.

Total heat of condensate before flashing

$= 2000 \times 171.4 = 342800$ kcal/hr.

Total heat of steam condensate after flashing

$= (2000 - x)$ 100.01 kcal/hr.

Total heat of flash steam $= x \times 639.11$ kcal/hr

$639.11x + (2000 - x)\ 100.04 = 342800$

$539.07x = 342800 - 200080 = 142720$ kcal/hr.

$x = 264.76$ kg. Ans.

27. The combustion reaction are :

$2C(s) + 2O_2(g) \rightarrow 2CO_2(g)$ ...(1)

$2H2(g) + O_2(g) \rightarrow 2H_2O(g)$ ...(2)

$C_2H_4(g) + 3O_2(g) \rightarrow 2CO_2(g) + 2H_2O(g)$ ...(3)

Adding equations 1 and 2 and subtracting from equation 3 gives

$2C(s) + 2H_2(g) \rightarrow C_4H_4(g)$

So heat of formation of ethylene = heat of combustion of carbon (solid) + heat of combustion of hydrogen (g) – heat of combustion of ethylene (g).

$= -2 \times 94.05 - 2 \times 57.796 + 316.19 = 12.498$ kcal/g. mole.

28. Let x, y and z be the amounts of a, b, c, to be mixed.

By overall mass balance $2000 = x + y + z$

By HNO_3 balance : $0.113x + 0.9y = 640$

By H_2SO_4 balance : $0.444x + 0.98z = 1200$

Solving the above three equations simultaneously, gives

$x = 152.4$ kg, $y = 691.8$ kg, $z = 1157.8$ kg.

29. Reaction is :

Tallow + caustic soda = sodium stearate + glycerine

$890 + 3 \times 40 = 3 \times 306 + 92$

Caustic soda required $= (3 \times 40)\dfrac{200}{890} = 26.96$ kg.

Glycerine produced = (92/890) × 200 = 20.6 kg.

30. Basic : 100 kg of free water.

Water associated with Na_2SO_4 = (180/142) × 32 = 40.66 kg.

Free water = 68 – 40.66 = 27.34 kg.

Glauber's salt present in 100 kg water

= (32 + 40.66) × 100/27.34 = 265.8 kg.

The final mother liquor contains 19.4 kg Na_2SO_4 per 100 kg water.

Water associated with Na_2SO_4 in solution = 180 × 19.4/142 = 24.6 kg.

Free water = 100 – 24.6; 75.4 kg.

Glauber's salt present in 100 kg free water

$$= \frac{(19.4 + 26.4)}{75.4} \times 100$$

= 58.33 kg.

Yield of Glauber's salt per 100 kg free water

= 265.8 – 58.33

= 207.47 kg Na_2SO_4, 10 H_2O

% yield of Glauber's salt

= (207.47/265.3) × 100 = 78.0)

SECTION—D

1. (a)

Solution by oxygen balance

Air × 0.21 + O_2 × 1 = (Air + O_2) 0.5

Let Air = 1 mole; Air = Moles of air, O_2 = Moles of oxygen.

Air/O_2 = 1.72 (0.50/0.29)

2. 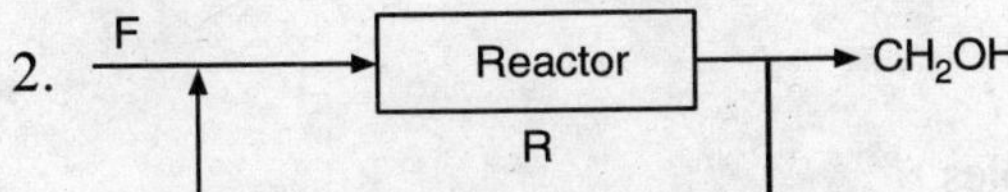

3200 kg/hr = 100 kg mole/hr.

Coin $(F + R)$ = 100/0.15 = 667; H_2 = 667 × 2 = 1334

$F + R$ = 2000 kg mole/ hr; F = 100 kg mole/hour (a)

So R = 1900 kg mole/hr (b).

3. (c)

4. $\Delta H = 3 * - 393.3 + 4 * -284.5 - (-393.3) = - 1924.6$ kJ/ mole.

$$\Delta E = \Delta H - \Delta n\ RT = -1924.6 - (3-6)\frac{8.314 \times 298}{1000}$$

= – 1932.03 kJ/mole.

5. (d)

6. Water volume = 0.1 m^3 = water mass = $\frac{0.1 \times 1}{0.001}$ = 100 kg.

Steam volume = 0.9 m³; steam mass $\frac{0.9}{1.7} = 0.53$ kg.

Total mass = 100 + 0.53 = 100.53 kg.

Dryness fraction = 0.53/100.53 = 0.00527

7. At triple point total pressures of solid and liquid ammonia should be equal to 1 atm. The temperature is obtained by trials.

At 200 K, total pressure is : 0.932 + 0.856 = 1.788 atm

At 190 K total pressure is : 0.3825 + 0.3472 = 0.73 atm

At 195 K, total pressure is : 0.5784 + 0.5762 = 1.14 atm

So temperature at triple point is nearly 195 K.

8. (a) Salt = 500 kg; water = 500 kg; salt separated = 400 kg.

$$\frac{\text{Salt}}{\text{Water}} \times 100 = \frac{400}{500} \times 100 = 80$$

9. (a) CO_2 = 1 mole; Nitrogen = $2 \times \frac{79}{21} = 7.5$ moles;

$$\text{Volume \% of } CO_2 = \frac{1 \times 100}{(1+7.5)} = 11.7$$

10. $2FeS_2 + \frac{11}{2}O_2 \rightarrow Fe_2O_3 + 4\,SO_2$

$$SO_2 + \frac{1}{2}\,O_2 \rightarrow SO_3$$

For 2 moles of FeS_2 to convert : $O_2 = \frac{11}{2}$ moles

$$N_2 = \frac{11}{2} \times \frac{79}{21} = 20.7 \text{ moles}$$

100 % excess is = 5.5 moles O_2; 20.7 moles N_2

For SO_3, oxygen required is = $4.0 \times \frac{0.8}{2} = 1.6$ moles

Final gas :			
SO_3	= 3.2 moles		6.49%
SO_2	= 0.8 moles		1.62%
O_2	= 5.5 – 1.6	= 3.9 moles	7.91%
N_2	= 20.7 + 20.7	= 41.4 moles	83.98%
		49.3	100.00

11. (d) The kinetic energy per molecule = kT where k = Boltzmann constant; T is temperature in Kelvin; So K.E is zero at $T = -273$°C (i.e. : 0 K)

12. (a) Let x is the amount of solids any y is the solvent

$$0.10\,(x + y) = x;\; x + y = 10x$$

$0.4\,[x + (y - 3000)] = x$ solving both simultaneously gives, $x = 400$, $y = 3600$.

13. (d) Initial water = 600 kg; dry solid = 400 kg.

Final $X = 0.20/\,(1 - 0.20) = 0.25$

Final water = 400 × 0.25 = 100 kg.

So water removed = 600 – 100 = 500 kg.

14. Done in the chapter on Fuels.
15. (b) One overall material balanced equation and three component balance equations.
16. (b)
17. (b) $F = B = 100$; $(100 + R)\,0.20 = 100$; $R = 400$ moles/HR.
18. 1000 = B + D

1000 * 0.03 = B × 0.05 + 0.0005 D

Solving $B = 596$ kg/HR; $D = 404$ kg/HR.

$F = E + D = E + 404$; $0.04\,F = 0.05\,E + 0.0005\,D$

Solving simultaneously gives : $F = 2000$, $E = 1596$

$E = R + D$; So $R = 1192$; $R/E = 1192/1596 = 0.75$

19. (d)
20. (a) Since no ethanol vapor in product, so 100% conversion.

SECTION—E

1. (c) 2. (a) 3. (b)

4. (b)

Solution : viscosity of water = 1 cP = 0.01 P

Viscosity of oil = 100 * w = 100 cP = 1 P = 0.1 Ps

5. (d)

Solution : Specific gravity of solution = 1.011

Density of solution in CGS units = 1.011 g/cubic cm. = 1011 g/liter

Weight of sulphuric acid present in solution per liter = 1011 * 0.0245 = 24.7695 g.

Normality = 24.7695/49 = 0.5055 Ans.

6. (b)

Solution : weight fraction of methanol = 0.64

Weight fraction of water = 0.36 (1 – 0.64)

Molecular weight of water = 18, Molecular weight of methanol = 32 (CH_3OH)

Mole of methanol = 0.64/32; Mole of water = 0.36/18

$xm = (0.64/32)/[0.64/32+0.36/18] = 0.50$

7. (d)

CHAPTER 2

Mechanical Operations

1. Sphericity = $\phi_S = \dfrac{6/D_P}{(S_P/D_P)}$; D_P = nominal diameter of particle; S_P = surface area of particle; V_P = volume of one particle.

2. Specific surface area of a mixture, $A_W = \dfrac{6x_1}{\phi_S P_P D_P} + \dfrac{6x_2}{\phi_S P_P D_P} + ... + \dfrac{6x_n}{\phi_S P_P D_P}$

   $x_1, x_2, ... x_n$ are mass fractions in a sample; n = number of components; D_P = average particle diameter; P_P = particle density ϕ_S = sphericity of particle.

3. **Calculating average particle size :**

   (*a*) volume surface mean diameter, $\bar{D}_S = 6/(\phi_S A_W P_P)$

   substituting for AW gives : $\bar{D}_S = 1/\Sigma(xi/D_{pi})$

   xi = weight fraction of ith component ; D_{pi} = average particle diameter.

   (*b*) Arithmetic mean diameter, $\bar{D}_N = \Sigma(Ni\ D_{pi})/NT$

   N_T = number of particles in entire sample; $\bar{D}_{pi}$ = average particle diameter.

   (*c*) Mass mean diameter, $\bar{D}_W = \Sigma xi\ \bar{D}_{pi}$

   (*d*) Volume mean diameter, $\bar{D}_V = 1/\Sigma xi/D_{pi}^{\ 3}$

4. **Screen analysis :** Tyler (British); Mesh number-clear opening, mm; 3-6.68; 5–3. 962; 7–2.794; 9–1. 981 ; 12–1. 397; 16 – 0. 991 ; 24 – 0. 701 ; 32 – 0. 495 ; 42 – 0. 351; 60 – 0.240 mm; 80 – 0. 175; 115 – 0. 124; 170 – 0. 088 ; 200 – 0.074; 325 – 0.044 mm.

5. **Mixers for non-cohesive solids :** Mixing is by slow speed agitation of the mass with an impeller, by tumbling, by centrifugal smearing and impact.

   (1) An internal screw mixer : consists of a vertical tank, containing a helical conveyor that elevates and circulates the material.

   (2) Ribbon blender : two counter acting ribbons are mounted on the same shaft, one moving the solid slowly in one direction and the other moving it quickly in the other.

   (3) Tumbling mixers : include ball mills, thumbing drums, both of which can handle dense slurries and heavy solids.

(4) Centrifugal smearing and impact are active in impact wheels in which a premix of the dry ingredients is fed continuously near the centre of a high speed spinning desk which throws it outward into a stationary casing.

6. **Mixtures for cohesive solids :** Eg. pastes, plastic materials, and rubber,

(*a*) Change : can mixers : blend viscous liquids or light pastes as in food processing or paint manufacture in a removable agitated can or vessel of 5 to 100 gallon in size

(*b*) Kneading machines: mix deformable or plastic solids by squashing the mass flat, folding it over and squashing it once more.

Two arm kneader : Handles suspensions, pastes and light plastic masses.

Dispenser : It works additives and colouring agents into stiff materials.

Masticator : It can disintegrate scrap rubber and compound the toughest plastic masses that can be worked at all. These are also called as intensive mixers.

Banbury mixer : This is a heavy duty two arm mixer in which the agitators are in the form of interrupted spirals. They compound rubber and plastic solids, masticate crude rubber, devulcanize rubber scrap and make water dispersions and rubber solutions.

MULLER MIXER : Mulling is a smearing or rubbing action similar to that in a mortar and pestle. In a large scale processing this action is given by the wide heavy wheels of the mixer.

7. **Size reduction :** Solids are broken by (*a*) compression (*b*) impact (*c*) attrition or rubbing and (*d*) cutting. Egs. for the above are : Nut cracker, Hammer, File and a pair of shears. Size reduction is one of the least energy efficient of all the unit operations.

Mechanical efficiency is from 25 to 60 percent (energy delivered to solids/energy input to the machine).

8. CRUSH LAWS:

(*a*) Bond's law: $P/m = K_b / \sqrt{D_p}$

where K_b is a constant; P = power ; m = mass flow rate : D_p = average particle diameter.

(*b*) Work index : gross energy requirement in Kilowatt hour per ton of feed needed to reduce a very large feed to such a size that 80% of the product passes a 100 μm screen. $K_b = 0.3162$ W_i; D_p in mm; P in KWs, m in TONS PER HOUR.

9. Equipments for size reduction :

(*a*) crushers (*b*) grinders (*c*) ultrafine grinders (*d*) cutting machines.

CRUSHERS : slow speed machines for coarse reduction of large quantities of solids. Types are: Jaw crushers, gyratory crushers, smooth-roll crushers, toothed roll crushers. First three operate by compression and can break large lumps of very hard materials.

Grinders : These are for intermediate duty. Product from a crusher is fed to a grinder for further reduction.

Commercial grinders are : hammer mills, impactors, rolling compression machines, attrition mills, and thumbing mills.

HAMMER MILLS and IMPACTORS : These contain high speed rotor turning inside a cylindrical casing. In hammer mill, the particles are broken by sets of swing hammers, pinned in a rotor disk. An impactor resembles a heavy duty pinned to a rotor disk. An impactor resembles a heavy hammer

duty hammer mill except it contains no screen.

Particles are broken by impact alone without rubbing action of hammer mill.

ROLLER MILLS : Solids are caught and crushed between vertical cylindrical rollers and a stationary anvil ring or bull ring. The rollers are driven at moderate speeds in a circular path.

ATTRITION MILLS : Particles of soft solids are rubbed between the grooved flat faces of rotating circular disks.

TuMBLING MILL : A cylindrical shell slowly turning about, a horizontal axis and filled to about one half its volume with a solid grinding medium forms a tumbling mill.

BALL MILL OR PEBBLE MILLS : Most of the reduction is done by impact as the balls or pebbles drop from the top of the shell.

Critical speed of the ball mill, $NC = \frac{1}{2\pi}\sqrt{\frac{g}{R-r}}$

where R = Radius of mill, r = radius of grinding element ; g = acceleration due to gravity.

Ultra fine grinders are : Fluid energy mills, Agitated mills, colloid mills.

10. **Screening :** Gyratory screen : Two screens one above the other are, held in a casing inclined at an angle between 16 and 30 deg. with the horizontal.

VIBRATING SCREENS : Screens that are rapidly vibrated with small amplitude are less likely to blind than are gyratory screens. The vibrations are generated mechanically or electrically.

$$\frac{B}{F} = \frac{x_D - x_F}{x_D - x_B}$$

F, *D*, *B* are mass flow rates of feed, overflow, and underflow. x_F, x_D, x_B are mass fractions of oversize material in *F*, *D*, *B*.

11. **FILTRATION :** Filtration is the removal of solid particles from a fluid by passing the fluid through a filtering medium or septum on which the solids are deposited.

FILTER PRESS : Contains a set of plates designed to provide a series of chambers or compartments in which solids may collect. The plates are covered with a plate filter medium such as canvas. Slurry is admitted to each compartment under pressure, liquor passes through canvas and out a discharge pipe, leaving a wet cake of solids behind.

SHELL and LEAF FILTERS : For filtering under high pressure than are possible in a plate and frame press, to economise labour or better washing of the cake.

AUTOMATIC BELT FILTER : Pressure filter that separates, compresses, washes, and automatically discharges the cake.

VACUUM FILTERS : Pressure filters are discontinuous but vacuum filters are continuous. Egs. Rotary drum filter, precoat filters, Horizontal belt filter.

CENTRIFUGAL FILTERS : Solids that form a porous cake can be separated from liquids in a filtering centrifuge.

FILTER AIDS : To increase the porosity of the cake to permit passage of liquor at a reasonable rate. Egs. diatomaceous silica, perlite, purified wood cellulose, other inert porous solids,

Pressure drop through filter cake: $\frac{dP}{dL} = \frac{150\mu u(1-\in)^2}{(\phi_S D_P)^2 \in^3}$

u is superficial velocity of filtrate.

12. MEMBRANE FILTERS : Microfiltration is used for particles in size range of 0.5 to 10 μm.

SECTION — A

1. The main objectives of crushing and griding are to obtain products to meet

(a) specific-size requirement (*b*) specific-surface requirement

(c) both (d) neither

2. Dressing of metalliferous ores prior to smelting results in

(a) savings in freight and total smelting cost

(b) reduction in losses of metal at the smelter

(c) neither

(d) both

3. Reduction of large pieces to intermediate or substantially smaller sizes is called :

(a) breaking, crushing or grinding (b) breaking, crushing or cutting

(c) crushing, grinding or comminution (d) crushing, grinding or disintegration.

4. Reduction to fine sizes or powders is called

(a) grinding and comminution (b) pulverization

(c) disintegration and dispersion (d) all the above.

5. The constants K_R, K_K and K_B in the laws of crushing depend on

(a) type of machine (b) feed material

(c) both (d) neither

6. The ways in which solids generally break in size reduction machines are

(a) compression and impact (b) attrition or rubbing

(c) cutting (d) all the above

7. Due to compression size reduction occurs in

(a) jaw crushers (b) gyratory crushers

(c) smooth-roll crushers (d) all the above.

8. An ideal crusher or grinder should

(a) have large capacity (b) consume less power per unit mass of product

(c) give highest % of desired product size (d) all the above.

9. Size reduction in grinders (intermediate and fine) is generally due to

(a) compression

(b) tearing and compression

(c) impact and attrition, sometimes combined with compression

(d) cutting action

10. Size reduction in toothed-roll crushers occurs due to
(a) compression
(b) tearing (shear), attrition, impact and compression
(c) cutting action
(d) impact and attrition

11. Ultrafine grinders (colloid mills) are
(a) hammer mills with internal classification
(b) fluid energy mills
(c) agitated mills and micronizers
(d) all the above.

12. Size reduction ultrafine grinders (colloid mills) is principally by
(a) compression (b) impact
(c) attrition (d) cutting action

13. Cutting machines are
(a) knife cutters (b) dicers
(c) slitters (d) all the above

14. Use of grinding aids results in
(a) finer products (b) higher production rate
(c) both (d) neither

15. With a given energy input, production rate
(a) decreases with increased fineness
(b) increases with increased fineness
(c) remains unchanged with increased fineness
(d) none of the above.

16. The energy consumption decreases with increase in
(a) size of feed (at constant reduction ratio) and size of product (at constant size feed)
(b) capacity of machine
(c) both
(d) neither

17. Tumbling mills refer to
(a) ball mills and pebble mills (b) rod mills and tube mills
(c) compartment mills (d) all the above

18. Length of a ball mill is
(a) 1.5 times more than its diameter (b) slightly more than its diameter
(c) equal to its diameter (d) less than its diameter

19. The tube mill
(a) is long in comparison with its diameter (b) uses smaller balls
(c) produces finer products (d) all the above.

20. Buhrstone mills are extensively used for grinding
(a) non-metallic ores (b) basic slag
(c) talc and soapstone
(d) saw dust, cork, pigments, spices, grains and cereals

21. For milling of cereals and other vegetable products and grinding wheat into flour, we may use
(a) roller mills (b) Buhrstone mills
(c) attrition mills (d) all the above

22. For making flours from starch, tapioca, potato and banana we may use
(a) ball mill (b) pebble mill
(c) hammer mill with internal classification (d) none of the above

23. Crushing rolls, rod mills and ball mills are suitable for size reduction of
(a) nonmetallk ores (b) metalliferous ores
(c) basic slag (d) mica and asbestos

24. Tube mills and balls mills with flint or porcelain balls are employed for size reduction of
(a) nonmetallic ores (b) gums
(c) rubber (d) none of these

25. Size reduction of talc, soapstone, carbonates and sulphates can be achieved by using
(a) ring-roll mills (b) compartment mill
(c) either (d) neither

26. Size of mica and asbestos can be reduced by using
(a) crushing rolls (b) rod mills
(c) gyratory crushers (d) hammer mills

27. Basic slag can be ground in
(a) ball mills (b) tube mills
(c) compartment mills (d) all the above

28. Gums, resins, waxes and moulding powders can be ground in
(a) hammer mill (b) cage mill
(c) either (d) neither

29. Steam-heated rollers and roll crushers are used to reduce size of
(a) refractory silicates (b) hard rubber
(c) gums (d) resins

30. Specific surface is the surface of a unit
(a) weight of materials (b) volume of materials
(c) either (d) neither

31. Limiting particle size is the size of the
(a) largest particle in the sample (b) smallest particle in the sample
(c) either (d) neither

32. Sauter mean diameter is nothing but the
(a) volume-surface mean diameter (b) arithmetic mean diameter
(c) mass means diameter (d) none of the above

33. The material received for separation is called
(a) feed or heading (b) feed or tailing
(c) feed or middling (d) feed or concentrate

34. The capacity of a screen
(a) has no relationship with the screen aperture
(b) is directly proportional to the screen aperture
(c) is inversely proportional to the screen aperture
(d) is directly proportional to the square of the screen aperture.

35. A gravity settling process in the removal of
(a) coarse particles from gases (gravity settling chamber)
(b) coarse particles from liquid (classifier)
(c) fine particles from liquids (sedimentation and thickening)
(d) all the above

36. The drag coefficient in hindered settling is
(a) equal to in free settling (b) greater than in free settling
(c) less than in free settling (d) none of the above

37. In sugar mills we use
(a) suspended batch basket centrifuge
(b) perforated horizontal-basket continuous centrifuge
(c) tubular-bowl centrifuge
(d) disc-bowl centrifuge.

38. For nonfriable crystals we recommend
(a) suspended batch basket centrifuge
(b) perforated horizontal-basket continuous centrifuge.
(c) tubular-bowl centrifuge
(d) disc-bowl centrifuge.

39. For separation of isotopes one uses
(a) disc-bowl centrifuge (b) ultra centrifuge
(c) either (d) neither

40. For transportation of grain, asphalt, crushed coal, ashes, gravel and sand to a short distance we may use a
(a) screw conveyor (b) ribbon conveyor
(c) flight conveyor (d) slat or drag conveyor

41. For transportation of sticky, gummy liquids like molasses, hot tar and sugar we may use a
(a) screw conveyor (b) ribbon conveyor
(c) flight conveyor (d) slat conveyor

42. Nonabrasive loose materials, like grain, food waste, garbage and coal can be transported to a short distance by a

(a) screw conveyor (b) ribbon conveyor
(c) flight conveyor (d) slot conveyor

43. Handling of ashes and similar materials can be achieved by a

(a) screw conveyor (b) ribbon conveyor
(c) flight conveyor (d) slat or drag conveyor

44. For transportation of cold, nonabrasive materials (granular, irregular shape, fines and similar materials) to a long distance we recommend

(a) belt conveyor (b) apron conveyor
(c) bucket elevator (d) none of the above

45. For transportation of materials which are lumpy, abrasive and hot we use

(a) belt conveyor (b) apron conveyor
(c) either (d) neither

SECTION—B

1. In size reduction machines, solids generally break in.....ways.

2. Jaw crushers and Gyratory crushers are.....crushers whereas crushing rolls (smooth roll crushers and toothed roll crushers) are called.....crushers.

3. Between the Blake jaw and the Doge jaw crushers,.....is more common.

4. The.....jaw crusher combines the principles of the Blake and the Dope jaw crushers.

5. (*i*) Hammer mills (impactors), (*ii*) Cage mills (disintegrators), (*iii*) roling-compression mills (bowl mills and roller mills), (*iv*) attrition mills and (*v*) tumbling/revolving mills (rod, ball, pebble, tube- and compartment mills) are called.....(intermediate and fine).

6. Size reduction in cutting machines takes place principally by.....

7. Stamp mills are well suited for crushing.....

8. Disintegrators exert a.....action. Size of.....materials like wood, mica and asbestos can be reduced by disintegrators.

9. Bradford breakers are widely used to reduce size of.....

10. .....crushers can be employed to reduce size of materials like coal, gypsum and ice.

11. In tumbling mills, rolling of balls or pebbles from top to bottom of the heap is called......and throwing of the balls through the air to the toe of the heap is called.....

12. A ball mill may have.....baffles.

13. Balls for ball mills are made from....., ..., ...or.....

14. Balls charge equal to .....% of the mill volume gives the maximum capacity.

15. Minimum-size balls capable of grinding the feed in a ball mill give.....efficiency.

16. α (half the angle of nip) is generally about.....

17. ..... is a measure of the grinding characteristics of a material.

18. Coarse material returned to a mill by a classifier is known as the..... .....

19. Grinding aids are used in the.....grinding.

20. Wear in dry crushing is.....when compared to wet crushing.

21. The products of the separation are called concentrate and tailing where.....refers to valuable part and....., to discarded product. If more than two products are made, the other is called an additional concentrate or a.....

22. The surface is a.....of the individual grain surfaces.

23. The surface shape factor K_a, for spheres is.....

24. The volume shape factor K_v for spheres is.....

25. The specific surface shape factor K_s for spheres is.....

26. Oldest dressing method was.....

27. .....is the operation dealing with separation of solid particles according to their densities.

28. .....means separation of materials into products characterized by difference in size.

29. In the treatment of sulphide ores,.....technique is without a competitor.

30. A tangent to the gas-liquid interface forms the.....angle (always measured through the more dense phase).

31. Separation of solid particles into two or more fractions on the basis of their velocities of flow through fluids is called.....

32. .....is a classification operation to determine the grain size distribution or sizing of particles in the range 5 to 10 microns or below the finest screen size.

33. Separation of two materials of different densities by passing dilute pulp over a table or a deck inclined about 2 to 5 degrees from the horizontal is called.....

34. The separation of a suspension into a supernatant clear fluid and a rather dense slurry containing a higher concentration of solid is called.....

35. A.....is a simple enlargement in a pipeline which reduces the velocity enough to permit the solids to settle out.

36. Separation of solids and fluids of different densities using a force greater than that of gravity is called.....

37. When the different rates of flow are used to separate materials of same density according to their sizes, the process is called.....

38. For collection of sulphuric acid mist, we use a.....

39. Opposite of.....is dispersion.

40. Opposite of size reduction is.....

41. For concentration of magnetite, ilmenite, chromite, monazite ores and for, removal of minor quantities of iron particles from ceramic raw materials....., can be used.

42. Materials like sulphur and coal are generally stored.....

43. Screw and helical flight conveyors are called.....

44. An apron conveyor's trough is made from..... or .....

45. .....elevators are used for vertical travel

SECTION—C

1. Write the Moh scale in increasing order of hardness.
2. Give approximate feed size and product size range for crushers, grinders, ultra fine grinders and cutting machines.
3. State Taggart's rule for selecting between a jaw crusher and a gyratory crusher.
4. Why are the gyratory crushers more widely used than the jaw crushers?
5. Write a simple formula for determination of capacity of jaw and gyratory crushers.
6. What is free crushing?
7. What is choke feeding?
8. Differentiate closed circuit grinding from open circuit grinding.
9. What are the main advantages of closed circuit grinding over batch or continuous open circuit grinding?
10. State Rittinger's law.
11. State Kick's law.
12. Define work index.
13. State Bond's law.
14. State a simple mathematical relationship from which all the laws of crushing can be derived.
15. What is the range of energy consumed per a tonne crushed, if idling loss is kept to minimum ?
16. Define reduction ratio.
17. A crusher crushes rock having a volume surface mean diameter of 0.2 m and discharges product of volume surface mean diameter of 0.04 m. To crush 3.5 kg/s, 7.0 kW of power is required. Using Rittinger's law, calculate Rittinger's constant. K_R.
18. Solve the above problem using Kick's law and calculate Kick's constant L_x.
19. Solve the same problem using Bond's law and calculate Bond's constant K_B (work index need not be taken into consideration).
20. Define crushing efficiency.
21. Define gape.
22. How can you determine capacity of crushing rolls ?
23. Define the angle of nip (2α).
24. What is the relationship between coefficient of friction μ and α for the particle to be crushed ?
25. What is the simple relationship existing between α and the radius of the roll r, the radius of the feed R and the radius of the largest particle in the product d (*i.e.*, half the clearance between the two rolls) ?
26. If the coefficient of friction u between rock and roll material is 0.4. What is the minimum radius of crushing rolls to reduce 0.06 m pieces of rock to 0.02 m ?
27. How does a hammer mill reduce size ?
28. What is a compartment mill ?
29. What is a rod mill ?
30. What is a pebble mill ?

31. What is the relationship between size of the ball and size of the feed (in a ball mill) ?
32. From what materials liners of ball mill are made ?
33. Give a simple relationship to determine critical speed of a ball mill.
34. In case of a ball mill, what should be the actual mill speeds as compared to critical for various cases ?
35. When is wet grinding preferred ?
36. Name the grinding aids used in cement industries.
37. When do you recommend grinding to be carried out in an inert atmosphere ?
38. What should be the moisture content of solids to be reduced in size ?
39. What is the significance of particle-size distribution?
40. List the methods for particle-size measurement.
41. Define mesh.
42. Explain the Tyler standard screen scale.
43. Write the various methods of indicating size fractions.
44. What are grizzlies ?
45. Write three sentences on trommels.
46. How do you calculate critical speed of rotation of a trommel ?
47. Write three sentences on vibrating screens.
48. Define effectiveness of screens.
49. Distinguish between the capacity and effectiveness of a screen.
50. Describe a reel in brief.
51. What do you mean by blinding of a screen ?
52. When is wet sieving employed?
53. Define sphericity ψ.
54. If ψ = sphericity of particle and D_p = equivalent diameter, what is the total surface area A ?
55. Give the graphical/numerical method of evaluating specific surface A_w.
56. If m = mass of particles, in a sample, ρ_m = density of particles and V_p = volume of one particle, what is the number of particles N in the sample?
57. Describe the various screen analysis plots.
58. How do you calculate the volume surface mean diameter D_s ?
59. What is the method of calculating arithmetic mean diameter D_N ?
60. What is the simplest method of calculating mass mean diameter D_w ?
61. What is the graphical /numerical method of finding number of particles N_w, in a mixture ? (Volume shape factor should be taken into consideration).
62. What is a Gate diagram ?
63. Why cumulative analysis for determining surface area is preferred over differential analysis ?
64. Define Selective Index as applied to concentration operations.
65. If lead recovery in a lead concentrate is 94% and gangue rejection in deleted pulp is 95%, find the selective Index.

66. What is flocculation ?
67. To prevent flocculation, what dispersing agents are commonly used ?
68. Define terminal velocity.
69. What do you mean by free settling ?
70. What is hindered settling ?
71. Define the spherical setting ratio both for laminar flow and turbulent flow.
72. Give a list of classifiers.
73. How can you determine capacity of a classifier ?
74. What is jigging ?
75. Explain what you mean by flotation and froth flotation.
76. Define collectors, promoters, frothing agents and modifying agents.
77. For concentration of coal, sulphides and gold by flotation process give the size range of particles.
78. Define Stoke's law of sedimentation by viscous resistance.
79. Give a list of sedimentation equipments.
80. What is a clarifier ?
81. What is a centrifuge ?
82. What is size enlargement ?
83. Describe a cyclone separator.
84. How does a Cottrell precipitator work ?
85. Define the angle of repose and the angle of slide.
86. Distinguish among bins, hoppers and silos.
87. List the mechanical conveyors under the division 'scrapers'.
88. Give a simple formula to calculate capacity of flight 'conveyors'.
89. List the mechanical conveyors under the division 'carriers.'
90. Name the common types of bucket elevators.

SECTION—D

1. A particle attains its terminal setting velocity when :
   (a) gravity force + drag force = buoyancy force
   (b) gravity force – drag force = buoyancy force
   (c) buoyancy force + drag force = gravity force
   (d) drag force = buoyancy force.

2. A fluid energy mill is used for :
   (a) cutting (b) grinding
   (c) ultra grinding (d) crushing

3. Calculate the surface volume mean diameter for the following particulate material. Show detailed calculations.

Size Range, μm	*Mass of particles in the Range*, gm
– 704 + 352	25
– 352 + 176	37.5
– 176 + 88	62.5
– 88 + 44	75.0
PAN	50.0

4. A suspension of uniform particles in water at a concentration of 500 kgs. of solids per cubic meter of slurry is settling in a tank. Density of the particles is 2500 kg/m^3 and terminal velocity of a single particle is 20 cm/S. What will be the setling velocity of suspension? Richardson and Zaki index is 4.6.

(a) 20 cm/s (b) 14.3 cm/s

(c) 7.16 cm/s (d) 3.58 cm/s

5. For geometrically similar baffled stirred tanks, the power number is known to remain constant at high Reynold numbers.

(a) let *p* to power supplied per unit volume of the fluid. *N* the revolutions per second of the agitation, ρ density of the fluid. μ the viscosity of the fluid, and *D* the diameter of the impeller. Then determine α, β, γ and δ in the following equation.

$$P = N^{\alpha}\, r^{\beta}\, \mu^{\gamma}\, D^{\delta}$$

(b) What is the effect of proude number *P*?

6. Match the following :

(1) Gyratory crusher	(a) shear force
(2) Hammer mill	(b) attrition
(3) Buhrstone mill	(c) compressive force
(4) Fluid energy mill	(d) impact

7. For separating particles of different densities the differential setting method uses a liquid sorting medium of density.

(a) Intermediate between those of the light and heavy ones.

(b) less than that of either one

(c) greater than that of either one

(d) of any arbitrary value.

8. At very low R.P.M. (N Re less than 5) the power required for agitation is proportional to :

(a) D (b) D^2

(c) D^3 (d) D^5

9. For crushing of solids, Rittinger's law states that the work required for crushing is proportional to

(a) the new surface created (b) the size reduction ratio

(c) the change in volume due to crushing (d) none of these

10. At low Reynolds numbers the power (*P*) required for agitating a fluid in a stirred tank becomes independent of inertial forces. In this limit, indicate which of the following relations is satisfied :

(a) $P_o \propto Re^{-1.0}$ (b) $P_o \propto Re^{0.0}$

(c) $P_o \propto Re^{0.50}$ (d) $P_o \propto Re^{1.0}$

ANSWERS

SECTION–A

1. c	2. d	3. b	4. d	5. c
6. d	7. d	8. d	9. c	10. b
11. d	12. c	13. d	14. c	15. a
16. c	17. d	18. d	19. d	20. d
21. d	22. c	23. b	24. a	25. a
26. d	27. d	28. c	29. b	30. c
31. c	32. a	33. a	34. b	35. d
36. b	37. a	38. b	39. b	40. a
41. b	42. c	43. d	44. a	45. b

SECTION–B

1. four 2. coarse, fine 3. blake 4. Universal 5. grinders.
6. cutting action 7. gold ores. 8. tearing, fibrous 9. coal
10. toothed-roll crusher 11. cascading, cataracting
12. horizontal.
13. C.I., forged steel, cast steel, alloy steel.
14. 50. 15. maximum. 16. 16°. 17. grindability
18. circulating load
19. dry. 20. less. 21. concentration, tailling middling.
22. summation. 23. $\pi\left(\frac{A}{D^2}\right)$ 24. $\frac{\pi}{6}\left(=\frac{V}{D^3}\right)$ 25. $\frac{K_a}{K_v}\left(=\frac{AD}{V}\right)$
26. hand sorting. 27. sorting. 28. sizing. 29. flotation. 30. contact.
31. classification. 32. elutraiation 33. tabling. 34. sedimentation
35. dust collector. 36. centrifugal 37. sizing. 38. venturi scrubber.
39. flocculation 40. size enlargement 41. magnetic separators
42. outdoors 43. scrapers. 44. wood, metal. 45. vertical.

SECTION-C

1. (*i*) talc (*ii*) gypsum (*iii*) calcite (*iv*) fluorite (*v*) apatite (*vi*) feldspar (*vii*) quartz (*viii*) topax (*ix*) corundum (*x*) diamond.

2.

	Feed size	Product size
Crushers	250-4 cm	upto 0.5 cm
Intermediate grinders	5-0.5 cm	40 mesh screen
Fine grinders	0.5-0.2 cm	200 mesh screen
Ultrafine grinders	0.5-0.02 cm	0.01 micron
Cutting machines	1-0.02 cm	0.5-0.01 cm

3. If the hourly tonnage to be crushed divided by the square of the gape expressed in cm. is less than 0.0184, use a jaw crusher, otherwise a gyratory.

4. Gyratory crushers crush continuously, most suitable for coarsest materials-power consumption is more steady have greater capacity per unit of discharge area than the reciprocating jaw crushers.

5. Capacity C (Tonnes/hour) = 0.087 LS

where L = length of the receiving opening, cm.

S = greater width of the discharge opening, cm.

The formula is accurate for gapes 10 to 60 cm.

6. In free crushing, the crushed products are removed after a short sojourn in the crushing zone. This method presents an excessive amount of fines.

7. In choke feeding, the crusher is filled so that it does not freely discharge the crushed products. This method increases fines, decreases capacity but may result in economy.

8. If the oversize material is returned to the crusher for recrushing, the operation is termed closed circuit grinding otherwise open circuit grinding.

9. Product is more uniform in size. Product specific surface is less. Production rate at a given limiting size is greater. Operation is economical.

10. $P/m = K_R(1/\bar{D}_{Sb} - 1/\bar{D}_{Sa})$

where P = power required, m = feed rate

$\bar{D}_{sp}$ = volume surface mean diameter of product

$\bar{D}_{sa}$ = volume surface mean diameter of feed

K_R = Rittinger's constant.

Assumption : Energy required for size reduction is proportional to the new surface created.

Validity of the law : Where energy input per unit mass of solid is not too great.

11. $$\frac{P}{m} = K_K \log \frac{\bar{D}_{sa}}{\bar{D}_{sc}}$$

K_K = Kick's constant

Assumption : Energy required for size reduction is proportional to the logarithm of the ratio between the initial and final diameters.

12. Work index is defined as the gross energy in kilowatt-hours needed to reduce a ton of very large feed to such a size that 80% of the product passess a 100-micron screen.

13. $$K_B\left(\frac{1}{\sqrt{D_{pb}}} - \frac{1}{\sqrt{D_{va}}}\right)$$

where P in Kilowatts, D in mm and $KB = \sqrt{100 \times 10^{-3} W_i}$

$= 0.3162\ W_i$ for dry crushing or wet grinding and

$K_B = 0.3162 \times \frac{4}{3} W_t$ for grinding.

14. $\frac{dL}{dE} = -CL^p$

Energy dE required to affect a small change dL is a simple power function of the size.

(*a*) $p = -2$ Rittinger's law

(*b*) $p = -1$ Kick's law

(*c*) $P = -3/2$ Bond's law

15. 0.3 to 1.5 kWh

16. It is the ratio of the maximum size of the particles in the feed and in the product.

17. $P/m = K_R(1/\bar{D}_{sb} - 1/\bar{D}_{sb})$

$7.0/3.5 = K_R(1/0.04 - 1/0.2)$

$K_R = 0.1$ K w.s.m./ kg.

18. $P/m = K_K \log(\bar{D}_{sa} / \bar{D}_{sa})$

$7.0/3.5 = K_K \log (0.2/0.04)$

$K_K = 1.243$(kW.s/kg)

19. $P/m = K_B\left(\frac{1}{\sqrt{D_{pb}}} - \frac{1}{\sqrt{D_{pa}}}\right)$

$7/3.5 = K_B(1/\sqrt{0.04} - 1/\sqrt{0.20})$

$K_B = 0.722$ kW.s.m$^{1/2}$ kg.

20. It is the ratio of surface energy created by crushing to the energy absorbed by the solid.

21. Gape is the greatest distance between the jaws or crushing surfaces.

22. Capacity of crushing rolls C (tons/hour)

$= 3.6\ uLDp\rho$

where u = peripherial velocity, mps

L = width of rolls, m

Dp = distance between rolls, m

ρ = density of materials, kg/m^3

If N = speed of rolls in rps and D = roll dia in m, then $u = \pi ND$ and $C = 3.6\ \pi NDLP_p\rho$

Actual capacity is 0.1 to 0.3 times of the theoretical.

23. It is the angle between the roll faces at the level where they will just take hold of a particle and draw it into the crushing zone.

24. $u \geq \tan \alpha$

25. $\cos \alpha = (r + d)/ (r + R)$

26. $\cos \alpha = (r + d)/ (r + R)$

$\tan \alpha = 0.4$, $\alpha = 21°48'$ and $\cos \alpha = 0.9285$

$$0.9285 = \frac{r + \frac{0.02}{2}}{r + \frac{0.06'}{2}} = r = 0.2497 = 0.25 \text{ m.}$$

27. A hammer mill reduces size due to impact and attrition between the feed particles, and the housing and grinding elements.

28. A compartment mill is a tumbling mill divided into two or more sections by perforated partions. Preliminary grinding takes place at one end and finishing grinding, at the discharge end.

29. A rod mill is a tumbling mill with rods arranged horizontally as grinding elements. It delivers more uniform and more granular products than other tumbling mills, thus minimising fines.

30. A pebble mill is a tube mill with flints or ceramic pebbles as the grinding medium and may be lined with ceramic or other non metallic liner

31. $D_b^2 = KD_P$ where D_b = ball diameter, m

Dp = size of coarse feed, m

K = grindability constant, 0.9 to 1.4 increasing in the increase order of hardness.

32. Liners of ball mill are made from alloy steel, C.I., rubber or ceramic.

33. $N_c = \sqrt{\dfrac{g}{R - r}}$

where Nc = critical speed, rps

g = acceleration of gravity, m/sec^2

R = radius of mill, m

r = radius of ball, m

34. 65 to 80% of critical

65 to 70% for wet grinding in viscous suspension.

70 to 75% for wet grinding in low viscous suspension and dry grinding of large particles upto 1.25 cm size. 5% may be more for unbaffled mills.

35. In the fine dry pulverizing, surface forces come into action to cause flocculation and cushioning with a less efficient use of energy. In such a case, wet grinding is preferred.

36. Vinsel resin, cod oil, beef tallow or aluminium stearates.

37. Many finely divided metal powders in suspension in air and many non-metallic materials like sulphur, starch, wood flour, cereal dust, dextrin, coal, pitch, hard rubber and plastics are potential explosion hazards. Maintenance of oxygen content below 12% in grinding systems (8% for sulphur grinding) or grinding in an inert atmosphere is recommended.

38. 3 to 4%

39. Particle-size distribution gives the most complete description of a powder.

40. Sieve separation (110 mm to 10 microns), elutriation (5 to 10 microns), electronic counting (0.6 micron), sedimentation (0.01 micron) and microscopic measurement (to below hundreths of a micron). Also, turbidity meter, sorption, permeability and X-ray measurements.

41. Mesh is a term stating the number of openings per linear inch of screen surface. The size of the opening depends on the thickness of the wire.

42. In this scale, the ratio of diameters of openings in two successive screens is $1:\sqrt{2}$. This is based on a 200-mesh screen with wire 0.0054 cm thick and with an opening of 0.0074 cm $1:\sqrt{2}$ is sometimes used for more careful work and research.

43.

First	Second	Third
Over size 60 mesh	+ 60 mesh	+60 mesh
Through 60 mesh on 120 mesh	– 60 mesh + 120 mesh	60/120 mesh
Under size	–120 mesh	120/200 mesh

44. Grizzlies are of rugged construction and are used for screening of large lumps; 2.5 cm and over. They consist of gratings made up of bars, usually built on slope (20 to 50 degrees from the horizontal) across which material is passed. Grizzlies are about 1.0 to 1.5 m wide and 2.5 to 3 m long.

45. Trommels are revolving screens of perforated sheet metal or wire screen cylindrical or conical in form. They are best suited for screening of materials 1.25 cm and over. They are open at one or both ends and the axis of the cylinder is horizontal or slightly inclined, so that the material is advanced by the rotation of the cylinder.

46. $N_c = \dfrac{1}{2\pi}\sqrt{g/R}$

Where N_c = critical speed, rps

g = acceleration of gravity m/sec^2

R = radius of trommel, m

The best operating speed is 33 to 45% of critical.

47. Vibrating screens are used where large tonnages are to be treated. The vibrating motion is imparted to the screen surface by means of cams, eccentric shafts, unbalanced flywheels or electromagnetic means. A complete screen may have single screening surface or it may be double or triple-decked.

48. Effectiveness of screen = Recovery × Rejection

$$= \frac{X_P(X_P - X_R)}{X_F(X_F - X_R)} = 1 - \frac{(1-X_P)(X_P - X_R)}{(1-X_P - X_R)}$$

Where X_P = mass fraction of desired material in product

X_F = mass fraction of desired material in feed

X_R = mass fraction of desired material in reject

49. The capacity of a screen is measured by the mass of material that can be fed per unit time to a unit area of the screen. The effectiveness of screen is a measure of the success of a screen in closely separating materials Both are opposing factors.

50. A reel is a revolving screen driven at a speed of 2 to 4 rps. The screening surface consists of silk bolting cloth supported by wire mesh. A reel is about 0.5 to 1.0 m in diameter and 1.5 to 25 m long. It is used in flour mills and for screening other light, dry nonabrasive materials.

51. A screen is said to be blinded when it is plugged with solid particles.

52. When the product contains very fine particles in large amounts, wet sieving is employed.

53. Sphericity $\psi = \dfrac{\text{Surface area of sphere having same volume as particle}}{\text{Surface area of particle}}$

$$= \frac{D_{avg}}{D_S}\left(\frac{1}{n}\right)$$

Where D_{avg} = average screen size

D_S = diameter of sphere having same volume as particle

$$n = \frac{\text{specific surface of particle}}{\text{specific surface of sphere}}$$

54. $A = N\,S_P = \dfrac{6M}{\psi \rho_P D_P}$

Where N = number of particles in the sample

S_P = surface area of one particle

M = mass of particles

ρ_P = density of particles

D_P = equivalent diameter of particles.

55. $Aw = \dfrac{6M}{\psi \rho_P} \displaystyle\int_0^{1.0} \frac{d\phi}{D_P}$

where ϕ = mass fraction of the sample that consists of particles in size larger than D_P.

56. $N = \dfrac{m}{\rho_P V_P}$

57. (*a*) Fraction plots, showing the mass fraction retained on screens in the series interval $\sqrt{2}$ as a function of the average particle diameter of logarithm of the average particle diameter in each fraction.

(*b*) Cumulative plots showing the mass fraction passing through screens as a function of the screen aperture or logarithm of the screen aperture.

(*c*) Log-log fractional plot showing the mass fraction retained on each screen in the series interval $\sqrt{2}$ as a function of the average particle diameter in each fraction.

58. $\bar{D}_S = \dfrac{6}{\psi A_W \rho_P}$

59. $\bar{D}_N = \dfrac{\int_0^{N_w} D_P dN}{N_w}$

where N_w = number of particles in a unit mass of sample.

60. $\bar{D}_W = \int_0^{1.0} D_P d\phi$

61. $N_B = \dfrac{1}{a\rho_P} \int_0^{1.0} \dfrac{d\phi}{D_p 3}$

where a = volume shape factor.

62. A gate diagram is a plot of cumulative percent by weight under-size vs. reciprocal diameter, the area beneath the curve represents the surface.

63. Unlike differential analysis, "the assumption that all particles in a simple fraction equal in size" is not needed for cumulative analysis.

64. Selective Index

$$= \sqrt{\frac{R_a J_b}{R_b \cdot J_a}} = \sqrt{\frac{R_a J_b}{(100 - R_a)(100 - J_b)}}$$

Where
R_a = recovery of a in $A = 100 - J_a$
R_b = recovery of b in $A = 100 - J_b$
J_a = rejection of a in $B = 100 - R_a$
J_b = rejection of b in $B = 100 - R_b$

65. Selective Index

$$= \sqrt{\frac{94 * 95}{6 * 5}} = 17.7$$

66. Agglomeration of individual particles into clusters (flocs) is called flocculation.

67. Silicates, phosphates or alkyl aryl sulphonic acids.

68. The maximum attainable velocity (constant velocity) the particle reaches in gravitational settling through a fluid is called the terminal velocity.

69. While settling through a fluid if the particle is at sufficient distance from the boundaries of the container and from other particles so that its fall is not affected by them, the process is called free settling.

70. While settling through a fluid, if the motion of the particles is impeded by the boundaries of the container and other particles, the process is called hindered settling.

71. $\left(\dfrac{D_A}{D_B}\right)^2 = \dfrac{\rho_B - \rho}{\rho_A - \rho}$ Laminar flow

$$\left(\frac{D_A}{D_B}\right) = \frac{\rho_B - \rho}{\rho_A - \rho} \quad \text{Turbulent flow}$$

72. Surface velocity – (spitzkasten), free settling —double cone—Gaycopneumatic—rake, cross-flow (Akins)—Hardinge countercurrent—bowl, spiral-vane classifiers.

73. Capacity of a classifier (tonnes of solid/hour)

$C = 3.6$ Aus ρ.

Where A = cross-sectional area, m^2.

u = rising velocity of fluid, M/sec.

s = % of the solids by volume in the suspension.

ρ = density of solids, kg/m^3.

74. Jigging is a concentration operation carried out in a fluid and its efficiency depends on differences in specific gravities of granular particles present. Jigging helps in stratification of the particles into layers of different specific gravities.

75. Flotation is an operation in which one solid is separated from another by floating one of them at or on the surface of a fluid. The size of solid particles range from 10 mesh to under 200 mesh.

76. Collectors and promoters are reagents which are adsorbed on the surface of the solid as very thin films and which increase the contact angle. Pine oil is an example of a collector whereas sodium xanthate, of a promoter. Collectors form multimolecular thick film and promoters, monomolecular thick film. Frothing agents prevent coalescence of air bubbles and thereby maintain a persistent froth. Hexyl alcohol and pine oil are the examples. Modifying agents activate or depress the adsorption of the filming agents. Copper sulphate is an example.

77. Coal : 10/14 mesh, Sulphides : 46/65 mesh

Gold : 100/150 mesh.

78. Depending upon specific gravity of the solid and viscosity of the fluid, fine particles (0.05 to 0.2 mm) settle at a velocity proportional to the square of the particle size, proportional to the apparent specific gravity of the settling particle and inversely proportional to the viscosity of the fluid medium. This law is known as Stoke's law of sedimentation by viscous resistance.

79. For batch : tray thickner, filter thickener.

For continuous : dust collector and cyclone separator

(those which work dry), baffle-plate scrubber and rotary-sprayer scrubber (those which work wet).

80. The equipments which helps in removal of traces of solids from a liquid is called a clarifier.

81. A centrifuge consists of a perforated or imperforate basket in which a mixture of solids (diameter under 0.5 micron) and liquid or a mixture of two liquids is rotated at high speed so that it is separated into its constituents by the action of centrifugal force.

82. The building up of solid masses or coarse particles from small particulates is called size enlargement.

83. A cyclone separator separates solid particles, (coarser than 30 microns), from gases. It is a vertical cylinder. The inlet stream is introduced tangentially near the top, giving the suspension a spinning motion in the cylinder. The centrifugal force acting on the particles tends to throw them radially to the sides of the cylinder as they spiral downward to a conical bottom where they are removed.

84. If a gas is subjected to a strong unidirectional electrostatic field, the gas becomes ionized and drift towards one electrode. If a finely divided solid or liquid is suspended in the gas, the particle becomes charged and drift towards the same electrode as the ionized gas.

85. Angle of repose is the angle of maximum slope measured from the horizontal at which a heap of loose solid material will stand without sliding. It is useful for determining the capacity of a bin or a pile. Angle of slide is the minimum angle measured from the horizontal at which any loose solid material will flow from rest. This angle is useful for determining spouting angles. It may be greater or lesser for a given material than the angle of repose.

86. A bin is not very tall and is fairly wide. A hopper is a small bin with a sloping bottom. A silo is tall and relatively small in diameter. These are used to store solids that are too volatile or too soluble to expose in outdoors.

87. Screw conveyors, ribbon conveyors, flight conveyors, slat or drag conveyors.

88. Capacity of flight conveyors (tonnes/hour)

$C = 3.6\ BDS\rho$

where B = width of flight, m

D = depth of flight, m

S = speed of conveyor, m/s

ρ = bulk density of material, kg/m^3.

89. Belt conveyors, apron conveyors, bucket elevators.

90. Centrifugal-discharge, positive-discharge and continuous-discharge.

SECTION—D

1. (b)

2. (c)

3. $\bar{D}_S = \dfrac{1}{\Sigma x_i / D_{Pi}}$

Avg. size, M.M.	weight fraction
0.528	0.10
0.264	0.15
0.132	0.25
0.066	0.30
0.044	0.20

$$\Sigma \frac{x_i}{D_{Pi}} = \frac{0.10}{0.528} + \frac{0.15}{0.264} + \frac{0.25}{0.132} + \frac{0.30}{0.066} + \frac{0.20}{0.044} = 11.743$$

$1/(\Sigma x_i/D_{Pi}) = 1/11.743 = 0.08516$ mm.

4. (b)

5. This question is done in Fluid Mechanics chapter.

6. 2—D; 3—A; 1—C; 4—B.

7. (a)

8. (c)

9. (a)

10. (a)

$P_O = P/(\rho N^3 D^5)$ – Power number; $Re = \rho N D^2/\mu$ – Reynold Number.

N is the impeller rotational speed and D is the impeller diameter.

11. The material crushed or ground is highest for :

(a) jaw crusher (b) rod mill

(c) ball mill (d) fluid energy mill **Ans.** (d)

12. For a particle settling in water at its terminal settling velocity, which of the following is true.

(a) Buoyancy = weight + drag (b) weight = Buoyancy + drag

(c) drag = buoyancy + weight (d) drag = weight **Ans.** (b)

13. The critical speed of the ball mill of radius R, which contains balls of radius r is proportional to :

(a) $(R - r)^{-0.5}$ (b) $(R - r)^{-1}$

(c) $(R - r)$ (d) $(R - r)^2$ **Ans.** (a)

CHAPTER 3

Fuels and Combustion

1. The world has reserves of natural gas are of the order of $98*10^{12}$ cubic meter. Total radiant energy of the sun is about $3*10^{26}$ watts. A nuclear power reactor mainly consists of: (a) the core (b) heat transfer system (c) radiation shielding and instrumentation and control system. Thermal or slow reactors operate with natural uranium enriched with U^{235}. Fast reactors use U^{235}, U^{233} and Pu^{239} (highly enriched fuels).
   Biomass is the organic material produced from solar energy by photo synthesis. Energy is obtained from biomass in the form of gas or by combustion. The kinetic energy of wind is used in water pumping, windmills and in navigation. Other energies are tidal powder, geothermal energy.
2. **Definitions :** Proximate analysis of a fuel indicate the moisture, volatile, matter, fixed carbon, and ash content of the fuels in terms of weight percentages. Ultimate analysis of a fuel means analysis in terms of weight percentages of carbon, hydrogen, oxygen, nitrogen, and sulphur. Combustion is the process by which heat is liberated from fuel by its high temperature reaction with an oxidant which is the oxygen of the air.
   Net calorific value (constant volume) is the quantity of heat evolved when a unit quantity of fuel is burnt at constant volume in oxygen saturated with water vapor, the original material and final products of combustion being at a reference temperature of 25°C and the water obtained from fuel being in the vapor phase/state.
   Ignition temperature of a fuel is the minimum temperature at which it ignites in an oxidising atmosphere without the help of any external source of fire.
   Flash point is the minimum temperature at which the fuel gives enough vapors causing a flash in the atmosphere when an external source of fire, is brought to it.
   Fire point is the minimum temperature at which the fuel gives enough vapors produce a continuous flame, instead of just a flash in momentary contact with a fire source. Fire point is higher than flash point.
   For high volatile liquids, the flas point and fire point are much lower than ignition temperature.
3. **Rank of a coal :** The position of coal in the metamorphism or degree of maturity.
   Carbonisation of coal is the process carried out by heating it out of contact with air, to a sufficiently high temperature so that coal undergoes decomposition and a yields a residue which is richer in carbon content than the original fuel.
   Gasification is the process of converting solid or liquid fuels into gaseous fuels. Carbonisation is a partial gasification process.

Caking : Some coals are capable of producing a lump or cake from a powdery state by the carbonisation process. These coals are called as caking coals and the phenoments is caking of coal.

Coking : When carbonisation is carried out at about 1000°C and the resultant lumpy product is sufficiently hard and strong, then this solid product is coke. The process is coking of coal.

Cracking : This is a thermal decomposition process carried out with or without catalyst in which large molecules break into smaller ones with rearrangement of atoms and groups. This is widely used in petrochemicals.

When cracking is performed in presence of hydrogen and suitable catalyst the process is called as hydrocracking.

Orsat analysis : The composition of gaseous fuels is determined in orsat analysis apparatus. The apparatus gives the % quantities of carbon dioxide, oxygen, carbon monoxide, and nitrogen.

Fuel gas : This is the gaseous product of combustion of fuels in ovens and furnaces.

1 kcal = 2.205 CHU = 3.968 BTU = 3088 ft – 1 b. = 4186.8 J = 0.00116 kWh

1 kWh = 1.341 H.P. = 859.845 kcal = 3412.14 BTU

1 H.P. = 0.7457 kW hour

1 W = 0.10197 kgf.m/s = 1.341* 10^{-3} H.P.

1 BTU/sq. ft. h.°F = 4.882 kcal/cu.m. h.°C

1 W/sq.cm.°C = 8598.5 kcal/sq.m. h.°C

1 poise = 1 g/cm. sec.

1 kcal/kg = 1.8 BTU/lb.

1 kcal/cubic m = 0.1124 BTU/cubic ft

1 BTU/cubic ft = 9490 kcal/cubic m

1 kg of oil is equivalent = 10,180 kcal

1 kg of coal is equivalent = 7000 kcal.

4. **Solid fuels :**

Natural solid fuels are : Wood, peat, lignite, or brown coal, bituminous coal and anthracite coal.

Major coal fields in India.

West Bengal : Raniganj; Bihar : Jharia, Bokaro, Santhai, Rajmahal.

Odisha : Talcher, Rampur; M.P. : Singrauli, Korba, Bisrampur, Pench valley; Maharashtra : Wardha Valley, Kamptee; A.P. : Singareni, Kottagudem; North-Eastern : Namchik, Khasi, Garo hills; Tamil Nadu : Neyveli (lignite); J & K Jammu (Anthracite); Rajasthan : Palana (lignite); Gujarat : Panandhro (Kutch, lignite).

Anthracite is the most mature and hard form of solid fuel.

5. **Origin of coal :** 1. Growth in situ, 2. Drift origin theory.

Majority of coal seams in the world are of the in situ origin.

There are four types of macrocomponents in coal. 1. Vitrain, 2. Clarain, 3. Durain, 4. Fusain.

6. **Hydrogenation of coal:** Bergius process. This is liquefaction of coal. This is used for production of liquid fuels from coal. Anthracite coals resist hydrogenation.

7. Processing of solid fuels:

1-coal cleaning (Reducing ash content)/coal washing.

classification of coal cleaning processes.

Dry processes	**Name**	**Basis of separation**
1	Pneumatic scales	Density and friction
2	Spiral separation	Density and separation
3	Berrisford process	Resilience, density.
4	Air-sand process	Density
Wet processes		
1	Dense or heavy medium Separators	Density, size, shape.
2	Jig washers	Density, size, shape.
3	Cyclone separators	Density, size, shape.
4	Trough washers or launchers	Density, size, shape.
5	Upward current classifiers	Density, size, shape.
6	Froth floatation process	Wettability and density.
7	Concentration tables(wilfly)	Density, friction
8	Oil agglomeration process	Wettability, size, density*

8. Coal carbonisation :

1. Low temperature carbonisation : Neyveli (Tamil Nadu) and Narasapur (A.P.) at 600°C. The products are : Semi coke, low temperature tar, liquor, crude low temperature spirit, and gas. The properties of gas are : Specific gravity = 0.60, calorific value = 6500 kcal/cubic. m; composition is CO_2 = 4.0, C_n H_n = 4.0, CO = 7.0, H_2 = 33.0, C_n H_{2n+2} = 45.0, N_2 = 7.0.

   High temperature carbonisation : This process is commercially practiced in coke ovens and to much lesser extent in gas retorts. The chief products are coke and gas. The temperature of the ovens is 1350°C–1450°C. Silica bricks are used at these high temperatures and in this atmosphere. Coke oven gas composition is : H_2 = 54.0, CH_4 = 28.9 CO = 7.4, N_2 = 5.6, CO_2 = 2.0, C_n H_m = 2.6 and O_2 = 0.4.

9. Liquefaction of solid fuels :

(1) Hydrogenation of coal (Bergius process)

(2) Solvent extraction of coal and hydrogenation of coal extracts.

(3) Refining of tar and oil obtained by carbonisation of various solid fuels and oil shales.

(4) Gasification of solid fuels into synthesis gas and conversion Tropsch process and methanol synthesis.

10. Liquid fuels :

Sources of liquid fuels : 1. Petroleum; 2. Coal tar; 3. Crude benzol; 4. Synthetic liquid fuels from coal, shale oil, alcohol, and oil from sands.

Petroleum means oil obtained from the earth. The world production in 1986 is 2.9 * 10^9 tons. The indigenous production of crude oil is

6.8 * 10^6 tons in 1970.

Origin of petroleum : 1. Embedment of organic matter in source beds; 2. Conversion of source material into fluid; 3. Migration of oil into the reservoir rock; 4. Secondary transformation of oil in the reservoir tank.

Petroleum production :

1. Rotary drilling system (derrick, drill column, rotating mechanism, mud circulation system)
2. Working platform.

   Composition of petroleum : Carbon : 83–87%; Hydrogen : 11–14%; Sulphur : 0.50–3%; Nitrogen : 0.1%; and Oxygen : 2–3%.

   Classification of petroleum : 1. Paraffin base crude; 2. Asphalt base crude; 3. Mixed base crude.

   Petroleum refining processes : 1. Physical separation processes (distillation, solvent extraction); 2. Breakdown processes (cracking, visbreaking, and coiling); 3. Rebuilding processes (reforming, alkylation, isomerisation and polymerisation).

11. Common fractions of petroleum :

1. Fuel gas, –162 to –42°C, used as refinery fuel.
2. Propane, –42°C, liquefied petroleum gas.
3. Butane, –12 to 0.50°C, LPG, component of gasoline.
4. Light naphtha, –1 to 150°C, straight run gasoline.
5. Heavy naphtha, 150 to 205°C, solvent component of jet fuel, catalytic reformer feed.
6. Kerosene, 205 to 260°C, fuel, illuminant.
7. Stove oil, 205 to 290°C, fuel.
8. Light gas oil, 205 to 315°C, fuel oil, disel fuel, adsorbent.
9. Heavy gas oil, 315 to 430°C, catalytic cracking feed.
10. Vacuum gas oil, 430 to 540°C, lubricating oils, catalytic cracking feed.
11. Residue, 595°C and above, heavy fuel oil, petroleum coke.

11. Important petroleum products :

1. Gasolines (motor gasoline, aviation gasoline);
2. Kerosene—in oil lamps, stoves, cookers,
3. Jet fuels—in jet liners and;
4. Diesel fuels
5. Fuel oils—used in boilers and furnaces
6. Lubricants—to protect metal surfaces from corrosion.

12. Testing of petroleum and petroleum products :

1. Specific gravity : Higher specific gravity means higher carbon content. Heavier oils have low calorific value.
2. Viscosity : The action of lubricating oils depend on viscosity.
3. Distillation range : This is of value in the evaluation of petroleum products.
4. Flash point and Fire point : Fire point is higher than flash point. The flash point and fire point give indications of range and nature of the boiling point curve.

5. Pour point and cloud point : Pour point is defined as the temperature, 5°F higher than that at which the oil ceases to flow when cooled and tested according to prescribed conditions. The cloud point is the temperature at which an oil becomes cloudy when it is cooled in a specified manner. Cloud point is higher than (5 to 6°F) pour point. A low pour point is desirable for handling.
6. Smoke point and char value : These criteria are determined to assess the burning quality of kerosene as an illuminator as fuel. The smoke point is the maximum flame height in millimeters to which kerosene will burn without smoking in a standard apparatus. The char value is the amount of char in milligrams for every kg of kerosene burned, which is formed on the wick of a standard lamp burning under prescribed conditions. The char value of high grade kerosene should not be more than 20 mg/kg.
7. Carbon residue : This is important for oils used in gas production in diesel engines, in burners and for lubricating oils used in motors.
8. Sulphur content : The sulphur content should be as low as possible. Sulphur is largely present in oils in the form of corrosive organic compounds. On combustion it gives of foul gases.
9. Ash: Some constituents of petroleum ash have very harmful effects (vanadium, sodium).
10. Calorific value : The calorific value of oils is determined by Bomb-calorimeter.
11. Octane number : Some additives are like—tetra methyl lead, tertiary butyl acetate. When knocking occurs, the power is wasted. A good gasoline should resist knocking.
12. Cetane number : There is a delay between ignition of diesel fuel into hot compressed air, and its ignition. The delay period is connected with the type of hydrocarbon. A n-paraffin is given 100 cetane number and an aromatic is given zero cetane number.
13. Aniline point : This is defined as the temperature at which a liquid just becomes completely miscible with an equal volume of aniline.
14. Diesel index = Aniline point in °F *Degree API/100

13. Petroleum Refineries in India upto 1987

1. Digboi (Assam);
2. RPC (Mumbai);
3. BPCL (Bombay);
4. HPCL-Visakhapatnam;
5. Guwahati (Assam);
6. Barauni (Bihar);
7. Koyali (Gujarat);
8. Cochin (Kerala);
9. Bongaigaon (Assam);
10. MRL (Chennai);
11. Haldia (West Bengal);
12. Madhura (U.P.);
13. Karnal (Harayana);
14. Mangalore (Karnataka) .

14. Liquid fuels from non-petroleum sources :

1. Coal tat and coal tar fuels;
2. Fisher Tropsch to give liquid fuel;
3. Motor benzol;
4. Alcohols

15. Fisher Tropsch Process : Based on catalytic hydrogenation of carbon monoxide.

$nCO + 2n\,H_2 \Rightarrow C_n\,H_{2n+2} + H_2O$
paraffin

$nCO + 2n\,H_2 \Rightarrow C_n\,H_{2n} + nH_2O$
olefin

16. Gaseous fuels :

Types of gaseous fuels :

1. Fuel gases occurring in nature (natural gas) methane from (coal mines).
2. Fuel gases made from solid fuels (producer gas, water gas, gases from other gasification processes, coal gas, blast furnace gas, wood gas).
3. Fuel gases made from petroleum (refinery gases, liquefied petroleum gas, gases from oil gasification processes).
4. Fuel gases made from permeation of organic wastes.
5. Hydrogen, acetylene.

17. Wobble number = Calorific value/Sp. gravity.

18. Composition of natural gas : (wet and sweet)

Methane = 77.28; Ethane = 11.18; Propane = 5.83; Butanes = 2.34; Pentanes = 1.18; Carbon dioxide = 0.80; Nitrogen = 1.39; Total = 100.00: Gross calorific value, kcal/cubic m is 11,200. Specific gravity is 0.741.

19. Water gas : Calorific value is : 2800 kcal/N. cubic. m.

This is obtained by gasifying solid source of carbon in superheated steam. Carburetted water gas is the enriched water gas with hydrocarbons.

20. Complete gasification of coal :

1. Lurgi process : This is a single stage fixed bed process gasifying a range of coarse or granular fuels.
2. Winkler process : This is fluidised bed process. Fluidised bed attains a temperature of 1000°C.
3. Koppers-Totzek process : Gasifies pulverised fuel carried in suspension in oxygen at approximately atmospheric pressure in presence of steam.
4. Texaco process : This is partial oxidation of heavy fuel oil.
5. Underground gasification of coal : Air and oxygen are used as the gasifying medium. Bituminous, sub-bituminous and moisture frown coal deposits are suitable for underground gasification.
6. Coal gas : Produced by high temperature carbonisation of coal in gas retorts and coke ovens.
7. Blast furnace gas : This is a combustible gas from blast furnaces of steel parts. The composition is : Carbon dioxide–11.0; carbon monoxide–27.0; hydrogen–2.0; nitrogen–60.0; calorific value is 870 kcal/N cubic. m; Sp. gravity = 1.01.
8. Gases from biomass : Biomass can be partially gasified by carbonisation which is initially pyrolysis. Biomass can be converted to fuel gas as well as synthesis gas for chemicals, fertilizers and liquid fuels.

9. Liquefied petroleum gases (LPG) : Out of the gaseous hydrocarbons the C_3 and C_4 components compounds can be liquefied at room temperature by the application of moderate pressure and therefore be conveniently stored and transported as liquids in high pressure vessels.
10. Continuous steam reforming of naphtha : A mixture of desulphurised naphtha vapors, surplus hydrogen and superheated steam enters the reactor from top. The reaction is carried out at 20 atms pressure and 840°C temperature the composition is carbon dioxide–16.0, carbon monoxide–11.0, methane–12.0, nitrogen–61.0.
11. Partial combustion process: The feed stock ranges from natural gas to high sulphur heavy oils.
12. Hydrogenation processes : The process of hydro-cracking has been used in the field of oil gasification.

21. Combustion process :

Types of combustion processes: 1. Combustion with stationary flame; 2. Surface combustion or flameless combustion; 3. Submerged combustion; 4. Pulsating combustion; 5. Combustion with explosion flame; 6. Pulsating combustion; 7. Slow combustion; 8. Combustion of solid fuels on grate; 9. Down jet combustion.

22. Spontaneous ignition temperature (SIT) : The minimum temperature to which a combustible is raised for bringing about rapid combustion or explosion.

23. Oil burners : 1. Vaporising type; 2. Atomising type.

Gas burners : 1. Aerated gas burners/atmospheric; 2. Air aspiration; 3. Independent; 4. Flameless combustion type.

QUESTIONS

SECTION—A

1. The number of macro components in coal according to Stopes are :

(a) Three (b) Four

(c) Five

2. Calorific value of a solid fuel is determined by :

(a) Junker's calorimeter (b) Bomb calorimeter

(c) Boys calorimeter

3. The difference in net and gross calorific value of a solid fuel is due to :

(a) Moisture content of the fuel (b) Difference in rate of heating

(c) Mineral matter of the fuel (d) Volatile matter of the fuel

4. Seylor's coal classification method is based on :

(a) Proximate analysis of the coal

(b) Ultimate analysis of the coal

(c) Calorific value and caking index of the coal

5. Coal washing is done to :

(a) Decrease the mineral matter (b) Decrease the moisture content

(c) Decrease the ash content (d) Increase caking properties

6. Weathering of coal occurs due to :

(a) Faulty mining process (b) High moisture content
(c) Improper storage conditions (d) None of these

7. Soft coke is primarily used as a :

(a) Blast furnace fuel, (b) Foundry fuel
(c) Domestic fuel

8. Temperature employed for high temperature carbonisation is :

(a) 600–700°C (b) 700–900°C
(c) 900–1000°C

9. The only commercial low temperature carbonisation plant in India is located at :

(a) Durgapur (b) Sindri
(c) Neyveli (d) Jharia

10. In 'Beehive Ovens' by-products are :

(a) Recovered, (b) Not recovered
(c) Partially recovered

11. The process of by-product recovery is called direct, indirect or semidirect depending on the recovery of :

(a) Ammonia (b) Tar
(c) Benzol

12. Coke ovens are built of :

(a) Silica bricks (b) Fire clay bricks
(c) Magnesite bricks

13. The principal component of coke oven gas is :

(a) Methane (b) Carbon monoxide
(c) Hydrogen (d) Nitrogen

14. Direct hydrogenation of coal is called :

(a) Bergius process (b) Fischer-Tropsch process
(c) Chance-process (d) Koppers-Totzek process

15. High pressure coal gasification is called :

(a) Winkler's process (b) Lurgi process
(c) Koppers-Totzek process (d) Tromp-process

16. In determining ash content by proximate analysis, the coal is heated at a temperature of :

(a) 925 ± 25°C (b) 750 ± 25°C
(c) 600 ± 25°C

17. For determining volatile matter, the coal is to be heated in absence of air for :

(a) 7 minutes at 750 ±25°C
(b) 7 minutes at 925 ±25°C
(c) Till constant weight is obtained at 750 ± 25°C
(d) Till constant weight is obtained at 925 ± 25°C

18. On carbonisation, the ash percentage of original coal :

(a) Increases, (b) Decreases

(c) Remains unaltered

19. Caking coal is necessarily a type of :

(a) Anthracite (b) Bituminous

(c) Lignite

20. Assam coal is abnormally high in :

(a) Moisture (b) Sulphur

(c) Phosphorus (d) Nitrogen

21. A typical yield of coke in a hyproduct oven is :

(a) 60% (b) 70%

(c) 75% (d) 80%

22. Conversion, of semicoke to coke is accompanied by :

(a) Rise in porosity of the mass

(b) Rise in density of the mass

(c) Rise in electrical conductivity of the mass

23. A measure of agglutinating property of coal is :

(a) Caking index (b) Swelling index

(c) Graveking assay (d) Raga index

24. In Bomb calorimeter, the combustion of coal is supported by :

(a) Air (b) Hydrogen

(c) Ozone (d) Oxygen

25. The copper calorimeter in bomb calorimeter is made polished and silvery white in order to :

(a) Prevent corrosion (b) Look more attractive

(c) Avoid radiation loss (d) None of these

26. Origin of petroleum is believed to be :

(a) From vegetable source (b) From animal source

(c) From both the sources (d) None of the above

27. The major component of LPG is :

(a) Carbon monoxide (b) Hydrogen

(c) Butane (d) Ethylene

28. Petroleum coke is obtained by the process of :

(a) Cracking (b) Reforming

(c) Stabilisation (d) None of these

29. The leading coal producing state in India is :

(a) Odisha (b) West Bengal

(c) Madhya Pradesh (d) Bihar

30. Octane number is measure of anti-knock properties, of :

(a) Diesel oil (b) Gasoline
(c) Kerosene (d) Jet fuel

31. Cetane number is a measure of anti-knock quality of :

(a) Diesel oil (b) Gasoline
(c) Kerosene (d) Jet fuel

32. The additive used to improve octane number is :

(a) Tetraethyl lead (b) Tetramethyl lead
(c) Butyl nitrate (d) Triethyl lead

33. The crude producing both wax and asphalt is called :

(a) Paraffin base (b) Asphalt base
(c) Mixed base. (d) Natural gasoline is

34. Natural gasoline is :

(a) A refinery product (b) Obtained freely in the nature
(c) A byproduct of coal carbonisation (d) Obtained by stripping natural gas

35. The catalyst used for catalytic cracking of gas oil to gasoline in refineries is :

(a) Synthetic oxides of alumina and silica (b) Reduced iron oxide
(c) Vanadium pentoxide (d) Finely divided nickel

36. A gaseous fuel is called 'Rich gas' if the calorific value is :

(a) Above 4000 kcal/nm^3 (b) Between 3500 and 4000 kcal/nm^3
(c) Between 3000 and 3300 kcal/nm^3

37. Producer gas is obtained by passing over a bed of incandescent carboneous fuel :

(a) Air (b) Mixture of air and steam
(c) Oxygen (d) Oxygen and steam

38. Use of steam in the producer blast increases the C.V. of the gas due to :

(a) Oxygen (b) Hydrogen
(c) Carbon monoxide (d) Methane.

39. Blue gas' is an alternative name of :

(a) Coke oven gas (b) Water gas
(c) Carburetted water gas (d) Natural gas

40. Methane content of the product gas is high in :

(a) Lurgi process (b) Winkler's process
(c) Koppers-Totzek process

41. The fertiliser plant, which adopted Koppers-Totzek process for coal gasification is located at :

(a) Sindri (b) Nangal
(c) Trombay (d) Talcher

42. Winkler's process of coal gasification is a :

(a) Fixed bed process (b) Fluidised bed process
(c) Moving bed process (d) Suspension bed process

43. Semet-Solvey process of oil gasification is a :

(a) Catalytic-cyclic process, (b) Non-catalytic-cyclic process

(c) Catalytic-continuous process

44. In Kopper-Totzek process, ash is normally removed as :

(a) Solid ash (b) Slagging ash

(c) Fly ash

45. Octane number of India gasoline is :

(a) 83 (b) 93

(c) 87 (d) 100

46. In orsat apparatus N_2 in flue gas is determined by absorbing in :

(A) Alkaline pyrogallol (b) Potassium hydroxide

(c) Ammoniacal cuprous chloride (d) None of these

47. Besides calorific value, bomb calorimeter is also used to determine :

(a) Sulphur (b) Nitrogen

(c) Hydrogen in coal

48. Preheated air in the process of combustion increases :

(a) Calorific value of the fuel (b) Thermal efficiency of the process

(c) Adiabatic flame temperature

49. The commercial application of liquefied petroleum gas is as :

(a) Industrial heating gas (b) Manufacture of synthesis gas

(c) Domestic heating gas (d) Additive for gasoline

50. In dry air, the volume ratio of N_2 to O_2 is :

(a) 3.76 (b) 0.27

(c) 0.79

51. Incomplete combustion of a fuel can be judged by :

(a) Smoke formation, (b) High CO content of the flue gas

(c) High O_2 content of the flue gas

52. Pulverised fuel is burnt in large boilers in order to provide :

(a) High calorific value of the fuel (b) Complete combustion

(c) Trouble free operation (d) Easy ash removal

53. A travelling grate stoker may bum efficiently :

(a) Caking coal (b) Non-caking coal

(c) High ash coal (d) Pulverised coal

54. In case of pulverised coal firing, the secondary air is supplied to :

(a) Carry the coal to the furnace

(b) Dry the coal

(c) Burn the coal efficiently around the burner

(d) Enrich primary air

55. The gross calorific value of a gaseous fuel containing no hydrocarbon or hydrogen will be same as that of net calorific value :

(a) True (b) False
(c) Unpredictable

56. A coal containing higher amount of volatile matter will have :

(a) Low ash fusion point (b) High calorific value
(c) High caking index (d) Very low ash content

57. A non-luminous flame while burning a gaseous fuel indicates :

(a) Complete combustion, (b) High CO content in the flue
(c) Excess O_2 in the flue (d) Incomplete combustion

58. Which of the following does not contribute to calorific value of a fuel :

(a) Carbon (b) Hydrogen
(c) Oxygen (d) None of these

59. The air requirement will be highest for combustion of 1 nm^3 of :

(a) Blast furnace gas (b) Water gas
(c) Producer gas (d) Coke oven gas.

60. On removal of hydrogen from coke oven gas, its calorific value :

(a) Increases, (b) Decreases
(c) Remains unchanged

SECTION—B

1. The theories behind origin of coal are.....and.....
2. .....method is used to determine moisture in peat.
3. In Seylor's classification, the coals are divided in four.....according to their carbon content.
4. The coal is difficult to wash, if it contains more.....mineral matter.
5. Improper storage of coal finally leads to.....
6. Carbonisation means heating of coal in.....of air.
7. High temperature carbonisation is practised mainly to cater the needs of.....
8. Product from a low temperature carbonisation process is called as.....
9. Compound coke ovens are fired by.....and.....gas.
10. Normal carbonisation time in a hyproduct oven is.....hours.
11. Gross calorific value of a coke oven gas is.....
12. A rough guide of the extent of carbonisation is.....in coke.
13. The only commercial Fischer-Tropsch plant in the world is located at.....
14. Higher the reactivity of coke.....is the CAB value.
15. Rank of a coal increases from.....to.....variety.
16. In a high temperature carboniser coal tar is a result of.....of primary tar.
17. Direct process of byproduct recovery ensures complete removal of NH_3 in form of.....

18. Washing characteristics of coal are ascertained by.....test.

19. During the process of carbonisation of caking coal, plastic layer is formed between a temperature of.....°C to.....°C.

20. A better caking and swelling properties of coal implies.....swelling index number.

21. Average ash content of Indian coal is.....

22. Benzol in the byproduct gas is removed by scrubbing with.....

23. The reservoir rocks containing petroleum has got high.....and.....

24. The first oil refinery in India was located at.....

25. The source of petroleum is mostly.....in nature.

26. A petroleum product is called sweet if it passes.....test.

27. The solvent used for dewaxing process is.....

28. Reformed gasoline is obtained by reforming of.....

29. Heavy furnace oil is obtained by a process of.....

30. The burning qualities of kerosene are obtained by.....test.

31. Chemical name of iso-octane is.....

32. Octane number of n-paraffins are.....

33. Aniline point is a measure of.....in the oil.

34. The total reserve of crude in India is approximately.....million tonnes, including the OFF shore reserves.

35. The largest oil refinery in India is located at.....

36. The principal component of natural gas is.....

37. The natural gas called wet, if the condensate amount is above.....

38. The calorific value of water gas is.....

39. Water gas shift reaction is a reaction between.....and.....

40. Blast furnace gas is an example of.....gas.

41. Water gas generators are operated on.....basis.

42. In gasification process, the reaction between CO_2 and C is called as.....reaction.

43. Synthesis gas is normally a variable mixture of.....or.....

44. Dust content of blast furance gas is.....

45. The combustible component of blast furance gas is.....

SECTION—C

1. What do you understand by petrography of coal ?

2. What is the basis for Saylor's classification of coal ?

3. What are the constituents determined by proximate analysis ?

4. What are the changes occur in coal due to weathering ?

5. What are the desirable characteristics of coal to be used for manufacturing metallurgical coke ?

6. What are the essential characteristics of a blast furnace coke ?

7. What are the byproducts recovered in a coal carbonisation plant ?

8. What are the demerits of 'Beehive' ovens ?

9. Give the approximate composition of coke oven gas.

10. What are the various dry coal cleaning processes ?

11. What are the various wet coal cleaning processes ?

12. What is Fraser and Yancey formula for calculating washing efficiency ?

13. What are the important micro components of coal ?

14. What is Goutal's formula for calculating gross calorific value from proximate analysis ?

15. What are the external factors responsible for conversion of the source material to petroleum ?

16. What are the various refinery processes carried out in an integrated oil refinery ?

17. What are the additives used to improve quality of a commercial gasoline?

18. Define octane number.

19. Define diesel index number.

20. Give the reactions involved in manufacturing liquid fuels from coal by Fischer-Tropsch synthesis.

21. What is the basis for classification of gaseous fuel ?

22. What is "semi water gas" ?

23. What is the requirement for producing 1 Nm^3 of carburated water gas ?

24. How to avoid clinkering problem in gasifiers ?

25. What are the essential requirements of the solid fuel to be gasified in a Winkler's plant ?

26. What are the points in favour of Kopper-Totzek process to be used for synthesis gas production ?

27. Define "Gross calorific value" at constant volume.

28. Give the calorific values of coke oven gas, producer' gas and water gas.

29. The flue gas analysis of a furnace burning coal shows, CO_2 = 12.7%, O_2 = 7.1%, N_2 = 80.2%. Calculate the percentage of excess air supplied.

30. A coal has got the following analysis. Calculate the C.V. of the coal using Dulong's formula. If this coal is used to manufacture producer gas whose C.V. is 1,550 kcal/nm^3, gas yield is

$$3{,}000 \ \frac{m^3}{\text{ton of coal}},$$ find the cold gas efficiency of the gas producer.

Analysis : C = 68%, H = 3.9%, S = 0.3%.

SECTION—D

1. Match the following :

(1) Catalytic reforming	(a) High speed diesel
(2) Smoke point	(b) Kerosene
	(c) Gasolime
	(d) Aromatics

2. The orsat analysis of a flue gas is :

CO_2 = 12.7%; O_2 = 7.1%; N_2 = 80.2%

Determine the percent excess air used in the comhustion. The nitrogen present in the flue gas is contributed by air only. This problem is similar to Problem-29 of section C.

3. Bituminous coal with a calorific value of 2000 kJ/kg is used for generateing steam in a boiler. How much coal has to be burnt to generate 1 MW of energy. Efficiency of combustion is 0.75. How much air is needed if 50% exess air is used. Assume that coal contains 67% carbon and 33% hydrogen.

4. Which of the following fuels has the highest caloricfic value per unit mass ?

(a) Coal (b) Kerosene

(c) Natural gas (d) Furnace oil

5. Match :

(1) Catalytic crecking	(a) Kerosene
(2) Catalytic reforming	(b) Gasoline
	(c) Aromatics
	(d) Diesel

6. Match the reactions with names :

(1) $CO + H_2O \rightleftharpoons CO_2 + H_2$	(a) Boudard reaction
(2) $CH_4 + H_2O \rightleftharpoons CO_2 + 3H_2$	(b) Water gas shift reaction
	(c) Water gas reaction
	(d) Steam reforming

7. Pure CO is mixed with 100% excess air and burnt. Only 80% of CO burns. The reactants are at 100°C and the products are at 300°C. Calculate the amount or heat added or removed per kg. mole of CO fed to the reactor. Data : Mean molal specific heats between 25°C and T°C (given below) in kJ/(kg.mole °K) are :

Gas	**T = 100°C**	**T = 300°C**
CO	29.22	3.61
CO_2	—	43.77
O_2	29.64	30.99
N_2	29.17	29.66

Standard heat of formatin at 25°C in kJ/kg. mole are :

$CO \rightarrow 11052.4$; $CO_2 \rightarrow -393514$

8. Match :

(1) LPG	(a) CO, CO_2, H_2
(2) Synthesis gas	(b) $C_2H_4; C_3H_8$
	(c) C_3H_8, C_4H_{10}
	(d) CO, H_2

SECTION—E

1. 6 g of carbon is burnt with an amount of air containing 18 g of oxygen. The product contains 16.5 g of carbon dioxide and 2.8 g of CO besides other constituents. What is the reactant ?
   (a) 100% (b) 95% (c) 75% (d) 20%
2. In a refinary, petroleum crude is fractionated into gas fraction, light ends, intermediate distillates, having heavy distillates, residues and byproducts. The group of products including gas oil, diesel oil and byproducts. The group to the fraction :
   (a) Heavy distillates (b) Intermediate dstillates (c) Light ends (d) Residues.
3. The order of preference for feedstock to a catalytic reformer is :
   (a) Catalytic naphtha—coking naptha—virgin naphtha
   (b) Cocking naphtha—virgin naphtha—catalytic naphtha
   (c) Virgin naphatha—catalytic naphtha—coking naphtha
   (d) Virgin naphatha—coking naphtha—catalytic naphtha.
4. The proper arranggement of petroleum fractions in the order of their boiling points is :
   (a) Lubricating oils diesel oils petrol LPG
   (b) Lubriccating oils petrol diesel LPG
   (c) Petrol lubricating oils diesel LPG
   (d) Petrol diesel LPG lubricating oil.

ANSWERS

SECTION—A

1. (b)	**2.** (b)	**3.** (a)	**4.** (b)	**5.** (a)
6. (c)	**7.** (c)	**8.** (c)	**9.** (c)	**10.** (b)
11. (a)	**12.** (a)	**13.** (c)	**14.** (a)	**15.** (b)
16. (b)	**17.** (b)	**18.** (a)	**19.** (b)	**20.** (b)
21. (c)	**22.** (c)	**23.** (a)	**24.** (d)	**25.** (c)
26. (c)	**27.** (c)	**28.** (a)	**29.** (d)	**30.** (b)
31. (a)	**32.** (a)	**33.** (c)	**34.** (d)	**35.** (a)
36. (b)	**37.** (b)	**38.** (c)	**39.** (b)	**40.** (a)
41. (d)	**42.** (b)	**43.** (b)	**44.** (b)	**45.** (c)
46. (d)	**47.** (a)	**48.** (c)	**49.** (c)	**50.** (a)
51. (b)	**52.** (b)	**53.** (b)	**54.** (c)	**55.** (a)
56. (b)	**57.** (a)	**58.** (c)	**59.** (d)	**60.** (a)

SECTION—B

1. In situ and Drift	**2.** Dean and Stark	**3.** Species
4. Inherent	**5.** Spontaneous ignition	**6.** Absence
7. Blast furnace	**8.** Soft coke	

9. B.F gas and coke-oven gas **10.** 14–16 **11.** 5020 kcal/nm^3
12. Percentage of volatile matter **13.** South Africa **14.** Lower
15. Lignite; Anthracite **16.** Secondary cracking **17.** Ammonium sulphate
18. Float and sink **19.** 350 to 450° **20.** Higher
21. 30% **22.** Wash oil **23.** Porosity, permeability
24. Digboi **25.** Organic **26.** Doctor's
27. Methyl ethyl ketone **28.** Heavy naptha **29.** Vis breaking
30. Smoke point **31.** 2, 2, 4 trimethyl pentane **32.** Low
33. Unsaturates **34.** 123 **35.** Mathura (U.P.)
36. Methane **37.** 50 g/nm^3 **38.** 2,800 kcal/Nm^3
39. CO and H_2O **40.** Lean **41.** Intermittent
42. Boudourd **43.** CO and H_2 **44.** 20–30 g/Nm^3
45. CO

SECTION—C

1. It is the study of coal components by visual means with or without the aid of microsocope.

2. Ultimate analysis, volatile matter and gross calorific value.

3. (a) Moisture (b) Volatile matter
(c) Ash content (d) Fixed carbon (by difference)

4. (a) Decrease in calorific value (b) Decrease in caking properties
(c) Increase in oxygen content (d) Increases in friability

5. (a) High caking index (b) Low sulphur content
(c) Low ash content (d) Moisture content (around 8%)

6. (a) High purity, (b) High strength
(c) High porosity (d) Low reactivity

7. (a) Tar (b) Ammonia
(c) Benzol (d) Napthalene

8. (a) No recovery of byproducts (b) Lower coke yield due to partial combustion
(c) Unsuitable for low grade coal (d) Lack of flexibility of operation

9. H_2 = 54%, CH_4 = 28%, CO = 7.4%, N_2 = 5.6%, CO_2 = 2% C_nH_m = 2.6%, and O_2 = 0.4%.

10. Pneumatic tables, spiral separators, Jig washers, cyclone washers, trough washers, double cone classifier, froth flotation and wilfly tables.

12. $$\eta = \left[\frac{\text{Recovery of clean coal}}{\text{(theoretical recovery at a given specific gravity)}} \times \frac{\text{(\% ash in feed coal – \% ash in clean coal)}}{\text{(\% ash in feed coal – \% ash in theoretical clean coal)}}\right]$$

13. Vitrinite, fusinite, micrinite ad liptinite.

14. $C_G = 82F + a, V$; C_G = gross cal. value in kcal/kg
F = fixed carbon, %; V = volatile matter, %,
a = A constant, which depends on the value of V.M.

15. (a) Anaerobic bacterical action
(b) Pressure of oven burden
(c) Low temperature cracking
(d) Irradiation by radioacitve materials associated with sedimentary rocks.

16. (a) Physical separation process
(b) Break-down process
(c) Rebuilding process

17. Anticracking combustion deposit modifier, antioxidants to prevent gum formation in metal..... and antirust agents.

18. It is the perecentage by volume of iso-octane in a mixture of n-heptane and iso-octane having the same knocking tendency as that of the sample under test.

19. It is given by the formula :

$$\text{D.I.N.} = \frac{\text{Aniline point (°F)} \times \text{API Gravity}}{100}$$

Numerically, it is usually 3 units higher cetane number.

20. $n\text{CO} + (2_n + 1)\ H_2 \rightarrow C_n\ H_2 n + n\ H_2O$ (paraffin)
$n\text{CO} + 2n\ H_2 \rightarrow Cn\ H_{2n} + n\ H_2O$ (olefin)

21. Wobbe index given by the relation Calorific value/(Sp. gravity)$^{0.5}$

22. While manufacturing water gas, a proportion of air is added to steam, so as to get a mixture of hydrogen, carbon monoxide, carbon dioxide and nitrogen in the outlet gas which on removal of CO and CO_2 can be used as a synthesis gas for ammonia manufacture.

23. Coke–0.45 kg; oil–0.3 kg; steam–0.65 kg and air–1.5 Nm^3.

24. By maintaining high steam to O_2 ratio.

25. Particle size 8 mm, medium to high reactivity, high ash fusion temperature, low moisture content.

26. Single stage gasification, low CH_4 content and absence of tar, capability to take high ash coal and simple design.

27. It is the amount of heat liberated by burning completely an unit quantity of fuel under constant volume condition in oxygen saturated with water vapour, the original matter and final product being brought down to 25°C and water being in liquid state.

28. Coke oven gas—5020 kcal/nm^3
Producer gas—1,260 kcal/nm^3
Water gas—2,800 kcal/nm^3

29. Since no 'CO' is present in the flue gas, the O_2 is entirely from the excess air supplied.
Basis : 100 k mol of dry flue gas
Total O_2 supplied = Total nitrogen/3.76
(Assuming the nitrogen is entirely from the air supplied)
= 80.2/3.76 = 21.35 k mol. Excess O_2 in flue = 7.1 k-mol.

$$\%\ \text{Excess air} = \frac{\text{Excess } O_2}{\text{Theoretical } O_2} \times 100 = \frac{7.1}{21.35 - 7.1} \times 100$$

$$= 49.8 \approx 50\%.$$

30. Gross C.V. of the coal by Dulong's formula

= 80.8 × 68 + 345 × 3.9 + 22.2 × 0.3 = 6842 kcal/kg

Gas yield = 3,000/1,000 = 3 m^3/kg. of fuel

$$\rightarrow \eta \text{ cold gas} = \frac{\text{Gas yield} \times \text{C.V. of the gas}}{\text{C.V. of the fuel}} \times 100$$

$$= \frac{3 \times 1,550}{6842} \times 100 = 67.9\%.$$

SECTION—D

1. **Ans :** 1 – d; 2 – b
2. **Ans :** Similar to Problem-29 Section C.
3. **Ans :** 2000 × 0.75 m = $10^8 \times 10^{-3}$ kJ/sec

   where m is the mass of coal in kg/s

   m = 667 kg/sec

   $$C + O_2 \rightarrow CO_2;\ H_2 + \frac{1}{2}O_2 \rightarrow H_2O$$

   Coal burnt = 667 × 0.67 = 500 kg/sec

   Carbon = 500 × 2/3 = 333 kg

   $$\text{Hydrogen} = \frac{500}{3} = 167 \text{ kg}$$

   $$\text{Oxygen for carbon} = 333 \times \frac{32}{12} = 888 \text{ kg}$$

   $$\text{Oxygen for hydrogen} = 167 \times \frac{16}{2} = 1336 \text{ kg}$$

   Total oxygen = 888 + 1336 = 2224 kg

   $$\text{Amount of air} = \frac{77}{23} \times 2224 \times 1.5 = 11168.3 \text{ kg/s.}$$

4. **Ans.** (e)
5. **Ans :** 1 – c; 2 – b
6. **Ans :** 1 – b; 2 – d
7. **Ans.** $CO + \frac{1}{2}O_2 \rightarrow CO_2$

   ΔH = – 393514 + 110524 = – 282, 990 kJ/k. mole

   Let CO = 1 kg. mole

   O_2 required for complete combustion = 0.5 kg. mole

O_2 consumed = 0.4 kg. mole.

Reactants are : CO = 1; O_2 = 1; N_2 = 3.76 kg. mole

Products are : CO = 0.2; O_2 = 0.6; N_2 = 3.76; CO_2 = 0.8

By Enthalpy balance of inputs = outputs

(1 × 29.22 ×(100 – 25) + 1 × 29.64 × (100 – 25) + 3.76 × 29.17 × (100 – 25) + 0.8 × 282, 990 + ΔH = 0.2 × 29.22 (300 – 25) + 29.64 × 0.6 × (300 – 25) + 376 × 20.17 × (300 – 25) + 0.8 × 43.77 × (300 – 25)

239, 032.4 + ΔH = 12,3647.7

ΔH = – 115.3847 kJ (heat to be removed).

8. Ans : 1 – c : 2 – d

SECTION—E

1. (b)

Solution : Limiting reactant is carbon (C).

Initially moles of carbon present = 6/12 = 0.5 mole

Carbon in product = 16.5/44 + 2.8/28 = 0.475 mole

Percent conversion = 0.475/0.50 ∗ 100 = 95% Ans.

2. (a)

3. (b)

4. (a)

❑

CHAPTER 4

Chemical Technology

1. Schematic Representations :

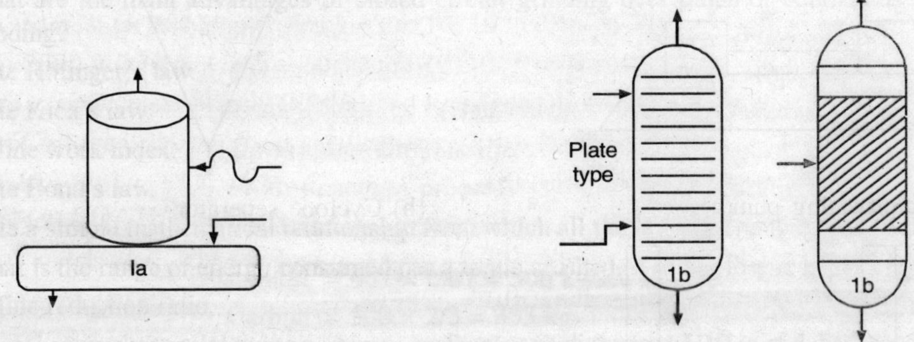

(a) Batch distillation (b) Continuous fractionator

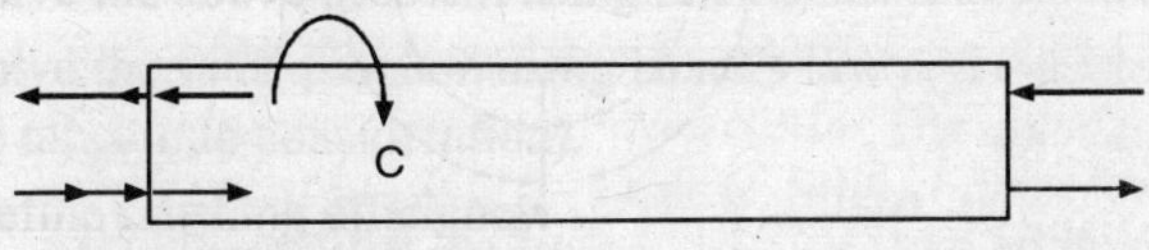

(c) Rotary dryer

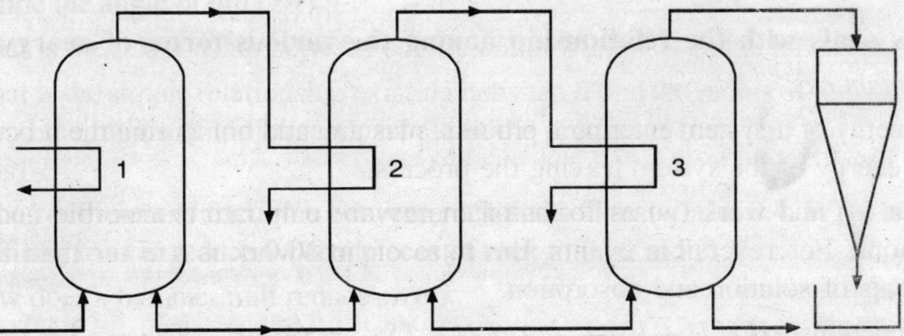

(d) Multiple effect evaporation

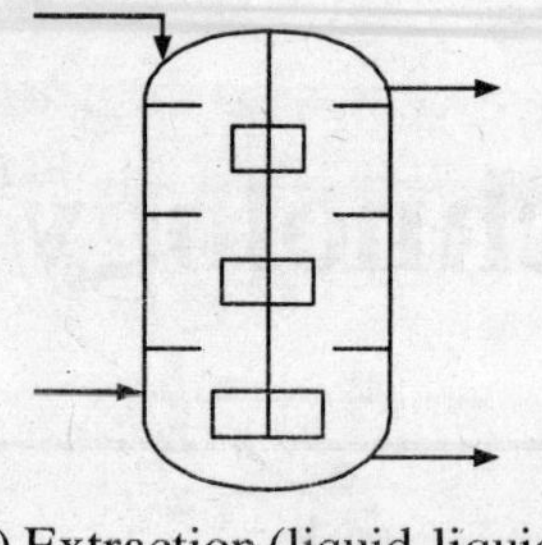
(e) Extraction (liquid-liquid)

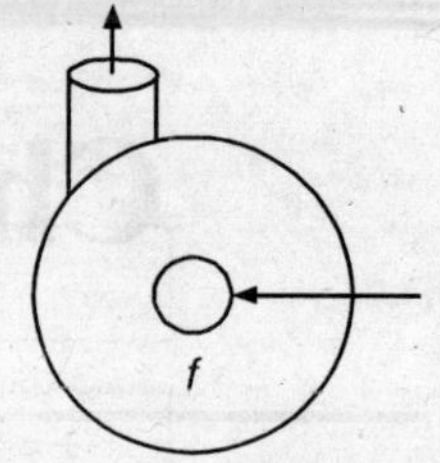

(f) Centrifugal pump

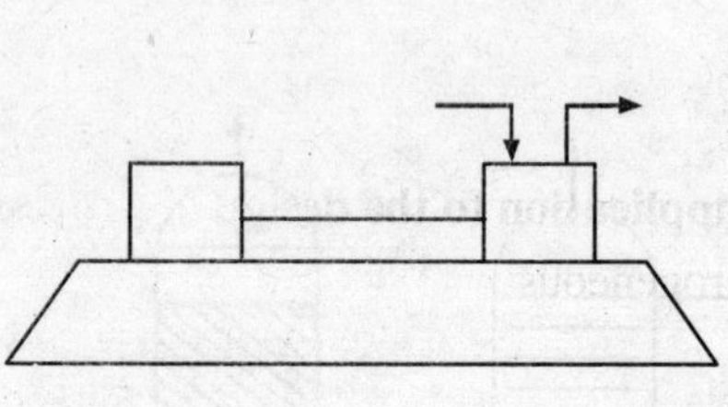
(g) Reciprocating pump

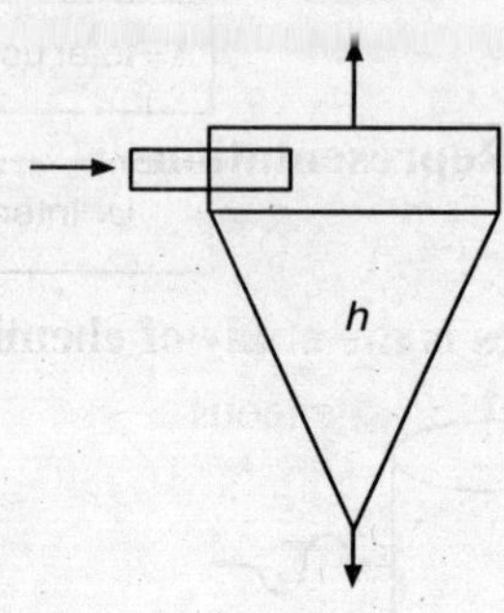

(h) Cyclone seperator

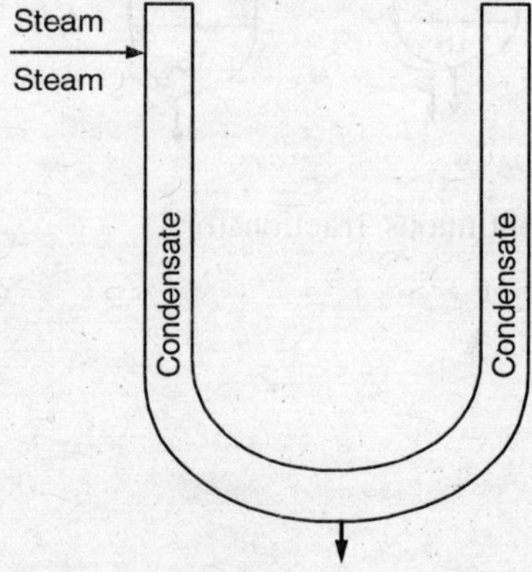

(i) Jacketed kettle

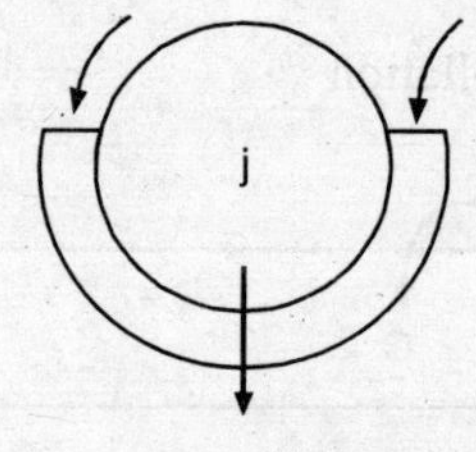

(j) Filter (rotary)

2. Thermodynamics deals with the relationship among the various forms of energy.

Laws of thermodynamics :

First law : Total energy of a system entering a process, plus any addition during the process must equal to the total energy of the system leaving the process.

Second law : Heat (q) and work (w) as forms of energy are only interchangeable under ideal reversible conditions. For reversible events, the ratio of heat/work is greater than for ideal conditions, e.g. heats of solution and absorption.

Thermodynamic functions : $H = U + PV$; $A = U - TS$; $G = H - TS$; $S = dq/T$;

H = Enthalpy; U = Internal energy;

PV = External works

Low temperature favours high conversion in an exothermic process.

High pressure favours high conversion in a reaction where there is a decrease in the number of total moles.

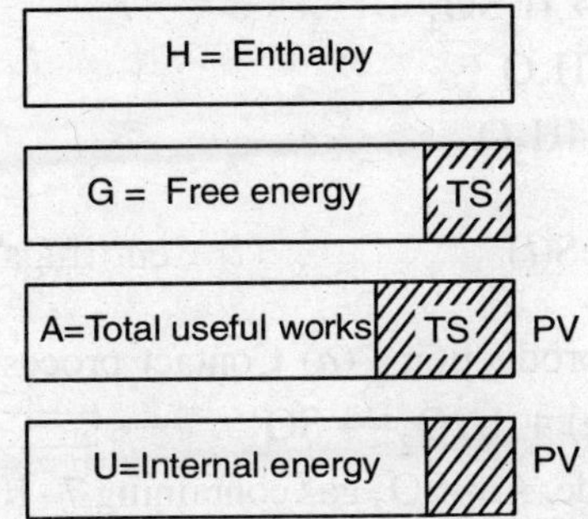

3. Reaction kinetics is the study of chemical kinetics and application to the design of processes.

Phases are : (1) Homogeneous (2) Heterogeneous

Mechanism : Molecular, chain, initiated, photochemical catalysis – catalytic, non-catalytic energy transfer – adiabatic, isothermal, intermediate.

Chemical equations : Simple reaction, parallel reaction, series reaction.

Effect of variables : Temperature, pressure, reactant concentration, flow pattern and transport properties.

Rate equations :

For $aA + bB \rightarrow R$; $-dcA/dt = r = kC_A^m\, C_B^n$

Molecularity is : $a + b$; order of reaction $= m + n$

For $A \rightarrow R$ and $A \rightarrow s$; $dcr/dt = k_1C_A{}^m$, $dcs/dt = k_2C_A^n$

$-dcA/dt = k_1C_A{}^m + k_2C_A^n$

For $A \rightarrow R \rightarrow S$, $-dcA\ dt = k_1C_a{}^m$; $dcR/dt = k_1C_A^m - k_2C_R{}^n$

Temp, effect on the rate constant (K) : $K = K_a e^{-E/RT}$

Where E = activation energy; R = gas constant; T = absolute temp; K_a = proportionality constant.

Classification of chemical reactors :

1. Simple batch homogeneous reactors
2. Semi batch reactors
3. Continuous homogeneous reactors

Design Equations :

For homogeneous systems : $Vr/F = \int_o^x dx/r = T$; where Vr is volume of reactor, F = Feed rate, r = Rate of reaction;

x = Conversion fraction, T = Holding time;

Heterogeneous systems (mostly catalytic systems) :

$W/F = \int_o^x dx/r$; W/F = kg catalyst/kg mole of feed/time;

Space velocity = F/Vr = Reciprocal of time units.

4. Inorganic Chemical Industries :

(*a*) Sulphuric acid :

Processes are : (1) Elemental sulphur (2) Hydrogen sulphide (3) Pyrites

(1) $S + O_2 \rightarrow SO_2 \rightarrow SO_3 \rightarrow H_2SO_4$

(2) $2H_2S + 3O_2 \rightarrow 2SO_2 + 2H_2O$

$4H_2S + 2SO_2 \rightarrow S_6(g) + 4H_2O$

(3) $FeS_2 \xrightarrow[1\text{ atm}]{1300°C} \frac{1}{2}\frac{S}{2} + FeS(l)$

Methods for sulphuric acid production : (*a*) Contact process ; (*b*) Chamber process

(*a*) $S + O_2 \rightarrow SO_2(g); SO_2(g) + 1/2O_2 \rightarrow SO_3$

Catalyst is vanadium pentoxide. Air–SO_2 gas containing 7–10% SO_2 and 11–14% O_2 is preheated by converter gas and sent to 1st stage reactors of steel. This is at 500–600°C. Conversion is 80% in first stage. The product is cooled at 300°C and sent to 2nd stage at 400–450°C giving 97% conversion of SO_2, SO_3 is dissolved in 98% H_2SO_4 giving Oleum.

(*b*) Fuel gases : Producer gas : $C + O_2 \rightarrow CO_2(g)$; $2C + O_2 \rightarrow 2CO$

$C + CO_2 \rightarrow 2CO$; $C + H_2O \rightarrow CO + H_2$; $CO + H_2O \rightarrow CO_2 + H_2$

Steam and air mixture is injected into the bottom of a water cooled jacketed steel furnace equipped with a rotating grate to remove fusible ash. Solid fuel is added on top of furnace.

Water gas

The correct ratio of steam, oxygen and coal is added in the reactor to yield a self sustaining reaction of zero heat releases.

C as coke or as coal : 58 kg; Air : 20 Nm^3; Steam : 80 kg for 100 cm^3 of water gas.

Coke oven gas :

From high temperature carbonization of coal.

H_2 = 60%, CH_4 = 25%, CO = 6%, CO_2 = 2%, impurities = 7%

Natural gas : 95% CH_4, C_2H_6 = 5%

Acetylene : $CaO + 3C \rightarrow CaC_2 + CO$; $CaC_2 + H_2O \rightarrow Ca(OH)_2 + C_2H_2$

5. Fertilizers : Major components are : Nitrogen phosphorous, potassium.

(*a*) Nitrogenous fertilizers : (1) Mixed fertilizers (2) Chemical nitrogen fertilizers (urea, calcium ammonium nitrate; ammonium sulphate nitrate, ammonium chloride).

Ammonia

$N_2 + 3H_2 \rightarrow 2NH_3 - 3.2$ kcal; Temp is 500–550°C.

Hydrogen is from synthesis gas, nitrogen from air, pressure is 100 to 1000 atms.

Nitric Acid : Oxidation of ammonia to NO. Oxidation of NO to NO_2. Absorption of NO_2 in water. Concentration of HNO_3. Pt-Rh alloy is catalyst, temp is 800°C.

Urea : $CO_2 + 2NH_3 \rightarrow NH_4COONH_2$; 170–190°C; 100–200 atm.

$NH_4COONH_2 \rightarrow NH_2CONH_2(\text{Urea}) + H_2O$

Side reaction : $2NH_2CONH_2CONH \rightarrow NH_2CONH_2 . H_2O$ (Biuret)

Ammonium nitrate : $NH_3 + NHO_3 \rightarrow NH_4NO_3$ –20.6 kcal

Phosphorous : $2Ca_3 (PO_4)_2 + 10C + 6SiO_2 \rightarrow P_4 + 6\ CaSiO_2 + 10\ CO$ in an electric furnace

Phosphorous acid : $Ca_3 (PO_4)_2 + 6HCL + 6H_2O \rightarrow 2H_3PO_4 + 3\ (CaSO_4, 2H_2O)$ gypsum

Tetraethyl lead : $2C_3H_5$ MgCl (Grignard reagent) + 2 C_2H_5Cl (Ethyl chloride) + Pb $\rightarrow 2MgCl_2$ + Pb $(C_2H_5)_4$.

6. **Chlor-alkali Industry :**

Soda Ash (Na_2CO_3) : Solvay process : Ammonia is dissolved in an NaCl solution and reacted with CO_2. A precipitate of $NaHCO_3$ is obtained. This is calcined to produce high purity Na_2CO_3. Chlorine, NaOH Electrolytic process : (1) Diaphram electrolytic process (2) Mercury electrolytic cell. Uses NaCl solution.

7. **Cement and Lime :** Cement is calcium aluminium silicate.

$CaCO_3 \rightarrow CaO + CO_2 + 44$ kcal

$CaO + Al_2O_3 + SiO_2 \rightarrow$ Cement

Lime : $CaCO_3(s) \rightarrow Cao(s) + CO_2$; Hydrated lime is :

$CaO + H_2O(l) \rightarrow Ca\ (OH)_2(s)$.

8. **Treatment of Water :**

Temporary hardness : Ca and Mg bicarbonates reduced by boiling to form precipitates.

Permanent hardness : Ca and Mg sulphates and chlorides removed by chemical reaction or ion-exchange.

Other impurities : Air, CO_2 and H_2S; insoluble matter, soluble color compounds, organism.

Conditioning methods : Lime soda process (hardness removal), filtration (insoluble matter), decolorization and deodorization (color removal, odor removal), chlorination (organisms). Ion-exchange (zeolite or synthetic resins for cation or anion exchange).

Fresh water from saline water :

(*a*) Evaporation or distillation
(*b*) Freezing or crystallization
(*c*) Solvent extraction
(*d*) Reverse osmosis
(*e*) Ion-exchange
(*f*) Electrodialysis.

9. **Oils, Soaps, Detergents :**

Fats and oils—mixtures of glycerides of fatty acids. Oils are all liquids and fats are all solids at normal temperature.

Waxes : Mixed esters of polyhydric alcohols other than glycerine. Solids at room temperature.

Extraction of oils : (1) Mechanical (2) Solvent extraction.

Solvents are : Petroleum cuts in hexane range, trichloroethylene.

Hydrogenation : Used in fat and oil to remove double bonds, raise melting point of fat and improve resistance to oxidation (Vanaspati).

$Oil + H_2 \xrightarrow{\text{Ni catalyst}} $ Hydrogenated oil/fat.

Soaps : Decrease surface tension and increase wettability.

Soap Production : Fat + $3H_2O \rightarrow$ Fatty acid + glycerine

Fatty acid + Na/K hydroxide $\rightarrow$ soap + water

Detergent : High molecular wt. Alcohol + $H_2SO_4 \rightarrow$ detergent

10. Fermentation : $C_{12}H_{22}O_{11} + H_2O$ invertase $\rightarrow 2C_6H_{12}O_6$

$C_6H_{12}O_6$ (starch)/Molasses zymase $\rightarrow 2\ C_2H_5O_2 + 2CO_2$

11. Chemicals Manufacture :

(*a*) **Ethylene oxide :**

$$CH_2 = CH_2 + \frac{1}{2}O_2 \xrightarrow{Ag_2O} C_2H_4O_2 - 29.2 \text{ kcal}$$

(*b*) **Vinyl acetates :**

$C_2H_2 + CH_3COOH$—4 to 5 atm, 250–300°C $\rightarrow CH_3COOH = CH_2 + 1/2O_2 \rightarrow CH_3COOCH = CH_2 + 1/2\ O_2 \rightarrow CH_3COOCH = CH_2 + H_2O$

(*c*) **Acetone :** $(CH_3)_2CHOH$ (Isopropanol) $\rightarrow CH_3COCH_3 + H_2$ $(CH_3)_2CHOH + 1/2O_2 \rightarrow CH_3COCH_3 + H_2O$

(*d*) **Phenol :** (1) Cumene + Air $\rightarrow$ Cumene hydroperoxide

Cumene hydroperoxide + Aqu. $H_2SO_4 \rightarrow$ [benzene ring–OH] (phenol) + Acetone

(*e*) Toluene + $11/2O_2 \rightarrow$ 150°C, Cobalt naphthenate $\rightarrow$ Benzoic acid + H_2O

[benzene ring–COOH]

(Benzoic acid)

$C_6H_5COOH + 1/2\ O_2 \rightarrow 220°C \rightarrow C_6H_5OH + CO_2 + H_2O.$

Cupric benzoate (Catalyst)

(*f*) Benzene + Cl_2 $\xrightarrow[\text{Fe/FeCl}_3 \text{ Catalyst}]{85°C}$ [benzene ring–Cl] Chlorobenzene + HCl

[benzene ring–Cl] + NaOH(aqu) $\longrightarrow$ [benzene ring–ONa] + HCl

[benzene ring–ONa] + HCl (aqu) $\longrightarrow$ [benzene ring–OH] + NaCl (aqu)

(g) Styrene :

$$\text{C}_6\text{H}_6\ (\text{Benzene}) + CH_2 = CH_2 \xrightarrow[C_2H_5Cl]{AlCl_3} C_6H_5CH_2CH_3\ (\text{Ethyl benzene})$$

$$C_6H_5CH_2CH_3 \xrightarrow[\substack{SnO\\FeO\\1500°F}]{} C_6H_5CH = CH_2\ (\text{Styrene})$$

(h) Phthalic Anhydride :

$$\text{Napthalene} + 4\tfrac{1}{2}\ O_2\ (\text{Air}) \longrightarrow \underset{400°C}{\overset{\text{Catalyst}}{V_2O_5}} \longrightarrow$$

$$C_6H_4(CO)_2O + 2CO_2 + 2H_2O$$

(i) Maleic Anhydride :

$$C_6H_6 + 4\frac{1}{2}O_2 \rightarrow V_2O_5 \rightarrow C_4H_2O_2 + 2H_2O + 2CO_2$$

$$C_4H_8\ (\text{Butene}) + 3O_2\ \text{Catalyst} \rightarrow C_4H_2O_3 + 3H_2O$$

(j) Nitrobenzene : $C_6H_6 + HNO_3\ (H_2SO_4) \rightarrow C_6H_5\,NO_2 + H_2O\ (H_2SO_4)$

(k) DDT-manufacture : Chloral $(3Cl_3CHO) + 2C_6H_5Cl \rightarrow (C_6H_4Cl)_2CHCl_3 + H_2O$

(l) BHC manufacture : $C_6H_6 + 3Cl_2 \rightarrow C_6H_6Cl_6$

12. Polymers :

Physical structure of polymers : (*a*) Linear polymers; (*b*) Cross-linked polymers; (*c*) Branched chain polymeric chemical reactions—(1) Polycondensation; (2) Addition reaction

(*d*) Modes of polymerization :

(1) Condensation polymerization methods (homogeneous polymerization, heterogeneous, bulk, solution, emulsion, suspension polymerization).

(2) Thermosetting (not melted, solvent insoluble)

(3) Elastomers (resistant solids with high flexural strength)

(4) Fibres (thread forming solids with high tensile strength)

(*e*) Polyethylene :

Ethylene $\rightarrow$ Al. Tetraethyl catalyst, 6 to 10 atm with $TiCl_4$ polyethylene.

(*f*) Polyvinyl chloride : 70°C, $nCH_2 = CH.\ Cl \rightarrow$ PVC. per sulphate is the catalyst.

(*g*) Phenol formaldehyde : Phenol + $XCH_2O \rightarrow$ Phenol formaldehyde acid solution catalyst, or alkaline catalyst temp is 130°C – 150°C

(*h*) Rayon : $(C_6H_9O_4.\ OH)x + 2NaOH \rightarrow (C_6H_9O4.\ ONa)x + xH_2O$

$(C_6H_9O_4.OH)x + xCS_2 \rightarrow (C_6H_9O_4\text{–}O\text{–}C_5SNa)x$

$(C_6H_9O_4\text{–}O\text{–}C_5SNa)x + xH_2SO_4 \rightarrow (C_6H_9O_4.\ OH)x + XCS_2 + xNaH_2SO_4$

(*i*) 6, 6-Nylon : Adipic Acid + Hexamethylene diamine → Nylon

(*j*) 6-Nylon : Caprolactrum + diamine → Nylon

(*k*) Dacron : Dimethyl terephthalate + Ethylene glycol (alkali catalyst) → Dacron + CH_3OH

13. Iron and Steel : $Fe_2O_3 + 3C \rightarrow 2Fe + 3CO$

14. Metallurgical Industries :

Aluminium : $6\ NaOH(Aqu) + Al_2O_3(s) \rightarrow 2Na_2AlO_3 + H_2O$

$Na_2AlO_3 + 3H_2O \rightarrow 3NaOH + Al\,(OH)_3$

$2Al(OH)_3 \rightarrow Al_2O_3 + 3H_2O$

$2Al_2O_3 + 3C \rightarrow 4Al + 3CO_2$

QUESTIONS

SECTION—A

1. Example of unit process :

(a) Alkylation (b) Drying

(c) Extraction (d) None

2. Example of unit operation :

(a) Combustion (b) Calcination

(c) Oxidation (d) Drying

3. Relative cost of complete process plants in India, taking : U.S.A, as 1.0 :

(a) 0.80 (b) 1.35

(c) 0.90 (d) 2.0

4. Percentage of sulphur average in pyrites :

(a) 10 (b) 20

(c) 35 (d) 45

5. Catalyst used in contact process for manufacture of sulphuric acid :

(a) Vanadium pentoxide (b) Iron oxide

(c) No catalyst (d) Platinum.

6. Calorific value of producer gas in kcal/cu. meter :

(a) 6000–14000 (b) 2500–2700

(c) 4500–8000 (d) 1200–1600

7. All synthesis gases contain :

(a) Nitrogen (b) Oxygen

(c) Carbon monoxide (d) Hydrogen

8. Calcium ammonium nitrate (CAN) fertilizer is produced by :

(a) Hindustan Steel Ltd., Rourkela (b) Hindustan Steel Limited, Bhilai

(c) Fertilizer Corp. of India Ltd. Sindri

9. Catalyst used in ammonia production :
(a) Platinum (b) Iron
(c) Iron oxide

10. Temperature used in ammonia reactor :
(a) 200–250°C (b) 500–550°C
(c) 300–350°C

11. Catalyst used in nitric acid production :
(a) Platinum (b) Rhodium
(c) Platinum-rhodium alloy

12. Pressure and temperature used in urea manufacture :
(a) 100–200 atm and 170–190°C (b) 200–300 atms and 100–150°C
(c) 100–200 atm and 400–500°C

13. Purpose of coating ammonium nitrate :
(a) To provide extra nutrient as fertilizer (b) To avoid explosion
(c) To reduce cost of production

14. Superphosphate is made by reacting phosphate rock with :
(a) Sulphuric acid (b) Phosphoric
(c) Nitric acid

15. The name of the process used for soda ash manufacture :
(a) Chamber process (b) Solvay process
(c) Contact process

16. The type of tower used in producing sodium bicarbonate in soda ash manufacture :
(a) Baffle tower (b) Plate tower
(c) Packed tower

17. Dual process of soda ash gives the following fertilizer :
(a) Ammonium sulphate (b) Ammonium nitrate
(c) Ammonium chloride

18. The material of construction used for ammonium chloride solution :
(a) Durmet-20 or rubber lined (b) Carbon steel
(c) Stainless steel

19. The material of construction for brine solutions :
(a) Rubber lined (b) Glass lined
(c) Nickel lined

20. Monazite sand contained uranium percent as :
(a) 1% (b) 0.5%
(c) 2–3% (d) 0.25%

21. Catalyst used in hydrogenation of oils :
(a) Copper (b) Silver
(c) Nickel (d) Platinum

22. Titanium dioxide is a pigment of colour :
(a) Blue (b) Black
(c) Yellow (d) White

23. Percentage of sugar in sugarcane is (on cane basis) :
(a) 9.5 to 15% (b) 15 to 20%
(c) 5 to 9.5%

24. One ton of pulp (dry basis) requires bamboo or wood in tons :
(a) 2.2 to 2.5 (b) 3 to 4
(c) 4 to 5

25. Natural gas contains mostly :
(a) Ethane (b) Ethylene
(c) Methane (d) Hydrogen

26. Amount of crude oil, India is importing from other countries in percentage :
(a) 70% (b) 50%
(c) 30%

27. Liquefied petroleum gas (LPG) contains mostly :
(a) Methane and ethane (b) Propane and butane
(c) Ethylene and hydrogen

28. Phthalic anhydride is manufactured using :
(a) Benzene (b) Naphthalene
(c) Toluene

29. Catalyst used in the manufacture of phthalic anhydride :
(a) Platinum (b) Iron oxide
(c) Vanadium pentoxide

30. An example of polycondensation reaction :
(a) Polythene (b) Phenol formaldehyde
(c) Polyvinyl chloride

31. One SBR (synthetic rubber) plant in India is :
(a) Delhi (b) Bareilly
(c) Kolkata

32. Butadiene is manufactured in India from :
(a) Benzene (b) Naphthalene
(c) Ethyl alcohol

33. Producer gas contains mostly :
(a) H_2, CO_2 (b) C_2H_2, CO_2, H_2
(c) CO, H_2, O_2 (d) CO, CO_2, N_2, H_2

34. Contact process :
(a) Obsolete or outdated

(b) Yields acid of lower concentration than chamber process
(c) Yields acid of higher concentration than chamber process

35. Mercury cell in caustic soda manufacture :
(a) Gives more concentrated NaOH than diaphragm cell
(b) Gives less concentrated NaOH than diaphragm cell

36. Gypsum's chemical name is :
(a) Calcium sulphate (b) Sodium sulphate
(c) Potassium sulphate

37. Oil is :
(a) Ester of alcohols other than glycerine
(b) A mixture of glycerides
(c) A mixture of glycerides of fatty acids

38. Wax is :
(a) A mixture of glycerides
(b) A mixture of glycerides of fatty acids
(c) A mixture of esters of polyhydric alcohols except glycerine

39. Solvent extracted oil :
(a) Is odourless (b) Has low free fatty acid content
(c) None of these

40. Permanent hardness of water is due to the presence of calcium and magnesium :
(a) Sulphates and chlorides (b) Bicarbonantes
(c) Carbonates

41. Soaps remove dirt by :
(a) Decreasing wettability (b) Supplying hydrophilic group
(c) Increasing the surface tension (d) None of these

42. Which is a detergent in the following ?
(a) Fatty acids (b) Fatty alcohols
(c) Alkyl benzene suphonates (d) Fatty esters

43. Bleaching of paper pulp is done by :
(a) Bromine (b) Activated clay
(c) Chlorine (d) Magnesium sulphate

44. Viscose rayon is a :
(a) Regenerated cellulose acetate (b) Cellulose nitrate
(c) Regenerated cellulose nitrate

45. High temperature carbonisation produces more :
(a) Coke (b) Tar
(c) Phenol (d) Ammonia

46. The chemical name of DDT is :
(a) Diethyl diphenyl trichloromethane
(b) Dichloro diphenyl trichloroethane
(c) Dichloro diphenyl trichloromethane

47. Molecular weights of polymers are in the range of :
(a) 10^3 to 10^7 (b) 10 to1000
(c) 10^5 to 10^9

48. Commercial name of polytetrafluoroethylene is :
(a) Dacron (b) Nylons
(c) Teflon (d) Bakelite

49. Glycerine comes as a byproduct in :
(a) Oil industry (b) Soap industry
(c) Sugar industry (d) Alcohol industry

50. Cellulose content of bamboo used for the manufacture of paper is :
(a) 5% (b) 10%
(c) 80%

51. In India most of the ethyl alcohol is manufactured from :
(a) Molasses (b) Rice
(c) Potatoes (d) Wheat

52. Polythene is a :
(a) Addition polymerisation product
(b) Condensation polymerisation product
(c) None

53. An example of elastomers :
(a) Polythene (b) SBR rubber
(c) Polyvinyl chloride (d) Rayon

54. Which of the following are a byproduct in the manufacture of sodium hydroxide using brine :
(a) Sodium bicarbonate (b) Chlorine
(c) Sodium carbonate (d) Graphite

55. The chemical formula of washing soda is :
(a) $NaHCO_3$ (b) $Na_2CO_3H_2O$
(c) $Na_2CO_310H_2O$

56. Bromine content in sea water in parts-per million (ppm) is around :
(a) 150 (b) 70
(c) 1000 (d) 2000

57. Carborundum consists mainly :
(a) Silicon carbide (b) Calcium carbide
(c) Boron carbide

58. Alum is manufactured from :

(a) Bauxite (b) Gypsum

(c) Galena

59. Salt cake's chemical formula is :

(a) $CaSO_4H_2O$ (b) Na_2SO_4

(c) $MgSO_4$

60. Margarine is a:

(a) Hydrogenated fat (b) Refined oil

(c) Maize oil

61. Leblanc process was used in the olden days to make :

(a) Caustic soda (b) Soda ash

(c) Bleaching powder

62. The suitable material for evaporator to concentrate sodium hydroxide :

(a) Nickel (b) Soda ash

(c) Bleaching powder

63. Dry chlorine is stored in a vessel made of :

(a) PVC (b) Nickel

(c) Iron or steel

64. Stainless steel contains :

(a) Copper (b) Aluminium

(c) Chromium and nickel

65. Urea autoclave is made of :

(a) Lead lined steel (b) Cast iron

(c) Stainless steel

66. Hydrochloric acid is stored in vessels of :

(a) Rubber lined steel vessel (b) Stainless steel

(c) Lead lined steel vessel

67. Aqueous nitric acid is stored in :

(a) Cast iron (b) Steel

(c) Stainless steel

68. Hydrochloric acid absorber is made of :

(a) Stainless steel (b) Mild steel

(c) Carbate

69. Packing material used in SO_3 absorber in contact process :

(a) Carbate (b) Chemical stoneware

(c) Mild steel

70. Presence of nickel in steel improves :

(a) Hardness (b) Wear resistance

(c) Corrosion resistance

71. Temporary hardness of water is reduced by :

(a) Adding chlorine (b) Filtering

(c) Boiling

72. Composition of sodium chloride in sea water in g/litre :

(a) 10 (b) 27

(c) 1 (d) 5

73. Concentration of sodium hydroxide produced in diaphragm cell in weight % is :

(a) 50% (b) 10%

(c) 20%

74. Vanillin is a kind of :

(a) Flavour (b) Insecticide

(c) Medicine

75. Main constituent of cotton seed oil is :

(a) Myristic acid (b) Palmitic acid

(c) Stearic acid (d) Oleic acid

(e) Linoleic acid

76. Percentage of oil in fish is :

(a) 10% (b) 20%

(c) 30% (d) 40%

77. Amount of bagasse available in cane sugar factory is (after consumption by the factory for heating etc.)

(a) Equal to the amount of sugar (b) Twice the amount of sugar

(c) Thrice the amount of sugar

78. Percentage of alcohol in bear is :

(a) 2 to 7% (b) 7 to 10%

(c) 10 to 15%

79. ($NaBr + Cl_2$) is a :

(a) Fire retardant (b) Insecticide

(c) Catalyst

80. Indigo dye gives colour :

(a) Black colour (b) Blue

(c) Green

81. Urea is :

(a) Potassic fertilizer (b) Nitrogeneous fertilizer

(c) Phosphatic fertilizer (d) Nutrient

82. Nitrogeneous fertilizer is graded based in its :

(a) NO_2 content (b) N_2 content

(c) HNO_3 content (d) N_2O_4 content

83. Potassic fertilizer is graded blased on its :

(a) K_2O content (b) K_2SO_4 content
(c) KNO_3 content (d) KCl content

84. Rock phosphate contains mainly :

(a) Dicalcium phosphate (b) Mono-calcium phosphate
(c) A mixture of metallic phosphates (d) Fluorapatite

85. Promoter in ammonia synthesis catalyst is :

(a) K_2O (b) V_2O_5
(c) U_2O_5 (d) SiO_2

86. 10-10-5 fertilizers mean that they contain :

(a) N_2, P_2O_5 and K_2O percentages are 10, 10, 5 respectively
(b) K_2O, P_2O_5 and N_2 percentages are 10, 10, 5 respectively
(c) P_2O_5, N_2 and K_2O percentages are 10, 10, 5 respectively

87. Fertilizer plants in Talcher (Odisha) and Ramagundam (Andhara) use :

(a) Coal for hydrogen (coal gasification) (b) Naphtha
(c) Coke oven gas (d) Electrolysis of water

88. NPK fertilizer is a :

(a) Complex fertilizer (b) Nitrogeneous fertilizer
(c) Mixed fertilizer (d) Trade name

89. The suitable fertilizer for paddy is :

(a) Potassium nitrate (b) Superphosphate
(c) Amonium sulphate (d) Urea

90. The fertilizer plant obtaining hydrogen from electrolysis of water is situated at :

(a) Sindri (b) Talcher
(c) Korba (d) Nangal

91. For cathodic protection of the buried steel pipelines, the following anodes are used :

(a) Magnesium and zinc anodes (b) Copper anode
(c) Brass anode (d) Nickel anode

92. Inhibitors used for minimizing corrosion of iron and steel in aqueous solutions :

(a) Chromates (b) Sulphates
(c) Chlorides (d) Carbonates

93. Cheapest material of construction used to store upto 75% NaOH :

(a) Copper (b) Carbon steel
(c) Aluminium (d) Stainless steel

94. Composition of chlorimet-2 alloy is roughly :

(a) 80% Ni, 13% Cr, 7% Fe (b) 63% Ni, 32% Mo, 3% Fe
(c) 54% Ni, 16% Mo, 5.5% Fe

95. Inconel composition is :

(a) 80% Ni, 13% Cr, 7% Fe (b) 91% Ni, 28% Mo, 5.5% Fe

(c) 60% Ni, 40% Fe

96. Monel alloy composition :

(a) 50% Ni, 40% Cr (b) 67% Ni, 30% Cu

(c) 30% Al, 65% Cu

97. Glass lined vessels are used for :

(a) Acids except hydrofluoric (b) Alkalies

(c) None

98. Fertilizers produced in Rourkela Steel Plant are :

(a) Phosphatic (b) Nitrogeneous

(c) Potassic (d) Mixed

99. Trade name of Fertilizer produced by Coramandal Fertilizers in Vizag is :

(a) Suphala (b) Sona

(c) Gromore

100. Location of one polystyrene plant :

(a) Rourkela (b) Chennai

(c) Hyderabad (d) Visakhapatnam

SECTION—B

1. Oleum contains more than 100 percent.....

2. The two processes used for production of sulphuric acid are.....process and.....process.

3. Acetylene is manufactured using.....and water as reactants.

4. Cryogenics deals with temperatures in the region from.....to.....

5. The three major components necessary in fertilizer are.....and.....

6. The two electroytic cell methods used for sodium hydroxide are.....cell and.....cell.

7. Portland cement is defined as finely ground.....and.....of varying composition.

8. To separate fresh water from sea water the operation used is.....

9. The nuclear energy equation based in U^{235} is.....

10. By terminology of industry oils are all.....and fats are.....at normal temperatures.

11. Antiknock compounds are added to petrol to.....

12. The raw materials for manufacture of vinyl chloride are.....

13. Phthalic anhydride is manufactured by the.....and.....

14. Maleic anhydride is manufactured using.....or.....as raw materials.

15. Massecuites is a terminology used in.....industry.

16. 9,6-Nylon is manufactured from.....and.....

17. One example of polyesters is.....

18. The iron ore mostly used in India to manufacture pig iron is.....

19. The raw materials for pig iron manufacture .....

20. The predominant ore of copper in India is.....

21. The pollution hazard in the manufacture of sodium hydroxide using mercury cell is.....

22. The catalyst used in fat splitting is.....

23. Plasticisers are added to the paints to.....

24. Varnish does not contain.....

25. Molasses is the starting material for the production of.....

26. Low temperature carbonization takes place at.....

27. In.....fertilizer, the nitrogen content is highest.

28. The amounts of nitrogen, phosphorous and potassium in fertilizers are expressed as ....,.....and.....

29. Amount of nitrogen in ammonium sulphate is.....%

30. Single superphosphate is produced by mixing ground.....rock with ....,

31. Triple superphosphate is manufactured by reacting finely ground..... rock with concentrated.....

32. Potassium schoenite is a double sulphate of.....and.....

33. Amount of nitrogen in urea is.....

34. Capacity of Ramagundam and Talcher fertilizer factories is.....thousand tons of urea per year.

35. Price of ammonium sulphate is around.....Rs/kg.

36. The trade name of sulphate process for manufacture of paper pulp is.....

37. Enzymes are organic catalysts used in.....reactions.

38. The two methods of polymerisation are (1) ..... and (2) .....

39. The raw materials required for sulphadiazine manufacture are.....

40. The drug used in contraceptives is.....

41. Most of the metals are corroded due to.....

42. Removal of a constitutent of an alloy by corrosion without apparent loss in volume but loss of strength is called as.....corrosion.

43. Fretting corrosion means.....

44. The two methods used for providing cathodic protection of metals to minimise corrosion are (1).....and (2).....

45. Wrought iron is used as material of construction for ..... and .....

46. Esterification means reaction of ..... and .....

47. This symbol

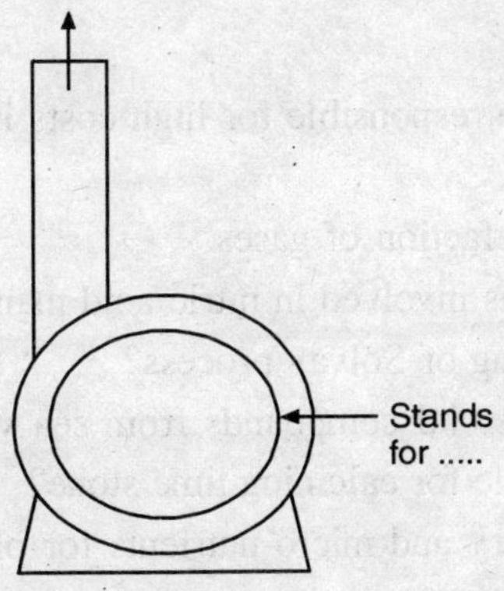

48. This symbol

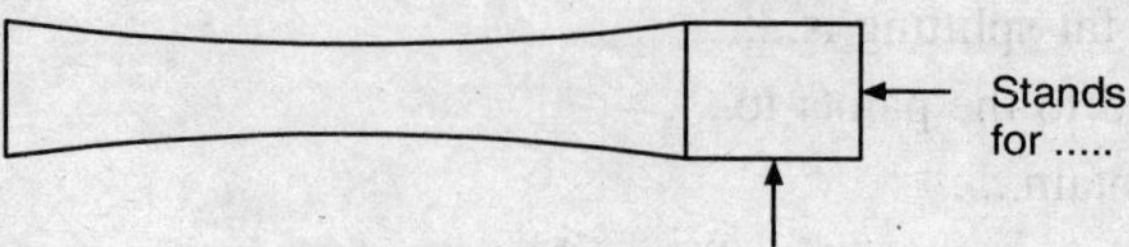

49. The fertilizer produced in Neyveli Lignite Corporation Ltd. is .....

50. The fertilizer used as cattle feed is.....

51. Minimum capacity of a cement plant is.....

52. Composition of uranium in pitchblend ore is.....

53. Composition of uranium in carnotite ore available in Jaduguda, Bihar is.....

54. Nuclear materials composition in monazite sand are.....

55. Location of one heavy water plant in India is .....

56. Hydrazine is mostly used as.....

57. Phthalic anhydride is used for.....and.....

58. Polythene is formed by.....polymerisation.

59. Phenol formaldehyde is formed by.....polymerisation.

60. In low pressure polyethylene manufacturing process, the temperature used is....., pressure is.....and catalyst is .....

61. The catalyst used in PVC manufacture is.....and temperature maintained is.....

62. .....and.....polymerisations are used in PVC manufacture.

63. Compounding of rubber means ....., ..., ..., ...and.....

64. Trade name of polychloroprene is.....

65. The biggest oil refinery in India is located at.....

66. Cold SBR is made at ..... C and ..... atms gauge pressure and hot SBR is made at ..... °C and ..... atms. gauge pressure.

67. The raw material for SBR Rubber in India is.....and in other countries.....

68. Thermoplastic meant that it.....

69. Thermosetting means that it.....

70. Stereospecific agents are.....

SECTION—C

1. What are the factors which are responsible for high costs in the Indian Chemical Industries?

2. List the pyrite ores.

3. List the methods used for liquefaction of gases.

4. What are the chemical reactions involved in nitric acid manufacture?

5. What is the principle of working of Solvay process?

6. What are the important commercial compounds from sea water?

7. What are the processes available for calcining lime stone?

8. What are the primary, secondary and micro-nutrients for plant growth?

9. List six nitrogeneous fertilizers.
10. What are the main raw materials for ammonia manufacture?
11. What are the methods of manufacture of ammonium sulphate?
12. List four industries in India which produce NPK fertilizers.
13. List compound and complex fertilizers.
14. List the chemical reactions taking place in Solvay process for the manufacture of soda ash.
15. List the various chemical reactions taking place in lead chamber process for the manufacture of sulphuric acid.
16. What are the constituents of lacquers and their uses?
17. What are the various constituents present in dry detergents and their purposes?
18. What are the products from crude petroleum after distillation?
19. List ten synthetic rubbers.
20. List the sequences in the manufacture of sulphuric acid by contact process.
21. How N, P and K from fertilizers help for plant growth?
22. Name four big manufacturing concerns of ammonium sulphate in our country.
23. What are the requirements of raw materials for one ton of cement?
24. What are the air pollutants?
25. Name eight food additives.
26. List four manufactures of polyethylene in our country.
27. List five manufacturers of polyvinyl chloride in our country.
28. Why multistage catalytic reactor is employed in converting SO_2 to SO_3?
29. Why single stage reactor is used in ammonia reaction instead of multistage reactor?
30. What is meant by hydrogenation in oil and fat industry?
31. What are the material requirements for hydrogenating 1 ton of cotton seed oil?
32. Name five manufacturers of Nylon in our country.

SECTION—D

1. Match :

(1) Naphtha	(a) Solvay process
(2) Sodium hydroxide	(b) Steam reforming
	(c) Mercury cell
	(d) DCDA process

2. A suitable material of construction to use with fuming sulphuric acid is :

(a) Carbon steel	(b) Stainless steel type-304
(c) Nickel	(d) Monel

3. Match :

(1) Global warming	(a) Unleaded gasoline
(2) Methyl tertiary butyl ether	(b) Nitrous oxides

(c) Chlorofluorocarbons

(d) Leaded gasoline

4. Match :

Na_2CO_3

(a) Haber process

(b) Solvay process

(c) Fischer-Tropsch process

(d) Castner-Kellner process

5. Match :

Prilling tower

(a) Cement

(b) Potassium chloride

(c) Urea

(d) Triple superphosphate

6. Match :

Extractive distillation

(a) Nitric acid

(b) Caustic soda

(c) Sulphuric acid

(d) Urea

7. Match :

Linear alkyl benzene sulphonate

(a) Detergents

(b) Perfumes

(c) Pharmaceuticals

(d) Refrigerants

8. Match :

V_2O_5

(a) Ammonia

(b) Sulphuric acid

(c) Polypropylene

(d) Refrigerants

9. Styrene-butadiene rubber is commercially manufactured by :

(a) Bulk polymerisation
(b) Suspension polymerisation
(c) Solution polymerisation
(d) Emulsion polymerisation

10. What are the catalysts employed in the following commercial reactor operations?

(a) Oxidation of SO_2
(b) Oxidation of ammonia
(c) Oxidation of C_2H_4
(d) Hydrogenation of vegetable oil
(e) Dehydrogenation of propanol

11. A biodegradable detergent is one which :

(a) Is manufactured using bio-technology
(b) Contains straight chain alkyl benzene
(c) Contains branch chain alkyl benzene
(d) Is easily decomposed by microorganisms

12. The most widely used coagulant for removing suspended impurities from water is :

(a) Bleaching powder (b) Chlorine

(c) Calcium sulphate (d) Alum

13. Explain in not more than 4 lines.

(a) In manufacture of sulphuric acid why is absorption of SO_3 done in concentrated sulphuric acid and not in water.

(b) A dilute solution of 20% nitric acid has to be concentrated to 80%. How is this done in Industry.

(c) A dilute solution of glycerine contains a soluble calcium salt as impurity. Why is the removal of calcium essential before concentration of this solution? How can this removal can be accomplished.

14. Hydrogenation of edible oils is done to :

(a) Decrease the number of unsaturated bonds

(b) Lower the melting point of oil

(c) Increase the thermal conductivity of oil

(d) Low pressure, low temperature

15. The preferred material of construction for storage tank for 98% sulphuric acid is :

(a) Aluminium (b) Lead

(c) Stainless steel (d) Mild steel

16. Match :

1. Titanium dioxide	(a) Detergents
2. Purified terephthalic acid	(b) Soaps
	(c) Paints and pigments
	(d) Synthetic fibres
	(e) Nuclear fuels
	(f) Synthetic rubber

17. Match :

1. Glycerine	(a) Rocket propellant
2. Ammonium chloride	(b) Soda ash
	(c) Battery grade MnO_2
	(d) Antibiotics
	(e) Soap
	(f) Sugar

18. A sample of well water contains 140 g/m^3 Ca^{2+} ions and 345 g/m^3 of Na^+ ions. The hardness of the sample of water, expressed in terms of equivalent $CaCO_3$ in g/m^3 (assuming atomic masses of Ca : 40 ; Na : 23 ; C : 12 ; and O : 16) is :

(a) 350 (b) 485

(c) 140 (d) 345

SECTION—E

1. Nylon-66 is so named because :

(a) The average degree of polymerization of the polymer is 1966

(b) The number of carbon atoms between two nitrogen atoms are 6

(c) The number of nitrogen atoms between two carbon atoms are 6

(d) The polymer was first synthesized in 1966

2. The catalytic converter for conversion of SO_2 to SO_3 by Contact process should have a feed with SO_2 content between :

(a) 2–5% (b) 7–10%

(c) 12–15% (d) 20–25%

3. The composition of fresh feed to the high temperature, high pressure, urea autoclave is :

(a) Excess liquid ammonia and liquefied CO_2

(b) Excess liquid ammonia and compressed CO_2

(c) Liquid ammonia and Excess compressed CO_2

(d) Compressed NH_3 gas and excess compressed CO_2

4. Identify the group in which all the polymers mentioned can be used to make fibres :

(a) Butadiene Copolymers, polyamides, urea aldehydes

(b) Cellulose derivatives, polyisoprene, polyethylene

(c) Cellulose derivatives, polyamides, polyurethenes

(d) Polypropylenes, polyvinyl chloride, silicons

5. The preferred Reacting System for oxidation of o-xylene to phthalic anhydride is :

(a) Jacketed liquid phase CSTR

(b) Jacketed steam heated multi tubular reactor

(c) Multi tubular reactor with cooling

(d) Multistage multi tubular reactor with inter stage cooling

6. Match :

Group 1	Group 2
P. Calcium Ammonium Nitrate	1. Fertilizer Industry
Q. $CaCl_2$- NaCl Liquor	2. Paper and Pulp Industry
	3. Soda Ash Industry

(a) P-1, Q-3 (b) P-1, Q-2

(c) P-3, Q-1 (d) P-2, Q-3

7.

Group-1	Group-2
P. Black Liquor	1. Petroleum Refining
Q. Activated Silica Alumina	2. Sugar Factory
R. Press Mud	

(a) P-1, R-2 (b) P-2, Q-1

(c) Q-1, R-2 (d) Q-2, R-1

8. In a Refinery, petroleum crude is fractioned into gas fraction, light ends, intermediate distillates, heavy distillates, residues and byproducts. The group of products including gas oil, diesel oil and heavy fuel oil belongs to the fraction.

(a) Heavy distillates (b) Intermediate distillates
(c) Light ends (d) Residues

9. The order of preference for feed stock to a catalytic reformer is :

(a) Catalytic Naphtha—Coking Naphtha—Virgin Naphtha
(b) Coking Naphtha—Virgin Naphtha/Catalytic Naphtha
(c) Virgin Naphtha—Catalytic Naphtha—Coking Naphtha
(d) Virgin Naphtha—Coking Naphtha—Catalytic Naphtha

10. Prilling tower is found in the flow sheet in the manufacturing of :

(a) Ammonia (b) Urea
(c) Super phosphate (d) Triple superphosphate

11. Multiple effect evaporators are used in the manufacture of :

P. Paper (a) P and Q
Q. Super phosphate (b) P and R
R. Sugar (c) P and S
S. Fats (d) R and S

12. Match the process in Group-I with the product in Group-II.

Group-I	Group-II
P. DCDA process	1. Sodium hydroxide
Q. Mercury cell	2. Sulphuric acid
	3. Sodium carbonate
	4. Nitric acid

(a) P-1, Q-4 (b) P-2, Q-3
(c) P-2, Q-1 (d) P-2, Q-1

13. Match the product in Group-I with the raw materials in Group-II

Group-I	Group-II
P. Urea	1. Ammonia and carbon dioxide
Q. Polyester	2. Dimethyl terephthalate and ethylene glycol
	3. Ammonia and carbon monoxide
	4. Hexamethylene diamine and adipic acid.

(a) P-1, Q-4 (b) P-3, Q-2
(c) P-3, Q-2 (d) P-1, Q-2

14. Match the product in Group-I with the name of the reaction in Group-II

Group-I	Group-II
P. Polyethylene	1. Condensation polymerization
Q. Nylon	2. Addition polymerization

R. Polystyreme

(a) P-1, Q-1, R-2
(b) P-2, Q-2, R-1
(c) P-1, Q-2, R-1
(d) P-2, Q-1, R-2

15. Match the process in Group-I with the catalyst used in Group-II

Group-I	Group-II
P. Sulphuric acid manufacture	1. Platinum
Q. Vegetable oil hydrogenation	2. Vanaium pentaoxide
	3. Iron
	4. Raney nickel

(a) P-3, Q-1
(b) P-2, Q-1
(c) P-2, Q-4
(d) P-4, Q-2

16. Which of the following is not employed in the commercial production of linear polyvinyl chloride ?

(a) Emulsion polymerization
(b) Suspension polymerization
(c) Additional polymerization
(d) Condensation polymerization

17. Which of the following is a detergent ?

(a) Benzene hexachloride
(b) Cellulose nitrate
(c) Polyvinly chloride
(d) Alkyl benzene sulphonate

18. The chief raw material for the commercial production of methanol is :

(a) Synthesis gas
(b) Formaldehyde
(c) Acetic acid
(d) Ethanol

19. Match the raw materials in Group-II with the product in Group-I

Group-I	Group-II
P. Nylon-66	1. Chlorodifluorothane
Q. Terylene	2. Dimethyl terepthalate and ethylene glycol
	3. Acetylene and hydrogen cyanide
	4. Hexamethylene diamine and adipic acid

(a) P-3, Q-4
(b) P-4, Q-3
(c) P-4, Q-2
(d) P-1, Q-2

20. Match the product in Group-I with the catalyst used for its production in Group-II

Group-I	Group-II
P. Nitric acid	1. Silver oxide
Q. Formaldehyde	2. Raney nickel
	3. Activated carbon
	4. Platinum-rhodium

(a) P-1, Q-2
(b) P-3, Q-4
(c) P-2, Q-3
(d) P-4, Q-1

21. Match the product in Group-I with the raw material in Group-II.

Group-I	Group-II
P. Caustic soda	1. Ammonia and sulphuric acid
Q. Soda ash	2. Sodium carbonate and slaked lime
	3. Salt and lime stone
	4. Salt and sulphuric acid

(a) P-2, Q-3 (b) P-1, Q-2
(c) P-3, Q-4 (d) P-4, Q-1

22. Match the feed in Group-I with the process in Group-II

Group-I	Group-II
P. Gas oil	1. Acetylene production
Q. Residual crude	2. Ethylene production
	3. Coking
	4. Cracking

(a) P-3, Q-4 (b) P-4, Q-3
(c) P-2, Q-3 (d) P-1, Q-4

ANSWERS

SECTION–A

1. (a)	2. (d)	3. (b)	4. (c)	5. (a)
6. (d)	7. (d)	8. (a)	9. (c)	10. (b)
11. (c)	12. (a)	13. (b)	14. (a)	15. (b)
16. (a)	17. (c)	18. (a)	19. (c)	20. (d)
21. (c)	22. (d)	23. (c)	24. (a)	25. (d)
26. (c)	27. (b)	28. (b)	29. (c)	30. (b)
31. (b)	32. (c)	33. (d)	34. (c)	35. (a)
36. (a)	37. (c)	38. (c)	39. (c)	40. (a)
41. (d)	42. (c)	43. (c)	44. (a)	45. (a)
46. (b)	47. (a)	48. (c)	49. (b)	50. (a)
51. (a)	52. (a)	53. (b)	54. (b)	55. (c)
56. (b)	57. (a)	58. (b)	59. (b)	60. (a)
61. (b)	62. (a)	63. (c)	64. (c)	65. (c)
66. (a)	67. (a)	68. (c)	69. (b)	70. (a)
71. (c)	72. (b)	73. (b)	74. (a)	75. (e)
76. (b)	77. (a)	78. (b)	79. (a)	80. (b)
81. (b)	82. (b)	83. (a)	84. (a)	85. (a)
86. (a)	87. (a)	88. (b)	89. (c)	90. (d)
91. (a)	92. (a)	93. (b)	94. (b)	95. (a)
96. (b)	97. (a)	98. (b)	99. (c)	100. (d)

SECTION–B

1. sulphuric acid (H_2SO_4).
2. contact, lead chamber.
3. calcium carbide, water.
4. absolute zero to 123°K
5. nitrogen, phosphorus and potassium.
6. diaphragm cell and mercury cell.
7. calcium aluminates and silicates.
8. reverse osmosis.
9. $_{92}U + {}_0n^1 \rightarrow X + Y + 2$ to $3\ {}_0n^1$.
10. liquid, solid.
11. slows down burning rate.
12. acetylene and hydrogen chloride.
13. oxidation of naphthalene and ortho-xylene.
14. benzene or *n*-butene.
15. Sugar
16. adipic acid and hexamethylene diamine.
17. decron.
18. hematite.
19. iron ore, coke and lime stone.
20. chalcopyrite.
21. mercury pollution in water.
22. zinc oxide.
23. gave elasticity and prevent cracking of the film.
24. a pigment.
25. ethyl alcohol.
26. 750 to 1100°F.
27. Urea.
28. N, P_2O_5 and K_2O.
29. 21
30. phosphate, sulphuric acid
31. phosphate, phosphoric acid
32. potassium and magnesium.
33. 46.0
34. 228.
35. 5.
36. Kraft process.
37. biochemical.

38. condensation polymerisation and addition polymerisation.
39. acetanilide, chlorosulphonic acid, alkali, 15% hydrochloric acid.
40. the drug used in contraceptives is mestranol.
41. chemical reactions.
42. selective corrosion.
43. corrosion between two surfaces accelerated by or resulting from the mechanical removal of a protective film of corrosion products.
44. (*i*) sacrificial anode method. (*ii*) impressed emf method.
45. alkalies and alkaline solutions.
46. alcohol and acid.
47. centrifugal pump.
48. jet ejector.
49. Urea
50. Urea.
51. 200 tons/day
52. 1 to greater than 10%
53. 0.1 to 0.5.
54. 0.25% uranium and 10 to 11% thorium oxide.
55. Fertilizer Corporation of India's Nangal unit.
56. As rocket fuel.
57. alkyd resins and plasticizers.
58. addition.
59. polycondensation.
60. 60 to 70°C, 6 to 10 atms gauge and aluminium alkyl as catalyst.
61. persulphate catalyst and 50°C.
62. emulsion and suspension.
63. addition of fillers, coloring agents, plasticizers, antioxidants and vulcanizing agents.
64. neoprene.
65. Mathura, U.P.
66. 5°C, 1 atm. gauge pressure; 50°C and 3 to 4 atms gauge pressure.
67. ethyl alcohol, ethylene.
68. can be repeatedly melted and useful for melt casting and spinning.
69. cannot be melted after forming.
70. ziegler catalysts, supported metal oxide catalysts and radiation.

SECTION—C

1. (*i*) Low capacity plants, often not operating in full capacity.

   (*ii*) Higher fixed capital investment since many of the plants are foreign built and/or Indian mechanical equipment is expensive.

(*iii*) Scarcity of raw materials, sometimes involving imports.
(*iv*) Technological problems particularly maintenance and repair.
(*v*) High cost of electric power, utilities and transportation.

2. (*i*) Chalcopyrities (copper pyrites) (sulphide ore)
(*ii*) Iron pyrites (sulphide ore)
(*iii*) Zinc sulphide (sulphide ore)
(*iv*) Lead sulphite (sulphide ore)
(*v*) Gypsum and selenite (sulphate ore)
(*vi*) Barite (sulphate ore)
(*vii*) Calestite (Sr SO_4) (sulphate ore)

3. (*i*) Exchange of heat with a colder stream.
(*ii*) Expansion through a throttle value (Joule-Thomson expansion) adiabatically.
(*iii*) Adiabatic expansion against a piston or in a turbine.

4.
$$NH_3 + 5/4O_2 \rightleftharpoons NO + 3/4H_2O(g)$$
$$2NO + O_2 \rightleftharpoons 2NO_2$$
$$3NO_2(g) + H_2O(l) \rightleftharpoons 2HNO_3(aq) + NO$$
$$2NO_2 \rightleftharpoons N_2O_4$$
$$2NO_2 + H_2O \rightleftharpoons HNO_3(aq) + HNO_2$$
$$2HNO_2 \rightleftharpoons H_2O + NO + NO_2$$

5. Ammonia is dissolved in sodium chloride solution and then reacted with carbon dioxide. A precipitate of sodium bicarbonate was obtained which is calcined to produce high purity sodium carbonate.

6. (*a*) Common salt (*b*) Magnesium compounds
(*c*) Potassium compound and (*d*) Bromine

7. (*a*) Vertical shaft (*b*) Moving bed
(*c*) Rotary kiln (*d*) Fluidized bed

8. Primary : carbon oxygen hydrogen, nitrogen, phosphorus, potassium
Secondary : calcium, magnesium sulphur.

9. (*i*) Ammonium chloride (*ii*) Ammonium nitrate (as CAN)
(*iii*) Ammonium sulphate (*iv*) Monoammonium phosphate
(*v*) Diammonium phosphate (*vi*) Urea

10. (*a*) Electric power (*b*) Natural gas
(*c*) Coke oven gas (*d*) Naphtha
(*e*) Fuel oil (*f*) Lignite
(*g*) Coke (*h*) Coal

11. (*a*) Recovery from coke oven gas by scrubbing the gas with sulphuric acid.

(*b*) Direct neutralisation of ammonia and sulphuric acid

$$2NH_3 + H_2SO_4 \rightarrow (NH_4)_2SO_4$$

(*c*) Using natural or byproduct gypsum and adopting the Merseburg process :

$$2NH_3 + CO_2 + H_2O \rightarrow (NH_4)_2CO_3$$
$$CaSO_4 . 2H_2O + (NH_4)CO_3 \rightarrow (NH_4)_2SO_4 + CaCO_2 + 2H_2O$$
$$CaSO_4 . 2H_2O + (NH_4)_2 CO_3 \rightarrow (NH_4)_2SO_4 + CaO_3 + 2H_2O$$

(*d*) Byproduct from caprolactum.

12. (*a*) Madras Fertilizers Ltd., Chennai
(*b*) Coramandal Fertilizers, Vizag
(*c*) ZACL, Goa
(*d*) IFFCO, Kandla

13. (*a*) Ammonium phosphate sulphate
(*b*) Diammonium phosphate
(*c*) Nitrophosphates
(*d*) Urea ammonium sulphate
(*e*) Mono-ammonium phosphate.

14. $CaCO_3 \rightarrow CaO(c) + CO_2(g)$
$C(s) + O_2(g) \rightarrow CO_2(g)$
$CaO(s) + H_2O(l) \rightarrow Ca(OH)_2(c)$
$NH_3(g) + H_2O(l) \rightarrow NH_4OH(aq)$
$2NH_3OH(aq) + CO_2(g) \rightarrow (NH_4)_2CO_3(aq) + H_2O(l)$
$(NH_4)_2CO_3 + CO_2 + H_2O \rightarrow 2NH_4HCO_3$
$NH_4HCO_3 \rightarrow NaCl \rightarrow NH_4Cl + NaHCO_3$
$2NHCO_3(c) \rightarrow Na_2CO_3(C) + CO_2(g) + H_2O(g)$
$2NH_4Cl(aq) + Ca(OH)_2(c) \rightarrow 2NH_3(g) + CaCl_2(aq)2H_2O(l)$
The overall reaction for the entire process is :
$CaCO_3 + 2NaCl \rightarrow CaCl_2 + Na_2CO_3.$

15. $2NO + O_2 \rightarrow 2NO_2$ —In all parts of the system
$SO_2 + H_2O \rightarrow H_2SO_3$ —In glover tower, chambers
$NO + NO_2 + H_2O \rightarrow 2H_2SO_4 \rightarrow 2ONO\ SO_2OH + H_2O$
—In Gay-Lussac tower, chambers (partly) reaction towers.
$H_2SO_3 + 2HNO_2 \rightarrow H_2SO_4 + 2NO + H_2O$
—glover tower, chambers, reaction towers.
$ONO\ SO_2OH + H_2O \rightarrow H_2SO_4 + HNO_2$
—glover tower chambers, reaction towers.

16. The constituents of lacquers and uses are :
(*a*) Film forming or basic materials to furnish water-proofness, hardness and durability.
(*b*) Pigments—to impart colour and to improve resistance to light.
(*c*) Solvents—to dissolve the film forming substances and to suspend the pigments.

(*d*) Diluents—to reduce viscosity and cost.

(*e*) Plasticizers—to remove film brittleness and to improve adherence.

17. (*a*) Surfactants : organic active, with suds regulators
Function : removal of oily soil, cleaning

(*b*) Builders : sodium triphosphate and/or tetrasodium
Pyrophosphate : for removal of inorganic soil, detergent building.
Sodium sulphate—filler with building action in soft water.
Soda ash—filler with some building action.

(*c*) Additives :
Sodium silicate—corrosion inhibitor with slight building action.
Carboxymethyl cellulose—antiredeposition of soil
Fluorescent dye—optical brightening
Tarnish inhibitors—preventation of silverware tarnish
Perfume and sometimes dye or pigment—aesthetic, improved product characeristics.

18. Products of crude petroleum after distillation.

(*a*) Natural gasoline and natural gas

(*b*) Light distillates : Motor gasolines, solvent naphthas, jet fuel, kerosene, light heating oils.

(*c*) Intermediate distillates : Heavy fuel oils, diesel oils, gas oils.

(*d*) Heavy distillates : Heavy mineral oils (medicinal), heavy floatation oils, waxes (candles, sealing, paper treating, insulating), lubricating oils (large range).

(*e*) Residues : Lubricating oils, fuel oils, petrolatum, road oils, asphalts, coke.

19. (*a*) Butadine rubbers,
(*b*) Isoprene rubbers synthetic
(*c*) Chloroprene rubbers
(*d*) Acrylic-butadiene rubbers
(*e*) Acrylate-butadiene rubbers
(*f*) Isobutylene-isoprene
(*g*) Nitrile-butadiene rubbers
(*h*) Nitrile-chloroprene rubbers
(*i*) Pyridine-butadiene rubbers
(*j*) Styrene-butadiene rubber
(*k*) Styrene-chloroprene rubbers
(*l*) Styrene-isoprene rubbers
(*m*) Ethylene-propylene-terpolymer rubber.

20. (*a*) Transportation of sulphur to the plant

(*b*) Melting of sulphur, if not shipped molten and often filtering of molten sulphur to remove traces of ash.

(*c*) Pumping and atomising of melted sulphur

(*d*) Drying of combustion air

(*e*) Burning of sulphur

(*f*) Recovery of heat from and cooling of hot SO_2 gas

(*g*) Purification of SO_2 gas by hot filtration

(*h*) Oxidation of SO_2 to SO_3 in converters

(*i*) Temperature control with heat transfer to secure high yields of SO_3

(*j*) Absorption of SO_3 in strong acid, 98.5 to 99% H_2SO_4

(*k*) Cooling of acid from absorbers

(*l*) Pumping of acid over absorption towers.

21. Nitrogen : Required during early stages of plant growth to promote development of stem and leaves.

 Phosphorous : Which stimulates early growth and accelerates seeding or fruit formation in later stage of growth.

 Potassium : Essential to the development of the starches of potatoes and grains, the sugar of fruits and vegetables and the fibrous material of plants; an ample supply of potassium in the soil helps to prevent disease and to lessen the effects of excessive nitrogen application.

22. (*a*) Fertilizer Corporation of India Ltd., Sindri

 (*b*) Fertilizer Corporation of India Ltd., Namrup

 (*c*) Fertilizers and Chemicals, Travancore Ltd. Alwaye

 (*d*) Gujarat State Fertilizers Co. Ltd. Baroda.

23. Clay = 0.1 to 0.3 ton; Limestone = 1.2 to 1.3 tons

 Gypsum = 0.03 to 0.05 ton; Coal = 0.25 to 0.4 ton

 Water = 3 tons, Electricity = 80 kWh.

24. Fallout particulate matter, suspended particulate matter, smoke, hydrocarbons, oxides of nitrogen, carbon monoxide, sulphur dioxide, pollen, odors.

25. (*a*) Calcium silicate (*b*) Benzoyl peroxide

 (*c*) Calcium citrate (*d*) Calcium chloride

 (*e*) Citric acid (*f*) Sorbic acid

 (*g*) Butylated hydroxy anisole (*h*) Invertase.

26. (*a*) ICL, Kolkata

 (*b*) Union Carbide, Trombay

 (*c*) Polyolefins Industries Ltd., Thane, Mumbai

 (*d*) Kasturbhai and Lalbhai Group, Koyali, Maharashtra.

27. (*a*) Shriram Vinyl and Chemical Industries, Ltd., Mumbai

 (*b*) National Organic Chemical Industries Ltd., Mumbai

 (*c*) Ahmedabad Manufacturing and Calico printing Co. Ltd., Mumbai :

 (*d*) Chemicals and Plastics Ltd., Chennai

 (*e*) Plastics Resins and Chemicals Ltd., Sahupuram, Tamil Nadu.

28. The kinetics of SO_2 reaction shows that the reaction is fast at high temperatures (500 to 600°C) but equilibrium conversion to SO_3 is low. So to have maximum conversion and fast reaction the 1st stage is maintained at 500–600°C (high temp) and 80% conversion is obtained. Before sending to second stage the gases are cooled and up to 97% conversion is obtained at 400–450°C reaction temperature in second stage.

29. The kinetics of ammonia reaction using nitrogen and hydrogen as raw materials is similar to SO_3 reaction. That is the reaction is fast at high temperatures but the equilibrium conversion is low. Since ammonia reaction requires high pressure vessels used for reactor, a multistage operation as

used with SO_2 oxidation is not economically feasible for ammonia production. So a one stage reactor is used.

30. Hydrogenation is a unit process which is used in the fat and oil industry to remove double bonds, raise its melting point of the fat and improve its resistance to rancid oxidation.

31. (*a*) Hydrogen = 50 MN^3

    (*b*) Nickel catalyst losses = 0.6 kg

    (*c*) Steam = 0.5 ton

    (*d*) Cooling water = 4 to 5 tons

32. (*a*) J.K. Synthetics (*b*) Nirlon

    (*c*) Modipom (*d*) Garware

    (*e*) Century Enka.

SECTION—D

1. 1 – b; 2 – c

2. (b)

3. 1 – c, 2 – a

4. (b)

5. (c)

6. (a)

7. (a)

8. (b)

9. (d)

10. (a) Vanadium pentoxide

(b) Iron oxide

(c) Silver oxide

(d) Nickel

(e) Copper or Brass.

11. (d)

12. (d)

13. (a) Heat of absorption or dilution is less in concentrated sulphuric acid. Oleums are made by absorption of SO_3 in conc sulphuric acid. Required strength sulphuric acid is made from oleum.

(b) Concentration by H_2SO_4. Rectification with 93% H_2SO_4 in silicon-iron or stone ware towers produces coc. HNO_3 and 70% H_2SO_4 which can be reevaporated to 93% H_2SO_4.

(c) To remove colour and dissolved salts, dilute glycerine is put through successive beds of anion and cation exchange resins. To avoid scaling in evaporators, calcium salts are removed.

14. (a)

15. (c)

16. 1 – c, 2 – d

17. 1 – c, 2 – b

18. (b)

Ans: Mol.wt. of $CaCO_3$ = 100

Mol.wt. of Na_2CO_3 = 106

$$145 + \frac{345 \times 100}{106} = 485$$

SECTION—E

1. (b)	**2.** (b)	**3.** (b)	**4.** (b)	**5.** (a)
6. (b)	**7.** (b)	**8.** (a)	**9.** (b)	**10.** (b)
11. (c)	**12.** (d)	**13.** (d)	**14.** (d)	**15.** (c)
16. (c,d)	**17.** (d)	**18.** (a)	**19.** (c)	**20.** (d)
21. (a)	**22.** (c)			

❑

CHAPTER 5

Fluid Mechanics

1. The branch of Engineering that deals with the behaviour of fluids is fluid mechanics. A fluid is a substance that does not permanently resist distortion. In manometer $Pa - Pb = g\ Rm\ (\rho a - \rho b)$; P = pressure; g = gravitation; Rm = level difference; ρ = density. Ideal fluid is incompressible and has zero viscosity. The effect of solid boundary on the flow is confined to a layer of the fluid immediately adjacent to the solid wall. This is called as boundary layer.
2. Rheological properties of fluids: Pseudoplastic, e.g. polymer solutions, starch, paints. Thixotropic, e.g. some polymer solutions, some paints. Newtonian, e.g. gases, simple liquids; Dialatent, e.g. cornflour, sugar solutions, starch in water, wet beachsand. Rheopectic, e.g. gypsum suspensions, clay suspensions.
3. Viscosity of a gas, $\mu = 0.00267\ (MT)^{1/2}/\sigma^2$; σ = molecular diameter, deg.A; μ = viscosity, cp; M = molecular weight; T = absolute temperature, deg. K. Kinematic viscosity (u) = μ/ρ; Reynolds number = DVp/μ. The approximate length of straight pipe necessary for completion of the final velocity distribution is : $xt/D = 0.05\ Re$ for laminar flow.
4. Equations of flow : Navier–Stokes equation (incompressible) $\rho(RHO)Dv/Dt = -\nabla P + u\nabla^2 V + \rho g$ Reynolds Number = Dvp/μ; Continuity Equation (mass balance): $d\rho/dt = -(\nabla.pV)$

   $D/Dt = d/dt + ud/dx + vd/dy + wd/dz$; Euler equation: for constant density and zero viscosity is: $\rho DV/Dt = \nabla P + \rho g$; Bernoullis Equation: $wp + Pa/\rho(RHO) + g\ Za + \alpha a\ Va^2/2 = \rho b/p + gZb + \alpha bVb^2/2 + hf$ where Wp is pump work; hf is friction head; $\rho(RHO)$ is pressure; ρ is density; Za is level from baser line; V = velocity of liquid: Hagen-Poiseuille Equation: $\Delta P = 32\ \mu L\ V/(gc\ D^*D)$; Friction factor in viscous flow = $16/N\ RE$; Frictional loss for sudden expansion: $hf = Ke\ Va^2/2g$ where $Ke = (1 - Sa/Sb)^2$.

5. **Compressible Fluids:** Mach Number = u/a where u is fluid velocity and a is sound velocity. $dm = \rho\ us$ = constant ; $\ln p + \ln u + \ln S$ = constant; $dp/p + du/u + ds/s = 0$ (continuity equation). Total energy balance equation: $m\ (ub^2 - ua^2)/2 + g(Zb - Za) + Hb - Ha = Q\text{-}ws$ Neglecting potential energy and shaft work: $Q/m = (Hb - Ha) + (ub^2 - ua^2)/2$

6. **Immersed Bodies:**

   The force in the direction of flow exerted by the fluid on the solid is called as Drag. Drag coefficient by Stokes law: CD = 24/Rep; Kozney-Carman equation $\Delta P/L = 150\ v_0\ \mu(1 - \varepsilon)^2/(\phi_S\ Dp^2\ \varepsilon^3)$ at low Reynolds Number.

Burke-Plummer equation: $\Delta P/L = 1.75\ p\ v_0^2\ (1 - \varepsilon)/(\phi_S^2\ Dp\varepsilon^3)$ at high Reynolds Number.

Combining both equations V_{om} (minimum fluidization velocity) = $g\ (p_P - p)\ \in^3 \phi^2_S\ Dp^2\ /(150\ \mu\ (1 - \in))$

Types of Fluidization:

1. Particulate fluidizaion: Characterized by a large but uniform expansion of the bed at high velocities.
2. Aggregative fluidization: The non-uniform nature of the bed is due to aggregation of the particles.
3. Bubbling fluidization: Particles do not stick together.

7. Expansion of fluidized beds: $g\ (1 - \in)\ (\rho_P - \rho) = \Delta P/L$

8. **Applications of Fluidization:**

   (a) Fluid bed catalytic cracking (b) Fluidized bed combustion of coal (c) Roasting of ores (d) Drying of fine solids (e) Adsorption of gases (f) Transportation of solids from point to point (g) Hydraulic or slurry transport (h) Pneumatic conveying.

9. **Optimum velocity in a pipe:** > 25 mm = V_{opt}. = $12\ m^{0.1}/\rho^{0.36}$ where m = mass flow rate in a pipe, lbs/second; ρ = density of fluid, lbs/cubic feet.

10. **Valves:** 1. Check valves (permit flow in one direction only); 2. Plug cocks and ball valves: a quarter turn of the system takes the valve from fuly open to fully closed. 3. Gate valves and globe valves: pressure drop is large.

11. **Pumps:** 1. Positive displacement pumps: a definite volume of liquid is trapped in a chamber which is alternately filled from the inlet and emptied at a higher pressure through the discharge.

12. **Reciprocating pumps:** Piston pumps, plunger pumps, and diaphragm pumps.

    Piston pumps: Liquid is drawn through an inlet check valve into the cylinder by the withdrawal of a piston and is then forced out through the discharge check valve on the return stroke.

    In a diaphragm pump, the reciprocating member is a flexibe diaphragm of a metal, plastic or rubber.

    Rotary pump: Gear pumps, lobe pumps, screw pumps, cam pumps, and vane pumps. Operate on clean, moderately viscous fluids such as light lubicating oil.

    Peristatic pumps: In biochemicals.

    Centrifugal pumps: The mechanical energy of the liquid is increased by centrifugal action.

12. **Flow meters:**

    Venturi meters: The pressure drop in the cone is used to measure the flow rate. $m = q * \rho = CvSb\ \sqrt{2(Pa - Pb)} \Big/ (\sqrt{1 - \beta^4})$ where m is flow rate, Sb = throat area; q = volumetric flow rate; Cv = venturi coefficient (0.98); $\beta = Db/Da$ where b is throat, a is pipe; P is pressure. This is expensive, occupies more space and power consumption is less.

    Orifice: It contains of an accurately machined and drilled plate mounted between two flanges with a hole concentric with the pipe in which it is mounted. Power consumption is high and of less cost.

    $Ua = Co\sqrt{(2(Pa - Pb)/\rho)} \Big/ \sqrt{1 - \beta^4}$ where $Co = 0.61$; Ua = velocity of fluid in the orfice; Co = orifice coefficient; Pa and Pb are pressures on the two sides of the orfice; ρ is density of flowing fluid; β is Db/Da as in venturi meter.

V-element meters: In these meters the flow is restricted by a V shaped indentation in the side of the pipe or by a metal wedge inserted in the pipe. Expensive but high accuracy.

Rotameters (area meters): Consists of devices in which the pressure is consant, or nearly so and the area through which the fluid flows varies with the flow rate. The area is related by calibraton to the flow rate.

Target meters: A sharp edged disk is set at right angles to the direction of flow .The drag force exerted on the disk by the fluid is measured. The flow rate is propotional to the square root of this force and to the fluid density. These are rugged and inexpensive.

Vortex shedding meters: Its body is designed to create when flow is turbulent a votex street in its wake.

Turbine meters: A bladed rotor is suspended axially in the flow stream and spins at a rate proportional to the fluid velocity.

Positive displacement meters: Flow is measured by counting the number of times the moving compartment is filled and emptied.

Ultrasonic meters: Transit type, Doppler shift type.

Insertion meters: A sensing element is inserted into the flow stream .This gives local velocity and from the insertion average velocity is calculated.

Thermal meters: These measure the flow rate by measuring the temperature rise in the fluid as it passes over a heating element.

Magnetic flow meters: The flow tube is lined with a non-conducting metal with two or more electrodes (metal) mounted flush with the liner wall. Electromagnetic coils surrounding the tube generate a uniform magnetic field within it. By Faraday's law of electromagneic induction, the motion of a conducting fluid through the magnetic field induces a voltage that is directly and linearly proportional to the velocity of the flowing fluid.

Coriolis meters: The fluid is passed through two U-shaped curved tubes that are vibrated at their natural frequency.This creates an alternating coriolis force that produces small elastic deformations in the tubes. From the magnitude of deformations the mass flow rate is calculated.

Pilot tube: measures local velocity. From local velocity average velocity is determined.

BLENDING AND MIXING:

13. **Impellers:** The types are: 1-propellers ; 2-turbines ; 3-high efficiency impellers.

    Power number $(Np) = P\, gc/(n^3\, Da^5\, \rho)$; where P is pressure, n = speed of impeller; Da is Diameter of impeller; ρ is density of liquid.

14. **Mixers:** Jet mixer. Circulation in large vessels is induced by one or more jets of liquid.

    If power input per unit volume is same in two vessels then, $n_0/n_1 = (Da_1/Da_2)^{2/3} = t_0/t_1$; where t is time, n= rotational speed, r/s; and Da is diameter of impeller.

15. **Kinemetics:** Deals with motion of particles without considering the forces causing motion.

16. Methods describing fluid motion are:

    1. **Eulerien method:** Velocity, pressure etc. are described at a point. A person stays at a point and observe the changes taking place in an element.
    2. **Lagrangian:** A particle is followed and its velocity, acceleration etc. are observed/described. Like a person sitting on an object and observing changes in it.

17. Steady flow: $\frac{\partial v}{\partial t} = 0$ (fluid properties do not change with time).

Unsteady state: Fluid characteristics change with time.

Uniform flow: Is that type of flow in which the velocity at any given time does not change with space $\left(\frac{\partial v}{\partial s}\right)_t$ = constant.

18. Laminar/viscous flow/streamline flow: In this the fluid particles move along well defined paths or streamlines. All the streamlines are straight and parallel. The particles move in layers sliding smoothly over the adjacent layer (no mixing of layers).

Turbulent flow: The fluid particles move in a zig-zag way. Because of zig-zig motion, eddies are formed. These eddies cause energy loss.

Re. No. (Reynolds Number) < 2000 is laminar flow.

> 4000 is turbulent flow.

Re. No. 2000 to 4000 is transition zone or buffer zone.

19. Compressible fluid means density varies: In compressible fluid means, density does not change. Rotational flow means in this the fluid particles while flowing along streamlines, also rotate about their own axis, e.g. rotation of earth around sun. Irrotational means the type of flow in which the fluid particles along streamlines do not rotate about their axis, e.g. moon rotates around earth.

20. Continuity equation: $\rho_1 A_1 V_1 = \rho_2 V_2 A_2$ (ρ = density; A = area, V = Velocity).

Equation of Continuity in 3 dimensions: $\frac{\partial \rho}{\partial t} + \frac{\partial}{\partial x}(\rho u) + \frac{\partial}{\partial y}(\partial v) + \frac{\partial}{dz}(\partial w) = 0$

21. Relations between streamline function (ψ) and velocity potential function (ϕ):

$$u = \frac{-\partial \phi}{\partial x};\ V = -\frac{\partial \phi}{\partial y};\quad u = \frac{-\partial \psi}{\partial y};\ V = -\frac{\partial \psi}{\partial x}$$

Hence $\frac{\partial \varphi}{\partial x} = \frac{\partial \psi}{\partial y};\ \frac{\partial \phi}{\partial y} = \frac{-\partial \psi}{dx}$

22. Euler equation of motion: $\frac{dP}{\rho} + gdz + \frac{vdv}{d\rho} = 0$; ρ is density, P = pressure, g = gravitation const. dz = level change; v = velocity.

Integrating $\frac{P}{\rho g} + \frac{v^2}{2g} + z$ = constant (Bernaulli equation).

23. OR $\left.\frac{P_1}{pg} + \frac{v_1^2}{2g} + z_1 + \omega \text{ (Pump)}\right\} = \frac{P_2}{\rho g} + \frac{v_2^2}{2g} + z_2 + h_L$ (friction head) (Bernaulli's equation)

24. Gauge pressure = Absolute pressure – atmospheric pressure.

25. Moment of inertias: 1. Rectangle: $bd^3/3$; 2. Triangle: $bh^3/12$; 3. Circle: $\pi d^4/64$;

4. Trapezium = $\left[\dfrac{a^2 + 4ab + b^2}{36(a+b)}\right]$

26. Loss of energy in pipes: $h_f = \dfrac{4flV^2}{d2g}$

Loss of head due to enlargement: $h_e = (V_1 - V_2)^2/(2g)$

Loss of head due to contraction: $h_c = 0.5\, v_2^2/(2g)$

Loss of head at entrance of pipe: $h = 0.5\, v^2/(2g)$

Loss of head due to bend in a pipe: $h_b = kv^2/2g$; Loss of head due to fittings = $kv^2/(2g)$;

Loss of head due to obstruction:

$$h_0 = \frac{v^2}{2g}\left[\frac{A}{C_C(A-a)} - 1\right]^2; C_C = \frac{A\text{ small}}{A\text{ big}};$$ A = area of pipe; a = area of obstruction.

27. In viscous flow, u (average)/u (maximum) = 0.5

In turbulent flow: v (average)/v (maximum) = 0.80

QUESTIONS

SECTION-A

1. For a streamline flow, velocity at a certain point is :

(a) Constant (b) A function of time only

(c) Constant, but depends on time (d) Constant and independent of time

2. The average velocity of a Newtonian fluid in a straight tube under viscous flow conditions equals to :

(a) The maximum velocity (b) 2/3rd the maximum velocity

(c) 50% of the maximum velocity (d) 1/3rd the maximum velocity.

3. The maximum velocity of a Newtonian fluid in a straight tube for turbulent flow conditions equals to :

(a) Average velocity (b) 1.2 times the average velocity

(c) 1.5 times the average velocity (d) 2 times the average velocity

4. A sewage sludge is non-Newtonian fluid of the type :

(a) Bingham plastic (b) Pseudo plastic

(c) Real plastic (d) Dilatant

5. A dilatant fluid is one :

(a) Whose apparent viscosity decreases with shear rate

(b) Whose apparent viscosity increases with shear rate

(c) Whose apparent viscosity remains constant with shear rate

(d) Which requires a threshold stress for initiation of flow.

6. Continuity equation for one-dimensional flow in x-direction of an incompressible fluid in unsteady state is :

(a) $\frac{\partial(\rho u)}{\partial x} = -\frac{\partial \rho}{\partial t}$ (b) $\frac{\partial u}{\partial x} = 0$

(c) $\frac{\partial(\rho u)}{\partial t} = 0$ (d) $\frac{\partial u}{\partial x} = -\frac{\partial \rho}{\partial t}$

7. Equivalent diameter in case of fluid flow in non-circular cross-section is given by :

(a) $\frac{\text{Flow cross-section}}{\text{Wetted-circumference}}$ (b) $4 \times \frac{\text{Flow cross-section}}{\text{Wetted-circumference}}$

(c) $\frac{\text{Flow cross-section}}{2 \times \text{Wetted-circumference}}$ (d) $2 \times \frac{\text{Flow cross-section}}{\text{Wetted-circumference}}$

8. For laminar flow of incompressible fluid in closed conduit of radius 'R', maximum shear stress occurs at :

(a) The wall (b) The centre

(c) A distance equal to $R/2$ from the wall (d) A distance equal to R/4 from the wall.

9. The relation between Fanning's friction factor (f) and Blasius friction factor (f_B) is :

(a) $f_B = 2f$ (b) $f_B = f$

(c) $f_B = 4_f$ (d) $f_B = 16f$

10. The relation between friction factor if Reynolds number (R_e) is given (for laminar flow conditions) :

(a) $f = 4/R_e$ (b) $f = 16/R_e$

(c) $f = R_e^{0.5}$ (d) $f = .0014 + R_e^{0.32}$.

11. Poiseulle's equation for pressure drop in laminar flow is given by :

(a) $32Lu\mu/g_cD^2$, (b) $16Lu\mu/g_cD^2$

(c) $L.u\mu/32g_cD^2$, (d) $Lu\mu/16g_cD^2$.

12. For ordinary flow conditions, transition from laminar to turbulent occurs at a Reynolds number of :

(a) 4000 (b) 6000

(c) 1500 (d) Between 2100-4000.

13. The dimension of kinematic viscosity is :

(a) L/θ^2 (b) L^2/θ^2

(c) L^2/θ (d) $L\theta$

14. Expression for Weber number is given by :

(a) $u\rho L/\mu$ (b) $u^2L\rho/\sigma\, g_c$

(c) $Fg_c/\rho u^2$ (d) gL/u^2.

15. Froude number represents the ratio of :

(a) Inertia and gravity forces (b) Inertia and viscous forces

(c) Gravity and viscous forces (d) Inertia and surface forces

16. The ratio of pressure force to inertia force is called .

(a) Cauchy number (b) Weber number

(c) Euler (d) Froude number

17. Flow of incompressible fluid with no shear is called :

(a) Creep flow (b) Streamline flow

(c) Potential flow (d) Boundary layer flow

18. In case of couette flow, the fluid flow is between two large flat parallel plates with :

(a) Both the top and the bottom plates fixed

(b) Both the top and the bottom plates moving

(c) Top plate fixed and the bottom moving

(d) Top moving and the bottom fixed

19. Head loss at the entrance of a straight pipe is given by :

(a) $u^2/2g_c$, (b) $0.5(u^2/2g_c)$

(c) $1.5(u^2/2g_c)$ (d) u^2/g_c.

20. In case of settling of a particle in a fluid medium, the Brownian movements is prominent in the particle size range of :

(a) 2–3 cm (b) 200–300 μ

(c) 2–3 μ (d) 20–30 μ

21. Drag coefficient for the settling of spherical particle in the Newton's range of flow is equal to :

(a) 0 .44 (b) $R_e/24$

(c) $24/R_e$ (d) $185/R_e^{0.6}$

22. For settling in the Stokes range, the value of Reynolds number is :

(a) $2 < R_e < 200$ (b) $200\ R_e < 500$

(c) $500 < R_e < 2100$ (d) $0 < R_e < 2$

23. A gravity settling chamber is effective in separating particles in the size range of :

(a) 0 – 5 μ (b) 5 – 10 μ

(c) > 74 μ (d) 10 – 10 μ

24. Separation of coarse particles from a slurry of fine particles is achieved by :

(a) Classifier (b) Thickener

(c) Decanter (d) Hydrocylone

25. A gravity decanter is used for the separation of :

(a) Two immiscible liquids of same density

(b) Two miscible liquids of same density

(c) Two miscible liquids of different density

(d) Two immiscible liquids of different density

26. Removal of fine solids from liquids is called :

(a) Classification (b) Sedimentation

(c) Decantation (d) Flocculation

27. When larger particles such as grains are subjected to fluidization, the corresponding bed is called:

(a) Teeter bed (b) Boiling bed

(c) Spouted bed (d) Slugging bed

28. The operating range of fluidization velocities for all particle sizes are :

(a) Almost the same (b) Much greater with small particles

(c) Greater with small particles (d) Greater with larger particles

29. For flow measurement in tubes and pipes in case of permanent installation, the best device is :

(a) Venturi meter (b) Orifice meter

(c) Pitot tube (d) Notch

30. In a hot-wire anemometer the resistance wire for conducting electrical current is made of :

(a) Nickel (b) Copper

(c) Momel (d) Tungsten

31. Friction losses in case of venturi meters are :

(a) About 2% of maximum velocity head (b) About 10% of maximum velocity head

(c) About 25% of maximum velocity head (d) About 50% of maximum velocity head

32. For velocity measurement in a large diameter pipe, the device used is :

(a) Rotameter (b) Pitot rube

(c) Notch (d) Venturi meter

33. Flow in an open channel is measured by a :

(a) Rotameter (b) Hooke's gauge

(c) Notch (d) Hot-wire anemometer

34. The density of the float material in a rotameter is :

(a) Equal to the density of the liquid replaced

(b) Heavier than the liquid it replaces

(c) Higher than the liquid it replaces

(d) Nothing in particular

35. In case of a standard venturi meter, the entrance cone angle is :

(a) Equal to the exit cone angle (b) Less than the cone angle at exit

(c) Greater than the cone angle at exit (d) Nothing in particular

36. The exit cone angle for a venturi meter is :

(a) 5 to 15° (b) 10 to 20°

(c) 21 ± 2° (d) 2 – 5°

37. Coefficient of discharge for square-edged circular orifices with $0.3 < \beta < 0.5$ is :

(a) 0.48 – 0.65 (b) 5 – 0.75

(c) 0.8 – 0.88 (d) 0.61 – 0.63

38. Automatic flow controller can be integrated to a :

(a) Venturi meter (b) Rotameter

(c) Orifice meter (d) Flow nozzle

39. For precise control of fluid flow rate, the best performance is obtained by a :
(a) Gate valve (b) Check valve
(c) Globe valve (d) None of these

40. The discharge pressure of compressor is generally :
(a) Between 2 atm to thousands of atm (b) Less than 2 atm
(c) Between 2 atm and 10 atm (d) Between 10 atm and 20 atm

41. For discharge of liquid against a pressure of 1500 kg/cm^2 or more, the recommended device is :
(a) Piston pump (b) Plunger pump
(c) Vane pump (d) Centrifugal pump

42. The most ideal device for the transportation of medium viscous lubricating oil is a :
(a) Plunger pump (b) Piston pump
(c) Centrifugal pump (d) Vane pump

43. Function of a volute in a centrifugal pump is to :
(a) Draw liquid to the pump casing
(b) Eliminate cavitation
(c) Convert the velocity head of the liquid to pressure head
(d) Minimize shock losses

44. Diffuser pumps are special type of liquid-transporting devices, where :
(a) Cavitation is minimised (b) Shock losses are minimised
(c) Disk friction is eliminated (d) Bearing losses are eliminated

45. In order to achieve a very high head (30 meters) in a centrifugal pump :
(a) Two or more impellers be mounted in series on a single shaft
(b) The impeller speed be increased
(c) The volute area be increased
(d) The number of vanes in the impeller be increased

46. Rotary vacuum pumps can lower the absolute pressure down to :
(a) 10 mm Hg (b) 1 mm Hg
(c) 0.01 mm Hg (d) 0.001 mm Hg

47. Sump pump used to drain shallow pits or sumps is a :
(a) Centrifugal pump (b) Diffuser pump
(c) Plunger pump (d) Single-stage vertical pump

48. Axial flow pumps are also called :
(a) Vertical pumps (b) Propeller pumps
(c) Diffuser pumps (d) Turbine pumps

49. Mixed flow is a characteristic of :
(a) Turbine pump (b) Diaphragm pump
(c) Sump pump (d) Piston pump

50. A liquid handling device that makes use of the momentum of one fluid to move another is called a :

(a) Vane pump (b) Acid egg

(c) Reciprocating pump (d) Jet pump

51. When the head pumped against is less than the head of the fluid used for pumping, the usual device is a :

(a) Blow case (b) Air lift

(c) Ejector (d) Injector

52. Liquid depth in an agitation tank is approximately equal to :

(a) Diameter of the tank (b) 2 × diameter of the tank

(c) 50% of the tank diameter (d) 75% of the tank diameter

53. Propeller agitators are used for liquids of:

(a) High viscosity (b) Low viscosity

(c) Medium viscosity (d) All viscosity ranges

54. Expression for power number in agitation is given by :

(a) $\dfrac{Pg_c \rho^2 D_t}{\mu^3}$ (b) $\dfrac{n^3 D^2 \rho}{Pg_c}$

(c) $\dfrac{Pg_c}{n^3 D^2 \rho}$ (d) $\dfrac{n^2 D^3 \rho}{Pg_c}$

55. In a rough qualitative way it may be said that 2 to 3 hp power per 1000 gal. of the liquid gives :

(a) Mild agitation (b) Thorough agitation

(c) Intense agitation (d) Solid suspension in liquid

56. Weber number is of significance in :

(a) Liquid-liquid mixing

(b) Gas dispersion in liquid

(c) Liquid dispersion in liquid

(d) Solid suspension in liquid

57. For blending of light pastes in paint industry, the device used is :

(a) Kneader (b) Disperser

(c) Masticator (d) Charge can mixer

58. Industrial paddle agitators turn at speed between :

(a) 20 and 100 rpm (b) 200 and 400 rpm

(c) 650 and 850 rpm (d) 2000 and 2500

59. Impeller diameter in case of turbine agitator is :

(a) 5% of less than the inside diameter of the vessel

(b) 30–50% of the inside diameter of the vessel

(c) 60–75% of the inside diameter of the vessel
(d) Nothing in particular

60. Mixing of plastic solids is generally achieved by :
(a) Kneading (b) Dispersing
(c) Dispersion (d) Intense mixing

61. The mechanism of mixing in a pan mixer is by :
(a) Kneading (b) Mixing-extrusion
(v) Dispersion (d) Mulling

62. A mixing device which operates under vacuum to deair clay or other materials is a :
(a) Ko-kneader (b) Muller-mixer
(c) Pug mill (d) Banbery mixer

63. In a mixer, the ratio of the product of longitudinal velocity of material and the mixer length of the diffusively in axial mixing (*uL/E*) is called :
(a) Froude number (b) Peclet number
(c) Stanton number (d) Weber number

64. Ribbon blenders are used for the mixing of :
(a) Heavy liquids (b) Dry powders
(c) Thick pastes (d) Liquids and solids

65. Fine light powders such as insecticides are blended by :
(a) Impact wheels (b) Pug mill
(c) Banbery mixer (d) Two-arm kneader

66. A screen set across a flow channel to remove dirt or rust from a flowing liquid is called a :
(a) Clarifier (b) Filter-thickener
(c) Strainer (d) Filler

67. Thickness of frames in plate frame filter press is usually :
(a) $\frac{1}{4}-2"$ (b) $\frac{1}{4}-8"$
(c) $\frac{1}{2}-6"$ (d) $1-10"$

68. To produce a thick suspension from a thin slurry, the device used is :
(a) Vacuum flitter (b) Edge filter
(c) Shell-and-leaf filter (d) Pressure filter thickener

69. Pressure differential across a continuous vacuum filter is of the order of :
(a) 5–10 cm of Hg (b) 10–20 cm of Hg
(c) 25–50 cm of Hg (d) 50–60 cm of Hg

70. In a common continuous rotary drum vacuum filter, vacuum is applied :
(a) Throughout the drum surface
(b) In the filtering zone only

(c) In the filtering and washing zones

(d) In the filtering, washing and drying zones

71. The drum for the vacuum filter turns with a speed of :

(a) 0.1–2 rpm (b) 2–10 rpm

(c) 4–6 rpm (d) 10–20 rpm

72. The function of a filter aid is to :

(a) Increase the filtration rate

(b) Increase cake porosity in case of a dense impermeable cake

(c) Remove the plugs from the septum

(d) Decrease the filtration pressure

73. The example of a commonly used filter aid is :

(a) Fuller's earth (b) Cellulose acetate

(c) Diatomaceous earth (d) Kyanite

74. Compressibility coefficient of the commercial compressible cake vary between :

(a) 0.1 and 1.0 (b) 0.2 and 0.8

(c) 0.2 and 0.7 (d) 0.4 and 0.8

75. In a washing type of plate and frame filter, the ratio of the rate of washing to the final rate of filtration is :

(a) 0.25 (b) 0.50

(c) 1.0 (d) 2.0

SECTION—B

1. Greater is the kinematic viscosity of the fluid,.....is the thickness of the boundary layer.

2. The shear stress *vs* the rate of shear curves are time dependent for.....and.....fluids.

3. Friction factor in case of pipe flow depends on.....and.....

4. The behavior of a mass of fluidized solids depends mainly on.....and.....

5. Slugging occurs in a bed which is.....and.....

6. Minimum fluidization velocity for a particular system depends on particle.....and.....and fluid.....and.....

7. The.....group gives a criterion from which the type of fluidization can be predicted.

8. In a fluidized bed, the total frictional force on the particles must be equal to.....

9. The uniformity of the gas fluidized bed is a function of the.....and.....of the solids.

10. Compared to gate valve, pressure drop is.....in globe valve.

11. A high speed rotary device that develop a maximum pressure of 2 atmosphere is called a.....

12. A compressor that takes suction at a pressure below atmospheric and discharges against atmospheric pressure is called a.....pump.

13. Commercial magnetic flowmeters cannot be used to measure velocity of hydrocarbons because their.....is too small.

14. Baffles are used in agitators to minimize.....and.....

15. The similarities taken into account in case of scale-up in design on the basis of similarity concept include.....,.....and.....similarities.

16. The significant dimensionless groups for calculation of power consumption in mixing are.....and.....

17. A sparker is used for dispersion of.....through.....

18. A ball mill without the balls resemble to a type of mixer called.....

19. For compounding rubber and plastic solids.....is used.

20. The finest control of the rate of flow is achieved by.....

21. Diaphragm valves are very much in use for the handling of.....fluids.

22. Friction loss in an union is.....compared to a 45° elbow.

23. Friction LOSS, in pipe fittings are generally expressed in terms of.....

24. .....is used for the measurement of small pressure differences in liquid.

25. Two liquid manometer is used for measuring.....

26. An instrument suitable for very high pressure measurement is.....

27. A differential pressure cell is used for.....

28. In a pitot tube the flow rate is obtained by measuring the difference.....

29. .....can be used for direct measurement of volumetric flow of liquids containing solid particles in suspension.

30. For accurate measurement of the flow of gases at low velocities......is used.

31. Ultra centrifuges are useful for separation of.....

32. Flocculation is achieved by the addition of.....or.....

33. Common surface active agents used in flocculation are.....,....., ...and.....

34. Compared to other types to pumps, maintenance cost for centrifugal pump is.....

35. For the production of very high vacuum as required for freeze dying or molecular distillation.....pumps are normally used.

36. Cartridge filters are called edge filters because.....

37. In a filter press, filtration is continued until.....

38. Various zones in a rotary drum vacuum filter are.....,.....and.....

39. When filtration is carried out by continuous increasing of the inlet pressure, the operation is.....

40. When filtration is carried out in a centrifuge, the driving force is.....

SECTION—C

1. For streamline flow between parallel plates what is the ratio of the maximum and the average velocity?

2. What is Prandtl one-seventh power law expression in case of turbulent flow in pipe?

3. Give the expression for pressure differential in case of an inclined manometer.

4. What is water hammer?

5. What is the most serious disadvantage in case of orifice meter?

6. What the main difference between an orifice meter and a pitot tube?

7. How can the range of a particular rotameter be increased?

8. What is Net positive Suction Head?

9. On what factors do the theoretical head developed by a centrifugal pump depend?

10. Name the following characteristics curves for centrifugal pumps plotted against capacity :

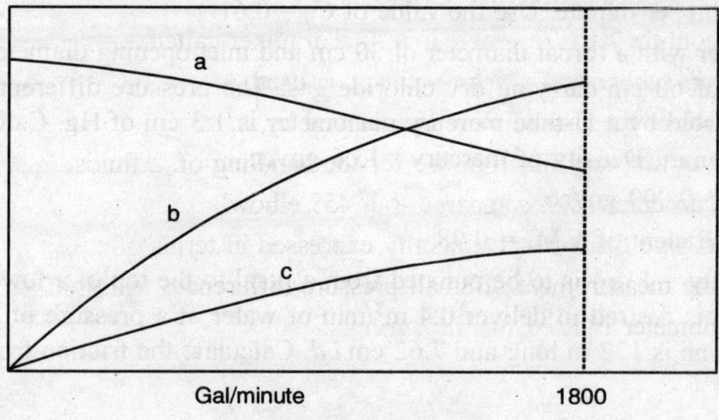

Fig. Gal/Minute

11. Which type of pumps are suitable for handling corrosive liquids and why?

12. What is a Mono pump?

13. What is cavitation?

14. What can be the range for maximum delivery pressure in case of a reciprocating compressor?

15. What are Hytor pumps?

16. What are the three chief advantages of fluidization from the point of view of transfer operations?

17. What is pneumatic transport?

18. What type of mixer is required to devulcanize rubber scrap and make water dispersion and rubber solution?

19. What are the main purposes of agitation of liquids?

20. What are the advantages of a shell-and-leaf filter over the plate-and-leaf filter over the plate-and frame filter?

21. On what factors does the capacity of a rotary filter depend?

22. What are the factors on which rate of filtration depend?

23. Give examples of some common filter aids.

24. When arc sand filters recommended?

25. What is Poiseuille's equation?

26. A pitot tube having a coefficient of 0.95 is inserted in the central line of a long smooth pipe of 25 cm diameter in which crude oil of density 0.9 gm/cc and viscosity 16.0 cP is flowing. Calculate the maximum velocity of oil in the pipe if the differential pressure in the manometer is 5 cm of water.

27. A pitot tube with a discharge coefficient of 0.9, when kept at the centre of a pipeline i.e., 8 cm, a differential pressure of 9.0 cm is recorded in a water manometer. The pipeline carries a light oil of specific gravity 0.8. Calculate the flow rate of oil in the line if the average velocity is 80% of the maximum.

28. Sulphuric acid of specific gravity 1.3 is flowing through a pipe 6 cm.*i.d.* An orifice of 1 cm diameter if fitted to the pipe and the differential pressure recorded by a mercury manometer is 10 cm. Assume that the leads to the manometer are filled with acid. Calculate the flow rate of acid in kilogram per minute. Use the value of $C_d = 0.61$.

29. A venturi meter with a throat diameter of 30 cm and inlet opening diameter of 60 cm is inserted in a pipeline of 60 cm carrying dry chloride gas. The pressure differential across the venturi meter as indicated by a U-tube mercury manometer is 1.3 cm of Hg. Calculate the flow rate of chlorine in kg/min. Density of mercury =13.6 gm/cc

Chlorine gas = 0.009 gm/cc.

Discharge coefficient of V.M. = 0.98.

30. Water (viscosity = 1 cp) is to be pumped from a pond to the top of a tower 1829 cm above the water level. It is desired to deliver 0.4 m^3/min of water at a pressure of 2.08 kg/cm^2 absolute. The delivery line is 122 m long and 7.62 cm.*i.d.* Calculate the friction from the equation

$$f = \frac{0.046}{R_e^{0.2}} \text{ (for turbulent flow).}$$

31. What is the cost at 18 p per kWh to pump oil of density 0.90 gm/cc and viscosity 30 cP through a 250 mm, i.e. pipe when the head loss due to friction is 813 m kgf/kg, The overall efficiency of the pump motor set is 50%. Pipeline is horizontal and rate of flow is 300 m^3/hr.

32. A bed of alumina (d_p = 246 microns) is to be fluidized with air at 350°C and 7 kgf/cm^2. The ultimate density of alumina particles is 3.7 gm/cc. The porosity of the static bed is 0.40. If the static bed is 4 meter deep, what is the minimum height of the fluidized bed?

33. A bed of sized pulverized coal is to be fluidized with a liquid petroleum fraction having a viscosity of 18 cp. The height of the static bed is 2 metres and the porosity is 0.40. The density of coal particles is 1.36 gm/cc and that of the liquid is 0.88 gm/cc. Calculate the pressure drop at the onset of fluidization. Minimum porosity for fluidization is 0.40.

34. What is the maximum dia. of a spherical particle of dust of density 2.5 gm/cc which will settle in the atmosphere (air density 1.225 × 10^{-3} gm/cc, kin. vis. 0.149 cm^2/sec.) in good agreement with Stokes' law.

35. What will be the settling velocity of a spherical steel particles 0.04 cm dia in an oil of sp. gr. 0.82 and viscosity 10 cp? The specific gravity of steel is 7.87.

36. A filler press with 24 frames each of 60 × 60 × 0.05 cm dimensions gets fill with cake in 2 hr while working at constant pressure. Washing of the like is done at the same pressure as filtration and the volume of wash water is equal to 10% of the volume of filtrate per cycle. Calculate the washing time if 0.05 m^3 of filter cake is formed per m^3 of filtrate collected. Neglect filter medium resistance.

37. A slurry is being filtered at constant pressure in a plate and frame filter using 10 frames, having a total filtration area of 10 m . The filter delivers 250 litres filtrate in 30 min. To increase filtration capacity 10 more identical frames are added to the filter. All other conditions being the as before, how long will it take to collect 500 lit. filtrate ? Initial filter resistance is found to be negligible.

SECTION—D

1. The shear stress-shear rate relationship for a liquid whose apparent viscosity decreases with increasing shear rate is given by

(a) $\tau y_x = -m \left|\frac{dv_x}{dy}\right|^{n-1} - \frac{dv_x}{dy}$ for $n < 1$ (b) $\tau y_x = -m \left(\frac{dv_x}{dy}\right)^n$, for $n = 1$

(c) $\tau_{yx} = -m \left|\frac{dv_x}{dy}\right|^{n-1} - \frac{dv_x}{dy}$, $n > 1$ (d) $\tau_{yx} = -m \frac{dv_x}{dy} + \tau_0$

2. A Newtonian fluid (ρ = density), μ = viscosity, is flowing with a velocity v in a tube of diameter D. Let DP be the pressure drop across the length L. For laminar flow DP is proportional to (1994 G).

(a) $L\rho V^2/D$ (b) $D\rho V^2/L$

(c) $L\mu V/D^2$ (d) $\mu V/L$

3. Hydraulic mean diameter for flow through packed bed of spherical par tides of size Dp with porosity $\in$ is : (1995 G)

(a) $Dp \frac{\in}{6(1-\in)}$ (b) $\frac{Dp(1-\in)}{6\in}$

(c) $Dp \frac{2}{3}\frac{(1-\in)}{\in}$ (d) $Dp \frac{2\in}{3(1-\in)}$

4. Use of Raschig Rings in place of crushed stones as packing in packed beds (other things being same)

(a) Increases pressure drop, increases surface area

(b) Increases pressure drop, decreases surface area

(c) Decreases pressure drop, increase surface area

(d) Decreases pressure drop, decreases surface area

5. A suspension of uniform particles in water at a concentration of 500 kg of solids per cubic meter of slurry is settling in a tank. Density of particles is 2500 kg/m^3 and terminal velocity of a single particle is 20 cm/sec. What will be the settling velocity of suspension? Richardson and Zaki index is 4.6.

(a) 20cm/S (b) 14.3 cm/S

(c) 7.16 cm/S (d) 3.58 cm/S

6. For an ideal fluid the Reynold's number is :

(a) 2100 (b) 100

(c) zero (d) infinity

7. Bed pressure drop in air fluidized bed of catalyst particles (ρ_q = 2000 kg/m^3, Dp = 0.05 cm) of 60 cm. bed depth and bed porosity of 0.5 expressed in cm of water (manometer) is :

(a) 90 (b) 60

(c) 45 (d) 30

8. Match :

(i) Venturimeter	(a) Static head
(ii) Wet gas meter	(b) Kinetic energy
	(c) Volumetric displacement
	(d) Fluid drag force

9. A pressure filter is operated in the constant rate mode to yield 10 m in the first ten minutes as the operating pressure increase from zero. In the next 20 minutes the filtration was continued at constant pressure, after which it was stopped.

(a) Estimate the total volume of filtrate obtained during filtration.

(b) Estimate the washing rate at the end of filtration. Also determine washing time if the volume of wash liquid equals the volume of filtrate obtained.

10. For laminar flow of fluid in a circular tube, h_1 is the convective heat transfer coefficient at a velocity V_1. If the velocity is reduced by half and assuming the fluid properties are constant, the new convective heat transfer coefficient is :

(a) 1.26 h_1 (b) 0.794 h_1
(c) 0.574 h_1 (d) 1.74 h_1

11. A binary mixture of 100 mm size having densities of 2 gm/cm^3 and 4 gm/cm^3 is to be classified by elutriation technique using water. Estimate the range of velocities that can do the job and recommend a suitable value.

12. A small model reactor is to be built for scale up studies of the behaviour I of a proposed large industrial stirred tank reactor having 100 times capacity. The bigger unit of 2 m diameter will have a liquid depth of 2 m. This will be fitted with a four bladed Rushton-turbine of 0.6 m diameter.

(a) Estimate the dimensions of smaller unit.

(b) For optimum stirrer speed of 330 RPM observed in the smaller model, what will be the recommended speed in the Industrial unit under the following conditions :

(i) Power per unit volume is kept constant

(ii) Reynolds number does not change.

(c) What design criteria would you recommend for this type of study?

13. For a sphere of density ρ_s and volume V placed in a fluid of density, match the following :

1. Weight	(a) $(\rho_s - \rho)/Vg$
2. Buoyancy force	(b) ρVg
	(c) $\rho_s V_g$
	(d) $V_g (\rho_s + \rho)$

14. (a) Air flows through a packed bed of powdery material of 1 cm depth at superficial gas velocity of 1 cm/S. A manometer connected to the unit registers a pressure drop 1 cm of water. The bed has a porosity of 0.40. Assuming that Kozney-Carman equation is valid for the range of study, estimate the particle size of the powder, ρ air = 1.23 kg/m^3; μ air = 1.8 * 10^{-5} kg/(m.S).

(b) The following data were obtained on a section of piping through which an incompressible viscous fluid is flowing. See figure given.

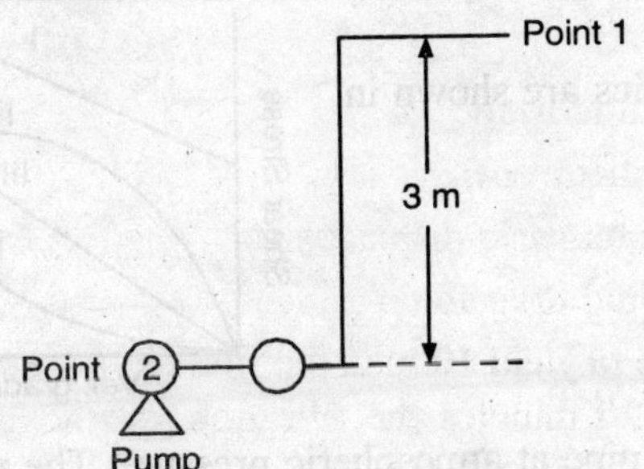

Point 1:
Pressure = 1.05×10^5 Pa
Cross sectional area = 5×10^{-4} m^2 elevation above point 2 = 3 m
Point 2:
Cross sectional area = $15 * 10^{-4}$ m^2
Pressure = $1.25 * 10^5$ Pa

Fluid velocity = 1 m/S; other data : Density of fluid = 1000 kg/m^3; power delivered by the pump: 7.5 W (assuming efficiency as 1).

Predict whether the flow is taking place from point 1 to 2 or from point 2 to1.

15. For an office meter the pressure recovery is..... than that of for a venturi meter.

16. As long as a pump is not used, a fluid will always flow from high pressure region to a low pressure regions. Reason, True or False.

17. The concentration and hydrodynamic boundary layers over a flat plate are of equal thickness, if Schmidt number is equal to unity. True/False (1994 G)

18. A tube of 0.05 sq. m cross-sectional area is packed with spherical particles upto a height of 0.25 m. Porosity of the bed is 0.35. It is desired to fluidize the particles with water (ρ = 1000 kg/m^3) $\mu = 10^{-3}$ Pa. Calculate the minimum velocity of fluidization given the Erguns equation.

$$\frac{-\Delta P}{L}\frac{D_P}{\rho_s V^2}\frac{\in^3}{(1-\in)} = \frac{150\mu(1-\in)}{D_p \rho_f V} + 1.75$$

Data : diameter of particles = 0.01 m; density of solid particles = 2600 kg/m^3.

19. For geometrically similar baffled stirred tanks, the power number is known to remain constant at high Reynolds number.

(a) Let P the power supplied per unit volume of fluid, N is the revolutions per second of the agitation, ρ the density of the fluid, μ the viscosity of the fluid and D the diameter of the impeller. Then determine α, β, γ and δ in the following equation.

$$P = N^{\alpha} \rho^{\beta} \mu^{\gamma} D^{\delta}$$

(b) What is the effect of Froude number on P?

20. A parabolic of radius R and density ρ_s is moving radially out in a centrifuge. The angular velocity of centrifuge is w. The density and viscosity of the fluid are ρ and μ. It is expected that Stokes law is valid. See figure below. Calculate velocity.

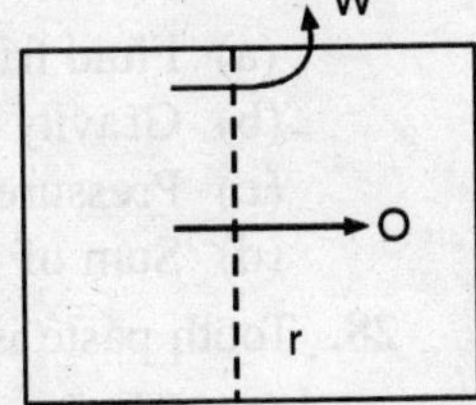

21. A particle attains its terminal settling velocity when,

(a) Gravity force + drag force = buoyancy force
(b) Gravity force – drag force = buoyancy force
(c) Buoyancy force + drag force = gravity force
(d) Drag force = buoyancy force

22. For a centrifugal pump the positive suction head is defined as :

(a) $h_{vs} + h_{ps}$ (b) $h_{vd} + h_{pd}$
(c) $h_{vs} + h_{ps} - P_s$ (d) $h_{vd} + h_{pd} - P_d$

23. Match the following :

The shear stress versus velocity gradient characteristics are shown in figure given :

I. — (a) (Newtonian)

II. — (b) (dilatent)

III — (c) (Bingham plastic)

IV — (d) (pseudo plastic)

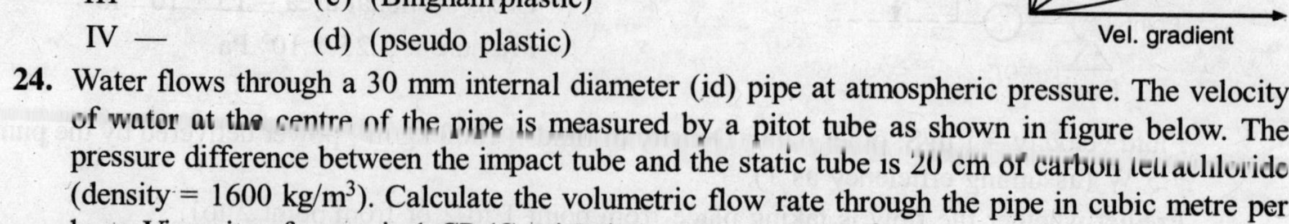

24. Water flows through a 30 mm internal diameter (id) pipe at atmospheric pressure. The velocity of water at the centre of the pipe is measured by a pitot tube as shown in figure below. The pressure difference between the impact tube and the static tube is 20 cm of carbon tetrachloride (density = 1600 kg/m^3). Calculate the volumetric flow rate through the pipe in cubic metre per hour. Viscosity of water is 1 CP (Centi Poise).

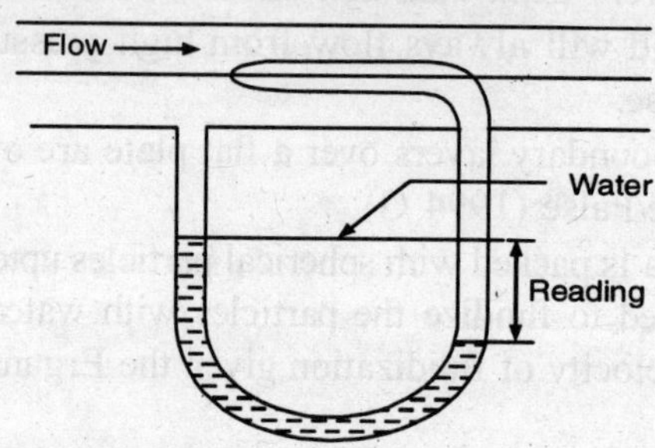

25. A piece of wood floats at the interface between layers of oil and water as shown in figure below. The specific gravity of wood and oil are 0.95 and 0.75 respectively. Determine the volume fraction of wood in the water layer. (Fig.)

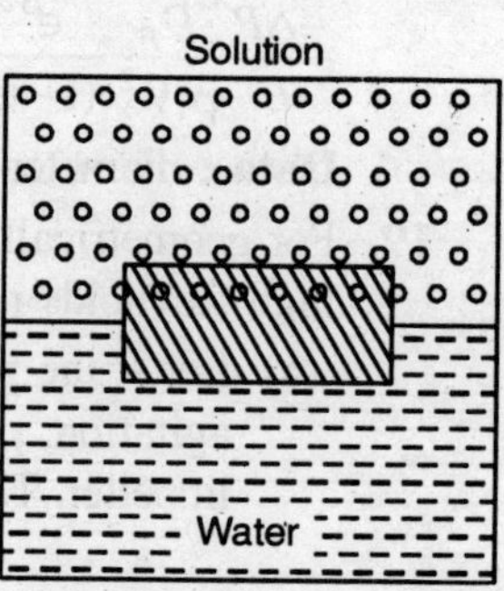

Fig.

26. In Filtration, If dt/dv = 9.3v + 8.5 and

Alpha (α) = Alpha O (α_0) * ΔP^S

and S = 0.30

Calculate the time for V to reach 3.50 V is Filtrate Volume.

27. Fuidized beds are formed when :

(a) Fluid friction is zero

(b) Gravity force is less than fluid friction

(c) Pressure forces equal gravity forces

(d) Sum of fluid friction and pressure forces is equal and opposite to gravity forces.

28. Tooth paste is a :

(a) Bingham plastic (b) Pseudo plastic

(c) Newtonian liquid (d) Dilatent

29. A Rotameter through which air at room temperature and atmospheric pressure is flowing, give a certain reading for a flow rate of 100 cc/s. If helium (molecular weight 4) is used and the rotameter shows the same reading, the flow rate is :

(a) 26 cc/S (b) 42 cc/S

(c) 269 cc/S (d) 325 cc/S

30. Match :

1. Gear pump
2. Air lift pump

(a) Suspension
(b) Concentrated sulphuric acid
(c) Viscous oil
(d) Toluene

31. Match:

1. 1/7th. power law
2. Hagen-Poiseuille equation

(a) Irrotational flow
(b) Turbulent flow
(c) Invicid flow
(d) Laminar flow in pipes

32. Match :

1. Saltation velocity
2. Compressible cake

(a) Filtration
(b) Fluidization
(c) Pneumatic conveying
(d) Screw conveyor

33. A tank full of water is open at the top with a hole near the bottom, the area of which is 0.25 m^2 and the cross-sectional area of tank is 1 m^2. If the height of water above the hole is maintained at 10 m, what is the volumetric flow rate out of the hole?

34. Water trickles by gravity over a bed of particles, each 1 mm diameter in 1 bed of diameter 6 cm and height 2m. The water is fed from a reservoir whose diameter, is much larger than that of the packed bed with water maintained at a height of 0.1 m above the top of the bed. The bed has a porosity of 0.31. Calculate the volumetric flow rate of water if its viscosity is 1.0 cP.

35. Ammonia at atmospheric pressure and 300° K with a bulk stream velocity of 10 m/S flows through a pipe id 25 cm. Calculate the pressure drop per 100 m length of the pipe and the power consumed. Friction factor $f = 0.079\ Re^{-0.25}$ in the turbulent regime. Viscosity of ammonia may be taken as 10.2×10^{-6} kg/m.S).

36. For laminar flow of fluid in a circular pipe of Radius R, the Hagen-Poiseuille equation predicts the volumetric flow rate to be proportional to :

(a) R (b) R^2
(c) R^4 (d) $R^{0.5}$

37. In the low Reynolds number region, the drag force on a sphere is proportional to:

(a) V (b) V^2
(c) V^4 (d) $V^{0.50}$

Where V is the terminal settling velocity of the sphere in a viscous medium.

38. The hydraulic diameter of an annulus of inner and outer radii R_i and R_o respectively is :

(a) $4\ (R_o - R_i)$ (b) $\sqrt{R_o - R_i}$
(c) $2\ (R_o - R_i)$ (d) $R_o + R_i$

39. For laminar flow of fluid through a packed bed of spheres of diameter d, the pressure drop per unit length of bed depends upon the sphere diameter is:

(a) d (b) d^2
(c) d^4 (d) $1/d^2$

40. A suspension of glass beads in Ethylene glycol has a hindered settling velocity of a single glass bead in Ethylene glycol is 17 mm/S. If the Richardson-Zaki hindered settling index is 4.5, then the volume fraction of solids in suspension is :

(a) 0.10 (b) 0.40
(c) 0.60 (d) None of these

41. In the figure below calculate the power required by the pump to deliver water at 3 m/S from a pond. The inner diameter of the pipe is 25 mm.

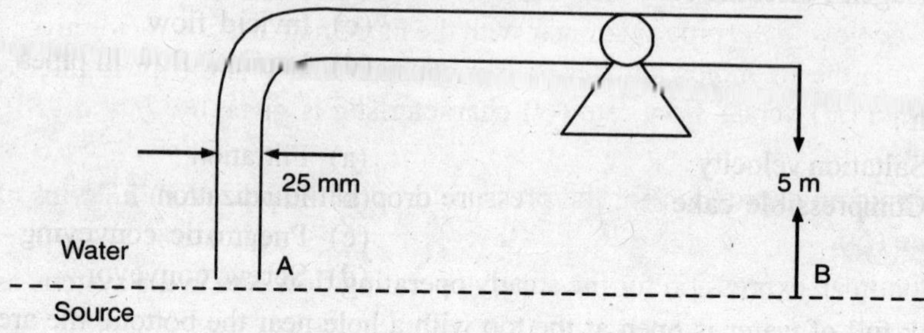

Neglect all losses in the pipe. Density of water is 1000 kg/m³.

42. In a delivery line for carbon tetrachloride at constant flow rate of 4×10^{-5} m³/S, the first 100 m long section is of 20 mm inside diameter smooth pipe followed by another 1000 m long section of 50 mm inside diameter smooth pipe as shown in figure below.

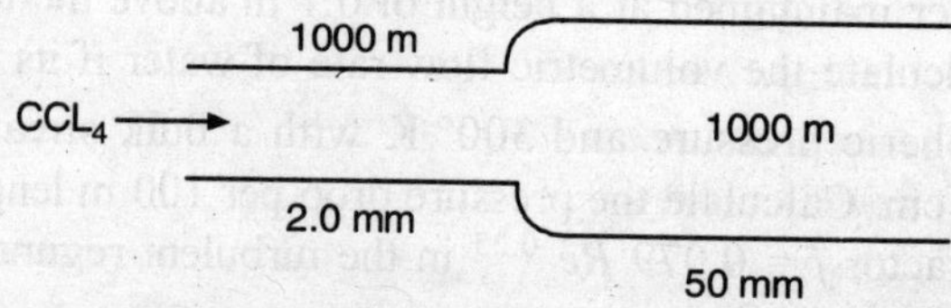

Estimate the pressure drop over the entire length of delivery line. Neglect the minor losses due to sudden enlargement of pipe diameter. For carbon tetrachloride : Viscosity = 10^{-3} Pa. s and density = 1500 kg/m³. For laminar flow, $f = 16/N_{Re}$. For turbulent flow $f = 0.079\ Re^{-0.25}$.

43. A rectangular piece of wood (density = 650 kg/m³) 75 mm × 75 mm × 2000 mm in size floats on water (density = 1000 kg/m³). How many kilograms of lead of density 11,200 kg/m³ must be put on top of the wooden piece so that it will float upright with 300 mm of its length above the water surface.

44. Boundary layer separation is characterized by one of the conditions given below, where Re. is the Reynolds number for the flow. Select the appropriate conditions.

(a) Re < 1, accelerating flow (b) Re > 1, accelerating flow
(c) Re < 1 decelerating flow (d) Re > 1, decelerating flow

45. A globe valve is most suitable for applications in which :

(a) The valve is required to be either fully open or fully closed
(b) Flow control is required
(c) The fluid contains dispersed particles
(d) One way flow is required

46. The dependence of volumetric flow rate (Q) on the pressure drop is given by : $\Delta P \propto Q^n$ for different flow regimes : Match the exponent n to the flow regime given below :

1. Laminar flow (a) $n < 0.50$

(b) $n = 0.50$

(c) $n = 1.0$

(d) $n > 1.0$

47. A fluid is pumped through a long horizontal pipe of diameter D and length L, using a centrifugal pump. The flow in the pipe is laminar with the Farming's friction factor given by $f = A/Re$ where $Re = DV\rho/\mu$ is the Reynolds number. A is a constant and V is the average velocity in the pipe. The pump head (H) versus flow rate (Q) characteristic is given by, $H = \alpha - \beta Q$ where α and β are constants.

(a) Obtain an expression for the pressure drop in the pipe (ΔP) in terms of the volumetric flow rate (Q).

(b) Obtain an expression for the steady operating flow rate in the system, assuming that the only pressure loss in the system is due to the pipe flow,

48. A round jet impinges on a flat plate and spreads out radially to form a film of decreasing thickness. The radius of the jet far from the stagnation point (O) is R, and the velocity is μ. Assume the flow to be inviscid and axisymmetric, and neglect gravity. Also assume that the velocity profiles are flat far from the stagnation point. The pressure outside the jet is atmospheric and the density of fluid (ρ) is constant.

(a) Obtain the velocity of the fluid in the film $v(r)$ and the thickness of film, $t(r)$ at any radial distance r far from stagnation point.

(b) Obtain the normal force acting on the plate.

49. A concentrated suspension of spherical Quartz particles in water settles under gravity. The particle diameter is $D_P = 10^{-5}$ m and the practices density is $\rho_P = 2650$ kg/m^3. The initial voltage in the suspension is $\in = 0.80$.

(a) Obtain the expression for the terminal velocity (u_t) of a single particle assuming Stoke's law to be valid.

(b) Find the initial setting velocity (v_s) of the particles in the suspension given $v_s = u_t \in^{4.6}$.

(c) Calculate the upward velocity of water in the suspension resulting from settling of particles for $\in = 0.80$.

SECTION—D

1. The velocity profile for a Bingham plastic fluid flowing (under laminar conditions) in a pipe is:

(a) Parabolic

(b) Flat

(c) Flat near the wall and parabolic in the middle

(d) Parabolic near the wall and flat in the middle.

2. A pipe has a porous section old length L as shown in the figure. Velocity at the start of the section is V_0. If the fluid leaks into the pipe through the porous section at volumetric rate per unit area is q/θ (X/L) 21, what will be their axial velocity in the pipe at any x? Assume incompressible one dimensional flow, i.e. no gradients in the radial direction.

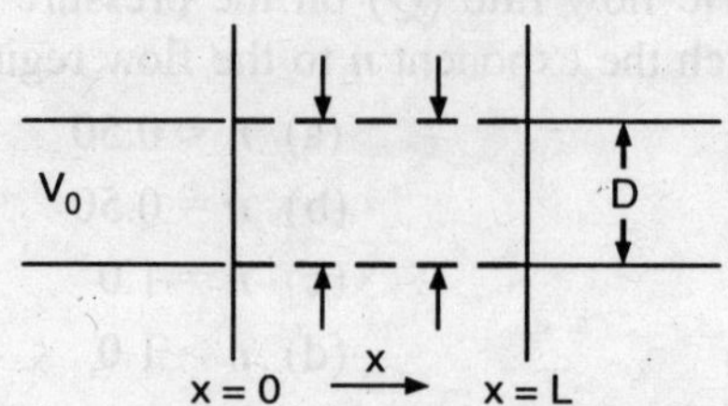

(a) $Vx = V_0 + qx^3/(L^2D)$ (b) $Vx = V_0 + qx^3/(3L^2)$

(c) $Vx = V_0 = V_0 + 2qx^2/(LD)$ (d) $Vx = V_0 + 4qx^3/(3LD)$

3. A centrifugal pump is used to pump water through a horizontal distance of 150 m and then raised to a overhead tank of 10 m above. The pipe is smooth with an id of 50 mm. What head (m. of water) must the pump generate at its exit (*E*) to deliver water at a flow rate of 0.001 cubic m^3/s. The fanning fraction factor is (*f*) 0.0062.

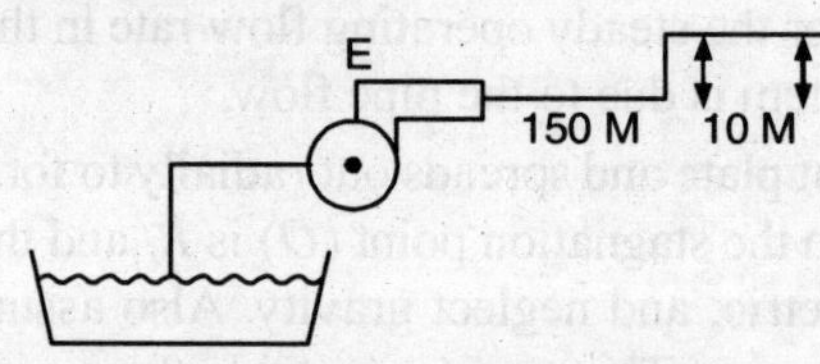

(a) 10 m (b) 11 m

(c) 12 m (d) 20 m

4. Match the following dimensionless numbers with appropriate ratio of forces:

Diamensionless number	Ratio of forces
P Froude number	1. Shear force/intert force
Q Reynolds number	2. Convective heat transfer/conductive heat transfer
R Friction factor	3. Gravitational force/viscous force
S Nusselt number	4. Inertia force/gravitational force
	5. Inertia force/gravitational force

(a) P–1, Q–2, R–5, S–3; (b) P–5, Q–4, R–3, S–2;

(c) P–5, Q–4, R–3, S–2; (d) P–3 Q–4, R–5, S–1;

5. A bed of spherical (SP.gr. = 2.0) of uniform size 1500 μm is 0.5 mm in diameter and 0.5 m high in the packed bed state, the porosdity may be taken as 0.40. Ergun equation for the above particle fluid system (in SI units) is given below:

$\Delta P/L = 375{,}000\ V_{om} + 10.96 \times 10^6\ V^2_{om}$ (SI units).

If water is to be used as the fluidizing medium, the minimum fluidization velocity V_{om} is:

(a) 12 mm/s (b) 16 mm/s

(c) 24 mm/s (d) 28 mm/s

6. In actual operation the above bed has a height of 1 m. What is the porosity of the fluidized bed?

7. The basic filtration equation is given as:

$dt/dv = \mu/(A\Delta P)\ (\alpha cV/A + Rm)$; where *V* is the volume of the filtrate; *A* is the filtration area; α is specific cake resistance μ is viscosity of the filtrate and *C* is the concentration of solids in slurry of feed.

In a 20 minute constant rate filtration, 5 cubic m of filtrate was obtained. If this is followed by a constant pressure filtration, how much more time in minutes will take for another 5 cubic m of filtrate to be produced? Neglect filtrate medium resistance, *Rm*. Assume incompressible cake.

(a) 10 (b) 20
(c) 25 (d) 30

8. The pressure differential across a vertical venturimeter is measured with the help of a mercury manometer to estimate the flow rate of water flowing through it. The expression for the velocity of water at the throat is:

(a) $(V_2^2 - V_1^2)/2g = h\,\rho m/\rho f$ (b) $(V_2^2 - V_1^2)/2g = h\,(\rho m - \rho w)/\rho f$
(c) $(V_2^2 - V_1^2)/2g = H + h\,(\rho m - \rho w)/\rho f$ (d) $V_2^2/2g = h\,(\rho m - \rho w)/\rho f$

9. The equivalent diameter for flow through a rectangle duct of width *B* and height *H* is:

(a) $HB/2\,(H + B)$ (b) $HB/(H + B)$
(c) $2HB/(H + B)$ (d) $4HB/(H + B)$

10. Match the system in Group-I with equipment to separate them in Group-II.

Group-I	Group-II
P. Gas–solid	1. Filter press
Q. Liqiuid–liquid	2. Cyclone
	3. Decanter
	4. Thickner

(a) P-1, Q-2 (b) P-2, Q-3
(c) P-3, Q-4 (d) P-4, Q-1

11. For a cyclone diameter of 0.2 m with tangent velocity of 15m/sec. at the wall,the separation factor is:

(a) 2250 (b) 1125
(c) 400 (d) 230

12. In constant pressure filtration:

(a) Resistence decreases with time (b) Rate of filtration is constant
(c) Rate of filtration increases with time. (d) Rate of filtration decreases with time

13. For the manometer set up shown in figure, the pressure diffrence $P_A - P_B$ is given by:

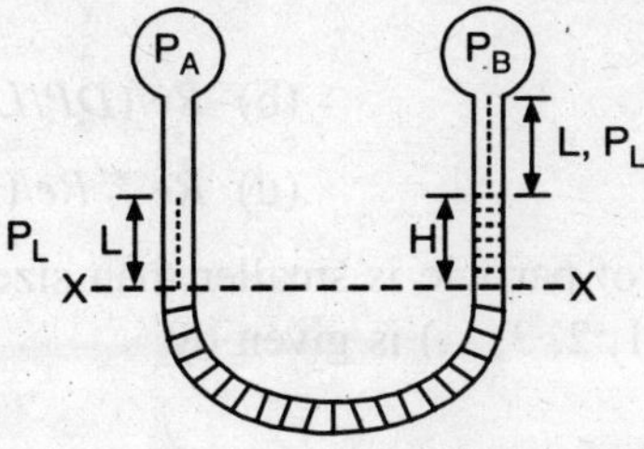

(a) $(\rho H - \rho_{air})g\,{*}H$ (b) $(\rho H - \rho L)g\,{*}H$
(c) $(\rho H - \rho L)g{*}\,H + (\rho L - \rho air)g{*}(L - H)$ (d) $(\rho H - \rho L)g{*}\,L + (\rho L - \rho_{air})g{*}H$

14. A conical tank with a bottom of cross sectional area, A is filled with water and is mounted on supports. What is the force *F* with which plate *X* must be pushed up to prevent water from leaking? Assume that the density of air is negligible as compared to the density of water ρ*L*.

(a) ρ*L V g* (b) ρ*LA H g*

(c) ρ*LVg*/2 (d) ρ*LV g*/3

15. Three piping net works as shown in the figure are placed horizontally. They are made using idential pipe segments and are subjected to the same pressure drop across them. Asuming no pressure losses at the junctions, the three networks are related as $Q_1 : Q_2 : Q_3$:

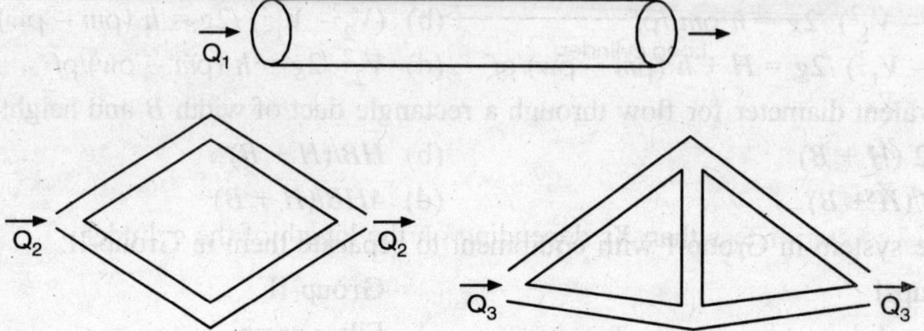

Figures: Q_1, Q_2, Q_3

(a) $1:\sqrt{3}:2$ (b) 1:2:3

(c) 1:2:2 (d) $1:\sqrt{2}:\sqrt{2}$

16. To keep power input constant for a stirred vessel operating under fully developed turbulent flow conditions (constant power number), if the impeller diameter is increased by a factor of:

(a) 1.2 to power of 1.5 (b) 1.2 power of 0.6

(c) 1.2 power of 0.67 (d) 1.2 power of 1.67

17. A centrifugal filtration unit operating at a rotational speed *w* has inner surface of the liquid (density ρ_L) located at a radial distance *R*, from the axis of rotation. The thickness of the liquid film is and if no cake is formed. The intial pressue drop during filtration is:

(a) $1/2\ w^2R^2\rho_L$ (b) $1/2\ w^2\delta^2\rho_L$

(c) $1/2\ w^2\delta\rho_L\ (2R + \delta)$ (d) $1/2\ w^2\ R\rho_L(R + 2\delta)$

18. The Kozney-Carman equation is written in terms of non-dimentional numbers, gives $\Delta P/\rho u^2$ proportional to:

(a) (*L*/*DP Re*) (b) *Re*/(*DP*/*L*)

(c) (*LDp*)/Re^2 (d) *Re* * *Re*/(*Dp*/*L*)

19. The cumulative mass fraction of particle is smaller than size *di* for collection of N_i particles of diameter *di* and mass *mi* (*i* = 1, 2, 3, ...) is given by

(a) $\sum_{i=1}^{j} N_i^3 d_i \ / \sum_{i=1}^{\infty} N_i d_i^3$ (b) $\sum_{1}^{j} N_i\ m_i\ d_i^3 / \sum_{i=1}^{\infty} N_i^{\ 3}\ m_i\ d_i$

(c) $\sum_{i}^{j} N_i\, m_i\, d_i^{\ 2} / \sum_{i=1}^{\infty} N^2{}_i\, m_i\, d_i$ (d) $\sum_{1}^{\infty} N_i\, m_i\, d_i \sum_{1}^{j} N_i\, m_i\, d_i$

20. A long cylinder and a sphere both of 5 cm diameter from the same porous material.The flat ends of cylinder are sealed. Both the cylinder and sphere are saturated with the same solution of sodium choride. Later both the objects are immersed for a short and equal interval of time in a large tank of water, which is well agitated. The fractions of salt remaining in the cylinder and sphere are *Xc* and *Xs* respectively. Which of the following statements is correct?

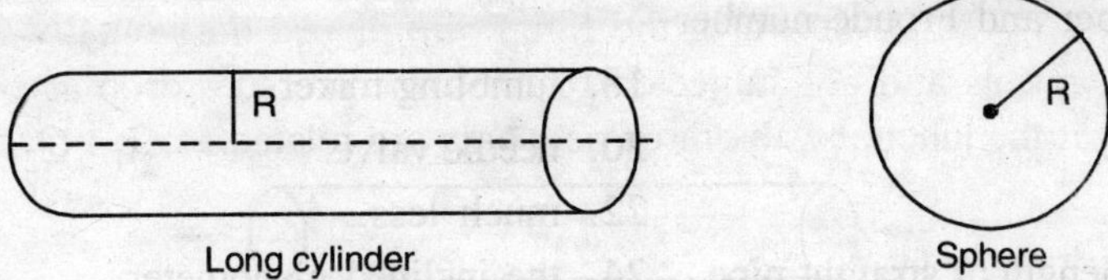

(a) $Xc > Xs$

(b) $Xc = Xs$

(c) $Xc < Xs$

(d) *Xc* is greater or less than *Xs* depending on the length of the cylinder.

ANSWERS

SECTION–A

1. (d	2. (c)	3. (b)	4. (a)	5. (b)
6. (b)	7. (b)	8. (a)	9. (c)	10. (b)
11. (a)	12. (d)	13. (c)	14. (b)	15. (a)
16. (c)	17. (c)	18. (d)	19. (b)	20. (c)
21. (a)	22. (d)	23. (c)	24. (a)	25. (d)
26. (d)	27. (c)	28. (a)	29. (a)	30. (d)
31. (a)	32. (b)	33. (c)	34. (b)	35. (c)
36. (a)	37. (d)	38. (b)	39. (c)	40. (a)
41. (b)	42. (d)	43. (c)	44. (b)	45. (a)
46. (c)	47. (d)	48. (b)	49. (a)	50. (d)
51. (b)	52. (a)	53. (b)	54. (c)	55. (d)
56. (c)	57. (d)	58. (a)	59. (b)	60. (a)
61. (d)	62. (c)	63. (b)	64. (b)	65. (a)
66. (c)	67. (b)	68. (d)	69. (c)	70. (d)
71. (a)	72. (b)	73. (c)	74. (b)	75. (a)

SECTION–B

1. greater.
2. thixotropic and the rheopectic.
3. Reynolds number and relative roughness.
4. particles size and nature of the fluid.
5. deep and narrow.
6. size, density, viscosity, density.

7. Froude
8. the effective weight of the bed.
9. size and surface properties.
10. more
11. blower.
12. vacuum.
13. electrical conductivity.
14. overtaxing and swirling.
15. geometrical, dynamic and kinematic.
16. Reynolds number and Froude number
17. gas, liquid,
18. tumbling mixer,
19. Banbury mixer.
20. needle valve.
21. corrosive.
22. much less.
23. the equivalent length of straight pipe.
24. the inclined manometer
25. small pressure differences in gases.
26. the Bourdon gauge.
27. remote recording of pressure difference.
28. between the impact and the static pressure.
29. rotameter.
30. hot wire anemometer.
31. colloidal particle.
32. an electrolyte or a surface active agent.
33. glue, lime, alumina and sodium silicate.
34. minimum.
35. diffusion.
36. most of the solids are removed at the periphery of the disks.
37. liquor no longer flow out the discharge (or the filtration pressure suddenly rises).
38. filtering, washing, dewatering and blow.
39. constant rate filtration.
40. the centrifugal pressure due to liquid.

SECTION–C

1. 1.5
2. $u/u_{max} = (y/r)^{1/7}$
3. r = radius of pipe, y = distance.
4. Water hammer is caused when water flowing in a long pipe is suddenly brought to rest by closing the valve.
5. Most of the pressure drop is not recoverable, (i.e., it is inefficient).
6. An orifice meter can measure the average velocity of the entire stream of fluid, while the pitot tube measures velocity at one point only.
7. The range can be increased by the use of floats of different densities.
8. It is the excess of the sum of velocity and pressure heads over the vapour pressure of the liquid at the suction.
9. It depends on :
   (a) The speed of the impeller.
   (b) The radius of the impeller.
   (c) The velocity of the fluid leaving the impeller.

10. (a) Head. (b) Efficiency. (c) BHP.

11. Gear pumps are suitable for a wide range of corrosive liquids, since these can he readily made in a variety of materials including plastics.

12. It is a positive acting rotary pump.

13. Dissolved gases contained in liquid may come out of solution during pumping giving rise to pockets of gas. This phenomenon is called cavitation.

14. Nearly 4000 atmospheres.

15. Hytor pumps are a group of vacuum pumps.

16. (a) Contacts fluid with all parts of the solid particles,
(b) Prevents particle segregation
(c) Minimizes temperature variation.

17. It is the case when the fluid velocity through a bed of solid is large enough to entrain all particles.

18. Banbury mixer.

19. The purpose are :
(a) Blending miscible liquids
(b) Dispersing gas in liquid
(c) Dispersing immiscible liquid in form of emulsion
(d) Suspending solid particles
(e) Promoting heat transfer between the liquid and a coil/jacket

20. The advantage are :
(a) Filtration under high pressure
(b) More effective washing
(c) Less labour cost.

21. (a) Characteristics of the feed slurry.
(b) Thickness of the cake.

22. The factors are :
(a) Area of filltering surface
(b) Pressure drop
(c) Resistance of the cake and the medium
(d) Properties of the cake the filtrate.

23. (a) Diatomaceous earth
(b) Asbestos
(c) Purified wood cellulose.

24. For removal of very small amounts of precipitate from large volumes of water.

25. $\Delta P = \dfrac{32\, Lu\mu}{gD^2}$

26. For the pitot tube

$$u_m = 0.95 \sqrt{\frac{2gc(P_s - P_o)}{\rho}}$$

$$P_s - P_o = 5 \times 1 = 5 \text{ gmf/cm}^2$$

$$u_m = 0.95\sqrt{\frac{2 \times 981 \times 5}{0.9}} = 99.20 \text{ cm/sec} = 99.20 \text{ cm/sec.}$$

27. Maximum velocity occurs at the centre of the pipe

$$P_s - P_o = 9 \text{ cm of water} = 9.0 \text{ gmf/cm}^2$$

$$u_{max} = 0.9\sqrt{\frac{2 \times 981 \times 9}{0.8}} = 133.7 \text{ cm/sec}$$

$$\frac{u_a}{u_{max}} = 0.8$$

So, $u_a = 0.8 \times 133.7 = 106.95$ cm/sec.

$$\text{Flow rate of oil} = \frac{\pi}{4}(8)^2 \times 106.95 \times 60 \times 1/10^3 = 322.5 \text{ lit./min.}$$

28. $P_A - P_B = \frac{g}{gc} R_m (\rho_A - \rho_B)$

$$= 1 \times 10(13.6 - 1.3) = 123 \text{ gmf/sec.}$$

$$\beta = 1/5 = 0.2$$

$$u_0 = \frac{Cd}{\sqrt{1-\beta^4}} \frac{2gc(P_A - P_B)}{\rho_\beta}$$

$$= \frac{0.61}{\sqrt{1-(0.2)^4}}\sqrt{\frac{2 \times 981 \times 123}{1.3}} = 263 \text{ cm/sec.}$$

Flow rate in kg/min

$$= \frac{\pi}{4}(1.0)2 \times 263 \times 1.3 \times 60 \times 1/10^3$$

$$= 16.1.$$

29. $P_A - P_B = \frac{g}{gc} R_m (\rho_A - \rho_B)$

$$= 1 \times 1.3\,(13.6 - 0.0009) = 17.67 \text{ gmf/cm}^2$$

$$\beta = \frac{30}{60} = 0.5$$

$$u_0 = \frac{0.98}{\sqrt{1-(0.5)^4}}\sqrt{\frac{2 \times 981 \times 17.67}{0.009}} = 1986.4 \text{ cm/sec.}$$

Flow rate of chlorine

$$= \frac{\pi}{4}(30)2 \times 1986.4 \times 0.009 \times 60 \times \frac{1}{10^3}$$

$$= 758.2 \text{ kg/min.}$$

30. Velocity in the pipe

$$= \frac{0.34 \times 10^6}{60 \times \frac{\pi}{4}(7.62)^2}$$

$$= 124.25 \text{ cm/sec.}$$

$$Re = Du\rho/\mu$$

$$= \frac{7.62 \times 124.25 \times 1}{1 \times 10^{-2}} = 94684$$

$$f = \frac{0.046}{(94684)^{0.2}} = 0.00465$$

31. Flow rate of oil

$$= \frac{300 \times 0.90 \times 1000}{3600} = 75 \text{ kg/sec.}$$

$$\eta W_p = 813 \times 75/76 = 802.3 \text{ hp}$$

$$W_p = 802.3/0.5 = 1604.6 \text{ hp} = 1180.19 \text{ kWh}$$

Cost of pumping

= Rs. 0.18 × 1180.19

= Rs. 212.43

32. $\in_{mf}$ = Minimum bed porosity = 1 – 0.356 (log246 – 1)

= 0.505

$$h_s(1 - \in pa)\, \rho s A = h_{mf}(1 - \in_{mf})\, P_s A$$

or,

$$hmf = hs\,\frac{(1 - \in pa)}{(1 - \in mf)} = \frac{400(1 - 0.40)}{(1 - 0.505)}$$

$$= 4085 \text{ cm.}$$

33. $Dp\ mf = g/gc\ hmf\,(1 - \in_{mf})\,(\rho s - \rho_f)$

Since, $\in pa = \in mf$

$h_{pa} = h_{mf}$

Hence, $\Delta P_{mf} = 1 \times 2 \times 100\,(1 - 0.4)\,(1.36 - 0.88)$

$= 57.6 \text{ gmf/cm}^2.$

34. $\rho_s = 2.5$ gm/cc

$$\frac{\mu}{\rho} = 0.149 \text{ cm}^2/\text{ sec. } r = 1.225 \times 10^{-3} \text{ gm/cc}$$

$$\mu = 1.225 \times 10^{-3} \times 0.149 = 0.183 \times 10^{-3} \text{ gm/cm sec.}$$

$$k = 3.3 = D_p \left[\frac{981 \times 1.225 \times 10^{-3} \times 2.5}{(0.183 \times 10^{-3})^{-2}}\right]^{1/3}$$

$$= 96.445 D_\mu$$

So, $D_p = 3.3/96.445 = 3.42 \times 10^{-2}$ cm.

35. $K = 0.04\left[\frac{981 \times 0.82(7.87 - 0.82)}{(10 \times 10^{-2})^2}\right] = 3.3$

$$u_f = \frac{981 \times (0.04)^2 (7.87 - 0.82)}{18 \times 10 \times 10^{-2}} = 6.148 \text{ cm/sec.}$$

Settling velocity = 6.148 cm/sec.

36. Volume of frame = volume cake

= 60 × 60 " 0.05 × 24 = 4320 cc.

Volume of filtrate = 4320/.05 = 86400 cc

= 86.4 lit.

For constant pressure filtration,

$$V^2 = 2k\,\theta \text{ ; } K = \frac{(86.4)^2}{2 \times 2}\frac{\text{lit.}^2}{\text{hr.}}$$

$$\text{Final rate of filtration} = K/V = \frac{(86.4)^2}{4 \times 86.4}$$

= 21.6 lit/hr.

$$\text{Rate of washing} = \frac{21.6}{4} = 5.4 \text{ lit/hr.}$$

Vol.of wash water = 86.4 × .10 = 8.64

$$\text{Washing time} = \frac{8.64}{5.4} = 1.6 \text{ hr.}$$

37. For constant pressure filtration,

$$V^2 = 2A^2 \frac{\Delta P}{\alpha W \mu} . \theta$$

For 1st filter,

$$V_1^2 = 2A_1^2\left(\frac{\Delta P}{\alpha W\mu}\right).\theta_1$$

For 2nd filter,

$$V_2^2 = 2A_2^2\left(\frac{\Delta P}{\alpha W\mu}\right)\theta_2$$

So ,
$$\frac{V_1^2}{V_2^2} = \frac{A_1^2\theta_1}{A_2^2\theta_2}$$

$$\frac{(250)^2}{(500)^2} = \frac{(10^2)}{(20)^2}.\frac{30}{\theta^2}$$

So,
$$\theta_2 = 30 \text{ min.}$$

Time taken = 30 minutes.

SECTION–D

1. (c)
2. (c)
3. (d)
4. (a)
5. (b)
6. (d)
   Solution Re $No = DV\rho/\mu$; $\mu = 0$ for ideal fluid (viscosity)
7. (b)
   Solution $\Delta P = L(1-\varepsilon)(\rho_s - \rho_f) = 60 \times 0.5 \times 2 = 60$ cm. of water $= \Delta P/\rho = 60/1 = 60$
8. (*i*) – (*b*); (*ii*) – (*c*)
9. Solution for filtration, general expression is,

$$\frac{dV}{d\theta} = \frac{A^2\Delta P}{\alpha\mu W(V+V_e)}; \text{ with } V_e = 0$$

$$\frac{dV}{d\theta} = \frac{A^2}{\alpha\, w\mu}\frac{\Delta P}{V} = K\frac{\Delta P}{V}; K = \frac{A^2}{\alpha\omega\mu};$$

At constant rate, $dV/d\theta$ = constant

So
$$DP = \frac{V}{K}\frac{dV}{d\theta} \quad ...(1)$$

At constant pressure, ΔP = constant

$$\frac{dV}{d\theta} = K\frac{\Delta P}{V}\text{; OR } V^2 = 2K\,\Delta P\theta$$

$$\theta = \frac{V^2}{2K\Delta P} + C\text{ ; at } \theta = 0\text{ ; } \therefore C = 0$$

For constant Rate, $\frac{dV}{d\theta} = \frac{10}{10} = 1 \text{ m}^3\text{/min.}$

Pressure developed at the end of constant rate $= \Delta P = \frac{V}{K}\frac{dV}{d\theta} = \frac{10}{K}$ (1)

Constant pressure operation $= \theta = \frac{V^2}{2K\Delta P} = \frac{V^2}{2K\left(\frac{10}{K}\right)}$

$= V^2/20 \text{ ; } \theta = 20 \text{ ; } V^2 = 400 \text{ m}^6 \text{ ; } V = 20 \text{ m}^3$

(*a*) Filtrate obtained $= 10 + 20 = 30 \text{ m}^3$

(*b*) $dV/d\theta = K\Delta P/V = K\,(10/K)\;1/20 = 1/2 \text{ m}^3\text{/min.}$

Washing Rate $= \frac{1}{1}\times\frac{1}{2} = \frac{1}{2}$ m^3/min (For filter press).

Total filtrate $= 10 + 20 = 30 \text{ m}^3$ = volume of wash liquid.

Washing time $= \frac{30}{1/2} = 60$ min.

10. (b)

11. More than fluidization velocity (use V_{om} of page 97)

12. Solution :

V | 1000 V | $Dt = 2$ mJ, $D_a = 0.6$ m, $h = 2$m, $h/Dt = 1$

$rpm = 330$

Volume $= \pi\, Dt^2\, h/4$

(*i*) $D_a = 0.06 \text{ ; } D_t = 0.2,\ h = 0.2$

$\therefore$ $h = Dt \text{ ; } \frac{2\pi}{1000} = \frac{\pi Dt^2}{4} Dt \text{ ; } Dt^3 = 8/1000$

$Dt = 0.2 \text{ m ; } h = 0.2 \text{ m ; } D_a = 0.3\ Dt = 0.06 \text{ m}$

$\frac{D_a}{D_t} = \frac{0.6}{2} = 0.3 \text{ ; Volume} = \frac{\pi Dt^2}{4}h = 1000 \text{ V}$

$$V \frac{\pi D_t^3}{4000} = \frac{\pi(2^3)}{4000} = \frac{2\pi}{1000}$$

(ii) (a) $$N_P = \frac{Pg_c}{n^3 D_a^5 P}; \frac{N_P \rho}{pg_c} = \frac{1}{330^3 (0.06)^5}$$

For large reactor, $$\frac{N_P \rho}{pg_c} = \frac{1}{n^3 D_a^5}$$

So $$\frac{1}{330^3 (0.06)^5} = \frac{1}{n^3 (0.6)^5}; n^3 = 330^3 \left(\frac{0.06}{0.6}\right)^5$$

$$n = 7.1 \text{ rpm}$$

(b) $$N_{Re} = n D_a^2 \rho/\mu \,;\, N_{Re}\, \mu/\rho = 330\,(0.06)^2$$

For larger one, $$\frac{N_{Re^{\mu}}}{\rho} = n D_a^2$$

$$330\,(0.06)^2 = n\,(0.6)^2 \,;\, n = 3.3 \text{ RPM}$$

13. $1 - c; 2 - a$

14. Solution (G) At (2), energy per unit mass :

$$\frac{P}{\rho * g} + \frac{g}{gc}\Delta Z + \frac{U^2}{2gc} = \frac{1.25 \times 10^5}{1000 \times 9.81} + 0 + \frac{1^2}{2 \times 9.81} = 12.79\text{m}$$

$$V_1 A_1 = V_2 A_2$$

$$V_1 * 5 \times 10^{-4} = 1 \times 15 \times 10^{-4} \,;\, V_1 = 3 \text{ m/s}$$

At point 1, $$\frac{1.25 \times 10^5}{1000 \times 9.81} + 3 + \frac{3^2}{2 \times 9.81}$$

Energy per unit mass $= 10.7 + 3 + 0.458 = 14.14$ m

Energy is supplied from fluid at point 2 to move to point 1, here flow is from (2) to 1.

15. less

16. Reason—Pressure gradient ; True

17. True.

Sc. $= \frac{\mu}{\rho D} = 1$; momentum diffusibility = mass diffusibility.

18. $$\frac{\Delta P}{L} \frac{D_P}{\rho V^2} \frac{\varepsilon^3}{1-\varepsilon} = \frac{150\mu(1-\varepsilon)}{D_p \rho f^V} + 1.75$$

$$A = 0.05 \text{ m}^2 \,;\, L = 0.25\text{m}$$

$$Dp = 0.01 \text{ m}; \rho s = 2600 \text{ kg/m}^3$$
$$\rho f = 1000 \text{ kg/m}^3 \; ; \mu = 10^{-3} \; PaS; = 10^{-3} \text{ kg/(m.s)}$$

Substituting the valves,

$$\left(\frac{2600-1000}{0.75}\right)\left(\frac{0.001}{1000V^2}\right)\frac{0.35^3}{(1.035)} = \frac{150\times10^{-3}(1-0.35)}{0.001\times1000V} + 1.75$$

Gives, $1.75V^2 + 0.00975V - 0.00422 = 0$

Solving $V = 0.046$ m/s

Assume that the particle moves only radially derive an expression for the radial velocity of the particle at any radial location r in the centrifuge.

19. Solution (*b*) $P = N^{\alpha} \rho^{\beta} \mu^{\gamma} D^{\delta}$

Putting in M, L, T dimensions

$$\frac{M}{MT} = (T)^{-\alpha}$$

$$ML^{-1} T^{-3} = T^{-\alpha-\gamma} M^{\beta+\gamma} L^{-3\beta+\partial-\gamma}$$

Comaring coefficients of left hand side and right hand side,

$$\beta + \gamma = 1 \;...(i) \quad \therefore \beta = 1 - \gamma$$
$$-\alpha - \gamma = -3 \quad ...(ii) \; ; \quad \therefore \alpha + \gamma = 3 \; ; \alpha = 3 - \gamma$$
$$3 - 3\gamma + \gamma + \delta = 1 \; ; \delta = 2(\gamma - 1) \; ; 3\beta + \gamma - \chi = 1 \quad ...(iii)$$

Let $\delta = 0 \; ; r = 1 \; ; \therefore \beta = 0$ and $\alpha = 2$

Putting $\beta = 0$, $r = 1$ in (*iii*), then $\delta = 0$

$$\therefore \quad P = N^2 \rho^0 \mu^1 D^0 = N^2 \mu$$

20. Ans. Use centrifuge equation.

21. Ans. *b* or *c*

22. h_{vs} = velocity head at suction; h_{vd} = velocity head at discharge; h_{ps} = pressure head at suction; h_{pd} = pressure head at discharge; Ps = vapour pressure of liquid at discharge temperature.

$$\text{NPRS} = h_{vs} + h_{ps} - \text{Ps. Ans (b)}$$

23. I – *c*; II – *d*; III – *a*; IV – *b*

24. Solution $\Delta P = h \rho g = 0.2 \times 1600 \times 9.81 = 3139.2$ N/m²

$$U_{max} = C_v \sqrt{2\Delta P/\rho} = 0.98 \sqrt{\frac{2\times3139.2}{1000}} = 2.455 \text{ m/s}$$

$$Re = \frac{DV\rho}{\mu} = \frac{0.03\times1000\times2.455}{1\times10^{-3}} = 73.66\times10^3$$

Avg. $U = U_{max}/2$; Volumetric flow

$$= \frac{\pi}{4}(0.03)^2 \times \frac{2.455}{2} \times 3600 = 3.12 \text{ m}^3\text{/h.}$$

25. Solution: V = Volume of wood

$$\rho_0 = 0.75 \; ; \rho \text{ wood} = 0.95$$

$$x = \text{fraction in water}$$

$$\text{Downward force} = V\rho \text{ wood } g$$

$$\text{Buoyancy force} = V\,x\,g + V\,(1-x)\,r_0\,g \text{ equating:}$$

$$V\,x\,g + V\,(1-x)\,g\,\rho o = V\rho \text{ wood } g$$

$$x + (1-x)\,0.75 = 0.95\,;\; x = 0.8$$

26. Solution:
$$\frac{dV}{dt} = \frac{A^2 DP}{Cv(V+V_c)}\;;\; Cv = \alpha\,w\,\mu$$

$$\frac{dt}{dV} = \frac{Cv}{A^2\Delta P}V + \frac{Cv}{A^2\Delta P}V_c = 9.3\text{ V} + 8.5$$

$$\frac{Cv}{A^2\Delta P} = 9.3;\; \therefore\; Ve = 8.5/9.3 = 0.914$$

$$\frac{\alpha w\mu}{A^2\Delta P} = 9.3\;;\; \alpha = \alpha_o\,\Delta p^s$$

$$\frac{\dfrac{\alpha_o(\Delta p_1)^s\,w\mu}{A^2\Delta P_1}}{\dfrac{\alpha_o(\Delta P_1)^s\,w\mu}{A^2\Delta P_2}} = \left(\frac{\Delta P_1}{\Delta P_2}\right)^{S-1} = \frac{9.3}{K};\left(\frac{1}{2}\right)^{0.3-1} = \frac{9.3}{K}$$

or
$$\frac{1}{\left(\frac{1}{2}\right)^{0.7}} = \frac{9.3}{K};\; K = 5.7248 = 5.725$$

$$\frac{W}{A^2\Delta P_1}V_e = 8.5\;;\; \frac{W}{A^2\Delta P_2}V_c = \frac{8.5}{2}\;;\; \Delta P_2 = 2\,\Delta P_1$$

New equation is $\frac{dt}{dV} = 5.725\,V + \frac{8.5}{2}$

$$t = 5.725\,\frac{V^2}{2} + 4.25\text{V; For } V = 3.5;\; t = 49.94\text{ sec.}$$

27. (d)

28. (b)

29. (c)

30. 1 – c; 2 – a

31. 1 – b; 2 – d

32. 1 – c; 2 – a

33. $V = \sqrt{2gh} = \sqrt{2 \times 9.81 \times 10} = 14$ m/sec.

$$\text{Flow rate } = V \times \text{area} = \frac{v \times \pi D^2}{4} = v \times 0.25 = 3.5 \text{ m}^3/\text{sec.}$$

34. Left for practice/as exercise.

35. Solution $T = 300°K$; $P = 1$ atm ; $V = nRT/P = \dfrac{1 \times 0.8205 \times 300}{1}$

$$V = 24.615 \text{ m}^3 \text{ /kg mol.}$$
$$\rho = 17/24.615 = 0.6906 \text{ kg/m}^3$$
$$D = 25\text{cm} \; ; \; V = 10 \text{ m/s}$$
$$\mu = 10.2 \times 10^{-6} \text{ kg/(m.s)} \; ; \; \rho = 0.6906 \text{ kg/m}^3$$

$$Re = \frac{0.25 \times 10 \times 0.6906}{10.2 \times 10^{-6}} = 169,265$$

$$f = 0.079 \text{ Re}^{-0.25} = \frac{0.079}{20.5} = 3.8948 \times 10^{-3}$$

$$f = \frac{g_c D \Delta P}{2v^2 Lp} \; ; \; \Delta P = \frac{2 f V^2 LP}{D_{gc}}$$

$$= \frac{2 \times 3.8948 \times 10^{-3} \times 10^2 \times 10^2 \times 0.6906}{9.81 \times 0.25}$$

$$\Delta P = 21.9 \text{kg/m}^2.$$

36. (c)

In Hagen–Poiseuille equation, $\Delta P = \dfrac{32 \mu l V}{g 4R^2}$; $V \propto R^2$

(*b*) Flow rate $= \pi R^2 V$ So q is ∞ to R^4.

37. (b)

38. (c)

$$\frac{4(\pi R_o^{\,2} - \pi Ri^2)}{2_p(R_o + R_i)} = \frac{4 \times \text{Flow cross section}}{\text{Wetted circumference}}$$

39. (d)

40. Ans (a)

41. $\dfrac{P_A}{\rho_A} + \dfrac{VA^2}{2g} + Z_A - F + W = Z_B + \dfrac{u_B^{\,2}}{2g} + \dfrac{P_B}{\rho_B}$

$$P_A/\rho_A = P_B/\rho_B \; ; \; V_A = 0 \; ; \; Z_A = 0 \; ; \; Z_B = 5$$

$$F = 0,\ u_B = 3\ ;\ \text{So } w = Z_B + \frac{u^2{}_B}{2g} = 5 + \frac{3^2}{2 \times 9.81} = 5.46 \text{ m}$$

$$= 17.93 \text{ ft. ; pressure} = 17.93 \times 62.4 = 1118.8$$

$$\text{H.P} = \frac{1118.8}{550} \times \frac{300}{2.54 \times 12} \times \left(\frac{2.5}{2.54 \times 12}\right)^2 \frac{\pi}{4} = 58.2 \text{ H.P}/550 \approx 0.11 \text{ HP}$$

42. Re. No. in 20 mm line $= \dfrac{20}{10 \times 100} \times \dfrac{4 \times 10^{-5} \times 1500}{(\pi \times 4 \times 10^{-4}/4)^{-3}{}_{10}} = 3819\ (\approx 4000)$

$$f = 0.079\ (3819)^{-0.25} = 0.010;\ V = \frac{4 \times 10^{-5}}{(\pi \times 10^{-4})} = 0.127$$

$$\Delta H = 4flv^2/(d2g) = \frac{4 \times 0.01 \times 1000 \times 0.127^2}{2 \times 10^{-2} \times 2 \times 981} = 1.644 \text{ m}$$

50 mm pipe gives viscous flow.

43. Similar to Problem 25.

44. (b)

45. (b)

46. (c)

47. (a) $f = \dfrac{g_c DH}{2V^2L} = \dfrac{A\mu}{DV\rho}\ ;\ H = \dfrac{2VLA\mu}{g_c D^2 \rho} = \alpha - \beta Q$

$$\Delta P = H\rho\ ;\ \frac{\Delta P}{\rho} = \frac{2VLA\mu}{gcD^2}; \frac{2VLA\mu}{g_c D^2 \rho}$$

$$DP = ; \frac{2VLA\mu}{{}_{gc}D^2}; \frac{2VLA\mu}{{}_{gc}D^2} = \rho(\alpha - \beta Q)$$

$$\Delta P = \rho\ (\alpha - \beta Q)$$

(b)

$$\Delta P/\rho = \alpha - \beta Q$$

$$\beta Q = \alpha - \Delta P/\rho$$

$$Q = \frac{\alpha}{\beta} = \frac{\Delta P}{\rho\beta}$$

$$Q = \frac{1}{\beta}\left(a - \frac{\Delta P}{\rho}\right)$$

48. Left for practice/as exercise.

49. $$u_t = \frac{d_p^2(\rho_p - \rho)}{18\mu}$$

(a) $$u_t = \frac{9.81\times10^{-10}(2650-1000)}{18\times0.80\times10^{-3}} = 0.1124 \text{ cm/s}$$

(b, c) $u_s = ut\in^{4.6} = 0.1124 \times 0.8^{4.6}$cm/S = 0.04 cm/s

SECTION–E

1. (d)
2. (d)

Sol. Initial volumetric flow rate, $Vo*A$ where A is cross section area = $D^2\pi/4$.

Let small increment in volumetric flow rate = dv in going through elemental distance dx.

$dv = \pi D\, dx\, q\, (x/L)^2$; $\sqrt{2}\, dv = \pi\, Dq^2x\, dx/L^2$; $vx - v_0 = \pi Dq\, x^3/(3L^2)$

$vx = v_0 + 4q\, x^3/(3L^2D)$; Here $vx = Vx\, \pi D^2/(L^2D)$ and $v_0 = V_0\, \pi\, D^2/4$.

3. (b)

Sol. $A = d^2\pi/4 = 0.001963$ m^2; velocity, $v = 0.001/A = 0.52$ m/sec.

Fractional loss, $hf = 4fl\, v^2/(2gd) = 4 \times 0.0062 \times 150 \times (0.51)^2 /(2 \times 9.81 \times 0.05) = 1.05$ m

Velocity head= $v \times v /(2 \times g) = 0.51^2 /(2 \times 9.81) = 0.01325$ m.

Total head = 10 + 0.05 + 0.0132 = 11.06 m. Nearly: 11m.

4. (c)
5. (b)

Sol. $150\mu\, (1 - \in) / (\phi S^2 DP\in^2) = 375{,}000$;$1.75\, P\, (1 - \in)/(\phi S\, DP) = 10.94 \times 10^6$

$(1 - \in) * 1000 * (2.5 - 1)g = 375.000V + 10.94 \times 10^6\, V * V$; Here $\in = 0.40$; Solving using algebraic equation gives; $V = 16.0266$ mm/second.

6. **Sol.** $\Delta p/L = (1 - \in)\, \rho_p - \rho)g$; $2/0.50 = (1 - 0.4)/(1 - x)$; $2 = 0.60/(1 - x)$ gives x as; 0.70 Ans.
7. (b)

Sol. $dt/dv = \mu/(A\Delta P)\, (\alpha c\, V/A + Rm)$; Since rate of filtration is constant, therefor, dt/dv is constant; $(dt/dv)\, 1 = (dt/dv)\, 2$; $T = 40$ minutes; Extra time needed = 40 – 20 = 20 minutes.

8. (b)

Sol. From Bernoulli equation; $P_1/\rho g + Z_1 + V_1^2/2g = P_2/\rho g + Z_2 + V_2^2/2g$.

$(P_1/\rho g + Z_1) - (P_2/\rho g + Z_2) = (V_2^2 - V_1^2)/2g$ ....(1)

Also $(P_1\backslash\rho g + Z_1) - (P_2/\rho g + Z_2) = h(pm/pw - 1)$ .... (2); puting in equation (1): $h\, (pm\text{-}pw/pw) = (V_2^2 - V_1^2)\, /2g$

9. (c)

Sol. Equivalent diameter = 4 rh; wherte rh is hydraulic radius, given by cross sectional area/wetted perimeter = S/lp; For rectangle duct,equivalent diameter is = $4\, rh = 4\, B*\, H/2\, (B + H) = 2\, BH\, (B + H)$.

10. (b)
11. (d)

Sol. Sepration factor, S is: $Vo*Vo/g * r$ where Vo is tangential velocity = 15 m/sec; r = radius =0.1 m; S = (15)* 15 /9.81*10* =230

12. (d)

13. (a)

Sol. Taking a datum line $X - X$, we have $PB + L *\rho L *g + H *g = L *\rho L *g + H * \rho_{air} *g + PA$; $PA - PB = H *g (\rho H - \rho_{air})$ Ans.

14. (b)

Sol. The force F which Plate X must be pushed up to prevent water from leaking is given by $(AH\rho L)g$.

15. (c)

Sol. $Q = (\pi\Delta P/L) *di*di/(128\mu)$; $Q \propto DP/L$; For $Q = 1$, pipe lenght (L) = L; For $Q = 2$, pipe length = $2L$. For $Q = 3$, pipe length = $2L$.

So the ratio is : $L : 2L : 2L = 1 : 2 : 2$ Ans.

16. (d)

Sol. Power number is, $Np = p/(n^3\ d^5\ p)$; For constant power,we have, $n^3\ d^5$ =constant. $n_2/n_1 = (d_1/d_2)\ 5/3$;here $d_2 = 1.2\ d_1$; $n_2 = n_1\ (d_1/d_2)^{5/3} = n_1/1.2^{5/3}$

17. (c)

Sol. For centrifugal filtration pressure drop is given by : $-\Delta P = 1/2\ \rho w^2\ (b^2 - x^2)$ where $b = R + \delta$ and $x = R$; For this case, puting the values, $-\Delta P = 1/2\ \rho w^2\ ((R + \delta)^2 - R^2) = 1/2\ w^2 \rho_L\ (2R\delta + \delta)^2$.

18. (a)

Sol. Kozney-Carman equation is:

$$\frac{\Delta P}{L} = \frac{150\ \mu v(1-\in)^2}{\phi_s^2\ D_p^2 \in^3} = \frac{150 v^2 (1-\in)^2 \rho}{(D_p v\rho/\mu)\ \phi_s^2 D_p \in^3}$$

$$\frac{\Delta P}{\rho v^2} = \frac{150\ (1-\in)^2 L}{\text{Re}\ \phi_s^2\ D_p^2 \in^3} = \frac{KL}{\text{Re}.D_p}; \quad \frac{\Delta P}{\rho v^2} \propto \left(\frac{L}{D_p}\right) \Big/ \text{Re}.$$

19. (b)

Sol. Cumulative mass fraction of particles = fraction of total weight of particles having a size smaller than a given screen opening. Hence required cumulative mass fraction is $\sum_{1}^{j} N_i m_i\ d_i^3 \sum_{i}^{\infty} N_i^3\ m_i d_i^3$

20. (c)

Sol. Diffusional area of cylinder = πDL; Diffussion area of $\pi DL >> \pi D^2$ (since $L >> D$). Sphere = $4\pi \dfrac{D^2}{4} = \pi D^2$. Hence More NaCl will diffuse. That is less fraction of salt will remain in the cylinder compared to the sphere. So $Xc < Xs$.

❑

CHAPTER

6 Heat Transfer

1. **Conduction:** If a temperature gradient exists in a continuous substance, heat can flow unaccompanied by observable motion of matter. This type of heat flow is conduction. Fourier's law is : $q = -KA\,dT/dx$; q = rate of heat flow, A = surface area, K = constant (thermal conductivity), T = temperature, x = distance, d = differential. This occurs in solids and in laminar flow.
2. **Convection:** This refers to the flow of heat associated with movement of a fluid. $Q = hA(Ts - Tf)$; q = rate of heat flow, h = convective heat transfer coefficient, A = area of heat transfer, Ts = surface temperature, Tf = bulk temperature of fluid.
3. **Radiation:** This is heat transfer of, energy through space by electromagnetic waves.; $Wb = \sigma T^4$; where Wb = rate of radiation energy emitted per unit area, σ = Stefan-Boltzmann constant, T = absolute temperature.
4. **Resistances in series:** $q/A = \Delta T/(BA/K_A + BB/K_B + BC/K_C + ...)$ where B is thickness of layer, K = thermal conductivity, ΔT =temperature difference, resistance = $x/(K{*}A)$; A = area, x = thickness of substance.
5. **Heat flow in a cylinder:** $q/A = q/(2\pi rL) = -K\,dT/dr$; integrating with limits of r_o, r_i and T_i, T_o gives:

   $Q = K(2\pi L)(T_i - T_o)/Ln(r_o/r_i) = KAL(T_i - T_o)/(r_o - r_i)\ Lm$
6. **One dimensional heat flow with constant temperature:** $dT/dt = \alpha dT^2/dx^2$; By separation variables method, the solution is:

   $(Ts - T)/(Ts - To) = 4/\pi\,(e^{-\omega_1 F_0}(\sin \pi x/2S) + 1/3\,e^{-9\omega_1 F_0}\sin(3\pi x/2S) + 1/5...)$ where $Fo = \alpha t/s{*}s$

   t = time, S = one half of slab thickness; $\omega_1 = \pi r^2/4$.
7. Overall heat transfer coefficients: $Q = UA\,\Delta T$ lm, where U = overall heat transfer coefficient, A = area, ΔT lm = log mean temperature difference,

   $U_o = 1/\{(Do/Di\ hdi) + (Do/Di\ Hi) + (xw/Km)(Do/DL) + 1/ho + 1/hdo\}$ where hdi and hdo are fouling factors; i = inside, o = outside.

   $Ui = 1/(1/hdi + 1/hi + (xw/Km)(Di/DL) + (Di/Doho)(Di/Dohdo)$
8. Magnitude of heat transfer coefficients:

Process	**h,w/sq.m.°C**
Steam (dropwise)	30,000–100,000 to convert from BTU/Hr.sq.ft. °F to
Steam (filmwise)	6,000–20,000 w/(sq.m.°C) Multiply By: 5.68

Boiling water	1700–50,000
Condensing organic vapors	1000–2000
Water (heating or cooling)	300–20,000
Oils (heating or cooling)	50–1500
Steam (super heating)	30–100
Air (heating or cooling)	1–50

9. Dimensionless numbers: (a) Prandtl number = $Cp\mu/K$; (b) Nusselt number = hD/K; (c) Fourier number (Fo) = $4KL/(Cp\ p\ D*D\ V)$; (d) Graetz number (Gz) = $m\ Cp/(K\ L)$; (e) Pecklet number = $Re * Pr$; (f) Reynolds number (Re) = $DV\ p/\mu$; (g) Stanton number (St) = $h/(Cp * G)$.

10. $Nu = hD/K = 0.023\ Re^{0.8}\ Pr^{0.33}\ \phi v$ where ϕv is viscosity ratio term.

11. Analogies: Reynolds analogy: $f = 0.046\ Re^{-0.20}$; $St = 0.023\ Re^{-0.20}$; From these, $h/(Cp\ G) = St = f/2$.

Coluburn analogy: $J_H = f/2 = (h/Cp\ G)\ Pr^{2/3}\ \phi v$; $J_H = f/2 = 0.023\ Re^{-0.2}$

12. Radiation $\alpha + \rho + \tau = 1$; ρ = replactivity; α = absorptivity; τ = emissivity, if $\alpha = 1$ then black body.

13. Laws of black body radiation: $wb = \sigma T^4$ (Stefan-Botzmann law) where wb = total emissive power, σ = constant; T = absolute temperature.

Planck's law: $w\ b, \lambda = 2\pi hc^2 \quad \lambda^{-5}/(e^{h_c/k\lambda T} - 1) = C_1\lambda^{-5}/(e^{C_2/\lambda T} - 1)$

λ = wavelength of radiation, T = absolute temperature, C_1 and C_2 are constants, $w\ b, \lambda$ = emissive power.

Wein's displacement law: $T\lambda_{max}$ = Constant, where C = 2890, λ is in micrometers, T is in Kelvin.

Kirchhoff's law: $w_1/\alpha_1 = w_2/\alpha_2$; where α = absorptivity, w = radiating power.

14. FG: correction factor: This is used since the flow is neither parallel nor counter current in multipass heat exchangers. FG is based on Z and ηH, where $Z = (Tha - Thb)/(Tcb - Tca)$ and $\eta H = (Tcb - Tca)/(Tha - Tca)$; Tha, Thb are hot side temperatures and Tca and Tcb are cold side temperatures. FG has lowest value of 0.50 and highest value of 1.0.

15. Shell side heat transfer coefficient: This is calculated using J factor.

$$h/(Cp\ G)\ (Cp\ \mu/K)^{2/3}\ (\mu a/\mu)^{0.14} = 0.20(DG/\mu)^{-0.40}$$

16. Choice of tube side fluid: (a) corrosive fluid—tube side (special metal tubes); (b) dirty fluids—tube side (easy for cleaning); (c) hot fluids—tube side; (d) very viscous fluids—on shell side.

17. Types of heat exchangers: (1) shell-tube type (2) plate type (3) extended surface (4) scraped surface exchangers (5) regenerators. Their special uses are: 1. to save metal, 2. used in dairy, food processing industries, 3. to conserve space and to reduce the cost of equipment, 4. evaporating viscous iquids, 5. passing cold and hot fluids alternately.

18. Codensers and vaporizers: 1. shell and tube condensers, 2. contact condenser (barometric), 3. dehumidifying condensers, 4. vaporizers.

19. Evaporation: In evaporation a solution is concentrated consisting of a non-volatile solute and a volatile solvent. For example, Salt + water; Sodium hydroxide + water.

20. Evaporators: The material to be evaporated flows in the tubes in a single effect evaporator, the vapor from the boiling liquid is condensed and discarded. To evaporate 1 kg of water from a solution calls for 1 to 1.3 kg of steam. Mutiple effect of evaporation is the mehod of increasing the evaporation of soluion per kg of steam by using a series of evaporators between steam supply and the condenser. Capacity is the number of kg of water evaporated per hour.

Economy is the number of kg vaporized per kg of steam fed to the unit.

21. Types of evaporators: 1. Long tube vertical evaporators (*a*) upward flow : useful for concentrating liquids that tend to foam.

(*b*) Falling film: concentrating heat sensitive materials like fruit juices and milk.

(*c*) Forced circulation: Moderately heat sensitive iquids are concentrated. Evaporating salting liquors.

2. Agitated film evaporators: Used for concentrating viscous solutions like gelatin, rubber, laton, fruit juices etc.

22. Boiling point of elevation, Duhring's rule: The boiling point of solution is higher than pure water. The increase in boiling point of solution over water is the boiling point of elevation of solution (BPE). The boiling point is found by Duhring rule. This rule states that the boiling point of a given solution is a linear function of the boiling point of pure water at same pressure. Thus if the boiling point of a solvent water is plotted against boiling point of solution at same pressure a straight line is obtained.

23. Overall coefficients in evaporators:

Type	**U, overall coefficient, w/(sq.m.deg.C)**
Long tube: Natural convection	1000–200
Forced convection	2000–5000
Agitated pan : 1 CP	2000
100 CP	1500
100 P	600

24. Enthalpy balance for single effect evaporator:

qS (heat from steam) = $mS\ \lambda S = (mf - m)\ Hv - mf\ Hf$; where Hv = enthalpy of vapor, Hf = enthalpy of feed; H = enthalpy of thick liquor, m = mass of thick liquor, mS = steam:

$q = UA\ \Delta T$; where U is overall heat transfer coefficient, A = area of heat transfer, Δt = temperature difference.

25. Mutiple effect evaporators: $Q = U_1 A_1 \Delta T_1 = U_2 A_2 \Delta T2 = U_3 A_3 \Delta T_3$; Normally $A_1 = A_2 = A_3$

Methods of feeding: Forward feed: concentration of liquid increases from first to the last. Requires a pump to the first effect and another pump from the last effect. Flow is in the direcion of decreasing pressure.

Backward feed: Dilute liquid is fed from the last effect and then pumped back to the first effect. This requires a pump for each effect. Used for viscous liquids.

Mixed feed: Dilute liquid enters an intermediate effect, flows in forward feed to the end of the series and then pumped back to the first effect. This eliminates some pumps.

26. Crystallization: This is the fomation of solid particles within a homogeneous substance, Magma. The two phase mixture of mother liquor and crystals of all sizes which occupies the crystallizer and is withdrawn, as a product. The birth of a new particle is called as Nucleation. The sequence of stages in evolution of a crystal are: cluster → embryo → nucleus → crystal.

Ostwald ripening is a small crystal dissolving in a solution until it disappears and the large one grows.

Solubility given by Kelvin equation is: in $\alpha = 4\ VM\sigma/(RTLx)$ where L = crystal size; α = ratio of super saturated solution/saturated solution; VM = molar volume of a crystal; σ = average interfacial tension between solid and liquid, x = number of ions per molecule of solute.

27. Stages of nucleation: 1. Primary nuclcation—birth of new particles within a solution. 2. Secondary nucleation—fluid shear nucleation; contact nucleation.

28. ΔL Law of crystal growth: The total growth of each crystal in the magma during the same time interval Δt is same $\Delta L = G\ \Delta T$.

29. General unsteady state heat conduction equation:

$$\frac{\partial^2 T}{\partial x^2}+\frac{\partial^2 T}{\partial y^2}+\frac{\partial^2 T}{\partial z^2}+\frac{q}{K}=\frac{1}{\alpha}\frac{\partial T}{\partial t}$$; q = heat generation, w/m^3; K = thermal conductivity;

$\alpha = K/(C_P\ \rho)$; ρ = density

C_P = Specific heat; Q (conduction heat transfer) = $\dfrac{-KAdT}{dx}$; T = temperature; x = distance; A = area.

30. General unsteady state heat conduction for a cylinder

$$\frac{\partial^2 T}{\partial r^2}+\frac{1}{r}\frac{\partial T}{\partial r}+\frac{1}{r^2}\frac{\partial^2 T}{\partial \theta^2}+\frac{\partial^2 T}{\partial z^2}+\frac{q}{K}=\frac{1}{\alpha}\frac{\partial T}{\partial t}$$

Here r = Radius; θ = angle;

31. General unsteady state heat conduction equation for a sphere:

$$\frac{d^2T}{dr^2}+\frac{2}{r}\frac{d_T}{d_r}+\frac{1}{r^2\sin\psi}\frac{d}{d\psi}\left(\sin\psi\frac{dt}{d\psi}\right)+\frac{1}{r^2\sin^2\psi}\frac{d^2T}{d\phi^2}+\frac{q}{K}=\frac{1}{\alpha}\frac{dT}{dt}$$

ψ, ϕ are angles; r = radius.

32. Means: Arithmetic mean = $(A_0 + A_i)/2$; $G.M = \sqrt{A_i\ .\ A_0}$

$$\text{H.M} = \frac{1}{A_0}+\frac{1}{A_i}=\frac{1}{A_{H.M}}$$

33. Heat flow in concentric spheres:

$$Q=\frac{4\pi(T_a-T_b)}{\left[\dfrac{1}{h_a r_1^2}+\dfrac{(r_2-r_1)}{kr_1r_2}+\dfrac{r_3-r_2}{kr_2r_3}+\dfrac{1}{h_b\ r_3^2}\right]}$$

34. Critical radius of insulation: $r_c = k/h$; $r_0 > r_c$.

r_c = critical radius; r_0 = outer radius.

35. Internal heat generation. Plane wall:

$T_{mese} = T_\omega + \dfrac{qL^2}{8K}$; T_ω = wall temp; q = heat; L = length of wall; K = thermal conductivity;

$T_\omega = T_\infty + \frac{qL}{2h}$; T_∞ = Ambient temperature

36. Hollow cylinder with heat generation:

$$T_0 = T_f + \frac{qr_i^2}{4K}\left[\left(\frac{2k}{hr_i} - 1\right)\left\{\left(\frac{r_0}{r_i}\right)^2 - 1\right\} + 2\left(\frac{r_0}{r_i}\right)^2 \ln\frac{r_0}{r_i}\right]$$

$T = T_i$ at $r = r_i$; $T = T_0$ at $r = r_0$; T_f = ambient temperature.

37. Sphere with internal heat generation max. temperature at centre of sphere: $T_e = T_0 + \frac{q.r_0^2}{6k}$

q = heat generation; $T = T_0$ at $r = r_0$; K = thermal conductivity.

38. Biot Number (B_i) = hL/K; h = heat transfer coefficient; L = length; K = thermal conductivity.

39. Extended surfaces and fins to increase heat transfer area. Used in cooling of electronic equipments, cooling of motors, cycle engines, compressors, refrigerators etc.

40. Transient heat conduction (Lumped):

$$\frac{T_0 - T_\infty}{T - T_\infty} = e\left(\frac{hA}{\rho_{VC}}\right)t;\ T_0 = \text{initial temperature;}$$

T_∞ = surface/ambient temperature; T = temp. at time t;
h = heat transfer coefficient; A = area; ρ = density; V = volume; C = specific heat.

41. Fourier number: $F_0 = \frac{\alpha t}{L_C^2}$ (Relative time)

L_C = Volume of body/surface area (characteristic length)

$$\frac{T - T_\infty}{T - T_\infty} = \exp(-B_i * F_o)$$

42. Continuity equation: $div\ (\rho V) = 0$; momentum equation (Navier – Stokes equation). Force balance

$$\rho\frac{DV}{Dt} = B - \Delta P + \mu\Delta^2 V$$

(1) (2) (3) (4)

(1) is mass * acceleration; (2) is body forces (gravitational; electrical, magnetic, centrifugal); (3) is pressure forces and; (4) is viscous forces.

Energy equation:

$$\rho * \frac{DE}{Dt} = K\left(\frac{\partial^2 T}{\partial x^2} + \frac{\partial^2 T}{\partial y^2} + \frac{\partial^2 T}{\partial z^2}\right) + \mu\phi;$$

(1) (3) (2)

(1) is mass; (3) is heat energy; (2) is viscous energy

43. To make dimensionless equations, two methods are available: (1) Rayleigh's method; (2) Buckingham method.

44. Reynolds Analogy: St. No. = $\dfrac{Nu}{Re.\,Pr} = \dfrac{C_f x}{2} = \dfrac{2T_S}{\rho u_\infty^2}$

$C_f x = \dfrac{0.664}{\sqrt{Re_x}}$; Coluburn analogy: $J_H = St_x\, Pr^{2/3} = \dfrac{Cfx}{2}$

Coluburn analogy: $St_x\, Pr^{2/3} = \dfrac{C_{fx}}{2} = JH$; *JH* is Coluburn factor, or *j* factor.

45. Turbulent boundary conditions:

$\dfrac{\delta}{x} = 0.381\ Re_x^{-0.20}$; $N_u = 0.037\, Re_L^{0.8}\ Pr^{0.33}$; $N_{uL} = \dfrac{h}{LK}$

46. Laws of Black body radiation:

(*a*) Plank's distribution law: $Eb_\lambda = \pi Ib_\lambda = \dfrac{2\pi C_1}{\lambda^2[\exp(C_2/\lambda T) - 1]}$ spectural distribution of Radiation intensity ($E_{b\lambda}$); λ = wavelength; T = temperature in Kelvin; C_1, C_2 are constants.

(*b*) Wein's law: $\dfrac{Ib_\lambda}{T^5} = \dfrac{2C_1}{(\lambda T^5)\exp(C_2/\lambda T)}$

(*c*) Rayleigh–Jeans formula: For $C_2/\lambda T << 1$

$\dfrac{Ib_\lambda}{T^5} = \dfrac{2C_1}{C_2}\dfrac{1}{(\lambda T)^4}$

(*d*) Wein's displacement law: $\dfrac{Eb\lambda_{max}}{T^5}$ = Constant

(*e*) Stefan–Boltzman's law:
Total emissive power of a surface = σT^4
T = temperature in Kelvin; σ = Stefan–Boltzman constant

47. Kirchhoff's law = $\dfrac{E(\text{emissive power})}{\alpha(\text{absorptivity})}$ same for all bodies at thermal equilibrium.

48. Radiation exchange between two bodies.

$$\omega_2 = \sigma(T_1^4 - T_2^4)\int_{A_1}\int_{A_2} \cos\phi_1 \cos\phi_2 \frac{dA_1\, dA_2}{r^2}$$

49. The equivalent $\in$ for two parallel grey planes is: $\overline{\in} = \dfrac{1}{\left(\dfrac{1}{\in_1} + \dfrac{1}{\in_2} - 1\right)}$

50. Radiation shields are used for insulation of cryogenic storage tanks. An example is measurement of temperature by a thermo-couple or thermometer shielded to minimize radiation effects.

51. Electrical analogy of radiation:

$$R = \frac{V}{I} = \frac{J_1 - J_2}{Q_{12}} = \frac{(J_1 - J_2)}{A_1 F_{12}(J_1 - J_2)} = \frac{1}{A_1 F_{12}} \text{ (Shape resistance).}$$

$$\text{Surface Resistance} = \frac{(1 - \in)}{\in A}$$

52. Heat exchangers.

$Q = UA\,\Delta T_m;\ U_i A_i = U_o A_o$

$$\text{For plane walls: } U = \frac{1}{[1/h_o + 1/h_i + L/K]}$$

$$\text{For cylindrical wall: } U_o = \frac{1}{\frac{1}{h_o} + \frac{r_o}{K}\ln\left(\frac{r_o}{r_i}\right) + \left(\frac{r_o}{r_i}\right)\frac{1}{h_i}}$$

$$U_i = \frac{1}{\frac{1}{h_i} + \frac{r_i}{K}\ln\left(\frac{r_o}{r_i}\right) + \left(\frac{r_i}{r_o}\right)\frac{1}{h_o}};\quad i - \text{inside};\quad o - \text{outside}$$

$$\text{LMTD} = \frac{(\Delta T_2 - \Delta T_1)}{\ln(\Delta T_2/\Delta T_1)}$$

Multi pass and cross flow heat exchangers: $Q = UA * F * \Delta T_{ln}$

F is correction factor (neither parallel flow nor counter flow).

53. NTU method of heat exchanger analysis: $\text{NTU} = \frac{UA}{C_{min}}$; C_{min} is smaller heat capacity of C_c (cold fluid) and C_h (hot fluid).

($\in$) Effectiveness = Actual heat transfer/Maximum possible heat transfer.

$$\text{For parallel heat exchanger, } \in = \frac{T_{co} - T_{ci}}{T_{hi} - T_{ci}} = \frac{1 - \exp\left[-NTU\left(1 + \frac{C_{min}}{C_{max}}\right)\right]}{\left(1 + \frac{C_{min}}{C_{max}}\right)}$$

For counter flow heat exchanger:

$$\in = \frac{1 - \exp\left[-NTU\left(1 - \frac{C_{min}}{C_{max}}\right)\right]}{\left\{1 - \left(\frac{C_{min}}{C_{max}}\right)[\exp(-NTU)]\left[\left(1 - \frac{C_{min}}{C_{max}}\right)\right]\right\}}$$

54. Condensation heat transfer: 1 – filmwise condensation; 2 – dropwise condensation (heat transfer rates larger, 5 to 10 times of filmwise).

Laminer filmwise condensation: $h_D = 0.725\left[\dfrac{g\rho_L(\rho_l - \rho_v)k_l^3\, h_{fg}}{\mu_l(T_{set} - T_s)D}\right]^{1/4}$

ρ = density of liquid; ρ_V = Vapor density;

K_l = thermal conductivity of liquid; D = outer diameter of tube; h_{fg} = latent heat; μ_l = viscosity of liquid.

55. Conduction shape factor: $Q = K\ \dfrac{N}{M}\ (T_1 - T_2) = K\ S(T_1 - T_2)$

S is shape factor = $\dfrac{N}{M}$; in a tubuler furnace, $T_1 - T_2 = M\ \Delta\ T$.

Total heat flow rate, $Q = N\ \Delta\ Q$; N is the number of heat flow tubes in the cross section.

QUESTIONS

SECTION—A

1. The constant of proportionality in Fourier's law is called as :

(a) Thermal conductivity (b) Thermal diffusivity

(c) Coefficient of heat transfer

2. The unit of thermal conductivity is :

(a) Kcal/hr m°C (b) Kcal/m² hr°C

(c) Cal/m² hr°C

3. If the rate of heat transfer varies with time, we call it :

(a) Steady state conduction (b) Unsteady state conduction

(c) Convection (d) Radiation

4. In case of steady state heat transfer, the temperature profile is :

(a) Linear (b) Parabolic

(c) Hyperbolic (d) Asymptotic

5. The expression for thermal diffusivity is :

(a) $\dfrac{\rho C_p}{K}$ (b) $\dfrac{K}{\rho C_p}$

(c) $\dfrac{C_p}{\rho K}$ (d) $\dfrac{\rho}{K C_p}$

6. Prandtle number is given by :

(a) $\dfrac{K}{C_p \mu}$ (b) $\dfrac{\mu}{K C_p}$

(c) $\dfrac{C_p \mu}{K}$ (d) $\dfrac{C_p}{\mu K}$

7. Grashof number is given by :

(a) $\dfrac{\mu^2}{\rho^2\beta g b^3 \Delta T}$ (b) $\dfrac{\rho^2\beta g b^3 \Delta T}{\mu^2}$

(c) $\dfrac{\beta g b^3 \Delta T}{\rho^2\mu^2}$

8. Nusselt's number is given by :

(a) DK/h, (b) K/hD,

(c) hD/K, (d) D/hK

9. Pick up the correct statement :

(a) Thermal conductivity of a gas at low density increases with increase of temperature.

(b) Thermal conductivity of a gas at low density increases with increase of temperature.

(c) Thermal conductivity of a gas at low density remains unchanged with change in temperature

10. With increase in temperature, thermal conductivities of most liquids :

(a) Increase (b) Decrease

(c) Remain unchanged (d) Unpredictable

11. Temperature remaining constant, with increase in pressure, thermal conductivities of gases :

(a) Increase (b) Decreases

(c) Remain unchanged (d) Unpredicatable

12. In case of liquid metals, ratio of thermal boundary layer to hydrodynamic boundary layer thickness is :

(a) 1.0 (b) > 1.0

(c) < 1.0.

13. Which area is used in case of heat conduction in a hollow cylinder :

(a) Logarithmic mean area (b) Arithmetic mean area

(c) Geometric mean area (d) None of these

14. Finns are provided on heat transferring surface in order to increase :

(a) Heat transfer area (b) Heat transfer coefficient

(c) Mechanical strength of the equipment (d) Temperature gradient.

15. Baffles are provided in heat exchangers to :

(a) Remove dirt (b) Increase heat transfer rate

(c) Increase heat transfer coefficient (d) Provide better mechanical spacing

16. Minimum baffle spacing is normally :

(a) 1/2 of shall diameter (b) 1/5th of shell diameter

(c) 1/3rd of shell diameter

17. 1 – 2 heat exchanger means :

(a) Single pass in shell side and double pass on tube side

(b) Single pass on tube side and double pass on shell side

18. Double pipe heat exchangers are preferred when :

(a) Overall heat transfer coefficient is low (b) Liquid is more viscous in nature

(c) Heat transfer area required is low (d) Liquid is corrosive

19. If ΔT_1 and ΔT_2 are the approaches of a heat exchanger, logarithmic mean temperature is given by the relation :

(a) $\Delta T_m = \dfrac{\Delta T_1 - \Delta T_2}{\ln\left(\dfrac{\Delta T_1}{\Delta T_2}\right)}$ (b) $\Delta T_m = \dfrac{\Delta T_1 - \Delta T_2}{\log\left(\dfrac{\Delta T_1}{\Delta T_2}\right)}$

(c) $\Delta T_m = \dfrac{\Delta T_2 - \Delta T_1}{\ln\left(\dfrac{\Delta T_1}{\Delta T_2}\right)}$ (d) $\Delta T_m = \dfrac{\Delta T_2 - \Delta T_2}{\log\left(\dfrac{\Delta T_1}{\Delta T_2}\right)}$

20. Correction of LMTD applied in case of :

(a) Parallel flow (b) Counter current flow

(c) Cross flow (d) None of these

21. Dittus-Boelter equation is applicable for :

(a) Fluid flowing in turbulent region (b) Fluid flowing in laminar region

(c) Fluid flowing under natural convection (d) None of these

22. In a heat exchanger one transfer unit means :

(a) Condition when the change in temperature of one stream numerically equals the average driving force.

(b) Section to the exchanger, where unit rate of heat is transferred

(c) That section of the exchanger, where the heat transfer area is unit

23. Floating heads are provided in heat exchangers to :

(a) Decrease pressure drop

(b) Facilitate easy fabrication and maintenance

(c) Avoid deformation of tubes due to thermal expansion

(d) None of these

24. In a shell and tube heat exchanger :

(a) The temperature drops in the fluid and the wall are proportional to individual resistances

(b) Temperature drops is inversely proportional to the resistance across which the drop occurs

(c) There exists no relationship between temperature drop and resistance

(d) Other external factors

25. Overall heat transfer coefficient largely depends on :

(a) The fluid whose individual coefficient is high

(b) The fluid whose individual coefficient is low

(c) Independent of the individual heat transfer coefficient

(d) Other external factors

26. Multipass heat exchangers are used :
(a) To obtain high rate of heat transfer (b) To obtain higher heat transfer area in a shorter space.
(c) To minimize pressure drop (d) For easy fabrication.

27. 'Fouling factor' is a :
(a) Factor of safety in heat exchanger design
(b) Dimensionless factor
(c) Conversion factor for individual heat transfer coefficient to insert overall heat transfer coefficient
(d) none of there

28. Heat transfer rate is maximum in :
(a) Laminar flow (b) Turbulent flow
(c) Co-current flow (d) Counter current flow

29. Nusselt number is the ratio of :
(a) Temperature gradient at the wall to that across th entire pipe
(b) Difference of temperature to the temperature gradient at the wall
(c) Heat flux at the wall to that across the entire pipe
(d) None of these

30. In case of a fluid flowing through a heated pipe, conduction is negligible in the turbulent core if the value of Prandtl number is :
(a) 0.1 (b) 0.5
(c) 0.6 (d) 1.0

31. Conduction will be negligible in the buffer zone, if the Pradtl number is :
(a) High (b) Low
(c) Very low (d) Very high

32. According to Reynold's analogy, '*j*' factor is given by :
(a) $j_H = (P_r)(S_t)^{2/3}$ (b) $j_M = (P_r)(S_t)^{3/2}$
(c) $j_H = (P_r)^{2/3}(S_t)$ (d) $j_H = (P_r)^{3/2}(S_t)$

33. In case of natural convention, Nusselt number is function of :
(a) *Re* and *Pr* (b) *Gr* and *Pr*
(c) *Gr* and *Re*

34. If the hydrodynamic and thermal boundary layers are indentical in case of a fluid flowing over heated plate, Pradtl number is :
(a) 1 (b) <1
(c) >1 (d) None of these

35. Prandtl number will be least in case of :
(a) Water (b) Viscous liquid
(c) Salt solution (d) Liquid metal

36. In case of heat transfer over a flat plate in laminar zone, if h_x is the local heat transfer coefficient and h_a is the average heat transfer coefficient over the entire length of the plate :
(a) $h_a = h_x$ (b) $h_a = 2h_x$
(c) $h_a = 0.5\ h_x$

37. If the tube surface is rough, heat transfer coefficient :

(a) Increases (b) Decrease

(c) Remains unaffected

38. The coefficient of thermal expansion (β) is given by the relation :

(a) $\dfrac{\left(\dfrac{\partial v}{\partial T}\right)p}{v}$ (b) $\dfrac{\left(\dfrac{\partial p}{\partial T}\right)v}{v}$ (c) $\dfrac{\left(\dfrac{\partial v}{\partial P}\right)T}{v}$

39. In case of a liquid flowing through a pipe, heat transfer in the laminar sublayer is mostly by :

(a) Convection (b) Decreases

(c) Eddies

40. If porosity increases, the thermal conductivity of a solid :

(a) Increases (b) Decreases

(c) Remains unchanged (d) Unpredicatable

41. For an opaque body sum of absorptivity and reflectively is :

(a) 0 (b) 1.0

(c) 0.5 (d) 0.8

42. As temperature changes, the radiant energy emitted by a black body :

(a) Changes (b) Remains unchanged (c) Unpredicatable

43. A black body is a substance, whose absorptivity is :

(a) Zero (b) Half (c) Unity

44. Stefan and Boltzman's law is applicable to :

(a) Grey body (b) White body (c) Black body

45. The range of electro-magnetic spectrum important for radiation is :

(a) 0.38 μ – 0.78 μ (b) 0.5 μ – 50 μ

(c) 100 μ – 1000 μ (d) 5 μ – 50 μ

46. In case of diffuse reflection, the value of absorptivity approaches unity :

(a) True (b) False

47. If the distance between the object emitting radiation and that receiving is '*d*'. Then the energy received is proportional to :

(a) $d^{1/2}$ (b) d

(c) d^2 (d) d^4

48. The interchange factor (F_{12}) between the grey bodies, where one is completely surrounded by the other is given by :

(a) $1/\in_1 + A_1/A_2\,(1/\in_2 - 1)$ (b) $\dfrac{1}{1/\in_1 + A_1/A_2(1/\in_2 - 1)}$

(c) $\dfrac{1}{1/\in_2 + A_2/A_1(1/\in_1 - 1)}$ (d) None of these

$\in_1 \cdot \in_2$ are emissivities of source and sink respectively.

49. Kirchhoff's law is applicable to :

(a) Monochromatic radiation (b) Total radiation
(c) Both *a* and *b* (d) Neither (a) nor (b)

50. According to Kirchhoff's law the ratio of total radiating power to the absorptivity of the body is dependent on :

(a) Temperature of the body (b) Nature of the body
(c) Wavelength of the radiation (d) None of these

51. Value of Stefan-Boltzmann's constant in kcal/m^2 hr °K^4 is :

(a) 0.1713×10^{-8} (b) 1.333×10^{-12}
(c) 4.876×10^{-8}

52. With increase in temperature the wavelength (λ_{max}) corresponding to maximum monochromatic emissive power :

(a) Increases (b) Decreases
(c) Remains unchanged

53. For a black body, the plot of (λT) against ($E\ \lambda/T^5$) is independent of temperature :

(a) True (b) False
(c) Only for a definite range of temperature

54. Solid angle subtended by the finite surface at the radiating element is :

(a) Called view factor
(b) Called angle of vision
(c) Called angle of incidence
(d) Dependent on the distance between the surfaces

55. Under the same condition of temperature difference filmwise condensation is more desirable than drop-wise condensation :

(a) False (b) True

56. Dropwise condensation occurs on a :

(a) Smooth surface (b) Oily surface
(c) Coated surface (d) Glazed surface

57. Compared to a superheated vapour, the film coefficient of saturated vapour is :

(a) Less (b) More
(c) Same

58. Presence of non-condensing gases in the vapours, decreases the film coefficient :

(a) True (b) False
(c) Partially true

59. Steam traps are used to :

(a) Remove condensate (b) Remove liquid from vapour
(c) Regulate pressure (d) None of these

60. Leidenfrost point is associated with :

(a) Condensation of vapour on a cold surface

(b) Evaporation of a solution

(c) Boiling of a liquid on a hot surface

(d) Exchange of heat between two fluids.

61. The important factor which controls the rate of bubble detachment from the solid surface is :

(a) Density of the liquid (b) Temperature of the hot surface

(c) Viscosity of the liquid (d) Interfacial tension

62. Liquid entrainer in evaporators are caused due to :

(a) High rate of evaporation (b) High rate of heat transfer

(c) Foaming of the solution (d) Low pressure in the evaporator

63. For concentrating fruit juice, the recommended evaporator to be used is :

(a) Falling film evaporator (b) Short tube vertical evaporator

(c) Long tube vertical evaporator (d) Agitated film evaporator

64. Terminal conditions and heat transfer area remaining same, the ratio of capacity between a triple effect evaporator and single effect evaporator is :

(a) 1 / 3 (b) 1

(c) 3

65. Boiling point elevation does not influence the economy of a multiple effect evaporator :

(a) True (b) False

66. In multiple effect evaporator, if the thick liquor is viscous, the feeding should be :

(a) Backward (b) Forward

(c) Mixed

67. Cane sugar juice is evaporated using :

(a) Short tube vertical evaporator (b) Long tube vertical evaporator

(c) Horizontal tube evaporator (d) Flash evaporator

68. Optimum number of effects in a multiple effect evaporator is decided by :

(a) Capacity required (b) Available floor area

(c) The terminal conditions (d) Cost analysis

69. For severely scaling liquids, the recommended evaporator is :

(a) Long tube vertical (b) Short tube vertical

(c) Agitated film (d) Horizontal tube

70. Condensation of a vapour under saturation condition means removal of :

(a) Latent heat (b) Sensible heat

(c) Super heat

71. Two identical blocks of copper and mild steel under same conditions of temperature will have same heat content :

(a) True (b) False

72. All conditions and dimensions remaining same, the ratio of velocity through the tubes of a four pass heat exchanger and single pass heat exchange is :

(a) 1/4 (b) 4
(c) 16 (d) 8

73. In a multipass shell and tube heat exchanger, the ratio of tube side return loss to velocity head ($V^2/2g$) is :

(a) 1 (b) 1/4
(c) 4 (d) 8

74. Preheating of feed into an evaporator :

(a) Reduces economy (b) Increases the heat transfer area
(c) Decreases the heat transfer area (d) Requires higher pressure of operation

75. Increasing liquor head in the evaporator :

(a) Increases the capacity (b) Decreases the capacity
(c) Increases the available temperature drop (d) Decreases the available temperature drop.

SECTION—B

1. Thermal conductivity is a..... of the conducting material and dependent on....

2. 1 kcal/hr m C equals to.....BTU/hr ft °F.

3. In steady state conduction, the only property of the substance which determines temperature distribution is.....

4. The equation for one dimensional unsteady state heat conduction is given by.....

5. Rate of heat transfer per unit area is called.....

6. In conduction heat flows in the direction of .....

7. An insulator should have..... thermal conductivity.

8. In forced convection heat transfer.....number and .....number appears.

9. In free convection heat transfer.....number and.....number appears.

10. Double pipe heat exchangers mostly find application in the field of.....

11. Fins installed on the heat transforming surface increases the.....

12. Natural convection is caused due to.....

13. Nusselt number is a measure of.....heat transfer coefficient.

14. The film heat transfer coefficient (h) in MKS unit is.....

15. For heat flow through a thick walled cylinder.....mean radius it used.

16. The two distance methods of condensation are.....and.....

17. A.....solid surface facilitates nucleate boiling.

18. For heating air by steam.....heat exchangers are recommended.

19. The unit of thermal diffusivity is .....

20. Peclet number is a product to.....and.....number.

21. Temperature of a furnace is measured by.....

22. Cooling of an air cooled IC engine takes place by.....

23. High rate of heat transfer is obtained in case of.....boiling.
24. Higher the temperature of the radiating body.....is the wavelength of radiation.
25. Agitated film evaporator is suitable for concentrating.....liquids.
26. Amount of water vaporization per hour in an evaporator is called.....
27. Boiling point elevation.....the capacity of an evaporator.
28. High rate of heat transfer in a viscous liquid may be obtained in a.....type evaporator.
29. .....rule states that the boiling point of a solution is a linear function of water at same pressure.
30. In evaporator, to separate liquid droplets from the vapour.....is used.
31. ......feed does not require any pumps between effects in multiple effect evaporator.
32. Cold feed should be fed.....in multiple effect evaporator.
33. The driving potential for a crystal growth is.....of the solution.
34. The first stage of crystal formation is called.....
35. The liquor left after removed of crystals from a solution is called.....
36. Swenson-Walker crystailiser operates on.....basis.
37. In heat exchangers, the baffle height is normally ....% of inside diameter.
38. Sensible heat of a hot industrial gas can be recovered by a.....
39. Thermal conductivity of a solid material is determined by.....method.
40. Thermal conductivity of a solid can be related to temperature according to the equation.....
41. In a shell and tube heat exchangers, the corrosive liquid is normally layer when Prandtl number is.....
42. In heat exchangers, the terminal point temperature differences between fluids are called.....
43. Thermal boundary layer is thicker than hydrodynamic boundary layer when Prandtl number is .....
44. The expression for logarithmic mean radius is.....
45. Energy radiated from a surface is proportion to the.....power of absolution temperature.
46. If the bubbles formed on a submerged hot surfaces, get absorbed in the mass of liquid, we call it a.....boiling.
47. The widely used insulating material for steam pipe is.....
48. Amount of heat available to a mass of cold water by condensation of 1 kg of saturated steam at 100°C is.....
49. The additional resistance to heat transfer by scale formation on the tubes of a heat exchanger is called.....
50. The ratio of momentum diffusivity to thermal diffusivity is called.....number.
51. An important application of mechanical recompression evaporation is in the production of.....
52. In a thermal recompression evaporator, the vapour is compressed by.....
53. A heat sensitive material should be concentrated by using.....
54. A smooth and clean surface will promote.....condensation.
55. Uncondensed gases and vapours from condensers are removed by.....

SECTION—C

1. What is the expression for heat transfer coefficient (h) for a fluid flowing inside a clean pipe under.

   (*i*) Turbulent flow (*ii*) Laminar flow.

2. What is the general form of Fourier's law of heat conduction through a cube ?
3. What is the expression for flow of heat through a thick walled cylinder ?
4. What is the equation of energy in general form ?
5. What is coefficient of thermal expansion of a fluid ?
6. Give the expression for overall heat transfer coefficient, based on outside area, for a clean fluid flowing through a tube.
7. A steam pipe 170 / 160 mm diameter is covered with two layers of insulation of thickness 30 mm (δ_1) and 50 mm (δ_2) respectively. Thermal conductivities of the pipe material and the insulating layers are 50, 0.15 and 0.08 kcal/m.hr.°C respectively. If the steam temperature is 300 °C and outermost surface temperature is 50°C, calculate the rate of heat loss per meter length of the pipe.
8. Define "fin effectiveness".
9. An aluminium plate of 400 mm × 400 mm × 4 mm, at 200°C is suddenly quenched by liquid oxygen at – 183°C.

   Find out the time required for the plate to reach – 70°C.

   h = 5000 kcal/m^{2}hr°C, specific heat (C_p) = 0.2 kcal/kg.°C. Density (ρ) = 3000 kg/m^3.

10. State the energy equation for thermal boundary layer in Cartesian coordinates in case of a fluid flowing over a flat plate.
11. Air at 20°C flows over a plate of 60 cm × 30 cm with a velocity of 29 m/sec. The critical Reynold's number is 5 × 10^5. Calculate the heat flow per hour from the surface of the plate assuming the flow is parallel to the 60 cm side. Temperature of the plate is 100°C. Properties of air at mean temperature (60°C) are as follows :

    ρ (Density) = 1.06 kg/m^3

    C_p = 0.24 kcal/kg.°C

    K = 0.025 kcal/m.hr.°C

    ν (kinematic viscosity) = 18.97 × 10^{-6} m^2/hr.

12. Calculate the heat transfer coefficient for water at 60°C, flowing through a tube of i.d. 0.625 cm with a velocity of 0.9 m/sec. Tube wall temperature is 40°C. At the mean temp. (*i.e.* 50°C)
13. What is wave theory of thermal radiation ?
14. What is quantum theory of thermal radiation ?
15. What is Plank's distribution law for monochromatic radiation of a black body ?
16. What are the various types of evaporators used industrially ?
17. A triple effect evaporator is concentrating a liquid that has no appreciable elevation in boiling point. The temperature of steam to the first effect is 108.3°C (5 psig). Vacuum in last effect in 26 inches Hg. (51.7°C). The overall heat transfer coefficients are 2440 in the first effect 1952 in second and 976 in third effect (MKS units). What are the approximate boiling temperatures of the liquid in the first and second effects ?
18. What is the law of Haüy of crystrallization ?

19. A 30% solution of Na_2CO_3 weighing 1000 kg is cooled slowly to 20°C. The crystals formed are Na_2CO_3, $10H_2O$. The solubility of Na_2CO_3 at 20°C is 21.5 parts of anhydrous salt per 100 parts water. During cooling 3% of the weight of original solution is lost by evaporation. What is the weight of $Na_2CO_2.10H_2O$ formed ?

20. What is Duhring's rule and what is its use in evaporation ?

21. An evaporator *i* to be fed with 10,000 kg/hr of solution containing 1 percent solution by weight. The feed is at a temperature of 38°C. It is to be concentrated to a solution of 1.5% solute in an evaporator operating at a pressure of 1 atm in vapour space. In order to carry out the operation, the heating surface is supplied with steam at 5 Psig (180.5°C). What is the weight of vapour produced and what is the total weight of steam required ? If overall heat transfer coefficient of the evaporator is (*U*) 1200 kcal/(hr.m^2°C), what is the surface required ?

SECTION—D

1. In thermal radiation for a black body :

(a) $\alpha = 1, \in \neq 1$ (b) $\alpha = 1, \in = 1$

(c) $\alpha \neq 1, \in \neq 1$ (d) $\alpha = 1, \in = 1$

2. At steady state the temperature variation in a plane wall, made of two different solids I and II is shown in the figure below :

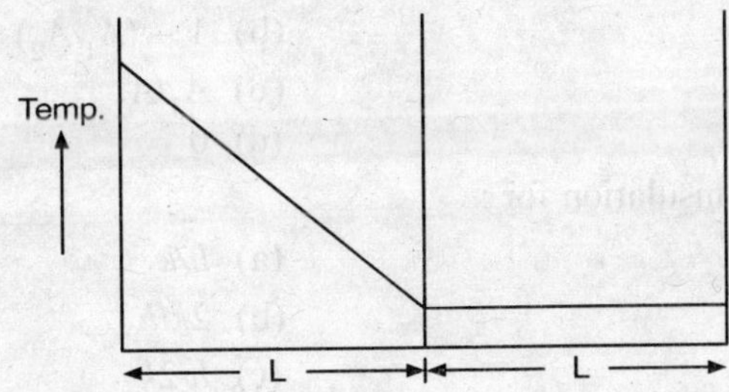

Then the thermal conductivity of the material 1:

(a) is smaller than that of II (b) is greater than that of II

(c) is equal to that of II (d) can be greater than or smaller than that of II.

3. The thermal radiative flux from a surface of emissivity = 0.4 is 22.68 kw/m^2. The approximate surface temperature (*K*) is :

(a) 1000 (b) 727

(c) 800 (d) 1200

4. For condensation of pure vapors, if the heat transfer coefficients in filmwise and dropwise condensation respectively are h_f and h_d then :

(a) $h_f = h_d$ (b) $h_f > h_d$

(c) $h_f < h_d$ (d) h_f

5. Water at the rate of 6800 kg/hr is to be heated from 311 K to 328 K in a double pipe heat exchanger by using another water stream in the annulus. The hot water enters the annulus counter currently at the rate of 3400 kg/hr, and at 366 K. The over-all heat transfer coefficient is 1420 W/(m^2.K), the outside diameter of the inner pipe is 0.115 m. Estimate the length of heat exchanger. Heat capacity of water is 4180 J/(kg.K).

6. A steel ball of 50 mm diameter is cooled by exposing to an air stream at 320 K. Under these conditions the convective heat transfer coefficient is 100 W/(m^2.K.). Estimate the time needed to cool the steel ball from 1120 to 520 K.

Properties of steel : Density = 8000 kg/m^3; Heat capacity = 450 J/(kg.K). Due to high thermal conductivity of steel there no temperature gradients within the ball.

7. It is proposed to reduce heat loss from a rectangular furnace wall by doubling its wall thickness. The temperature of the hot surface of the wall is 723 K and it loses heat from the other side exposed to air at 308 K. In case –1, the temperature of the wall surface exposed is 453 K. Estimate the reduction in heat loss due to the doubling of wall thickness. Neglect radiation losses and assume 1–D conduction. Also assume that K (thermal conductivity of furnace wall) and convective heat transfer coefficient (h) to be constant.

8. A metal wire of 0.01 m diameter and the thermal conductivity of 200 W/(m.K) is exposed to a fluid stream with convective heat transfer coefficient. 1000 W/(m^2.K). The Biot number is :

(a) 5.6 (b) 0.0125

(c) 3.5 (d) 0.0035

9. A small black body with a surface area A_1 having no concavities is surrounded by a large black surface of area A_2. Match the view factors.

1. F_{21} (a) 1
2. F_{22} (b) $1 - (A_1/A_2)$

(c) A_1/A_2

(d) 0

10. Critical thickness of insulation for :

1. Sphere (a) h/k
2. Cylinder (b) $2k/h$

(c) $h/2k$

(d) k/h

11. The maximum in the emissive power of a surface at a temperature T_1 occurs at a wavelength of λ_1. If the surface temperature is halved, the maximum in themselves power would occur at a wavelength of 0.5 λ_1. True or False and state the reason.

12. Saturated steam at 130°C is flowing through a steel pipe of 0.021 m inside diameter and 0.027 m outside diameter. The pipe is insulated outside with a 0.038 m thick insulation. The ambient air outside the insulation is at 27°C. Calculate :

(a) The rate of heat loss per meter length of tube

(b) Overall heat transfer coefficient based on inside surface area of steel pipe.

Date : Thermal conductivity of steel = 45 W/(m.K)

Thermal conductivity of insulation = 0.064 W/(m.K)

Convective heat transfer coefficient outside the insulation = 11 W/(m^2.K)

13. Saturated steam at 6.9 × 10^4 Pa pressure and 90°C saturation temperature condenses on a vertical pipe of 0.025 m outside diameter and 0.3 m length. The average condensing heat transfer coefficient on the tube is 1200 W/(m^2.K)

Other data : outside surface temperature of pipe = 86°C

Enthalpy of staturated steam = 2659 kJ/kg.

Enthalpy of condensate = 375 kJ/kg.

Viscosity of condensate at film temperature = 3.24×10^{-4} Pa.s.

Assume the flow of condensate is laminar calculate the rate of steam condensation. Check whether the flow is laminar.

14. An asbestos pad, square in cross-section measures 0.05 m on a side and increases linearly to 0.1 m on side at type other end. The length of the pad is 0.15 m.

If the small end is held at 600 K and larger end at 300 K, what be the heat flow rate if the other four sides are insulated. Assume one directional heat flow. Thermal conductivity of asbestos is 0.173 W/(m.K)

15. The non-dimensional temperature gradient in liquid at the wall of a pipe is :

(a) The heat flux
(b) The Nusselt number
(c) The Prandtl number
(d) The Schmidt number

16. When the ratio of Grashof number to the square of the Reynolds number is one the dominant mechanism of heat transfer is :

(a) Free convection
(b) Entry length problem in laminar forced convection (developing thermal boundary layer)
(c) Mixed convection (both free and forced)
(d) Forced convection

17. Match :

(*i*) Graetz number	(a) Heat transfer in creeping flow
(*ii*) Rayleigh number	(b) Thermally developing flow
	(c) Product of Grashof and Prandtl numbers
	(d) Product of Reynolds and Prandtl numbers.

18. A hot horizontal plate is exposed to air by keeping :

(a) The hot surface facing up.
(b) The hot surface facing down.

Heat transfer to ambient air is primarily by natural convection. In which of the above cases, heat transfer coefficient is higher and why ? Answer in three or four lines.

19. Choose correct answer.

Saturated steam is condensed as a film on a vertical plate of height 50 cm. If the local heat transfer coefficient at a distance of 50 cm from the top is 7200 W/(m^2.°C), the average heat transfer coefficient over the entire plate in W/m^2.°C is :

(a) 5400
(b) 9600
(c) 7200
(d) 14400

20. A solid cube of 30 cm at an initial temperature of 1000 K is kept in vacuum at absolute zero temperature. Calculate th time required to cool it to 500 K. The material has the following properties. Density = 2700 kg/m^3; specific heat = 0.9 KJ/(kg.K). Emissivity = 0.10; Stefan-Boltzmann constant, $\sigma = 5.669 \times 10^{-8}$ W/(m^2.K^4)

21. A pipe of 20 mm inner diameter and 30 mm outer diameter is insulated with 35 mm thick insulation. The thermal conductivity of the insulation is 0.15 W/(m^2.°C) and the convective heat transfer coefficient of outside air is 3 W/(m^2.°C). The surface temperature of Bare pipe is 200°C and the ambient air temperature is 30°C. The heat transfer resistance of the metal pipe can be neglected.

(*i*) Comment with reasoning about the heat transfer rates with and without insulation.

(*ii*) If the same insulating material is used, what is the minimum thickness above which there is reduction in heat loss as compared to the bare pipe.

(*iii*) For optimum design what conductivity of insulating material do you suggest for the conditions given in the problem

22. In a counter-current heat exchanger which has been in service for quite time, due to formation of scale, heat transfer rate is reduced to 85% of its original value based on clean surface.

Assuming that the terminal temperature of fluids are same in both cases and the effective heat transfer area does not change appreciably due to scale formation, determine the overall fouling factor if cleaned overall coefficient is 500 W/(m^2. K).

23. A long iron rod initially at a temperature of 20°C has one end dipped in boiling water (100°C) at time $t = 0$. The curved surface of the rod is insulated so that heat conduction is one dimensional in the axial direction. The temperature at a distance of 100 mm from the dipped end becomes 40°C at $t = 200$ s. The same temperature is achieved at a distance of 200 mm from dipped end at time.

(a) $t = 283$ s (b) $t = 356$ s

(c) $t = 400$ s (d) $t = 800$ s

24. The radiation heat flux from a heating element at a temperature of 800°C in a furnace maintained at 300°C is 8 kW/m^2. The flux when the element temperature is increased to 1000°C for the same furnace temperature is :

(a) 11.2 kW/m^2 (b) 12.0 kW/m^2

(c) 14.6 kW/m^2 (d) 16.5 kW/m^2

25. If the Prandtl number is greater than the Schmidt number :

(a) The thermal boundary layer lies inside the concentration boundary layer.

(b) The concentration boundary layer lies inside the termal boundary layer.

(c) The thermal and concentration boundary layers are of equal thickness.

(d) The hydrodynamic boundary layer is thicker than the other two.

26. In the lower portion of a spray tower, urea pellets (diameter = 2 mm) are cooled by air at a temperature T_w = 20°C. The pellets are falling at their terminal velocity, u_t = 6.9 m/s. The temperature within the pellets is uniform at all times, and the initial temperature of the pellets is 80°C. The heat transfer coefficient for a pellet falling in air is h = 208 W/(m^2.K)

(a) Obtain an expression for the change of temperature of a pellet with time.

(b) Calculate the height of tower for a pellet to cool to 60°C.

SECTION—E

1. In Joule experiments and insulated container contains 20 kg of water initially at 25°C. It is stirred by an agitator which is made to turn by a slowly falling body weighing 40 kg, through a height of 4 m. The process is operated 500imes. The acceleration due to gravity is 9.8 m/sq. sec. Neglecting heat capacity of agitator, the temperature to water in °C is:

(a) 40.5 (b)34.4 (c) 26.8 (d) 25

2. Three solid objects of the same material and of equal mass—a sphere, a cylinder (length = diameter) and a cube are at 500°C initially. These are dropped in a quenching bath containing a large volume of cooling oil each attaining the bath temperature evenly. The time required for 90% change of temperature is smallest for : (a) Cube; (b) Cylinder; (c) Sphere; (d) Equal for all the three.
3. A dilute aqueous solution is to be concentrated in an evaporator system. High pressure steam is available. Multiple effect evaportor system is employed because :
   (a) Total heat transfer area of all the effects is less than that in a single effect evaporator system.
   (b) Total amount of vapour produced per kg of feed steam in a multiple effect is much higher than in a single effect.
   (c) Boiling point of elevation in a single effect system is much higher than in any effect in multifactor system.
   (d) Heat transfer coefficient in a single effect is much lower than in any effect in a multi-effect system.
4. The units of resistence to heat transfer are :
   (a) J/(sq.m.K) (b) J/(m.K)
   (c) W/ (sq.m.K) (d) sq.m.K/w
5. A process system of dilute aqueous solution flowing at the rate of 10 kg/s. is to be heated. Steam condensate at 95°C is available, for heating purpose, also at a rate of 10 kg/s. A 1-shell and tube heat exchanger is available. The best arrangement is :
   (A) Counter flow with process stream on shell side.
   (B) Counter flow with process stream on tubbiest.
   (C) Parallel flow with process stream on shell side.
   (D) Parallel flow with process steam on tubbiest.
6. The inner wall of a furnace is at a temperature of 700°C. The composite wall is made of two substances 10 and 20 cm. Thick with thermal conductivities of 0.05 and 0.1 W/m.°C. respectively. The ambient air is at 30°C. and the heat transfer coefficient between the outer surface of wall and air is 20 W/sq.m.°C. The rate of heat loss from the outer surface in W/sq.m. is
   (a) 165.4 (b) 167.5
   (c) 172.8 (d) 175
7. Steam is to be condensed in a shell and tube heat exchanger of 5 m long with a sheel diameter of 1 m. Cooling water is to be used of removing the heat. Heat transfer coefficient for the cooling water whether on shell side or tube side is same. The best arrangement is :
   (a) Vertical heat exchanger with steam on tubel side,
   (b) Vertical heat exchanger with steam on shell side,
   (c) Horizontal heat exchanger with steam on tube side,
   (d) Horizontal heat exchanger with steam on shell side.
8. A fluid is flowing inside the inner tube of a double pipe heat exchanger with diameter "*d*". For a fixed mass flow rate, the tube side heat transfer coefficient for turbulent flow condition is proportional to :
   (a) $d^{0.8}$ (b) $d^{0.2}$
   (c) $1/d$ (d) $d^{-1.8}$

9. Air is to be heated by condensing steam. Two heat exchangers are available. (1) a shell and tube heat exchanger and (2) a finned tube heat exchanger. Tube side heat transfer area is equal in both cases. The recommended arrangement is :
   (a) Finned tube heat exchanger with air inside and steam inside.
   (b) Finned tube heat exchanger with air outside and steam inside.
   (c) Shell and tube heat exchanger with air inside tubes and steam shell side.
   (d) Shell and tube heat exchanger with air on shell side and steam inside tubes.
10. For a given ambient air temperature with increase in thickness of insulation of a hot cylindrical pipe, the rate of heat loss from the surface would :
    (a) Decrease (b) Increase
    (c) First decrease and then increase (d) First increase and then decrease.
11. In forced convection the Nusselt number is a function of :
    (a) Re & Pr (b) Re & Gr
    (c) Pr & Gr (d) Re & Sc
12. For an ideal black body :
    (a) = 1 (absorptivity) (b) Reflectivity = 1
    (c) Emissivity = 0 (d) Transmissivity = 1
13. The left face of a one dimensional slab of thickness 0.2 m is maintained at 80°C and the right face is exposed to air at 30°C. The thermal conductivity of slab is 1.2 W/(m.K) and the heat transfer coefficient from the right face is 10 W/(sq.m.h.). At steady state the temperature of the right face in °C is : 60º. Find time :
    (a) 555 hrs. (b) 55.5 hrs.
    (c) 0.55 hr. (d) 0.15 hr.
14. It is desired to concentrate a 20% salt solution (20 kg of salt in 100 kg of solution) to a 30% salt solution in an evaporator. Consider a feed of 300 kg/min at 30°C. The boiling point of solution is 110°C. The latent heat of vaporization is 2100 KJ/kg. The specific heat of solution is 4 KJ/(kg. K). The rate at which heat has to be supplied in KJ/min to the evaporator is :
    (a) $3.06*10^5$ (b) $6.12\ 10^5$
    (c) $7.24 * 10^5$ (d) $9.05 * 10^5$
15. Hot water (0.01 cubic. m/min) enters the tube side of co-current shell and tube heat exchanger, at 80°C and leaves at 50°C. Cold oil (0.05 cubic. m/min of density 800 kg/cubic.m and specific heat of 2 kJ/kg.K. enters at 200°C. The log mean temperature difference in °C approximately is :
    (a) 32 (b) 37
    (c) 45 (d) 50
16. Multiple effect evaporators are commonly used in the manufacture of :
    P-paper; Q-super phosphate; R-sugar; S-Fats.
    (a) P and Q (b) P and R
    (c) P and S (d) R and S

ANSWERS

SECTION–A

1. (a)	2. (a)	3. (b)	4. (a)	5. (b)
6. (c)	7. (b)	8. (c)	9. (a)	10. (b)
11. (b)	12. (b)	13. (a)	14. (a)	15. (c)
16. (b)	17. (a)	18. (c)	19. (a)	20. (c)
21. (a)	22. (a)	23. (c)	24. (a)	25. (b)
26. (a)	27. (a)	28. (b)	29. (a)	30. (c)
31. (a)	32. (c)	33. (b)	34. (a)	35. (d)
36. (b)	37. (a)	38. (a)	39. (b)	40. (b)
41. (b)	42. (a)	43. (c)	44. (c)	45. (b)
46. (a)	47. (c)	48. (b)	49. (c)	50. (a)
51. (c)	52. (b)	53. (a)	54. (b)	55. (a)
56. (b)	57. (c)	58. (a)	59. (a)	60. (c)
61. (d)	62. (c)	63. (a)	64. (b)	65. (a)
66. (a)	67. (b)	68. (d)	69. (d)	70. (a)
71. (b)	72. (b)	73. (c)	74. (c)	75. (b)

SECTION–B

1. porosity, temperature
2. 0.672
3. Thermal conductivity
4. $\dfrac{\partial t}{\partial \theta} = \alpha \dfrac{\partial^2 t}{\partial x^2}$
5. heat flux
6. temperature drop
7. low
8. Reynold, Prandtl
9. Pradtl, Grashof
10. refrigeration
11. heat transfer rate
12. density difference of air
13. convective
14. kcal/m^2.hr.°C
15. logarithmic
16. dropwise, filmwise
17. rough
18. finned tube
19. m^2/hr
20. Reynold number, Prandtl number
21. radiation pyrometer
22. natural convection
23. nucleate
24. shorter
25. viscous
26. capacity
27. reduces
28. agitated film
29. Duhring's
30. entrainment separator
31. forward feed
32. backwardly
33. super saturation
34. nucleation

35. mother liquor
36. continuous
37. 75
38. regenerator/recuperator
39. Lee's
40. $K = a + bT$
41. tube
42. approach
43. less than unity
44. $(r_1 - r_2)/\left(\ln \frac{r_1}{r_2}\right)$
45. fourth
46. pool
47. 85% mangesia and glass wool
48. 537 kcal
49. fouling factor
50. Prandtl
51. distilled
52. steam jet ejector
53. vacuum
54. film
55. vent tubes.

SECTION—C

1. (*i*) $h = 0.023 \text{ K/D} \left(\frac{DV\rho}{\mu}\right)^{0.8} \left(\frac{Cp\mu}{K}\right)^{0.4}$

for Re > 10,000.

(*ii*) $h = 1.86 \frac{K}{D}\left[\left(\frac{DG}{\mu}\right)\left(\frac{Cp\mu}{K}\right)\left(\frac{D}{L}\right)\right]^{0.33} \left(\frac{\mu}{\mu_s}\right)^{0.11}$

Re < 2,100

2. $\frac{\partial t}{\partial \theta} = \frac{K}{C_p \rho}\left(\frac{\partial^2 t}{\partial x^2} + \frac{\partial^2 t}{\partial y^2} + \frac{\partial^2 t}{\partial z^2}\right)$; t = temperature; θ time

3. $g = \frac{K\bar{A}_L(T_4 - T_0)}{r_o - r_i}$, where $\bar{A}_L = \frac{2\pi L(r_o - r)}{\ln(r_o - r_i)}$

i, refers to inside condition

o, refer to outside condition

L = Length of the cylinder.

4. $\underset{(i)}{\frac{\partial}{\partial t}\rho\left(U + \frac{1}{2}v^2\right)} = -\underset{(ii)}{\left[\Delta.\rho v\left(U + \frac{1}{2}v^2\right)\right]} \underset{(iii)}{\nabla q.}$

$+ \underset{(iv)}{\rho\,(u.g)} - \underset{(v)}{\nabla\,.\,Pv} + \underset{(vi)}{[\nabla\,(\tau.\,v)]}$

(*i*) Rate of gain of energy/unit volume

(*ii*) Rate of energy input/unit volume by convection

(*iii*) Rate of energy input/unit volume by conduction

(*iv*) Rate of work done on fluid/unit volume by gravitational forces.

(*v*) Rate of work done of fluid/unit volume by pressure forces.

(*vi*) Rate of work done on fluid/unit volume by viscous forces.

5. The cofficient of thermal expansion (β) of a fluid may be defined as the fractional increase in volume, at constant pressure of the fluid per unit rise in temperature.

Mathematically, $\beta = \left(\frac{\partial u}{\partial T}\right)_P / v$

6. $$U_i = \frac{1}{\frac{1}{h_t} + \frac{D_i}{D_L}\frac{x_w}{K_m} + \frac{D_i}{D_o h_o}}$$

7. Given data:

$r_1 = 80$ mm, $T_1 = 300°C$,

$K_1 = 50$

$r_2 = 85$ mm; $K_2 = 0.15$

$r_3 = 115$ mm; $K_3 = 0.08$

$R_4 = 165$ mm $T_4 = 50°C$

$L = 1000$ mm = 1 mt.

According to the formula, heat lost/hr

$$= \frac{(T_1 - T_4)2\pi L}{\frac{1}{K_1}\log\frac{r_2}{r_1} + \frac{1}{K_2}\log\frac{r_3}{r_2} + \frac{1}{K_3}\log\frac{r_4}{r_3}}$$

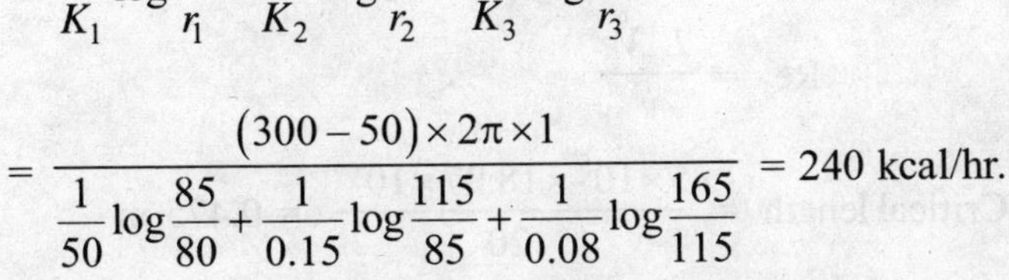

$$= \frac{(300-50)\times 2\pi \times 1}{\frac{1}{50}\log\frac{85}{80} + \frac{1}{0.15}\log\frac{115}{85} + \frac{1}{0.08}\log\frac{165}{115}} = 240 \text{ kcal/hr.}$$

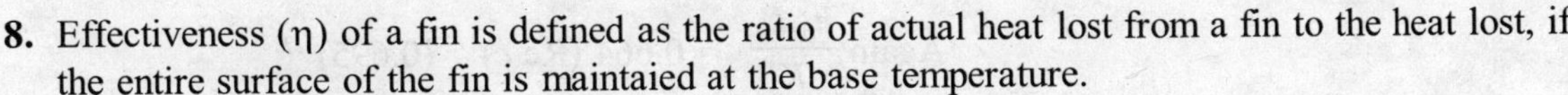

8. Effectiveness (η) of a fin is defined as the ratio of actual heat lost from a fin to the heat lost, if the entire surface of the fin is maintaied at the base temperature.

9. The formula to be used is,

$$\frac{T - T_a}{T_i - T_a} = \exp.\ \frac{hAt}{CV_P}$$

where, $T = -70°C$, $T_i = 200°C$; $T_a = -183°C$

Area (A) = 2 × 0.4 × 0.4 = 0.32 m² (neglecting the ends)

Volume (V) = 0.4 × 0.4 × 0.004 = 64 × 10^{-5} m³

$$\frac{-70-(-183)}{200-(-183)} = (e)\frac{-5000\times 0.32}{3000\times 0.2\times 64\times 10^{-5}} \times t$$

or, $0.295 = (e) - 4180\,t$

or $-4180\,t = \ln(0.295) = -1.22$

$$t = \frac{1.22}{4180} \times 3600 = 1.05 \text{ secs.}$$

10. $v_x \dfrac{\partial T}{\partial x} + v_y \dfrac{\partial T}{\partial y} = \alpha \dfrac{\partial^2 T}{\partial y^2}$

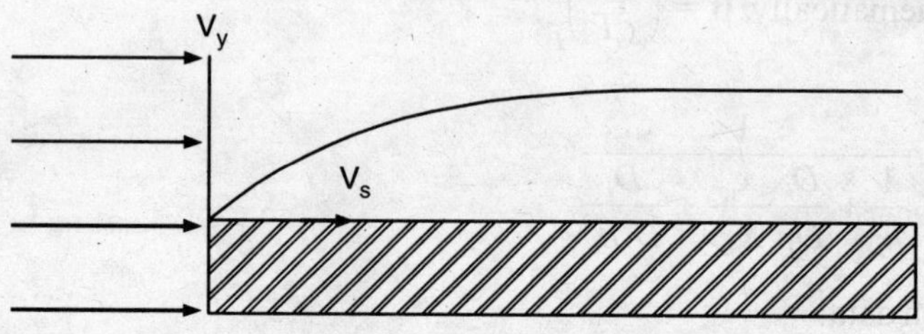

Assumptions :

(*a*) Heat conduction in x-direction is neglected comparead to y-direction

(*b*) Heat generated due to internal friction of the fluid is also neglected.

11. Prandtl number

$$(P_r) = \frac{C_p \mu}{K} = \frac{C_p \nu \rho}{K}$$

$$= \frac{0.24 \times 18.97 \times 10^{-6} \times 3600 \times 1.06}{0.025} = 0.695$$

Critical Reynold's number,

$$\text{Re}_e = \frac{L_c.V}{\nu}$$

$$\text{Critical length} = \frac{5 \times 10^5 \times 18.97 \times 10^{-6}}{20} = 0.475 \text{ m}$$

$$\text{Again } \frac{h_a.L_e}{K} = 0.664\,(\text{Re}\ c)^{1/2}\,(0.695)^{1/3}$$

h_a = average heat transfer coefficient in laminar flow.

$$h_a = \frac{0.664 \times 0.025}{0.475}(5 \times 10^5)^{1/2}\,(0.695)^{1/3}$$

$= 21.8$ kcal/m^2.hr.°C

Heat flow $= 21.8 \times (0.63 \times 0475)\,(100 - 20)$.

$= 248\ 5$ kcal/hr.

12. Prandtl number,

$$(P_r) = \frac{C_p\mu}{K} = \frac{1 \times 2.167}{0.54} = 4.01$$

Reynold's number

$$(R_e) = \frac{DV\rho}{\mu} = \frac{(0.625 \times 10^{-2})(0.9)1000}{2.167/3600}$$

$$= 9345.$$

$$h = \frac{0.023 C_p \rho V}{(R_e)^{0.2}(P_r)^{0.66}} = \frac{0.020 \times 1 \times 1000 \times 0.9 \times 3600}{(9345)^{0.2}(4.01)^{0.66}}$$

$$= 4790 \text{ kcal/m}^2\text{hr°C}.$$

13. According to the Wave theory, the species believed to be filled up by a hypothetical medium called 'ether'. Radiation occurs because of the propagation of electromagnetic waves through the ether. The electromagnetic waves travel though the medium of ether at the speed of light. When the electromagnetic waves emitted by a hot body is intercepted by a second body; a part of it gets absorbed by the second body and gets converted to stored energy by the absorbing body.

14. According to quantum theory, when the temperature of a solid body is raised, some of the constituting atoms and molecules are raised, to excited state of higher energy. These atoms and molecules, after emitting energy in form of electromagnetic radiations, return to the lower energy level. This emitted energy in form of successive and discrete quantities are called quanta, radiated at all frequencies.

15. At a given wavelength, the radiant energy emitted by a black body increases as the temperature is increases. According to this law, the monochromatic emissive power of a black body is given by the relation,

$$(E\lambda)_b = C_1 \frac{\lambda^{-5}}{\left(\exp\frac{C_2}{T\lambda}\right) - 1}$$

where $C_1 = 37.404 \times 10^{-6}$ erg. cm²/sec.

$C_2 = 1.4387 \times 10^{-2}$ m.K.

16. Evaporator types are :

(1) Apparatus heated by direct fire. (2) Apparatus with heating medium in jackets, double walls etc. (3) Vapour heated evaporators with tubular heating surfaces.

(a) Tubes horizontal (vapour inside tubes)

(b) Tubes vertical.

1. Standard type, 2. Basket type, 3. Long tube type

4. Forced circulation type.

17. The total temperature drop is : 108.3 – 51.7 = 56.6°C.

The temperature drops are approximately inversely proportional to the overall heat transfer coefficients.

$dt_1 : dt_2; dt_3 = \frac{1}{U_1} : 1/U_2 : 1/U_3$

That is $U_1\, dt_1 = U_2\, dt_2 = U_3 dt_3$

From this it follows by trial and error the temperature drops nearly,

$dt_1 = 12°C,\ dt_2 = 15°C;\ dt_3 = 29.6°C,$

Boiling temperature in first effect = 108.3 – 12 = 96.3°C.

Boiling temperature in second effect = 96.3 – 15 = 81.3°C

18. Although the relative development of the different faces of two crystals of the same material may be widely different, the interfacial angles of corresponding faces of the two crystals are all equal and that is a characteristic of that substance.

19. Molecular weight of Na_2CO_3 = 106;

Molecular weight of $Na_2CO_3\ 10\ H_2O$ = 286.2

The equation governing material balance is :

$$C = R\left[\frac{100w_0 - S(H_0 - E)}{100 - S(R-1)}\right]$$

C = Weight of crystals in final magma

$$R = \frac{\text{Molecular weight of hydrated solute}}{\text{Molecular weight of anhydrous solute}}$$

= 286.2/106.0 = 2.70

S = Solubility (parts by weight of anhydrous solute per 100 parts by weight of solvent) of material at final temperature.

w_0 = Weight of anhydrous solute in original batch

= 0.30 × 10000 = 3000 kg.

H_0 = Total weight of solvent in batch at the beginning of the process.

E = Evaporation during the process

= 10,000 × 00.3 = 300 kg.

$H_0 - E$ = 10,000 – 3000 – 300 = 6,700 kg.

$$C = 2.70\left[\frac{100 \times 3000 - 21.5 \times 6700}{100 - 21.5(2.70 - 1.0)}\right] = 6636 \text{ kg.}$$

20. For strong solutions the boiling point of elevation cannot be calculated from any known law. According to this rule of pure water at the same pressure, a straight line results. For a given concentration of solution at different pressure, different straight lines are obtained for different concentrations. The boiling point of elevation is required to calculate the temperature drop in the evaporator which is used for heat transfer area (surface) calculations.

21. F = 10,000 kg/hr : point of elevation is neglected. $X_F = 0.01$.

Total solids in feed = 10,000 × 0.01 = 100 kg

Total water in feed = 10,000 × 100 = 9900 kg

$t_F = 38°C,\ h_F = 38.1$ kcal/kg (from steam tables)

$X_1 = 0.015$

Total solids in thick liquor = 100 kg

L = 100/0.015 = 6670 kg

t_L = 100°C, h_F = 100.8 kcal/kg (from steam tables)

V = 10,000 – 6670 = 3330 kgs, t_v = 100°C.

H = 644 kcal/kg (from steam tables), t_s = 108.3°C.

H_s = 647.4 kcal/kg (from steam tables), t_e = 108.3°C.

h_e = 109.2 kcal/kg (from steam tables).

By energy transferred to the heating surface :

1000 (38.1) + 647.4 S = 3330 × 644 + 6670 × 100.8 + 109.2 S

S = 4530 kg/hr.

The heat transferred to the heating surface :

4530 × (647.4 –109.2) = 2,438.046 kcal = Q

$UAdT$ = 1200 A × (108.3–100) = 2,438,046

1200 × 8.3 × A = 2438,046, A = 244.8 sqm. **Ans.**

SECTION—D

1. (d) Where ε emissivity and α of absorptivity. Absorptivity is one and emissitiy is one for black body.
2. (a) $dt_1 + dt_2 + dt_3 = \frac{qL_1}{AK_1} + \frac{qL_2}{AK_2} + \frac{qL_3}{AK_3}$ and dt_2 is less than dt_1.
3. (a) $q = 22680 = 0.4 * 5.67 \times 10^{-8}\ T^4 = \varepsilon\, b\, T^4$

   T = 1000 K
4. (c)
5. q = 6800 * (328 – 311) 4180 = 3400 (366 – T_0) 4180

   T_0 = 332 K

   Δt_1 = 332 – 311 = 21; Δt_2 = 366 –328 = 38

   Δt_{1m} = 28.7; $q = 4832 \times 10^5$ = 1420 × 3600 × 3.142 × 0.11 * L × 28.7

   = $UA\ \Delta t$ lm giving L = 9.115 m
6. $\frac{T - Ta}{T_i - T_a} = e^{-hAt/(e\rho v)}$

   Here T = 520; T_i = 1120; T_a = 320.

   h = 100; $A = 4\pi r^2 = 4\pi \times (0.05)^2 / 4$; t = time, in seconds;

   ρ = density = 8000 kg/m^3;

   C = specific heat = 450;

   V = volume = $\frac{4}{3}\pi r^3 = \frac{4}{3}\pi\,(0.05)^3 / 8$

   t = 8301.2 seconds = 2.306 hr.
7. Similar to Q : D – 12, Q = UA ΔT and Similar to Q : D – 21.
8. (b) Biot Number = hL / K; V = volume of material. $\pi D^2 / 4 * L$

9. 1 – c; 2 – d.
10. 1 – b; 2 – c.
11. False. *wb*, $\lambda = C_1\, \lambda^{-5} / (e^{C_2/\lambda T} - 1)$
    λ = wavelength, T = Temperature.
12. (a) $U_o = \dfrac{1}{\dfrac{1}{h_o} + \dfrac{x_w}{k_m}\dfrac{D_o}{D_L} + \dfrac{x_{ins}\cdot}{k_{ins}}\dfrac{D_{ins}\cdot}{D_{L_{ins}}} + \dfrac{D_o}{h_i\, D_i}}$

    $h_o = 11$; $x_w = 0.006$; $k_m = 45$; $D_o = 0.027$
    $D_L = (0.027 + 0.021) / 2$; $x_{ins} = 0.038$
    $k_{ins} = 0.064$; $D_{ins} = 0.065$; $D_{L_{ins}} = (0.065 + 0.027) / 2$
    $D_o = 0.027$; $D_i = 0.021$; $h_i = 5678$
    $U_o = 1.162$ W/(m^2.°C)
    $q/A = U_o\, \Delta t = 1.162 \times (130 - 27) = 119.7$ W/m^2.

    (b) $U_i = \dfrac{1}{\dfrac{1}{h_i} + \dfrac{x_m}{k_w}\dfrac{D_i}{D_L} + \dfrac{1}{h_o}\dfrac{D_i}{D_o} + \dfrac{x_{ins}}{k_{ins}\cdot}\dfrac{D_{ins}}{D_{L_{ins}}}}$

    $h_i = 5678$; $x_w = 0.006$; $k_w = 45$; $D_i = 0.027$;
    $D_L = 0.024$; $h_o = 11$; $D_i = 0.027$; $D_o = 0.065$;
    $x_{ins} = 0.038$; $k_{ins} = 0.064$; $D_{ins} = 0.065$; $D_{L_{ins}} = 0.046$
    $U_i = 1.24$ W/(m^2.°C)

13. $h\left(\dfrac{\mu^2}{kf^3 \rho f^2 g}\right) = 1.47\left(\dfrac{4T_b}{\mu f}\right)$

    $h = 12000$ W/(m^2.c); $\mu_f = 3.24 \times 10^{-4}$ Pa.s.
    $\rho_f = (0.0373 / 62.4)\ 1000$ kg/m^3; $g = 9.81$ m/s^2
    $k_f = 0.0136 \times 1.7308$ watt/(m.K)

    Substituting τ_b (kg/m.hr.) is $\dfrac{0.0060319 \text{ kg}}{\text{m.s}}$

    $= \dfrac{217 \text{ kg}}{\text{m.hr.}} = \dfrac{m}{\pi D_o}$

    $m = 21.7 \times 3.142 \times 0.025 = 1.705$ kg/hr.
14. $q = -KA\, dT/dX$; $K = 0.173$
    at $x = 0$; area $(A) = 0.0025$ m^2
    at $x = 0.15$; $A = 0.01$ m^2; $A = 0.0025 + 0.05\, X$

    $q = 0.173\,(0.0025 + 0.05x)\left(\dfrac{600 - 300}{0.15}\right)$

    at $x = 0.15$; $q = 3.46$ Watt.

15. (*b*) *hD* / *k* is Nuisselt number.
16. (*c*).
17. (*i*) – *a*, (*ii*) – *b*
18. In case of (*a*) it is more. Air after getting heated by plate leaves immediately for next cool air to come without forming and film resistance.
19. (*b*) 9600

The average coefficients is $\frac{4}{3}$ times the local coefficient at the bottom of the tube.

20. The answer is similar to Question no : 6 but here loss is by radiation.
21. (*i*) $U_o = \dfrac{1}{\dfrac{1}{h_i}\dfrac{D_o}{D_i} + \dfrac{x_w}{k_m}\dfrac{D_o}{D_L} + \dfrac{1}{h_o}}$

With insulation : x_w = 0.035 m; k_w = 0.15;
D_o = 0.065 m; D_L (log mean diameter) = 0.0453 h_o = 3
U_o comes as 1.497; Heat loss/m^2 = $U\Delta t$
= 1.497 × (200 –30) = 255 W/m^2
Without insulation, U_o = 3; Heat loss/m^2 = 3 × 170 = 510 W/m^2.
(*ii*) Thickness of insulation at 10 m.m, heat loss is 415.1 W/m^2; at 20 mm, heat loss is 340 W/m^2 at 35 mm, heat loss is 255 W/m^2.
(*iii*) *k* should be as less as possible to have less heat loss.

22. $U_i = \dfrac{1}{\dfrac{1}{h_{di}} + \dfrac{x_w}{k_m}\dfrac{D_i}{D_L} + \dfrac{D_i}{D_o h_o}}$

$$500 = \frac{1}{\underset{\downarrow \atop A}{\dfrac{1}{h_i}} + \underset{\downarrow \atop B}{\dfrac{x_w}{k_m}\dfrac{D_i}{D_L}} + \underset{\downarrow \atop C}{\dfrac{D_i}{D_o h_o}}}$$

$$A + B + C = \frac{1}{500} = 0.002$$

$$U_i = \frac{1}{\dfrac{1}{h_{di}} + \dfrac{1}{h_i} + \dfrac{x_w}{k_m}\dfrac{D_i}{D_L} + \dfrac{D_i}{D_o h_o} + \dfrac{D_i}{D_o h_{Do}}}$$

h_{di}, h_{Do} are dirt factors. Neglecting h_{Do}.

$$500 \times 0.85 = \frac{1}{500 \times 0.85} = 0.002353$$

$$\frac{1}{h_{di}} = 0.002353; \; h_{di} = 425 \text{ W/m}^2\text{.K}$$

23. Use unsteady state heat conduction equation (Item–6 on P–136).
24. (d) $8 = k\,(1073^4 - 573^4)$; $Q = K\,(1273^4 - 573^4)$; $Q = 16.54$ KW/m^2
25. (a)
26. Hint : (a) $\frac{T(t)-T(\infty)}{T(i)-(\infty)} = e\; hAt/\,(PVCP)$ time (t) comes at $T = 60°C$; (b) Height of Tower = t * 6.9 m.

SECTION—E

1. **Ans. (b). Solution:** Heat lost by falling body = heat gained by water; $500*mb*g*h = mw*Cp*(T-25)$; where mb is mass of body; mw is mass of water, h = height; Cp = specific heat of water; T = temperature.
   500*40*9.8*4 = 20*4180* (T-25); T = 34.4°C. Ans.
2. **Ans :** (a) **(Solution :** A cube has largest surface area when length = diameter of cylinder (breadth and width).
3. **Ans.** (b) 4. **Ans.** (d) 5. **Ans.** (a)
6. **Ans.** (a)
   (Solution: $Q = T_1 - T_2/\,(R_1 + R_2 + R_3) = T_1 - T_2/\,(L_1/\,K_1A + L_2/K_2\,A + 1/hA)$; $q = Q/A = T_1 - T_2/(L_1/K_1 + L_2/K_2 + 1/h) = 700 - 30/(0.10/0.05 + 0.20/0.10 + 1/20) = 670/4.05 = 167.5$ W/sq.m)
7. **Ans.** (b)
8. **Ans.** (b) **(Solution.** $h_L = 3.52\; V/D$; $h_L \infty D^{-0.20}$)
9. **Ans.** (b) 10. **Ans.** (d) 11. **Ans.** (a)
12. **Ans.** (a)
    Solution : An ideal black body absorbs all radiations falling on it. So absorptivity is one.
13. **Ans.** (d)
    Solution : For unsteady state energy balance, we have the relation : $q(t)/q\;(0) = e^{-mt}$
    i.e. $[(T(t) - T(\infty)]/T(i) - T(\infty) = \exp\,(-h\,At/pV\,Cp)$; Here $T(i) = 80°C$; $T\,(\infty) = 30°C$; $T(t) = 60°C$; $h = 50$; $A = 4\pi\;(0,\,1) * 0.1$ sqm, $p = 3000$ kg/sq. m; $Cp = 0.4$ KJ/kg.K = 400J/kg.K; $V = 4\pi/3 * (0.1)^3$ cubic m. 60 – 30 (90 – 30) = cx (– 50 * 40 * (0.10) * (0.1)/ 3000*400* $4\pi/3$ * $(0.1)^3$ * t;
    t = 554.4 seconds = 0.154 hr. Ans.
14. **Ans. (a) Solution :** Overall material balance : $F = L + V$; $F * xF = L*xL + V*yV$;
    since $yV = 0$; 300*0.20 = L*0.30; L = 200 kg/min; $V = F - L$
    = 300 – 200 = 100 kg/min. Heat supply rate :
    $F * CpF\;(TBF - TF) + V * \lambda V$ = 300*4* (110 – 30) + 100*2100 = 3.06*10^5 KJ/min.
15. **Ans.** (a)
    Solution. QH = 0.01 cubic. m/min; Qc = 0.05 cubic. m/min.
    ρ_c = 800 kg/cubic. m; Cpc = 2000J/kg.; ρ_H = 1000 kg/cubic.m;
    CpH = 4.18 KJ/kg.; Heat balance for the case is:
    $mH\;CpH\;(TH1 - TH2) = mc\;Cpc\;(Tc1 - Tc2)$; $mH = QH\;\rho H$ and $mc = Qc\;\rho c$;
    1000*0.01*4.18*1000*(33 – 323) = 0.05*800*2000*(293 – $Tc2$); $Tc2$ = 36.75°C. DT1 = 80 – 20 = 60; $DT2$ = 50 – 35.675 = 14.325; $DT\,Lm = DT1 - DT2/\,Ln$ (DT1/DT2) = 60 – 14.325/Ln (60/14.325) = 32°C.
16. **Ans.** (c)

❑

CHAPTER 7

Mass Transfer

1. Five Thumb Rules of chemical engineering are:
   (a) Material balance
   (b) Energy balance
   (c) Momentum balance
   (d) Equilibria relations
   (e) Cost/economy relations.
2. **Unit Operations:** In unit operations no chemical reactions but only physical change, e.g. drying, crushing, filtration, evaporation, etc.
3. **Unit Processes:** Chemical reactions are present, e.g. nitration, polymerization.
4. Mass Transfer Operations are classified as:
   (a) Direct contact of miscible phases (thermal diffusion, centrifugation)
   (b) Contact of miscible phases separated by membranes (dialysis, gas permeation)
   (c) Contact of immiscible phases (distillation, evaporation, extraction).
5. Material balance is based on law of conservation of mass.
6. Average molecular weight = $(Wa + Wb + Wc + ... + Wn)/(Wa/Ma + Wb/Mb + ... + Wn/Mn)$.
7. Mole fraction of A is = $Wa/Ma/(Wa/Ma + Wb/Mb + ... + Wn/Mn)$
8. Ideal gas law is $PV = nRT$ where P is pressure, V is volume, n = no of moles, R is gas constant and T is Temp.
9. One gram mole of ideal gas at 0°C and 760 mm Hg pressure occupies 22.41 litres volume. One pound mole of gas occupies 359 cu.ft at a pressure of 760 mm Hg and 32°F. For ideal gases, volume % = pressure % = mole %.
10. Steady state, unsteady state: If the operating conditions do not vary with time then it is steady state, if they vary with time, then unsteady state.
11. **Equilibriun:** If a system undergoing changes is left to itself, it will eventually reach a state where no further action takes place.
12. Driving force, gradient: $\Delta C/\Delta x$ is concentration gradient, where C is concentration, and x is distance.
    ΔC is (Concentration Difference) driving force.
13. Weight fraction of A = $Wa/(Wa + Wb +... + Wn)$ where Wa is mass of A.
14. Phase rule: $F = C - P + R$, F = degrees of freedom, C = no. of components, P = number of phases, R = number of external variables.
15. Relative volatility, Alpha (α) = $y_1/x_1/y_2/x_2$.
16. Diffusion is the movement of molecular species through a mixture due to some physical driving force.

17. Concentration diffusion is due to concentration difference driving force, e.g. distillation, absorption, etc.

18. Thermal diffusion is due to temperature difference as driving force, e.g. isotopes separation, etc.

19. Pressure Diffusion: Pressure difference as driving force, e.g. UF6 isotopes separation, etc.

20. Diffusivities of gases are nearly 10000 times greater than diffusivities of liquids.

21. Net rate of mass transfer: $N_A = -D_{AB}\dfrac{dc_A}{dz} + Y_A(N_A + N_B)$. Where $-D_{AB}\dfrac{dc_A}{dz}$ is due to diffusion and $Y_A(N_A + N_B)$ is due to bulk flow. N_A = molar flux of A, N_B = molar flux of B, z = distance, D_{AB} = diffusion coefficient.

22. For equal molar flow ($N_A = -N_B$), $N_A = D_{AB}/RTL\ (P_{A1} - P_{A2})$, where N_A is molar flux of A, D_{AB} is diffusion coefficient, P_{A1} and P_{A2} are partial pressures, R is gas constant. T is temp and L is distance.

23. For stagnant case ($N_B = 0$), $N_A = D_{AB}Pt/RTL\ PBM\ (P_{A1} - P_{A2})$, where Pt is the total pressure, PBM is $(P_{B1} - P_{B2})/Ln(P_{B1}/P_{B2})$ and $P_{A1} + P_{A2} = Pt$ and $P_{A2} + P_{B2} = Pt$.

24. Fick's first law of diffusion is: $N_A = -D_{AB}\dfrac{dc_A}{dz}$.

25. Fick's second law is unsteady state diffusion in a medium given by: $dc_A/dt = D_{AB}\, d^2c_A/dz^2$. Where c_A is concentration of A, t is time and z is distance.

26. Stefans equation for diffusion of a liquid into air: Stagnant case $N_A = D_{AM}\ Pt/RTZ\ (P_{A1} - P_{A2})/PBM = \rho L/M\ dz/d\theta\ DAM = PBM\ RT\ \rho L\ (Z_1^2 - Z_2^2)/2Ma\ Pt\ (P_{A1} - P_{A2})$.

ρL = density of liquid, Z_1 = final air gap distance, Z_2 = initial air gap distance, θ = time, Pt = total pressure, M = molecular weight of liquid, P_{A1}, P_{A2} are partial pressures of liquid.

27. Analogy equations: Mass, $N_A = -D_{AB}\dfrac{dc_A}{dz}$; Momentum, $Tgc = -\mu du/dz$. And heat is $q = -k/CP\rho\ d/dz\ (Cp\rho T)$ and corresponding momentum, equation is $Tgc = -\mu/\rho\ d/dz\ (u\rho)$ in viscous flow. The analogy is the coefficients have the same units.

28. Reynolds analogy equation is: $h/CP\rho U \propto NRe^{-0.25} = f/2$; Re = Reynolds number; Cp = Sp. heat, ρ = density, u = velocity, f = friction factor.

29. Chilton – Colburn analogy is: $J_h = 0.0\ 23Re^{-0.2} = J_D = f/2$

$J_h = h/CP\rho U\ (Cp\mu/K)^{0.67}$; $J_D = \dfrac{h_D PBM}{uP}\left(\dfrac{\mu}{\rho D}\right)^{0.67}$.

h = heat transfer coefficient, Cp = specific heat of liquid, ρ = density of liquid, u = velocity of liquid, μ = viscosity of liquid, k = thermal conductivity of liquid, K_d = mass transfer coefficient, PBM = inert component pressure, u = velocity of fluid, P = total pressure, d = diffusion coefficient.

30. Overall theories of Mass Transfer:

(1) Film theory (interphase mass transfer).

(2) Penetration theory ($K_L = 2\text{sqrt}(D_{AB}/\pi\theta)$)

(3) Surface renewal theory ($KL_{Avg} \propto D_{AB}^{0.50}$).

31. (1) Film penetration method (a combination of film, penetration and surface renewal).

(2) Convective transport theory (includes bulk transport with surface renewal).

(3) Wason-Ulwerya theory: Modified surface theory.

Film theory is still used for design of equipment.

32. Diffusion coefficients of gases by Chapmon-Enskog theory is: $D_{AB} = \frac{0.00407 / r_{AB}^2}{\Omega}$ square root $(T(1/M_A + 1/M_B)$ where T is temperature in degrees R, M_A and M_B are molecular weights of A and B, r_{ab} is the molecular separation of collision of $A = (r_a + r_b/2)$; Ω given as a function of KT/E_{AB} and E_{AB} = square root $(E_a + E_b)$; E_a, E_b are energies of molecular interaction).

33. The Gilliland equation for diffusion coefficient of gases is, $D = 0.0043$ $T^{1.5}$sqrt$(1/M_A + 1/M_B)$/ $[p(V_A{}^{1/3} + V_B{}^{1/3})]$ D is in cm²/s; where T is absolute lumperature in Kelvin. M_A and M_B are molecular weights, p = total pressure in atmospheres, V_A and V_B are molecular volumes.

34. The equation for diffusion coefficients of gases by Hirshfelder, Bird and Spotz is given as $D_{AB} = 0.0009297\ T^{3/2}\ (1/M_A + 1/M_B)^{1/2}\ f(KT/E_{AB})$. Where $R_{AB} = (r_A + r_B)/2$, K is the Boltzmann constant and $f(KT/E_{AB})$ is a collision function given in tables, p is pressure, T is temperature.

35. The diffusion coefficients of liquids by Wilke is $D_{AB} = 9.4 * 10^{-8}\ (\psi_B\ M_B)^{1/2}\ T/\mu V_A^{0.60}$ where ψ_B is association parameter, M_B is molecular weight of B, T is temperatures in kelvin and V_A is molar volume of solute A.

36. Distillation is a method of separating the components of a solution which depend upon the distribution of substances between a gas and liquid phase, by providing heat.
Relative Volatility is given as alpha $(\alpha) = y/(1 - y)\ /\ x/(1 - x)$.

37. $y - x$ compositions (vapor, liquid) are obtained by Raoult's law and Dalton's law (ideal solutions) $p_A = P_A * X_A : p_B = P_B * X_B$; $Y_A = p_A/(p_A + p_B)$. The boiling temperature is, when total pressure becomes equal to $p_A + p_B$ (ideal solution/ Raoult's law) and ambient pressure.
$(\alpha) = P_A/P_B$.

38. **Minimum boiling azeotrope:** Carbon disulphide-acetone; Ethanol-water; for positive deviations from ideality; $B.\ P$ is minimum at azeotrope composition.

39. **Maximum boiling azeotrope:** HCl-water; acetone-chloroform; negative deviations from ideality; at azeotropic composition ($y = x$) maximum boiling temperature.

40. Vapour pressure equations (can be used for calculating T-X-Y); $\text{Log}_{10}\ P = A - B/(T + C)$; P is vapour pressure of liquid; T = temperatures of liquid; A, B, C are constants.

41. **Flash distillation:** $F = L + V$, $FXF = LXL + VYV$; solving both; $Y = (F/V)\ XF - (L/V)X$; slope is $-L/V \rightarrow$ at X_F; F = Feed; L = Liquid (bottom); V = Vapor; Y = Vapor composition; X = Liquid composition.

42. **Differential or simple distillation:** Rayleigh's equation $\ln(Bo/Bl) = \int dx/(y^* - x)$; Bo = Initial number of moles of liquid; Bl = final no of moles; y^* = theoretical (equilibrium's) vapor composition.

43. McCabe – Thiele method's Assumptions:
(a) Equal molar latent heats of vaporization and condensation.
(b) No heat loss from column.
These two assumptions give constant molar overflow and underflow.

44. (a) Upper operating line equation in a continuous distillation column
$Yn + 1 = (R/R + 1)\ Xn + Xd/(R + 1)$
(b) Lower operating line equation in a continuous distillation column
$Ym + 1 = (L/L - w)\ Xm - (w/L - w)Xw$
(c) Feed operating Line: $y = (q/q - 1)X - XF/(q - 1)$; $Q = (Hv - H_F)/(HV - HL)$;

45. Optimum reflux ratio = 1.2 to 1.5 Rm; $R = L/D$.

46. For α (alpha) is near to 1 then an entrainer is used, to form low boiling azetrope for separation. This is called as azetropic distillation, e.g. Acetic acid-water using butyle acetate; Ethanol-water using benzene.

47. Extractive distillation: For α (alpha) is near to 1, a solvent is added to alter α (alpha). The added solvent is of low volatility, e.g. Toluene-isoctane using phenol.

48. For packed towers: N TO G = ∫ $dy/(y^* - y)$; Z = N TO G * H TO G.

H TO G = V/KYA.

49. Steam distillation: This is for insoluble liquids. Each component exerts its own vapor pressure. $Wa/Wb = MaPa/(MbPb)$.

50. At total reflux with total condenser and a reboiler:

$(n + 1)$ = Log [$Xd/(1 - Xd)$ / $Xb/(1 - Xb)$] / log (alpha) avg (Fenske-Underwood equation).

51. H.E.T.P is height equivalent of a theoretical plate used to express packing height in a number of theoretical plates.

52. Absorption is defined as the separation of one or more constituents of a gas mixture by contacting with a liquid solvent, e.g. Removal of ammonia vapor from air-ammonia vapor mixture by contacting with chilled water as a solvent.

53. Stripping is reverse of absorption. In this solute diffuses from liquid to gas, e.g. Seperation of benzol vapors from wash oil using steam.

54. Solubility of gas increases with pressure in a liquid and solubility of a gas in liquid decreases with temperature.

55. Raoult's law is for ideal solution; $p = Px$ where p is partial pressure, P = vapor pressure; x = mole fraction of solute in liquid phase, $y = mx$ in mole fractions; y = mole fraction of solute in gas phase.

56. Henry's Law (Non-ideal Solution): $P = mx$.

57. **Balances in distillation:**

Upper operating line in enthalpies (Ponchan Savarit) method:

V/Hv and Yv; L/hL and XL; $D/hD - QCD$, Xd are in straight line

$F = D + B$; $FXF = DXD + BXB$

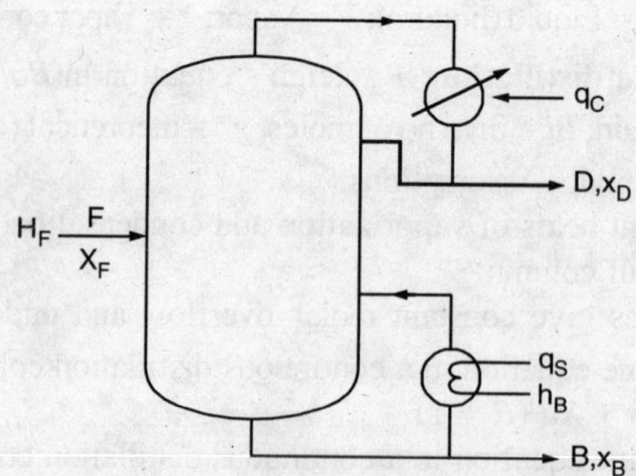

Lower operating line:

L / h_L and XL; V/HV and YV and $W(B)/hB - QSB$ and XB, are in straight line.

Feed operating line:

$F/XH, hf; D/hD - QCD; W/XB, hB - QSB$ are in straight line

$QCD = qc/D; QsB = qs/B$

58. Balances in absorption:

$L1 + G2 = G1 + L2$

$L_S X_1 + GSY2 = GSY1 + LSX2$

$Y2 - Y1 = (LS/GS)(X2 - X1)$

This is operating line: $X = x/(1 - x)$; $Y = y/(1 - y)$;

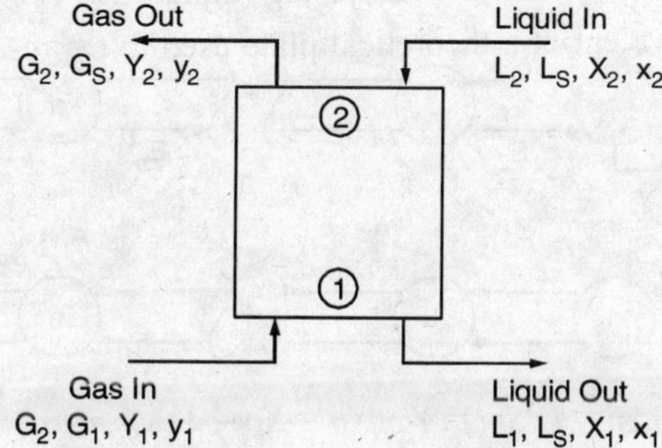

59. In absorption, operating line is above the equilibrium curve. In stripping or desorption, operating line is below the equilibrium curve. When operating line touches equilibrium curve (or minimum slope), it gives minimum solvent and infinite number of plates. Co-current absorber is used when a pure gas is dissolved in a liquid.

60. In a packed absorber: $Z = N\ TO\ G * H\ TO\ G$.

$N\ TO\ G = \int dy/(y - y^*)$; $H\ TO\ G = G/KYA$.

61. Absorption factor (A) = $L/(mG)$. Theoretical stages and *NTUs* are calculated using absorption factor.

$N\ TO\ G = \ln[(y2 - mxl/y1 - mx)(1 - 1/A) + 1/A]/(1 - 1/A)$

62. Equipments used in gas liquid operations:

(a) Agitated vessels; (b) Bubble cap tray towers; (c) perforated plate towers; (d) Grid tray towers; (e) Tourills; (f) Wetted wall towers; (g) Spray towers; (h) Baffle towers; (i) Packed towers.

The various packing are: Raschig ring, Lessing ring, Partition ring, Bert Saddle, Intalox Saddle, plate towers have: High pressure drop, not used in vacuum, used when suspended solids are present, used for large diameters, permit withdrawal of intermediate gas or liquid and cooling coils, economical in large sizes.

63. Liquid-liquid extraction: This is the separation of components of a liquid solution (miscible) by contacting with an insoluble liquid called as a solvent, e.g. Acetic acid-water are separated by adding milk as solvent. Temperature increases the solubility of components increase with increase of temperature. The various systems examples are: one pair partially soluble (acetic acid-water-benzene); Two pairs partially soluble (chloro benzene-water-methyl-ethyle ketone); one solid two partially soluble liquids (aniline, naphthalene; iso-octane). Solvent is selected based on high: selectivity, distribution coefficient, insolubility of solvent, recoverability, density, difference and physical and chemical properties.

64. Methods of extraction: 1. Batch (stage wise); 2. Continuous cocurrent, Counter current).

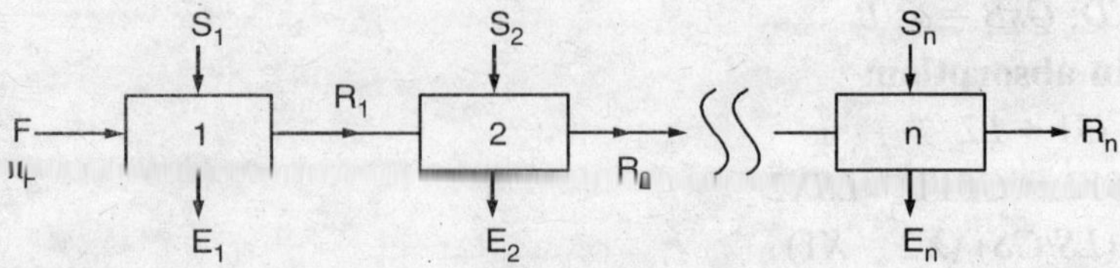

F – Feed; R – Raffinate; E – Extract; S – Solvent

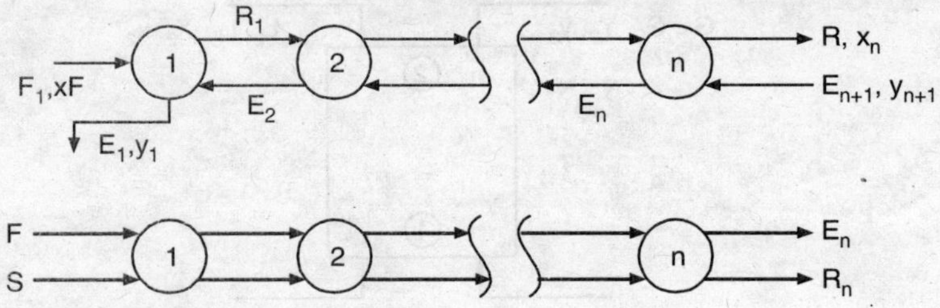

(3) For insoluble liquids: $-D/Sm = Ym(Xn - Xn - 1)$

65. Packed extraction Towers:

$Z = N\ TO\ R * H\ TO\ R$; $N\ TO\ R = \int dx/(x - x^*)$; $H\ TO\ R = R / KRA$.

66. Contacting and seperation equipment in extraction:

(a) Spray tower (b) Mixer-settler (c) Packed towers (d) Baffle towers (e) Perforated plate towers (f) Pulse columns (g) Centrifugal extractors (h) Scheibell extractor.

67. Psychrometry: Psychrometry deals with any vapor carried by a dry gas. Humidity deals with water vapor carried by dry air. Humidity is defined as the kg of water vapor per kg of dry air $(Y) = (18/29)pA/(1 - pA)$; Molar humidity is the kg moles of water vapor per kg mole of dry air = $pa/(1 - pa)$. Saturation humidity is the humidity when the air is saturated with water vapor = $(18/29)PA/(1 - PA)$.

Where Pa is vapor pressure of A (water vapor); Relative saturation or relative humidity = 100 pa/Pa where pa is partial pressure of A and Pa is V.P of A; Percentage saturation or percent absolute humidity = 100 $pA/(1 - pA)$ / $PA/(1 - PA)$; Humid heat is the kilocalories of heat necessary to increase the temperature of one kg of dry air plus the water vapor it contains by one degree centigrade = $0.24 + 0.45Y$.

Humid volume is defined as the volume in cubic meters of one kilogram of dry air plus whatever vapor contains at one atmosphere pressure and at air temp (t) = Vy.

$22.414(t + 273)/29*273 + 22.414(t + 273)Y/18*273$ L/g of dry air.

$2.81*10^{-3} + 4.561 * 10^{-3} (t + 273)$ m^3/kg of dry air;

Dew point is the temperature to which the air-water vapour mixture has to be cooled at constant humidity to make the mixture completely saturated. Dry bulb temperature is the temperature shown by vapor-gas mixture (ambient temp.)

Adiabatic saturation temperature is the temperature at which the sensible heat given up by gas in cooling equals to the latent heat required to evaporate the added liquid.

Wet bulb temperature is the steady state temperature attained by a small amount of liquid evaporating into a large amount of unsaturated vapor gas mixture.

Enthalpy is the heat content of one kg of dry air plus the heat content of water vapor along with it; $H = Cs(tg - tr) + Y\lambda$ kcal. kg dry air, $Cs = 0.24 + 0.45\ Y$ (Humid heat).

68. Lewis relation is: $Y - Yas/(t - tas) = - Cas/\lambda as$ (adi. sat. temp)

$(Y - Yw)/(t - tw) = - hg/Ky\lambda w$ (wet bulb temperature). These two are equal if $hg/Ky = Cs$ (psychrometric ratio). $hG/KyCs$ is Lewis relation.

69. The various methods for humidity measurement are: (a) Direct chemical method (b) Hygrometer method (c) Sling psychomoter (d) Dew point methods.

70. Calculations: Material balance and Energy balance.

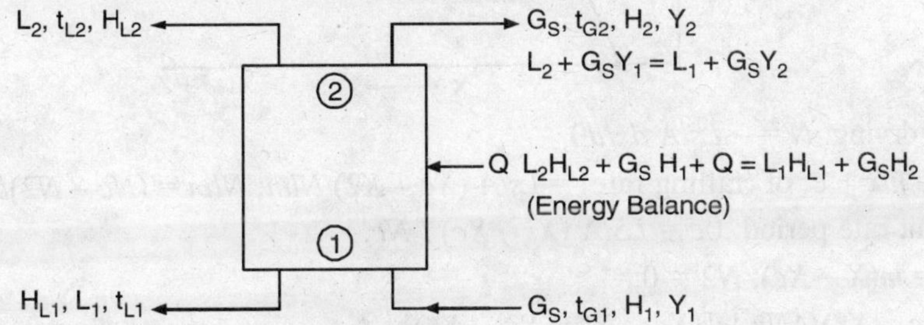

$L2 + GS\ Y1 = L1 + GsY2$ (Material balalnce)

$L2HL2 + GsH1 + Q = L1HL1 + GSH2$ (Energy balance)

In packed tower: $Z = NTU{*}HTU$; $NTU = \int dh/(H^* - H)$; $HTU = Gs/Kya$.

In adiabatic humidifier: $NTU = \ln[(Yas - Y1)/(Yas - Y2)]$; $HTU = Gs/Kya$.

$NTU = \mathrm{Ln}[(tg1 - tas)/(tg2 - tas)]$; $HTU = GsCs/hGa$;

In humidification, operating line is below equilibrium curve. In dehumidification, operating line is above equilibrium curve. Humidification means water cooling and dehumidification deals with air cooling.

71. The various equipment's used in air cooling or water cooling are:

(a) Natural circulation– (1) atmospheric (2) natural draft

(b) Mechanical draft– (1) Forced draft cooling tower; (2) Induced draft cooling tower;

(c) Spray tower; (d) Spray chambers; (e) Spray ponds.

72. Drying: Removal of moisture from a substance by contacting with a hot gas is termed as drying, e.g. drying of wet clothes by contact of ambient hot air.

Moisture content, dry basis is kg of moisture per kg of dry solid. Moisture content—wet basis is the kg of moisture per kg of wet solid. Equilibrium moisture. Moisture content of a substance which is at equilibrium with the partial pressure of moisture in the gas vapor mixture.

Bound moisture is the moisture content of a substance which exerts an equilibrium vapor pressure less than that of pure liquid at the same temperature.

Unbound moisture is the moisture content of a substance which exerts an equilibrium vapor pressure equal to that of pure liquid at the same temperature.

Free moisture is the moisture contained by a substance in excess of the equilibrium moisture.

73. Rate of drying curve is a plot of kg of moisture evaporated / (hr.sq.m) (N) versus kg moisture / kg dry solid (X).

$$N = -\ (L_s / A)\frac{dx}{d\theta}$$

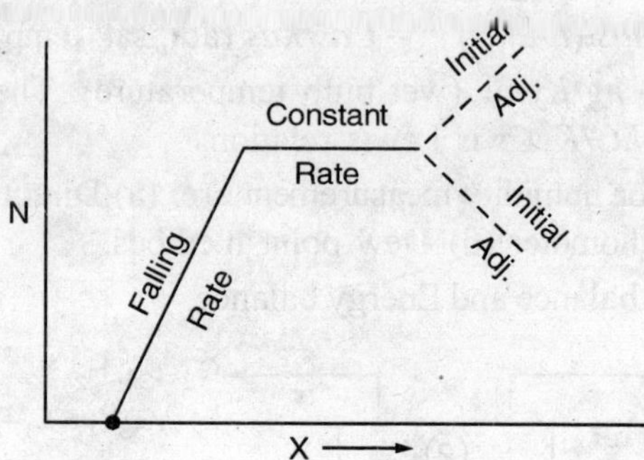

74. Times of drying: $N = -\ Ls/A\ dx/d\theta$

(2) If $N = mx + c$; of (falling rate) $= Ls/A\ (Xc - X2)/Nlm$; $Nlm = (Nc - N2)/\ln (Nc/N2)$.

In constant rate period: $\theta c = LS/A\ (Xi - Xc)\ /\ Nc$.

(3) If $N = m(X - Xc)$; $N2 = 0$;

$tf = Ls(Xc - X^*)/ANc\ \ln[(Xc - X^*)/(X2 - X^*)]$.

75. The various theories of movement of moisture are: (a) Diffusion; (b) Capillary movement; (c) Vapor diffusion; (d) Pressure diffusion; (e) Percolation theory.

76. The heat relations are: $Nc = q/\lambda s$; $q = hc(tg - ts)$; $hc = 0.01G^{0.8}$ or $hc = 0.37G^{0.37}$ where G = mass velocity of air.

77. The various equipment's for batch drying are:

(a) Tray dryer; (b) Truck dryer; (c) Through circulation dryer; (d) Vacuum shelf dryer; (e) Agitated pan dryer, x vacuum rotary dryer.

78. Continuous drying:

Gas
G_S, Y_1, t_{G1}, H_1 ← G_S, Y_1 ← G_S, Y_2, t_{G2}, H_2
Solid
L_S, t_{L1}, X_1, H_{L1} → L_S, t_{L1} → L_S, X_2, t_{L2}, H_{L2}
2

Material balance: $LsX_1 + GsY_2 = LsX_2 + GsY_1$

Enthalpy of solid, $HL = CL(tL - tR) + XCA(tL - tR) + \Delta HA$

Enthalpy balance: $Ls^*HL_1 + GsH_2 = LsHL_2 + GsH_1$.

79. The various equipment used in continuous drying are: (a) Thermal dryers; (b) Turbo dryer; (c) Through circulation; (d) Rotary dryers – direct heat, Counter flow- direct heat- parallel flow; Indirect heat;Counter current flow, Direct-indirect; (e) Drum dryers; (f) Pneumatic dryer; (g) Spray dryer.

80. Leaching (Solid-liquid operation). This is the removal of one or more constituents of a solid mixture by contact with a liquid solvent like hexane. The equilibrium data in leaching is: *N Vs X* and *N Vs Y*; $N = b/(a + c)$;

X = weight fraction of C in solution on b free basis.

Y = weight fraction of C in slurry or solid on b free basis,

b = kg of insolubles, C = kg of solute, w = kg of solvent.

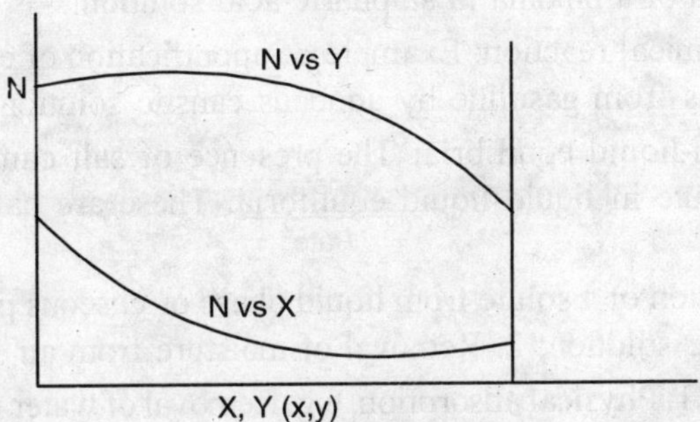

80. Material Balance:

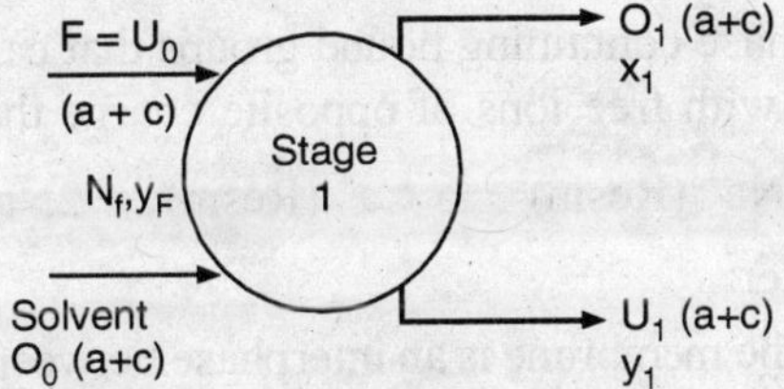

Solute balance (C): $Fy_f + Oox_0 = O1x_1 + U1y_1$

Solvent balance(a): $F(1 - yf) + O_0(1 - x_0) = O1(1 - Xl) + Ul(1 - y1)$

$(a + c)$ balance: $UO + F = O1 + U1$;

For constant underflow, the number of ideal stages are:

$K = \log [(Xb - Xb^*)/(Xa - Xa^*)]/\log[(Xb - Xa)/(Yb - Ya)]$

Number of stages by variable underflow (counterflow) are calculated by graphical method.

81. Various equipments used in leaching are:

(A) Batch process: (1) Heap leaching; (2) Percolation tanks; (3) Shanks system; (4) Percolation in closed vessels; (5) Filter press leaching and; (6) Agitated vessels.

(B) Continuous process: (1) Leaching during grinding; (2) Agitated vessels; (3) Thickners; (4) Classifiers; (5) Rotocol; (6) Kennedy extractor; (7) Miag wheel; (8) Hilderbrandt extractor; (9) Bollmann extractor; (10) Bonotto type extractor.

82. Crystallization is defined as the formation of solid particles from homogeneous phase, e.g. formation of sodium chloride crystals from brine solution; Sugar crystals from sugar solution. Various laws of crystallization:

1. Law of Haüy (the angles).
2. Gibbs-Wulff theorem: The crystal of the same material maintains the same shape.

3. Survival theory: The large the value of super saturation the greater the survival of smaller crystals.
4. ΔL law of crystal: The total growth of each crystal in the magma during the same time interval Δt is same and $\Delta L = G\Delta t$ where G is growth rate of crystal.
5. Miers theory: Below supersaturation curve no nucleation occurs and thereby no crystallization.

83. Mass transfer with chemical reaction: Examples: Absorption of carbon dioxide in carbonate solution, Absorption of ammonia in sulphuric acid solution.

84. Extraction with chemical reaction: Examples: Saponification of esters of faty acids; Extraction of odorous mercaptains from gasoline by aqueous caustic solution.

85. Salt effect in liquid-liquid equilibria: The presence of salt can raise or lower the distribution coefficient of a solute in liquid-liquid equilibria. These are called as salting out or salting in respectively.

86. Adsorption: Separation of a solute from liquid phase or gaseous phase by a solid, e.g. 1. Removal of colour from sugar solution; 2. Removal of moisture from air by silica gel (solid).

Types of adsorption: 1. Physical adsorption, e.g. Removal of water from gasoline; 2. Chemisorption: Chemical reaction takes place, e.g. Manufacture of ammonia from nitrogen and hydrogen. Adsorption equilibria are: 1. Langmur; 2. Brunauer-Emmett-Teller (BET); 3. Freundlich isotherm.

87. Ion-exchange: A solid phase containing bound groups that carry an ionic charge either positive or negative in conjunction with free ions of opposite charge that can be displaced.

Example: Ca^{++} (aq) + $2Na^{+}$ (Resin) $\rightleftarrows$ Ca^{+} $(Resin)_2$ + $2Na^{+}$ (aq)

Use: Purification of water.

88. Membrane separation: The membrane is an interphase between the two bulk phases. The membrane is selective to one of the species. One bulk phase is enriched in one of the species while the other is depleted of it. The driving forces are: Chemical potential and pore size, e.g. separation of air; desalination of water.

89. Gas separation membranes: These separate the components a gas mixture. Gas membranes work because individual gases differ in their solubility and diffusivity through non-porous polymers. Some membranes operate by sieving, Knudsen flow or chemical composition. $J = P\,\Delta P$; J = flux of component; P = Permeability; ΔP = Pressure gradient.

90. Froth floatation: This is an important industrial operation to separate finely ground valuable minerals from associated gangue. The chemistry components used are: Collectors, frothers, activators, depressants and PH. Flotation and agglomeration are the processes based on the adhesion to air of some particles from pulp and the simultanous adhesion of other particles to water. Proper adhesion is effected between gas bubbles and small particles in such a way that the specific gravity of the mineral-air associations is less than that of the pulp so that they rise in that pulp. The floating mineralized froth is then mechanically separated from the pulp.

91. Zone melting: This is the technique of repeated passage of liqid zones in one direction through a solid. A main object of zone refining is to concentrate impurities at one or both ends of the solid charge so as to purify the remainder, e.g. Purifying germamium for transistors.

92. Adductive crystallization: In this the material added to the feed causes formation of a crystalline solid phases. This is analogous to absorption among the vapor-liquid processes, e.g. *m* and *p*-cresols are separated by this method (B.P.s are 396.3°F and 395.6°F).

93. Electrophoresis: Depends upon the differeuces in the rate of a molecular process whereas most of the other unit operations depend on differences in some equilibrium property to accomplish separation.

94. Chromatographic refining: Dependent on the components distributing themselves differently between the two phases, one mobile and the other stationary. The mobile phase may be gas or liquid. The stationery phase may be: non-vollitile solvent adsorbed on the surface of a solid, solid acting as a carrier or a solid phase such as adsorbent, molecular sieve, ion-exchange resin, etc.

95. Pervaporation: Pervaporation membranes are non-porous. The membranes used show-preferential permeation rates to certain substances over others, e.g. Separation of ethanol-water azeotropes.

96. Osmosis: Solvent can pass through membrane but the solute can not. In reverse osmosis, solutes pass through the membrane and solveut is enriched.

Dialysis is the separation of substances of different molecular sizes in a liquid solution by virtue of their different rates of diffusion through a semipermeable membrane, e.g. blood purification.

Electrodialysis: In this ions are removed from water by passing through a semipermeable membranes that are impervious to water, e.g. production of drinking water.

Donnan dialysis: This is used to exchange ions between two solutions, e.g. softening of hard water.

QUESTIONS

SECTION—A

1. Range of diffusivity of gases at atmospheric pressure in cm^2/sec:

(a) > 1 (b) < 1

(c) > 10

2. Range of diffusivity of liquids in cm^2/sec:

(a) $1*10^{-5}$ (b) 0.1

(c) 1

3. For diffusivities in liquids the equation by the following is available:

(a) Hirschfelder (b) Gilliland

(c) Wilke and Chang

4. At the plait point the two phases in liquid-liquid extraction become:

(a) One (b) Two

(c) Three

5. Wet bulb and adiabatic saturation temperature become equal when psychrometric ratio is:

(a) Equal to one (b) Greater than one

(c) Less than one

6. Schmidt number is given by:

(a) $\mu/\rho D$ (b) $\rho D/\mu$

(c) $\rho/\mu D$

7. Prandtl number is given by:

(a) $\mu/C_p K$ (b) $C_p\mu/K$

(c) K/Cpu

8. Lewis number is:
(a) *SC/Pr* (b) *Pr/SC*
(c) None

9. Pick out the right statement:
(a) Mass transfer coefficient is same as diffusion coefficient.
(b) Mass transfer coefficient accounts for all types of flow/diffusion.
(c) None of these.

10. Film theory was developed by:
(a) Higbie (b) Fick
(c) Reynolds

11. Surface Renewal theory was developed by:
(a) Higbie (b) Danckwerts
(c) Prandtl

12. The dimensionless group in mass transfer, similar to Prandtl number in heat transfer is:
(a) Sherwood number (b) Lewis number
(c) Schmidt number

13. The dimensionless group in mass transfer, similar or Nusselt number in heat transfer is:
(a) Sherwood number (b) Lewis number
(c) Schmidt number

14. Lewis relation is:
(a) $h_G/K_y = 1$ (b) $h_G/K_y = C_s$
(c) $K_y/h_G = C_S$

15. Dehumidification means:
(a) Contacting warm vapour-gas mixture with cold liquid
(b) Contacting hot liquid with cold vapour-gas mixture
(c) None

16. Humidification means:
(a) Contacting warm vapour-gas mixture with cold liquid
(b) Contacting hot liquid with cold vapour-gas mixture
(c) None

17. Which is the correct figure for counter-current absorber? *A* is operating line and *B* is the equilibrium curve.

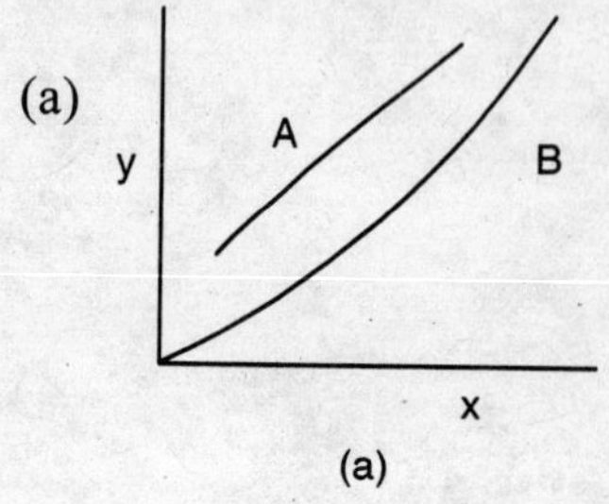

(a)

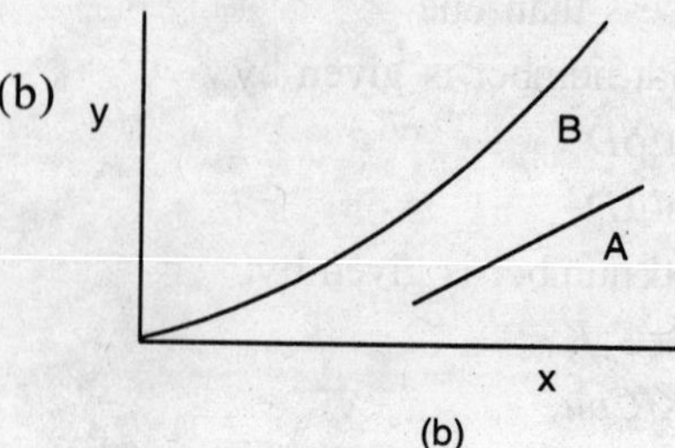

(b)

18. Which is the correct minimum operating line?

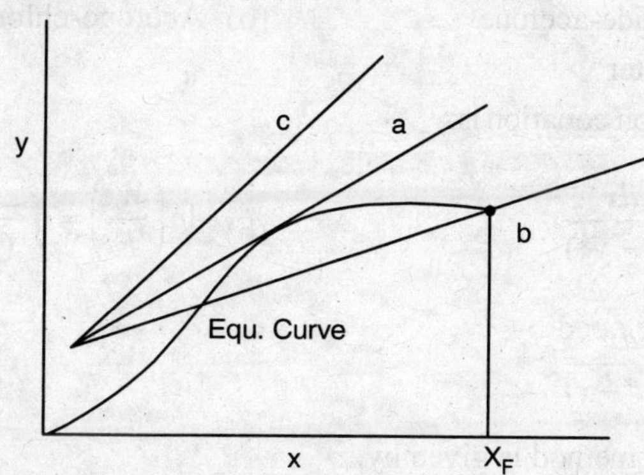

19. Which is the operating line of cocurrent absorber? Where *A* is a equilibrium curve and *B* is the operating line.

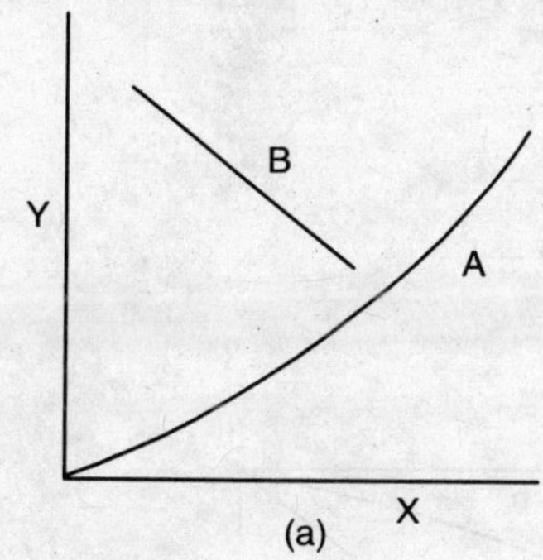

(a)

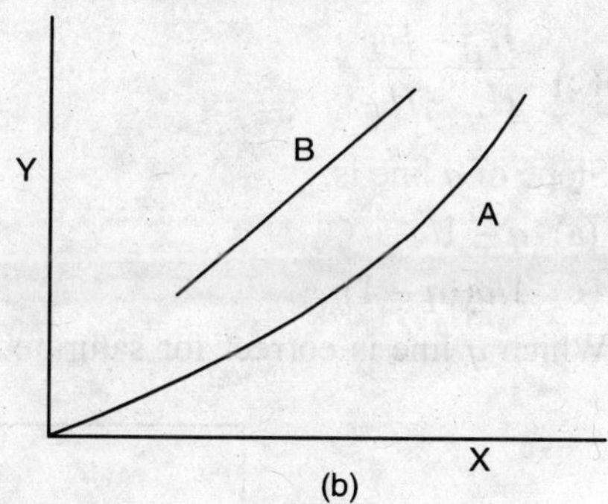

(b)

20. Absorption factor is:

(a) mG/L (b) L/mG

(c) G/mL

21. *N to G* in absorption is given by for dilute mixtures:

(a) $\int_{y_2}^{y_1} \frac{dy}{(y - y^*)}$ (b) $\int_{y_2}^{y_1} \frac{dy}{(y^* - y)}$

(c) $\int_{y_2}^{y_1} (y - y^*)dy$

22. Relative volatility is given by:

(a) $\frac{x/(1-x)}{y/(1-y)}$ (b) $\frac{y/(1-y)}{x/(1-x)}$

(c) $\frac{yx}{(1-y)/(1-x)}$

23. An example of minimum boiling azeotrope:

(a) Carbon-disulphide-acetone (b) Acetone-chloroform

(c) Iso-butanol-water

24. Differential distillation equation is:

(a) $\ln\left(\frac{F}{W}\right) = \int_{x_W}^{x_F} \frac{dx}{(x - y^*)}$ (b) $\ln\left(\frac{W}{F}\right) = \int_{x_W}^{x_F} \frac{dx}{(y^* - x)}$

(c) $\ln\left(\frac{F}{W}\right) = \int_{x_W}^{x_F} \frac{dx}{(y^* - x)}$

25. q in McCabe-Thiele method is given by:

(a) $\frac{H_G - H_F}{H_G - H_L}$ (b) $\frac{H_F - H_G}{H_G - H_L}$

(c) $\frac{H_G - H_L}{H_G - H_F}$

26. Slope of q line is:

(a) $q - 1/q$ (b) $q/q - 1$

(c) $1/q((q - 1)$

27. Which q line is correct for saturated vapour feed?

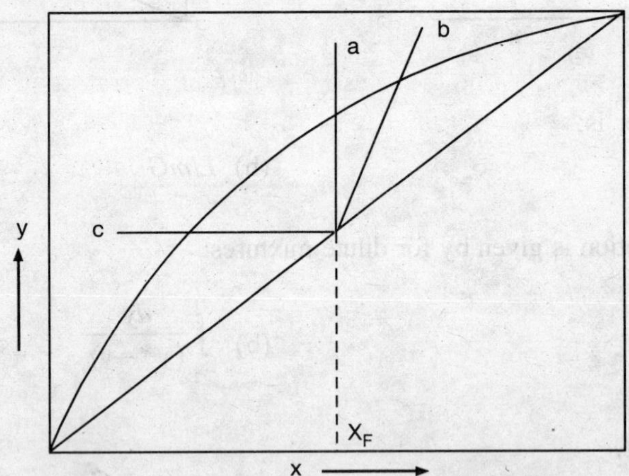

28. For acetic acid-water separation by distillation the entrainer used is:

(a) Benzene (b) Butyl acetate

(c) Hexane

29. An example of adsorption:

(a) Removal of moisture from gasoline (b) Removal of oil from cake

(c) Separation of ammonia from coke-oven gas using chilled water.

30. The equation that comes in adsorption equilibria is:

(a) Gibb's equation (b) Fick's law

(c) Freundlich equation

31. Which figure is correct for counter-current stripper ? *A* is equilibrium curve and *B* is the operating line.

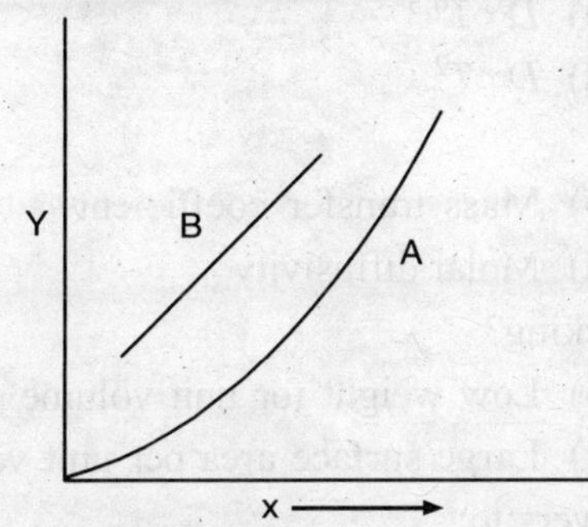

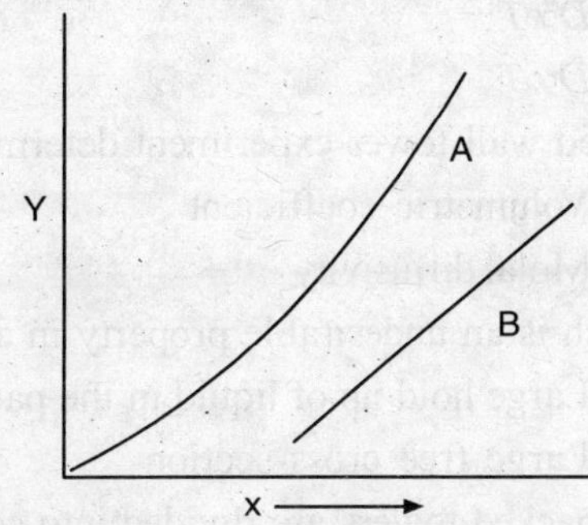

32. An example of drying is:

(a) Removal of water from clothes by keeping in hot air

(b) Removal of water from gasoline

(c) Separation of water from ethanol-water mixture.

33. Equilibrium moisture is:

(a) This is the moisture content of a substance when at equilibrium with a given partial pressure of the vapour.

(b) Moisture contained by a substance which exerts an equilibrium vapour pressure less than that of the pure liquid at the same temperature.

(c) Moisture contained by the substance which exerts an equilibrium vapour pressure equal to that of the pure liquid at the same temperature.

34. Decoction refers to:

(a) Use of solvent at its boiling temperature (b) Use of solvent as a vapour

(c) Use of cold solvent

35. An example of leaching is:

(a) Separation of benzene-toluene mixture (b) Removal of oil from oil cakes

(c) Removal of colour from gasoline

36. A practical application of reverse osmosis is:

(a) Desalination of sea water (b) Recovery of sodium chloride from sea water

(c) No practical application

37. A commercial application of thermal diffusion:

(a) Separation of helium isotopes for scientific work

(b) Separation of gaseous components in natural gas

(c) Separation of benzene-toluene liquid mixture.

38. Soret effect accounts:
(a) Temperature change due to concentration difference in heat transfer
(b) Concentration change due to temperature difference in mass transfer
(c) Convection heat transfer in heat conduction problem.

39. The diffusivity in a binary gas mixture is related to the temperature (T) as:
(a) $D \propto T^{1.5}$ (b) $D \propto T^{0.5}$
(c) $D \propto T$ (d) $D \propto T^2$

40. Wetted wall tower experiment determines:
(a) Volumetric coefficient (b) Mass transfer coefficient
(c) Molal diffusivity (d) Molar diffusivity

41. Which is an undesirable property in a tower packing?
(a) Large hold up of liquid in the packing (b) Low weight for unit volume
(c) Large free cross-section (d) Large surface area per unit volume

42. Tall packed towers are divided into series of towers to:
(a) Reduce liquid hold up (b) Avoid flooding
(c) To avoid channelling (d) Reduce the overall pressure drop

43. Boiling point diagram is effected by:
(a) Ambient pressure (or system pressure) (b) Ambient temperature
(c) Ambient humidity

44. In azeotropic mixture, the equilibrium vapor composition is:
(a) Less than liquid composition (b) Same as liquid composition
(c) More than liquid composition (d) Not dependent on pressure.

45. In batch distillation with constant reflux, overhead composition of light component:
(a) Decreases with increase of time (b) Increase with time
(c) Does not vary with time

46. Rose oil is extracted from Rose leaves using:
(a) Extractive distillation (b) High pressure distillation
(c) Steam distillation (d) Vacuum distillation

47. Which of the following methods assume constant molal vaporisation and overflow:
(a) Ponchan-Savarit method (b) McCabe-Thiele method
(c) Ellis method (d) Enthalpy-concentration method

48. The following one provides maximum contact surface for a liquid vapour system:
(a) Wetted wall column (b) Sieve plate column
(c) Packed tower (d) Bubble cap tower

49. At total reflux the number of plates is:
(a) Minimum (b) Maximum
(c) In between maximum and minimum

50. Rayleigh's equation is used for:
(a) Flash vaporisation (b) Differential distillation
(c) Equilibrium distillation (d) Continuous rectification

51. A ternary composition in triangular coordinates comes:
(a) In the corners of the triangle (b) On the sides of the triangle
(c) Inside the triangle

52. To extract oil seeds the following equipment is used:
(a) Centrifugal extractor (b) Packed column
(c) Bollman extractor (d) Pulse column

53. Milk powder is made from milk using:
(a) Tray dryer (b) Spray dryer
(c) Freeze dryer (d) Vacuum dryer

54. Detergent solution is dried to powder using:
(a) Tray dryer (b) Vacuum dryer
(c) Spray dryer (d) Drum dryer

55. For gas liquid contact, the pressure drop is the least in the following equipment:
(a) Bubble cap tower (b) Wetted wall tower
(c) Perforated tray tower (d) Grid try tower

56. In the following operations, steady state equimolal counter diffusion is encountered in:
(a) Liquid-liquid extraction (b) Liquid-solid operation (leaching)
(c) Binary distillation (d) Absorption

57. Sticky material is dried in a:
(a) Rotary dryer (d) Tray dryer
(c) Fluidised bed dryer (d) Spouted bed dryer

58. The use of flights in a rotary dryer is:
(a) To increase the residence time of solid
(b) To reduce the residence time of solid
(c) To lift and shower the solids for drying action of the gas
(d) To increase mechanical strength of the dryer

59. The optimum reflux ratio is around (where R_m is the minimum reflux ratio):
(a) 5 R_m (b) 2.5 R_m to 3 R_m
(c) 1.2 to 1.5 R_m (d) 1.5 to 2 R_m

60. Radioactive nuclear waste is treated in:
(a) Bollman extractor (b) Pulsed column extractor
(c) Mixer settler extractor (d) Sieve plate extraction column

61. The equipment used to disperse liquid in a continuous gas phase:
(a) Pulse column (b) Packed column
(c) Sieve plate column (d) Bubble cap column

62. The following method is used for design calculations for multicomponent distillation:
(a) McCabe-Thiele method (b) Ponchon-Savarit
(c) Tray-to-tray calculations (d) Enthalpy-concentration

63. Thiele-Geddes method is used in:
(a) Multicomponent distillation (b) Binary distillation
(c) Multicomponent absorption (d) Multicomponent absorption

64. Benzene-cyclohexane mixture is separated using:
(a) Ordinary distillation (b) Extractive distillation
(c) Extraction (d) Absorption

65. Removal of oxygen from water is done using:
(a) Ion-exchange (b) Absorption
(c) Absorption (d) Distillation

66. Langmuir equation comes in:
(a) Absorption (b) Distillation
(c) Absorption (d) Extraction

67. Hypersorption means:
(a) Moving bed process for absorption (b) Fixed bed process for absorption
(c) Moving bed process absorption (d) Fixed bed process absorption

68. Dialysis is used for:
(a) Separation of sea water into fresh water (b) Refining of beet sugar
(c) Refining of gasoline (d) Water purification

69. Electrodialysis used:
(a) In the manufacture of sodium hydroxide
(b) In the manufacture of sulphuric acid
(c) Refining of gasoline
(d) Refining of beet sugar

70. An example of absorption with chemical reaction:
(a) Removal of ammonia from coke oven gas using chilled water
(b) Removal of CO_2 from gas using alkaline solution
(c) Removal of benzol vapours from coke oven gas using wash oil

71. In the system of SO_2-air-water, the controlling phase/phases for absorption are:
(a) Gas phase (b) Liquid phase
(c) Both gas and liquid phase

72. Which packing is the costliest out of the following:
(a) Raching rings, ceramic (b) Bed saddles, ceramic
(c) Intalox saddles, ceramic (d) All rings, ceramic

73. Reynolds' number in an agitated tank is:
D = diameter of impeller, N = speed of impellar, ρ = density, μ = viscosity.

(a) $DN\rho/\mu$ (b) $DN^2\rho/\mu$
(c) $N\rho\mu/D$.

74. Power number is:
(a) $PgC/(N^3D_a^5\rho)$ (b) $N^3D_a^5\rho/(PgC)$
(c) $KN^3D_a^5$

75. Equipment used for leaching gold ores:
(a) Bollman extractor (b) Pachuca tanks
(c) Centrifugal extractor (d) Kennedy extractor

76. Pug mill is used for:
(a) Size reduction to any size desired (b) Crushing
(c) Mixing (d) Drying

77. Collection efficiency in electrostatic precipitator:
(a) 85–90% (b) 50–60%
(c) 30–50% (d) 70–80%

78. An example where gaseous diffusion is used:
(a) Separation of air into N_2 and O_2 (b) Separation of uranium isotopes
(c) Separation of ammonia and air (d) Separation of helium isotopes

79. Extraction tower used for desulphurization of gasoline:
(a) Rotating disc extractor (b) Bollman extractor
(c) Packed extraction column (d) Plate type extraction column

80. Equipment used in extraction of penicillin:
(a) Packed extraction column (b) Plate type extraction column
(c) Schiebel extraction column (d) Centrifugal extractor

81. Peclet number is:
(a) Re.No*PrNo. (b) Pr.No/Re.No.
(c) Re.No/Pr.No (d) Pr.No./(Re.No.*SC No.).

82. The theory that is used now a days for design to calculate mass transfer coefficients is:
(a) Penetration theory (b) Surface renewal theory
(c) Film theory (d) Film penetration mode

83. *L/mG* ratio used in absorption normally for economical operation is:
(a) 1.0 to 15 (b) 2 to 3
(c) 3 to 4 (d) 1.25 to 2.0.

84. Refractory bricks are dried in:
(a) Rotary dryers (b) Truck dryers
(c) Tray dryers

85. Silica gel is dried using (large amount):
(a) Tray dryer (b) Rotary dryer
(c) Through circulation dryer

86. For drying of paper and cloth:
(a) Tray dryers (b) Truck dryers
(c) Tunnel dryers (d) Cylinder dryers

87. To dry wheat on large scale:
(a) Fluidised bed dryers (b) Spouted bed dryers
(c) Truck dryers

88. Bonotto type extractor is used in:
(a) Solid-liquid extraction (leaching) (b) Liquid-liquid extraction

89. Spray dryers arc designed upto sizes of (maximum):
(a) 20 ft dia and 50 ft high (b) 40 ft dia and 100 ft high
(c) 10 ft dia and 25 ft high.

90. For drying sugar in large quantity:
(a) Truck dryer (b) Tray dryer
(c) Rotary dryer (d) Fluidized bed dryer

91. For hydrogenation in Vanaspati industry:
(a) Sieva tray column (b) Bubble cap tray tower
(c) Agitated vessel (d) Grid tray

92. Separation is obtained in extraction if selectivity is (desirable):
(a) Less than one (b) Equal to one
(c) Greater than one

93. $D_{AB} = D_{BA}$ in:
(a) Equimolar counter diffusion (b) Stagnant diffusion case

94. Diffusion coefficient of water into air at 26°C nearly: in sq. cm/sec is:
(a) 1.5 (b) 2.2
(c) 0.26 (d) 0.53

95. Normally the solubility of a gas:
(a) Decreases with temperature increase (b) Increases with increase of temperature
(c) No effect of temperature

96. Operating line slope in absorbers and strippers is:
(a) G_s / L_s (b) L_s / G_s
(c) L_s / mG_s (d) G_s / mL_s

97. Fenske-Underwood equation is used to calculate the theoretical number of plates in distillation at:
(a) Minimum reflux (b) Total reflux
(c) Operating reflux

98. In open steam distillation:
(a) Reboiler is not used (b) Reboiler is used

99. "Clathrates" are used in the following operation:
(a) Absorption (b) Drying
(c) Adsorption (d) Crystallization

100. "Adducts"are used in the following operation:
(a) Crystallization (b) Drying
(c) Adsorption (d) Distillation

SECTION—B

1. .....is a unit operation in which a soluble component of a gas mixture is dissolved in a liquid.
2. The inverse of absorption is called as.....or.....
3. Steady state equimolar counter diffusion equation is.....
4. Steady state diffusion equation for stagnant case is.....
5. Fick's 1st law is.....
6. Liquid mass transfer coefficient by Higbie's penetration theory is.....
7. The analogy relation of mass, heat and momentum is.....
8. The number of transfer units equation in gas absorption based on gas phase in mole fraction units is.....
9. The height of transfer unit in absorption is.....
10. An example of absorption with chemical reaction is.....
11. Psychrometry deals with the determination of the properties of.....mixtures.
12. Fick's second law equation is.....
13. Absolute humidity is.....
14. Relative saturation is given by the equation.....
15. Percentage saturation is given by the equation.....
16. Equipment used for water cooling are called as.....
17. Raoult's law is given by the equation.....
18. Henry's law is given by the equation.....
19. Liquid-liquid extraction means.....
20. The two types of absorption are (1).....and (2).....
21. The theories that are existing to explain drying are (1).....(2).....(3).....(4).....
22. Leaching means.....
23. Equipment used in continuous leaching are.....
24. Dialysis means.....
25. Another name of reverse osmosis is.....
26. The various types of diffusion based on gradients are.....
27. For absorber, both equilibrium and operating line will be straight for.....
28. Operating velocity in a packed tower is usually.....the flooding velocity.
29. Cocurrent absorbers are usually used when the gas to be dissolved in the liquid is a.....
30. Steam distillation is used to separate.....
31. Psychrometry deals with the determination of the properties of.....
32. Absorption accompanied by a chemical reaction provides three major advantages over that of physical absorption. They are.....

33. As bubble diameter increases, the terminal velocity increases/decreases.

34. Separation process used for clarifying Rayon-Spinning baths is.....

35. Separation process used in separating water soluble oils in grinding operations is.....

36. Flame photometry is used for determination of compositions of.....

37. For solids and metals analysis.....spectroscopy is used.

38. Polarimetry is used for measuring concentrations of.....solutions.

39. Electrical conductivity method is used for analysis of.....solutions.

40. Dielectric constant and loss factor method is used for measuring small amount of in

41. Oscillometry is used for the analysis of small amount of.....in.....

42. Paramagnetism method is used for composition measurement of.....and.....

43. Distillation in packed columns is limited to column diameters less than.....and columns heights less than.....

44. Weeping in sieve trays means.....

45. Dumping in a vapour-liquid contacting equipment means.....

46. Flooding occurs in a try when.....

47. Fish is dried using.....drying.

48. Plotting $(1 - a)/a$ versus $(1 - b)/b$ where a = weight fraction solvent in solvent rich-phase and b is the weight fraction of diluent in equilibrium reffinate phase, is a correlation for.....

49. In packed columns flooding determines the minimum.....

50. Overall gas phase mass transfer coefficient in terms of individual coefficients is given by $1/K_G =$ .....

51. Stefan's equation is.....

52. Gilliland equation for diffusion coefficients of gases is given as.....

53. In wetted wall towers mass transfer area is known /unknown to calculate mass transfer coefficients experimentally.

54. Mass transfer coefficients are also called as.....

55. Blasius equation relates.....and Reynold's number.

56. J_H in analogy of heat, mass and momentum transfer is given by.....

57. Lewis relation under analogy of heat, mass and momentum transfer is.....

58. How Higbie's penetration theory differs from film theory is.....

59. How Danckwert's surface renewal theory differs from penetration theory is that.....

60. Diffusion by definition is the.....

61. Isotopes of methane are separated using.....diffusion.

62. The diffusivities of gases are approximately.....times greater than the diffusivities of liquids in the same units.

63. For non-ideal solution.....law is used.

64. Fenske equation is.....

65. Open steam is used when one component is.....in the solution to be separated.

66. Ellis method in distillation assumes.....

67. Fenske-underwood equation is useful for calculating theoretical number of plates at.....conditions.
68. The driving forces the cause diffusion are.....
69. In adiabatic humidification cooling NTG is given by.....and HTG is given by.....
70. One example of membrane used in dialysis is.....
71. Cascade means.....
72. Ergun's equation for pressure drop in cylindrical packed beds is.....
73. Elution means.....
74. Drying refers generally.....
75. Elutriation refers to.....

SECTION—C

1. List the various equations available for calculating the diffusion coefficients of gases.
2. List the equipment used in absorption.
3. List the equipment used in liquid-liquid extraction.
4. What are the methods available for humidity measurement?
5. Match the following:

(*i*) Fick's first law	(a) $dC_A/dt = D_{AB}\dfrac{d^2CA}{dz^2}$
(*ii*) Fick's 2nd law	(b) $q = -KAdT/dx$
(*iii*) Fourier's law	(c) $\tau = -\mu dv/dx$
(*iv*) Newton's law	(d) $q = \epsilon\sigma T^4$
(*v*) Boltzmann equation	(e) $J_A = -D_{AB}\dfrac{dC_A}{dz}$

6. What are the equipment used for gas dispersion ?
7. What are the equipment used for liquid dispersion ?
8. Define dew point and wet bulb temperature.
9. Define gas absorption and give two examples.
10. What are the factors that are considered for selection of a solvent in absorption?
11. What are the equipment used in liquid-liquid extraction?
12. A wet solid is to be dried from 80 to 50% moisture (on wet basis). Calculate the moisture to be evaporated for 1000 kg of dried product.
13. List the driers used for batch drying.
14. List the equipments used for continuous drying.
15. What are the characteristics of ideal liquid solutions ?
16. Calculate the diffusion coefficient of oxygen (*A*) assuming that the non-diffusing gas is a mixture of methane (*B*) and hydrogen (*C*) in the volume ratio 2 : 1. The diffusivities are estimated to be $D_{O_2} - H_2 = 0.690$. $D_{O_2} - CH_4 = 0.184$ sq.cm/sec.

17. Oxygen (A) is diffusing through carbon monoxide (B) under steady state conditions with carbon monoxide nondiffusing. The total pressure is 1 atm. and the temperature is 0°C. The partial pressure.of oxygen at two planes 0.2 cm apart is respectively, 100 and 5mm Hg. The diffusivity for the mixture is 0.185 sq.cm/sec. Calculate the rate of diffusion of oxygen in g moles/sec through each square centimeter of the two planes.
18. In a mixture of benzene vapour (A) and nitrogen gas (B) at a total pressure of 800 mm Hg and a temperature of 60°C, the partial pressure of benzene is 100 mm Hg. Express the benzene concentration in absolute humidity.
19. What are the psychrometric relations for the system air (1) water (2) at 1 atm pressure ?
20. Derive the time of drying in constant rate period.
21. A glass tube of 1 mm bore was filled with acetone to within 23 cm of the top and maintained at a temperature of 18.8°C in gentle current of air. The position of liquid surface was measured with a travelling microscope. After 80 minutes the level had fallen by 0.148 cm.
    Calculate the diffusivity of acetone in air.
    Data: V.P. of acetone at 18.8°C = 170 mm Hg
    Barometric pressure = 7655 mm Hg;
    Density of acetone = 0.79 g/c.c.
    Molecular weight of acetone = 58.1.
22. Derive an expression for the rate of molecular diffusion of gas A through a stagnant layer of another gas B.
23. Ammonia gas is diffusing through a layer of stagnant air 2.5 mm thick at a constant rate. Conditions are fixed so that the gas contains 50% ammonia by volume at one boundary of the stagnant layer. The ammonia diffusing to the other boundary is quickly absorbed and its concentration is negligible at the plane. The temp. is 20°C and the pressure is atmospheric and D value at these conditions is 0.18 sq. cm/sec. Calculate the rate of diffusion of ammonia through air layer in g moles/(hr.cm).
24. The coefficient of diffusion of HCl gas in air is 0.145 cm^2/sec at 20°C and 1 atmosphere pressure. What will be the diffusion coefficient in g moles/(cm.sec.) at the same conditions?
25. Derive Stefan's equation.
26. What are the various models (theories) for mass transfer?
27. A gas mixture at equilibrium has the following composition at 2 atm total pressure and 75°F.
    CH_4 – 60%, C_2H_6 – 20%, Propane – 8%, *n*-butane – 6%, *n*-pentane – 6%.
    Calculate the composition of equilibrium solution. Vapour pressures are very high, 41.5, 8.84, 2.33, 0.65 respectively in atmospheres.
28. A coal gas to be freed to its benzene by scrubbing with wash oil. Vapour pressure of benzene is 100 mm Hg at the operating temperature of 80°F. The column operating pressure is 800 mm Hg. Calculate the equilibrium curve relation.
29. The feed to a distillation column is 10,000 kg/hr of alcohol and water, containing 30% alcohol by weight. The distillate product contains 80% alcohol by weight and bottoms product contains 5% alcohol by weight. Calculate the amounts of distillate and bottom product in kg/hr.
30. Compare packed and plate towers.

31. Calculate vapour-liquid equilibrium data using ideal solution law at constant pressure of 1 atm, for mixture of heptane and octane at 110°C. The data:

t°C	P_A mm Hg	P_B mm Hg
98.4	760	333
105	940	417
110	1050	484

32. For ideal binary mixture (solution) derive relative volatility is equal to the ratio of vapour pressures.

33. Draw the ternary diagrams for the following systems in liquid-liquid extraction giving one example each (1) one pair partly soluble (2) two pairs partially soluble (3) 1 solid, two partially soluble liquids.

34. An air-water vapour mixture at 1 atm pressure has a dry bulb temperature of 148°F and a wet bulb temperature of 88°F. If latent heat is 1043 BTU/lb. and saturation humidity is 0.018 in lb of water/lb dry air. Calculate the psychrometric ratio (hG/K).

35. Draw the equilibrium curves and operating lines for dehumidification, humidification and adiabatic humidification operations.

36. Derive the expression for wet bulb depression as a function of the psychrometric ratio.

37. Under constant drying conditions, a wet solid is dried from 30% to 4%. The time taken is 4 hours. All are on dry basis. The equilibrium moisture is 2%. Critical moisture content is 10%. How long it takes to dry to 7% ?

Assume the falling rate period is linear, *i.e.* rate of drying is proportional to free moisture.

38. Dry air at atmospheric pressure and 62°C is blown through liquid ethyl ether under adiabatic conditions. To what temperature will the liquid be ultimately cooled ? Vapour pressure of ether is:

Tem., °C	– 20	– 10	0	10	20
V.P. mm Hg.	63	111.8	184.9	291.8	942.4

Latent heat of vaporisation of ether: 92 cal/g.

Specific heat of ether = 0.47 cal/(g.°C)

Specific heat of dry air = 0.238.

39. What are the factors to be considered for selection of solvent in liquid-liquid extraction?

SECTION—D

1. A gas bubble at a pressure of P_g is passed through a solvent with a saturation vapour pressure of *Pa*. If the time of passage of the bubble is ling and air is insoluble in solvent, the mole fraction of solvent in the bubble will be equal to.....

2. The concentration and hydrodynamic boundary layers over a flat plate are of equal thickness of Schmidt number is equal to unity, True/False and Reason.

3. 1000 kg of wet solid are to be dried from 60% to 20% moisture (by weight). The mass of moisture removed in kg is

   (a) 520 (b) 200

   (c) 400 (d) 500

4. In distillation $Q = 1$ for saturated liquid feed; $Q = 0$ for saturated vapor feed; Q is < 1 for superheated vapor feed and Q is > 1 for sub-cooling feed. True or False.

5. In gas-liquid contact operation the number of ideal stages $N = (x_a - x_b)/(x_b - x_b^*)$. This is true when the stripping factor is (S):

(a) $S > 1$ (b) $S < 1$

(c) $S = 1$ (d) $S = \infty$

6. For absorbing a sparingly soluble gas in a liquid.

(a) Gas side mass transfer coefficient should be increased.

(b) Liquid side mass transfer coefficient should be increased.

(c) Liquid side mass transfer coefficient should be decreased

(d) Mass transfer coefficients must be kept constant.

7. Match:

(i) Solutropic system (a) Crystallization

(ii) Equilibrium moisture content (b) Distillation

(c) Extraction

(d) Drying

8. It is necessary to dry a batch of 160 kg of a wet material from 30% to 5% moisture content under constant rate and falling rate period. The falling rate is assumed to be linear. Calculate the total drying time considering an available drying surface of 1 m²/40 kg of dry solid. A constant drying flux of 3×10^{-4} kg/m²s is given

$x_c = 0.20$ kg moisture/kg solid; $x^* = \dfrac{0.05 \text{ kg moisture}}{\text{Solid}}$

9. The equilibrium data for the distribution of solute τ between feed solvent A and the extracting solvent B is given below.

X, kgs τ/A	0	0.1	0.2	0.30	0.40	0.45
Y, kgs τ/B	0	0.096	0.17	0.252	0.275	0.28

In a counter current operation 10 kg/s of the feed containing 20% τ in A (w/w) is treated with pure solvent so as to reduce the solvent extraction to 50% of feed.

Calculate the solvent rate for operation with 2.5 times the minimum.

10. Name the crystal system whose unit cell has three non-equivalent edges and only one non-ninety degree angle.

11. Absorption towers are operated under conditions of:

(a) Low pressure, high pressure

(b) High pressure, high, temperature

(c) High pressure, low temperature

(d) Low pressure, low temperature

12. A counter current plate absorber is to be installed for scrubbing of an air mixture containing 5% ammonia by volume. The scrubber is fed with water containing 0.002 mole NH_3 per mole water. The scrubbing water flows at a rate of 1.0 mole water per mole air. It is necessary to absorb 85%

of ammonia present in the gas by operating the absorber at 20°C, $K = 0.80$ (mole ammonia/mole of air)/(mole ammonia/mole water). Calculate the concentration of NH_3 in the outgoing liquid and estimate the number of stages necessary for this operation.

13. 100 kg of a solution containing 50% by a weight of a salt dissolved in it is cooled. 40 kg of anhydrous salt is separated out. The solubility of salt at the lower temperature in kg/100 kg of water is:

(a) 80 (b) 50
(c) 40 (d) 20

14. Diffusion coefficient in a binary gas mixture at low pressures varies with pressure as:

(a) P (b) P^2
(c) $1/P$ (d) Independent of P

15. According to film theory for equimolar counter diffusion, the mass transfer coefficient is given by..... .

16. The Reynolds analogy for mass transfer is given by.....and is applicable when Schmidt number is.....

17. Sherwood number for flow in a pipe can be expressed as the ratio of the concentration gradient at.....to the overall concentration difference.

18. If an insoluble gas is passed through a volatile liquid placed in a perfectly insulated container, the temperature of the liquid will increase. True/False.

19. For laminar flow over a flat plate of length L, the local a mass transfer coefficient at a distance L from the leading edge is 1.5×10^{-2} m/s. Then the average mass transfer coefficient for the plate is 2×10^{-2} m/s. True or False.

20. The Z component of the total mass flux of component A in a binary mixture of A and B is given by: $-D_{AB}\dfrac{d_{CA}}{dZ}$. True or False.

21. The diffusion coefficient varies as a function of temperature for:

(a) Gases: (A) $T^{0.5}$
(b) Liquids: (B) $T^{1.5}$
(C) $1/T$
(D) T

22. Match:

(a) Schmidt number (A) Momentum diffusivity
(b) Lewis number (B) Thermal diffusivity/mass diffusivity
(C) Momentum diffusivity/mass diffusivity
(D) Mass diffusivity/thermal diffusivity.

23. Air at 30°C and 150 kPa in a closed container is compressed and cooled. It is found that the first droplet of water condenses at 200 kPa and 15°C. Calculate the percent relative humidity of the original air. The vapour pressure of water at 15°C and 30°C are 1.7051 kPa and 4.246 kPa respectively.

24. What is the significance of reflux in distillation operation ? Explain in one or two sentences.

25. Simplify the Fick's law of mass diffusion for equimolar counter diffusion for a binary system.

26. In a gas mixture of hydrogen and oxygen, steady state equimolar counter diffusion is occurring at a total pressure of 100 kPa and a temperature of 20°C. If the partial pressures of oxygen at two planes 0.01 m apart and perpendicular to the direction of diffusion are 15 kPa and 5 kPa respectively and the mass diffusion flux of oxygen in the mixture is 1.6×10^{-5} k mole/m^{2}s, calculate the molecular diffusivity for the system.

27. In distillation under minimum reflux conditions, number of theoretical stagesa would be:

(a) One
(b) Minimum
(c) Maximum
(d) None of these

28. A stream of air at 100 kPa pressure and 300°K is flowing on the top surface of a thin flat sheet of solid naphthalene of length 0.2 m with a velocity of 20 m/sec. The other data are:

Mass diffusivity of naphthalene vapour in air = 6×10^{-6} m^2/s

Kinematic viscosity of air = 1.5×10^{-5} m/s

Concentration of naphthalene at the air-solid napthalene interface = 1×10^{-5} mole/m^3.

Calculate:

(*a*) Average mass transfer coefficient over the plate

(*b*) The rate of loss of naphthalene from the surface per unit width.

Note: For heat transfer over a flat plate, convective heat transfer coefficient for laminar flow can be calculated by the equation:

$Nu = 0.664\ \text{Re}_L^{1/2}\ P_r^{1/3}$. You may use analogy between mass and heat transfer.

29. What relationship does any point on an operating line for absorption in a packed bed represent ?

30. Associate the following dimensionless groups with heat transfer, mass transfer, and momentum transfer.

(a) Sherwood number
(b) Prandtl number
(c) Nusselt number
(d) Schmidt number

31. Carbon disulphide is to be absorbed from a dilute gas mixture of $CS_2 - N_2$ into a pure non-volatile oil at atmospheric pressure in a counter current absorber. The mole fraction of CS_2 in the inlet gas stream is 0.05 and the flow rate of gas stream, G is 1500 k mol/hr. The equilibrium relation is given by $y = 0.50\ x$ where x is the mole fraction of CS_2 in liquid stream.

It is desired to reduce the mole fraction of CS_2 in the exit gas stream to 0.005.

(a) Calculate the minimum value of L/G where L is the liquid flow rate, K mole/hr.

(b) Derive the equation for operating line if L/G is equal to 1.5 times the minimum value.

32. Oxygen is diffusing in a mixture of oxygen–nitrogen at 1 std. atm 25°C. Concentration of oxygen at planes 2 mm apart are 10 and 20 volume % respectively. Nitrogen is non-diffusing. Derive the appropriate expression to calculate the flux of oxygen. Define units of each term clearly.

33. A binary distillation column is operating under conditions specified below.

Feed rate = 350 k mole/hr; overhead product rate = 150 k mole/hour;

Mole fraction of more volatile component in overhead = 0.97

Mole fraction of more volatile component in bottom = 0.02

Bottom product rate = 200 k mole/hr reflux ratio = 3.5

In the stripping section it was found that the mole fraction of the volatile component in the vapour leaving the plate is 0.33, while its mole fraction in the liquid coming to some plate is 0.25. Assuming constant molal counter flow, determine whether the feed is vapor, liquid or partially vaporized.

34. A mineral containing 20% elemental sulphur is to be leached with hot gas oil in which sulphur is soluble to the extent of 10 parts in 100 parts of oil by weight. The oil is recycled over the batch of ground material till no further dissolution of sulphur takes place. On drainage, the solid materials retain the solution to the extent of one tenth the weight of undissolved solid (sulphur and gangue). No preferential adsorption takes place. Calculate the equilibrium compositions and quantities of solid and liquid phases if 50 kg oil is used for leaching 100 kg of feed.

35. Agranular solid with dry bulk density of 1.6 g/c.c is dried in a batch dryer with air with adiabatic saturation temperature of 25°C, and dry bulb temperature of 65°C. The solids containing 0.5 kg water/kg dry solid are in trays of 62 cm × 62 cm × 1.25 cm size, insulated so that heat and mass transfer occurs to the top surface only. The solids are to be dried to a final moisture content of 0.02 kg water/kg dry solid. Air passes over the Pan with a miss velocity of 5900 kg/(hr. m^2). Equilibrium moisture content is zero $h = 0.0175\ G^{0.8}$. Calculate drying time.

36. A granular material containing 40% moisture is fed to a counter current rotary dryer at a temperature of 20°C and is withdrawn at 35°C containing 5% moisture. The air supplied contains 0.006 kg water vapor/kg dry air at 110°C and leaves at 38°C. The dryer handled 200 kg of dry solid per hour. Assuming radiation losses amount to 5 kcal/kg dry air used. Determine the weight of dry air supplied per hour and humidity of air leaving. Latent heat erf water = 585 kcal/kg; Sp. heat of 7dried material = 0.21 cal/(g.°C); specific heat of dry air = 0.238 cal/(g.°C); specific heat of water vapour = 0.480 cal/g.°C.

37. What is wet bulb approach? How does it effect the performance of a cooling tower?

38. What conclusion can be reached from Lewis relationship for thermal and molecular diffusivities?

39. Acetone is evaporated into a stream of nitrogen gas and the mixture of acetone vapor and nitrogen flows through a duct of cross-section 0.08 m^2. The pressure and dry bulb temperature at one point in the duct are 800 mm Hg and 45°C and at this point, the average velocity is 4.0 m/s. The wet bulb temperature being 28°C. Calculate the mass flow rate of acetone vapor carried by duct. The data are λ_{acetone} = 403,000 J/kg, h_c/K_y = 1700 J/kg.k; vapor pressure of acetone = 50 mm Hg.

40. A crushed oil seed containing 55% oil by weight is to be extracted at the rate of 4000 kg/hr using 6,000 kg/hr of *n*-hexane containing 5% oil by weight as solvent. A counter current two stage system is employed. The oil seed will retain 1 kg of solution per oil free cake. Estimate % oil recovery.

41. Roasted copper ore containing the copper as cupric sulphate ($CuSO_4$) is to be extracted in a series of extractors with water as solvent. The system is loaded hourly with 10 tons of gangue, 1.2 tons of sulphate and 0.5 ton water in ore. The strong extract solution contains 0.07 mass fraction of $CuSO_4$ and 98% $CuSO_4$ initially loaded into the system is to be recovered. If each ton of gangue retains 2 tons of solution, compute the number of stages. Find the amount of fresh solvent.

42. Acetone is to be recovered from an aqueous solution containing 20% of acetone by means of kerosene as solvent. The distribution of acetone in water and kerosene follows the relationship, $X = 6.45\ Y$ where X = kg acetone/kg water and Y = kg acetone/kg kerosene. Find the minimum amount of solvent for one kg of feed.

43. In extraction of acetic acid from an aqueous solution with benzene in packed column of height 1.37 m and cross-sectional are of 0.0045 sq. m; Acetic acid concentration in inlet phase = 0.69 kg mole/m^3; Acetic acid concentration in outlet water phase = 0.684 kg mole/m^3.

Flow rate of benzene phase = 0.02 mm/hr.

Inlet benzene concentration = 0.0039 kg mole/cu.m.

At equilibrium C benzene = 0.0247 C water. Calculate mass transfer coefficient and height of transfer unit based on extract phase.

44. A solution containing 35% $MgSO_4$ and 65% water is cooled to 15°C, while cooling 10% of the total water evaporated. Determine the amount of crystals obtained in kg per kg of original mixture. Given at 15°C, the concentration of mother liquor is 24.0% anhydrous $MgSO_4$ and 76% water. From the solubility curve of $MgSO_4$ - water system, the crystals obtained are $MgSO_4$, 7 H_2O at 15°.

45. A crystallize? is to be used to produce 1000 kg/hr of $FeSO_4$, H_2O crystals by the cooling of a saturated solution which enters the crystallizer at 49°C. The slurry leaving the crystallizer will be at 27°C cooling water enters the crystallizer jacket at 20°C and leaves at 25°C. The heat transfer coefficient for the crystallizer is 171.1 kcal/(hr. m.°C). (a) Estimate the cooling water requirement in kg/hr

(b) Determine the heat transfer area of the crystallizer.

Data: At 49°C, the saturation solution contains 140 parts of $FeSO_4$, $7H_2O$ per 100 parts excess water. At 27°C, the saturated solution contains 74 parts $FeSO_4$ $6H_2O$ per 100 parts excess water. Average specific heat of initial solution = 0.70 kcal/(kg °C). Heat of solution of $FeSO_4$, $7H_2O$ at 18°C = – 4400 kcal/kg mole of Fe SO_4, $7H_2O$ (mole. wt 278).

46. A vacuum crystallizer is used to produce 20,000 kg $FeSO_4$, $7H_2O$ per hour. The feed is a solution containing 40.0 parts $FeSO_4$ per 100 parts total water, and enters the crystallizer at 72°C. The crystallizer vacuum is such that as to produce a crystallizer temperature of 27°C. What is the amount of feed ?

Data: Saturated solution at 27°C contains 30.5 parts $FeSO_4$ per 100 parts total water. Enthalpy of liquor at 27°C = – 50.6 kcal/kg. Enthalpy of feed solution = 26.0 kcal/kg.

47. Light oil is absorbed from a mixture of light oil vapour and air by means non-volatile absorption oil. The gas entering the absorber contains 2.25 % by volume of light oil and the scrubbed gas leaves the absorber containing 0.18% of light oil by volume. The molecular weight of light oil is 80. The average molecular weight of air is 29.0. The lean oil entering the absorber contains 0.5% by weight of light oil and the rich oil leaving the absorber contains 5% by weight of light oil. The equilibrium curve is $Y = 0.65\ X$ where Y is kg of light oil per kg of light oil free air, X = kg of light oil per kg of light oil free absorption oil. Calculate (*a*) kg of light oil free air to kg of light oil free absorption oil (*b*) number of theoretical plates?

48. A soluble gas is absorbed in water using a packed tower. The equilibrium relationship is $Y^* = 0.06X^*$. Terminal conditions are:

	top	bottom
x	0	0.08
y	0.001	0.009

If the individual height of transfer unit based on liquid phase and gas phase respectively are $H_x = 0.24$ m and $H_y = 0.36$ m, determine the height of packing.

49. What are the various laws in crystallization?

50. A liquid containing 30 mole % toluene, 40 mole% ethyl benzene and 30 mole % water is subjected to continuous flesh distillation at a total pressure of 0.5 atm. Vapour pressure data in mm Hg is given below. Calculate the vapour composition.

Temp. °C	E.B	Toluene	Water
60	78.6	139.5	149.4
70	113.0	202.0	233.0
80	160.0	290.0	355.0

51. Sodium chloride at 18°C is diffusing through a stagnant layer of water 1.5 mm thick when the concentrations at the opposite ends of the film are 25 and 15 wt. % respectively. The density of 25 and 15 w % solutions are 1025 and 1012 kg/m^3 respectively. Determine the diffusivity of NaCl through water. The rate of diffusion of NaCl is 1.76×10^{-6} k.mol./ (m^2.s).

52. Ammonia from an air-ammonia stream containing 5% ammonia by volume is absorbed by sulphuric acid in a counter-current wetted wall column, 15 mm internal diameter, *j* 800 mm long. The inlet gas rate is 0.15 kg mole/hr and exit gas contains 05% NH_3 by volume. The average temperature is 20°C and the pressure is normal. The change in acid concentration may be neglected, (*a*) Calculate the value of absorption coefficient kg (*b*) If the value of heat transfer coefficient is 50.0 kcal/(hr. m^2.°C), estimate K_G by equating J_D and J_H.

Data for air: $C_p\ \mu/k = 0.74$; $C_P = 0.24$ for NH_3 – air; Sc. No = 0.60; value of P_{BM} is taken as unity.

53. A feed solution contains 100 moles of benzene-toluene mixture having 70 mole % benzene. One third of feed is vaporized. The total pressure is one atmosphere. Calculate the distillate composition (vapor) and bottom composition using differential distillation.

$\alpha_{avg.} = 2.50$

54. The vapour pressure of toluene and water are given below in mm Hg at different temperatures.

Temp. °C	70	80	90	100
Water	245	350	510	760
Toluene	205	280	400	550

Determine the steam distillation temperature of toluene at atmospheric pressure and kilograms of water per kilogram of toluene in the distillate assuming no condensation takes place in the still.

55. Show that vapor liquid equilibrium for a binary system can be expressed by the relation $y = \dfrac{\alpha x}{1 + (\alpha - 1)x}$ where y is vapour composition in mole fraction and x is liquid composition in mole fraction and α is relative volatility.

56. Find the number of plates at minimum reflux ratio for Benzene-Toluene system ($\alpha_{avg} = 2.5$) having vapour composition of 0.95 mole fraction of benzene and liquid composition of 0.05 mole fraction of benzene.

57. (A) Removal of colour in sugar solutions:

(a) Absorption (b) Adsorption

(c) Desorption (d) Boiling

(B) The curve showing the variation of pressure with temperature for a given amount of adsorption is called as.....

(C) The curve showing the effect of.....on the extent of adsorption at a given pressure of the adsorbate is called as adsorption.....

58 The reason for preferring packed towers over plate tower in distillation practice is that the packed towers operation gives:

(a) Low pressure drop and high hold up

(b) High pressure drop and low hold up

(c) Low pressure drop and low hold up

(d) High pressure drop and high hold up.

59. When a multistage tower uses a minimum reflux ratio it implies:

(a) Infinite trays and maximum reboiler heat load

(b) Infinite trays and minimum reboiler heat load

(c) Minimum trays and minimum reboiler heat load

(d) Minimum trays and maximum reboiler heat load.

60. 990 kg/hour of an aqueous solution has a solute concentration of 0.10 kg solute/kg water. It is treated with 600 kg/hour of a pure immiscible organic solvent in a 2 stage mixer-settler system. If the organic and aqueous streams having each stage are in equilibrium such that

$$\frac{\text{Kg solute on organic phase}}{\text{Kg solvent}} = 2\left(\frac{\text{Kg solute on aueous phase}}{\text{Kg water}}\right)$$

Calculate the kg of solute extracted per kg of solvent if counter current operation is employed.

61. A liquid mixture containing 50 mole % each of benzene and toluene at 40°C is to be continuously flash vaporized to vapourise 60 mole% of the feed. The residual liquid product contains 35 mole% benzene. If enthalpies per mole of feed, the liquid product and the vapour product are respectively 5, 2 and 39 kJ/mole.

(*a*) Calculate the heat added in kJ per mole of vapour product.

(b) Represent the process on $H - x - y$ diagram.

62. 160 kg of wet solid is to be dried from an initial moisture content of 25% to a final moisture content of 6%. Drying test shows that the rate of drying is constant at 3×10^{-4} kg water/m^{2}s in the region 0.2 to 0.4 kg water/kg dry solid. The drying rate falls linearly in the range of 0.01 to 0.2 kg water/kg solid. If the equilibrium moisture content is 0.01 kg water/kg dry solid, calculate the time of drying. The drying surface is 1 m^2/30 kg dry weight.

63. In binary distillation the separation of components is easier if the relative volatility (a) is:

(a) $\alpha >>1$ (b) $\alpha << 1$

(c) $\alpha = 1$ (d) None of these

64. Molecular diffusivity of a liquid:

(a) Increases with temperature (b) Decreases with temperature

(c) May decrease or increase with temp (d) Independent of temperature

65. For turbulent mass transfer in pipes, the Sherwood number depends upon the Reynolds number as:
(a) $Re^{0.33}$ (b) $Re^{0.53}$
(c) $Re^{0.83}$ (d) Re

66. In distillation column the number of bubble caps per tray primarily depends on the:
(a) Allowable liquid velocity (b) Allowable gas velocity
(c) Feed composition (d) Allowable gas and liquid velocity

67. In a binary distillation column, if the feed contains 40 mole % vapor, the *q* line will have a slope of:
(a) 15 (b) – 0.60
(c) –1.5 (d) 0.60

68. According to film theory of mass transfer, the mass transfer coefficient is proportional to:
(a) D (b) D^2
(c) $D^{0.5}$ (d) $1/D$

69. One mole feed of a binary mixture of a given composition is flesh vapourized at a fixed *P* and *T*. If Raoult's law is obeyed, then changing the feed composition would effect.
(a) The product composition but not the fraction vaporised.
(b) The product composition as well as the fraction vaporised.
(c) The fraction vaporized but not the product composition.
(d) Neither the product composition nor the fraction vaporised.

70. An equimolar mixture of benzene and toluene is contained in a piston/cylinder arrangement at a temp. *T*. What is the maximum pressure below which this mixture will exit as a vapor phase rule ? At the given *T*, the vapor pressures of benzene and toluene are 1530 and 640 mm Hg. Assume Raoult's law holds .

71. Calculate the time required to reduce the moisture content of a solid from 0.66 to 0.25 kg moisture/kg dry solid. The rate of drying N: kg. water evaporated/ (m^2.s) is given as:
$N = 0.0015 \times$ for $X \leq 0.3$
$= 0.00045$ for $X > 0.3$
The drying surface is 0.025 m^2/kg dry solid.

72. A feed of know-composition (binary mixture of constant relative volatility) is to be distilled in a continuous fractionating column, consisting of a partial condenser, one plate and a reboiler. The feed enters the reboiler from which a bottom product is continuously withdrawn. The liquid reflux from the partial condenser is returned to the plate. The distillate composition x_D is 0.80 and the reflux ratio is 2.
(*a*) What is the slope of the operating line on *x*-*y* plot and its intercept on *y* axis ?
(*b*) Using the McCabe-Thiele method, Qualitatively locate on the *x*-*y* plot:
(*i*) The composition of the streams leaving the plate and
(*ii*) The bottom product composition. (do not use graph paper)

73. The mass transfer coefficient for a solid sphere of radius '*a*' dissolving in a large volume of quiescent liquid, in which *D* is the diffusivity of solute is:
(a) D/a (b) $D/2a$
(c) Proportional to $\sqrt{D}$ (d) Dependent on the Reynolds number

74. In distillation column design, the McCabe -Thiele procedure is inadequate and a Ponchon-Savarit procedure is needed when:

(a) Saturated feed is not used

(b) An azeotrope forms

(c) The latent heats of vapourisation of the more and less volatile components are greatly different

(d) A total condenser is used

75. At intially at 101.3 kPa and 40°C and with a relative humidity of 50% is cooled at constant pressure to 30°C. The cooled air has:

(a) A higher dew point
(b) A higher absolute humidity
(c) A higher relative humidity
(d) A higher wet bulb temperature

76. 100 moles of an acetonitrile-nitromethane mixture is differentially distilled in a batch still at a pressure of 70 kPa. The feed contains 74 mole % acetonitrile. Distillation is continued till the liquid left behind in the still contains 32 mole % acetonitrile. The vapour-liquid equilibrium for the system at this pressure are correlated as:

$$y^* = 1.05x + 0.13 \text{ for } 0.3 \le x \le 0.52$$

$$y^* = 0.77x + 0.28 \text{ for } 0.52 \le x \le 0.80$$

where x, y^* refer to the mole fractions of acetonitrile in liquid and equilibrium vapor respectively. Find the average composition of distillate product.

77. A pure gas is bubbling through a column of liquid at a sufficiently slow rate that the bubbles are well separated from each other. The bubbles are spherical are 3 mm in diameter and rise with a constant velocity of 0.24 m/s. At any given time ten bubbles of gas can be seen in the liquid, the liquid volume being 3.14×100^{-4} m^3. The liquid contains no dissolved gas to start with, and may be assumed to be well mixed at all times. The mass transfer coefficient for this situation is given by Higbies penetration theory as:

$$K_L = \sqrt{\frac{4D}{\pi t^*}} \quad \text{where } D \text{ is diffusivity of the gas in liquid} = 2 \times 10^{-9} \text{ m}^2/\text{s}$$

and t^* = time of contact between liquid elements and the gas, in seconds. Determine the time it takes for the liquid to attain 50% of the saturation concentration.

SECTION—E

1A. The diffusion coefficient in m^2/s of acetic acid in Benzene (liquid in liquid) is:

(a) $2.09 * 10^{-4}$
(b) $2.09 * 10^{-5}$
(c) $2.09 * 10^{-9}$
(d) $2.09 * 10^{-12}$

1B. Na_2SO_4, $10H_2O$ crystals are formed by cooling 100 kgs of 30% by weight of aqueous solution of Na_2SO_4. The final concentration of the solute is 10%. The weight of the crystals is:

(a) 20
(b) 32.2
(c) 45.35
(d) 58.65

2. H_2S is being absorbed in a gas absorbed unit. The height of the transfer unit based on the gas side is 0.4 m. The equilibrium data is given by $Y = 1.5\ X$. The bulk concentration of H_2S has to be reduced from 0.05 to 0.001 mole fraction in the gas side. The height of the tower (in meters) corresponding to an operating line are given by $y = 5x + 0.001$ is:

(a) 2.0 (b) 1.56

(c) 1.0 (d) 0.56

3. The Reynolds number of the liquid is increased 100 fold for a laminar falling film used for gas liquid contacting. Assuming penetration theory is applicable, the fold-increase in the mass transfer coefficient (Kc) for the same system is:

(a) 100 (b) 10

(c) 5 (d) 1

4. A solid is being in the linear drying rate regime from moisture content X_0 to X_f. The drying rate is zero at $X = 0$ and the critical moisture content is the same as initial moisture, X_0. The drying time for $M = Ls/\text{Arc}$ is:

(a) $M\ (X_0 - X_f)$ (b) $M\ (X_0/X_f)$

(c) $M\ Ln(X_0/X_F)$ (d) $MX_0 \ln (X_0/X_f)$

5. The following plot gives the saturated humidity (He) versus temperature (T).

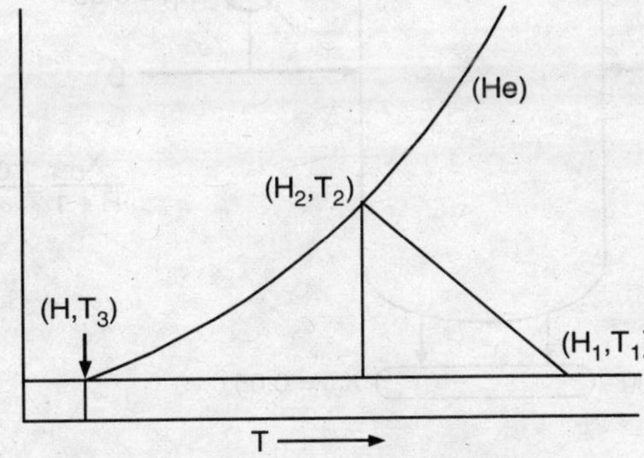

Line joining (H_1, T_1) and (H_2, T_2) is the constant enthalpy line. Choose the correct one from among the Temp T, alternatives A, B, C and D.

(a) T1-dew point Temp; T2-dry bulb temp; T3-wet bulb Temp

(b) T1-dew point Temp; T2-wet bulb Temp; T3-dry bulb temp

(c) T1-wet bulb Temp; T2-dry bulb temp; T3-dew point temp

(d) T1-dry bulb Temp; T2-wet bulb Tempo; T3-dew point temp

6. Compound A is extracted from a solution of ($A + B$) into a pure solvent S. A co-current unit is used for the liquid-liquid extraction. The inlet rate of the solution containing A is 200.0 moles of B/hr. m². The equilibrium data is represented by $Y = 3X^2$ Where Y is in mole of the maximum percentage extraction achieved in the unit is:

(a) 25% (b) 50%

(c) 70% (d) 90%

7. In Distillation column sizing calculations by short cut methods, match the following

P. Underwood Equation	1. No. of Real Trays
Q. Fenske Equation	2. Column Diameter
R. Gilliland Equation	3. Minimum No of ideal trays
S. Vapour Velocity at flooding	4. Actual no of ideal trays
	5. Minimum Reflux Ratio
	6. Tray Efficiency.

(a) P-1, Q-3, R-4, S-6
(b) P-2, Q-5, R-1, S-3
(c) P-5, Q-3, R-6, S-2
(d) P-5, Q-3, R-4, S-2

8. A distillation column separates 10,000 kg/hr of a benzene-Toluene mixture as shown in figure below. In the figure x_F, x_D, x_W represent the weight fraction of benzene in the feed, distillate and residue respectively.

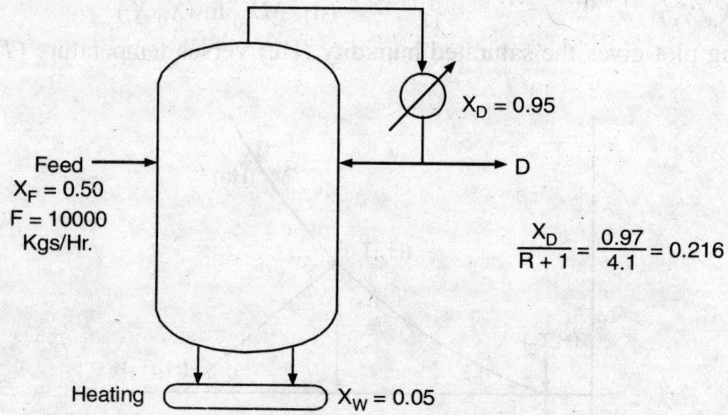

The reflux ratio is:

(a) 0.50 (b) 0.60
(c) 1.0 (d) 2.0

9. Acetone is to be removed from air in an isothermal dilute absorber using pure water as solvent. The incoming air contains 5 mole % of acetone ($Y_{in} = 0.05$). The design equation to be used for obtaining the number of trays (N) of the absorber is: $N + 2 = 6 \ln (Y_{in} / Y_{out})$. For 98% recovery of acetone the number of trays required are:

(a) 1 (b) 8
(c) 9 (d) 10

10. The boiling points for pure water and pure toluence are 100°C and 110.6°C respectively. Toluence and water are completely immiscible in each other. A well agitated equimolar mixture of toluence and water is prepared. The temperature at which the above mixture will exert a pressure of one standard atmosphere is:

(a) Less than 100° C (b) 100° C
(c) Between 100 and 110° C (d) 110.6° C **(Ans. c)**

11. At a total pressure of one standard atmosphere exerted by vapors of Toluence and water, the mole fraction of water *Xw* in the vapor phase satisfies.

(a) $O < Xw < 0.50$ (b) $Xw = 0.50$
(c) $0.5 < Xw < 1.0$ (d) $Xw = 1.00$

12. In a distillation operation, what is the effect of temperature on the reflux stream (given below) on the condenser and reboiler loads?

Reflux conditions (*i*) Reflux stream is completely liquid reflux and is at its bubble point (*ii*) Reflux stream is below its bubble point.

(a) Condenser and reboiler loads are same in both cases.
(b) Reboiler load is the same in both cases but condenser load is high in case (*ii*).
(c) Condenser load is same in both cases that reboiler load is higher in case (*ii*).
(d) Both condenser and reboiler loads are higher in case (*ii*) as compared to case (*i*).

13. At equilibrium the concentrations of water in vapour phase (C^*) in kg/m^3 of air space. And the amount of related by $C^* = 0.0067$ m; to maintain dry conditions in a room of air space, 100 m^3 containing 2.2 kg of water vapour initially, 10 kg of dry silica gel is kept in room. The fraction of initial water remaining in the air space after a longtime (during which the temperature is maintained constant) is m:

(a) 0.0 (b) 0.20
(c) 0.40 (d) 1.0

14. In liquid-liquid extraction, 10 kg of a solution containing 2 kg of solute *C* and 89 kg of solvent *A* is brought into contact with 10 kg of solvent *B*. Solvents *A* and *B* are completely immiscible in each other, whereas solute *C* is soluble in both the solvents. The extraction process attains equilibrium. The equilibrium relationship between the two phases is $Y^* = 0.90\ X$ where Y^* is kg of *C*/kg of *B* and *X* is kg of *C*/kg of *A*. Choose the correct answer:

(a) The entire amount of *C* is transferred to solvent *B*.
(b) Less than 2 kg but more than 1 kg of *C* is transferred to Solvent *B*.
(c) Less than 1 kg of *C* is transferred to *B*.
(d) No amount of *C* is transferred to *B*.

15. A 25 cm * 25 cm * 1 cm flat sheet is weighing 1.2 kg, during intial conditions. It took 1500 seconds for the weight of sheet to reduce to 1.05 kg. Another 1 m * 1 m * 1 cm flat sheet of same material is dried from one side only. Under the same drying conditions time required for drying in seconds from its initial weight of 19.2 kg to 17.6 kg is: second

(a) 1000 (b) 1500
(c) 2000 (d) 2500

16. A distillation column with *N* plates is being operated under normal conditions. At same point in time, the operation is shifted to total reflux condition (i.e no product and residue are being with drawn and feed to the column is stopped) at the new steady state.

(a) Composition of vapors and that of liquid do not vary through out the column.
(b) Reboiler load and condenser load are minimum.
(c) The top and bottom compositions are unchanged with and without total reflux
(d) The top and bottom compositions correspond to the maximum enrichment available

17. An aqueous solution of methanol is to be distilled in a tray column. High pressure steam is available as a source of heat. For a given reflux ratio and overhead composition, two options are being explored (*i*) a reboiler is used (*ii*) no reboiler is used but steam is fed directly to the bottom of the column. As compared to option (*i*) in option (*ii*).

(a) Less number of trays are required.
(b) Composition of residue remains unchanged.
(c) More number of trays are required but the residue composition remains unchanged.
(d) More number of trays are required and the residue composition is more dilute in methanol.

18. In a tray column, separating a binary mixture with non-ideal stages, one of the following statement is true.

(a) Point efficiency can exceed 100%
(b) Murphree efficiency can not exceed 100%
(c) Murphree efficiency can exceed 100%
(d) Both Murphree and point efficiency can exceed 100%

19. the ratio of the diffusion coefficient in a gas to that of liquid is in the order to:

(a) 10^{+5} (b) 10^{-5}
(c) 10^{-2} (d) 10^{+2}

20A. In the triangular diagram, represented below for a batch separation process, as stream *F* is mixed with a solvent *B* to produce products *R* and *E*.

Substances *A* is the carrier liquid and *C* is the solute to be extracted. The amount of *B* and *E* are 1 kg and 1.20 kg respectively. The ratio of *R*/*E* is:

(a) 1.285 (b) 2
(c) 0.751 (d) 2.5

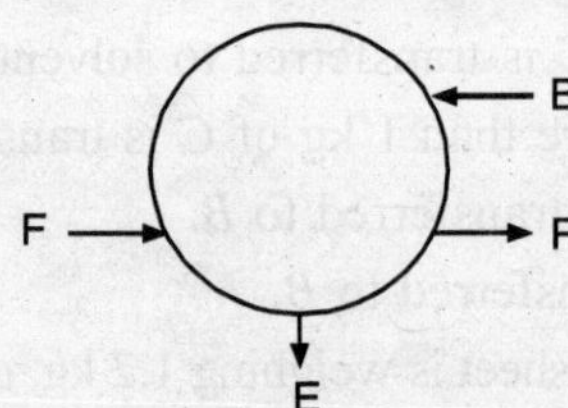

20B. For a two phase feed where 80% of feed is vaporized under column conditions, the feed line slope in McCabe-Thiele method for distillation column design is:

(a) –1/4 (b) +1/4
(c) –4 (d) 2.5

21. A liquid mixture of benzene and Toluene is in equilibrium with its vapor at 101 kPa and 373°K. The vapor pressures of benzene and toluene at 373°K respectively are 156 and 63 kPa. Assuming that the system obeys Raoult's law. The mole fraction of benzene in liquid is:

(a) 0.650 (b) 0.410
(c) 0.065 (d) 0.04

22. Two solid discs of benzoic acid (molecular weight =122) of equal dimensions are spinning separately in large volumes of water and air at 300°K. The mass transfer coefficients of benzoic acid in water and air are $0.9*10^{-5}$ and $0.47*10^{-2}$ m/s respectively. The solubility of benzoic acid in water is 3 kg/m^3. And the equilibrium vapour pressure of benzoic acid in air is 0.04 kPa. Then the disc:

(a) Dissolves faster in air then in water
(b) Dissolves faster in water then in air
(c) Dissolves at same rate in both water and air
(d) Does not dissolve in either water or in air

23. Match the variation of mass transfer coefficient given by theory in Group-I with the appropriate relation in Group-II.

Group-I	Group-II
P. Film theory	1. X(alpha) *DAB*
Q. Penetration theory	2. X(alpha) $DAB^{2/3}$
R. Boundary layer theory	3. X(alpha) $DAB^{1/2}$

(a) P-1, Q-2, R-3 (b) P-2, Q-1, R-3
(c) P-1, Q-3, R-2 (d) P-3, Q-2, R-1

24. A binary gas mixture of solute and carrier gas is treated in a counter current gas absorption column containing ideal trays using a solvent. The compositions of *y* and *x* are mole fractions of solute in gas and liquid respectively. Also *V* and *L* are the molar flow rates of gas and liquid respectively. Assuming that the carrier gas in the solvent and the vapor pressure of solvent is very low at the given conditions of the column. Further the gas and liquid streams are sufficiently dilute that *L* and *V* may be assumed to be constant throughout the column. The equilibrium relation is given by: $y^* = mx$ where *m* is a +ve constant.

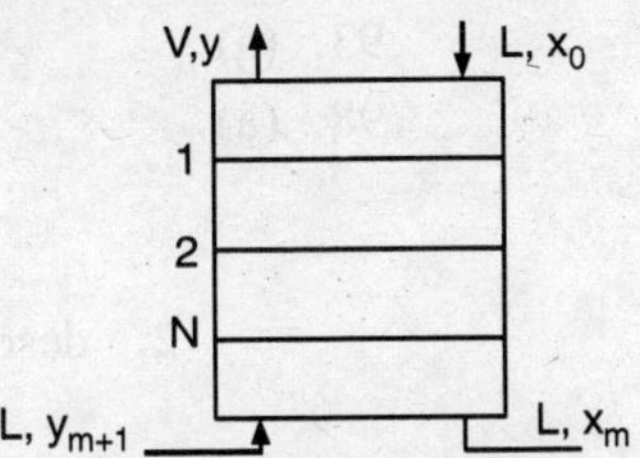

For any value of *m*, the change in liquid composition across a tray is independent of the tray location. If

(a) $L/V = 1$ (b) $mL/V = 1$
(c) $m/(LV) = 1$ (d) $L/(mv) = 1$

25. Under the correct condition corresponding to part (a), the number of ideal trays in the column is given by:

(a) $N = Xo - XN/[Xo - (Yn + 1/m]$ (b) $N = Xo - (Yn + 1/m)/(Xo - Xn)$
(c) $N = Xn - Xo/(Yn + 1/m) - XN]$ (d) $N = (Y n + 1/m) - X_n/(XN - Xo)$

ANSWERS

SECTION–A

1. (b)	2. (a)	3. (c)	4. (a)	5. (a)
6. (a)	7. (b)	8. (a)	9. (b)	10. (a)
11. (b)	12. (c)	13. (a)	14. (b)	15. (a)
16. (b)	17. (a)	18. (a)	19. (a)	20. (b)
21. (b)	22. (b)	23. (a)	24. (c)	25. (a)
26. (b)	27. (c)	28. (b)	29. (a)	30. (c)
31. (b)	32. (a)	33. (a)	34. (a)	35. (b)
36. (a)	37. (a)	38. (b)	39. (a)	40. (b)
41. (a)	42. (c)	43. (a)	44. (b)	45. (a)
46. (c)	47. (b)	48. (a)	49. (a)	50. (b)
51. (c)	52. (c)	53. (b)	54. (c)	55. (a)
56. (c)	57. (b)	58. (c)	59. (c)	60. (b)
61. (b)	62. (c)	63. (a)	64. (b)	65. (a)
66. (c)	67. (a)	68. (b)	69. (a)	70. (b)
71. (c)	72. (b)	73. (b)	74. (a)	75. (b)
76. (c)	77. (a)	78. (b)	79. (a)	80. (d)
81. (a)	82. (c)	83. (d)	84. (b)	85. (c)
86. (d)	87. (b)	88. (a)	89. (b)	90. (c)
91. (c)	92. (c)	93. (a)	94. (c)	95. (a)
96. (b)	97. (b)	98. (a)	99. (d)	100. (a)

SECTION—B

1. absorption

2. desorption or stripping

3. $N_A = \dfrac{D_{AB} P_t}{RTZ}(y_1 - y_2)$

4. $N_A = \dfrac{D_{AB} P_t}{RTZP_{BM}}(PA_1 - PA_2)$

5. $N_A = -D_{AB}\dfrac{dC_A}{dZ}$

6. $K_L = \dfrac{2}{\sqrt{\pi}}\sqrt{\dfrac{D}{t}}$

7. $J_M = J_H = f/2,\ J_M = \dfrac{K_G}{G_M}\left(\dfrac{\mu}{\rho D}\right)^{2/3}$

$$J_H = \frac{h}{C_P G}\left(\frac{CP\mu}{K}\right)^{2/3}$$

8. $N_{OG} = \displaystyle\int_{y_2}^{y_1}\frac{dy}{(y - y^*)} + \frac{1}{2}\ln\frac{1 - y_1}{1 - y_2}$

9. $G/[K_{Ga}P_t(1 - y)m]$

10. SO_2 absorption in alkali solution

11. gas-vapour.

12. $\partial C_A/\partial\theta = D\partial^2 C_A/\partial x^2$

13. mass of vapour = mass of gas

14. $100p_A/P_A$ is where P_A is vapour pressure at dry bulb temperature of mixture, p_A = partial pressure.

15. $100y/Y_S$ where y is absolute humidity and Y_S is saturation humidity.

16. natural circulation towers, mechanical draft towers, spray chambers and spray ponds.

17. $P^* = PX$ where P^* = partial pressure, P = vapour pressure and X = mole friction in liquid phase.

18. $y^* = mx$, y^* = mole friction of solute in gas phase, x = mole fraction of solute in liquid, m is a constant.

19. The separation of the constituent of a liquid solution by contact with another insoluble liquid.

20. Physical adsorption, chemical adsorption.

21. Liquid diffusion, capillary movement, vapour diffusion, pressure.

22. Preferential solution of one or more constituents of a solid mixture by contact with a liquid solvent.

23. 1. Agitated vessels, 2. Thickners. 3. Hydrocyclones
4. Classifiers 5. Kennedy extractor 6. Bollman extractor
7. Continuous horizontal extractor.

24. The separation of solutes of different molecular sizes in a liquid solution by use of a suitable membrane.

25. Ultrafiltration.

26. Concentration diffusion (ordinary diffusion), pressure, thermal diffusion and forced diffusion.

27. dilute solution and isothermal operation.

28. half.

29. pure substance.

30. high boiling substances from non-volatile impurities.

31. gas-vapour mixtures.

32. 1—great capacity of solvent for the solute gas,
2—an increased rate of absorption and
3—increased utilization of stagnant zones of the liquid phase.

33. increases.

34. froth flotation.

35. froth flotation.

36. alkali metals.

37. emission

38. sugar and similar

39. water purity

40. water, organic systems or solids.

41. salt, water.

42. oxygen, oxides of nitrogen

43. 0.6 meter, 6 meters.

44. Phenomenon of liquid passing through the perforations countercurrent to vapour flow.

45. Is the bubble cap tray analogy to weeping, but occurs due to excessive liquid flow.

46. when the pressure drop through a tray exceeds the liquid head available le in the down comer.

47. freeze.

48. lines in liquid-liquid extraction

49. possible diameter,

50. $1/K_g = 1/k_g + m/k_L$

51. $D_{AB} = \dfrac{P_{BM} RT \rho_L (Z_2 - Z_1^2)}{2Pt(p_{A1} - p_{A2}) M \theta}$.

52. $D = 0.0043 \dfrac{T^{3/2}}{P(V_A^{1/3} + V_B^{1/3})} \sqrt{\dfrac{1}{M_A} + \dfrac{1}{M_B}}$

where V_A, V_B are atomic volumes, M_A and M_B are molecular weights, P is total pressure and T is absolute temperature, D is cm²/sec.

53. known.

54. film coefficients.

55. Friction factor and Reynolds number,

56. $\left(\dfrac{h}{CP_G}\right)\left(\dfrac{C_p \mu}{K}\right)^{0.67}$

57. $h_D = \dfrac{h}{C_p \rho}$ where ρ = density.

58. that in many situations the time of exposure of a fluid to mass transfer is short so that the concentration gradient of film theory (which is steady state) would not develop.

59. The time of exposure is not constant but varying lengths of time.

60. diffusion by definition is the movmeent of a molecular species due to some physical driving force through a mixture.

61. thermal diffusion

62. 10,000 times

63. Henry's law

64. $N_{m+1} = \log\left[\dfrac{X_D/(1-X_D)}{X_W/(1-X_W)}\right] / \log \alpha_{avg}$.

65. Water.

66. enthalpy concentration lines of liquid and vapour are straight and parallel.

67. total reflux conditions.

68. (*a*) concentration (*b*) temperature (thermal) (*c*) force.

69. $NTG = \ln\left(\frac{Y'_{as} - Y'_1}{Y'_{as} - Y'_2}\right), HTG = G_s / K_{ya}$

70. cellulose acetate membranes.

71. A group of stages interconnected so that the various streams flow from one to the other.

72. $\frac{\Delta P}{Z} \frac{g_e \in^3}{(1-\in)G^2} \frac{p\rho_G}{} = \frac{150(1-\in)}{R_e} + 1.75$

73. the desorption of the adsorbed solute hy a solvent.

74. the removal of moisture from a substance.

75. to when the soluble material is largely on the surface of an insoluble solid and is merely washed off by the solvent.

SECTION—C

1. (1) Gilliland equation
(2) Chapman-Enskog equation (using Lennard's potentials)

2. (*a*) Packed towers (*b*) Plate towers (*c*) Spray towers
(*d*) Cyclone scrubbers (*e*) Stirred absorbers (*f*) Wetted wall columns.

3. (*a*) Spray towers, (*b*) Plate towers, (*c*) Packed towers
(*d*) Pulse columns (*e*) Podbielniak extractor (centrifugal extractors)

4. (a) Dew point methods (*b*) Sling psychrometer (*c*) Hygrometer
(*d*) Gravimetric method (*e*) Electric hygrometers.

5. (*i*) e (*ii*) *a* (*iii*) *b*
(*iv*) *c* (*v*) *d*

6. (*a*) Agitated vessels, (*b*) Tray towers (bubble cap trays, perforated, sieve trays), (*c*) Proprietary trays.
(*a*) Venturi scrubbers (*b*) Wetted wall towers (*c*) Spray towers
(*d*) Spray chambers (*e*) Baffle towers (*f*) Shower trays
(*g*) Packed towers.

8. Dew point: This is the temperature at which a vapour gas mixture becomes saturated at constant total pressure out of contact with liquid. Wet bulb temperature: It is the steady temperature reached by a small amount of liquid evaporating into a large amount of unsaturated vapour-gas mixture.

9. Definition: Gas absorption is an operation in which a gas mixture is contacted with a liquid for the purpose of preferentially dissolving one or more components of the gas and to provide a solution of these in liquid.
Examples: (1) Removal of ammonia gas from coke oven gas using water as solvent.
(2) Removal of benzol vapours from coke oven gas using wash oil

10.(*a*) Gas solubility: The gas solubility should be high, thus increasing the rate of absorption and decreasing the quantity of solvent required.

(*b*) Volatility: The solvent should have low vapour pressure.

(*c*) Corrosiveness: The materials of construction required for the equipment should not be unusual or expensive.

(*d*) Cost: The solvent should be inexpensive so the losses are not costly and should be readily available.

(*e*) Viscosity: Low viscosity is preferred for reasons of rapid absorption rates improve flooding characteristics, low pressure drops and good heat-transfer characteristics.

(*f*) Miscellaneous: The solvent should be, non-toxic non-flammable and chemically stable and should have a low freezing point.

11. (*a*) Mixers (*b*) Agitated vessels (*c*) Flow mixers
(*d*) Spray towers (*e*) Packed towers (*f*) Perforated (sieve plate) towers
(*g*) Agitated counter current extractors (*h*) Centrifugal extractors.

12. Initial moisture content

$$\frac{0.80}{1-0.80} = 4.00 \text{ kg/kg dry solid}$$

Final moisture content

$$= \frac{0.05}{1-0.05} = 0.0527 \text{ kg water/kg solid.}$$

Kg dry solid in product = 1000 (0.95) = 950 kg

Moisture to be removed 950 (4 – 0.0527) = 3750 kg Ans.

13. (1) Direct driers: Truck dryer, Agitated Pan dryers, Vacuum rotary dryers. Through circulation dryer.

(2) Indirect: Vacuum shelf dryers, Agitated Pan dryers, Vacuum rotary dryers.

14. (*a*) Tunnels dryers (*b*) Turbo-type dryers (*c*) Through circulation dryers
(*d*) Rotary dryers (*e*) Throcirculation rotary dryers (*f*) Drum dryers
(*g*) Spray dryers (*h*) Fluidised bed dryers (*i*) Spouted bed dryers.

15. (a) The average intermolecular forces of attraction and repulsion in the solution are unchanged on mixing the constituents.

(*b*) The volume of the solution varies linearly with composition.

(*c*) There is neither absorption nor evolution of heat in mixture of the constituents. In the case of gases dissolving in liquids, however, this criterion should not include the heat of condensation of the gas to the liquid state.

(*d*) The total vapour pressure of the solution varies linearly with composition expressed as mole fractions.

16. $$D_{AE} = \frac{1}{y'_B / D_{AB} + y'C / D_{AC}} = \frac{1}{0.667/0.184 + 0.333/0.690}$$

= 0.244 sq. cm/sec.

$y_b' = 2/3, y'c = 1/3$

17. D_{AB} = 0.185 sq. cm.sec.

P_t = 1.0 atm, Z = 0.2 cm.

R = 82.06 (cu. cm) (atm)/(g. mole) (°K),

T = 276°K, P_{A_1} = 100/760 = 0.1317 atm.

P_{A_2} = 50/760 = 0.0658 atm, P_{B_1} = 1 – 0.1317 = 0.8683 atm.

P_{A_2} = 1 – 0.0658 = 0.934 atm.

$$P_{BM} = \frac{P_{B_1} - P_{B_2}}{\ln(P_{B_1}/P_{B_2})} = \frac{0.8683 - 0.9342}{\ln(0.8683/0.9332)} = 0.901 \text{ atm.}$$

$$N_A = \frac{D_{AB}P_t}{RTZ\,P_{BM}}(P_{A_1} - P_{A_2})$$

$$= \frac{0.185(1.0)(0.1317 - 0.0658)}{82.06 * 273 * 0.2 * 901}$$

$= 3.01*10^{-6}$ g moles benzene/sec. sq. cm.

18. $Y = y_A/y_B = p_A/p_B$ = 0.125/0.875

$$= \frac{100}{700} = 0.143 \text{ mole benzene/mole nitrogen.}$$

$$Y' = Y\,M_A/M_B = 0.143\frac{78.05}{28.08} = 28.97$$

19. M_1 (molecular weight) = 18.02

M_2 (molecular weight) = 28.97

$$Y' = 0.622\frac{p_{\text{water}}}{1 - p_{\text{water}}} \text{ kg water vapour/kg air.}$$

$$Y'_S = 0.622\frac{p_{\text{water}}}{1 - p_{\text{water}}} \text{ kg water vapour/kg air at saturation}$$

$v_H = (0.0252 + 0.405\ Y')\ (t_G + 460)$ cu. ft mixture/lb air

$C_S = 0.24 + 0.45\ Y'$ BTU for mixture/(lb air) °F)

t_0 = 0°C, 32°F.

λ_0 = 1075.8 BTU / 1b. latent heat of vaporization of at 0°C.

$H' = (0.24 + 0.45\ Y')\ (t_G - 32) + 1075.8\ Y'$ BTU/lb of dry air plus containing water vapour.

20. Time of drying in constant rate period:

$$N = \frac{-L_s dX}{A d\theta}, \theta = \int_0^\theta d\theta = \frac{Ls}{A}\int_{x_2}^{X_1}\frac{dX}{N}$$

at constant rate period $N = N_C$, X_1 and X_2 are greater than X_C.

$$\theta_C = \frac{L_S(X_1 - X_2)}{AN_C}$$

where L_S = weight of dry solid; A = drying surface, cross-sectional area: X = moisture content of solid on dry basis; θ = time, Nc = constant rate of drying.

21. Using Stefan's equation,

$$D_{AB} = P_{BM}\frac{RT\rho L}{2Pt\theta}(Z_2^2 - z_1^2)$$

$$p_{A1} = 170;\ p_{B1} = 765.5 - 170 = 595.5 \text{ mm Hg}$$

$$p_{A2} = 0;\ p_{B2} = 765.5 - 5 \text{ mm Hg}$$

$$P_{BM} = \frac{765.5 - 595.5}{\ln\frac{765.5}{595.5}} = 680 \text{ mm Hg}$$

θ = 80 minutes, R = 82.06. T = 291.8 K, z_3 = 2.448 cm

z_1 = 2.300 cms, P_t = 765.5, P_{A1} = 170, p_{A2} = 0 substituting the values,

$$D_{AB} = \frac{680 \times 82.06 \times 291.8(2.448^2 - 2.300^2)}{2(765.5)/760(170/760)80 \times 851.1}$$

= 337.44 cm²/hr.

22. $N_A = -\frac{Dy_A}{dz} + (N_A + N_B)y_A$

For stagnant case $N_B = 0$

$$N_A = -\frac{dy_A}{dz} + N_A * y_A, N_A dz = \frac{-Ddy_A}{(1 - y_A)}$$

Integrating between the limits, 0 to Z

$$N_A = \int_o^z dz = -\int_{y_{iA}}^{Y_{oA}} \frac{Ddy_A}{(1 - y_A)}$$

$$N_A = \frac{D}{z}\ln\left(\frac{1 - y_{iA}}{1 - y_{oA}}\right)$$ where D is diffusion coefficient,

z = film thickness, y_{iA} and Y_{oA} are mole fractions on the sides of the film.

23. $N_A = \frac{D_{AB}P_t}{RT_z}\ln\left(\frac{1 - y_{iA}}{1 - y_{oA}}\right)$

D_{AB} = 0.18, T = 293° K, R = 82.06

$$N_A = \frac{0.18 * 1}{82.06 * 293}\ln\frac{1}{(1 - 0.5)} * 3600$$

= 0.0748 g moles/ (hr. cm²).

24. Using ideal gas law,

$$P_1V_1/T_1 = P_2V_2/T_2$$

$$V_2 = \frac{V_1T_2}{T_1P_2} = \frac{22.41*293}{273*1}$$

$$= 24.05 \text{ L/g mole; } D \text{ in g moles/(cm.sec)}$$

$$= \frac{D \text{ in cm}^2\text{/sec}}{\text{(C.C.s /g mole)}} = 0.145/(24.05*1000)$$

$$= 0.603*10^{-6} \text{ g moles/(cm. sec.)}$$

25. For stagnant cases, $N_A = \dfrac{D_{AB}P_t(P_{A1} - P_{A2})}{RTZP_{BM}}$ and rate of diffusion is $N_A = \dfrac{\rho_A}{M}\dfrac{dZ}{d\theta}$.

Equating both and integrating between the limits gives, Stefan's equation

$$D_{AB} = \frac{\rho_A RTP_{BM}(Z_2^2 - Z_1^2)}{2M_AP_t\theta(P_{A1} - P_{A2})}.$$

26. (*a*) Stagnant film model (film theory)

(*b*) Penetration model—by Higbie 1935

(*c*) Surface renewal model—by Danckwerts 1951

(*d*) The film penetration model—Toor and Marchello

(*e*) Convective transport theory—Kishinevskiii

(*f*) Modified surface renewal theory of Danckwerts—by Wason-Ulwerya.

27.

Component	V.P. at 75°F atm P	P.P atms P^*	mole fraction $= P^*/P$ x in liquid
methane	Very high	2*0.6 = 1.2	1.2/V.P = 0
ethane	41.5	0.4	0.4/41.5 = 0.0097
propane	8.84	0.16	0.018
n-butane	2.33	0.12	0.052
n-propane	0.65	0.12	0.084
			0.264

Remaining liquid is = 1 – 0.264 = 0.736 solvent oil.

28. V.P. of benzene at 80°F = 100 mm Hg for ideal solution,

$P^* = Px$, $P = 100/760 = 0.1315$ atm.

$P^* = 0.1315x$; $Pt = 800/760 = 0.053$ atm.

$y^* = P^*/P_t = 0.1315x/1.053 = 0.1250x$

So equilibrium line is $y^* = 0.1250x$.

29. $F = D + W$ (overall balance)

$$D = F - W = 10{,}000 - w,\ F_{xF} = D_{xD} + W_{xW} \text{ (component balance)}$$

$$10{,}000*0.30 = 0.80*D + 0.05\ W$$

$$3000 = 0.8\ (10000 - w) + 0.05\ w,\ w = 6667 \text{ kg/hr.}$$

$$D = 10{,}000 - 6667 = 3337 \text{ kg/hr.}$$

30. Packed towers	Plate towers
1. Smaller pressure drop	1. High pressure drop
2. Useful for vacuum operation and for and for corrosive fluids.	2. Allows installation of cooling coils.
3. Moderate diameter	3. Large sizes are more economical than packed type.
4. Not useful if suspended solids are present in fluid stream.	4. Permits withdrawal of intermediate gas or liquid Streams.
5. Not favoured for large towers due to distribution problem.	
6. High cost in large size.	

31. at 110°C; $P_t = P_A x + P_B (1 - x) = x(P_A - P_B) + P_B$

$$x = P_t - P_B/(P_A - P_B) = \frac{760 - 484}{1050 - 484} = 0.487 \text{ mole fraction}$$

$$y = P_A x/P_t = 1050*0.487/760 = 0.674 \text{ mole fraction.}$$

32. $P_A^* = P_A x;\ P_B^* = P_B (1 - x)$.

If the vapour is ideal, $P_t = P_A^* + P_B^* = P_A x + P_B(1 - x)$

$$y^* = P_A^*/P_t = P_A x/P_t;\ 1 - y^* = P_B^*/P_t = P_B(1 - x)P_i$$

$$\alpha = \frac{y^*/(1 - y^*)}{x/(1 - x)} = P_A / P_B.$$

33. (1) A is diluent, B is solvent, C is solute

Examples: Water, benzene, acetic acid.

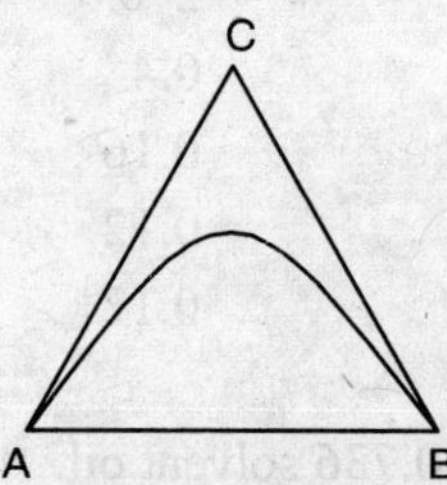

(2) Example: Chlorobenzene-water-methylethyl ketone.

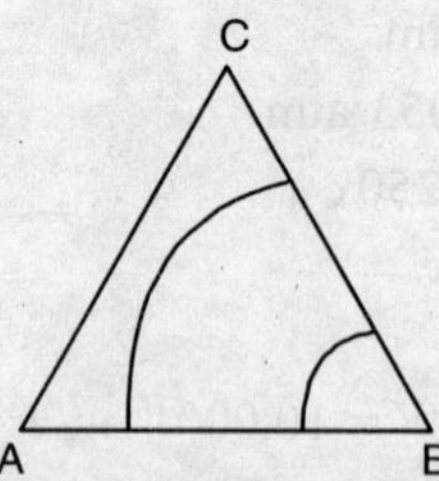

(3) Example: Aniline-napthalene-iso-octane.

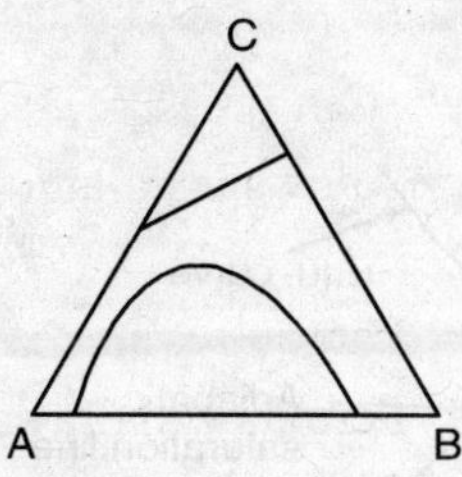

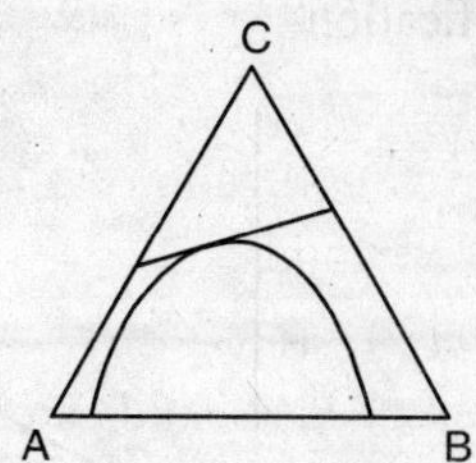

34. $t_G - t_W = \dfrac{\lambda_{\omega} K_y}{h}(Y_{W'} - Y')$

t_G = dry bulb temperature, °F (148 °F)

t_W = wet bulb temperature, °F (88 °F)

λ_W = latent heat of water in BTU/lb (1043)

Y_W = saturation humidity (0.031)

Y' = air humidity (0.018)

$$h_G/K = \frac{\lambda_{\omega}(Y'_W - Y')}{t_G - t_W} = \frac{1043(0.031-0.018)}{148-88}$$

$$= 0.227.$$

35. Dehumidifaction: H-enthalpy of vapour gas mixture.

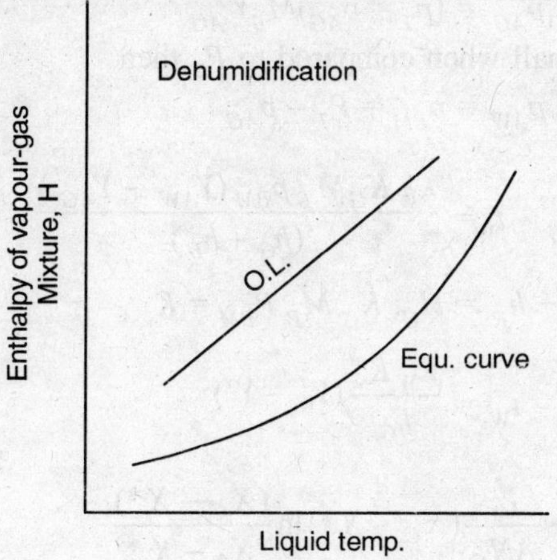

Humidification: Enthalpy of vapour gas mixture

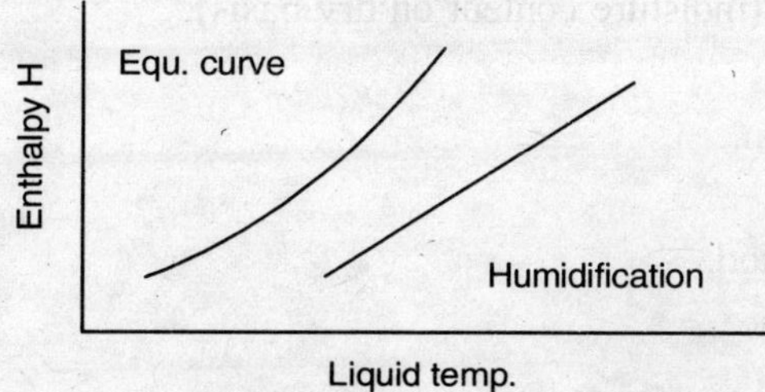

Adiabatic humidification

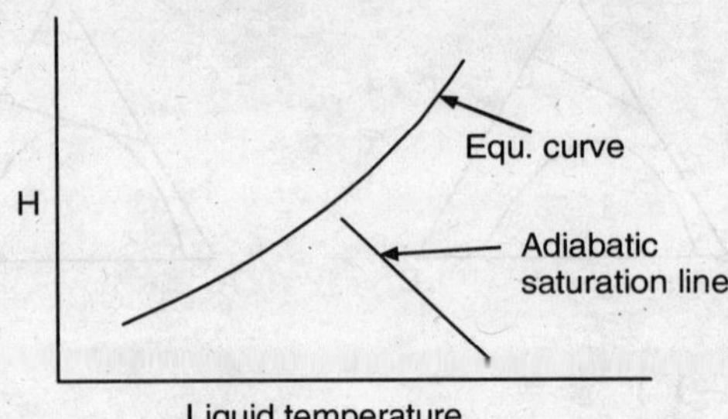

36. $q = \lambda_\omega W$ ...(1)

By mass transfer: $W = M_A S K_G (P_{AW} - P_{AG})$ ...(2)

By heat transfer: $q = (h_G + h_R) S (t_G - t_W)$ ...(3)

Substituting Eq. (3) and Eq. (2) in Eq. (1), gives

$$t_G - t_W = \frac{\lambda_W M_A K_G}{(h_G + 4_R)}(P_{AW} - P_{AG}) \quad ...(4)$$

Since $$Y_A' = W_A/W_B = P_A/(P_T - P_A)\frac{M_A}{M_B} \quad ...(5)$$

$$M_A p_{AW} = (P_T - p_{AW}) M_B Y'_{AW}$$

$$M_A p_{AG} = (p_T - p_{AG}) M_B Y'_{AG}$$

If p_{AW} and p_{AG} are small when compared to P_T then

$$P_T - p_{AW} = p_{BM} = P_T - p_{AG}$$

so Eq. (4) becomes, $$t_G - t_W = \frac{\lambda_W K_G M_B p_{BM} (Y_{AW}' - Y_{AG}')}{(h_G + h_B)}$$

$$h_G + h_R = H_G;\ K_G M_B P_{BM} = K_v$$

$$t_G - t_W = \frac{\lambda_W K_v}{h_G}(Y_W' - Y')$$

37. $$\theta = \frac{L_S}{AN_C}(X_i - X_C) + \frac{L_s}{AN_C}(X_C - X^*)\ln\frac{(X_C - X^*)}{X_2 - X^*}$$

X_2 = final condition (moisture content on dry basis)

X_t = initial condition (moisture content on dry basis)

L_S = dry solid weight;

X_C = critical condition,

A = area of drying,

N_C = critical rate period,

X^* = equilibrium value.

$$4 = \frac{L_S}{AN_C}(0.3-0.10) + \frac{L_S}{AN_C}(0.10-0.02)\ln\left[\frac{0.1-0.02}{0.04-0.02}\right]$$

$$L_S/AN_C = 13.1 \text{ for } X_2 = 0.07$$

$$\theta = 13.1\ (0.30 - 0.10) + 13.1\ (0.10 - 0.02)\ \ln\left[\frac{0.10-0.02}{0.07-0.02}\right]$$

$$= 3.11 \text{ hr.}$$

38. For adiabatic saturation condition

$$\frac{H-H_s}{t-t_s} = -C_s/\lambda_S$$

$$C_S = 0.238 + 0.47{*}0 = 0.238$$

$$H_S = 18\ P_A/MEE(1-P_A)$$

$$\frac{18P_A}{74(1-P_A)(69-t_S)} = 0.238/92.$$

By trial and error, $t_s = 10$ deg. C

MEE = Molecular wt. of ethyl ether (74)

H = humidity of entering air = 0

H_S = humidity of air after saturation

P_A = vapour pressure of ethyl ether

C_S = humid heat

λ_S = latent heat of vaporisation at saturation.

39. (*a*) Selectivity: β = weight fraction *A* in *R** weight fraction of *C* in *E*/*wt* fraction *C* in *R** *wt* fraction *A* in *E*, should be more than one.

(*b*) Distribution coefficient: *y**/*x* greater than one is desirable since less solvent will be required.

(*c*) Insolubility of solvent: Solvent should be less soluble in diluent but more soluble with solute.

(*d*) Recoverability: The solvent should be easily recoverable from the solute.

(*e*) Density: A difference in densities of the saturated liquid phases is necessary for separation.

(*f*) Interfacial tension: The larger the interfacial tension the more readily will coalescence of emulsions occur, but the more difficult will the dispersion of one liquid in the other be. Coalescence is usually of greater importance and interfacial tension should be therefore high.

(*g*) Chemical reactivity: The solvent should be stable chemically and inert towards the other components of the system and toward the common materials of construction.

(*h*) Viscosity, vapour pressure and freezing point: These should be low for ease in handling and storage.

(*i*) The solvent should be non-toxic, non-flammable and of low cost.

SECTION—D

1. Ps/Pg

2. True. $\frac{\delta u}{\delta c} = \text{Sc.No}^{1/3} \cdot \delta u$ = thickness of velocity boundary layer,

   δc = Thickness of conc. boundary layer.

3. (d) 1000 kg wet solid, $1000 \times \frac{60}{100} = 600$ kg moisture; = 1000 – 600 = 400 kg; Dry solid

$$X_i = \frac{60}{100-60} = 1.50 \text{ (initial moisture content on dry basis).}$$

$$X_f = \text{(final moisture content on dry basis)} = \frac{20.0}{100-20} = 0.25$$

4. True. Water removed = 400 (1.5 – 0.25) = 500 kg
5. (c)
6. (a) Since gas is less soluble in liquid, so gas side film will be controlling.
7. (i) – (a); (ii) – (d)
8. $Q = \frac{L_S}{AN_C}(X_I - X_C) + \frac{L_S}{AN_C}(X_C - X^*)\ln\left(\frac{X_C - X^*}{X_2 - X^*}\right)$

$$\frac{L_S}{A} = 40;\ N_C = 3 \times 10^{-4};\ X_1 = \frac{0.30}{1-0.30} = 0.43;$$

$$X^* = 0.05;\ S_C = \frac{0.20}{1-0.20} = 0.25;\ X_2 = \frac{0.05}{1-0.05} = 0.0526$$

$$Q = \frac{4.0}{3*10^{-4}}(0.43-0.25) + \frac{40}{3*10^{-4}}(0.25-0.05)*\ln\left(\frac{0.25-0.05}{0.0526-0.05}\right)$$

   Q = 139, 464 secs; = 38.74 hrs.

9. Feed = 100 kg/s; A = 10 × 0.80 = 8 kg/s

   Initial X = 0.20; X final = 0.20 × 0.05 = 0.01

$$\frac{Y_n^* - 0}{0.20-0.01} = \frac{A}{B}; Y_n^* \text{ at } X = 0.20 \text{ is } 0.17$$

$$0.17/0.19 = 8/3;\ B = 8.94 \text{ kg/S}$$

   2.5 times of min B = 8.94 × 2.5 = 22.35 kg/S

10. Rhombic
11. (c)

12.

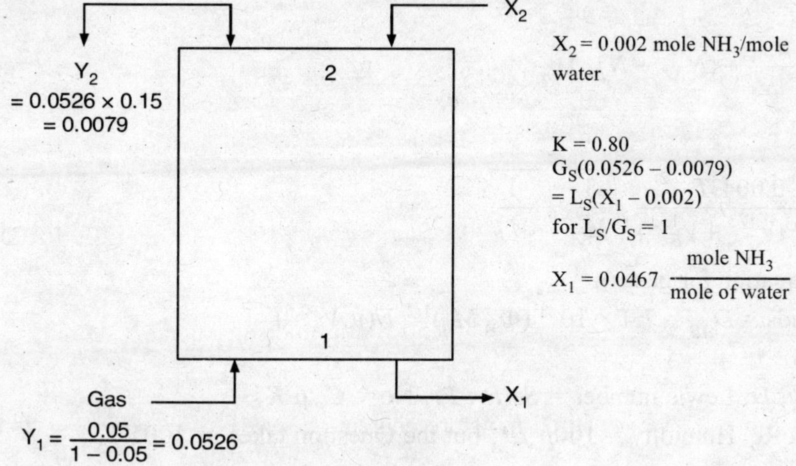

Assuming $L/G = 1$; $L/mG = 1/0.8 = 1.25$

No. of stages for constant K value (m):

$$\frac{Y_{N_p+1} - Y_1}{Y_{N_p+1} - mX_0} = \frac{A^{N_p+1} - A}{A^{N_p+1} - 1} = \frac{0.0526 - 0.0079}{0.0526 - 0.8 \times 0.002}$$

$$= \frac{1.25^{N_p+1} - 1.25}{1.25^{N_p+1} - 1}; \text{ by trial and error } N_{p+1} = 5$$

13. (d) 20

Salt = 50.0 kg; Solvent = 500 kg; salt remaining = 50 – 40 = 10 kg

Solubility in kg/100 kg water = $\frac{10}{50} \times 100 = 20.$

14. $C(1/P)$

For binary gas mixtures at low pressure, D_{AB} is inversely proportional to pressure, increases with increasing temperature and is independent of composition for a given gas-pair.

15. $D_{AB}Pt/(RTZ)$

$$N_A = \frac{D_{AB}\, p_t}{RTZ}(P_1 - P_2) = K_y(P_1 - P_2)$$

16. $\frac{h_D}{u}\left(\frac{C_{BM}}{C_T}\right) = \frac{R}{\rho u^2} = \frac{f}{2}$, one. C_{BM} = Concentration of non-diffusing component , C_T = overall concentration.

17. Surface

18. False

When a gas is soluble in a liquid then there may be decrease or increase of temperature of the solution.

19. False

20. False

$$N_A = -D_{AB}\frac{d_{CA}}{dZ} + (N_{AZ} + N_{BZ})x_A$$

21. *a—B; b—D.*

$$(a)\ D_{AB} = \frac{0.0043T^{3/2}}{P(V_A^{1/3} + V_B^{1/3})}\sqrt{\frac{1}{M_A} + \frac{1}{M_B}}$$

(Gilliland equation for gases).

(*b*) For liquids = $D_{AB} = 7.4 \times 10^{-8}\ (\Phi_B M_B)^{1/2}\ T/(\mu\ V_A^{0.6})$

22. *a—C; b—B*

Sc. No. = $\mu/\rho D$; Lewis number = Sc/Pr; Pr. No = $C_P\,\mu/K$

23. Normally % Re. Humidity = $100p_A/P_A$ but the Question takes as $100Y/Y_S$.

$$YS \text{ (Saturation humidity) at } 30°C = \frac{4.25}{150-4.25} * \frac{18}{29} = \frac{4.25}{145.75} \times \frac{18 \text{ kg water vapour}}{29 \text{ kg dry air}}$$

$$Y \text{ at } 15°C \text{ is} = \frac{1.71}{200-1.71} \times \frac{18}{29} = \frac{1.71}{198.29} \times \frac{18}{29}$$

$$\% \text{ Rel. humidity} = \left(\frac{1.71}{198.29}\right) / \left(\frac{4.25}{145.75}\right) \times 100 = 29.6\%$$

24. Reflux is used for enriching top and bottom products. Increasing reflux ratio will increase the Quality of products, decreases in number of stages but increases the diameter of column heating and cooling loads and pump capacities.

25. $N_A = (N_A + N_B)y_A - D_{AB}\dfrac{\partial_{yA}}{\partial Z}$; for $N_A = -N_B$

$$N_A = -D_{AB}\frac{\partial_{yA}}{\partial Z} = \frac{-D_{AB}}{RT}\frac{dp_A}{d_Z}; \text{ integrating gives.}$$

26. $N_A = \dfrac{D_{AB}}{RTZ}(P_{A1} - P_{A2})$; 1atm = 101.33 kPa

15 kPa = 0.148 atm; 5 kPa = 0.0493 atm; 1 atm = 10.332 kg/m^2

$$R = \frac{0.082 \text{ lit. atm}}{\text{g mole. °k}} = \frac{0.082\text{m}^3 \text{ atm}}{\text{kg mole. °k}}; 1 \text{ kg/m}^2 = 9.807 \text{ Pascals}$$

$$1.6 \times 10^{-5} = \frac{D_{AB} * (0.148 - 0.0493)}{0.082 \times 293 \times 0.01}; D_{AB} = 3.895 \times 10^{-5} \text{ m}^2/\text{s}.$$

27. Infinite.

28. Hint: Use mass transfer and heat transfer analog.

29. Operating line represents material balance in the absorber. Any point on the operating line represents material balance relation at the height of bed.

30. Sherwood number—mass transfer
Prandtl number— heat transfer
Nusselt number — heat transfer by convection
Schmidt number — mass transfer

31.

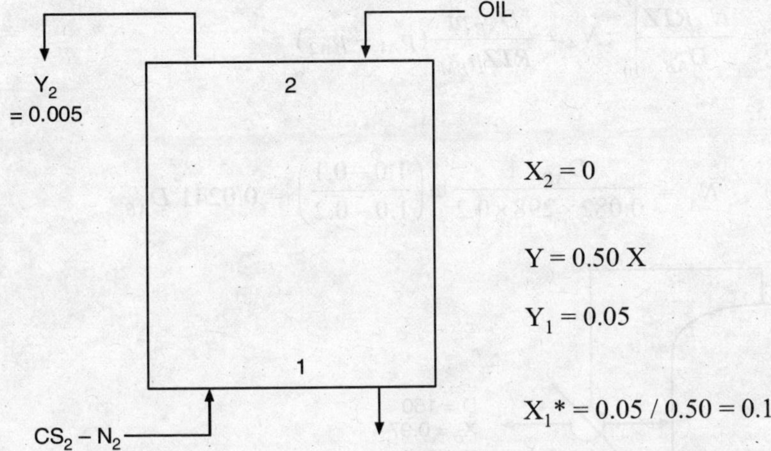

$X_2 = 0$

$Y = 0.50\ X$

$Y_1 = 0.05$

$X_1^* = 0.05 / 0.50 = 0.1$

$G = 1500$ K mole/hr; $G_S = 1500 \times 0.95 = 1425$ K mole/hr.

$$L_S\left(\frac{0.1}{1-0.10}-0\right) = G_S\left(\frac{0.05}{0.95}-\frac{0.005}{0.995}\right); \frac{L_S}{G_S} = 0.4284$$

$$L_S = 1425 \times 0.4284 = 610.5 \text{ K mole/hr.}$$

(*b*) $L_S (X_1 - X_2) = G_S(0.0476)$. Assuming $X_2 = 0$

$$L_S = 1.5 \times 610.5 = 915.8 \text{ K mole/hr}$$

$$951.8\ X_1 = 1425 \times 0.0476;\ X_1 = 0.0741$$

Operating points: $Y_1 = 0.526;\ X_1 = 0.0741$

$Y_2 = 0.00508;\ X_2 = 0,\ Y = mX;\ m = 0.0476/0.0741 = 0.6423$

32. For stagnant case:

$$N_A = \frac{D_{AB}Pt}{RTZP_{BM}}(p_{A1} - p_{A2});\ \text{T is °K} = 273 + 25 = 298\text{°K}$$

P_t = total pressure (atm); p_{A1} = partial pressure of A in atm. at point 1;

p_{A2} = partial pressure of A in atm. at point 2; R = gas constant, $\frac{\text{lit. atm}}{\text{g mole. °K}}$

Z = thickness of film, cm.

$$P_{BM} = (p_{B2} - p_{B1})/\ln (p_{B2}/p_{B1})$$

Using

$$N_A = (N_A + N_B)\frac{p_A}{p_t} - \frac{D_{AB}}{RT}\frac{dp_A}{dZ}$$

For stagnant case, $N_B - 0$

$$N_A = N_A \frac{p_A}{p_t} - \frac{D_{AB}}{RT}\frac{dp_A}{dZ}; N_A\left(1 - \frac{p_A}{RT}\right) = \frac{D_{AB}}{RT}\frac{dp_A}{dZ}$$

$$P_t \frac{dp_A}{(P_t - p_A)} = \frac{N_A RTdZ}{D_{AB}}; \text{ Integrating}$$

$$-p_t \ln(p_t - p_A)\Big|_{p_{A1}}^{p_{A2}} = \frac{N_A RTZ}{D_{AB}}\Big|_0^Z; N_A = \frac{D_{AB}\,pt}{RTZp_{BM}}(p_{A1} - p_{A2})$$

Using the given data,

$$N_A = \frac{D_{AB} * 1}{0.082 \times 298 \times 0.2} \ln\left(\frac{1.0 - 0.1}{1.0 - 0.2}\right) = 0.0241\, D_{AB}$$

33.

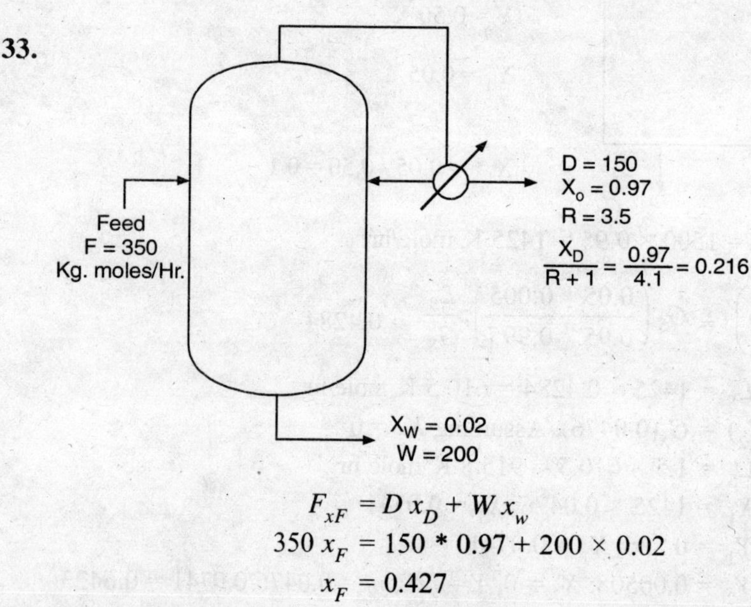

$$F_{xF} = D\, x_D + W\, x_w$$

$$350\, x_F = 150 * 0.97 + 200 \times 0.02$$

$$x_F = 0.427$$

By graph to have a stage of $y = 0.33$ and $x = 0.25$ the feed is to be partially vapourized.

34.

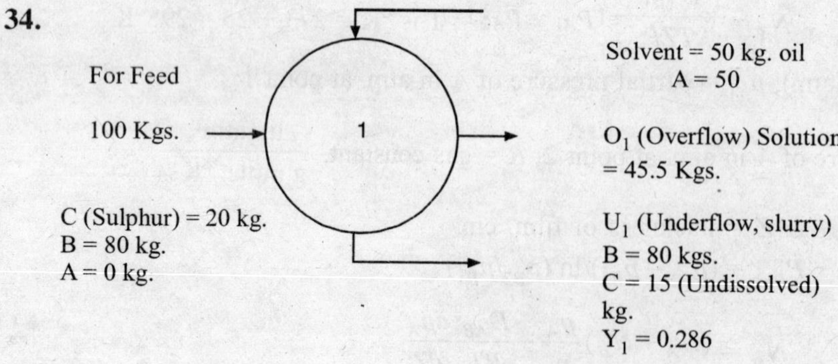

50 kg of solvent can dissolve 5 kg of sulphur.

Undissolved $C + B = 15 + 80 = 95$ kg

$A + C$ in $U_1 = 95/10 = 9.5$ kg

by $(A + C)$ balance: $50 + 20 = 9.5 + 15 + 0_1$

$0_1 = 45.5$

By C balance: $20 = 0_1 x_1 + U_1 y_1 = 45.5\, x_1 + 24.5\, y_1$

By A balance: $50 = 01\,(1 - x_1) + U_1\,(1 - y_1)$

$50 = 45.5\,(1 - x_1) + 24.5\,(1 - y_1)$

Using $x_1 = y_1$ (for no adsorption)

$x_1 = y_1 = 0.286$

35. $h = 0.0175 * 5900 = 18.2$ kcal/ (hr. m² °C)

$Q = hA\Delta T = 18.2\,(65 - 25)\,62 \times 62/10^4 = 258.9$ kcal/hr.

Amount of water evaporated $= Q/\lambda_S = 258.9/585 = 0.443$ kg/hr.

Total dry solid $= 62 \times 62 \times 1.25 \times 1.6/1000 = 7.69$ kg

Total water removed $= 7.69\,(0.50 - 0.02) = 3.6912$ kg

Time of drying $= 3.6912/0.443 = 8.33$ hr

36. $L_S = 500$; $X_1 = 0.40/(1 - 0.40) = 0.667$

$X_2 = 0.05/(1 - 0.05) = 0.053$;

Enthalpy of entering air $= (0.238 + 0.48 \times 0.006)\,(110 - 0) + 585 * 0.006 = 30.23$ kcal/kg dry air

Enthalpy of leaving air

$= (0.238 + 0.48\,Y_1)\,(38 - 0) + 585\,Y_1 = 9.12 + 603.24\,Y_1$

Enthalpy of entering solid $= 0.21\,(20 - 0) + 0.667 \times 1.\,(20 - 0) = 17.54$

Enthalpy of leaving solid $= 0.21\,(35 - 0) + 0.053\,(35 - 0) = 9.205$

By material balance: $200 * 0.667 + G_S * 0.006 = G_S\,Y_1 + 200 \times 0.053$ ...(1)

By enthalpy balance: $200 \times 17.54 + G_S \times 30.23$

$= (9.12 + 603.24\,Y_1)\,G_S + 9.205 \times 200 + 5 \times G_S$ ...(2)

Solving equations (1) and (2), $G_S = 5797$ kg dry air/hr

$$Y_1 = 0.0272 \frac{\text{kg water vapor}}{\text{kg dry air}}$$

37. The difference between exit liquid temperature and the entering air wet bulb temperature is called wet bulb temperature approach. It is a measure of driving force available for diffusion at the tower end of the equipment. In the design of cooling towers, this is ordinarily specified to be from 2.5 to 5°C. This effects the size of tower and leaving temperature of gas and liquid. On the performance of water cooling tower this effects the leaving water temperature.

38. $\dfrac{h_G}{k_y C_s} = 1$ is Lewis relation. $\dfrac{h_G}{k_y C_s} = \left(\dfrac{SC}{Pr}\right)^{0.567} = Le^{0.567}$

The thermal and molecular diffusivities are equal or if $Sc = Pr$; the Lewis number is (Le) is one.

39. $t_G - t_w = \dfrac{\lambda_w(Y_w - Y)}{h_G / k_y}; Y_w = \dfrac{50}{(800-50)}\dfrac{58}{28} = 0.1381$

$$45 - 28 = \frac{403,000}{h_G / k_y}; Y = 0.0664$$

Flow rate = 0.08 × 4 = 0.32 m³/sec. at 318 °K = 12.26 g mole/sec.

$$\text{Acetone} = 0.01226 \times 0.0664 \times \frac{28}{58} = 0.000395 \text{ kg mole/s.}$$

$$G = 0.000395 \times \frac{3600}{0.08} \times 58 = 1025.7 \text{ kg. acetone/hr. m.}$$

40. O_3 x_3 → (2) → O_2 x_2 → (1) → O_1, x_1

← y_2, U_2 ← (2) ← U_1, y_1 ← (1) ← U_0 (Feed), Y_F

Solids = 1800 Kg/hr

Oil in feed = 400 – 4000 = 2200 kg/hr.

Underflow = 1800 kg/hr

By overall material balance $U_0 + O_3 = O_1 + U_2$

$$O_3 = 6000 \text{ kg/hr.}$$

$$2200 + 6000 = O_1 + 1800; O_1 = 6400 \text{ kg/hr}$$

Overall balance for stage 1: $U_O + O_2 = O_1 + U_1$

$$200 + O_2 = 6400 + U_1; U_1 = 1800; \text{ giving}$$

$$O_2 = 6000 \text{ kg/hr.}$$

Oil balance for stage 1: $6400\, x_1 + 1800\, y_1 = 6000\, x_2 + 2200$ ...(1)

Oil balance for stage 2: $6000 \times 0.05 + 1800\, y_1 = 1800\, y_2 + 6000\, x_2$ ...(2)

Solving equation (1) and (2): Using $x_1 = y_1$ and $x_2 = y_2$ gives $x_1 = 0.3567$; $y_1 = 0.3567$; $x_2 = 0.1208$

$$\% \text{ oil recovery} = \frac{2200(\text{oil in feed}) - 217.44(\text{oil lost})}{2200/100}$$

$$= 90.13\ \%$$

Oil lost = $1800\, y_2 = 1800 \times 0.1208 = 217.4$ kg/hr.

41.

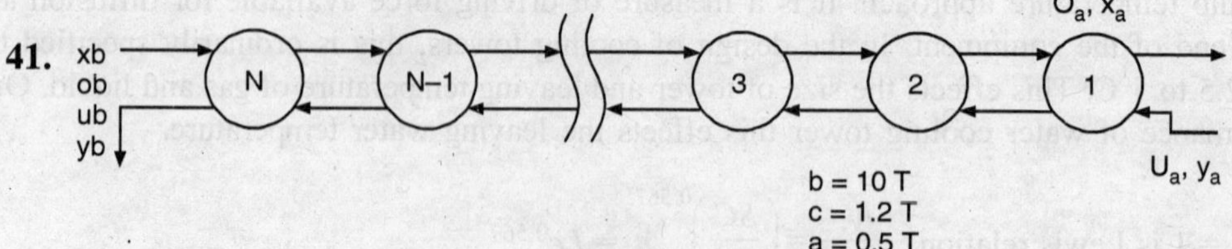

Fresh solvent:

By overall $(a + c)$ balance: $O_b + U_a = O_a + U_a$; $O_b = 1700$

$$= 20,000 + O_a$$ ...(1)

C balance: $1700 \times 0.706 = 20000 \times 0.0012 + O_a \times 0.07$ ...(2)

(Since $y_b = 1200 \times 0.02/20{,}000 = 0.0012$)

Solving equations (1) and (2) gives O_b (solvent) = 36,800 kg/hr.

Number of stages:

$$\text{Leached ore } y_b = 24/20{,}000 = 0.0012$$
$$N_y b = 10000/20000 = 0.50;\ x_b = 0;\ N_x b = 0$$
$$y_a = 1200/1700;\ N_{ya} = 10{,}000/1700 = 5.88$$
$$x_a = 0.07;\ N_{xa} = 0$$

Leaving first stage, rest have constant under flow.

X_2, O_2 → (1) → $X_2 = 0.07$

U_1, $Y_1 = 0.07$ ← (1) ← $y_2 = 0.706$

By $(a + c)$ balance: $O_2 = U_1 + O_a - U_a = 2000 + 16{,}800 - 1700$

$= 35{,}100$

By C balance: $35100\, x_2 + 200 = 20000 \times 0.07 + 16800 * 0.07$

$$x_2 = 0.0392$$

Leaving stage 1; number of stages for constant underflow is:

$$n = \frac{\log[(x_b - y_b)/(x_a - y_a)]}{\log[(x_b - x_a)/(y_b - y_a)]} = \frac{\log[(0 - 0.0012)/(0.0392 - 0.04)]}{\log\left(\dfrac{0 - 0.0392}{0.0012 - 0.07}\right)}$$

$n = 5.77$. Total stages are: $n + 1 = 6.77$ **Ans.**

Or another relation is

$$\frac{y_F - y_N}{y_F - mx_{m+1}} = \frac{(R/mE)^n - (R/mE)}{(R/mE)^n - 1}, \text{ here } m = 1$$

$$\frac{0.07 - 0.0012}{0.07 - 0} = \frac{(35100/20000)^n - (35100/20{,}000)}{(35100/20000)^n - 1}$$

By trial and error $n = 7$. Total stages are = 8 $(n + 1)$.

42. $$(A/B)_{\min} = \frac{X_F/6.45(y_1^*) - Y_{n+1}(0)}{X_F - X_N}$$

$$X_F = \frac{0.20}{1 - 0.20} = 0.25;\ X_N = \frac{0.04}{1 - 0.04} = 0.0417$$

$$A = 0.80 \text{ kg in 1 kg of feed; } \frac{0.8}{B_{\min}} = \frac{0.25/6.45 - 0}{0.25 - 0.0417}$$

$$= 0.1863;\ B_{MN} = 4.29 \text{ kgs.}$$

43.

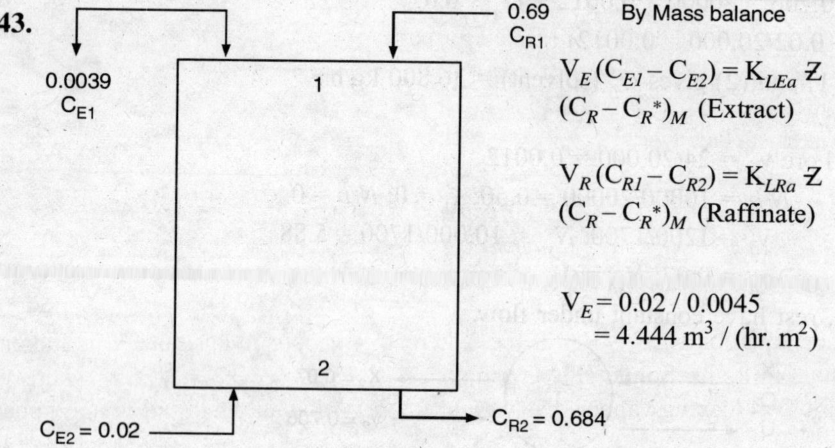

By Mass balance

$V_E(C_{E1} - C_{E2}) = K_{LEa} Z (C_R - C_R^*)_M$ (Extract)

$V_R(C_{R1} - C_{R2}) = K_{LRa} Z (C_R - C_R^*)_M$ (Raffinate)

$V_E = 0.02 / 0.0045 = 4.444 \text{ m}^3 / (\text{hr. m}^2)$

$V_E = 4.444 * 800/78 = 45.6$ kg mole.hr.

$$(C^* - C_E)_M = \frac{(C_{E2}^* - C_{E2}) - (CE_1^* - C_{E1})}{\ln\left(\dfrac{C_{E2}^* - CE_2}{CE_1^* - CE_1)}\right)}; C_{E1}^* = mCR_1\ ,\ C_{E2}^* = m\ CR_2$$

$$(C_E^* - C_E)_M = 0.01494 \text{ giving } K_{LE}a * 1.37 \times 0.01494 = 45.6 * 0.0039$$

$$K_{LE}a = 8.69 \text{ kg mole/ (hr . m}^3. \Delta C)$$

$$HTLE = V_E/K_{LE}a = 45.6/8.69 = 5.25 \text{ m}$$

44. For 1000 kg of original solution, total water = 0.65 × 1000 = 650 kgs

Water evaporated is = 0.10 × 650 = 65 kg

The molecular weight of Mg SO_4 and $MgSO_4$ and $MgSO_4$, $7H_2O$ are 120.4 and 246.5 respectively.

The total $MgSO_4$, $7H_2O$ in the batch is $1000 \times 0.35 \times \dfrac{246.5}{120.4} = 716.6$ kg

Free water = 1000 – 716.5 – 65 = 218.4 kg

In 100 kgs of mother liquor (15°C) there is = 24 × 246.5/120.4 = 49.1 kg $MgSO_4$, 7 H_2O

Free water at 15°C in mother liquor = 100 – 49.1 = 50.9 kg

The $MgSO_4$, $7H_2O$ in mother liquor is = $\dfrac{49.1}{50.9} \times 218.4 = 210.7$ kg

Amount of crystals obtained = 716.6 – 210.7 = 505.9 kg.

45. The yield is calculated from the solubility data.

At 49°C, the saturated solution contains 140 parts $FeSO_4$, $7H_2O$ per 100 parts excess water. The yield is therefore 140 – 74 = 66 parts per 100 parts excess water or 66 parts per 240 parts of initial solution. The total amount of feed, solution = $1000 \times \dfrac{240}{66} = 3635$ kg/hr. The cooling requirement is calculated using given data neglecting heat of dilution of 18°C to 27°C.

Heat of crystallization of $FeSO_4$, $7H_2O$ = – 4400/278 = – 15.83 kcal/kg of crystal.

$$\text{Heat removed } (q) = 3650 \times 0.70 \times (49 - 27) + 1000 \times 15.38$$
$$= 72{,}040 \text{ kcal/hr.}$$

$$\text{Cooling water required} = \frac{72,040}{(25-20)\times 1\times 60} = 240 \text{ kg/min}$$

$$\text{Amount of heat transfer surface } (A) = \frac{72040}{171.1\times 13.8}$$

$$= 30.5 \text{ m}^2$$

where $$\Delta t_m = \frac{\Delta t_1 - \Delta t_2}{\ln(\Delta T_1/\Delta T_2)} = \frac{(49-25)-(27-20)}{\ln[(49-25)/(27-20)]} = 13.8$$

46. Total material balance equation is: $F = V + L + C$ where V is vapour in kgs/hour, L is liquor in kg/hr and C is crystals in kg/hr. Solute ($FeSO_4$) balance is: $Fx_F = Vy + Lx_L + Cx_c$ where y is weight fraction of solute in leaving vapour, x is weight fraction of solute in liquid or solid phase.

$$x_F = \frac{40}{100+40} = 0.236; \; x_L = \frac{30.5}{100+30.5} = 0.234$$

The crystals are $FeSO_4$, $7H_2O$. The molecular weight of $FeSO_4$ is 151.9 and the molecular weight of $FeSO_4$, $7H_2O$ is 278.0.

$$x_e = 151.9/278.0 = 0.546$$

Substituting in overall balance and solute balance equations

$$F = V + L + 20{,}000; \; 0.28\,F = 0.2346L + 0.546 \times 20{,}0000.$$

Energy balance equation is: $Fh_F = VH + Lh_L + Ch_C$

Neglecting boiling point of elevation, enthelpy of vapour (H) as that of saturated steam at 27°C which is 611.1 kcal/kg.

$$F \times 26.0 = V \times 611.1 + \text{L} \times -1.33 + 20{,}000 \times -55.6$$

Solving the above three equations,

$$F = 93,\,264 \text{ kg/hr}; \; L = 67{,}302 \text{ kg/hr}; \; V = 5962 \text{ kg/hr}.$$

47. $$y_1 = \frac{0.18\times 80}{0.18\times 80 + 99.82\times 29} = 0.004950$$

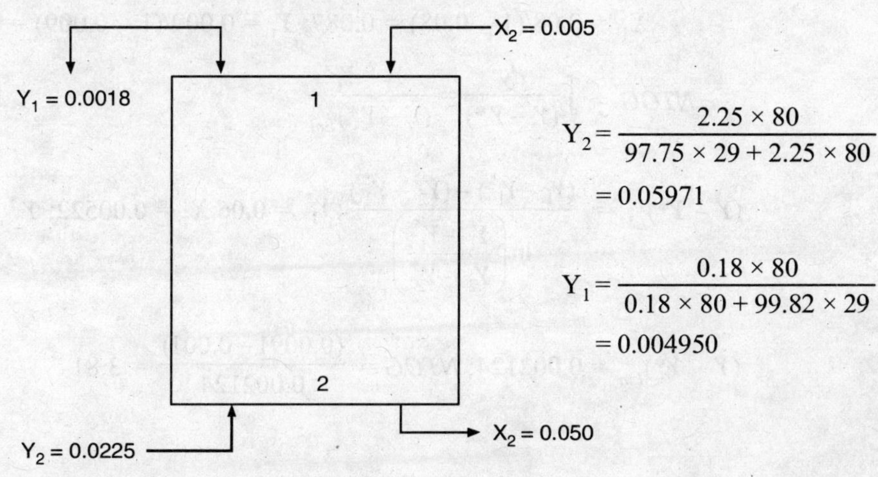

$$Y_2 = \frac{2.25\times 80}{97.75\times 29 + 2.25\times 80}$$

$$= 0.05971$$

$$Y_1 = \frac{0.18\times 80}{0.18\times 80 + 99.82\times 29}$$

$$= 0.004950$$

$$Y_2 = \frac{2.25 \times 80}{97.75 \times 29 + 2.25 \times 80}$$

$$= 0.05971$$

$$Y_1 = \frac{0.18 \times 80}{0.18 \times 80 + 99.82 \times 29}$$

$$= 0.004950$$

$$y_2 = \frac{2.25 \times 80}{97.75 \times 29 + 2.25 \times 80} = 0.05971$$

$$Y_2 = \frac{0.05971}{1 - 0.05971} = 0.0635; \; Y_1 = \frac{0.00495}{1 - 0.00495} = 0.004975$$

$$X_2 = \frac{0.05}{1 - 0.05} = 0.05263; \; X_1 = \frac{0.005}{1 - 0.005} = 0.005025$$

By solute balance:

$$G_S(0.0635 - 0.004975) = L_S\,(0.05263 - 0.00\,5025)$$

$$G_S/L_S = 0.8134$$

(*b*) Since equilibrium curve and operating lines are straight.

$$N = \frac{\log[(Y_1 - Y_1^*)/(Y_2 - Y_2^*)]}{\log[(Y_1 - Y_2)/(Y_1^* - Y_2^*)]} \text{ here Y1} = 0.004975$$

$$Y_1^* = 0.005025 \times 0.65 = 0.0032663$$

$$Y_2 = 0.0635; \; Y_2^* = 0.65 \times 0.5263 = 0.03421$$

$$N = 4.47 \text{ stages.}$$

48. The composition in the ratios are: taking top as (2) and bottom as (1): $X_2 = 0$; $Y_2 = \frac{0.001}{1 - 0.001} =$ 0.001

$$X_1 = 0.08/(1 - 0.08) = 0.087; \; Y_1 = 0.009/(1 - 0.009) = 0.0091$$

$$NTOG = \int \frac{dy}{(Y - Y^*)} = \frac{Y_1 - Y_1}{(Y - Y^*)_{Lm}}$$

$$(Y - Y^*)_{LM} = \frac{(Y_1 - Y_1^*) - (Y_2 - Y_2^*)}{\ln\left(\dfrac{Y_1 - Y_1^*}{Y_2 - Y_2^*}\right)}; \; Y_1^* = 0.06\,X_1 = 0.00522; \; Y_2^* = 0$$

$$(Y - Y^*)_{Lm} = 0.002124; \; NTOG = \frac{(0.0091 - 0.001)}{0.002124} = 3.81$$

$$\frac{G_S}{L_S} = \frac{X_1 - X_2}{Y_1 - Y_2} = \frac{0.087 - 0}{0.0091 - 0.001} = 10.74; \text{Assuming } \frac{G}{L} \text{ is near to } G_S/L_S;$$

$$HTOG = H_y + \frac{mG}{L} H_x = 0.36 + 0.06 \times 10.74 \times 0.24 = 0.5150$$

$$Z = NTOG \times HTOG = 3.81 \times 0.515 = 1.96 \text{ m}$$

49. **1. Law of Haüy:** When crystals are allowed to form without any interference from other crystals or outside bodies, though the relative sizes of the faces and edges of various crystals of the same material may be different and angles made by corresponding edges of the same material are equal.

2. Gibbs-Wulff theorem: If a crystal grows at constant temperature in equilibrium with mother liquor, the crystal maintains same shape provided the surface energy per unit area of face is constant.

3. Survival theory: The larger the value of supersaturation, the greater the survival of smaller crystals. The number of nuclei per contact is proportional to supersaturation.

4. ΔL law of crystal growth: The total growth of each crystal in the magma during the same interval Δt is same.

5. Miers theory: The greater the degree of supersaturation, the better for nuclei forming. If supersaturation cross certain range the nucleus formation is highly rapid.

50. Total pressure = 380 mm Hg. = $P_T^O x_T + P_{EB}^O x_{EB} + P_w^O$

at 70°C : 113 × 0.4 + 202 × 0.3 + 233 = 338.8

So boiling point is slightly above 70°C. Interpolating linearly at 73°C.,

127.1 × 0.4 + 228.4 × 0.30 + 269.6 = 389 (near to 380)

$$y_T = 68.52/389 = 0.176$$

$$y_{EB} = 50.84/389 = 0.131$$

$$y_w = 269.6/389 = \frac{0.693}{1.000}$$

51. This is steady stae diffusion of A through non-diffusing B. N_A = const. $N_B = 0$

$$N_A = \frac{D_{AB}}{Zx_{BM}} \left(\frac{\rho}{M}\right)_{avg} (x_{A1} - x_{A2}); A \to \text{NaCl}, B - \text{water}$$

$$M_A = 58.5; M_B = 18.02$$

$$x_{A1} = \frac{0.25/58.5}{(0.25/58.5 + 0.75/18.02)} = 0.09313$$

$$x_{A2} = \frac{0.15/58.5}{(0.15/58.5 + 0.85/18.02)} \quad 0.05155$$

At 25 wt. % NaCl Mol. wt. = 1/0.045894 = 21.8 where 0.045894 is (0.25/58.5 + 0.75/18.02).

At 15 wt. % NaCl Mol. wt $= \dfrac{1}{0.15/58.5 + 0.85/18.02}$

$$= 20.1; \left(\frac{P}{M}\right)_{avg.} = \left(\frac{1025}{21.8} + \frac{1012}{20.1}\right)/2$$

$$x_{BM} = (x_{B2} - x_{B1})/\ln(x_{B2}/x_{B1}) = \frac{(0.94845 - 0.90687)}{\ln(0.94845/0.90687)} = 0.9275$$

$$N_A = 1.75 \times 10^{-6} = \frac{D_{AB} \times 48.4(0.09313 \quad 0.05155)}{0.0015 \times 0.9275}$$

$$D_{AB} = 1.215 \times 10^{-9} \text{ m}^2\text{/s.}$$

52. Entering air concentration $= \dfrac{0.05}{(1-0.05)} = 0.05263$

Leaving air concentration $= \dfrac{0.005}{1-0.005} = 0.005025$

Air flow rate $= 0.15 \times 0.95 = 0.1425$ kg mole/hr.

NH_3 lost $= N_A = 0.1425\ (0.05263 - 0.005025) = 0.006784$

$$N_A = K_G A\ (y - y^*)L_m : A = \pi DL = \frac{3.143 \times 15}{100 \times 10} \times \frac{800}{1000} = 0.0377 \text{ m}^2$$

$$y_1 - y_1^* = 0.05263; y_2 - y_2^* = 0.005025$$

$$(y - y^*)L_m = \frac{(y_1 - y_1^*) - (y_2 - y_2^*)}{\ln[(y_1 - y_1^*)/(y_2 - y_2^*)]} = 0.02027$$

So $K_G = 8.873$ kg. mole/(hr. m²)

(*b*)
$$J_H = \left(\frac{h}{C_P G}\right)\text{Pr}^{2/3} = J_D = \frac{Sh}{\text{Re}.Sc} Sc^{2/3} = \frac{K_G d\mu\rho D_{AB}(Sc)^{2/3}}{D_{AB} dG\mu}$$

$$\frac{50}{0.24}[0.74]^{2/3} = k_G\,\rho\,(Sc)^{2/3} = K_G \times 1.4 \times (0.61)^{2/3}$$

$$K_G = 169.3 \text{ kg mole/hr. m}^2$$

53.
$$\ln\left(\frac{w_0}{w_1}\right) = \frac{1}{(\alpha - 1)}\left[\ln\left(\frac{x_0}{x_1}\right) + a\ln\left(\frac{1 - x_1}{1 - x_0}\right)\right]$$

$$\alpha = 2.5; x_0 = 0.70; x_1 = ?$$

$$\ln\left(\frac{100}{66.7}\right) = \frac{1}{1.5}\ln\left(\frac{0.70}{x_1}\right) + 2.5\ln\left(\frac{1 - x_1}{0.30}\right)$$

By trial and error $x_1 = 0.63$; y_V average is calculated by material balance.

$$Vy_V + Lx_L = Fx_F$$

$$33.3\, y_V + 66.7 \times 0.630 = 100 \times 0.70; y_v = 0.8402$$

54. For boiling: $P_A + P_B = P_t = 760$ mm Hg.

where P_A is vapour pressure of toluene in mm Hg and P_B is vapour pressure of water in mm Hg at 80°C; $P_A + P_B = 350 + 280 = 630 \neq 760$; at 85°C; assuming linear interpolation,

$$P_A = 340, P_B = 430; P_A + P_B = 770$$

at 84.5°C, $P_A = 334; P_B = 422; P_A + P_B = 756$

(near to 760). So Boiling point it 84.5°C.

$$w_a/w_b = M_a P_a/(M_b\, P_b) = \frac{92}{18} \times \frac{334}{422} = 4.045$$

where w_a = weight of Toluene and w_b = weight of water in kg.

55. $\alpha = \dfrac{y/(1-y)}{x/(1-x)}$ for binary system

Rearranging $y = \alpha x\,(1-y)/(1-x)$, collecting y terms and x terms gives, $y = \dfrac{\alpha x}{1+(\alpha-1)x}$.

56. Using Fenske-Underwood equation:

$$\frac{x_D}{1-x_D} = a_{avg}^{N+1}\left(\frac{x_w}{1-x_w}\right); \text{ here } x_D = 0.95;\ x_w = 0.05$$

$$N+1 = \log\left[\left(\frac{x_D}{1-x_D}\right)\frac{(1-x_w)}{x_w}\right] / \log \alpha_{avg}$$

$$N+1 = \log\left(\frac{0.95}{0.05} * \frac{0.95}{0.05}\right) / \log(2.5) = 5.43$$

57. Ans. A – b; B – isostere; C – isochore

58. Ans. (c)

59. Ans. (a)

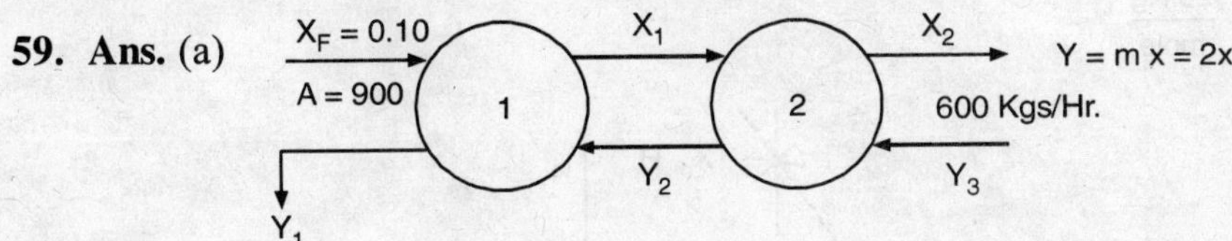

60. Ans.

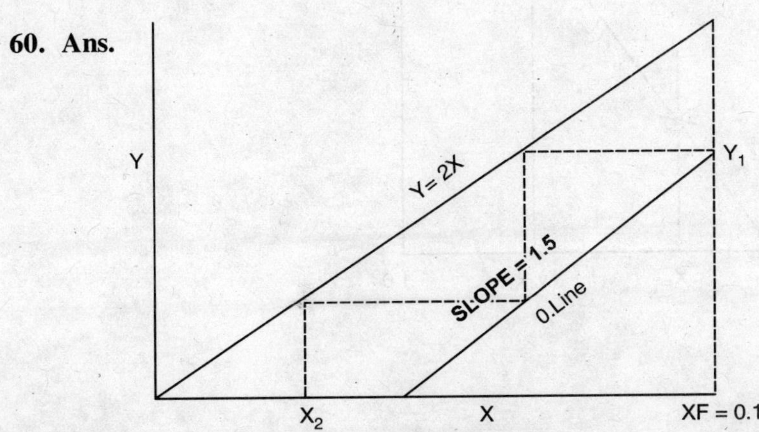

$$\text{Solute in feed} = 990 \times \frac{0.1}{1.1} = 90 \text{ kg}; A = 990 - 90 = 900 \text{ kg}.$$

Operating line point X_F, $Y_1 = 0.10$, Y_1; X_2, $Y_3 = X_2$, 0

Slope of operating line: $\dfrac{A}{B}$

By overall balance: $900\,(X_2 - 0.1) = 600\,(Y_1 - 0)$

$$\frac{A}{B} = \frac{900}{600} = 1.5$$

0.L is moved up and down so that two stages are obtained giving Y_1 and X_2.

61. Ans.

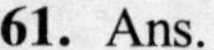

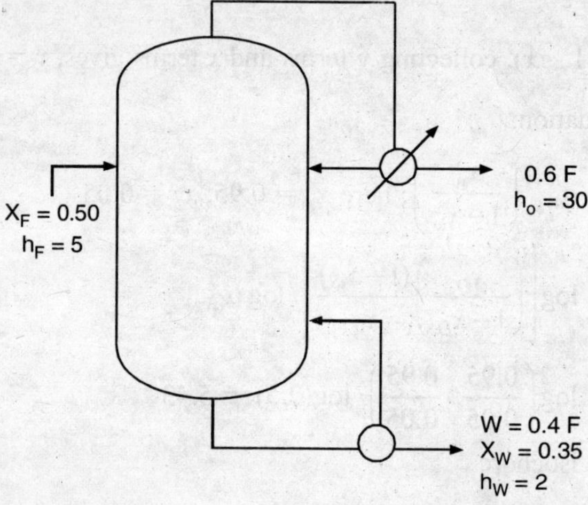

(a) $13.8F/0.60F = \dfrac{23 \text{ Kg}}{\text{mole}}$

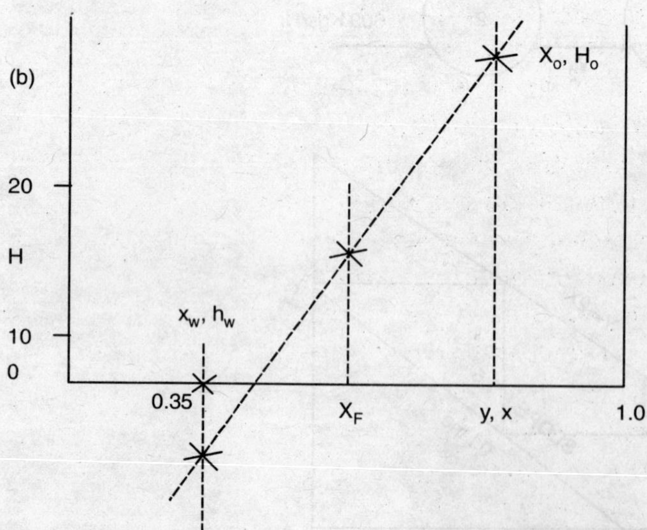

By enthalpy balance $F \times h_F + Q = 0.6\ F \times 30 + 0.4\ F \times 2$

$$F \times 5 + Q = 18F + 0.8\ F$$

$$Q = 13.8\ F$$

By component balance $F \times 0.5 = 0.6\ F \times x_D + 0.4\ F * 0.35$

$$x_D = 0.60$$

62. $X_1 = 25/75 = 0.333;\ X_2 = 6/94 = 0.0638$

$$N_C = 3 \times 10^{-4}\ \text{kg/(m}^2\ .\ \text{s)};\ X_c = 0.20;\ X^* = 0.01$$

$$\frac{A}{L_S} = 1/30;\ L_S = 160 \times 75/100 = 120\ \text{kgs};\ A = 120 \times \frac{1}{30} = 4\ \text{m}^2$$

$$\theta = \theta_C + \theta_f$$

$$= \frac{L_S}{AN_C}\left[(X_1 - X_C) + (X_C - X^*)\ln\left(\frac{X_C - X^*}{X_2 - X^*}\right)\right]$$

$$= \frac{120}{4 \times 3 \times 10^{-4}}\left[(0.333 - 0.20) + (0.2 - 0.01\ln\left(\frac{0.2 - 0.01}{0.0638 - 0.01}\right)\right]$$

$$= 37273\ \text{seconds} = 10.35\ \text{hours.}$$

63. (a)
64. (a)
65. (b)
66. (b)

67. (c) $q = \frac{H_G - H_f}{H_G - H_L}$; here $H_f = 0.4\ H_G + 0.6\ H_L$

So, $q = 0.6$; slope of q line is $\frac{q}{q-1} = -\ 1.5$

68. (a)
69. (a)
70. $p_B = 1530 \times 0.50 = 765;\ p_T = 640 \times 0.5 = 320$

Total pressure = 765 + 320 = 1085 mms. Hg

71. $\theta = \frac{-L_S}{A}\int_{0.66}^{0.30}\frac{dX}{0.00045} - \frac{L_S}{A}\int_{0.30}^{0.25}\frac{dX}{0.0015X}$

$$= \frac{1}{0.025}\left[\frac{0.36}{0.0045} + \frac{1}{0.0015}\ln\frac{0.30}{0.25}\right]$$

$$= \frac{1}{0.025}[800 + 666.7 * 0.1823] = 36861.5\ S$$

= 10.24 hrs

72. (a) Intercept $= \dfrac{x_D}{R+1} = \dfrac{0.8}{3} = 0.27$

$$\text{Slope} = \frac{0.80-0.27}{0.80-0} = 0.66$$

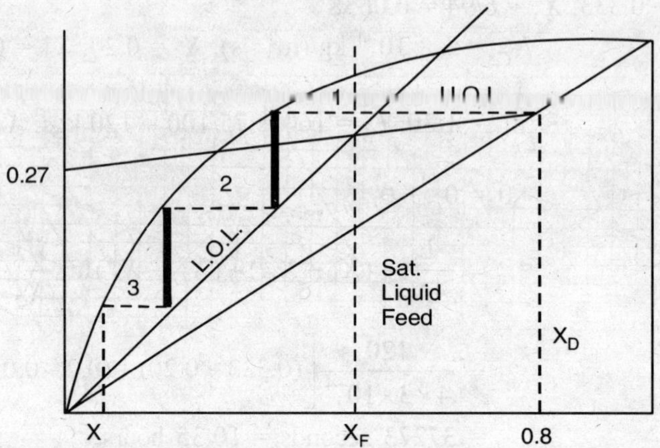

There are total 3 stages 1 (partial condenser + 1 (plate) + 1 (Reboiler).

Assuming Sat. liquid feed.

x_w = bottom product composition. By changing lower operating line a total of 3 stages are drawn/obtained by trials, giving x_w.

73. (c)

74. (c)

75. (c)

76. $\ln(F/w) = \int_{x_w}^{x_f} \dfrac{dx}{(y-x)}$

$$\ln(100/w) = \int_{0.32}^{0.52} \frac{dx}{(1.05x+0.13-x)} + \int_{0.52}^{0.74} \frac{dx}{(0.77x+0.28-x)}$$

$$= \frac{\ln}{0.05}(0.05x+0.13)\Big|_{0.32}^{0.52} + \frac{1}{0.23}\ln(0.28-0.23x)_{0.52}^{0.74}\Big| = 3.7$$

$$W = 0.67 \text{ moles; } D = 100 - 0.67 = 99.33 \text{ moles}$$

By solute balance:

$$100 \times 0.74 = 0.67 \times 0.32 + 99.33\, x_D$$

$$x_D = 0.743 \text{ Ans.}$$

77. $N_A = K_L\,(y^*-y)$; $N_A/(y^*-y) = 1$; So : $K_L = 1$

From this t^* is calculated.

HUMIDITY CHART

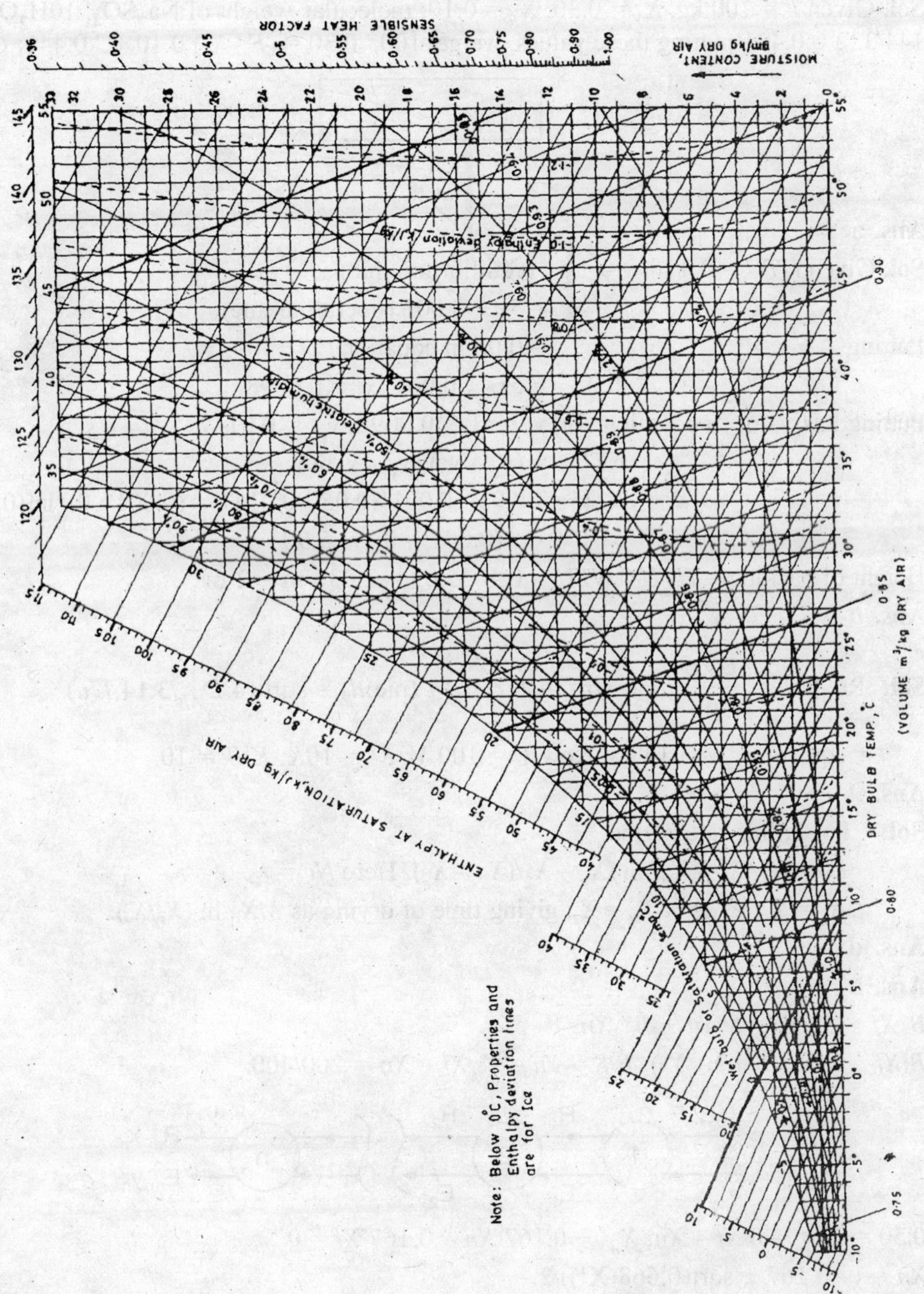

SECTION—E

1A. Ans. a

1B. Ans. d

Sol. Given $F = 100$ kg, $X_Y = 0.30$; $X_F = 0.10$; molecular weight of Na_2SO_4, $10H_2O = 322$; $Xc = 144/322 = 0.441$ putting these values, we get $100^* \; 0.30 = (F - C)\; 0.10 + C\; 0.441$; $C = 58.65$ kg.

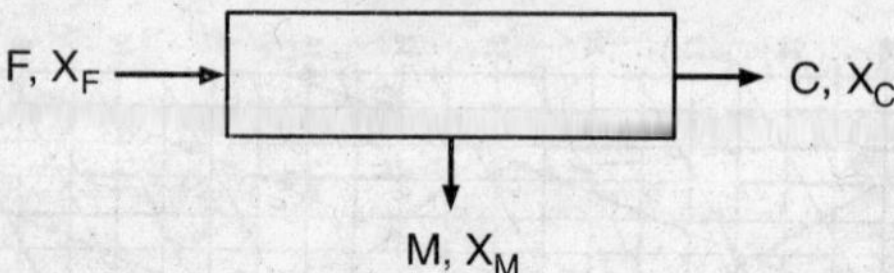

2. Ans. a

Sol. Given $HTU = 0.4$ m; $y = 1.5x$ (equilibrium line)

$$y = 5x + 0.001 \text{ (operating line)} \quad \ldots (2)$$

Putting $y_1 = 0.05$ in equation (2)

$$0.05 = 5xi + 0.001;\; xi = 0.0098$$

Putting this value in equation (1), $y_{e2} = 1.5{*}0.0098$; $y_{e2} = 0.0147$

$$NTU = (y_1 - y_2)/(y_1 - y_2)\text{lm}$$

$$NTU = 0.05 - 0.001/(0.05 - 0.0147) - (0.001 - 0)/\ln[(0.05 - 0.0147)/(0.001 - 0)] = 5.90$$

Height of packing = $HTU * NTU = 0.40{*}5.09 = 2.036$ m = 2 m

3. Ans. *b*

Sol. Re. No. = $\dfrac{DV\rho}{\mu}$; $k = \text{sqrt}\,(4\,D_{AB}/3.14\,(\text{pie})tl) = \text{sqrt}\,(4\,D_{AB}/3.14\,L/u)$

$= \text{sqrt}\,(4\,D_{AB}\,u/3.14\,L)$; Now $U = 100\,V$; $K^1 = 10K$; $K^1/k = 10$.

4. Ans. *d*

Sol. For falling rate period:

$Ls/ANC\,(X_c - X_e)\ln(X_c - X_2/(X_2 - X_e)$. Here $Nc = r_c$.

$X_c = X_0$, $X_e = 0$; $X_2 = X_f$; giving time of drying as $MX_0 \ln(X_0/X_f)$.

5. Ans. d

6. Ans. b

Sol. $B{*}Xf + S{*}\,Ys = S{*}Yn + B{*}\,Xn$.

$B(Xf - Xn) = S(Yn - Ys)$; $B/S = Yn - Ys/Xf - Xn = 200/400$.

B, F R₁ R₂
X_F 1 2 n R_n, x_n
S E₁ E₂ E_n, y_n

$0.50 = 3X^2_n - 0/Xf - Xn$; $X_n^{\,2} + 0.167\,Xn - 0.167\,Xf = 0$

$Xn = (-\,0.167 \pm \text{sqrt}(0.668\ \text{Xf}))/2$.

$Xf = 0$; $Xn = 0$;

$Xf = 0.10$; $Xn = 0.0457$; $(Xf - Xn)/XF = 0.543$

$Xf = 0.30$; $Xn = 0.1403$; $(XF - Xn)/XF = 0.532$

7. Ans.c

(a) P-1, Q-3, R-4, S-6

(b) P-2, Q-5, R-1, S-3

(c) P-5, Q-3, R-6, S-2

(d) P-5, Q-3, R-4, S-2

8. Ans. b

Sol. Reflux $= R = L_0/D$; $R + 1 = (L_0 + D)/D$. Mass balance around condenser (top one) is: $8000 = L_0 + D$; $R + 1 = 8000/D$.

Mass balance around column $= F = D + W$.

$FXf = DXD + WXw = DXD + (F - D)\ Xw$

$F(Xf - Xw) = D(Xd - Xw)$; $D = F(Xf - Xw)/\ (XD - Xw)$.

$= 10{,}000\ (0.50 - 0.05)/(0.95 - 0.05) = 5000$

$R + 1 = 8000/5000 = 1.60$; $R = 0.60$

9. Ans. c

Sol. $Y_{in} = 0.05$; For 98% Recovery of acetone, $Y_{out} = 0.05 - 0.98 * 0.05 = 0.001$

Number of trays by equation is given by: $N + 2 = 6 \log (Y_{in}/Y_{out})$

$N + 2 = 6 \log (0.05/0.001) = 6 \log 50 = 10.19$.

$N = 8.19$; This is more than 8 trays. So $N = 9$ are required.

10. Ans. c

Sol. The mixture will exert a pressure of 1 atmosphere ie: at that point the system will boil. Now boiling point of equimolar mixture is given by $½*100 + ½*110.6 = Tbp = 105.3°$ C

i.e: $100 < Tb.p < 105.3$

11. Ans. c

Sol. Since the boiling point of water is 100°C and that of Toluence is : 110.6°C, hence the mole fraction of water is greater than 0.50 in vapor. Hence $0.50 < Yw < 1.0$.

12. Ans. d

Sol. In condition (*i*) the reflux is liquid at its bubble point and in condition (*ii*) the reflux is liquid below its bubble point. So more condenser load is required in condition (*ii*).

When the reflux is liquid below its bubble point, the temperature of the tower decreases, requiring more reboiler load.

Hence both condenser and reboiler load are more in condition (*ii*) compared to condition (*i*).

13. Ans. c

Sol. $C^* = 0.0667$ m, $Ls = 100\ m^3$; $S_s = 10$ kgs; $Co = 16/\text{ARC} = 0.022\ kg/m^3$; $C1^* = 0.0667$ m; $m_0 = 0$

$Ls\ (Co - C1) = Ss\ (m1 - mo)$

$100\ (0.022 - 0.0667\ m1) = 10\ m1 = 0.131$ kg water kg absorbent

Kg water absorbed = 0.131*10 = 1.31 kgs

Kg water remaining = 2.2 – 1.31 = 0.89

Fraction of water remaining = 0.89/2.2 = 0.40

14. Ans. a

Sol. In feed kg of C/kg A = 2/8 = 1/4 = X.

Y^* = kg C/kg B; kg C/kg B = 0.225

0.225 kg of C required 1 kg B

2 kg of C requires ? = 2/0.225* 1 = 8.89 kg B

Solvent B, Provided = 10 kg

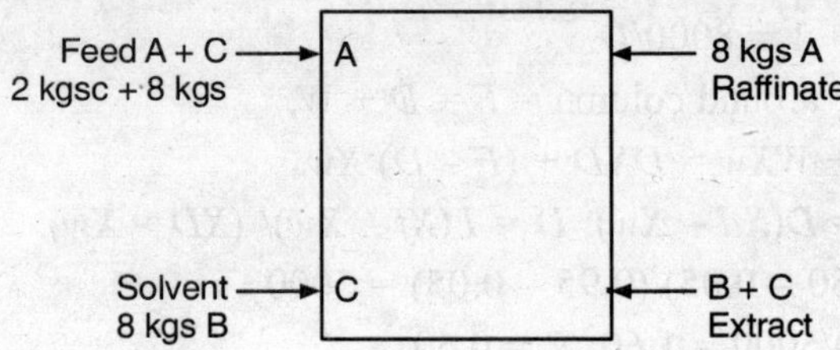

So all of C will be transferred to solvent B.

15. Ans. a

Sol. Equations for constant drying conditions is: $\theta c = Ls/\text{Arc}\ (X_1 - Xc) = 1500 = 2*1.2/\text{Arc}\ (X_1 - Xc)$ (1)

And for next material = t = 1.6/ Arc($X1 - Xc$) ... (2)

Eq. (1)/Eq. (2) = 1500/t = 2*1.2/1.6; t = 1000 seconds.

16. Ans. d

Sol. As reflux ratio is increased for the same number of plates in the column, the compositions of top and bottom products increase.

17. Ans. a

Sol. This is open steam distillation where one component is water (methanol + water). Less number of trays are enough. Reboiler is considered as an extra tray.

18. Ans. b

19. Ans. a

20A. Ans. a

Sol. $F + B = R + E$ ...(1)

$F*FM = MB*B$; $F/B = MB/FM$ = 5.4/3.1 = 1.74

Using Eqn. (1): 1.74 B + B = R + E; 2.74 = 1.2 + R; R = 1.54

So, R/E = 1.54/1.2 = 1.285 Ans.

20B. Ans. a

Sol. $q = Hg - H/(Hg - HL)$; Here $Hf = 0.8\ HG + 0.2\ HL$

So q = 0.2; Slope of q line: $q/q - 1 = -1/4$.

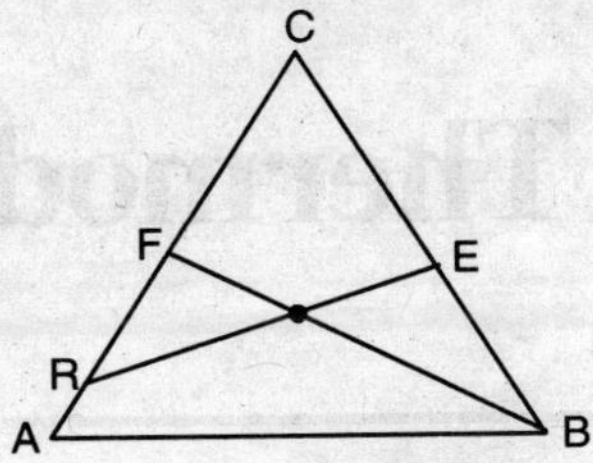

21. Ans. b

Sol. V.P. of benzene = 156; V.P. of Toluene = 63

$Pt = 101 = 156 * Xb + 63\ (1 - Xb)$

$38 = 93\ Xb$; $Xb = 38/93 = 0.410$.

22. Ans: b

23. Ans. c

24. Ans. d

25. Ans. c

❑

CHAPTER 8

Thermodynamics

1. First law of thermodynamics: Although energy assumes many forms,the total quantiy of energy is constant, and when energy disappears in one form it appears simulaneously in other forms, $dE = dQ + dW$; or $E = Q + W$; where E is internal energy, Q = heat added, W = work done. The thermodynamic properties are: Temperatue,pressure,and density.
2. Equilibrium denotes a static conditiont: In thermodynamics it means not only the absence of change but the absence of any tendency towards a change on a macroscopic scale.
3. Phase rule: This is $F = C - P + 2$; F = degrees of freedom, C = number of components, P = number of phases.
4. Reversible process: A process is reversible when its direction can be reversed at any point by an infinitesimal change in external conditions.
5. Constant pressure process: $dQ = d(nU) + Pd(nV)$ and $H = U + PV$
6. Heat capacities: $C_V = (dQ/dT)V = (dU/dT)V$; $C_P = (dQ/dT)P = (dH/dT)P$; $C = dQ/dT$
7. Mass balance at steady state: $m = u_1 A_1/V_1 = u_2 A_2/V_2 = u A/V$ where u = velocity, m/s, A = area, sq.m.; V = specific volume, cubic m/kg; m = kg/second.
8. Energy balance at steady state: $\Delta H + \Delta u^2/(2gc) + g/gc\ \Delta Z = Q + Ws$
9. $Cp = Cv + R$; $PV = RT$ (ideal gas)

   For ideal gases: isothermal: $Q = RT \ln(P_2/P_1)$; Isobaric process: $Q = Cp\ dT$; Isochoric process (constant volume): $Q = Cv\ dT$.
10. Adiabatic process: $r : Cp/Cv$; $W = RT_1 /(r - 1) \{(P_2/P_1) - 1)$; Poytropic process: PV^γ = constant; Work done = $RT_1/(\gamma - 1)\{(P_2/P_1)^{\gamma - 1/\gamma} - 1\}$
11. Van der Waal's equation of state: $P = RT/(V - b) - a/(V^*V)$; Virial equation of state: $Z = PV/(RT) = 1 + B/V + C/(V^*V)$

    Theorem of corresponding states: All fluids when compared at the same reduced pressure and reduced temperature have approximately the same compressibility factor,and all deviate from the ideal gas behaviour to about the same degree. Pitzer correlation for compressibility factor ($Z = PV/RT$): $Z = Z^0 + wZ^1$
12. **Second law of thermodynamics:**

    Statement - 1: No apparatus can operate in such a way that its only effect is to convert heat absorbed by a system completely into work done by the system.

    Statement - 2: No process is possible which consists solely in the transfer of heat from one temperature level to a higher one. Carnot cycle efficiency: $\eta = 1 - Tc/Th$.

13. Entropy: There exists a property called entropy which is an instrinsic property of a system, functionally related to the measurable coordinates which characterize the system. For a reversible process, $dS = dQ$ rev./T

14. Third law of thermodynamics: Absolute entropy is zero for all perfect crystaine substances at absolute zero temperature.

15. Thermodynamic relations: $H = U + PV$; $A = U - TS$; $G = H - TS$; $dU = TdS - P\,dv$; $dH = TdS + vdP$; $dA = PdV - SdT$; $dG = vdP - SdT$; $(dT/dV)S = (dP/dS)V$; $(dT/dP)S = (dV/dS)P$; $(dP/dT)V = (dS/dV)T$; $(dV/dT)P = -(dS/dP)T$.

16. Woller diagram: The most common diagrams are: temperature/entropy; Pressure/enthalpy; and enthalpy/entropy (Mollier diagram).

17. Rankine cycle, Carnot cycle is a reversible cycle. Actual operations are not reversible. These are represented by Rankine cycle.

18. Otto engine/cycle, diesel engine/cycle come in internal combustion engine.

19. Refrigeration: This implies the maintenance of a temperature below that of the surroundings. This requires continuous absorption of heat at a low temperature level, usualy accompanied by evaporation of liquid in a steady flow process. The vapor formed may be returned to its original liquid. In a steady flow process .The vapor formed may be returned to its original liquid state for reevaporation in either of two ways. Most commonly it is simply compressed and then condensed. Alternately it may be absorbed by a liquid of low volatility, from which it is subsequenty evaporated at higher pressure.

Coefficient of performance = Heat absorbed at low temp/Net work (W); $w = !Qc!/W = TC/(TH - TC)$

Vapor compression cycle: A liquid evaporating at constant pressure (1——2) provides a means of heat absorption at a low constant temperature.The vapor produced is compressed to a higher pressure and is then cooled and condensed with rejection of heat at a higher temperature level. Liquid from the condenser- returns to its original pressure by an expansion process. In principle this is carried out in an expander, from which the work is obtained. In practice this is accomplished by throttling through a partly open valve.

Choice of refrigerant: Toxicity, flammability, cost, corrosion properties and vapor pressure. Main one is vapor pressure of refrigerant. The refrigerants are: Ammonia, methyl chloride, carbon dioxide, propane and other hydrocarbons. Recent ones are: Hydrogenated chlorofluorocarbons ($C\,Cl_3$ F- trichlorofluoromethane, $C\,Cl_2\,F_2$-dichlorodifluoromethane)

Heat pump: This is a reversible heat engine for heating houses and commercial buildings during winter and cooling them during the summer.

20. Liquefaction of gases is done by: (1) Heat exchange at constant pressure; (2) Expansion process; (3) Throttling process.

21. Models for vapor-liquid equilibrium: (a) Raoult's law (b) Henry's law (c) Modified Raoult's law (using activity coefficients) (d) K-value correlation ($Ki = yi/xi$).

22. Solution thermodynamics: $S = (dG/dT)P, x$; $V = (dG/dP)T, x$; $H = G + TS$; At equilibrium for phases $ui^{\alpha} = ui^{\beta} = ... = ui^{\pi}$.

Margulus equations: $!n\,\gamma_1^{\infty} = A12$ ($xl = 0$); Ln $\gamma_2^{\infty} = A21$ ($x_2 = 0$); $fl = y_1\,P$ (f is fugacity, y is vapor composition in mole fracion, P is total pressure); $f_2 = y_2P$.

Gibbs–Duhem equation for binary systems is: $x_1\, d \ln \gamma_1/dx_1 + x_2\, d \ln \gamma_2/dx_2 = 0$ (constant T.P). Enthalpy of a solution: $H : x_1 H_1 + x_2 H_2 + \Delta H$; where ΔH is enthalpy of mixing; Heat of solution: when solids or gases are dissolved in liquids, the heat effect is called the heat of solution, and is based on dissolution of one mole of solute.

23. Chemical reacion equilibria: $K = \text{Exp.}(-\Delta G/RT)$; $d \ln K/dT = \Delta H/R\, T$; Or Ln $(K/K1) = \Delta H/R\, (1/T - 1/T1)$

$$\Delta H° = \Delta H°0 + R\int \Delta C°p / R\, dT;\ \Delta S° = \Delta S°0 + R\int_{T_0}^{T} \Delta Cp / R\, dT;\ \Delta S°0 = (\Delta H°0 - \Delta G°0)/T0$$

24. Fuel cells: is a device in which a fuel is oxidized elecrochemically to produce electric power. It has the characteristic of a battery in that it consists of two elecrodes separated by an electrolyte. The reactants are not stored in cell but are fed continuously and the products are withdrawn coninuously, e.g. Fuel—Hydrogen–oxygen

Hydrogen, methane, butane, methanol as anode and oxygen or air as cathode.

25. Equilibrium: The equilibrium state of a closed system is that state to which the total Gibbs energy is a minimum with respect to all possible changes at the given temperature and pressure. At constant temperature and pressure, ΔG and its first and second derivatives must be coninuous functions of $x1$, and the second derivative must be everywhere positive.

26. Virial equation of state: $Z = 1 + B'p + C'P^2 + D'P^3 + ...$

$$Z = 1 + \frac{B}{V} + \frac{C}{V^2} + \frac{D}{V^3} + ...$$

27. Entropy change of an ideal gas: $\dfrac{\Delta S}{R} = \displaystyle\int_{T0}^{T} \frac{C_p^{ig}}{R}\frac{dT}{T} - \ln\left(\frac{P}{P_0}\right)$

28. Uses of refrigeration: 1. Air conditioning; 2. Preservation of foods; 3. Gas liquefaction; 4. Ice manufacture and 5. Hydration of gases.

29. Methods of refrigeration: 1. Vapor compression; 2. Vapor absorption.

30. Coefficient of performance: $w = \dfrac{W}{(Q_e)} = \dfrac{(TH - TC)}{\text{Pressure}}$

31. The various refrigerants used are: 1. Ammonia; 2. Methyl chloride; 3. Carbon dioxide; 4. Propane; 5. Hydrocarbons; 6. Halogeneted hydrocarbons.

32. Ozone depletion is caused by chlorine compounds.

33. Liquefaction (of gases) methods: 1. By heat exchange at constant pressure; 2. Expansion process; 3. Throttling.

34. Gas liquefaction processes: 1. Linde liquefaction; 2. Claude liquefaction process.

35. Internal energy of ideal gases (microscopic view):

$U = KT^2\left(\dfrac{\partial \ln Z}{\partial T}\right)_V$; $Z = \Sigma g_i e^{-Ei/(kT)}$ where E_i is energy level; g_i is degeneracy of the level; k is Boltzmann constant; T is temperature; V = specific volme; U = internal energy.

36. Thermodynamic properties and statistical mechanics:
$A = -kT \ln Z$; A = Helmholtz energy.

37. Lattice model: Since the liquid state is in some sense intermediate between the crystalline state and the gaseous state, so there are two types of approaches. The first considers liquids to be gas like; a liquid is pictured to be dense and highly non-ideal gas whose properties can be described by some equation of state, e.g. van der Waals is the best known example. An equation of state description of pure liquids can be extended to liquid mixtures.

The 2nd approach is a liquid to be solid like, in a Quasi-Crystlline state, where the molecule do not translate fully in a chaotic manner as in a gas, but where each molecule tends to stay in a small region, a more or less fixed position in space about which it vibrates back and forth.

38. Excess Gibbs free energy (solution thermodynamics):

$$\frac{d(G^E/RT)}{dx_1} = \ln\left(\frac{\gamma_1}{\gamma_2}\right); \quad \underset{x_1 \to 0}{Lt}\left[\frac{G^E}{x_1 x_2 RT}\right] = \ln \gamma_1^{\infty}; \quad \underset{x_1 \to 1}{Lt}\left[\frac{G^E}{x_1 x_2 RT}\right] = \ln \gamma_2^{\infty}$$

QUESTIONS

SECTION—A

1. Thermodynamics deals with:
(a) Calculation of rates of chemical processes
(b) Mechanism of reactions
(c) Transformation of energy from one kind to another
(d) None of these

2. A vessel containing a volatile liquid in contact with its vapours constitutes a:
(a) Open system (b) Closed system
(c) Isolated system (d) None of these

3. Which of the following are extensive properties?
(a) Density (b) Viscosity
(c) Free energy (d) Entropy
(e) Refractive index

4. Which are the intensive properties among the following?
(a) Surface tension (b) Chemical potential
(c) Heat capacity (d) Entropy

5. Which of the following statements are true?
(a) Heat and work are state functions
(b) All the natural processes are irreversible
(c) In a closed system the energy and matter cannot be exchanged with the surroundings
(d) Enthalpy and entropy are state functions

6. The coefficient of thermal expansion is given by:

(a) $\frac{1}{V}\left(\frac{\partial V}{\partial T}\right)_P$ (b) $-\frac{1}{V}\left(\frac{\partial T}{\partial P}\right)_T$

(c) $\left(\frac{\partial V}{\partial T}\right)_H$ (d) none of these

7. Joule-Thompson coefficient is given by:

(a) $\left(\frac{\partial H}{\partial T}\right)_T$ (b) $\left(\frac{\partial T}{\partial P}\right)_H$

(c) $\left(\frac{\partial P}{\partial T}\right)_H$ (d) None of these

8. Mark the reversible processes in the following:

(a) Precipitation of CdS by H_2S from Cd NO_3 solution

(b) An ideal gas expands against constant pressure

(c) Vaporization of water at 100°C and 1 atm pressure

(d) Solution of $BaNO_3$ in water at 40°C

9. Work and heat are examples of:

(a) Point functions (b) Path functions

(c) State fucntions (d) None of these

10. Joule-Thomson experiment is:

(a) Isothermal and isobaric (b) Isobaric and isenthalpic

(c) Isenthalpic and adiabic (d) Isothermal and isentropic

11. The specific heat at constant pressure (C_P) is given by:

(a) $\left(\frac{\partial E}{\partial V}\right)_P$ (b) $\left(\frac{\partial V}{\partial T}\right)_P$

(c) $\left(\frac{\partial H}{\partial T}\right)_P$ (d) $\left(\frac{\partial H}{\partial V}\right)_P$

12. The specific heat at constant voluem (C_V) is defined by:

(a) $\left(\frac{\partial E}{\partial P}\right)_V$ (b) $\left(\frac{\partial H}{\partial T}\right)_V$

(c) $\left(\frac{\partial E}{\partial T}\right)_V$ (d) None of these

13. The expansion of a gas into vacuum is:

(a) Reversible process (b) Irreversible process

(c) None of these

14. The pressure and temperature relationship in an adiabatic system is represented by:

(a) PV = constant (b) PV^n = constant

(c) PV^Y = constant (d) None of these

15. Mark the correct statement:

(a) Reversible work is less than irreversible work in an isothermal system

(b) Isothermal work is more than the aidabatic work under reversible conditions

(c) The heat exchange measured at constant pressure is equal to the increase in internal energy

16. Heat absorbed is equal to internal energy for:

(a) Constant pressure process (b) Constant volume process

(c) Adiabatic process (d) Polytropic process

17. A polytropic process is described by:

(a) PV = constant (b) PV^{Y} = constant

(c) PV^{n} = constant (d) none of these

18. The change in enthalpy when HCl gas is dissolved in wated is termed as:

(a) Heat of formation (b) Heat of solution

(c) Heat of dilution (d) None of these

19. The direction of a particular spontaneous process is ascertained from the law of thermodynamics:

(a) Zeroth law (b) First law

(c) Second law (d) Third law

20. The value of R in cal/g, mole °K is:

(a) 8.314 (b) 82.05

(c) 0.7302 (d) 1.987

21. An isometric process is carried out under:

(a) Constant volume (b) Constant temperature

(c) Constant pressure (d) None of these

22. The units of entropy of a system are:

(a) cal (b) cal/°K

(c) cal/°C (d) None of these

23. For an irreversible isolated system ΔS:

(a) $= 0$ (b) < 0

(c) > 0 (d) None of these

24. The entropy change for one mole liquid ethyl chloride (B.P. = 12°C) if its heat of vaporization is 90 cal/g:

(a) 10 (b) 20.33

(c) 15 (d) – 20.33

(e) 0

25. For a reversible adiabatic process ΔS is:

(a) < 0 (b) > 0

(c) $= 0$ (d) None of these

26. For expansion of one mole of ideal gas from V_1 to V_2, Δ_S:

(a) $\ln\left(\frac{V_1}{V_2}\right)$ (b) $R \ln\left(\frac{V_1}{V_2}\right)$

(c) $R\left(\frac{V_1}{V_2}\right)$ (d) $R \ln\left(\frac{V_2}{V_1}\right)$

27. For a spontaneous process, ΔG is:

(a) Negative (b) Zero

(c) Positive (d) None of these

28. At equilibrium Gibbs free energy, at constant T and P is:

(a) Maximum (b) Zero

(c) Minimum (d) None of these

29. ΔS for an isolated system is:

(a) $\frac{\Delta A}{T}$ (b) $-\frac{\Delta A}{T}$

(c) $\frac{\Delta G}{T}$ (d) $-\frac{\Delta G}{T}$

30. In an irreversible isolated process ΔS:

(a) $= 0$ (b) < 0

(c) > 0 (d) None of these

31. For vaporization process of an ideal gas if molar volume of liquid is negligible compared to that of vapour, Clausius-Clapeyron equation is:

(a) $\frac{d \ln P}{dT} = \frac{RT}{H_{vap}}$ (b) $\frac{d \ln P}{dT} = \frac{H_{vap}}{RT^2}$

(c) $\frac{dP}{dT} = \frac{H_{vap}}{RT^2}$ (d) $\frac{dP}{dT} = \frac{H_{vap}}{R \ln T}$

32. Molar entropy of vaporization at normal boiling point in J/°K mole is:

(a) 80 (b) 90

(c) 200 (d) 92

33. Mark the correct statement (subscript : *s*-solid *l*-liquid, *g*-gas):

(a) $S_s > S_l < S_g$ (b) $S_s > S_l > S_g$

(c) $S_s < S_l > S_g$ (d) $S_s < S_l < S_g$

34. Gibbs-Duhem equation for an isothermal and isobaric process for a binary system is:

(a) $n_1\mu_1 + n_1\mu_2 = 0$ (b) $n_2\mu_2 + n_2\mu_2 = 0$

(c) $n_2 d\mu_2 + n_2 d\mu_1 = 0$ (d) $n_1 d\mu_1 + n_2 d\mu_2 = 0$

35. A partial molar property ($\bar{J}$) is defined as:

(a) $\left(\frac{\partial J}{\partial ni}\right)_{V,P,nj}$ (b) $\left(\frac{\partial J}{\partial ni}\right)_{T,P,nj}$

(c) $\left(\frac{\partial J}{\partial nj}\right)_{T,nj}$ (d) None of these

36. For a *n*-component system *G* at constant *T* and *P* is:

(a) $\sum \frac{n_i}{\mu_i}$ (b) $\sum \frac{\mu_i}{n_i}$

(c) $\Sigma ni\ \mu i$ (d) None of these

37. For transfer of a component *A* from phase *P* to phase *Q* is:

(a) $\mu_P < \mu Q$ (b) $\mu_P = \mu Q$

(c) $\mu_P > \mu Q$ (d) None of these

38. The chemical potential for an ideal gas is:

(a) $Rd \ln P$ (b) $RTd \ln P$

(c) $\frac{R}{T} d \ln P$ (d) $RPd \ln T$

39. Activity is defined as:

(a) $\ln \frac{f_1}{f_i^0}$ (b) $\ln \frac{f_i^0}{f}$

(c) $\frac{f_1}{f_i^0}$ (d) None of these

40. Units of fugacity are:

(a) atm./mole (b) atm.

(c) atm./°K (d) None of these

41. The chemical potential for real gas is expressed as:

(a) $Rd \ln f$ (b) $RTd \ln f$

(c) $\frac{R}{T} d \ln f$ (d) $RPd \ln T$

42. Gibbs phase rule is given as (for a non-reacting system):

(a) $P = C + F + 2$ (b) $P + 2 = C + F$

(c) $P + C = F + 2$ (d) $P + F = C + 2$

43. At triple point the number of degrees of freedom is:

(a) 2 (b) 1

(c) 0 (d) None of these

44. If *R* is the number of distinct chemical reactions that the components undergo, Gibbs phase rule is:

(a) $P + F = C + 2 = R$ (b) $P + F = C + 2 - R$

(c) $P + F = C + R - 2$ (d) $P - F = C + R + 2$

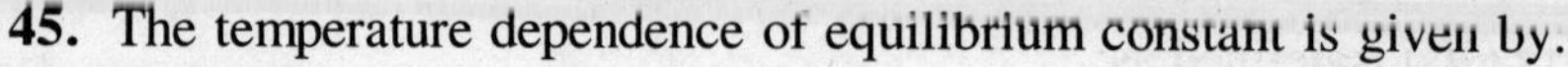

45. The temperature dependence of equilibrium constant is given by:

(a) $\frac{dK}{dT} = \frac{\Delta H}{RT^2}$ (b) $\frac{d \ln K}{dT} = \frac{1}{RT^2}$

(c) $\frac{d \ln K}{dT} = \frac{\Delta H}{RT^2}$ (d) None of these

46. The free energy change (ΔG) and equilibrium constant (K) for a chemical reaction are related by:

(a) $\Delta G - RT \ln K$ (b) $\Delta G = -T \ln K$

(c) $\Delta G = -RTK$ (d) $\Delta G = -RT \ln K$

47. The number of degrees of freedom for two miscible non-reacting species in vapour liquid equilibrium forming as azeotrope:

(a) 2 (b) 0

(c) 1 (d) 3

48. The number of degree of freedom for a gaseous system consisting of CO, CO_2, H_2, H_2O and CH_4 in a chemical reaction (no. distinct reactions = 2).

(a) 1 (b) 2

(c) 3 (d) 4

(e) 0

49. For a reaction $A(g) + B(g) \rightarrow 2C(g)$ the equilibrium constant K_p is:

(a) $RT\frac{P_c^2}{p_a p_b}$ (b) $RT \ln\frac{Pc^2}{p_a p_b}$

(c) $-RT \ln\frac{Pc^2}{p_a p_b}$ (d) None of these

50. The equilibrium constant at two different temperatures is given by:

(a) $\frac{Kp_2}{Kp_1} = \frac{\Delta H}{R}\left[\frac{1}{T_1} - \frac{1}{T_2}\right]$ (b) $\frac{Kp_2}{Kp_1} = \Delta H\left[\frac{1}{T_2} - \frac{1}{T_1}\right]$

(c) $Kp_2 = \frac{1}{R}\left[\frac{1}{T_1} - \frac{1}{T_2}\right]$ (d) None of these

51. In a Carnot cycle if heat is absorbed at temperature T_2 (low) and heat is rejected at T_1 (high) the coefficient of performance is (COP) is:

(a) $\frac{T_1}{T_2}$ (b) $\frac{T_2}{T_1}$

(c) $\frac{T_1}{T_1 - T_2}$ (d) $\frac{T_1 - T_2}{T_1}$

(e) $\frac{T_1 - T_2}{T_2}$

52. If in absorption refrigeration, heat is taken from a source at T_1 and the surrounding temperature is T_2 and provides refrigeration at T_R, the maximum COP is:

(a) $\frac{T_R}{T_2 - T_R} \times \frac{T_1}{T_1 - T_2}$ (b) $\frac{T_R}{T_2 - T_R} \times \frac{T_1 - T_2}{T_1}$

(c) $\frac{T_R}{T_2 - T_R} \times \frac{T_2 - T_2}{T_1}$ (d) None of these

53. For a reversible process at constant pressure and temperature the free energy change is:

(a) Greater than zero (b) Less than zero

(c) Equal to zero

54. Viral equation of state is:

(a) $PV = RT$ (b) $PV = A + B/V = C/V^2 + ...$

(c) $(P + a/V^2)(V - b) = RT$

55. Avogadro's number in molecules/gm mole is:

(a) 6.024×10^{23} (b) 6.024

(c) 1.987 (d) 82.06

SECTION—B

1. Write the missing terms.

(a) $dE = dq - ?$ (b) $G = H - ?S$

(c) $? = E + PV$ (d) $A = ? - TS$

2. In an isolated system neither the......nor the ......can be exchanged with the surroundings across the boundary.

3. ...... is used to get exact differential from an inexact differential by multiplication of the later.

4. The work done by the expansion of a gas into vacuum is......

5. The relationship between C_p and C_V for an ideal gas is......

6. The first law of thermodynamics states that......

7. Mathematically the first law of thermodynamics is represented by......

8. Enthalpy of a system is given by the equation......

9. dq and dW are not exact differentials because......

10. Heat of reaction can be found using a......

11. Heat of formation of a compound is the heat of reaction from the corresponding......

12. The integral heat of solution is that heat exchange encountered when......

13. An equation of state for a gas relates...... and is given by in the form......

14. Compressibility factor is......

15. Reduced temperature, reduced pressure and reduced volume are given by......

16. The virial equation of state is given by......

17. Redlich-Kwong equation of state is......

18. Entropy is a measure of......

19. The entropy change for a reversible process is represented mathematically......

20. A system is said to be at equilibrium if the entropy of the system has reached...... value.

21. Clausius-Clapeyron equation is given by......

22. Clausius-Clapeyron equation finds application in

(a) ...... (b) ......

(c) ......processes

23. Approximate vapour pressure-temperature data can be obtained knowing latent heat of vaporization and vapour pressure at one temperature from......equation.

24. Trouton's rule for estimation of ΔS_{vap} at normal boiling point (T_b) is......

25. Gibbs free energy per mole for a pure substance is......

26. The fundamental property relation for single-phase fluid systems, applicable to systems of constant or variable composition is......

27. $\left(\frac{\partial G}{\partial ni}\right)_{T,P,nj}$ is......

28. For a n-component system at constant pressure and temperature, the Gibbs-Duhem equation is......

29. At equilibrium, the chemical potential of a material in different phases in contact with each other is......

30. For a real gas chemical potential is expressed as......

31. A solution is said to be ideal if it obeys......law.

32. Fugacity coefficient of a solution is defined as......

33. Chemical potential for a component i in a mixture of real gas is expressed as......

34. Activity coefficient of a non-ideal solution is defined as......

35. The equations used for the prediction of activity coefficient from experiments are (1)......(2)......(3)......

36. The various forms of Gibbs-Duhem equation for a binary system at constant temperature and pressure can be expressed as......

37. The criterion for equilibrium of a chemical reaction such as $aA + bB + cC + ... \rightarrow rR + sS + tT +$ ......is......

38. Refrigeration means......

39. The working fluid in refrigeration is called......

40. The ton of refrigeration capacity means.....

41. Two main methods used for refrigeration are.....

42. The energies associated with a process are (*a*)......and (*b*)......

43. The energy balance equation for a flow system is......

44. Free energy function is given by......

45. Van't Hoff equation is given by......

46. Zeroth law of thermodynamics deals with......

47. Nitration reactions are exothermic/endothermic type.

SECTION—C

1. Calculate α(coefficient of thermal expansion) for an ideal gas.
2. Calculate β(compressibility) for an ideal gas.
3. Write the expression for work done under isothermal and adiabatic conditions for a reversible process, for an ideal gas.
4. Define heat of a reaction.
5. What do you understand by heat of combustion?
6. What is the law of corresponding states?
7. Write short notes on accentric factor.
8. How is the standard heat of reaction evaluated at a temperature T which is above 25°C?
9. State second law of thermodynamics.
10. What is known as thermodynamic equation of state?
11. What are the formulae used to calculate the entropy change under:
    (a) Reversible isothermal volume change
    (b) Heating or cooling of a substance
    (c) Simultaneous heating and expansion of an ideal gas
    (d) Pressure and temperature are changed.
12. What are the mathematical definitions for Gibbs free energy (G) and Helmholtz free energy (A)?
13. What equation is called Gibbs-Helmholtz equation?
14. What are the Maxwell's equations?
15. Prove $dG = - SdT + Vdp$, from $G = H - TS$.
16. Derive an equation for free energy change with pressure at constant temperature.
17. Define chemical potential.
18. What is the physical significance of chemical potential?
19. What is known as Gibbs-Duhem equation?
20. If μi and μi^0 are chemical potentials of component i in a mixture and pure component i respectively, and x_i is the mole fraction of i in the mixture, how are they related?
21. State Raoult's law and Henry's law.
22. What are the standard states for gases and liquids in the definition of chemical potential?
23. What is known as Lewis-Randall rule?
24. Give the equations used for the predicting activity coefficient from experimental data.
25. What do you understand by excess property?
26. How the relibility of isothermal vapour-liquid equilibria data of binary solutions can be tested?
27. Define equilibrium constant of a chemical reaction. How this constant is related when the equilibrium concentrations are expressed in terms of activity (ai), partial pressure (p_i) or molarity (moles per unt volume, (C_i)?
28. Describe briefly a refrigeration cycle?
29. What are the important characteristic of a refrigerant? Give a few examples of refrigerants.
30. What are the units in: (*i*) Vapour compression refrigerator and; (*ii*) Absorption refrigerator.

31. How the performance of a refrigeration machine is reported?
32. What is a heat pump?
33. One mole of liquid 'A' is converted reversibly into vapour at its boiling point 90°C by supplying heat. The vapour expands against the pressure of 1 atm. The heat of vaporisation is 300 J/g. Calculate q, w, ΔE and ΔH for the process. The molecular weight of A is 100.
34. A well insulated vessel is divided into two parts by a membrane which separates 1 kg of water from 0.01 kg of fairly strong sulphuric acid. The vessel is sealed and is at a constant temperature of 40°C throughout. The membrane is punctured and after mixing is complete, temperature is 42°C. What was the change in the total energy for the system comprising the vessel and the contents?
35. Two moles of an ideal gas are held by a piston under 5 atm and at 300 K. The pressure is suddenly released to 0.5 atm and the gas is allowed to expand isothermally. Calculate w, q, ΔE and ΔH for the system.
36. The formation of SO_3 from SO_2 and oxygen.

$$SO_2(g) + \tfrac{1}{2}O_2(g) \rightarrow SO_3(g)$$

is exothermic by 97000 J at 298°K and 1 atm measured in a bomb calorimeter. What is the heat of reaction?
37. Calculate the enthalpy change involved when reacting liquid water with SO_3 to form H_2SO_4 at 25°C and 1 atm.

Heats of formation for : $H_2SO_4(l) = -\ 193750$ cal.

$H_2O(l) = -\ 68370$ cal, $\quad SO_3(l) = -\ 104200$ cal.
38. When hydrogen is burnt in oxygen, 204.5 kJ heat is liberated per mole of the explosion product measured at 27°C. Calculate the maximum $CH_2O(g) = 24$ J/K.
39. Calculate the change in entropy in heating to 50°C one mole of a gas which occupies a volume of 23.64 m^3 at 150°C if its final volume is 26.51 m^3. C_V for the gas is 2.98 cal/mol °K.
40. The temperature dependence of C_p within the interval from 298 to 1500 K for gaseous benzene is expressed by the expression.

$C_p = 8.1 + 112.78 \times 10^{-3}T - 71.31 \times 10^{-6}T^2$

Find the change in entropy for two moles of benzene when it is heated from 27°C to 273°C.
41. Calculate the change in entropy as a result of the reaction of incomplete oxidation of methane

$$CH_4 + \frac{3}{4}O_2 \rightarrow \frac{1}{6}C_6H_6(g) + \frac{3}{2}H_2O(g)$$

conducted in standard conditions at 25°C.

$(\Delta S^\circ{}_{298})_g$, cal/mol °K: $CH_4 = 44.5 \quad O_2 = 49.0$

$C_6H_6 = 64.34 \quad H_2O = 45.11$
42. Five moles of an ideal gas are compressed isothermally at 400 K from 1 atm. to 5 atm. Calculate the free energy change for the process.
43. One mole of an ideal gas expands isothermally at 30°C into an evaluated vessed so that the pressure drops from 10 to 1 atm. Calculate the ΔG and ΔS.

44. The latent heat of vaporization of a liquid is 10,000 cal/mole. The molar volume of liquid is 20 mL and that of its vapour is 30 litres at 200°C(B.P.) and 1 atm. What would be the change in the boiling point of the liquid at 200°C if the atmospheric pressure is changed by 5 mm?

45. For the reaction $A(g) + 2B(g)$ at 27°C and 1 atm. $K_p = 0.0161$ atm. Calculate K_C and K_X for this reaction.

46. Determine whether the following reaction is possible at 25°C.

$$A \rightleftharpoons B + 2C + D$$

		A	B	C	D
Given $\left[\Delta H^0_{298}\right]_{form}$	in Kcal/mol	– 108	– 35	– 4.0	+ 5
$\left[\Delta S^0_{298}\right]$	in Cal/mol °K	148	83.0	70.0	64

47. Calculate the equilibrium constant for the reaction below at 227°C,

$$A + B \rightleftharpoons C$$

Given $\left[G^0 - H^0_0\right]/T$, Cal/mol °K	– 41.0	– 44	– 56
$\left[H^0 - H^0_0\right]/T$, Cal/mol °K	8.0	7.0	11.0
$\left[\Delta H^0\right]_{formation}$ Kcal/mol °K,	– 58	– 26.0	– 90.0

48. What are the various energies associated with a system?

49. What is third law of thermodynamics?

SECTION—D

1. A system undergoes a change from a given initial state to a given final state either by an irreversible process or by a reversible process. Then:

(a) ΔS_i is always > ΔS_R
(b) ΔS_i is sometimes > ΔS_R
(c) ΔS_i always < ΔS_R
(d) ΔS_i is always = ΔS_R

When ΔS_i and ΔS_R are entropy changes of the system for the irreversible and reversible process respectively.

2. The change in Gibbs free energy for vaporisation of a pure substance is:

(a) Positive
(b) Negative
(c) Zero
(d) May be positive or negative

3. The molar excess Gibbs free energy (G^b) for a binary liquid mixture at T and P is given by : $(G^b/RT) = A\, x_1 x_2$ where A is constant. The corresponding equation for ln r_1 where r_1 is activity coefficient of component 1 is:

(1) Ax_2^2
(a) Ax_1
(2) Ax_2
(b) Ax_1^2

4. A feed at 128 K consisting of flue gas (CO_2, O_2 and N_2) and air is passed through a bed of pure carbon. The two reactions that occur both go to completion.

$CO_2(g) + C(s) \rightarrow 2CO(g)$; $\Delta H_R{}^0$ at 298 K = 170 kJ/Mol

$O_2(g) + 2C(s) \rightarrow 2CO(g)$; $\Delta H_R{}^0$ at 298 K = 220.4 kJ/Mol

The combustor is adiabatic and product gases exit at 1298 K. Calculate the required moles of CO_2 per mole of O_2 in the feed stream, so that the net heat generated is zero and the bed temperature remains constant at 1298 K.

Data : Mean molar heat capacities, CPM

Substance	CPM, kJ/(k. Mol. K)
C	0.02
O_2	0.03
CO	0.03
CO_2	0.05

5. Pressure is an extensive property. Yes/No, state reason?

6. The equation $dU = TdS - pdV$ is applicable to infinitesional change occurring in:

(a) An open system of constant composition

(b) A closed system of constant composition

(c) An open system with changes in composition

(d) A closed system with changes in composition

7. The theoretical minimum work required to separate one mole of liquid mixture at 1 atm, containing 50 mole% each of *n*-heptane and *n*-octane into pure compounds each at 1 atm.

(a) $-2RT \ln 0.5$ (b) $-RT \ln 0.5$

(c) $0.5\,RT$ (d) $2\,RT$

8. Identify the extensive and intensive properties from the given list.

(a) Chemical potential (b) Entropy

(c) Fugacity (d) Enthalpy

(e) Activity coefficient

9. At the inversion point, the Joule-Thompson coefficient is:

(a) Positive (b) Negative

(c) Zero

10. (a) Find ΔH^0_{298} for the reaction

$C_3H_8(g) + 5O_2(g) \rightarrow 3CO_2(g) + 4H_2O(l)$

If the standard enthalpies of formation ΔH_f^0 of

$C_3H_8(g) + 5O_2(g) \rightarrow 3CO_2(g) + 4H_2O(l)$

If the standard enthalpies of formation ΔH_f^0 of

$C_3H_8(g) + CO_2(g)$ and $H_2O(l)$ are – 125.5, – 393.3, and 284.5 kJ/mole respectively at one atmosphere.

(b) What is the value of ΔE^0_{298} for the above reaction?

11. Choose correct answer.

An insulated rigid vessel contains a mixture of fuel and air. The mixture is ignited by a minute spark. The contents of vessel experience:

(a) Increase in temperature and pressure and energy.

(b) Decrease in temperature, pressure and energy.

(c) Increase in temperature and pressure but no change in energy.

(d) Increase in temperature and pressure but decrease in energy.

12. The first law of thermodynamics takes the form $W = -\Delta H$ when applied to:

(a) A closed system undergoing a reversible adiabatic process.

(b) An open system undergoing an adiabatic process with negligible changes in kinetic and potential energies.

(c) A closed system undergoing a reversible constant volume process

(d) A closed system undergoing a reversible constant pressure process.

13. A reversible heat transfer demands:

(a) The temperature difference causing heat transfer tends to zero.

(b) The system receiving heat must be at a cosntant temperature.

(c) The system transferring out heat must be at a constant temperature.

(d) Both interacting systems must be at a constant temperature.

14. Which among the following relations is/are valid only for reversible process under/by a pure substance?

(a) $dQ = dU + dW$

(b) $TdS = dU + dW$

(c) $TdS = dU + pdV$ (reversible process)

(d) $dQ = PdV + dU$ (reversible process)

15. When a system executes an irrversible cycle?

(a) $\oint \frac{dQ}{T} < 0$ (b) $\oint dS > 0$

(c) $\oint dS = 0$ (d) $\oint \frac{dQ}{T} > 0$

16. The relationship $\left(\frac{\partial T}{\partial P}\right)_H = 0$ holds good for:

(a) An ideal gas at any state (b) A real gas at any state

(c) Any gas at its critical state (d) Any gas at its inversion point

17. During the change of phase of a pure substance:

(a) $dG = 0$ (b) $dP = 0$

(c) $dH = 0$ (d) $dU = 0$

18. The figure below shows a thermodynamic cycle undergone by a certain system. Find the mean effective pressure, in N/m^2.

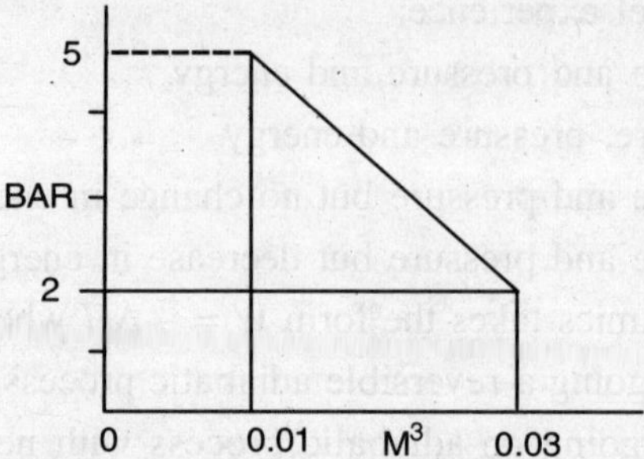

19. A vertical cylinder with a freely floating piston contains 0.1 kg air at 1.2 bar and a small electrical resistor. The resistor is wired to an external 12 volt battery. When a current of 1.5 amps is passed through the resistor for 90 seconds, the piston sweeps a volume of 0.01 m^3. Assume

(*i*) Piston and the cylinder are insulated and

(*ii*) Air behaves as an ideal gas with C_V = 700 J/kg. K. Find the rise in teperature of air.

20. Figure below shows a reversible heat engine E_R having heat interactions with three constant temperature systems. Calculate the thermal efficiency of the heat engine.

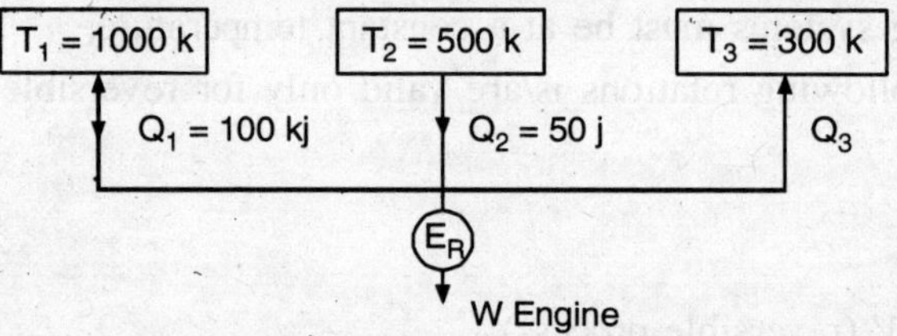

21. Air expands steadily through a turbine from 6 bar, 800 K to 1 bar, 520 K. During the expansion, heat transfer from air to surroundings at 300 K is 10 kJ/kg air neglect the change in kinetic and potential energies and evaluate the irreversibility per kg of air. Assume air to behave as an ideal gas with C_P = 1.0 kJ/(kg K) and R = 0.3 kJ/(kg K).

22. In the abvoe Problem (21) find the actual work and miaximum work per kg. air.

23. A vessel of volume 1.0 metre cu. contains a mixture of liquid water and steam in equilibrium at 1.0 bar. Given that 90% of the volume is occupied by steam, find the dryness fraction of the mixture. Assume at 1.0 bar, v_f = 0.001 m^3/kg and v_g = 1.7 m^3/kg.

24. A rigid insulated cylinder has two compartments separated by a thin membrane. While one component contains one kilo mole nitrogen at a certain pressure and temperature, the other contains 1 kg mole carbon diodxide at the same temperature and pressure. The membrane is ruptured and the two gases are allowed to mix. Assume that the gases bahave as ideal gases. Calculate the increase in entropy of the contents of the cylinder. Universal gas constant is 8314.3 J/(kg mole °K).

25. In the vicinity of triple point the vapour pressure of liquid and solid ammonia are respectively given by

$$\ln P = 15.16 - 3063/T \quad \text{and} \quad \ln P = 18.70 - 3754/T$$

where P is in atmospheres and T is in kelvin. What is the temperature at the triple point?

26. A change in state involving a decrease in entropy can be spontaneous only if:

(a) It is exothermic (b) It is isenthalpic

(c) It takes place isothermally (d) It takes place at constant volume

27. A Carnot cycle consists of the following steps:

(a) Two isothermals and two isentropics (b) Two isobarics and two isothermals

(c) Two isochorics and two isobarics (d) Two isothermals and two isochorics

28. It is desired to bring about a certan change in the state of a system by performing work on the system under adiabatic conditions.

(a) Amount of work needed is path dependent

(b) Work done cannot bring about such a change of state

(c) The amount of work needed is independent of path

(d) More information is needed to conclude anything about the path dependence or otherwise of the work needed.

29. If the heat of solution of an ideal gas is negative then its solubility at a given partial pressure varies with the temperature as:

(a) Solubility increases as temperature increases

(b) Solubility decreases as temperature increases

(c) Solubility is independent of temperature

(d) Solubility increases or decreases with temperature depending on Gibbs free energy change of solution.

30. One mole of nitrogen at 8 bar and 600 K is contained in a piston-cylinder arrangement. It is brought to 1 bar isothermally against a resisting pressure of 1 bar. The work done in joules by gas is:

(a) 30554 (b) 10373

(c) 4988.4 (d) 4364.9

31. For water at 300°C, it has a vapour pressure of 8592.7 kPa and fugacity 6738.9 kPa under these conditions, one mole of water in liquid phase has a volume of 2.28 cubic cm and has in vapour phase is 39.1 cubic cm. Fugacity of water in kPa at 900 kPa will be:

(a) 6738.9 (b) 6753.5

(c) 7058.3 (d) 9000

32. Heat capacity of air can be approximately as $C_p = 26.693 + 7.365 \times 0.001 \times T$ where C_p is in joules/(mole K) and T is in K. The heat given off by one mole of air when cooled at 1 atmospheric pressure from 500°C – 100°C is:

(a) 10.73 kJ (b) 16.15 kJ

(c) 18.11 kJ (d) 18.33kJ

33. A solid metallic ball weighing 5 kg has an initial temperature of 500°C. 40 kg of water initially at 25°C is contained in a perfect insulated tank. The metallic block is brought into contact with water. Both of them come to equilibrium. Specific heat of block material is 0.4 kJ/kg K ignoring the effects of expansion and contraction and also the heat capacity of the tank, the total enthalpy change in kJ/kg K is:

(a) 1.87 (b) 0

(c) 1.26 (d) 3.91

34. The following heat engine produces power of 10,000 kW. The heat engine operates between 800 K and 300 K. It has a thermal efficiency equal to 50% of that of the Carnot engine for the same temperatures. The rate at which heat is absorbed from the hot reservoir is:

(a) 100,000 kW (b) 160,000 kW
(c) 200,000 kW (d) 320,000 kW

35. A steam turbine operates with a superheated steam flowing at 1 kg/s. The steam is supplied at 41 bar and 500°C and discharges at 1.01325 bar and 100°C.

Data ; 41 bar, 500°C. Enthalpy = 3443.9 kJ/kg; Entropy = 7.0785 kJ/kg K

41 bar, 251.8°C. Enthalpy of saturated steam = 2799.9 kJ/kg

Entropy of saturated steam = 6.058 kJ/kg/K

1.01325 bar, 100°C. Enthalpy of saturated vapour = 2676 kJ/kg

Enthalpy of saturated liquid = 419.1kJ/kg

Entropy of saturated vapour = 7.3554 kJ/kg K.

Entropy of saturated liquid = 1.3069 kJ/kg. K.

The maximum power output in kilowatt will be:

(a) 644.0 (b) 767.9
(c) 871.3 (d) 3024.8

36. At 60°C vapour pressures of methanol and water are 84.562 kPa and 19.953 kPa respectively. An aqueous solution of methanol at 60°C exerts a pressure of 39.223 kPa; the liquid phase and vapour phase mole fractions of methanol are 0.1686 and 0.5714 respectively. Activity coefficient of methanol is:

(a) 1.572 (b) 1.9398
(c) 3.389 (d) 4.238

37. One kg of saturated steam at 100°C and 1.01325 bar is contained in a rigid walled vessel. It has a volume of 1.673 cubic m. It cools to 98°C the saturation pressure, is 0.943 bar. One kg of water vapour under these conditions has a volume of 1.789 cubic m. The amount of water vapour condensed in kg is:

(a) 0.0 (b) 0.065
(c) 0.10 (d) 1.0

38. The latent heat of condensation, kJ/kg under these conditions is:

(a) 40732 (b) 2676
(c) 2263 (d) 540

39. As pressure approaches zero, the ratio of fugacity to pressure (f/P) for a gas approaches:

(a) Zero (b) Unity
(c) Infinity (d) An indeterminate value

40. A perfectly insulated container of volume, V is divided into two equal halves by a partition. One side is under vacuum while the other side has one mole of an ideal gas (with constant heat capacity) at 298 K. If the partition is broken, the final temperature of the gas in the container :

(a) Will be greater than 298 K (b) Will be 298 K
(c) Will be less than 298 K (d) Can not be determined

41. If the average specific heat of all the gases/vapour is 40 J/mole the maximum teperature rise of the exhaust gases in °C would be approximately equal to:

(a) 1225 (b) 1335
(c) 1525 (d) 1735

42. A vessel of volume 1000 cubic m contains air which is saturated with water, vapour. The total pressure and temperature are 100 kPa and 20°C, respectively. Assuming that the vapour pressure of water at 20°C is 2.34 kPa, the amount of water vapour in kg in the vessel is approximately :

(a) 17 (b) 20 (c) 25 (d) 34

43. The number of degrees of freedom for an azeotropic mixture in a two component vapour liquid equilibria are:

(a) Zero (b) One
(c) Two (d) Three

44. A car tyre of volume 0.057 cubic m is inflated to 300 kPa at 300 K. After the car is driven for ten hours, the pressure in the tyre increases to 330 kPa. Assuming the air is an ideal gas and C_V for air is 21 J/mol K. The change in the internal energy of air in the tyre in J/mol. is:

(a) 380 (b) 630
(c) 760 (d) 880

45. A gas obeys $P(V - b) = RT$. The work obtained from reversible expansion of isothermal of one mole of this gas from an intial molar volume V_f is:

(a) $RT \ln (V_f/V_i)$ (b) $RT \ln \left(\frac{V_f - b}{V_i}\right)$

(c) $RT \ln \left(\frac{V_f}{(V_i - b)}\right)$ (d) $RT \ln \left(\frac{V_f - b}{V_i - b}\right)$

46. A cycle engine exchanges heat with two reservoirs maintained at 100 and 300°C respectively. The maximum work in Joules that can be obtained from 1000 J of heat extracted from the hot reservoir is:

(a) 349 (b) 651
(c) 667 (d) 1000

47. The vapour pressure of water is given by, $\ln P^{sat} = A - \frac{5000}{T}$ where A is a constant, P^{sat} is vapour pressure in atm and T is tmeperature in °K. The vapour pressure of water in atm at 50°C is approximately as:

(a) 0.07 (b) 0.09 (c) 0.11 (d) 0.13

48. At standard conditions:

$N_2 + 2O_2 \rightleftharpoons 2NO_2 \rightarrow (1)$; $\Delta G^0 = 100$ kJ/mol

$NO + \frac{1}{2}O_2 \rightleftharpoons 2NO_2$; $\Delta G^0 = -35$ kJ/mol. The standard free energy of formation of NO_2 in kJ/mole is:

(a) 15 (b) 30 (c) 85 (d) 170

ANSWERS

SECTION–A

1. (c)	2. (a)	3. (c) and (d)	4. (a) and (b)	5. (b) and (d)
6. (a)	7. (b)	8. (b) and (c)	9. (b)	10. (c)
11. (c)	12. (c)	13. (b)	14. (c)	15. (b)
16. (b)	17. (a)	18. (b)	19. (c)	20. (d)
21. (a)	22. (b)	23. (c)	24. (b)	25. (c)
26. (d)	27. (a)	28. (c)	29. (d)	30. (c)
31. (b)	32. (d)	33. (d)	34. (d)	35. (b)
36. (c)	37. (c)	38. (b)	39. (c)	40. (b)
41. (b)	42. (d)	43. (c)	44. (b)	45. (c)
46. (d)	47. (a)	48. (d)	49. (c)	50. (a)
51. (e)	52. (b)	53. (c)	54. (b)	55. (a)

SECTION–B

1. (a) dW (b) T (c) H (d) E
2. Energy, mass
3. Integrating factor
4. Zero
5. $C_p - C_v = R$
6. energy can never be created nor destroyed but chagnes from one form to another.
7. $dq = dW + dE$
8. $H = E + PV$
9. q and W are not state functions but path functions
10. bomb calorimeter
11. elements
12. a specified amount of pure solute is added to a specified amount of pure solvent.
13. (a) pressure, molar or specific volume and temperature
    (b) $f(PVT) = 0$
14. $Z = PV/RT$
15. $T_r = T/T_C$; $P_r = P/P_C$; $V_r = V/V_C$
16. $PV/RT = 1 + B/V + C/V^2 + ...$
    A, B, C, etc. ......are called virial coefficients.
17. $T = \dfrac{RT}{V-b} - \dfrac{a}{T^{1/2}\,V(V+b)}$

    where a and b are constants.
18. molecular disorder existing in the system.

19. $dS = dq/T$; T is absolute temperature.
20. maximum.
21. $\frac{dP}{dT} = \frac{\Delta H}{T\Delta V}$; ΔV = volume chagne from one phase to other phase.
22. (a) vapour liquid equilibria
    (b) solid-liquid equilibria
    (c) solid-vapour equilibria.
23. Clausius-Clapeyron.
24. $\Delta H_{vap}/T_b$
25. Chemical potential.
26. $d(nU) = Td(nS) - Pd(nV) + \mu_i dn_i$
27. μ_i
28. $n_i d\,\mu_i = 0$
29. equal.
30. $\mu = \mu° + RT \ln f$
    where $\mu°$ is standard chemical potential at unit fugacity ($f° = 1$ atm) and the gas is behaving ideally.
31. Raoult's law
32. f/p where f is fugacity and p is pressure.
33. $\mu i = \mu i° + RT$ in $(fi/fi°)$
34. Activity coefficient is the ratio of activity and mole friction.
    or $Y_i = a_i/x_i = \frac{f_i}{f_i^0 x_i}$ $\left[\text{Since } a_i = \frac{f_i}{f_i^0}\right]$
35. (*i*) Van Laar (*ii*) Margules and (*iii*) Wilson's equation
36. (a) $x_1 d \ln p_1 + x_2 d \ln p_2 = 0$
    (b) $x_1 d \ln x_1 + x_2 d \ln x_2 = 0$
    (c) $x_1 d \ln y_1 + x_2 d \ln y_2 = 0$
    (d) $x_1 = \left(\frac{\partial \ln p_1}{\partial x_1}\right) T, P, X_2 = x_2\left(\frac{\partial \ln p_2}{\partial x_2}\right) T, P, X_2$
37. $\Sigma v_i \mu_i = 0$
    where v_i is the stoichiometric coefficient of species i.
    v_i is positive for products and negative for reactants.
    For the reaction given this may be expressed as
    $S(r\,\mu R + s\,\mu s + t\,\mu_T - a\,\mu_A - b\,\mu_B - c\,\mu_C) = 0$
38. maintenance of temperature below that of the surroundings.
39. refrigerant.
40. the capability of absorbing the heat evolved in freezing one ton of water at 0°C, in one day.

41. (a) vapour compression refrigiration and
    (b) absorption refrigiration
42. heat, work
43. $U_1 + P_1V_1 + Z_1 + \dfrac{u_1^2}{2g_c} + q - W = U_2 + P_2V_2 + Z_2 + \dfrac{u_2^2}{2g_c}$
44. $G_T{}^0 - H_0{}^0/T$
45. $d \ln K/dT = (\Delta H^\circ/R_T{}^2)$
46. equilibrium.
47. exothermic type.

SECTION—C

1. $PV = RT$ for one mole of ideal gas

   Differentiation gives $PdV + VdP = RdT$; $(\partial V/\partial T)_P = R/P$

   $$= \frac{1}{V}\left(\frac{\partial V}{\partial T}\right)_P = \frac{R}{VP} = \frac{1}{T} = T^{-1}$$

2. $\left(\dfrac{\partial V}{\partial P}\right)_T = -\dfrac{V}{P}$; $\beta = -\dfrac{1}{V}\left(\dfrac{\partial V}{\partial P}\right)_T = \dfrac{V}{VP} = \dfrac{1}{P} = P^{-1}$
3. $W_{\text{isothermal}} = nC_V(T_2 - T_2)$
4. Heat of reaction is defined as the amount of heat absorbed or evolved at a given temperature when the reactants have combined to form the products as represented by the balanced chemical equation.

   or

   Heat of reaction = Σ enthalpy of products – Σ enthalpy of reactants
5. The enthalpy change when one mole of a substance is burnt in oxygen, is defined as the heat of combustion.
6. All gases, when compared at the same reduced temperature and reduced pressure have the same compressibility factor and all deviate from ideal gas behaviour to about the same degree.
7. The accentric factor (W) is defined by Pitzer as

   $W = -\log_{10}(P_r{}^{\text{sat}})\ Tr - 0.7^{-1.0}$

   where $P_r{}^{sat}$ is reduced vapour pressure and T_r is reduced temperature. W can be determined for any fluid from T_c, P_c and a single vapour pressure measurement made at $T_r = 0.7$. The parameter 'W' is used to find out the compressibility factor.
8. The standard heat of reaction at any temperature T is given by

   $$\Delta \overset{0}{\underset{T}{H}} = \Delta \overset{0}{\underset{R}{H}} + \Delta \overset{0}{\underset{298}{H}} + \Delta \overset{0}{\underset{P}{H}}$$

   where ΔH^0 = standard heat of reaction at any temperature T.

$\Delta \overset{0}{\underset{R}{H}}$ = enthalpy change of the reactants for cooling from T to 25°C = $\sum\limits_{\text{reactants}} n \int_T^{298} \overset{0}{\underset{p}{C}}\, dT$

$\Delta \overset{0}{\underset{298}{H}}$ = standard heat of reaction at 25°C

$\Delta \overset{0}{\underset{P}{H}}$ = enthalpy change of the prodcuts for heating from 25°C to T.

$$= \sum_{\text{products}} \left(n \int_{298}^{T} C_P dT \right)$$

9. Any process which consists solely in the transfer of heat from one temperature to a higher one is impossible.

or

No apparatus can operate in such a way that its only effect in system and surroundings is to convert heat absorbed by a system completely into work.

10. This is the equation relating E, V, J and P and is given by

$$\left(\frac{\partial U}{\partial V}\right)_T = T\left(\frac{\partial P}{\partial T}\right)_V - P$$

and is true for all substances under all conditions

(a) $\Delta S = -\, nR \ln (P_2/P_1)$

(b) $\Delta S = nC_p \ln (T_2/T_1)$

(c) $\Delta S = nR \ln (V_2/V_1) + nC_V \ln (T_2/T_1)$

(d) $\Delta S = nR \ln (V_2/V_1) + nC_V \ln (T_2/T_1)$

12. $G = H - TS$, $A = E - TS$.

13. $$\left[\frac{\partial (G/T)}{\partial T}\right]_P = -\frac{H}{T^2}$$

$$\left[\frac{\partial (A/T)}{\partial T}\right]_V = -\frac{E}{T^2}$$

14. $(\partial T/\partial V)_s = -\,(\partial P/\partial S)_V$

$(\partial T/\partial P)_s = (\partial V/\partial S)_P$

$(\partial P/\partial T)_V = (\partial S/\partial V)_T$

$(\partial V/\partial T)_P = -\,(\partial S/\partial P)_T$

15. $G = H - TS$; $dG = dH - TdS$; $H = E + PV$

$dH = dE + PdV + VdP - SdT - TdS$,

But $dE = dq - dw = TdS - dW$

So $dG = TdS - dw + PdV + VdP - VdP - SdT - TdS$

Assuming that the only work done is by expansion

$W = VdP$, So $dG = -\,SdT + VdP$

16. $dG = -SdT + VdP$

    At constant temperature, $dG = VdP$

    If the state of the system is changed from 1 and 2

    $$\int_1^2 dG = \int_1^2 VdP; G_2 - G_1 = \Delta G = \int_1^2 \frac{RT}{P} dP$$

    Since for an ideal gas $PV = nRT$; So $\Delta G = nRT \ln \frac{P_2}{P_1}$.

17. $$\mu_i = \left(\frac{\partial U}{\partial ni}\right)_{S,V,n_j} = \left(\frac{\partial G}{\partial ni}\right)_{T,P,n_j}$$

    $$= \left(\frac{\partial H}{\partial ni}\right)_{S,P,n_j} = \left(\frac{\partial A}{\partial ni}\right)_{T,V,n_j}$$

18. The chemical potential of a component of a phase is the amount by which the capacity of the phase for doing work (other than work of expansion) is increased per unit amount of substance added for an infinitesimal addition at constant pressure and temperature.
19. $-SdT + VdP - \Sigma n_i \, d\,\mu_i = 0$
20. $\mu_i - \mu_i^0 = RT \ln xi$.
21. **Raoult's law:** Partial pressure of a component i, equal to the product of its mole fraction x_i, in the solution and the pure component vapour pressure P_i^0

    $$P_i = P_i^0 x_i$$

    Henry's law: If very small amounts of a substance (say B) are added isothermally to a liquid (A) with which B does not react. It is found that the partial pressure of the component B in an equilibrium vapour phase increases linearly with the mole fraction x_B

    $P_B = K_2 x_2$ where K_2 is constant.
22. Standard state for the gases is the gas at 1 atm. For the liquid solutions, the standard state of solvent is the pure component.
23. $$\overset{\Delta}{f_i} = f_i x_i$$

    where $\overset{\Delta}{f_i}$ is the fugacity of component i in the solution

    f_i is the fugacity of the pure component i

    x_i is the mole fraction of component i is the solution.

24. **Van Laar equation:** $$\ln \gamma_1 = \frac{B/T}{\left(1 + A\frac{x_1}{x_2}\right)^2}$$

    $$\ln \gamma_2 = \frac{AB/T}{(A + x_2/x_1)^2}$$

Margules equation: $\ln g_1 = bx_2^2 + Cx_2^2$

$$\ln g_2 = bx_1^2 + C(0.5 + x_2)\, x_1^2.$$

25. An excess property is the difference between an actual property and the property that would be calculated for the same conditions of T, P and x by the equations for an ideal solution. Thus by definition

$$M^E = M - M^{\text{ideal}}$$

where M^E is called the excess solution property.

26. The equation $x_1\left(\dfrac{\partial \ln P_1}{\partial x_1}\right)_{T,P,x_2} = x_2\left(\dfrac{\partial \ln P_2}{\partial x_2}\right)_{T,P,x_2}$

is used to test the experimental isothermal vapour-liquid composition data for binary solutions. The quantity

$x_1\left(\dfrac{\partial \ln P_1}{\partial x_1}\right)_{T,P,x_2}$ plotted against $x_2\left(\dfrac{\partial \ln P_2}{\partial x_1}\right)_{T,P,x_2}$

should yield a straight line going through the origin, with a slope of unity if the data are reliable.

27. Consider the reaction

$$aA + bB \underset{Kr}{\overset{Kf}{\rightleftharpoons}} mM + nN$$

Equilibrium constant $(K_{eq}) = \dfrac{Kf}{K_b} = \dfrac{[M]^m[N]^n}{[A]^a[B]^b}$

[] indicates concentrations.

The equilibrium concentrations [] can be expressed in terms of partial pressure (p_i), mole fraction (x_i) and moles per unit volume (c_i). Thus

$$K_P = \frac{p_M{}^m P_N{}^n}{P_A{}^a P_B{}^b};\; K_c = \frac{C_m{}^m C_N{}^n}{C_A{}^a C_B{}^b};\; K_x = \frac{x_M{}^m x_N{}^n}{x_A{}^a x_B{}^b}$$

and $K_P = K_C(RT)^{\Delta V} = K_x(P)^{\Delta V}$

where $\Delta v = (m + n) - (a + b)$.

28. In refrigeration, since low temperature must be maintained, it is necessary to absorb heat continually at low level. This necessitates a continuous process. One way of absorbing heat at a low temperature is through the evaporation of a liquid whose properties are such that, at the pressure of evaporation,the boiling temperature is low. For continuous operation, the evaporated fluid must be returned to its original liquid state so that it can again absorb heat at the low level. The complete series of steps through which the fluid passes cnstitutes the refrigeration cycle.

29. **Characteristics:**

1. Should be non-toxic, non-corrosive, non-explosive and of low cost.
2. The material should not have a subatmospheric vapour pressure at the temperature in the refrigerator coils.
3. The vapour pressure at the condenser temperature should not be unduly high.

Examples :

(a) Dichlorodifluoromethane (Freon-12)

(b) Ammonia (c) SO_2 (d) CO_2 (e) CCl_4

30. (*i*) Vapour cmpression refrigeration:

(a) Compressor (b) Condenser

(c) Expansion valve (d) Refrigerator

(*ii*) Absorption refrigeration:

(a) Absorber (b) Pump (c) Degenerator

(d) Condenser (e) Expansion valve (f) Refrigerator

31. The performance of a refrigeration machine is reported in terms of the coefficient of performance (COP).

$$\text{COP} = \frac{\text{Refrigeration obtained}}{\text{Work required}}$$

32. Heat pump is a device for heating commercial buildings etc. during the winter months and cooling them during the summer.

33. The process is $A(l) \to A(\text{vapour})$

Heat supplied for conversion of 1 mole of A into its vapour

$= q =$ Heat of vaporization × Mol. wt. of liquid

$= 300 \times 100 = 30000$ J/mole.

Work done $= w = \int PdV = \int_{V_l}^{V_v} dV$; P is constant (= 1 atm).

$= PV = P(V^v - V^l)$

$V^v > V^l$ (assumed).

So $w = PV^V = RT = 8.314 \times (273 + 90)$

$= 8.314 \times 363 = 3018$ J/mole

$\Delta E = q - w = 26982$ J/mole

$\Delta H = \Delta E + b\Delta P = q = 30{,}600$ J/mole

34. For this process; Heat exchange (q) = 0

Work done (W) = 0, So change in total energy

$\Delta E = q - W = 0$

35. Since the pressure is released suddenly the expansion is irreversible.

Hence, $n = 2$ moles; $P_1 = 5$ atm; $P_2 = 0.5$ atm

$T_1 = T_2 = 300$ K

(*i*) Work done : $= nRT(1 - P_2/P_1)$

$= 2 \times 8.314 \times 300(1 - 0.5/5) = 4489.56$ J

(*ii*) Since the gas expands isothermally D$T = 0$; So, $\Delta E = 0$

$\Delta H = \Delta E + \Delta(PV) = \Delta E + \Delta(RT) = 0$

(*iii*) $q = w + \Delta E = 4489.56 + 0 = 4489.56$ J

36. $\Delta H = \Delta E + \Delta(pV) = \Delta E + p(\Delta V) = \Delta E + \Delta ng(RT)$

$\Delta n_g = 1 - \left(1+\frac{1}{2}\right) = -\frac{1}{2}$, $q = \Delta E$ under constant volume

So $\Delta H = -97000 - \frac{1}{2}(80314)\ 2(98) = -98240$ J

37. $H_2(g) + 2O_2(g) + S(s) \rightarrow H_2SO_4 \qquad \Delta H = -193750$

$H_2(g) + \frac{1}{2}O_2 \rightarrow H_2O(l) \qquad \Delta H = -68370$

$S(s) + \frac{3}{2}O_2(g) \rightarrow SO_3(l) \qquad \Delta H = -104{,}200$

Subtracting the last two equations from the first

$H_2O(l) + SO_3(l) \rightarrow H_2SO_4(l); \qquad \Delta H = -21{,}180$ cal

So the enthalpy change = – 21180 cal.

38. The reaction is $H_2(g) + \frac{1}{2}O_2(g) \rightarrow H_2O(g)$

The amount of heat liberated is used to raise the temperature of the product

$$\Delta H_{\text{reaction}} = \int_{T_1}^{T_2} nC_i dT; 240500 = \int_{300}^{T_2} 24.0\, dT$$

$240500 = 24(T_2 - 300)$, $T_2 = 10{,}320.83$ °K

39. $\Delta S = C_V \ln(T_2/T_1) + R \ln(V_2/V_1)$

$= 2.98 \times 2.303 \times \log_{10}(323/288) + 1.987 \times 2.303 \log(26.51/23.64)$

$= 0.57$ cal/mol °K

40. $\Delta S = S_{T_2} - S_{T_1} = \int_{T_1}^{T_2} C_p \frac{dT}{T}$

$T_2 = 1000$ °K, $T_1 = 300$ K

$$\Delta S = S_{1000} - S_{300} = \int_{T_1}^{T_2} (8.1 + 112.78\times10^{-3}\, T - 71.31\times10^{-6} T^2)\frac{dT}{T}$$

$$= 8.1 \ln\frac{1000}{300} + 112.78 \times 10^{-3}(1000 - 300) - \frac{71.31\times10^{-6}}{2}(1000^2 - 300^2)$$

$= 9.76 + 78.4 - 32.1 = 36.54$ cal/mol K

For two moles C_6H_6, $\Delta S = 73.08$ cal/K

41. For a reaction of the type

$V_1A_1 + V_2A_2 + \ldots\ldots = V_1'A_1' + V_2'A_2' + \ldots\ldots$

$\Delta S = (V_1'S_{A_1}' + V_2'S_{A_2}' + \ldots\ldots) - (V_1S_{A_1} + V_2S_{A_2} + \ldots\ldots)$

So, in the present problem

$$\Delta S_{298}{}^0 = \left(\frac{1}{6}\times 64.34 + \frac{3}{2}\times 45.11\right) - \left(44.5 - \frac{3}{4}\times 49.0\right)$$

$78.3 - 81.25 = -2.95$ cal/K

42. Number of moles = 4

$T = 400$ K, $P_2 = 5$ atm, $P_1 = 1$ atm

$\Delta G = nRT \ln (P_2/P_1) = 4 \times 1.987 \times 400 \ln (5/1) = 3116.7$ cal.

43. $\Delta G = nRT \ln (P_2/P_1) = 1 \times 1.987 \times 303 \ln (1/10)$

$(\Delta S)_{system} = -nR \ln (P_2/P_1) = 1 \times 1.987 \ln (1/10) = -4.575$ cal.

$(\Delta S)_{surroundings} = -q$ system$/T = 0$

(as $q = \Delta E + w = 0 + 0 = 0$)

$(\Delta S)_{universe} = (\Delta S)_{system} + (\Delta S)_{surr} = 4.575 + 0 = 4.575$ cal.

44. $\Delta H_{vap} = 10{,}000$ cal/mole; 1 lit. atm = 24.21 cal.

$$\frac{dP}{dT} = \frac{\Delta H_{vap}}{T(V^V - V^l)} = \frac{10000}{473 \times 29980} = 7.052 \times 10^{-4} \text{ cal/lit K}$$

$$7.052 \times 10^{-4} \frac{\text{cal}}{\text{lit K}} = \frac{0.7052}{24.2100} \text{ atm/ K}$$

$= (0.0291 \times 760)$ mm/K $= 22.13$ mm/K

So $dT/dP = 0.045$ ...... deg/mm

So change in B.P. $= 0.045 \times 5 = 0.225$ K

45. The reaction is $A(g) \rightarrow 2B(g)$, $\Delta v = 2 - 1 = 1$

$$RT = 0.082 \left(\frac{\text{m}^3\text{atm}}{\text{kg mole K}}\right) \times 300\,(\text{K}) = 24.6$$

$K_p = K_C(RT)^{\Delta v} = 0.160$

$K_c = K_P/(RT)^{\Delta v} = 0.160/24.6 = 6.50 \times 10^{-3}$ kg mol/m^3

$K_X = K_P(P)^{-\Delta v} = 0.160(1)^{-1} = 0.160$

46. $\Delta H^\circ_{298} = -35 - 4 \times 2 + 5 + 108 = 70.0$ kcal

$\Delta S^\circ_{298} = 83 + 2 \times 70 + 64 - 148 = 139$ cal/K

$\Delta G^\circ_{298} = \Delta H^\circ_{298} - T\Delta S_{298}$

$= 70{,}000 - 298 \times 139 = 28578$ cal.

Since ΔG is positive the reaction is impossible at 298 K or 25°C.

47. $\Delta(G^\circ - H_0^\circ) / T = -56 - (44 + 41) = 29.0$ cal/mol K

$\Delta(G^\circ - H_0^\circ) / T = 11 - (8 + 7) = -4.0$ cal/mol K

$\Delta H^0_{T500°K}$, kcal/mol $= -90 - (-58 - 26) = -6$ kcal $= -6000$ cal.

$$\Delta H_0^\circ = \Delta H_T^\circ - \Delta\left(H^\circ - \frac{H_0^\circ}{T}\right) T$$

$$\text{where } \Delta\left(\frac{H^\circ - H_0^\circ}{T}\right) = \Delta\left(\frac{H^\circ - H_0^\circ}{T}\right)_{\text{products}}$$

$$-\Delta\left(\frac{H^\circ - H_0^\circ}{T}\right)_{\text{reactant}}$$

So $\dfrac{\Delta H_0^\circ}{T} = \dfrac{-6000}{500} - (-4) = -8.0$

and $R \ln K_P = -\left[\Delta\left(\dfrac{G^\circ - H_0^\circ}{T}\right) + \dfrac{\Delta H_0^\circ}{T}\right]$

So, $\log K_P = \dfrac{1}{2.303 \times 1.987}(29 - 8) = 4589$ or $K_P = 2.576 \times 10^{-5}$ **Ans.**

48. (a) Internal energy (b) Flow or pressure energy
(c) Kinetic energy (d) Potential energy
(e) Surface energy (f) Magnetic energy

49. The third law states that the entropy of perfect crystalline substance is zero at absolute zero temperature.

SECTION—D

1. (a)

2. (c) Molar or specific Gibbs energy which for a pure species does not change during a phase transition such as melting, vaporization and sublination.

3. (a)

4. Overall reaction :
$CO_2 + 3C + O_2 \rightarrow 4CO(g) - 50.4$ kJ/K mole by heat balance:
$50.4 = 1298 \times 4 \times 0.03 - CO_2 \times 0.05 \times 128 - 128(3 \times 0.02 + O_2) - 99.4 = -6.5\, CO_2 - 3.9\, O_2$
For $O_2 = 1$ mole; CO_2 is nearly 14 moles.

5. Pressure in an intensive property, since it does not depend on the amount of material making up the system.

6. (b)

7. (b)
Entropy of mixing $= -R\, \Sigma x_i \ln x_i = -R[0.5 \ln 0.5 + 0.5 \ln 0.5]$; work $= T\, \Delta s = -RT \ln 0.5$

8. (a) Intensive
(b) Intensive
(c) Intensive
(d) Extensive
(e) Intensive

9. (c) Joule-Thompson coefficient is $\left(\dfrac{\partial T}{\partial P}\right)_H$. The process is isenthalpic and adiabatic.

10. $\Delta H = 3 \times -393.3 + 4 \times -284.5 - (-125.5) = -2192.4$ kJ
$\Delta H = \Delta E + \Delta W$; $\Delta W = \Delta n\, RT$
$\Delta n\, RT = (3 + 4 - 1)\, 8.314 \times 298 = 14864.4$ kJ
$\Delta E = \Delta H - \Delta W = -2192.4 - 14864.4 = -17057.8$ kJ

11. (c)

12. (d)

13. (a)

14. (c) and (d)

15. (d)

16. (d)

17. (a)

18. $\int PV/\delta V = \dfrac{\frac{1}{2}(5-2)(0.03-0.01)}{(0.03-0.01)} = 1.5$ bar

Effective pressure = 2 + 1.5 = 3.5 bar

19. $dQ = dE + dW$; $dW = pdV = 1.2 \times 0.01 = 0.012$ m^3 atm

= 1200 Joules

$dE = mC_V dT = 0.1 \times 700 \times dT = 70\ dT$ Joules

$dQ = VI \times 90 = 12 \times 1.5 \times 90 = 1620$

$1620 = 70\ dT + 1200$; $dT = 6$ K

20. For a reversible engine.

$$\frac{Q_1}{T_1} + \frac{Q_2}{T_2} - \frac{Q_3}{T_3} = 0$$

$$\frac{100}{1000} + \frac{50}{500} - \frac{Q_3}{300} = 0;\ Q_3 = 60 \text{ Joules}$$

$$\%\ \text{efficiency} = \frac{100 \times \text{work done}}{\text{Heat input}} = \frac{100\,(150-60)}{100+50} = 60\%$$

21. $\Delta S = C_p \ln\left(\dfrac{T_2}{T_1}\right) - R\ \ln\,(P_2/P_1)$

$$= 1 \times \ln\left(\frac{520}{800}\right) - 0.300 \ln\,(1/6) = 0.1068\ \frac{\text{kJ}}{\text{kg. K}}$$

22. $W = \dfrac{RT_1}{r-1}\left[1 - \left(\dfrac{1}{6}\right)\dfrac{0.43}{1.43}\right] = 232.5$ Joules/kg air

Maximum work = 232.5 + Unavailable work

$$= 235.5 + 0.1068 \times \left(\frac{800+520}{2}\right) = 306 \text{ J/kg air}$$

23. Steam volume = 0.9 m^3; $v_f = 0.001$ m^3/kg; $v_g = 1.7$

Steam amount = 0.9/1.7 = 0.53 kg

Water = 0.1/0.001 = 100 kg

Dryness fraction = 100 × 0.53/(100 + 0.523) = 0.527%

24. Increase in Entropy of mixing of ideal gas behaviour is = $R\ Sy_k \ln\left(\frac{1}{y_k}\right)$

$y_1 = 0.5$, $R = 8314.3$ J/(kg mol K)

$$= 8314.3\left[0.50\ln\left(\frac{1}{0.5}\right)+0.5\ln\left(\frac{1}{0.5}\right)\right] = 5763 \text{ J/(kg mole K)}$$

25. At triple point total pressures of solid and liquid ammonia should be equal to 1 atm.

The temperature is obtained by trials.

At 200 K, total pressure is = 0.932 + 0.856 = 1.788 atm

At 190 K, total pressure is = 0.382 + 0.347 = 0.73 atm

At 195 K, total pressure is = 0.5784 + 0.5762 = 1.14 atm

So temperature at triple point is nearly 195 K.

26. (a)

27. (a)

28. (c)

29. (c)

30. (b)

Work done in an isothermal process = $n\,RT\ln(V_2/V_1) = 1 \times 8.314 \times 600 \times \ln(8/1) = 10373$ Joules

31. (b)

$f_i = f_i^{sat}$ exp ($Vl(P\text{-part})/RT$) = 6738.9 exp (25.28 × (9000 – 8592.70/(8314 × 573.15) = 6753.7

32. (c)

C_p = 26,693 + 7.333375 × 0.001 T J/mole. K. Heat rejected during the process for one mole is:

$$Q = \int_{T_1}^{T_2} C_p dT = \int_{T_1}^{T_2} (26.693 + 7.375\times 0.001\,T)\, dT = (26.693\ T + 7.365 \times 0.001\ T{*}T{*}/2)\Big|_{T_1}^{T_2};$$

$T_1 = 500 + 273 = 773$ K; $T_2 = 273 - 100 = 173$ K

Q = 26.693(173 – 773) + 7.365 * 0.001 * (173 *173 – 773 *773)/2) = – 18.11 kJ

Negative sign shows that heat is rejected.

33. (b)

Let equilibrium temperature = T_f; Heat loss = heat gain; mb $Cpb(500 - T_f) = mw\ Cpw(T_f - 25)$;
$5{*}0.4{*}(500 - T_f) = 40{*}4.18\ {*}(T_f - 25)$; $T_f = 30.61$°C; $T_f = 303.61$ K

Initial temperature of block = 500 + 273 = 773 K; Initial temperature of water = 25.0 + 273 = 298 K.

$$\Delta S = \int_{773}^{303.61} C_p / T\ dT + \int_{298}^{303.61} C_p / T\ dT = Cpb \ln T\Big|_{773+}^{303.61} Cpw \ln T\Big|_{298}^{303.61} = 0.4$$

ln (303.61/773) + 4.13 ln (330.61/298) = 0.3738 + 0.0779 = – 0.2958 (nearly zero)

34. (d)

Power produced by engine = work done

(W) = 100,000 kW; H.E. = 50% of Carnot cycle or engine = 1/2 (Carnot engine)

$- 1/2\ (1 - \frac{T_2}{T_1}) = 1/2 \left(1 - \frac{300}{600}\right) = 5/16$; H.E. = W/Q_2; $Q_2 = 100{,}000\ T_1 *16/5 = 320{,}000$

35. Work done by turbine = $W_T = h_1 - h_2 = 3443.9 - 2676 = 767.9$ kW.

36. $x_i = 0.1686$; $y_1 = 0.5714$; Now, $P = 39.233$ KPa; $P_i^{sat} = 84.562$ KPa
$\gamma_i = y_i\ P/(x_i\ P_i^{sat}) = (0.5714 *0.39233)/(0.1686 *84.562) = 1.572$

37. $V_1, T_1 = 100°C$; $V_2 = V_1 + x *V_{fg}$
$1.789 = 1.673 + x*1.789$
$V_2, T_2 = 98°C$; $x = 0.065$

38. From steam tables.

39. For f/P ratio, since P tends to zero, fugacity tends to P.
So $f/P = 1$

40. One portion of the container is vacuum, and the other is filled with gas and since the container is perfectly insulated, thus the wall separating the two portions when removed,then free expansion takes place. The free expansion prduces cooling. So the final temperature of gas in the container will be less than 298 K. This also can be explained by adiabatic process, using, $T_2/T_1 = (V_1/V_2)^{\gamma-1}$ where γ is greater than one.

41. $\Delta H_{298}° + \Delta H_P° = \Delta H$; C_P of product = 2*40 + 1*40 + 7.52*40 = 420.8;

$$T = 298.15 - \frac{\Delta_{298}^{\ 0}}{C_p} = 298.15 - (-730{,}000/420.8) = 2032.94°K = 1735°C$$

The above question's data is : one mole of methane undergoes complete combustion in a stoichiometric amount of air. The reaction proceeds
$CH_4 + 2O_2 \rightarrow CO_2 + 2H_2O$.
Both the reactants and products are in gas phase. $\Delta H_{298}° = -730$ kJ/mole of methane.

42. Humidity of air = 2.34/(100 – 2.34) *18/29 = 0.01487 kg water/kg dry air.

$$\text{Humid volume} = \left(\frac{0.01487}{18} + \frac{1}{29}\right) *22.41\ \frac{293}{278}\ \frac{1.013*10^5}{1000} = 0.8602$$

Kg of dry air = 1000 * 1/0.8602 = 162.5 kg
Amount of water vapour = humidity * kg of dry air
= 0.01487 *162.52 = 17.28 = 17 kg.

43. The system consists of two non-reacting species in two phases. If there are no azeotropes, by phase rule, F(degrees of freedom) = $C - P + 2$ where C is number of components, P = number of phases, Here $P = 2$, $C = 2$, $F = 2 - 2 + 2 = 2$.
Since azeotrope, one more relation is $x_i = y_i$;
So $F = 2 - 1 = 1$

44. (b)
$PV = RT$ for ideal gas,
Here $P_1 = 300$ kPa, $T_1 = 300$ K; $P_2 = 330$ kPa; T_2 = ?
For constant volume process, $P_1/P_2 = T_1/T_2$

$300/330 = 300/T_2$; $T_2 = 330$ K; Change in internal energy $= dU = C_V\, dT$;
$dU = 21\ (330 - 300) = 630$ J/mol.

45. (d)

The given equation is :
$(V - b)\ P = RT$ work done,

$$W = \int_{V_1}^{V_2} PdV = \int_{V_i}^{V_f} RT/(V-b)\, dV = RT \ln (V - b) \int_{V_i}^{V_f} = RT[\ln (V_f - b) - \ln (V_i - b)]$$

$$RT \ln \left(\frac{V_f - b}{V_i - b}\right)$$

46. (a)

Efficiency of the cycle $= 1 - \dfrac{T_2}{T_1} = 1 - 373/573 = 200/573$

Maximum work obtained = 1000 *200/573 = 349.04

47. (d)

$\ln P^{sat} = A - \dfrac{5000}{T}$ where A is a constant, P^{sat} is pressure of water.

$T = 100°C = 373$ K; V_P of water, $P^{sat} = 1$ atm; $\ln 1 = A - \dfrac{5000}{373}$

$A = \dfrac{5000}{373}$; $\ln P^{sat} = 5000\left(\dfrac{1}{373} - \dfrac{1}{T}\right)$; at $T = 50°C$

$P^{sat} = 0.13$ atm.

48. (c)

$N_2 + 2O_2 \rightarrow 2NO_2$; $\Delta G° = 100$ kJ/mol; ...(1)

$NO + \frac{1}{2} O_2 \rightarrow NO_2$; $\Delta G° = -35$ kJ/mol. ...(2)

Multiplying by 2, the above equation becomes :

$2NO + O_2 \rightarrow 2NO_2$; $\Delta G° = -70$ kJ/mol. ...(3)

Substracting eqn. (3) from eqn. (1) gives :

$N_2 + 2O_2 \rightarrow 2NO_2$

$\Delta G° = 100$ kJ/mole; $2NO + O_2 \rightarrow 2NO_2$;

$\Delta G° = -70$ kJ/mole

$N_2 + O_2 \rightarrow 2NO$; $\Delta G° = 170$ kJ/mole

$N_2 + O_2 \rightarrow 2NO$; $\Delta G° = 170$ kJ/mole

Thus for one mole of NO, $\Delta G° = 85$ kJ/mole.

❑

CHAPTER 9

Chemical Reaction Engineering

1. $G^0 = -RT \ln K$; G^0 = free energy change; K = equilibrium constant.
2. Homogeneous: Single phase reaction.
3. Heterogeneous reaction: More than one phase.
4. Variables effecting reaction in homogeneous systems: Temperature, pressure, and composition.
5. Reaction rate, $ri = dNi/(V\, dt)$; moles i/vol. of fluid. time; for catalyst, $ri = dUi/(V\, W.dt)$ = moles formed/(mass of solid* time)
6. Series reactions are: $A \rightarrow R \rightarrow S$
7. Parallel reactions: $A \rightarrow R$ and $A \rightarrow R$
   $\searrow S$ $A \rightarrow S$
8. Molecularity: This is the number of moles involved in the reaction. This is an integer.
   Order of reaction: This is the sum of the powers in the rate equation.
9. Rate of reaction: $aA + bB + cC + \ldots \rightarrow$ ....

   $r_A = KC_A^a\, C_B^b\, C_C^c\, C_D^d \ldots$; $r_A = K\, p_A^a\, p_B^b\, p_C^c \ldots$ where C is concentration, p is partial pressure.
10. Free radicals: $H_2 + Br_2 \rightarrow 2HBr$ is explained as:

    $B^*r_2 \rightleftarrows 2Br$ (intiation and termination)

    $Br + H_2 \rightleftarrows HBr + H^*$ (propagation)
    $H^* + Br_2 \rightarrow HBr + Br^*$ (propagation)
11. Enzyme reactions: $A \rightarrow R$; A + Enzyme $\rightleftarrows$ (A^* Enzyme)*
    (A^* Enzyme)* $\rightarrow R$ + Enzyme.
12. Arrhenius law; $K = K_O\, e^{-}\, E/RT$ (variation of rate constant with temp.)
13. Collision theory: $K \infty T^{1/2}\, e^{-E/Rt}$
14. Transition state theory: $K \infty T\, e^{-E/RT}$
15. $r_i = 1/V\, (dN_i/dt) = dC_i/dt = 1/RT\, (dp_i/dt)$ (rate of reaction)
16. Integral method of analysis: $-\int dC_A/f(C_A) = Kt$
17. First order: $-\ln(1 - X_A) = Kt = -\ln(C_A/C_{A0})$; X_A is mole fraction of A and t is time.

18. 2nd order reaction: $2A \rightarrow$ Products, the conversion is:
$1/C_A - 1/C_{AO} = Kt$;

19. 2nd order reaction: $A + B \rightarrow$ Products, the conversion is:

$$\ln\left(\frac{1-X_0}{1-X_A}\right) = C_{AO}\,(M-1)\,Kt = (C_{BO} - C_{AO})\,Kt;\ M = C_{BO}/C_{AO}$$

20. For $A + 2B \rightarrow$ Products (if second order), the concentration equation is:

$$\frac{1}{C_A} - \frac{1}{C_{AO}} = 2Kt\ (M = 2) \text{ where } M = C_{BO}/C_{AO}.$$

21. For 3rd order reaction. $A + B + D \rightarrow$ Products, the relation is:

$$\frac{1}{(CAO - CBO)\,(CAO - CDO)} \text{Ln}\,(CAO/CA) + \frac{1}{(CBO - CDO)\,(CBO - CAO)} \text{Ln}\,(CBO/CB)$$

$$+ \frac{1}{(CDO - CAO)\,(CDO - CBO)} \text{Ln}\,(CDO/CD) = Kt$$

22. $A + 2B \rightarrow R;\ DCA/dt = -K\ CA\ CB^2;\ \ 1/CA^2 - 1/CAO^2 = 8Kt\ \ (M = 2)$

23. For $A + 2B \rightarrow R;\ r_A = -KCA\ CB^2$; Rate equation is: $1/CA^2 - 1/CAO^2 = 2Kt\ (M = 1)$

24. nth order: $CA^{1-n} - CAO^{1-n} = (n-1)\ Kt;\quad n \neq 1.$

25. Zero order reaction: $-rA = dCA/dt = K$; the concentration equation is:
$CAO - CA = Kt = CAO * XA.$

26. Overall order of irreversible reactions from half life:

$$t_{1/2} = \frac{2^{n-1} - 1}{K(n-1)} CAO.$$

27. Irreversible reactions from half life in parallel:

$$A \xrightarrow{K_1} B;\ \ A \xrightarrow{K_1} S;\ -r_A = -dVA/dt = K_1CA + K_2CA$$

$r_R = dCR/dt = K_1\ CA;\ r_S = dCS/dt = K_2\ CA$:

$$\text{Solving; } \frac{CR - CRO}{CS - CSO} = K_1/K_2;\ \text{Ln}\,(CAO/CA) = (K_1 + K_2)\ t.$$

28. Irreversible reaction in series:

$$A \xrightarrow{K_1} R;\ \ R \xrightarrow{K_2} S;\ dCA/dt = rA = -K_1CA;\ rR = dCr/dt = K_1\ Ca - K_2\ CR.$$

$r_S = dC_S/dt = K_2\ CS$; from equation of rA . $CA = CAO\ \ e^{-K_1}\,t$; from rR equation replacing CA and solving gives:

$$C_R = CAO\ K_1\left(\exp\frac{-K_1\ t}{K_2 - K_1} + \frac{\exp - K_2\ t}{K_1 - K_2}\right);\ \text{using } CAO = CA + CB + CC;$$

$$CS = CAO\left(1 + \frac{K_2}{K_1 - K_2}\exp - K_1\, t + \frac{K_1 \exp - K_2\, t}{K_2 - K_1}\right)$$

29. For first order reversible reactions:

$A \underset{K_2}{\overset{K_1}{\rightleftarrows}} R$; $Kc = K$ = equilibrium constant; $M = CRO/CAO$;

$dC_R/dt = -\,dCA/dt = CAO\; dXA/dt = K_1\;\; CA - K_2\;\; CR$; $-$ Ln $(1 - YA/XAO)$

$$= -\,\mathrm{Ln}\left(\frac{CA - CAO}{CAO - CAe}\right) = \frac{M+1}{M + X\,Ae}K_1\;\; t;$$

30. Variable volume batch reactor;
$V = VO\,(1 + EA\;XA)$; $EA = (V_{XA} = 1 - V_{XA} = O)/V_{XA} = 0$;
$-\,r_A = CAO\,(1 + EA\;XA)\; d \times A/dt = dCA/dt$;

31. Temperature and reaction rate: Arrhenius law: $K \propto e^{-E/RT}$
Collision or transition state theory: $K \propto T^m\, e^{-E/RT}$;

32. Ideal batch reactors: $t = NAO \int_0^{X_A} d \times A/(-r\;A)V) = -\int_{CAO}^{CA} dCA/-rA$

33. Steady state mixed flow reactor: $T = 1/S = V/VO = V\;CAO/FAO$:
$= CAO\;\; XA/-\,rA = (CAO - CA)/-\,r_A$.

34. Steady state plug flow reactor: Input = output + disappearance + accumulation (zero)
$F = (F_A + dF_A) + (-\,r_A)\; dV$; $dF_A = FAO * dX_A = r_A * dV$

$$T = V/Vo = CAO\int_0^{XAf} dXA/(-rA) = -\int_{CAO}^{CAF} dCA/-\,rA.$$

35. Heat of reaction: $\Delta H_{r2} = \Delta H_{r1} + \Delta\alpha\;(T_2 - T_1) + \Delta\beta/2\;(T_2^{\,2} - T_1^{\,2}) + \Delta r/3\;(T_2^{\,3} - T_1^{\,3})$

36. Equilibrium constant: $AG° = -RT \ln K$
Equilibrium conversion: Ln $(K_2/K_1) = -\,\Delta H_r/R\;(1/T_2 - 1/T_1)$

37. The thermodynamic, equilibrium constant is uneffected by the pressure of the system, pressure or absence of inerts, or by kinetics of the reaction but is effected by the temperature of the reaction. The equilibrium concentrations of material and equilibrium conversion of reactants are influenced by pressure, inerts, or by the kinetics of the reaction. For $K >> 1$ (equilibrium constant) shows, the reaction is irreversible. For $K << 1$ indicates the reaction will not proceed to appreciable extent. For an increase in temperature, equilibrium conversion rises for endothemic reaction and drops for exothermic reaction. Increase of pressure in gas reactions, conversion rises when the number of moles decreases with reaction, conversion drops when the number of moles increases with reaction. A decrease in inerts for all reactions acts in the way that an increase in pressure acts for gas reactions.

38. Residence time distribution: (a) F – curve (step input); (b) C – curve (impulse input (c) E – the age distribution of fluid leaving a vessel. dF/dt.

39. Fluid-particle reactions: The models are: (1) Progressive conversion model (2) Unreacted core model (represents reality).

1. The reactant gas enters and reacts through the particle at all times) most likely at different rates at different locations within the particle. Thus solid reactant is converted continuously and progressively through particle.
2. Developed by Yagi and Kunii (1955). The steps are:
   (a) Diffusion of gaseous reactant *A* through the film surrounding the particle to the surface of the solid.
   (b) Penetration and diffusion of *A* through the blanket of ash to the surface of the unreacted core.
   (c) Reaction of gaseous *A* with solid at the reaction surface.
   (d) Diffusion of gaseous products through the gas film back into the main body of fluid.
   Equation for (a) is: $t/T = 1 - (rc/R)^2 = XB$
   Equation for (b) is: $t/T = 1 - 3\,(1 - XB)^{2/3} + 2\,(1 - XB)$
   Equation for (c) is: $t/T = 1 - rc/R = 1 - (1 - XB)^{1/3}$

40. Experimental methods for finding rates:
(a) Differential (flow) reactor; (b) Integral (plug flow) reactor; (c) Mixed reactor; (d) Batch reactor.

For (a) $(-r_A')$ avg. = $FAo\,(X_A \text{ out} - X_A \text{ in})/w$; For (b): $w/FAo = \int_O^{XA} dxA/(-rA)$

For (c): The equation is $w/FAo = XA$ out/$-rA$ out.
For recycle reactor: $w/FAo = (R + 1)\, dXA/-rA'$

41. Fluidized models:
(a) Dispersion and tanks in series; (b) RTD models; (c) Contact time distribution models; (d) Two region model; (e) Hydrodynamic flow model.

42. Catalyst deactivation occurs due to; (a) Decay reactions; (b) Pore diffusion; (c) Form of surface attack by poision; (d) Pore mouth blocking, reactvation.

42. Elementary reactions: In these rate equation corresponds to stoichiometric equation. In non-elementary reaction there is no correspondence between rate and stoichiometry.

43. Equilibrium means: From thermodynamics, if a system is in equilibrium with surroundings, $(\Delta G) > 0$. From statistical mechanics point of view the state of the system that has the overwhelmingly great probability of occuring. From chemical kinetics, the system is at equilibrium if the rates of change of all the forward and reverse elementary reactions are equal.

44. A reaction is zero order when the rate of reaction is independent of concentrations of materials. Thus: $-r_A = dC/dt = K$ (a constant).

45. Three types of ideal reactors:
(a) Batch reactor: (b) Plug flow reactor (c) Mixed flow reactor:

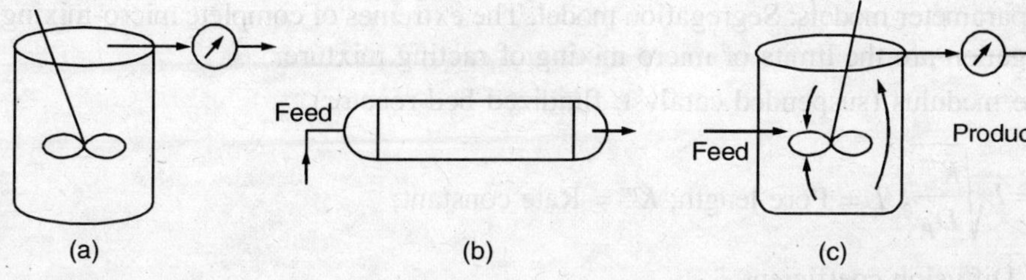

46. Material balance for the reactor:

$$\left.\begin{matrix}\text{Rate of reactant}\\ \text{flow into the}\\ \text{reactor}\end{matrix}\right\} - \left.\begin{matrix}\text{Rate of reactant}\\ \text{flow from the}\\ \text{reactor}\end{matrix}\right\} \mp \left.\begin{matrix}\text{Rate of reactant}\\ \text{loss due to chemical}\\ \text{reaction}\end{matrix}\right\} = \begin{matrix}\text{Rate of accumutation}\\ \text{of reactant in reactor}\end{matrix}$$

For non-isothermal reactions, the energy balance is:

$$\left.\begin{matrix}\text{Rate of heat}\\ \text{flow into the}\\ \text{reactor}\end{matrix}\right\} - \left.\begin{matrix}\text{Rate of heat}\\ \text{flow from the}\\ \text{reactor}\end{matrix}\right\} \mp \left.\begin{matrix}\text{Rate of heat}\\ \text{change in the}\\ \text{reactor}\end{matrix}\right\} = \begin{matrix}\text{Rate of heat accumutation}\\ \text{in the reactor}\end{matrix}$$

47. Series–parallel reactions:

$$A + B \xrightarrow[K_1]{} R \quad OR \quad A \xrightarrow{K_2 + B} R + \xrightarrow{K_2 + B} S \xrightarrow{K_3 + B} T$$

$$R + B \xrightarrow[K_2]{} S \quad r_A = \frac{dCA}{dt} = -K_1 C_A C_B; r_B = \frac{dCB}{dt} = -K_1 C_1 C_B - K_2 C_R C_B$$

$$S + B \xrightarrow[K_3]{} T \quad r_R = \frac{dCR}{dt} = K_1 C_A C_B - K_2\, C_R C_B$$

$$r_S = \frac{dC_S}{dt} = K_2 C_R C_B$$

$$\frac{r_R}{r_A} = \frac{d_{CR}}{d_{CA}} = -1 + \frac{K_2 C_R}{K_1 CA};$$ This is a first order O.D.E.

The solution is: $\dfrac{C_R}{CAO} = \dfrac{K_1 C_A}{1 - K_2 / K_1}\left[\left(\dfrac{CA}{CAO}\right)^{K_2/K_1} - \left(\dfrac{CA}{CAO}\right)\right] + \dfrac{C_{RO}}{C_{AO}}\left(\dfrac{CA}{CAO}\right)^{K_2/K}$

$$\frac{C_R}{CAO} = \frac{CA}{CAO}\left(\frac{C_{RO}}{CAO} - \ln\frac{CA}{CAO}\right) \text{ for } \frac{K_2}{K_1} = 1.$$

48. One parameter model: Single parameter to account for non ideality of reactor. This parameter is evaluated by analysing the RTD determined from a tracer test. Examples are: Reactor dead volume, fraction of fluid by passing the reactor. Other examples are: tanks in series model and dispersion model.

Two parameter models: Here adjustable parameters are two. 1. Volume of the reactor; 2. Flow rates (e.g. CSTR modeled using by passing and dead space).

Zero parameter models: Segregation model. The extremes of complete micro-mixing and complete segregation are the limits of micro mixing of racting mixture.

49. Thiele modulus (suspended catalyst, fluidized bed reactor),

$M_T = L\sqrt{\dfrac{K^m}{D_P}}$; L = Pore length; K^m = Rate constant;

D_P = Diffusion coefficient

50. Design for Multiple Reactions (Parallel "x"):

(a) $A \xrightarrow[K_1]{} R(\text{desired}); A \xrightarrow[K_2]{} S(\text{unwanted})$

$$r_R = \frac{dC_R}{dt} = K_1 C_A^{a_1}; r_S = \frac{dCs}{dt} = K_2 C_A^{a_2}$$

$$\frac{r_S}{r_R} = \frac{dC_s}{dC_R} = \frac{K_2}{K_1} C_A^{a_2 - a_1}$$ For this ratio to be small for $a_1 = a_2$:

(1) By changing temperature of the operation, (2) By using a suitable catalyst.

(b) $A + B \xrightarrow{K_1} R(\text{desired}); A + B \xrightarrow{K_2} S(\text{unwanted}).$

$$r_R = \frac{dCR}{dt} K_1 C_A^{a_1} C_B^{b_1}; \frac{dC_s}{dt} = K_2 C_A^{a_2} C_B^{b_2}$$

$$\frac{r_S}{r_R} = \frac{dCS}{dC_R} C_A^{a_2 - a_1} C_B^{b_2 - b_1}; \frac{d_{CS}}{dC_R}$$ is to be minimum for maximum C_R.

For reactions in parallel the concentration level of the reactants is the key to proper control of product distribution. A higher reactant concentration favors the reaction of higher order, a low concentration, level has no effect on the product distribution for reactions of same order.

51. Solid – catalyzed reactions:

Spectrum of kinetic regimes:

1. Surface kinetics or what happens at the surfaces; interior or exterior of the particle. This involves adsorption of reactant A onto the surface, reaction on the surface or desorption of product back into gas stream.
2. Pore diffusion resistance which may cause the interior of the particle to be starved for reactant.
3. Temperature gradients within the particle. This is caused by lot of heat. Release or absorption during reaction.
4. Film ΔT between the outer surface of the particle, and the main gas stream. For example, the particle has uniform temperature throughout but hotter than surrounding gas.
5. Film diffusion resistance or concentration gradients across the gas film surrounding the particle.

Rate equation for surface kinetics of $A + B \rightleftarrows R + S$, occuring in pressure of inert carrier material U, the rate expression when adsorption of A controls is:

$$-r_A'' = \frac{K(P_A - P_R P_S / kP_B)}{(1 + K_A P_R P_S / kP_B + K_B P_B + K_R P_r + K_S P_S + K_U P_U)}$$

For desorption of R:

$$-r_A'' = \frac{k(P_A P_B / P_S - P_R / K)}{(1 + K_A P_A + K_B P_B + KK_R P_A P_B / P_S + K_S P_S + K_U P_U)}$$

For diffusion resistance combined with surface kinetics:

Single cylindrical pore, first order reaction: A → Product

$$-r_A'' = -\frac{1}{S}\frac{dN_A}{dt} = K''C_A; C_A/C_{Ai} = \frac{\cosh m(L-x)}{\cosh mL}$$

$m = \sqrt{\frac{2K''}{Dr}}$; K'' is rate constant; D = diffusion coefficient;

r = Radius

Experimental methods for finding rates:

Differential reactor: $\frac{w}{FA_O} = \frac{X_{A\,\text{out}} - X_{A\,\text{in}}}{(-r_A')_{\text{out}}}$

X = mole fraction of component A;

w = weight of catalyst; FA_O = feed flow rate; $(-r_A')_{\text{out}}$ = Rate of reaction based on outlet composition

Integral reactor: $\frac{w}{FAO} = \int_0^{X_A} \frac{dX_A}{(-r_A)}$

Mixed flow Reactor: $\frac{w}{FAO} = X_{A\,\text{out}} / -r_{A\,\text{out}}$

Recycle Reactor: $\frac{w}{FAO} = (R+1)\int_{(R/R+1)\,xAf}^{X_Af} dX_A / -r_A'$

Batch Reactor: $\frac{1}{CA_O} = \int_0^{X_A} \frac{dX_A}{-r_A} = \frac{V}{W}\int \frac{dXA}{-r_A'}$; V = Volume of gas.

Minimum fluidizing velocity:

$$150(1 - E_{mf})\, u_{mf} + 1.75\, u_{mf}^2\, dp = E_m\, (dp)^2$$

Terminal velocity, $u_t = \left[\frac{18}{dp^2} + \frac{0.591}{dp^{1/2}}\right]^{-1}$

For irregular shape particles: $u_t = \left[\frac{18}{dp^2} + \frac{2.33 - 1.744\phi_S}{dp^{1/2}}\right]^{-1}$

Batch reactor: No input or output until the reaction is completed.

Continuous reactor: Composition at any point does not change with time.

Semi-batch: One stream is sent as steady flow and the other stream is sent as batch.

In plug-flow reactor/batch reactor: $A \xrightarrow{K_1} R \xrightarrow{K_2} S$

$$\frac{dCA}{dt} = -k_1 C_A; \frac{dC_R}{dt} = -K_2 C_R + K_1 CA;$$

$$\frac{C_A}{CAO} = e^{-K_1 t}; \frac{C_R}{CA_O} = \frac{K_1}{(K_2 - K_1)}(e^{-K_1 t} - e^{-K_2 t})$$

$$\frac{dCS}{dt} = K_2 CR; C_S = C_{AO} - C_A - C_R;\ \text{Feed has no R and S.}$$

$$\frac{CR\max}{CA_O} = \left(\frac{K_1}{K_2}\right)^{K_2/(K_2 - K_1)}$$

In mixed flow reactor: $A \xrightarrow{K_1} R \xrightarrow{K_2} S; \ \frac{CR\max}{CAO} = \frac{1}{\left[(K_2/K_1)^{1/2} + 1\right]^2};$

For plug flow/batch reactor: $A + B \xrightarrow{K_1} R; R + B \xrightarrow{K_2} S$

$$r_A = \frac{dCA}{dt} = -K_1 C_A C_B; r_B = \frac{d_{CB}}{dt} = -K_1 C_A C_B - K_2 C_R C_B;$$

$$r_R = \frac{dCR}{dt} = K_1 C_A C_B - K_2 C_R C_B; \frac{r_R}{r_A} = \frac{d_{CR}}{dCA} = -1 + \frac{K_2 CR}{K_1 CA}$$

Solution is: $\frac{CR}{CAO} = \frac{1}{1 - K_2/K_1}\left[\left(\frac{CA}{CAO}\right)^{K_2/K_1} - \left(\frac{CA}{CAO}\right)\right] + \frac{CRO}{CAO}\left(\frac{CA}{CAO}\right)^{K_2/K_1}; \frac{K_2}{K_1} \neq 1$

$$\frac{CR}{CAO} = \frac{CA}{CAO}\left(\frac{CRO}{CAO} - \ln\frac{CA}{CAO}\right); \frac{K_2}{K_1} = 1$$

For missed flow reactor: $C_R = \frac{CA(CAO - CA + CR_O)}{CA + (K_2/K_1)(CAO - CA)}$

The rate equation for mass transfer of A (absorption only)

$$-r_A = \frac{1}{(1/kAga + HA/k_{Ala})} * (P_A - H_A C_A); P_A = H_A C_A$$

Rate equation for mass transfer and reaction

A(gas) + bB(liquid) $\rightarrow R$ (product) instantaneous reaction)

$$-r_A = \frac{1}{S}\frac{dNA}{dt} = \frac{\frac{D_{Bl}}{D_{Al}}\frac{C_B}{b} + \frac{B_A}{H_A}}{\frac{1}{H_A K_{Ag}} + \frac{1}{K_{Al}}}$$

Fluid–particle reaction:

2 ZnS (s) + $3O_2$ (g) $\rightarrow$ 2 ZnO (s) + $2SO_2$ (g)

Models used are: (1) Progressive conversion model (PCM) and (2) Shrinking un-reacted core model (SCM)

On first one, gas enters the solid and reacts throughout the particle. Solid reactant is converted continuously.

In second model, here reaction is first at outer skin of the particle. The zone of reaction moves into the solid, leaving behind converted material (ash). At any time there exists unreacted core which shrinks in size during reaction.

Models for building fluidized beds:

1. Dispersion and tanks in series model
2. RTD model
3. Contact time distribution models
4. Two region models
5. K.L. model for batch fluidized bed.

Circulating fluidized bed: When gas velocities are higher than batch fluidized beds.

Turbulent bed: At higher gas velocities the batch fluidized bed (BFB) tranforms into turbulent fluidized bed.

Fast fluidized bed (F.F): $U_{TB-FF} = \dfrac{1.53\sqrt{(\rho_S - \rho_g)\, g\, dP}}{\rho_g}$

Pneumatic conveying (PC):

$$\frac{U_{FF} - PC}{\sqrt{g\, dP}} = 21.6\left(\frac{G}{\rho a\, U_{FF} - PC}\right)^{0.542} (dP)^{0.315}$$

QUESTIONS

SECTION—A

1. The rate of a chemical reaction is expressed as:

(a) Moles per unit time (b) Moles/(unit volume) (unit time)

(c) Moles/(unit volume) (d) None of these

2. Rate of a chemical reaction is function of:

(a) Temperature only (b) Pressure only

(c) Composition only (d) Temperature, pressure and composition

3. Arrhenius theory states that rate constant k is proportional to:

(a) T (b) $T_e^{-E/RT}$

(c) $e^{-E/RT}$ (d) $T^{1/2}e^{-E/RT}$

4. Integral method of analysis of kinetic data is used for:

(a) Testing specific mechanism (b) Large amount of accurate data

(c) Scattered data (d) None of these

5. Differential method of analysis is preferred when we have:

(a) To develop a rate equation to fit data (b) Scattered data

(c) To test specific mechanism (d) None of these

6. An ideal plug flow reactor has:

(a) Uniform mixing (b) Axial dispersion

(c) Flat velocity profile (d) None of these

7. The composition in the reactor and at the exit of the reactor is same in a:

(a) Tubular reactor (b) Back mix reactor
(c) Batch reactor (d) Semi-batch reactor

8. A liquid '*A*' decomposes according to the following reaction scheme:

$$A \xrightarrow{K_1} R; \quad A \xrightarrow{K_2} S; \quad A \xrightarrow{K_3} T$$

The rate $(-r_A)$ is given by:

(a) $(K_1 + K_2 + K_3)\, C_A$ (b) $K_1\, C_A$
(c) $(K_1 + K_2)\, C_A$ (d) $(K_2 + K_3)\, C_A$

9. The fractional conversion of a first order reaction in a given time depends on initial concentration as below:

(a) Rises (b) Decreases
(c) Remains constant (d) None of these

10. 1 litre/sec of gaseous reactant *A* is introduced into a mixed reactor of volume 2 litres. the space time is:

(a) 2 sec (b) $\frac{1}{2}$ sec
(c) 1 sec (d) None of these

11. For any given duty and for all positive reaction orders,the size of mixed reactor is:

(a) Higher than plug flow reactor (b) Smaller than plug flow reactor
(c) Same as plug flow reactor

12. The units of activation energy are:

(a) cal (b) kcal
(c) cal/g mol (d) None of these

13. The rate equation r_B for the reaction $A \xrightarrow{K_1} B \xrightarrow{K_2} C \xrightarrow{K_3} D$ is given by:

(a) $K_1 C_B$ (b) $(K_1 + K_2 + K_3)C_B$
(c) $(K_1 + K_2)C_B$ (d) $K_1 C_A - K_2 C_B$

14. The point selectivity of product *B* in the reaction

$A \xrightarrow[K_2]{K_1} B$ is:

$\swarrow$
C

(a) $\dfrac{K_2}{K_1}$ (b) $K_1 + K_2$
(c) $\dfrac{K_1}{K_2}$ (d) $K_1 - K_2$

15. Thermodynamic equilibrium constant in a system is affected by:

(a) Both pressure and temperature (b) Pressure, temperature and inerts
(c) Temperature only (d) Both pressure and inerts

16. For an increase in pressure in gas reactions, when the number of moles decreases conversion:

(a) Decreases (b) Increases
(c) Unaffected

17. The response curve for a step input signal from a reactor is called:

(a) E-curve (b) C-curve
(c) F-curve (d) I-curve

18. The response curve equation for a pulse input in a mixed flow reactor is:

(a) $1 - e^{-\theta}$ (b) $\pm e^{-\theta}$
(c) $1 + e^{+\theta}$ (d) $1 + e^{-\theta}$

19. A vessel in which the fluid enters and leaves solely by plug flow is called:

(a) Closed vessel (b) Open vessel
(c) Open-closed vessel (d) None of these

20. The dispersion number for a plug flow reactor is:

(a) 1 (b) ∞
(c) 0 (d) None of these

21. The dispersion number for a CSTR is:

(a) ∞ (b) 0
(c) 1 (d) None of these

22. The flow in a real packed bed can be approximated:

(a) Plug flow model (b) Mixed flow model
(c) Dispersion model (d) Tanks-in-series model

23. The variance of C-curve in a tanks-in-series model consisting of n tanks is given by:

(a) $\frac{1}{n}$ (b) n
(c) n^2 (d) 1

24. Unreacted core model represents the following reaction:

(a) Manufacture of CS_2 from elements (b) Roasting of sulphide ores
(c) Sodium cyanide from sodium amide (d) Coal particle combustion

25. When a fluid solid (non-catalytic) is found to be highly temperature sensitive then the controlling step is:

(a) Gas film diffusion (b) Chemical reaction
(c) Ash diffusion (d) None of these

26. The relationship between the time needed to achieve the same fractional conversion for particles of different sizes (R), when chemical reaction control is:

(a) $t \propto R^{1.5}$ (b) $t \propto R^3$
(c) $t \propto R$ (d) $t \propto R^2$

27. Due to the presence of a catalyst, the equilibrium constant:

(a) Increases (b) Decreases
(c) Unaffected

28. The physical significance of effectiveness factor of a catalyst pellet is that it is a measure of:
(a) Gas film resistance (b) Pore diffusion resistance
(c) Chemical reaction resistance (d) None of these

29. Thiele modulus is defined as:
(a) $\frac{k}{D}$ (b) $\sqrt{\frac{k}{D}}$
(c) $\frac{D}{K}$ (d) $L\sqrt{\frac{k}{D}}$

30. If T is Thiele modulus, effectiveness factor for a first order reaction is:
(a) $\frac{\tan T}{T}$ (b) $\frac{\tan(hT)}{T}$
(c) $\frac{T}{\tan T}$ (d) $\frac{\tan(hT)}{\tan T}$

31. Pore diffusion resistance in a catalyst is cosidered negligible if Thiele modulus is:
(a) > 1 (b) < 1
(c) < 0.5 (d) > 0.5

32. Knudson diffusion is affected by:
(a) Pressure (b) Temperature
(c) Temperature and pressure (d) None of the above

33. Bulk diffusin in catalyst pore is:
(a) Directly proportional to pressure (b) Inversely proportional to pressure
(c) Not affected by pressure

34. Knudson diffusion is proportional to:
(a) T (b) $T^{1/2}$
(c) $1/T$ (d) T^2

35. Brunauer-Emmet-Teller method uses the extension of the following isotherm:
(a) Freundlich (b) Tempkin
(c) Langmuir (d) None of these

36. The gas usually used in B.E.T method is:
(a) Oxygen (b) Helium
(c) Nitrogen (d) Hydrogen

37. Addition of promoter to a catalyst improves:
(a) Surface area (b) Sensitivity
(c) Porosity (d) None of these

38. The catalyst for oxidation of SO_2 is:
(a) Cobalt oxide (b) Alumino-iron oxide
(c) Vanadium pentoxide (d) None of these

39. The production of trichloro-benzene from benzene and chlorine gas belongs to the type of:
(a) Parallel reaction (b) Elementary
(c) Consecutive (d) None of these

40. The units of frequency factor in Arrhenius equation are:
(a) Dimensionless (b) Same as rate constant
(c) Same as activation energy

41. Calcination of lime stone is usually carried out industrially in:
(a) Fixed bed reactor (b) Fluidized bed reactor
(c) Moving bed reactor (d) None of these

42. Thermal cracking of hydrocarbons belongs to the type of reaction:
(a) Heterogeneous catalytic (b) Heterogeneous non-catalytic
(c) Homogeneous catalytic (d) Homogeneous non-catalytic

43. Gas-phase oxidation of SO_2 in chamber process is carried out industrially by:
(a) Non-catalytic reaction (b) Homogenous catalytic reaction
(c) Heterogenous catalytic reaction

44. The heats of adsorption in physical adsorption are of the order of:
(a) 10000 cal/g, mole (b) 100000 cal/g, mole
(c) 1000 cal/g, mole (d) None of these

45. The effectiveness of solid catalysts for reaction depends upon:
(a) Physical adsorption (b) Physical and chemical adsorption
(c) Chemical adsorption (d) None of these

46. Space time for a plug flow reactor (for a first order reaction) is:
(a) $K \ln \dfrac{C}{C_0}$ (b) $\dfrac{1}{K} \ln \dfrac{C_0}{C}$
(c) $\dfrac{1}{K} \dfrac{C_0}{C}$ (d) $K \ln \dfrac{C_0}{C}$

47. A reaction of the type $A \xrightarrow{+B, K_1} R \xrightarrow{+B, K_2} S \xrightarrow{+B, K_3} T$ is:
(a) Series reaction (b) Parallel reaction
(c) Series-parallel reaction (d) None of these

48. For the reaction $N_2 + 3H_2 \rightarrow 2NH_3$, increase in pressure causes:
(a) Higher conversion (b) Lower conversion
(c) No effect

49. For endothermic reactions maximum equilibrium conversion is obtained at:
(a) Lowest possible temperatures (b) Highest possible temperatures
(c) Medium temperatures

50. For exothermic reversible reactions when the temperature is maximum—attainable conversion:
(a) Increases (b) Decreases
(c) No effect

51. Model suggested for a fluidized bed is:
(a) Plug flow (b) Dispersed plug flow
(c) Tank-in-series model (d) Bubbling bed model

52. In a reaction where the particle changes in size, the gas phase rate constant in Stoke's regime is proportional to:

(a) d_p (b) d_p^2

(c) $\frac{1}{d_p}$ (d) None of these

53. Promoter added to iron catalyst in NH_3 synthesis is:

(a) Silica gel (b) Alumina

(c) Sodium oxide (d) None of these

54. Reactor where there is exchange of heat with the surroundings but with sizeble temperature variations is:

(a) Isothermal reactor (b) Adiabatic reactor

(c) Non-adiabatic reactor (d) None of these

55. Catalyst usd in the alkylation of hydrocarbons is:

(a) Nitric acid (b) Nitric oxide

(c) Phosphoric acid (d) Sulphuric acid

56. The equilibrium composition for a reversible reaction with different catalysts is:

(a) Different (b) Same

57. The mechanism of catalysis is explained by:

(a) Collision theory (b) Thermodynamics

(c) Activated complex theory (d) None of these

58. The function of a carrier in catalyst is to:

(a) Improve activity (b) Improve performance

(c) Provide large surface area (d) None of these

59. The type of reactor used for hydrogenation of vegetable oils is:

(a) Plug flow (b) Homogeneous catalytic

(c) Slurry reactor (d) None of these

60. The best combination of reactors for an exothermic reaction is:

(a) Plug flow followed by CSTR (b) CSTR is series

(c) CSTR followed by plug flow reactor (d) None of these

61. Most suitable reactor for autocatalytic reaction is:

(a) Plug flow (b) CSTR

(c) Recycle reactor (d) CSTRs in series

62. Decrease in the size of the temperature interval:

(a) Increases the uncertainty in E (b) Decreases the uncertainty in E

(c) No effect

SECTION—B

1. A reaction is said to be elementary reaction if......

2. The dimensions of a *n*th order reaction rate constant are......

3. A given reaction is ......temperature sensitive at......temperature (choose from, more, less, high, low).

4. A reaction in which one of the......acts as catalyst is called......reaction.

5. The material balance equation used for the design of all reactors is......

6. Performance equations in flow reactors interrelate (1) ......, (2) ......, (3) ...... and (4) ......

7. For any particular duty and for all positive reaction orders the ratio of mixed reactor to plug flow reactor......with order.
8. Activation energy for a reaction is obtained from the...... of the Arrhenius plot.
9. The intercept of the Arrhenius plot is called......
10. The yield of product B in the reaction $A \xrightarrow[K_2]{K_1} B$ is......

C

11. The overall selectivity of product B in the reaction $A \xrightarrow[K_2]{K_1} B$ is......

C

12. For an increase in temperature equilibrium conversion......for endothermic reaction.
13. Presence of inerts......equilibrium conversion of reactants (affects/does not affect).
14. K (equilibrium constant) >> 1 indicates that reaction is......
15. The experimental technique used to study the non-ideal flow reactors is called......
16. The response curve for a pulse input from a reactor is called......
17. The single parameter models proposed for description of non-ideal flow are (1)...... and (2)......
18. The dispersion number for a reactor or vessel is given by......
19. Non-ideal flow occurs in reactors due to......
20. The basic differential equation representing the dispersion model is......
21. The relationship between exit age description function (E) and internal age distribution function (I) is......
22. The relatively slow moving portions of the fluid or stagnant fluid region in a reactor is called......
23. The models used to represent non-catalytic fluid-solid reactions are (1)...... and (2)......
24. Carbon particle combustion is represented by......model.
25. The roasting fo iron pyrites (FeS_2) is represented by the equation......
26. Among a number of steps involved in a reaction, the rate controlling step is the one......
27. The resistance involved in the combustion of a carbon particle are......
28. The design equation for a tubular reactor packed with a catalyst of weight W is......
29. A catalyst is defined as the material......
30. The resistances involved in a catalytic (gas-solid) reaction are......
31. Effectiveness factor of a catalyst pellet is defined as......
32. The technique employed for the measurement of surface area of catalysts is......
33. Two kinds of diffusion occurring in pores of catalyst are (a)......, (b)......
34. The method used for the determination of pore volume and porosity of a catalyst is......
35. A substance is said to be catalyst poison if......
36. The empirical relation proposed for Freundlich isotherm is......
37. The rate of reaction for a catalytic reaction is expressd as......

38. A catalyst is said to be negative catalyst if......

39. Catalytic reactions are divided into (1)...... and (2)......

40. The catalyst used in the gas phase oxidation of SO_2 in chamber process is......

41. The types of adsorption taking place in catalysts are (1)...... and (2)......

42. The type of adsorption studies useful in the determination of catalyst surface area and pore size distribution is......

43. The factors which increase the importance of diffusion in a catalyst are (*i*)......, (*ii*)...... and (*iii*)......

44. The reason for limiting the temperature in the SO_2 oxidation is......

45. Space time of *N* number of CSTRs connected in series (assuming first order reaction) is......

46. The reaction $C_6H_6 \xrightarrow{+HNO_3} C_6H_5NO_2 \xrightarrow{+HNO_3} C_6H_3(NO_2)_3$ belongs to the......type of reaction.

47. Multiple reactions are classified into (*i*)...... and (*ii*)......

48. The variation of equilibrium constant with temperature is given by the equation......

49. A higher temperature favours the reaction of activation energy (choose from lower, higher)

50. Disperesed plug flow is defined as......

51. In a reaction where the particle size changes, the regime where $K_g \propto 1/dp$ is called......

52. A catalyst reduces......over which the reactants must pass to form products.

53. A reactor is said to be ideal when......can be exactly calculated from the knowledge of mixing patterns.

54. Fraction of the material that stays longer than the mean residence time is called......and is represented by the equation......

55. Freundlich equation for adsorption isotherm is given by......

56. For a reaction $A(g) \rightarrow B(g)$ over a catalyst bed, the global rate of reaction is defined as......

57. Catalyst in Petroleum industry is regenerated by......

58. An autothermal reactor is one......

59. Rate of a reaction is independent of concentration for......order reaction.

60. For a reaction the type $A \rightarrow 2R$ increase in pressure......conversion (increases/decreases).

61. It is desirable to monitor the......during initial stages and the......during the final stages of the reaction (choose from products or reactants) to study any reaction.

SECTION—C

1. Given the reaction $2NO_2 + \frac{1}{2}O_2 \rightarrow N_2O_5$. What is the relation between the rate of formation and disappearance of the three components of the reaction?

2. What are the various laws proposed to predict the temperature dependency of rate constant?

3. How is the temperature dependence expressed in the above theories?

4. For a gas reaction the rate is reported as

$$-\frac{dp_A}{dp} = Kp_A{}^2 \text{ atm/hr.}$$

what are the units of rate constant?

5. What are the methods used to interpret kinetic data from a batch reactor?
6. A simple autocatalytic reaction is represented by $A + B \rightarrow B + B$. Starting with a small concentration of B, show in a plot how the rate will rise as B is formed.
7. If the rate of a gaseous reaction is measured by noting the change in partial pressure of one of the components, write down the rate equation in terms of partial pressure 'p' assuming the gaseous mixture is an ideal gas.
8. Let S be the fractional change in volume of the reactor system between no conversion and complete conversion of reactant A. What is the value of 'S' for reaction $A \rightarrow 3B$, starting with 50% inerts?
9. Write down the performance equation for a batch, plug flow and stired tank reactor.
10. Define space time and space velocity in flow reactors.
11. What are the design equations for a batch, plug flow and stirred tank reactor, in which the volume of reacting mixture changes proportionately with conversion. Let S be the fractional change in volume of the reactor system between no conversion and complete conversion for a reactant 'A'.
12. What do you understand by Arrhenius' plot for a reaction?
13. What is the relation between initial concentration (C_0) and final concentration C_4 for a system of 4 equal size mixed reactors in series? The reaction is the first order and t is of residence time.
14. Show in a diagram the effect of temperature of equilibrium conversion for endothermic and exothermic reactions.
15. What are the theories of reaction rate proposed?
16. What are the signals used usually to study non-ideal flow by stimulus response technique?
17. Define mean (t) and variance σ^2 for a C(concentration) and t (time) curve.
18. Write the unsteady state equation representing mass transfer (diffusion) with chemical reaction (first order).
19. Show graphically the response curve of a pulse and step-input in a plug flow reactor.
20. How does the response appear for a step input and pulse input in a back mix reactor?
21. What is the main characteristic of a linear system?
22. What are the types of regions used in the constructin of mixed models?
23. What are the steps involved in the roasting of iron pyrites (FeS_2) as represented by unreacted core model?
24. Show graphically the progress of reaction of a single solid particle with surrounding fluid as represented by unreacted core model in a plot of rc/R and t (time for complete reaction) when chemical reaction is controlling step.
25. What are the expressions relating to fractional conversion (X_B) and fractional time for complete conversion (t/τ) under conditions of gas film diffusion control, ash diffusion control and chemical reaction control?
26. What are the step involved in the cobustion of carbon particle in air (as represented by shrinking core model).

27. Give a pictorial representation of various resistances involved in an unreacted core model.
28. Give a few industrial examples of fluid-solid contacting reactors.
29. What is meant by Knudsen diffusion? Give the equation for the same.
30. What are the methods of preparing the catalysts?
31. What are various types of catalyst poisons?
32. How are the height of a catalytic unit (HCU) and height of reactor unit (HRU) defined?
33. Write down the energy balance equation for a batch reactor with an exothermic reaction and heat exchange. Explain all the terms involved.
34. How is effective diffusivity in a catalyst pellet related to molecular diffusivity and Knudsen diffusivity?
35. Show graphically the design equations for a mixed reactor and a plug flow reactor.
36. A reaction is represented by $A + B \xrightarrow{K_1} R$ and $R + B \xrightarrow{K_2} S$. what is the rate equation for the disappearance of B?
37. State the distinguishing characteristic of each of the following reactions. Single, multiple, elementary and non-elementary.
38. Show the effect of inerts (increase and decrease) on equilibrium conversion of a gaseous reaction $A \rightarrow 2R$ graphically.
39. What are the types of reactors used for the study of kinetics of catalytic reaction?
40. Write the B.E.T. equation used for the measurement of surface area of catalysts.
41. Show graphically the cncentration variation for a plug flow reactor (with distance) and for series of CSTRS (with stages).
42. What are the catalysts used for the following reactions:
    (a) Water gas reaction
    (b) Oxidation of O–xylene to phthalic anhydride
    (c) Hydrogenation of vegetable oil
    (d) Fischer-Tropsch process.
43. What are the factors decided in the design of a reactor?
44. What are the main differences between physical and chemical adsorption?
45. What are the advantages of a fluidized bed reactor?
46. What is the equation of Langmuir Isotherm?
47. At 500 K the rate of a bimolecular reaction is ten times the rate at 400 K. Find the activation energy for this reaction.
48. The pyrolysis of ethane proceeds with activation energy of about 70 kcal. How much faster is the decomposition at 600°C than 500°C?
49. A liquid 'B' decomposes by first order kinetics and in a batch reactor 60% B is converted in 10 minutes. How much longer would it take to reach 80% conversion?
50. After 8 minutes in a batch reactor, reactant (C_{A0} = 1 mole/litre) is 80% converted; afte 18 minutes, conversion is 90%. Find a rate equation to represent this reaction.
51. Two litres/min of liquid containing R and S(C_{RO} = 0.03 mole/litre; C_{SO} = 0.02 mol/litre; C_{sf} = 0.04 mol/litre, C_{Tf} = 0.05 mol/litre). Find the rate of reaction of R, S and T.

52. The reactor setup shown in figure consists of three plug flow reactors in two parallel branches. Branch 'R' has two reactors of 40 and 50 litres volume respectively. Branch 'S' has a reactor of volume of 30 litres. What fraction of the feed should go to branch R?

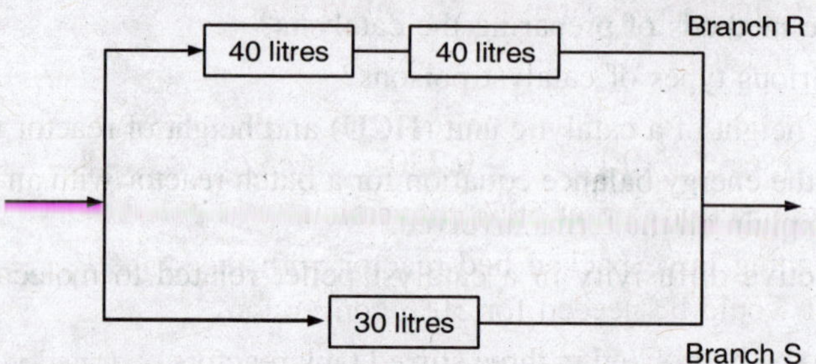

53. A gaseous reaction is given by $E \rightarrow B + C + D$. The initial pressure of the system of 420 mm Hg. Pressure of the system after 85 sec in 662 mm Hg. What are the pressures of components E, B, C and D after 85 sec.?

54. Consider a gas phase reaction $2A \rightarrow B + 2C$ with unknown kinetics. If a space velocity of 2/min is needed for 75% conversion of A in a plug flow reactor, find the corresponding space time and mean residence time of the fluid in the reactor. Calculate the same for a mixed reactor also.

55. Sulfuryl chloride vapours are heated in a closed vessel for 45 min at 320°C and 1 atm initial pressure. The dissociation of SO_2Cl_2 to SO_2 and Cl_2 is a first order reaction. The reaction velocity constant is 0.0013 min^{-1}. Calculate the percentage decomposition of SO_2Cl_2 and the time required to decompose 80% of SO_2Cl_2.

56. A homogeneous liquid phase reaction $B \rightarrow C; -r_A = KC_B^2$ takes with 50% conversion in a mixed reactor. What will be the conversion if this reactor is replaced by one 6 times as large—all else remaining unchanged?

57. Two samples of cracking catalysis have particle densities 1.126 and 0.962 g/c.c. respectively as determined by mercury displacement. The true density of solid material in each case in 2.37 g/c.c. The surface area of the first sample is 467 m^2/g and that of the second is 372 m^2/g. Find the difference in their mean pore radii.

58. A homogeneous reaction $B \rightarrow 4R$ has a reported rate at 200°C with first order rate $-r_B$ = $(0.05/\text{min})C_B$. What size of plug flow reactor operrating at 200°C and 4 atm can produce 75% conversion of feed consisting of 2 g moles of pure B per hour?

59. In an experiment to determine the pore volume and catalyst particle porosity the following data were obtained on a sample of catalyst.

Mass of catalyst sample placed in chamber = 100 g

Volume of helium displaced by sample = 50 c.c.

Volume of mercury displaced by sample = 100 c.c.

Calculate pore volume, density of the catalyst and void fraction of the catalyst particle.

60. A dilute aqueous solution A is to be hydrolyzed continuously at 30°C. At this temperature the rate equation is $r = 0.2\ C_A$ g. moles/c.c. min. The feed rate is 1 litre/min. of solution, with initial concentration of A of 2.0×10^{-4} g moles/c.c. Calculate the conversion:

(a) If one litre vessel was used as a steady-tank flow reactor

(b) If two 5 litres vessels were used in series and

(c) If a tubular flow reactor of 10 litre volume was used.

61. The following kinetic data on the reaction $A \rightarrow B$ are obtained in an experimental bed reactor using various amounts of catalysts and fixed rate.

F_{AO} = 10 kg moles/hr.

W (kg catalyst)	1	2	3	4	5
X_A	0.122	0.230	0.325	0.41	0.48

(a) Find the reaction rate at 50% conversion.

(b) In designing large packed bed reactor with feed rate F_{AC} = 500 kg moles/hr, how much catalyst would be needed for 50% conversion?

62. A liquid is to be hydrolized in three stirred tank reactors operated in series. Suppose that each has a volume of 2 litres and the feed rate to the first reactor is 500 c.c./min. compute the percent hydrolysis accomplished in the three reactors.

The rate of hydrolysis is r = 0.16 g. moles/(c.c.) (min).

63. The concentration readings in table below represent a continuous response to a pulse input into CSTR operating at steady state. Determine the average residence time of the fluid and (t) curve.

Time t(sec)	0	100	200	300	400	500	600	800	1000
Tracer concentration g/c.c.	0.0	6.5	12	12	10	5	25	0.0	0.0

64. The reduction of iron ore of density ρ_B = 4.6 g/c.c. and size R = 5 mm by hydrogen can be approximated by the unreacted core model with no water vapour present. The stoichiometry of the reaction is

$$4H_3 + Fe_2O_3 \longrightarrow 4H_2O + 3Fe$$

with rate approximately proportional to the concentration of hydrogen in the gas stream. The first order rate constant has been measured to be $K = 2.0 \times 10^5\ e^{-2400/RT}$ cm/sec. Find out the time required for complete conversion assuming chemical reaction is rate controlling step.

65. The reduction of a ore takes place according to the equation

$$A(g) + B(s) \longrightarrow C(s) + D(g)$$

A is passed at atmospheric pressure at high flow rate. The temperature was 4500°C. The effective diffusion coefficient is 4×10^{-2} cm²/sec. The particle size is 0.1 mm. Assuming that diffusion through product is controlling. Calculate the time required for complete conversion. The density of the particles is given by 5 g/c.c. Molecular wt of A is 2.

SECTION—D

1. For the gaseous reaction $2A \rightarrow B$ the feed consists of 50 mole % A and 50 mole % inerts. The expansion factor is:

(a) 1 (b) – 0.5

(c) – 0.25 (d) 0

2. To maximise the formation of R in simultaneous reactions:

$A + B \rightarrow R,\ r_R = 2C_A^{0.5}C_B^{2}$

$A + B \rightarrow S$, $r_S = 1.5\ C_A C_B$ we should have

(a) Low C_A, low C_B (b) Low C_A, high C_B
(c) High C_A, low C_B (d) High C_A, high C_B

3. For an ideal mixed flow reaction (CSTR) the exit age distribution $E(t)$ is given by:

(a) A Dirac delta function (b) A step function
(c) A ramp function (d) None of the above

4. A first order reaction $A \rightarrow B$ occurs in an isothermal porous catalyst pellet of spherical shape. If the concentration of A at the centre of the pellet is much less than at the external surface, the process is limited by:

(a) Diffusion within the pellet (b) Reaction
(c) External mass transfer (d) None of the above

5. The rate constant of a chemical reaction increases by 100 times when the temperature is increased from 400 K to 500 K. Assuming the transition state theory is valid the value of E/R is:

(a) 8897 K (b) 9210 K
(c) 8764 K (d) 8621 K

6. For a first order reactin taking place in an isothermal batch reactor, 80% of liquid reactant is converted to product in 15 minutes. Calculate space velocity required to achieve same conversion in a plug flow reactor and in a back mix flow reactor.

7. For a given conversion and a first order reaction, the volume required for a mixed reactor is......than that for a plug flow reactor.

8. Two mixed reactors of unequal size are available for producing a specified product, formed by a homogeneous second order reaction. To achieve maximum production rate the smaller reactor should be placed in series before the larger reactor. True or False and reason.

9. Liquid A decomposes in a batch reactor by a second order kinetics. If 50% of A is converted in 5 minute run, how long could it take to reach 75% conversion. For the same conversion, the holding time required in a batch reactor is always equal to space time required in a plug flow reactor. True or False and reason.

10. In a CSTR 50% conversion is obtained for a homogeneous, isothermal, liquid phase, irreversible second order reaction. What is the conversion if the reactor volume is five times the original all else remaining unchanged.

11. A homogeneous gas phase decomposition reaction $4A \rightarrow B + 7S$ takes place in an isothermal plug flow reactor. The reaction rate is $-r_A = K_1 C_A$ where $K_1 = 0.17S^{-1}$; feed concentration of $A(CA_0)$ = 0.1 mole per m^3. Feed flow rate (F_{AO}) = 0.17 mole/s. Determine the size of reactor in order to achieve 50% conversion.

12. The mechanism for the decomposition of aceteldehyde (CH_3CHO) into CH_4 and CO in presence of Iodine is:

$$CH_3CHO + I_2 = CH_3I + HI + CO \qquad \text{(Slow)}$$

$$CH_3I + HI \longrightarrow CH_4 + I_2 \qquad \text{(Fast)}$$

Then the rate of disapearance of acetaldehyde is equal to $KC_{CH_3}C_{HI}$ and HI acts as catalyst. True or False and Reason.

13. Match the items:

1. $A \rightarrow B \rightarrow C$	(a) Rate of reaction of A shows a maximum with time
2. $A + B \rightarrow 2B + C$	(b) C_B shows a maximum with time
	(c) C_C shows a maximum with time
	(d) C_B continuously decreases with time

14. (a) The Thiele modulus for a first order isothermal reaction for a flat plate geometry catalyst is found to be 2. Calculate the catalyst effectiveness factor?

(b) A gaseous reactant diffuses through a gas film and reacts on the surface of a non-porous spherical catalyst particle. The rate of surface reaction is K_1C_S where C_S is the reactant concentration in the catalyst surface. The reaction rate constant $(K_1) = 0.83 \times 10^{-4}$ m/s and the gas film mass transfer coefficient $(k_m) = 1.66 \times 10^{-4}$ m/s. Derive the reaction rate expression in terms of bulk gas phase concentration (C_G).

15. (a) (i) The temperature dependence of a chemical reaction rate in general can be written as:

$$K = K_0T^m \exp(-E/RT)$$

What is the value of m from collision theory, transition state theory and from Arrhenius law?

(ii) For every 2 minutes one reactor volume of feed is being treated at specified conditions in a continuous reactor. What is the space velocity?

(b) Choose the correct answer. For multiple reactions:

$$2A \rightarrow R$$

$$2S \rightarrow S$$

the number of moles of S present when the number of moles of A and R are 0.3 and 0.5 respectively (initially 2 moles of A are only present) are:

(a) 0.125 (b) 0.175

(c) 0.535 (d) 0.350

(c) Sketch (rough) diagram of exist distribution $E(t)$ and cumulative age distribution function, $F(t)$ for plug flow and perfectly mixed flow vessels and indicate the mean residence time and the maximum values of $E(t)$ and $F(t)$ in each case.

16. The observed rate using a cube of 2 cm. Catalyst pellet is 10^{-8} mole/s cm^3 of pellet, calculate the value of first order reaction rate constant. The effective diffusivity is 0.01 cm^2/sec.

17. A reversible liquid phase reaction (endothermic) is to be carried out in a plug flow reactor. For minimum reactor volume, it should be operated such that the temperature along the length.

(a) Decreases (b) Increases

(c) Is at the highest allowable temperature throughout

(d) First increases and then decreases

18. Residence time distribution of an ideal CSTR is:

(a) $\frac{1}{T}$ exp. $(-t/T)$ (b) T exp. $(-t/T)$

(c) exp. $(-t/T)$ (d) $\frac{1}{T}$ exp. $(-t/T)$

19. The decomposition of A into B is represented by the endothermic reaction $A \rightleftharpoons 2B$. To achieve maximum decomposition it is desirable to carry out the reaction:

(a) At high P and high T
(b) At low P and high T
(c) At low P and low T
(d) At high P low T

20. In the gas phase reaction:

$2A \rightarrow B$ is carried out in an isothermal plug flow reactor. The feed contains 800 mole % A and 20 mole % inerts. If the conversion of A at the reactor exit is 50% then CA/CA_0 at the outlet of reactor is:

(a) 2/3
(b) 5/8
(c) 1/3
(d) 3/8

21. For a first order reaction (isothermal) in a porous catalyst, the effectiveness factor is 0.30. The effectiveness factor will increase if the:

(a) Catalyst size is reduced or the catalyst diffusivity is reduced.
(b) Catalyst size is reduced or the catalyst diffusivity is increased
(c) Catalyst size is increased or the catalyst diffusivity is reduced
(d) Catalyst size is increased or the catalyst diffusivity is increased

22. The rate of heterogeneous catalytic reaction $A(g) + B(g) \rightarrow C(g)$ is:

$$-r_i = \frac{kk_t P_A P_B}{1 + K_A P_A + K_C P_C}$$

where K_A and K_C are the adsorption equilibrium constants. The rate controlling step for this reaction is:

(a) Adsorption of A
(b) Surface reaction between adsorbed A and B in the gas phase
(c) Surface reaction between adsorbed A and B
(d) Surface reaction between A in gas phase and adsorbed B.

23. For a second order liquid phase reaction $A \rightarrow$ products, occuring in an isothermal plug flow reactor, the conversion is 50%. What will be the conversion, if the plug flow reactor is replaced by a continuous stirred tank reactor of the same volume.

24. Ethanol is manufactured by the vapour phase hydration of Ethylene according to the reaction:

$$C_2H_6(g) + H_2O(g) \rightleftharpoons C_2H_5OH(g)$$

The reactor operates at 400 K and 2 bar and the feed is a gas mixture of ethylene and steam in the ratio (mole) 1 : 3. The equilibrium constant is 0.55.

Estimate the composition (mole%) of the equilibrium mixture. Assume ideal gas behaviour and take $f_1^{\,0}$ = 1 bar, where $f_1^{\,0}$ is the standard state fugacity of component i.

25. Liquid A decomposes in a batch reactor by zero H order kinetics. The initial concentration of A is 0.5 k mol/m^3 and for a reaction time of 1200 s, the conversion is 40%. Assuming isothermal conditions.

(a) Determine the rate constant for this reaction
(b) What will be the conversion for reaction time of 3600 s.

26. The liquid phase parallel reaction.

$$r_S = K_1 C_A \qquad K_1 = 0.3\ \text{s}^{-1}$$
$$r_S = K_2 \qquad K_2 = 0.3\ \text{k mol/m}^3\ \text{s.}$$

is conducted in an isothermal plug flow reactor. The inlet concentration of A is 2.0 k. mol/m^3. No products are present in the feed. If the conversion of A is 80% then determine the exit concentration of R.

27. The first order exothermic irreversible liquid phase reaction $A \rightarrow B$ is to be conducted in an adiabatic CSTR. The desired conversion is 70% and the reactor is to be operated at 353 K. For the data given below, find:

(a) The required volume of the CSTR (b) The feed temperature

Data: The rate constant at 353 K = 0.02 s^{-1}

Inlet volumetric flow rate = 0.01 m^3/s; Density of inlet stream = 1000 kg/m^3; Inlet concentration of A = 2.0 k. mole/m^3; Heat of reaction (independent of temperature) = – 50,000 kJ/K. mol. of A reacting. Average heat capacity of inlet and outlet streams = 2 kJ/(kg K).

28. For reaction $P + 2Q \rightarrow 3R$; molar rate of consumption of P is:

(a) Double of that of Q

(b) Same as that of Q

(c) Half of that of Q

(d) 2/3 of that of Q

29. Molecularity of an elementary reaction:

$P + Q \rightarrow R + S$ is:

(a) 1 (b) 2

(c) 3 (d) 4

30. Second order consecutive irreversible reactions $A \xrightarrow{K_1} B \xrightarrow{K_2} C$ were carried out in a constant volume isothermal batch reactor with different initial feed compositions. Reactor temperature was same in all the cases. In an experiment where the ratio of concentration of B to that of A in initial feed was less than 0.50, the concentration of B increased first, reached a maximum and then declined with time. However for all the experiments where this concentration ratio was 0.50 or above, conc. of B decreased monotonically with time, right from the beginning. What is the ratio of the two rate constants (K_1/K_2).

(a) 1/4 (b) 1/2

(c) 2 (d) 4

31. A spherical porous catalyst particle of radius R is subjected to reactant A which reacts to form B by a zero order surface reaction $A \rightarrow B$. Film mass transfer resistance is negligible and pore diffusion of A is rate controlling. The effectiveness factor of the catalyst is reported as 7/8, which of the following statements is true?

(a) Inner catalyst core of radius $R/8$ does not participate in reaction.

(b) Inner catalyst core of radius $R/2$ does not participate in reaction.

(c) Inner catalyst core of radius $7R/8$ does not participate in reaction.

(d) Effectiveness factor for a zero order reaction cannot be 7/8 as it must always be 1.

32. A liquid phase first order, reversible reaction $A \rightarrow B$ is carried out in a CSTR. Molar densities of A and B are same. Other things (such as space time, flow rate, temperature) remaining same, a feed of pure A to the reactor results in 40% conversion of A, while a feed of pure B results in

50% conversion of B. Estimate the reaction equilibrium constant. Assume steady state operation in both the cases.

33. Steady state plug flow reactor (PFR) data for isothermal irreversible reaction: $A \rightarrow B$ is as shown in the table. Reactor space time was 10 seconds in both the cases.

$\rightarrow$ Conc. of A(k mol/m^3)

	In Feed	In Product
Case I	1	0.50
Case II	2	0.555

If the reaction is known to be of non-zero integer order, find the reaction order and rate constant.

34. A particular metal reacts with a certain liquid and the product passes into solution. Three non-porous solid spheres of same metal and of diameters 10, 20 and 30 mm respectively were placed in a very large stirred pool of reactive liquid at the same time. After an hour, it was found that the pool had only two spheres of diameter 10 mm and 20 mm respectively. After another hour, the pool had only one sphere of diameter 10 mm. This sphere also disappeared after another hour. Explain these observations through appropriate derivation, using a more likely rate controlling step out of the following two:

(*i*) Film mass transfer (*ii*) Surface reaction

35. For a series of reactions: $A \xrightarrow{K_1} B \xrightarrow{K_2} C$ having $K_1 << K_2$, the reaction system can be approximated as:

(a) $A \xrightarrow{K_1} B$ (b) $A \xrightarrow{K_2} B$

(c) $A \xrightarrow{K_2} C$ (d) $A \xrightarrow{K_1} C$

36. An elementary liquid phase decompostion reaction; $A \xrightarrow{K_1} 2B$ is to be carried out in a CSTR. The design reaction equation is:

(a) $KT = \dfrac{X_A}{(1-X_A)}$ (b) $KT = \dfrac{X_A(1+X_A)}{(1-X_A)}$

(c) $KT = \dfrac{X_A}{(1-X_A)^2}$ (d) $KT\,C_{AO} = \dfrac{\dfrac{X_A}{(1-X_A)}}{(1-X_A)^2}$

37. Find a mechanism that is constant with rate equation and reaction given below.

$2A + B \rightarrow A_2B; -r_A = KC_A C_B$

(a) $A + B \rightleftharpoons AB;\quad AB + A \rightarrow A_2B$

(b) $A + B \rightarrow AB;\quad AB + A \rightarrow A_2B$

(c) $A + A \rightarrow AA;\quad AA + B \rightarrow A_2B$

(d) $A + A \rightleftharpoons AA;\quad AA + B \rightarrow A_2B$

38. A CSTR is to be designed in which an exothermic liquid phase reaction of first order of the type: $A \rightarrow R$ is taking place. The reactor is to be provided with a jacket in which coolant is flowing.

Following data is given C_{AO} = 5 K mol/cubic. m; X_A = 0.501 Feed temperature = Reactor temperature = 40°C; Rate constant at 40°C = 1 min.$^{-1}$; ΔH = 40 kJ/mole; P = 1000 kg/cubic. m; C_P = 4 J/g°C; q = 0.001 m^3/min (P and C_P are same for the reactant and product streams). The amount of heat to be removed is:

(a) 2/3 kW (b) 1 kW
(c) 5/3 kW (d) 4 kW

39. A liquid phase reaction is to be carried out under isothermal conditions. The reaction rate as a function of conversion has been determined, experimentally. What choice of reactor or combination of reactors will require the minimum overall reactor volume, if a conversion of 0.9 is desired.

(a) CSTR followed by PFR
(b) PFR followed by a CSTR
(c) CSTR followed by a PFR followed by CSTR
(d) PFR followed by a CSTR followed by PFR

40. The following gas phase reactions are carried out isothermally in a CSTR:

$A \rightarrow 2R$; $r_1 = K_1 P_A$; K_1 = 20 mol/(sec cubic m bar)

$A \rightarrow 3S$; $r_2 = K_2 P_A$; K_2 = 40 mol/(sec cubic m bar)

Total pressure = 1 bar; F_{AO} = 1 mol/sec: Feed is pure A. What is the maxmium possible value of F_R(mol/sec)?

(a) 1/3 (b) 1/2
(c) 2/3 (d) 2

41. The volume of a CSTR required for a fractional conversion of A equal to 0.30 due to the first reaction is:

(a) 0.011 (b) 0.21
(c) 0.275 (d) 0.375

42. A step input tracer test is used to explore the flow pattern of fluid through a vessel of total volume equal to 1 cubic. m. having a feed rate of 1 cubic m/minute.

identify in each curve in Group-I, a suitable flow model from the list given under Group-II.

Group-I	**Group-II**
P curve-1	1. PFR and CSTR are in series
Q curve-2	2. CSTR with dead space
	3. PFR and CSTR in series and dead space

(a) P – 4; Q – 3 (b) P – 4; Q – 1
(c) P – 2; Q – 3 (d) P – 2; Q – 1

43. The following isothermal kinetic data are obtained from a basket type mixed flow reactor for a porous catalyst. Determine the role of pore diffusion and external mass transfer processes.

Pellet diameter	**Leaving conc. of the reactant**	**Spinning rate of basket**	$(-r_A)$
1	1	high	2
2	1	low	1
2	1	high	1

(a) Strong pore diffusion control and mass transfer not controlling
(b) Both pore diffusion and mass transfer not controlling
(c) Both pore diffusion and mass transfer controlling
(d) Mass transfer controlling

44. For an ideal gas mixture undergoing a reversible gaseous phase chemical reaction, the equilibrium constant:
(a) Is independent of pressure
(b) Increases with pressure
(c) Decreases with pressure
(d) Increases/decreases with pressure depending on the stoichiometric coefficients of the reaction.

45. The rate of ammonium synthesis for the reaction
$N_2 + 3H_2 \rightleftharpoons NH_3$
is given by $r = 0.8\, P_{N_2}^{\ 0.5}\ \ P_{H_2}^{\ 1.5} - 0.6\ \ P_{NH_3}^{\ 2}$. If the reaction is represented as, $1/2_{N_2} + 3/2H_2 \rightleftharpoons NH_3$. The rate of ammonia synthesis is:
(a) $r = 0.8\, P_{N_2}^{\ 0.5}\ \ P_{H_2}^{\ 1.5} - 0.6\, P_{NH_3}$
(b) $r = 0.8\, P_{N_2}\ \ P_{H_2}^{\ 3} - 0.6\, P_{NH_3}\ P_{NH_3}$
(c) $r = 0.5\, (0.8\, P_{N_2}\ \ P_{H_2}^{\ 3} - 0.6\ \ P_{NH_3}^{\ 2})$
(d) $r = 0.5\, (0.8\, P_{N_2}^{\ 0.5}\, P_{H_2}^{\ 1.5} - 0.60\, P_{NH_3})$

46. An endothermic aquous phase first order irreversible reaction is carried out in an adiabatic plug flow reactor. The rate of reaction:
(a) Is maximum at the inlet of the reactor
(b) Goes through a maximum along the length of the reactor
(c) Goes through a minimum along the length of the reactor
(d) Is maximum at the exit of the reactor

47. A first order gaseous phase reaction is catalysed by a non-porous solid. The kinetic rate constant and the external mass transfer coefficient are K and K_g respectively. The effective rate constant (K_{eff}) is given by:
(a) $K_{eff} = K/K_g$
(b) $K_{eff} = (K + K_g)/2$
(c) $K_{eff} = (K\,K_g)^{1/2}$
(d) $1/K_{eff} = 1/K + 1/K_g$

48. For a packed bed reactor, the presence of a long tail in the residence time distribution curve is an indication of:
(a) Ideal plug flow
(b) By pass
(c) Dead zone
(d) Channeling

49. The following gas phase reaction is taking palce in a plug flow reactor:

$$A + \frac{1}{2}B \rightarrow C$$

A stoichiometric mixture of A and B at 300 K is fed to the reactor. At 1 meter along the length of the reactor the temperature is 360 K. The pressure drop is negligible, and the ideal gas behaviour can be assumed. Identify the correct expression relating the concentration at the inlet (C_{AO}), concentration of a at 1 m (C_A) and the corresponding conversion of $A(X)$.
(a) $C_A = 1.2\, C_{AO}\,(1 - X)/(1 - 0.33\, X)$
(b) $C_A = 1.2\, C_{AO}(1 - X)/(1 - 0.5\, X)$
(c) $C_A = 0.83\, C_{AO}\,(1 - X)/(1 - 0.33\, X)$
(d) $C_A = 0.83\, C_{AO}\,(1 - X)/(1 - 0.5\, X)$

50. A second order liquid phase reaction $A \rightarrow B$ is carried out in a mixed flow reactor operated in a semi-batch mode (no exit stream). The reactant A at concentration C_{AF} is fed to the reactor at volumetric flow rate of F. The volume of the reacting mixture is V and the density of the liquid mixture is constant. The mass balance for A is:

(a) $d(V\,C_A)/dt = -F(C_{AF} - C_A) - KC_A^2\,V$ (b) $d(VC_A)/dt = F(C_{AF} - C_A) - KC_A^2V$

(c) $d(VC_A)/dt = FC_A - KC_A^2V$ (d) $d(VC_A)/dt = FC_{AF} - KC_A^2V$

51. For an isothermal second order aqeous phase reaction $A \rightarrow B$. The ratio of the time required for 90% conversion to the time required for 45% conversion is:

(a) 2 (b) 4

(c) 11 (d) 22

52. An isothermal aqueous phase reaction reversible type $P \rightleftharpoons R$ is to be carried out in a mixed flow reactor. The reaction rate in k mole/cubic m. hr. is given by $r = 0.5\ C_P - 0.125\ C_R$. A stream containing only P enters the reactor. The residence time required (in hours) for 40% conversion of P is:

(a) 0.80 (b) 1.33

(c) 1.60 (d) 2.67

53. A pollutant P degrades according to first order kinetics. An aqueous stream containing P at 2 K mol/cubic. m and volumetric flow rate of 1 cubic m/hour requires a mixed flow rate of reactor of volume V to bring down the pollutant level to 0.50 k mole/cubic m. The inlet concentration of the pollutant is now doubled and the volumetric flow rate is tripled. If the pollutant level is to be brought down to the same level of 0.50 k mole/cubic m, the volume of the mixed flow reactor should be increased by a factor of:

(a) 7 (b) 6

(c) 3 (d) 7/3

54. Consider a reversible exothermic reaction in a plug flow reactor. The maximum and minimum permissible temperatures are T_{max} and T_{min} respectively. Which of the following temperature profiles (T) will require the shortest residence time to achieve the desired conversion?

55. An irreversible aqueous phase reaction $A + B \rightarrow P$ is carried out in an adiabatic mixed flow react or a feed containing 4 k mol/m^3 of each A and B entering the reactor at 8 m^3/hour. If the temperature of the exit stream is never to exceed 390 K, what is the maximum feed temperature allowed?

Data: Heat of reaction = – 50 kJ/mol, density of reactant mixture = 1000 kg/m^3; specific heat of reacting mixture = 2 kJ/(kg. K).

The above data can be assumed to be independent of composition and temperature.

(a) 190 (b) 290

(c) 390 (d) 490

56. Pick the wrong design guideline for a reactor in which the reaction $A \rightarrow R$ (desired) and $A \rightarrow S$ (undesired) are to take place. The ratio is: $r_R/r_S = (K_1/K_2)\ C_A^{a-b}$

(a) Use high pressure and eliminate inerts when $a > b$

(b) Avoid recycle when $a > b$

(c) Use batch reactor or plug flow reactor when $a > b$

(d) Use CSTR with high conversion when $a > b$

ANSWERS

SECTION–A

1. (b)	2. (d)	3. (c)	4. (a) and (c)	5. (a)
6. (c)	7. (b)	8. (c)	9. (c)	10. (a)
11. (a)	12. (c)	13. (d)	14. (c)	15. (c)
16. (b)	17. (c)	18. (b)	19. (a)	20. (c)
21. (a)	22. (c)	23. (a)	24. (b)	25. (b)
26. (c)	27. (c)	28. (b)	29. (d)	30. (b)
31. (c)	32. (b)	33. (b)	34. (b)	35. (c)
36. (c)	37. (b)	38. (c)	39. (c)	40. (b)
41. (c)	42. (d)	43. (b)	44. (c)	45. (c)
46. (b)	47. (c)	48. (a)	49. (b)	50. (b)
51. (d)	52. (c)	53. (b)	54. (c)	55. (d)
56. (b)	57. (c)	58. (c)	59. (c)	60. (c)
61. (c)	62. (a)			

SECTION–B

1. Its rate equation corresponds to its stoichiometric equation.
2. $(\text{time})^{-1}$ $(\text{concentration})^{1-n}$.
3. more, low.
4. Products, autocatalytic.
5. $\begin{pmatrix}\text{Rate of}\\ \text{reactant}\\ \text{flow in}\end{pmatrix} = \begin{pmatrix}\text{Rate of}\\ \text{reactant}\\ \text{flow out}\end{pmatrix} + \begin{pmatrix}\text{Rate of reactant}\\ \text{loss due to}\\ \text{chemical reaction}\end{pmatrix} + \begin{pmatrix}\text{Rate of}\\ \text{accumlation}\\ \text{of reactant}\end{pmatrix}$
6. (a) rate of reaction, (b) extent of reaction
   (c) reactor volume (d) feed rate
7. Increase.
8. Slope.
9. Frequency factor.
10. Fraction of reactant *A* converted to *B*.
11. Ratio of yield of *B* and $C = \dfrac{X_B}{X_C}$, where 'X_B' denotes fraction of *A* converted to *B*.

12. rises.
13. affects.
14. practically irreversible
15. stimulus-response technique.
16. *C*-curve.
17. (a) dispersion model (b) Tanks in series model
18. D/uL.
19. Channelling, recycling or by creation of stagnanat regions.
20. $\dfrac{\partial C}{\partial \theta} = \left(\dfrac{D}{UL}\right)\dfrac{\partial^2 C}{\partial Z^2} - \dfrac{\partial C}{\partial Z}$
21. $E = -\dfrac{dI}{d\theta}$
22. Dead water regions.
23. (a) unreacted core model (b) continuous reaction model
24. unreacted core model (shrinking core).
25. $4FeS_2(s) + 11O_2(g) \longrightarrow 8SO_2(g) + 2Fe_2O_3(s)$
26. which is the slowest of all.
27. Gas film resistance and chemical reaction resistance.
28. $\dfrac{W}{F_{AO}} = \int_{X_A \text{ in}}^{X_A \text{ out}} \dfrac{dX_A}{(-r_A)}$
29. which affects the rate of a reaction but is neither a product or reactant.
30. (a) gas film resistance for reactants
    (b) pore diffusion resistance for reactants
    (c) surface phenomenon resistance
    (d) pore surface diffusion resistance for products
    (e) gas film resistance for products.
31. $E = \dfrac{\text{(Actual rate within pore of the catalyst)}}{\text{(Rate if not slowed by pore diffusion)}}$
32. Brunauer-Emmett-Teller technique.
33. (a) molecular diffusion (b) Knudsen diffusion
34. Helium-Mercury method.
35. If it lowers the activity of the catalyst.
36. $v = \beta(p)^{1/n}$

    where v volume of gas adsorbed

    β, n = specific constants for system of gas of surface
37. moles/(times) (wt. of catalyst).

38. The catalyst reduces the rate of the reaction.
39. (a) homogeneous catalytic (b) heterogeneous catalytic reaction
40. Nitric oxide (NO)
41. (a) physical adsorption (b) chemisorption
42. physical adsorption
43. (a) an active catalyst surface (b) large particle size
    (c) small pore diameter.
44. The equilibrium yield is adversely affected by temperature.
45. $N \text{ reactors} = \frac{N}{K}\left[\left(\frac{C_0}{C_N}\right)^{1/N} - 1\right]$
46. series.
47. series and parallel reactions.
48. $\frac{d(\ln K)}{dT} = \frac{\Delta H_r}{RT^2}$
49. higher, lower.
50. plug flow on which is superimposed some degree of backing or inter-mixed, (the degree of which is independent of position within the vessel).
51. Stokes regime
52. potential energy barrier
53. the residence time or spread in the residence time.
54. Hold back (H); $H = \int_t^{\infty} I\, d\theta$
55. $V = K_p^{1/n}$ where V is the volume of gas adsorbed, p, the pressure of the gas, K and n are constants.
56. Rate of reaction per unit volume of catalyst bed in terms of the temperature and concentration of A in the bulk gas stream.
57. passing steam/air (to burn C to CO or CO_2).
58. which is self-supporting in its thermal energy requirements.
59. zero
60. decreases
61. products; reactants

SECTION—C

1. $-\frac{1}{2}\frac{d(NO_2)}{dt} = -\frac{2d(O_2)}{dt} = \frac{d(N_2O_5)}{dt}$
2. Arrhenius law, activated complex theory, collission theory.

3. Arrhenius law: $Ke^{-E/RT}$: collission theory: $T^{1/2}e^{-E/RT}$
   Activated complex theory: $Te^{-E/RT}$
4. (atm.)$^{-1}$ (hr)$^{-1}$.
5. (a) Integral method (b) Differential method

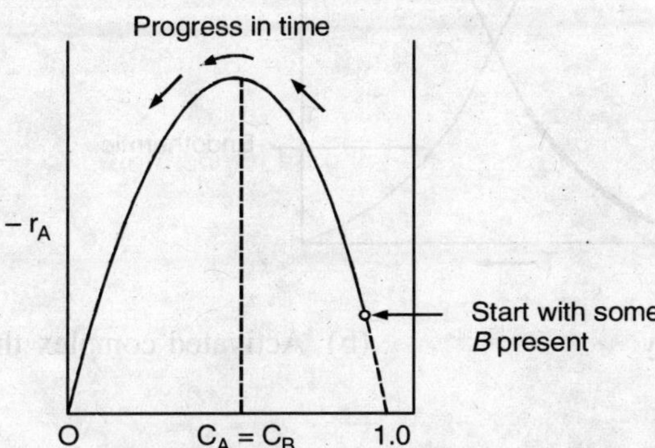

6. Figure
7. $r = \dfrac{1}{RT}\dfrac{dp_i}{dt}$
8. 2
9. batch $t = \displaystyle\int_0^{XA} \frac{dXA}{-r_A}$;

   Plug flow: $\dfrac{V}{F} = \displaystyle\int_0^{XA} \frac{dXA}{-r_A}$

   Stirred tank: $\dfrac{V}{F} = \dfrac{X_{Af} - X_{A\text{in}}}{(-r_A)_f}$
10. Space time $(\tau) = \dfrac{\text{Reactor volume}}{\text{Volumetric feed rate}}$ (time units)

    Space velocity $(S) = \dfrac{1}{\tau} = \dfrac{\text{Volumetric feed rate}}{\text{Reactor volume}}$ (time^{-1})
11. Batch reactor: $t = C_{AO}\displaystyle\int_0^{XA} \frac{dXA}{(1+\delta SX_A)(-r_A)}$

    Plug flow reactor: $t = C_{AO}\displaystyle\int_0^{XA} \frac{dXA}{(1+\delta SX_A)}(-r_A)$

    Back mix reactor: $t = \dfrac{C_{AO}X_A}{(1+X_A)(-r_A)}$
12. Plot of ln K (rate constant) *vs* $1/T$ (°K) is called Arrhenius plot.
13. $C_C/C_A = (1 + K\tau)^4$.

14.

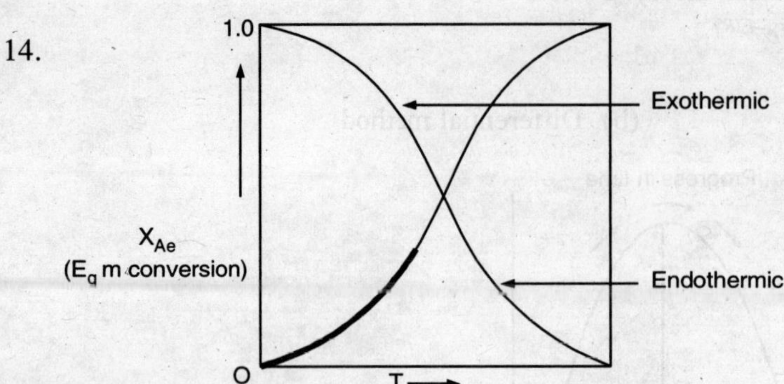

15. (a) Arrhenius theory (b) Activated complex theory
16. Step; pluse.

17. Mean $(\bar{t}) = \dfrac{\int_0^\infty tCdt}{\int_0^\infty Cdt}$

Variance $(\sigma^2) = \dfrac{\int_0^\infty s^2Cdt}{\int_0^\infty Cdt} - \bar{t}^2$

18. $\dfrac{\partial C_A}{\partial t} = D\dfrac{\partial^2 C_A}{\partial x^2} - KC_A$

19.

Input
Response
O → t
Response to step input

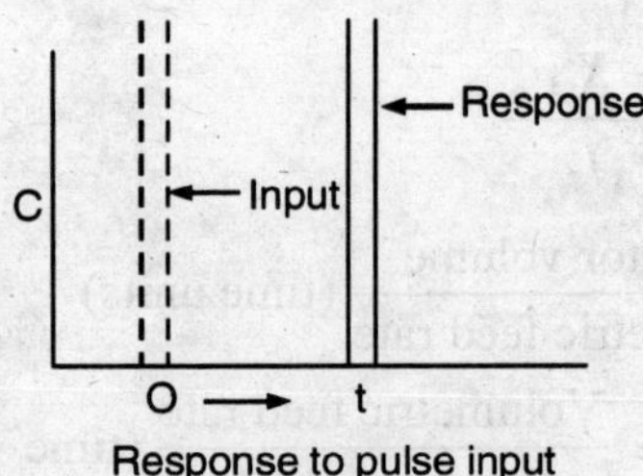

Response to pulse input

20. If a number of linear processes are occurring simultaneously in a system.

21.

F
$1 - e^{-\theta}$
θ
Response to step input.

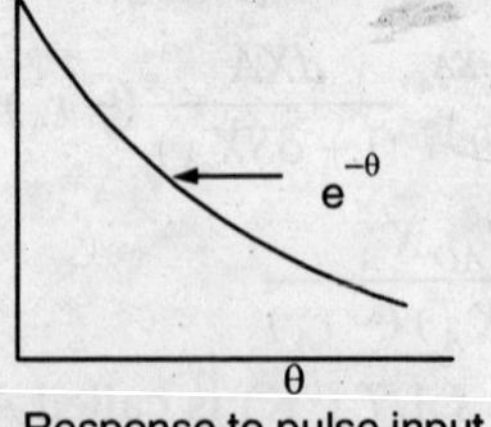

Response to pulse input.

their overall effect is also a linear process.

22. Plug flow regions, backmix flow regions, dispersed plug flow regions, dead water regions.
23. (a) Diffusion of O_2 through the gas film surrounding FeS_2 particle.
    (b) Diffusion of O_2 through the Fe_2O_3(ash) to the surface of unreacted core.
    (c) Reaction of O_2 with FeS_2
    (d) Diffusion of SO_2 through Fe_2O_3 back to the surface of the solid.
    (e) Diffusion of SO_2 through the gas film.
24.

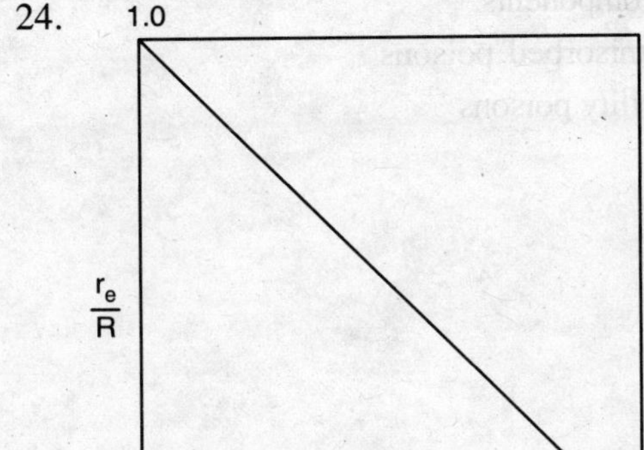

25. Gas film diffusion control: $\frac{t}{\tau} = X_B$

    Ash diffusion control; $\frac{t}{\tau} = 1 - 3(1 - X_B)^{2/3} + 2(1 - X_B)$

    Chemical reaction control = $\frac{t}{\tau} = 1 - (1 - X_B)^{1/3}$.
26. 1. Diffusion of O_2 through the gas film
    2. Reaction between carbon particle and O_2
    3. Diffusion of CO_2 through gas film.
27.

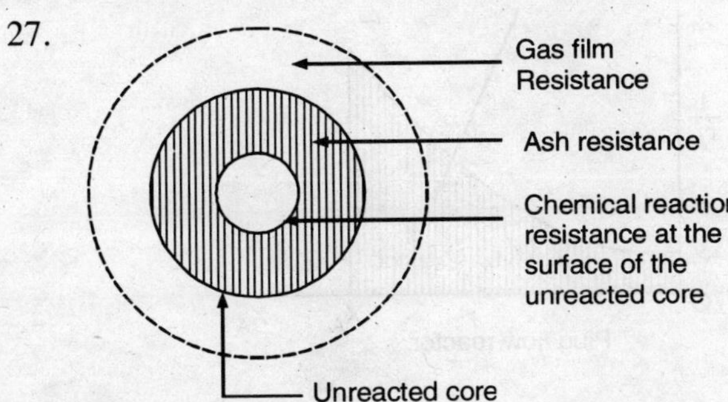

28. Blast furnance, fluidized bed reactors.
29. If the pore size is small in comparison with the mean free path, collisions with the pore wall control with process. The diffusivity under this condition is called Knudsen diffusivity (D_k).

$$D_k = \frac{2r}{3}\sqrt{\frac{8K_BT}{\pi m}}, \; r \to \text{pore radius}$$

$m \to$ mass of the diffusing molecule.

30. Precipiation, gel formation, mixing of catalyst components.
31. (a) Deposited poison (b) Chemisorbed poisons
(c) Selectivity poisons (c) Stability poisons
(e) Diffusion poisons
32. $HCU = \dfrac{Z}{\int \dfrac{dp_{gA}}{(p_t - p_g)_A}}$; $HRU = \dfrac{Z}{\int \dfrac{dp_{gA}}{(p_g - P_g)_A}}$

where p_g = partial pressure in gas phase
p_i = partial pressure at catalyst gas interface
P_g = partial pressure of a given component when the reaction mixture is at equilibrium
Z = depth of the catalyst in a reactor.

33. $\underset{(1)}{mCvdT} = \underset{(2)}{(-\Delta H)V_{Br}d\theta} - \underset{(3)}{Qsd\theta}$

where (1) $\to$ net energy added to the reaction mixture in time $d\theta$
(2) $\to$ energy evolved due to the amount of reaction occurring in time $d\theta$
(3) $\to$ heat transfer to surroundings in time $d\theta$.

34.

$1/De$	$= 1/D_M$	$+ 1/D_K$
Effective diffusivity	Molecular diffusivity	Knudsen diffusivity

35.

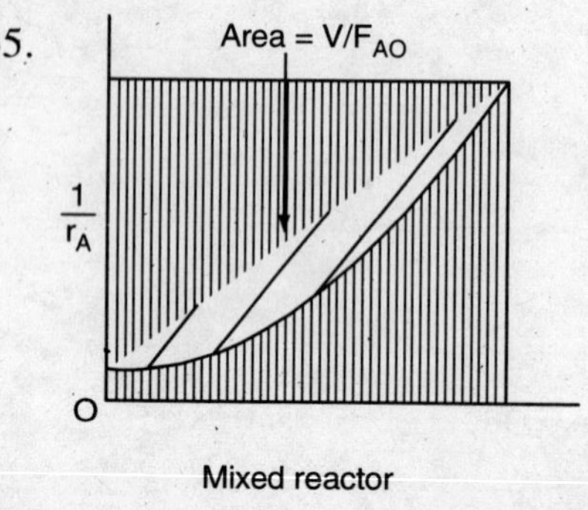

Mixed reactor

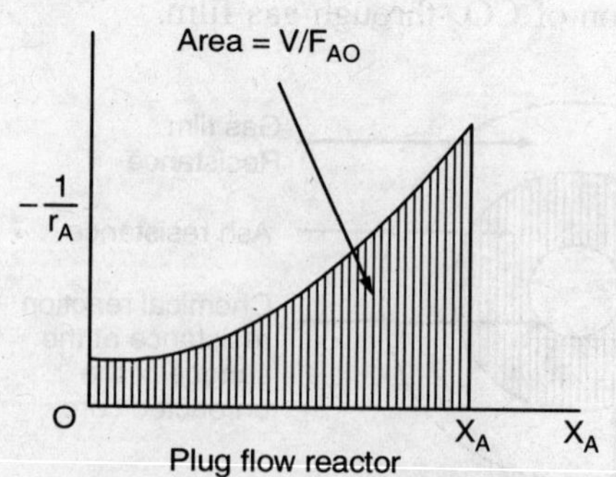

Plug flow reactor

36. $-r_B = -dO_B/dt = -K_1C_AC_B - K_2C_RC_B.$

37. Single reactions: A single stoichiometric equation and single rate equation are chosen to represent the progress of the reaction

Multiple reactions: More than one stoichiometric equation and more than one kinetic expression are needed to represent the progress of the reaction.

38.

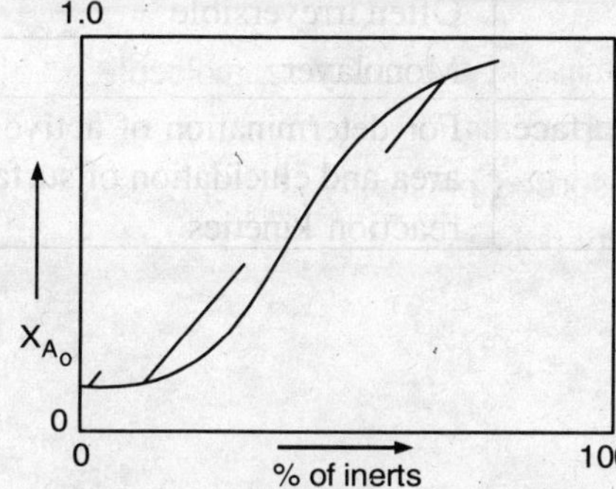

Elementary reactions: In these reactions the rate equation corresponds to stoichiometric equation.

Non-elementary reaction: There is no correspondence between stoichiometry and the rate equation.

39. (a) Differential (flow) reactor (b) Integral (plug flow) reactor

(c) Mixed reactor (d) Batch reactor

40. $p/V(p_0 - p) = (1/V_{me}) + p(c - 1)/V_m C_{po}$

where C is constant

V is volume of gas adsorbed at pressure P.

V_m is volume required to form a monomolecular layer.

41.

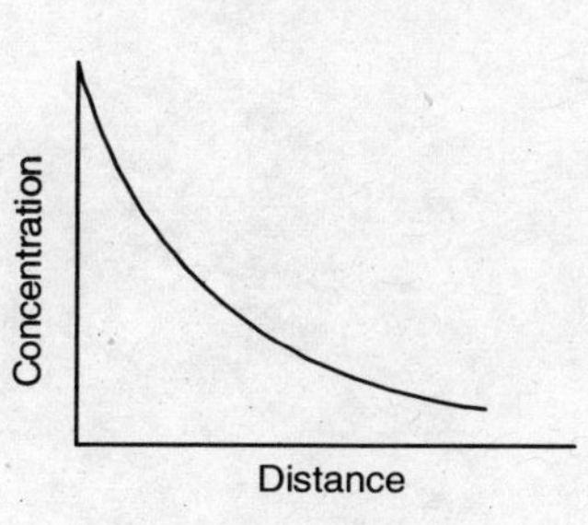

Plug flow reactor

Concentration

1 2 3 4

Stages

CSTR in series

42. (a) Iron oxide-magnesia (b) Vanadium pentoxide

(c) Nickel (d) Iron oxide and cobalt thoria.

43. $\Delta G = nRT \ln (P_2/P_1) = 1 \times 1.987 \times 303 \ln (1/10)$

$(\Delta S)_{\text{system}} = - nR \ln (P_2/P_1) = 1 \times 1.987 \ln (1/10) = - 4.575$ cal.

$(\Delta S)_{\text{surroundings}} = - q$ system$/T = 0$

(as $q = \Delta E + w = 0 + 0 = 0$)

$(\Delta S)_{\text{universe}} = (\Delta S)_{\text{system}} + (\Delta S)_{\text{surr}} = 4.575 + 0 = 4.575$ cal.

44.

Parameter	Physical	Chemical
Temp. range	Low temperature	Generally high temperature
Heat of adsorption	Low	High
Rate	Very rapid, low E	High E
Reversibility	Highly reversible	Often irreversible
Coverage	Multilayer	Monolayer
Importance	For determination of surface area, pore size	For determination of active centre area and elucidation of surface reaction kinetics

45. (a) Uniform temperature hot spots
(b) Can use small sizes of catalyst particles
(c) Regeneration of catalyst can be combined.

46. $\theta = v/vm = K_p/(1 + K_p)$
$K \rightarrow$ adsorption equilibrium constant
$v_M \rightarrow$ volume of gas adsorbed when all the active sites are covered.
$v \rightarrow$ volume of gas actually adsorbed,
$\theta \rightarrow$ fraction covered by adsorbed by molecules

47. $K_{500}/K_{400} = 10 = \dfrac{e^{-E/R(500)}}{e^{-E/R(400)}}$

$\ln 10 = E/R(1/400 - 1/500)$
or $E = 9148$ cal/mole

48. $K = K_0 e^{-E/RT}$; $K_{600}/K_{500} = e^{+E/R(1/773 - 1/873)}$
$= e^{+70{,}000/1.987(1/773 - 1/873)}$
$= e^{5.2102} = 185.014$ times.

49. For first order reaction
$\ln(1 - X_A) = K_t$
$-\ln(1 - 0.6)/ - \ln(1 - 0.8) = K(10)/Kt$;
$t = -\ln 0.2 \times 10/ - \ln 0.4 = 17.56$ min.

50. Assume that order of reaction is n,
For nth order reaction it can be proved that
$C_A^{1-n} - C_{AO}^{1-n} = (n - 1)\,Kt$

$$\frac{C_{AO}^{1-n}(1 - X_1)^{1-n} - C_{AO}^{1-n}}{C_{AO}^{1-n}(1 - x_2)^{1-n} - C_{AO}^{1-n}} = \frac{(n-1)\,Kt_1}{(n-1)\,Kt_2} = \frac{t_1}{t_2}$$

substituting the values we have

$$\frac{(1 - 0.8)^{1-n} - 1}{(1 - 0.9)^{1-n} - 1} = \frac{8}{18} = \frac{4}{9} \quad \text{or} \quad \frac{(0.2)^{1-n} - 1}{(0.1)^{1-n} - 1}$$

By trial and error it can be shown that $n = 2$

so the rate equation for the reaction is $\frac{-dC_A}{dt} = KC_A^{\,2}$

51. $-r_R = C_{RO} - C_{Rf}/\tau = C_{RO} - C_{Rf}/(V/v) = (0.3 - 0.03)/(2/2)$
$= 0.27$ mole/litre – min.
$-r_S = C_{SO} - C_{sf}/\tau = 0.02 - 0.04/1$
$= -0.02$ mole/litre – min.
$-r_T = C_{TO} - C_{Tf}/\tau = O - 0.05/1 = 0.05$/mole/litre-min.
Thus R is disappearing while S and T are being formed.

52. Branch R consists of two reactors; hence it may be considered to be a single reactor of volume

$$V_R = 40 + 50 = 90$$

Now for reactors in parallel V/F must be identical if the conversion is to be the same in each branch.
Hence $(V/F)_R = (V/F)_S$
$F_R/F_S = V_R/V_S = 90/30 = 3$
Hence three-fourths of the feed must be fed to branch R.

53. From the analysis of the method of total pressure data we have the following equations.
For a reaction of the type

$$aA + bB + \longrightarrow rR + sS + ...$$

For reactants $p_A = P_{AO} - a/\Delta n(\pi - \pi_0)$
For products $p_R = p_{RO} + r/\Delta n(\pi - \pi_0)$
where p_{AO}, p_{RO} indicate initial pressures,
$\Delta n = (r + s + ...) - (a + b + ...)$
π = pressure of the system at any times
π_0 = initial pressure of the system
Here $n = 3 - 1 = 2$; $\pi_0 = 420$ mm Hg
π_{85} sec $= 662$

So $P_E = 420 - \frac{1}{2}(662 - 420) = 299$ mm Hg

$P_B = 0 + \frac{1}{2}(662 - 420) = 121$ mm Hg

$P_C = 0 + \frac{1}{2}(662 - 420) = 121$ mm Hg

$P_D = 0 + \frac{1}{2}(662 - 420) = 121$ mm Hg

54. (a) Space time for plug flow reactor

$$= 1/\text{space velocity} = 1/2 \text{ min.} = V/v0$$

(b) Space time for mixed reactor also is same as that in plug flow

$$= 1/2 \text{ min} = V/vo$$

(c) Mean residence time in mixed reactor = i_{mixed}

$$\bar{t} = V/v_O(1 + \in_A X_A) = \frac{V/v_0}{(1+\in_A X_A)}$$

$$\in_A = \frac{3-1}{1} = 2$$

$$\text{So} - t_{\text{mixed}} = \frac{1/2}{(1+0.9\times 2)} = 0.5/2.8 = 0.1785 \text{ min.}$$

(d) Mean residence time in the plug flow reactor:

Since the gas reacts progressively as it passes through the plug flow reactor, it expands correspondingly—not immediately on entering and not all at one time as it leaves—in which case $T_{\text{plug flow}}$ = 0.1785 min. to 1/2 min.

55. For a first order reaction it is known that

$$-\ln(1 - X_A) = Kt$$

where X_A is fractional conversion, K is rate constant and t is time of reaction.

So we have here, $-\,ln\,(1 - X_{SO_2Cl_2}) = Kt$

t = 45 min; K = 0.0013 min^{-1}

(a) % conversion of $X_{SO_2Cl_2}$ at 45 min = 5.68

(b) Time required for 80% conversion = 1238 min.

56. For a mixed reactor the performance equation is

$V/F_{BO} = X_{Af} - X_{Ai}/(-r)_B$

For I case, $V_1/F_{BO} = X_{Af_1} - X_{Ai}/ + (KC_B{}^2)$

$= 0.5/ + [KC_{BO}(1 - X_{A1})^2]$ ...(1)

For II case, $V_2/F_{BO} = 6V_1/F_{BO} = X_B{}^2/KC_{BO}(1 - X_B{}^2)$ ...(2)

X_{B_2} has to be evaluated.

Dividing equation (2) by equation (1) and substituting the values we have

$6 = (1 - X_{A1})^2 X_{A2}/X_{A1}(1 - X_{A2})^2$

$= 0.25\, X_{A2}/0.5(1 - X_{A2})^2$

or $12 = X_{A2}/(1 - X_{A2})^2$; By trial and error X_{A2} = 75%

57. From the true density of the catalyst the helium volume displaced per gram can be calculated

$V_H = 1/\rho_s$; so $V_H = 1/\rho_{si}$ = 0.4219 c.c./g

Volume displaced by mercury for each sample per gram is,

V_{Hg1} = 1/1.126 = 0.888, V_{Hg2} = 1/0.962 = 1.0395

Pore volume (V_g) = (volume displaced by mercury – volume displace by Helium)/weight of the catalyst.

V_{g1} = (0.888 – 0.4219)/1 = 0.4661

V_{g2} = (1.0395 – 0.4219)/1 = 0.6176

Pore radius (r) = $2V_g/S_g$

So $r_1 = 2 \times V_{g1}/S_{g1} = 2 \times 0.4661/467 \times 10^4$

$= 19.96 \times 10^{-8}$ cm

$r_1 = 19.96$ A; $r_2 = 33.2$ A

The difference $= r_2 = r_1 = 13.24$ A.

Note: Helium-Mercury method is used for the determination of pore volume. Volume displaced by Helium gives the volume occupied by the solid material. Mercury does not fill up the pores. The mass of the catalyst divided by volume displaced by mercury gives the density of porous particles.

58. The reaction is $B \longrightarrow 4R$ with $(-r_B) = (5/\text{hr})C_B$

The volume of plug flow reactor is given by:

$$V = F_{BO}\int_0^{X_B}\frac{dX\ B}{-rB} = F_{BO}\int_0^{X_A}\frac{dXB}{KC_B}$$

At constant pressure, $C_B = C_{BC}(1 - X_B/(1 + \in_B X_B))$

$$\text{Hence, } V = F_{BO}/KC_{BO}\int_0^{X_B}\frac{1+\in_B X_B}{1-X_B}\,dX_B$$

Integrating we obtain

$$V = \frac{F_{BO}}{KC_{BO}}(1+\in_B)\ln\frac{1}{1-X_B} - \in_B X_B$$

The individual terms are

$F_{BO} = 4$ kg moles/hr, $K = 0.05\ \text{min}^{-1}$

$$C_{BO} = \frac{p_{BO}}{RT} = \frac{4\ \text{atm}}{\left(82.06\ \frac{\text{c.c. atm}}{\text{g. mol K}}\right)(473\ \text{K})}$$

$= 1.0305 \times 10^{-4}$ g. moles/c.c.

$\in_B = 4 - 1/1 = 3$, $X_A = 0.75$

Hence the volume of the reactor is

$$= \frac{2\ (\text{g moles/min})}{(0.05/\text{min})\left(1.0305\times10^{-4}\ \frac{\text{g moles}}{\text{c.c.}}\right)}\left[(1+3)\ln\frac{1}{0.25} - 3(0.75)\right]\frac{(4/60)\,10^4}{1.0205\times0.05}$$

$= 1.2789 \times 10^4$ c.c. $= 1.2789\ \text{m}^3$.

59. Pore volume = (volume of mercury displaced/g – helium displacement volume per gram)

Hence $V_g = 100 - 50/100 = 0.5$ c.c./g

$$\text{True density of the catalyst} = \frac{\text{Weight of catalyst}}{\text{Helium displacement volume}}$$

$\rho_s = 100/50 = 2.0$ g/c.c.

Porosity $= V_g\rho_S/(V_g\rho_s + 1) = (0.5 \times 2/(0.5 \times 2 + 1)) = 0.5$

60. (a) For a stirred tank reactor of 10 litres capacity

$V_s/F = (C_0 - C)/r = (C_C - C)/KC$

or $x = (C_0 - C)/C_0 = K(V_s/F)/(1 + K(V_s/F))$

$$= \frac{0.2(10000/1000)}{1+0.2(10000/1000)} = \frac{0.2\times10}{1+0.2\times10} = 0.667$$

(b) For two 5 litre reactors in series (CSTRs) residence time ($\bar{t}$) is equal in the reactors.

$\bar{t} = 5000/1000 = 5$ min.

For 2 reactors in series (CSTRs) (for a first order reaction).

$$\text{Exit conversion } x = 1 - \frac{1}{(1+Kt)^2} = 1 - \frac{1}{(1+0.2\times5)^2} = 0.75$$

(c) For a tubular flow reactor of 5 litre capacity

$\bar{t} = 1000/100 = 10$ min.

$$\text{and } \frac{V_P}{F} = \bar{r} = C_0\int_0^X \frac{dx}{r} = C_0\int_0^X \frac{dx}{KC_0(1-x)}$$

$$= -\frac{1}{K}\ln(1-x)$$

or $x = 1 - e^{-Kt} = 1 - e^{-0.2\times10} = 0.8647$.

61. For heterogeneous catalytic reactions in a plug flow reactor the design equation is:

$$\frac{W}{F_{AO}} = \int_0^{X_A} \frac{dX_A}{(-r_A)} \qquad ...(1)$$

Assuming first order reaction, $-r_A = KC_A = KC_{AO}(1 - X_A)$ where C_{AO} is initial concentration of A and is a constant value. So, $-r_A = K'(1 - X_A)$ where $K' = KC_{AO}$ a constant substituting for $-r_A$ in equation (1) and integrating

$$\frac{W}{F_{AO}} = \frac{1}{K'}\ln(1 - X_A)$$

or $$K' = -\ln(1 - X_A)/(W/F_{AO})$$

values of K' are calculated for the given conditions given below:

S.No.	W/F_{AO}	$-\ln(1 - X_A)$	K'
1.	0.1	0.13	1.300
2.	0.2	0.2614	1.306
3.	0.3	0.3973	1.310
4.	0.4	0.527	1.320
5.	0.5	0.658	1.307
			1.309 (average value)

The value of K' is found to be constant. Hence the assumption of first order is correct.

(a) Reaction rate at 50%

$-r_A = K'(1 - X_A) = 1.309\ (1 - 0.5)$

$= 0.6545\ \dfrac{\text{kg moles}}{\text{hr kg of catalyst}}$

(b) Weight of catalyst needed (W) at 50% conversion and

$(F_{AO} = 500 \text{ kg moles/hr}) = F_{AO} \times -\dfrac{1}{K'} \ln (1 - X_A)$

$= 500\ X - \dfrac{1}{1.309} \ln (1 - 0.5)$

$= 264.76$ kg of catalyst.

62. 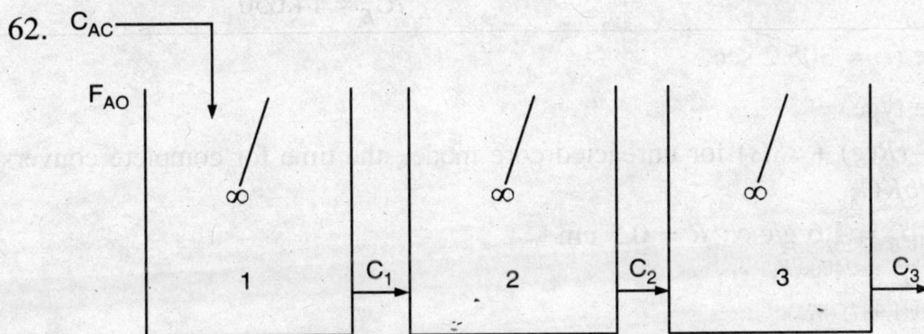

The design equation for a stirred tank reactor is

$V/F_{AO} = \bar{t} = (C_0 - C)/(-r)$

$-r = KC$; So, $\bar{t} = (C_0 - C)/KC$ or $C = C_c/(1 + K\bar{t})$

Hence for tank '1' here $\bar{t} = 2000/500 = 4$ min

$C_1 = C_0/(1 + K\bar{t})$ or $C_1 = 0.6097\ C_0$

For tank '2', $C_2 = C_1/(1 + K\bar{t})$ or $C_2 = 0.6097\ C_1$

For tank '3', $C_3 = C_2/(1 + K\bar{t})$ or $C_3 = 0.6097\ C_2$

or $C_3 = (0.6097)^3\ C_0$

Fraction hydrolysed at the end of three reactors

$= 1 - C_3/C_0 = 1 - (0.6097)^3\ C_0/C_0 = 0.773$ or 77.3%

63. $F_{(t)} = \dfrac{\sum_0^t C_E}{\sum_0^\infty C_E}$; $\bar{t} = \dfrac{\sum_{t=0}^{t=\infty} tC_E}{\sum_{t=0}^{t=\infty} C_E}$

The data work up is shown in table.

Time, t(sec)	$C_E(g)$/c.c.	$t_0 C_E$	$F(t)$	tC_E
0	0.0	0.0	0	0
100	6.5	6.5	0.135	650
200	12	18.5	0.385	2400
300	12	30.5	0.635	3600
400	10	40.5	0.844	4000
500	5	45.5	0.940	2500
600	2.5	48.0	1.0	1500
800	0.0	48.0	1.0	0.0
1000	0.0	48.0	1.0	0.0
$C_E = 48$			$tC_E = 14.650$	

Mean residence time (t) = 305.2 sec.

64. For a reaction of the type

$aA(g) + bB(s) \longrightarrow rR(g) + sS(s)$ for unreacted core model, the time for complete conversion (T) of B is $T = \rho_B R/bKC_{Ag}$

In the present case, $\rho_B = 4.6$ g/c.c.; $R = 0.5$ cm

$b = 1/4$; $K = 2.0 \times 10^5\, e^{-24000/R_S T}$

$= 2.0 \times 10^5 \times e^{-24000/(1.987)\,(873)}$

$= 2.0 \times 10^5 \times 9.8 \times 10^{-7} = 0.196$ cm/sec.

$C_{Ag} = p/R_g T = 1/(82.06)\,(273 + 600)$

$= 1.396 \times 10^{-5}$ gm moles/c.c.

$$T = \frac{4.6 \times 0.5}{(1/4)\,(0.190)\,(1.396) \times 10^{-5} \times 2} = 16.8 \times 10^5 \text{ sec}$$

65. For a reaction of the type

$aA(g) + bB(s) \longrightarrow rR(g) + sS(s)$ for unreacted core model, the time required for complete converstion when diffusion through the product is controlling, is given by $T = \rho_B R^2/6bD_e C_{Ag}$

In the present problem

$\rho B = 5.0$ g/c.c.; $R = 0.01$ cm

$b = 1$; $D_e = 4.0 \times 10^{-2}$ cm²/sec

$C_{Ag} = p/RT = 1/820.6\,(273 + 450)$

$= 1.69 \times 10^{-5}$ g moles/c.c.

$$\text{So, } T = \frac{5.0 \times (0.01)^2}{6 \times 1 \times 4.0 \times 10^{-2} \times 1.69 \times 10^{-5} \times 2}$$

= 61.637 sec.

SECTION—D

1. (c)

Let V be volume of reaction mixture.

Volume of $A = 0.5\ V$; volume of inerts $= 0.5\ V$

$0.5\ V$ of a A gives $0.25\ V$ of B ($\because 2A \rightarrow B$)

$\therefore$ Final volume $= 0.25\ V$ of $B + 0.5\ V$ of inerts $= 0.75\ V$

$$\text{Expansion factor} = \frac{\text{Final volume} - \text{Initial volume}}{\text{Initial volume}}$$

$$= \frac{0.75V - V}{V} = -0.25$$

2. (a)

3. (d)

The exit age distribution for a CSTR is given by $e^{-\theta}$

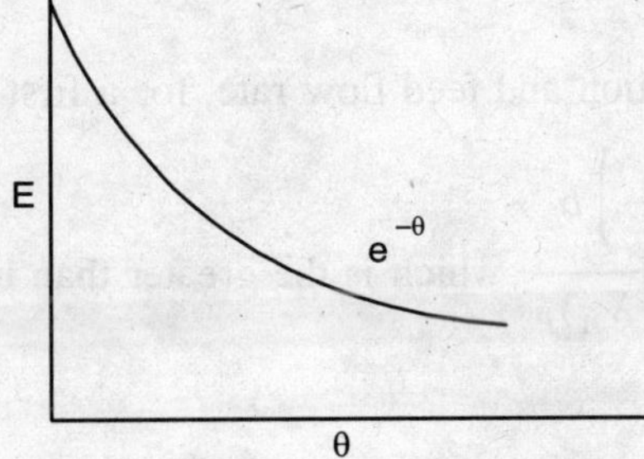

4. (a)

5. (b)

$$\frac{K_2}{K_1} = e^{-E/R\left(\frac{1}{T_2} - \frac{1}{T_1}\right)};\ K \propto e^{-E/RT}$$

$$\frac{K_2}{K_1} = e^{-E/R\left(\frac{1}{T_1} - \frac{1}{T_2}\right)};\ \ln\left(\frac{K_2}{K_1}\right) = \frac{E}{R}\left(\frac{1}{T_1} - \frac{1}{T_2}\right)$$

$$\ln 100 = \frac{E}{R}\left(\frac{1}{400} - \frac{1}{500}\right);\ \frac{E}{R} = 9210$$

6. The design equation for a batch reactor is:

$$t = \int_{C_{AO}}^{C_A} \frac{dC_A}{-r_A};\ C_A = C_{AO}(1 - X_A)$$

$$-r_A = KC_A \text{ (First order)}$$

$$t = \int_{CA}^{CA_0} \frac{dC_A}{KC_A} = \frac{1}{K} \ln C_A \Big|_{C_A}^{C_{AO}} = \frac{1}{K} \ln\left(\frac{C_{AO}}{C_A}\right)$$

Substituting the data, $15 = \frac{1}{K} \ln \frac{C_{AO}}{C_{AO}(1 - X_A)} = \frac{1}{K} \ln \frac{1}{(1 - 0.8)}$

$K = \ln 5/15 = 0.107$

Plug flow reactor: $\tau = \frac{1}{5} = C_{AO}\int_0^{X_A} \frac{d_{XA}}{-r_A} = C_{AO}\int_0^{X_A} \frac{d_{XA}}{K_{CA}}$

$$= \frac{C_{AO}}{C_{AO}}\int_0^{X_A} \frac{dX_A}{K(1-X_A)} = -\frac{\ln}{K}(1-X_A)\Big|_0^{0.8}$$

$$\tau = \frac{\ln}{K}(1-X_A)\Big|_0^{0.8} = \frac{\ln 5}{K};\ S = \frac{K}{\ln 5} = \frac{\ln 5}{15 \ln 5} = 0.067 \text{ min}$$

Back mix reactor:

$$\tau = \frac{1}{S} = \frac{C_{AO}X_A}{-r_A} = \frac{C_{AO}X_A}{KC_A} = \frac{C_{AO}X_A}{KC_{AO}(1-X_A)}$$

$$= \frac{X_A}{K(1-X_A)} = \frac{0.8}{K\times 0.2} = \frac{4}{K};\ S = \frac{K}{4} = \frac{0.107}{4} = 0.026/\text{minute}$$

7. Ans. Greater

For identical feed composition and feed flow rate, for a first order reaction,

$$\frac{V_{\text{mixed reactor}}}{V_{\text{plug flow reactor}}} = \frac{\left(\frac{X_A}{1-X_A}\right)b}{-\ln(1-X_A)_P}$$ which is the greater than unity for any conversion.

8. Ans. True

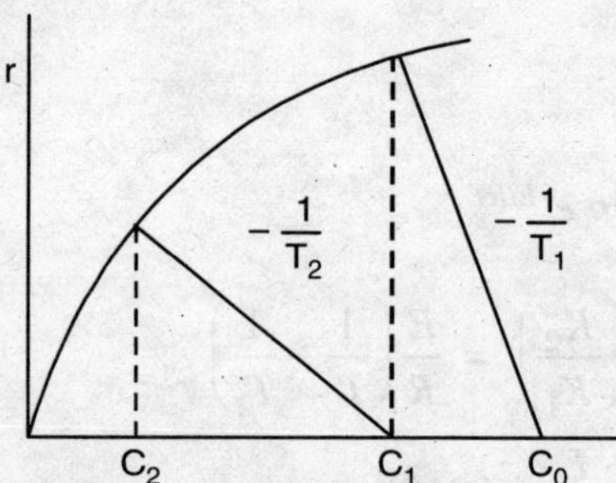

For reactions whose order is greater than unity, the plot $-\frac{1}{r}$ *vs*. X (conversion) is concave. In order to attain maximum production, the given system of reactors should be arranged so that the concentration of the reactant is kept as high as possible. It is possible only when the smaller reactor is placed before the larger reactor. As seen from figure higher concentration is obtained if smaller reactor is kept before big reactor since $\tau_1 < \tau_2$.

9. Ans. True

The design equation for a batch reactor is given by:

$$t = C_{AO}\int_0^{X_A} \frac{dx_A}{-r_A};\ C_A = C_{AO}(1-x_A)$$

$$-r_A = KC_A^2 = KCA_O^2(1-x_A)^2;\ t = CA_0\int_0^{X_A}\frac{dx_A}{KCA_0^2(1-x_A)}$$

$$= \frac{1}{KCA_0}\left(\frac{x_A}{1-x_A}\right)$$

$$t_{0.75} = \frac{1}{KCA_0}\left(\frac{0.75}{1-0.75}\right) = \frac{3}{KCA_0} \quad \text{...(1)}$$

$$t_{0.50} = 5_{\min} = \frac{1}{KCA_0}\frac{0.5}{(1-0.5)} = \frac{2}{KCA_0} \quad \text{...(2)}$$

Eqn. (2) divided by Eqn. (1) gives, $t_{0.75} = 5 \times \dfrac{3}{2} = 7.5$ min.

Time required for 75% conversion is $7\frac{1}{2}$ minutes. The design equation for a plug flow reactor is given by, τ (space time) $= C_{A0}\int_0^{X_A}\dfrac{dx_A}{-r_A}$ same as the holding time required for a batch reactor.

10. The design equation for a CSTR, assuming that the expansion factor is zero and the initial conversion is zero is given by:

$$\frac{V}{F_{AO}} = \frac{X_{Af}-0}{(-r_A)_x} = \frac{X_{Af}}{KCA^2f} = \frac{X_{Af}}{K[C_{AO}(1-X_{Af})]^2}$$

For $V = V_1$; $\dfrac{V_1}{FA_O} = \dfrac{0.50}{(-r_A)f} = \dfrac{0.5}{KCA_O^2(1-0.5)^2}$ ...(1)

For $V = V_2$; $\dfrac{V_2}{FA_O} = \dfrac{X_2}{(-r_A)f} = \dfrac{X_2}{KCA_O^2(1-X_2)^2}$ ...(2)

Eqn. (2) divided by Eqn. (1); $\dfrac{V_2}{V_1} = \dfrac{X_2}{(1-X_2)^2}\dfrac{1}{2} = 5$

$10 = \dfrac{X_2}{(1-X_2)^2}$; $10 \times 2^2 - 21 \times 2 + 10 = 0$; $X_2 = 1.37$ or 0.73

Since conversion cannot be greater than unity $X_2 = 0.73$ or 73%.

11. $4A \longrightarrow B + 7S$

$-r_A = 0.17\ C_A$; $K_1 = 0.17\ s^{-1}$;

$C_{AO} = 0.1$ mol/m^3; $F_{AO} = 0.17$ mol/s.

Percentage conversion to be achieved = 50% = X_A

The design equation for plug flow reactor is:

$$\frac{V}{F_{AO}} = \int_0^{X_A} \frac{d_{XA}}{-r_A} = \int_0^{X_A} \frac{dX_A}{KC_A};$$

At constant pressure

$$C_A - CA_O \frac{(1-X_A)}{1+\in_A \times A}; V = \frac{FA_0}{KC_{AO}}\left[(1+\in_A)\ln\frac{1}{(1-X_A)} - \in_A X_A\right]$$

$$\in_A - \frac{8-4}{4} - 1$$

$$\therefore \quad V = \frac{0.17}{0.17 \times 0.1}\left[(1+1)\ln\frac{1}{(1-0.5)} - 1 \times 0.5\right]$$

$V = 10[2 \ln 2 - 0.5] = 8.86$ m^3

Volume of the reactor = 8.86 m^3.

12. Ans. False

The first step is slower and hence this becomes rate controlling step. The rate of disappearance of CH_3CHO depends on the concentration of CH_3CHO and I_2.

13. Ans. 1 – b; 2 – d

14. Ans. (a) Effectiveness factor (E) = $\frac{\tan h}{h}$ where h is Thiele modulus.

$$E = \frac{\tan h_2}{2} = 0.482$$

(b) Liquid a decomposes in a batch reactor by a second order kinetics. If 50% of A is converted in a five minute run, how long would it take to reach 75% conversion.

The design equation for a batch reactor is

$$t = CA_0 \int_0^{X_A} \frac{dX_A}{-r_A}; -r_A = KC_A^2 = KC_{A0}^2 (1-X_A)^2$$

$$t = \frac{CA_0}{KC_{A0}^2}\int_0^{X_A} \frac{dX_A}{(1-X_A)^2} - \frac{1}{KC_{A0}}\left(\frac{1}{1-X_A} - 1\right)$$

$$= \frac{1}{KC_{A0}}\left(\frac{X_A}{1-X_A}\right) = \frac{1}{KC_{A0}}\left(\frac{0.5}{1-0.5}\right) = \frac{1}{KC_{A0}} = 5 \text{ minutes}$$

$$\therefore \quad \text{For 75\% conversion, } t = 5\left(\frac{X_A}{1-X_A}\right) = 5\left(\frac{0.75}{1-0.75}\right) = 15 \text{ minutes.}$$

15. (a) (i) $K = K_0\, T^m\, e^{-E/RT}$

Transition state theory $m = 1$

Collision theory $m = 1/2$

Arrhenius theory $m = 0$

(ii) Space velocity = $\dfrac{\text{(Volume of feed entering at specified conditions/time)}}{\text{Volume of reactor}}$

$= \dfrac{V/2}{V} = \dfrac{1}{2}\ \text{min}^{-1}$

(b) Ans. (d)

$2A \rightarrow R$

$2A \rightarrow S$

Initial moles of $A = 2$; at time t, moles $= 0.30$ moles of $R = 0.5$

For one mole of R, two moles of A are consumed

$\therefore$ 1 mole of A is consumed for R

No. of moles of A consumed for $S = 2 - 1.3 = 0.70$

For one mole of S, two moles of A are consumed

$\therefore$ 0.35 moles of S are present at time, t

(c) Plug flow: Perfectly mixed flow

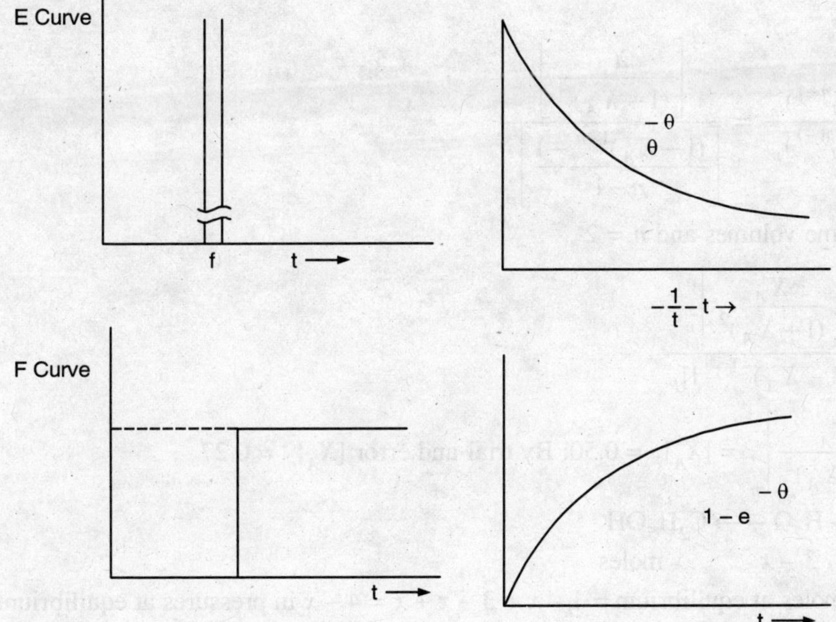

16. The observed rate in a catalyst pellet is given by,

$W_{AS} = 4\pi\, r\, D_{\text{eff.}}\, C_{AS}\, [1 - KR \text{ Coth } KR]$

where $K = \sqrt{\dfrac{K''a}{D_{\text{eff.}}}}$; $R = \dfrac{3 \times \text{Volume of particle}}{\text{Surface area}}$ for a non-spherical particle.

For cube $R = \dfrac{\partial V_P}{S_P} = \dfrac{3S^3}{6S^2} = \dfrac{3S}{6}$

For $S = 2$ cms, $R = 3 \times \dfrac{2}{6} = 1$ atm

$W_{AS} = 10^{-8}$ moles/c.c. S (given) C_{AS} = Surface conc.

$10^{-8} = 4\pi \times 1 \times 0.1\ C_{AS}\,[1 - K \text{ Cot } h\ k]$

or $(1 - K \cot h\ k) = 10^{-8}/(C_{AS} \times 0.04\ \pi)$

C_{AS} (Surface concentration is required but not given in the problem).

If C_{AS} is given, k has to be determined by trial, from which k'' (first order rate constant) can be determined.

17. (d)

18. (d)

19. (c)

20. (b)

21. (b)

22. (a)

23. $$\frac{(TC_{AO}^{\ n-1})_m}{(TC_{AO}^{\ n-1})_p} = \frac{\left[\dfrac{X_A}{(1-X_A)^m}\right]}{\left[\dfrac{(1-X_A)^{1-n}-1}{n-1}\right]_P}$$

For same volumes and $n = 2$

$$1 = \frac{\left[\dfrac{X_A}{(1-X_A)^2}\right]_m}{[(1-X_A)^{-1}-1]_P}$$

$$\left[\frac{X_A}{(1-X_A)^2}\right]_m = [X_A]_P = 0.50; \text{ By trial and error } [X_A]_m = 0.27$$

24. $C_2H_4 + H_2O \longrightarrow C_2H_5OH$

$1 - x \quad 3 - x \qquad x$ moles

Total moles at equilibrium = $1 - x + 3 - x + x = 4 - x$ in pressures at equilibrium.

$$0.25 = \frac{\dfrac{2x}{(4-x)}}{2\left(\dfrac{1-x}{4-x}\right)\left(\dfrac{3-x}{4-x}\right)^2};\ x = 0.268$$

mole % of C_2H_5OH at equilibrium = $\left(\dfrac{x}{4-x}\right) \times 100 = 7.18$

25. (a) $-dCA/dt = k$, $CA_0 - C_A = kt$; $C_{AO} = 0.50$

$C_A = 0.5 * 0.60 = 0.30$

$0.5 - 0.3 = 1200\ k$; $K = 0.167 \times 10^{-3}$ k mol./(m^3 . S)

(b) For $t = 3600\ S$; $C_A = -0.10$ which is not possible. So conversion is 100%.

26. $-r_A = -\dfrac{dC_A}{dt} = K_1 C_A + K_2$; $\dfrac{dC_R}{dt} = K_1 C_A$

$$\frac{dC_S}{dt} = K_2$$

$\dfrac{dC_A}{dt} + K_1 C_A = -K_2$; solving gives

$$C_A = \left[CA_0 + \frac{K_2}{K_1}\right] e^{-K_1 t} - \frac{K_2}{K_1}$$

$$\frac{dC_R}{dt} = -K_2 + K_1 C_{A0}\ e^{-k_1 t} + K_2 e^{-k_1 t}$$

$$C_R = \frac{K_2}{K_1} - K_2 t - C_{A0} e^{-k_1 t} - \frac{K_2}{K_1} e^{-k_1 t}$$

For 80% conversion. $C_A = 0.2 \times 2.0 = 0.40$

Using C_A equation, $t = 2.93$ seconds

C_R at this $t = 2.93$ is 0.717

27. (a) $\dfrac{V}{F_{A0}} = \dfrac{X_A}{-r_A} = \dfrac{0.70}{0.02 \times 2 \times 0.30}$

$-r_A = k_1 C_A = K_1 C_{A0}(1 - x_A)$

$\dfrac{V}{F_{A0}} = \dfrac{0.70}{0.012}$; $F_{A0} = 0.01$; $V = 0.6$ m^3;

(b) By energy balance:

$0.01 \times 1000 \times T_f \times 2 + 50{,}000 \times 0.02 \times 0.6 \times 1.4 - 0.01 \times 1000 \times 353 \times 2 = 0$, $T_x = 311$ K.

28. (c)

29. (b)

30. (b)

31. (a)

32. $\dfrac{1}{X_{Ae}} K_1 t = \dfrac{1}{X_{Be}} K_2 t$; $K_e = \dfrac{K_1}{K_2} = \dfrac{X_{Ae}}{X_{Be}} = \dfrac{0.40}{0.50} = 0.80$

33. $\tau = \int_{C_{A0}}^{C_{Af}} \frac{d_{CA}}{-r_A}; -r_A = KC_A^{\ n}$

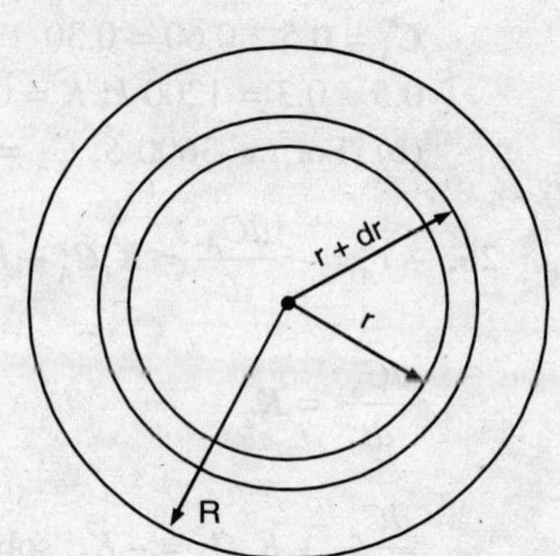

$$= \frac{1}{K}\int_{C_{AO}}^{C_{AF}} C_A^{\ -n}\, d_{CA} = \frac{-C_A^{\ -n+1}}{(-n+1)}\Bigg|_{C_{AO}=1,2}^{C_{AF}=0.50,\,0.555}$$

$1 - 0.5^{-n}X^{0.5} = 2^{-n} \times 2 - 0.555^{-n} \times 0.555$

So $n = 1$ by trial for

$$n = 1,\ \frac{1-0.5}{0.5} = \frac{2}{2} - \frac{0.555}{0.555}$$

$$\tau = \int \frac{d_{CA}}{r_A};\ r_A = K_{CA}$$

$$\tau = \frac{1}{k}\ln\left(\frac{C_{Af}}{C_{AO}}\right);\ 10 = \frac{1}{k}\ln\left(\frac{0.50}{1}\right)$$

$k = \ln(0.5)/10$

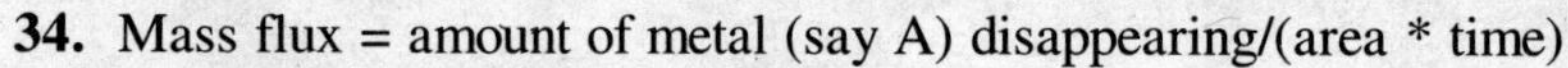

34. Mass flux = amount of metal (say A) disappearing/(area * time)

From data mass flux, N_A = constant

Taking a differential volume disappearing in time dt, the radii being r and $r + dr$. Any time dt, volume disappearing there is:

$$\frac{4}{3}\pi\,[(r+dr)^3 - r^3] = \frac{4}{3}\pi * 3r^2dr \text{ (neglecting small terms)} = 4\pi\, r^2\, dr$$

Surface area = $4\pi r^2 = 4\pi(r + dr)^2$ (neglecting smaller terms)

$$N_A = \frac{4\pi\, r^2 dr\, \rho}{4\pi\, r^2\, dt} = \rho\frac{dr}{dt}$$

Integrating, $N_A \times t = pr$ + constant from Initial condition: $r = r_1$ at $t = 0$

$$N_A = \frac{\rho}{t}(r - r_1);\ N_A = K_A(x_{A1} - x_{A2})$$

Assuming $x_{A2} = 0$, at bulk of fluid and

$$x_{A1} \approx x_A;\ N_A = K_A x_A;\ K_A x_A = \frac{\rho}{t}(r - r_1)$$

So film mass transfer controlling.

35. (d)

36. (a)

$$V = \frac{F_{AO}\, X_A}{-r_A} = \frac{VO\, C_{AO} X_A}{KC_{AO}\,(1 - X_A)}$$

$$\frac{V}{V_O} = \frac{X_A}{K(1-X_A)}; T = \frac{X_A}{K(1-X_A)}; TK = \frac{X_A}{(1-X_A)}$$

37. (d)

38. (c)

The amount of heat removed is:

0.001 * 5 * 0.5 * 40 kJ/sec./80 = 5/3 kW.

39. (d)

40. (c)

$\frac{r_1}{r_2} = \frac{1}{2}$; r_1 ?; $(r_1 + r_2) = \frac{1}{3}$; $\frac{r_1}{r_A} = \frac{1}{3}$; For $X_A = 2$; 1 mole of *A* produces 2 moles of *r*. So maximum possible value of $r = \frac{2}{3}$

41. (b)

$A \rightarrow 2R; A \rightarrow 3S; r_1 = K_1P_A; r_2 = K_2P_A$

$V = \frac{F_{AO} * X_A}{-r_1}; -r_1 = K_1P_A = 20 * 0.70 * 1 = 14; V = 2 * 0.30/14 = 0.021$ cubic m.

42. (a)

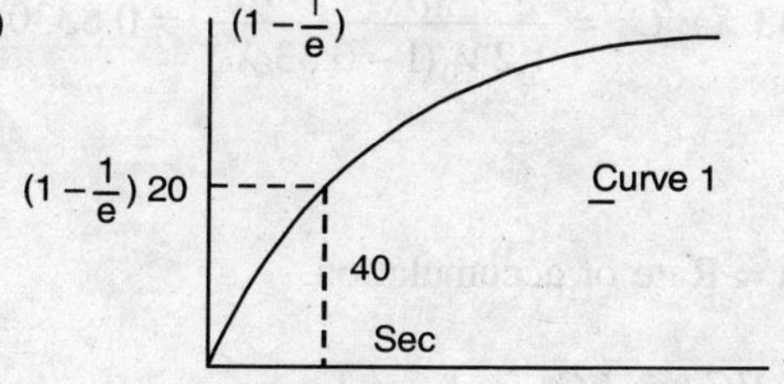

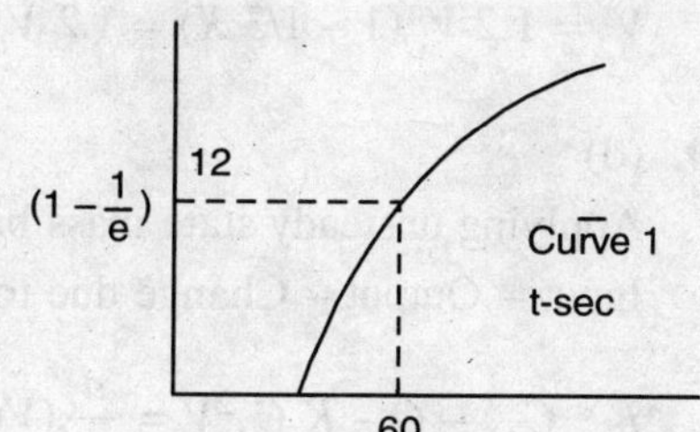

43. (a)

44. (d)

The equilibrium constant is given by: $K_P = \frac{P_C^{\ c}\ P_D^{\ d}}{P_A^{\ a}\ P_B^{\ b}}$

Thus equilibrium constant increases/decreases with pressure depending on the stoichiometric coefficients of the reactions.

45. (a)

$N_2 + 3H_2 \rightleftharpoons 2NH_3; r = 0.8\ P_{N_2}^{\ 1/2}\ P_{H_2}^{\ 3/2} - 0.6\ P_{NH_3}^{\ 2}$

when the reaction is represented by: $\frac{1}{2}H_2 + \frac{3}{2}H_2 \rightleftharpoons NH_3$

$r = 0.8\ P_{N_2}^{\ 1/2}\ P_{H_2}^{\ 3/2} - 0.6\ P_{NH_3}$

46. (a)

For endothermic aqueous phase first order irreversible reaction carried out in a adiabatic PFR. The rate of reaction is maximum at the inlet of the reactor.

47. (d)

Kinetic rate constant = K; external mass transfer coefficient = K_g

$K_{eff} = 1/(1/K + 1/K_g)$; $1/K_{eff.} = 1/K_g + 1/K$

48. (c)

For a packed bed reactor, the presence of a long tail in the residence time distribution curve, The long tail represents the dead zone.

49. (c)

$A + \frac{1}{2}B \rightarrow C$; A and B are at 300 K, PFR is used.

$$C_A = \frac{N_A}{V} = \frac{N_{AO}(1-X)}{V_2}$$

For gas phase reaction at constant pressure,

$$\frac{V_1}{T_1} = \frac{V_2}{T_2};\ V_2 = V_1 * \frac{T_2}{T_1} = V_1 * \frac{6}{5} = 1.2\ V_1;\ V_1 = V_0\ (1 + X);\ \in = \delta Y_{AO} = -\frac{1}{2} * \frac{2}{3} = -\frac{1}{3};$$

$$V_2 = 1.2\ V_O(1 - 1/3\ X) = 1.2\ V_0\ (1 - 0.33\ X):\ C_A = \frac{N_{AO}(1-X)}{1.2\,V_0(1-0.33\,X)} = 0.8330$$

50. (d)

Applying unsteady state mass balance:

Input – Output – Change due to reaction = Rate of accumulation.

$$F * C_{AF} - O - K\,C_A^2 V = \frac{d}{dt}(VC_A);\ \frac{d}{dt}\ (VC_A) - FC_{AF} - KC_A^2\ V$$

51. (c)

$$A \rightarrow B; -r_A = KC_A^2 = KC_{AO}^2(1 - X_A)^2; -\frac{dC_A}{dt} = KC_A^2;\ \text{integrating:}$$

$$\frac{1}{C_A} = Kt + \text{constant; at } t = 0,\ C_A = C_{AO};\ \frac{1}{C_A} - \frac{1}{C_{AO}} = Kt = \frac{1}{C_{AO}}\left(\frac{X_A}{-X_A}\right)$$

Here $X_{A1} = 90\%$ and $X_{A2} = 45\%$. Taking the ratio of the above equation, susbtituting the conversions,

$$\frac{\frac{1}{C_{AO}}\left(\frac{0.90}{1-0.90}\right)}{\frac{1}{C_{AO}}\left(\frac{0.45}{1-0.45}\right)} = Kt_1/K\,t_2 = t_1/t_2 = 11$$

52. (c)

$$V = \frac{FP_0\ X_P}{-r_P} \text{ evaluated at } X_P;\ V = \frac{C_{PO}\ X_P\ v_o}{(0.5\ C_P - 0.125\ C_R)}$$

$t = C_{PO}\ X_P/(0.5\ C_{PO}\ (1 - X_P) - 0.125\ C_{PO}X_P)$

$T = C_{PO}\ X_P/(0.5\ C_{PO}\ (1 - X_P) - 0.125\ C_{PO}\ X_P)$; $T = X_P/(0.5\ (1 - X_P) - 0.1250\ X_P)$

at $X_P = 0.40$, $T = 1.6$ hrs. $(0.40/(0.5 * 0.6 - 0.125 * 0.40))$

53. For mixed flow reactor, $V/V = (C_{AO} - C_A)/(KC_A)$; $C_{AO} = 2$ mole/m^3

$C_{AO} = 0.5$ K mol/m^3; $V = 1$ m^3/hr; $V = V_1$

For first case, $V_1/1 = (2 - 0.50)/(K * 0.50) = 3/K$

For second case, $V_2 = 3\ V_1 = 3$ m^3/hr.; $C_{AO2} = 2C_{AO1}$ $\dfrac{V}{V_2} = \dfrac{C_{AO2} - C_{A2}}{KC_{A2}}$; $\dfrac{V_2}{3} = \dfrac{4 - 0.5}{3/V_1 * 0}$

$\dfrac{V_2}{V_1} = 7$ **Ans.**

54.

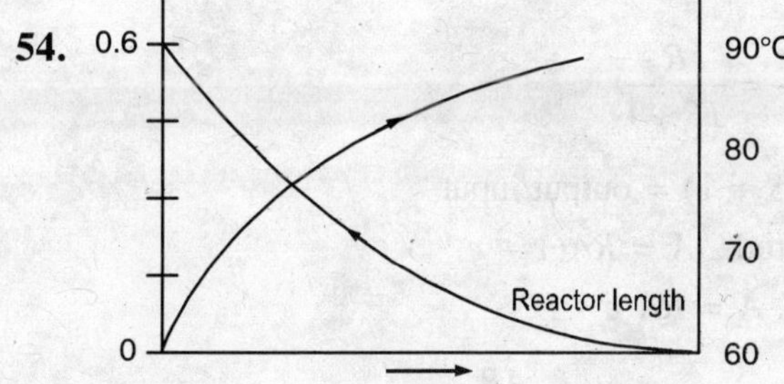

55. (b)

$C_P = 2$ kJ/kg K + 1000 kg/m^3 + 1/4000 m^3/mole = 0.5 kJ/kg mole K

For complete conversion, $X = 1$

$-\Delta H_r - C_P\Delta T$; $-(-50) = 0.5\ (390 - T)$; $T = 290$ K

56. (d)

$A \to R$ (desired); $A \to S$ (undesired)

Selectivity, $S = r_R/r_S = K_1/K_2\ C_A^{\ a-b}$

For $a > b$, $a - b = +$ ve.

For high sensitivity, we have to make C_A as possible. We can use high pressure and eliminate inerts. We can use use batch or PFR. We can use CSTR with high conversion. We can use recycle. Since we have to opt wrong design guide lines, hence avoiding recycle is the option.

❑

CHAPTER 10

Instrumentation and Process Control

1. Laplace transforms of commonly used functions:
$L(t^n) = n!/S^{n+1}$; $L(e^{at}) = 1/(S - A)$; $L(\sin at) = 1/(S^2 + a^2)$;
$L(\cos at) = S/(S^2 + a^2)$; $L(u(t - a)) = e^{-aS/S}$; $L(\text{unit impulse}) = 1$;
$L(t^n f(S)) = (-1)^n d^n x(S)/dS^n$; $L(e^{bt} \sin at) = a/((s - b)^2 + a^2)$
$L(t \sin at) = 2aS/(S^2 + a^2)$; $L(t \cos at) = (S^2 - a^2)/(S^2 + a^2)$
$L(e^{bt \cos at}) = (S - b)/(S - b)^2 + a^2)$; $L(f^2(t)) = S^2 f(S) - Sf(0) - f(0)$
$L(e^{-at} t^n) = n!/(S + a)^{n+1}$; $L(\text{step of magnitude } A) = A/S$

2. First order transfer function: $\dfrac{y(S)}{x(S)} = \dfrac{\text{Output}}{\text{Input}} = \dfrac{R}{TS+1}$

2nd order transfer function: $R/(T^2 S^2 + 2\xi TS + 1)$ = output/input
Step response of 1st order system of magnitude, $A = RA(1 - e^{-t/T})$
Impulse response of 1st order of magnitude, $A = RA\, e^{-t/T}/T$

Sinusoidal response of 1st order of magnitude: $y(t) = \dfrac{AR}{\sqrt{(w^2 T^2 + 1)}} \sin(\omega t + \phi)$

$\phi = -\text{Tan}^{-1}(\omega T)$
Linear or ramp response: $y(t) = Kt - KT + KT\, e^{-t/T}$

3. Automatic control is the maintenance of a desired value by measuring the existing value, comparing with desired value and using this difference to intiate action for reducing the difference. Examples are: Heat exchanger, exothermic CSTR, home heating by gas supply, water heater. In heat exchanger hot fluid temperature is controlled by cold fluid flow rate and vice versa. In exothermic reactor (CSTR) constant temperature in CSTR is maintained by changing cooling fluid flow rate. In water heater hot water temperature is maintained by changing heat supply. The main uses are: 1. improves the uniformity of the products, 2. decreases human errors, 3. economical in some processes.

4. Block diagrams: Represent the various units present in a controlled process. Analysis is done with the help of block diagrams. These are like flow diagrams in process control. The figures used in drawing block diagrams are:

(a) Dynamic function: $y \rightarrow \boxed{} \rightarrow x$; $x = f(y, t)$; t = time

(b) Algebraic function: $z \rightarrow \otimes \rightarrow x$; $x = z - y$ (with y entering the summing junction from below)

5. The process variables in process control are:

(a) Controlled variable: This is that variable which directly or indirectly indicates the form or state of the product.

(b) Manipulated variable: It is that variable which is selected for adjustment by the controller so as to maintain the controlled variable at the desired value.

(c) Load variable: All other variables of independent type except controlled variable and manipulated variable.

6. Process decrees of freedom: These are the number of independent variables in a process, $n_p = n_v - n_e$ where n_p is process degrees of freedom, n_e is number of equations, n_v is number of variables.

7. Dead time (Transport lag): Dead time is defined as any definite time delay between two related actions. Dead time is found in chemical reactions, when a definite time is to be waited for the reaction to begin.Transportation lag is due to transport of fluid. This is given by: Volume of fluid, in pipe/flow rate of fluid. The units are time units.

8. (a) For single tank, for linear system the transfer functions are:

$h'(S)/q_i'(S) = R/(TS + 1)$; $q_0'(S)/q_i'(S) = 1/(TS + 1)$

(b) for CSTR with first order chemical reactions:

$$\frac{C_o'(S)}{C_i'(S)} = \frac{F/(F + KV)}{V/(F + KV)S + 1)} = + \frac{R}{TS + 1}$$

(c) For non-linear single tank, $q = c\sqrt{h}$; $\dfrac{H(S)}{Q_1(S)} = \dfrac{2h^{1/2}}{(A2\sqrt{h_S}\, S + C)} = \dfrac{R}{TS + 1}$

For mercury in glass thermometer, $y(S)/x(S) = 1/[(mC/hA)S + 1]$

(d) For single tank $C(S)/m(S) = R/(TS + 1)$; $C(S)/u(S) = 1/(TS + 1)$

(e) For noninteraction tanks; $H_2(S)/Q_1(S) = + \dfrac{1}{T_1S + 1}\dfrac{R_2}{T_2S + 1} = \sum \dfrac{K_i}{(T_iS + 1)}$

For manometer: $\dfrac{h(S)}{P(S)} = \dfrac{R_2}{T_1 T_2 S^2 + (T_1 + T_2 + A_1R_2)S + 1}$

(f) Interacting: $\dfrac{H_2(S)}{Q_1(S)} = \dfrac{R_2}{T_1 T_2 S^2 + (T_1 + T_2 + A_1R_2)S + 1}$

For manometer: $h(S)/P(S) = \dfrac{1/2P}{(L/2g)S^2 + (16\,\mu L/\rho D^2 g)S + 1}$

9. Parts of a control system are: (a) Process; (b) Measuring instrument; (c) Automatic controller; (d) Final control element; (e) Transmission.

10. Various control actions: The automatic controller determines the error and generates counteraction to produce the smallest error. The method by which this the automatic controller generates this counter action is mode of control or control action.

P-action: $m(t) = K_C\, e(t) + M_S$; 1-action: $m(t) = \int e/T_i\, dt$

D-action: $m(t) = T_d\, de/dt$; *P–I*-action: $m(t) = \frac{K_C}{T_i}\int edt + K_C e$

P–D-action: $m(t) = K_C T_d\, de/dt + K_C e$; *P–I–D*: $m(t) = K_C e + K_C Td\, de/dt + \frac{K_C}{T_i}\int edt$

On–Off control: In this the manipulated variable is changed to either maximum value or minimum value depending on the error, whether it is greater than zero or less than zero.

Floating control: The manipulated variable changes at a constant rate in one direction when the deviation is positive and in the reverse direction when the deviation in negative.

P has large maximum deviation. Offset is present. Stabilizing time is less. *P–D* has smallest maximum deviation because *D* allows high K_C values.

Offset is half of *P*-control. Stabilising time is the smallest.

P–I: Offset is almost zero. Highest maximum deviation. Stabilizing time is increased by integral control.

P–I–D-action: Next smallest deviation. Offset is zero because of *I*-action.

Stabilization time is increased by *I*-action.

11. Optimum controller settings; methods: (a) Continuous cycling method; (b) Damped oscillation method; (c) Reaction curve (Ziegler-Nichols method); (d) Error integral; (e) Frequency response method.

12. Various controlling methods: (a) Feedback control; (b) Feed forward control; (c) Cascade control; (d) Proportioning or ratio control; (e) Optimal control; (f) Adaptive control; (g) Stochastic control.

13. Conversion of block diagrams:

(a) Negative feedback: $y(S)/x(S) = \frac{\pi_f}{1+\pi_t}$

where π_f = Product of transfer functions between *y* and *x*.

π_t = Product of transfer functions in the entire loop.

(b) Positive feedback: $\frac{\pi_f}{1-\pi_t} = y(s)/x(s)$

14. Responses of a 2nd order system:

$\frac{y(S)}{x(S)} = \frac{R}{T^2S^2 + 2\xi TS + 1}$; Solutions depend on values of ξ.

Step response values: $\xi = 1$; $y(t) = 1 - \left(1+\frac{t}{T}\right)e^{-t/T}$

$\xi > 1$; $y(t) = 1 - e^{-\xi t/T}\,(\cos h\sqrt{\xi^2 - 1}\, t/T + \frac{\xi}{\sqrt{\xi^2-1}} \sin h\sqrt{\xi^2 - 1}\, t/T)$

$\xi < 1$; $y(t) = 1 - \frac{e - \zeta t/T}{\sqrt{1-\xi^2}} \sin(\sqrt{1-\xi^2}\ t/T + \tan^{-1}\frac{\sqrt{1-\xi^2}}{\xi})$

Impulse response: $\xi = 1; y(t) = \frac{t}{T^2} e^{-t/T}$

$$\xi > 1; y(t) = \frac{e^{-\xi T/T}}{T\sqrt{\xi^2 - 1}} \left[\frac{\xi}{\sqrt{\xi^2 - 1}} \sin h\sqrt{\xi^2 - 1}\frac{t}{T} \right]$$

$$\xi < 1 \,; y(t) = \frac{1}{T(\sqrt{1 - \xi^2})} e^{\xi t/T} \sin\left((\sqrt{1-\xi^2})t/T\right)$$

For step responscs: Overshoot = $\exp(-\pi\xi/\sqrt{(1-\xi^2)})$

Decay ratio $\exp(-2\pi\xi/\sqrt{1-\xi^2})$; Period of oscillation = $\sqrt{1-\xi^2}/T = w$

Natural period of oscillation = $w_n = 1/T$

15. (a) Frequency response: Step and impulse responses are used upto 2nd order. For higher order, frequency response is used.

(a) First order: A.R. $= \frac{1}{\sqrt{1 + w^2 T^2}}$; $\phi = -\tan^{-1}(wT)$;

(b) Characteristic equation is the denominator of closed loop transfer function. $\frac{G(S)}{1 + G(S)}$; $G(S)$ is open loop transfer function. $G(S)$ is the product of all elements in the block diagram (Entire Loop).

(c) Ruth stability method uses characteristic equation, of form:

$b_n S^n + b_{n-1} S^{n-1} + \ldots\ldots + b_1 S + b_0 = 0$

Arranging as: $b_n \quad b_{n-2} \quad b_{n-4}$

$b_{n-1} \quad b_{n-3} \quad b_{n-5}$

Coefficients of 3rd row are: $(b_{n-1} * b_{n-2} - b_n * b_{n-3})/b_{n-1}$;

$(b_{n-1} * b_{n-4} - b_n * b_{n-5})/b_{n-1}$.

The coefficients of fourth row are calculated by cross multiflying the coefficients of 3rd row and second row. The system is stable if:

(1) All the coefficients of the characteristic equation are of the same sign.

(2) If all the coefficients of the first colunmn of the matrix are of similar sign.

(d) Second order system: $AR = \frac{1}{((1 - w^2 T^2)^2 + (2\xi wT)^2)^{1/2}}$

$\phi = \tan^{-1}(-2\xi\, wT/(1 - w^2 T^2))$

(e) Transportation lag: $AR = 1$; $\phi = -w * T * (180/3.143)$ degrees.

(f) *P*–controller: $AR = K_C$; $\phi = 0$ degrees.

(g) P–I: $AR = K_C\sqrt{1+(1/wT_i)^2}$; $\phi = \tan^{-1}(-1/wT_1)$

(h) $S : AR = w$; $\phi = 90°$

(i) $(TS + 1)$: $AR = \sqrt{1+w^2 T^2}$; $\phi = \tan^{-1}(wT)$

(j) $P\text{–}D : AR = K_C(1+w^2T_d^{\;2})^{1/2}$; $\phi = \tan^{-1}(wT_d)$

(k) P–I–D: $AR = K_C\left[\left(wT_d - \dfrac{1}{wT_i}\right)^2 + 1\right]^{1/2}$; $\phi = \tan^{-1}\left(wT_d - \dfrac{1}{wT_i}\right)$

(l) $AR_0 = AR_1 * AR_2 * \ldots\ldots AR_n$

$f_0 = \phi_1 + \phi_2 + \phi_3 \ldots\ldots + \phi_n$

16. Various plots of frequency response are: (a) Bode plots: *AR* vs. log *w*, ϕ vs. log *w*); (b) Polar plots (Real part vs. imaginary part i.e. magnitude versus phase angle). (c) Nichols plot (magnitude ratio vs. phase margin).

17. Root locus diagram: This is for stability analysis. This is plotting the roots of open loop transfer function with controller settings.

(a) If roots are real, then response is non-oscillatory; (b) if roots are complex and real parts are positive then response is increasing sinusoid; (c) if roots are complex and real parts are negative then response is oscillatory; (d) when the roots are imaginary then the system is unstable.

18. Nyquist stability method: Open loop transfer function is used for this stability method. It is the polar plot of frequency response of open loop transfer function.The amplitude is the radius vector and the phase angle is degrees clockwise from right handed axis.

Rule 1: If the polar plot passes –1. Point for $P = 0$ cases then it is unstable. P = number of roots with positive real parts.

Rule 2. If $N = P$ is satisfied then also stable. N = number of counter clockwise arrows minus number of clockwise arrows obtained by drawing radial line from –1 point to cut polar curve and marking the arrows towards increasing frequency.

19. Hurwitz-Mikhalov method: The characteristic equation, is arranged as:

$a_0 S^n + a_1 S^{n-1} + a_2 S^{n-2} + \ldots + a_{n-1} S + a_n = 0$

Arranging the elements as:

The determinants as marked should be greater than or equal to zeroes for stable.

$$\begin{matrix} a_1 & a_0 & 0 & 0 \\ a_3 & a_2 & a_1 & a_0 \\ a_5 & a_4 & a_3 & a_2 \\ a_7 & a_6 & a_5 & a_4 \end{matrix}$$

20. Closed loop frequency response values from open loop frequency response values:

$$M = \left[\frac{1}{\left(1+\dfrac{\cos\phi}{A}\right)^2 + \left(\dfrac{\sin\phi}{A}\right)^2}\right]^{1/2};\ \theta = \tan^{-1}\left[-\frac{\sin\phi/A}{1+\dfrac{\cos\phi}{A}}\right]$$

A and ϕ are for openloop, θ and M are for closed loop. Decibles = 20 $\log_{10}$ (AR)

21. (a) Lumped parameter systems:

Examples: mixer, CSTR. Response is a function of time only.

Ordinary differential equations. Phase lag is limited.

(b) Distributed parameter systems: Examples: tubular reactor, heat exchanger (shell and tube) partial differential equations. Phase lag has no limit. Response is a function of position and time.

22. (1) Self operated controllers: Examples: (a) pressure regulator (b) float level controller (c) thermostatic valve.

(2) Nozzle pneumatic controller: Principle: Flapper nozzle is a displacement type. It works by converting controlled variable and set point into corresponding displacements and converting these displacements in order to produce a pneumatic signal (air pressure output).

(3) Force type pneumatic controller: Principle: The diaphragms produce a force representing the error. This is also called as stack controller.

23. The various control valve plugs: (a) piston type (b) V-port type (c) parobolic type (d) popett type. Various control valves (automatic) are: (a) single seat valve (b) double seat valve (c) gate valve (d) weir valve (e) rotating plug valve (f) butterfly value (g) louvers.

24. (1) Parts of an instrument: (a) Primary or sensing element; (b) Secondary element; (c) Manipulating element; (d) Functioning element.

(2) Static characteristics of an instrument: These are range, span, accuracy static error, reproducibility, sensitivity, dead zone and drift.

(3) Dynamic characteristics: These are response, dynamic error, time lag, fidelity.

25. Transmission is of two types: (a) Pneumatic transmission. Example is pneumatic transmission of temperature through pressure bulb.

(b) Electric transmission – example is displacement of a variable using a differential transformer.

26. Modern control theory: This is developed in 1960 by Poincare. This uses state variables, state vector and state space. This can be used for linear, non-linear, time varying, multi–inputs and outputs etc. Laplace transforms can be used for linear and time invariant systems. The state input equation is $\dot{X} = A X + BM : \dot{X} = dX/dT$; t = time

X = state vector; A and B are constant matrices; M = input vector: State output is: $0 = CX + DM$; 0 is output vector; C and D are constant matrices.

27. Feed forward controller transfer function is: $G(S) = -G_1 S/G_p(S)$

28. Principles of working of temperature measuring instruments and ranges of operation.

(a) Mercury in glass thermometer: Volumetric expansin of mercury with increase of temperature. – 30°C to 350°C.

(b) Constant volume gas thermometer: Pressure of a gas increases with temperature increase (at constant volume). H_2 : O – 100°; N_2 : above 100°C; He : below 0°C.

(c) Bimetalic thermometer: When two dissimilar metal strips are wlded together having different coefficients of thermal expansion, if one end is fixed and the other end is free and if the fixed end is exposed to high temperature, then the free end deflects. This deflection is calibrated to temperature. Invar, brass, nickel alloys are used 0 to 400°C.

(d) Vapor pressure thermometer: – 20 to 300°C; The vapor pressure increases with increase of temperature. The pressure is calibrated to temperature by a Bourdon tube.

(e) Thermocouples: – 20 to 1600°C; when two dissimilar metal wires are connected at both ends and if one end is kept at a low temperature and the other end at a high temperature, then an emf in milli volts is generated in the circuit. This emf is calibrated to tempeature. Cu-constantan: 100–600°C; Chromel-Alumel: 800–1100°C.

(f) Iron–constantan: 0 to 600°C; pt – pt + Rh : 600 – 1600°C.

Resistance thermometers: – 200 to 650°C; Resistance of a metal wire increases with increase of temperature. Cu, Ni, platinum wires are used.

(g) Thermistors: In these, resistance decreases with increase of temperature. 6 to 300°C; mixtures of metallic oxides like manganese, cobalt, copper, iron, nickel and their salts.

(h) Radiation pyrometer: 800 to 2000°C; uses Stefan–Boltzmann law of radiation, $q = \sigma T^4$; The radiation from the object is focussed on a thermofile (number of thermocouples) by an optical system.

The thermofile generates emf (mVs) based on the amounts of radiation. The emf is calibrated to temperature.

(i) Optical pyrometer: 800–2000°C; uses monochromatic light (red color, 0.65 μ).

Method–1: Lamp light intensity is matched with object light. The lamp intensity is calibrated to temperature.

Method–2: Lamp light intensity is kept constant. An optical wedge (varying thickness) is moved to match object light and lamp light. The wedge thickness is calibrated to temperature.

(j) Photo-electric pyrometer: 800–16000°C; photovoltaic call gives an emf based on the amount of radiation falling on it. Photo emissive tube produces current proportional to the amount of radiation falling on it. The emf or current is calibrated to temperature.

29. Composition measuring instruments:

(a) Mass spectrometer: In analysing isotopes, inorganic gases, vapors. A gas is bombarded with electrons giving positive ions. The radius of circular path depends on mass and velocity of ion.

(b) X–ray diffraction: Dyes, filters in rubber, selection of clay particles, battery manufacture. In this one beam of X–ray is passed through the sample cell to a flore scent screen. The other beam is passed through an aluminium alternator and then to second sample cell; to the same screen. For the two beams to have the same intensity, the aluminium thickness is changed. The aluminium thickness is calibrated to composition.

(c) Polarograph: For analysing non-ferrous alloys, steels, ores etc. This is used for electrolytic materials. Mercury is dropped into the solution containing the sample. Potential supply is given to the electrodes from 0 to 5 volts in steps. Due to attraction of ions to the electrodes, the current rises in steps. The voltage of each step shows the presence of the elements and height of step shows concentration.

(d) Thermal conductivity cells: For CO_2 analysis, hydrogen in nitrogen, oxygen analysis, air mixtures, water vapor in air. The heat conducted from a heated filament depends on the gas and its composition. The resistance of the filament changes with gas and its composition. This variable resistance is measured by a Wheatstone bridge circuit.

(e) Psychrometer: Measures dry bulb temperature and wet bulb temperature of air. From these humidity, percent saturation etc. can be obtained from psychrometric chart.

(f) Hygrometer: Expansion and contraction of materials like wood, hair etc. with, humidity is used. This instrument when exposed to the air sample gives—percent saturation or relative percent saturation on the dial.

(g) P.H. meter: When two solutions of different hydrogen concentration are separated by a then glass membrane, then an emf in milli volts is generated. This emf is related to the concentration of solution. For measuring acidity, basicity of a solution.

(h) Refractometer: Monocromatic light rays (from sodium lamp) when passed through a liquid are refracted. The amount of refraction depends on the characteristics of a fluid. Benzene + tolune solution; ethanol + water etc.

(i) Conductivity: The electrical conductivity of a solution changes with composition. Used: Water analysis, salt solutions.

(j) Chromotograph: The sample gas is carried by a carrier gas like nitrogen.The components of the sample are separated by raising the temperature of the column gradually. Variation of thermal conductivity with gas composition is used. Examples: Air analysis, pollued gases analysis etc.

(k) Calorimetry: Refers to the determination of a substance from its ability to absorb visible light.

(l) Spectroscopic methods: Atoms of a substance are exited and thereby their configuration is changed. Then they either absorb or radiate energy at unique frequencies.

Line spectra are caused due to absorption or emission of discrete wavelengths of radiation by atoms.

The absorption spectroscopy is based on when X-ray radiation or infrared radiation or ultra-violet radiation is passed through an unknown material, some frequencies of radiation are absorbed. In the emission spectroscopy, the sample of unknown substance is placed in a flame or an arc. This gives emission of radiation which is characteristic of each substance.

In fluorescence spectroscopy, the substance is excited by absorption of light. The excitation of a substance caused emission of light which is related to the composition of a substance.

30. Bode-Stability method:

1. If *AR* overall crosses the point 1 at ϕ overall is – 180° (phase) then it is unstable.
2. At *AR* overall = 1, if phase (ϕ) overall exceeds – 180°, then it is unstable.

31. Signal flow graphs: This method is developed by S.J. Mason for complicated systems, signal flow graphs can be used to get relations among system variables. The elements of signal flow graphs are: node, branch, transmittance.

32. Mason's gain formula:

Overall gain $= T = \dfrac{1}{D}\sum_{K} T_K\, D_K$ where D is determinant of graph; T_K = transmittance or path gain of Kth forward path

ΣM_a = Sum of all different loop gains; $\Sigma M_f M_C$ = Sum of gain products of all possible combinations of two non-touching lops; $\Sigma M_d\, M_e\, M_f$ = Sum of gain products of all possible of three non-touching loops; D_k = Cofactor of Kth forward path determinant of the graph with the loops touching the Kth forward path removed.

33. Obtaining response equation from state equation:

$\dot{X} = AX + BM$ (State equation)

$$X(t) = X(t_0)\, e^{A(t-t_0)} + \int_{t_0}^{t} e^{A(t-\tau)}\, BM(\tau)\, d\tau$$

$\phi(t) = e^{At}$; t_0 = intial time, τ = time

$\phi(t) = e^{At}$; $\times (t) = e^{At} \times (0)$

34. Discretization of continuous equation

$X[(K + 1)T] = \phi(T) \times (KT) + \phi(T)\, M(KT)$

$t = KT$; K is integer ($K = 0, 1, 2, \ldots n$); T = Sampling period

$$\theta(T) = \int_0^T \phi(T-\tau)\,.\, Bd\tau;\ \phi(T) = e^{AT}$$

35. Conversion of pulse data to frequency response data:

$$G(w_j) = \frac{\Sigma y(t) \cos \omega t\,(dt) - j\Sigma y(t) \sin wt\, dt}{\Sigma x(t) \cos \omega t\, dt - j\Sigma x(t) \sin \omega t\, dt}$$

where $x(t)$ = is input, $y(t)$ is output, ω is frequency in radians/time and dt is time interval. $G(w_j)$ is a complex quantity.

$$AR = (\text{Real part}^2 + \text{imaginary part}^2)^{1/2};\ \text{phase angle} = \tan^{-1}\left(\frac{\text{Img. part}}{\text{Real part}}\right)$$

36. Optimal control: The design criteria used in optimal control are: (a) Time optimal (b) Based on performance index like economy, attaining target, output's variation etc. In mathematical terms:

(a) is: $P = \int_{t_0}^{t_f} dt = t_f - t_0$

(b) $P = \int_{t_0}^{t_f} [f(x, t), x(t), t_0]\ dt$ where t_0 is initial time and t_f is final time.

37. Controllability conditions: $[BAB\ A^2\ B\ \ldots\ A^{n-1}\ B]$ A, B are matrices. The above one should be non-singular (not equal to zero).

38. Observability condition: $\left[\, C' A' C' (A')^2\ C' \ldots (A')^{n-1} C' \,\right]$ should be non-singular. " ' " is transpose.

39. Sampled data control systems:

$X[(K + 1)T] = \phi(T)X(KT) + \theta(T)\, M(KT)$

$\theta(t) = C \times (KT) + DM(KT)$

where K is an integer, $\theta(t)$ is output vector, T = Sampling period; M = Control signal (scalar).

$$\phi(T) = L^{-1}\,(SI - A);\ \theta(T) = \int_0^T \phi(T-\tau)d\tau$$

40. Lyapunov stability method: The second metod of Lyapunov is applicable to non-linear, any order and time varying systems.

2nd method of Lyapunov: Rule 1: $dx/dt = f(x, t)$ where $f(0, t) = 0$ for all t. If there exists a scalar function $F(x, t)$ having continuous first partial derivatives and satisfying the condition given below:

(a) $F(x, t)$ is positive definite; (b) $F(x, t)$ is negative definite then the equilibrium state at the origin is uniformly asymptotically stable. Example: Consider:

$$\dot{x}_1 = x_2 - x_1(x_1^2 + x_2^2) \text{ and } \dot{x}_2 = x_1 - x_2(x_1 + x_2)$$

The origin is $x_1 = 0$; $x_2 = 0$; To check for stability by Rule 1:

defining $f(x) = x_1^2 + x_2^2$ which is positive definite, then the derivative of $F(x)$ with time is:

$$\dot{F}(x) = 2x_1\dot{x}_1 + 2x_2\dot{x}_2 = 2x_1[x_2 - x_1(x_1^2 + x_2^2)] + 2x_2[-x_1 - x_2(x_1^2 + x_2^2)] = -2(x_1^2 + x_2^2)$$

Which is negative definite. So $F(x)$ is a Lyapunov function and the system is stable.

Rule 2: Of a system is described by $\dot{x} = f(x, t)$ where $f(0, t) = 0$ for all $t \geq 0$ and there exists a scalar function $F(x, t)$ having continuous first partial derivatives and satisfying the conditions given below:

(a) $F(x, t)$ is a positive definite (b) $F(x, t)$ is negative definite

(c) $F[f(t, x_0, t_0)\ t]$ does not vanish;

Example: $\begin{bmatrix} \dot{x}_1 \\ \dot{x}_2 \end{bmatrix} = \begin{bmatrix} 0 & 1 \\ 1 & -1 \end{bmatrix}\begin{bmatrix} x_1 \\ x_2 \end{bmatrix}$; Assumning $F(x) = x'\,R\,x$ and R is from: $A'R + RA = -1$;

$$\begin{bmatrix} 0 & +1 \\ 1 & -1 \end{bmatrix}\begin{bmatrix} R_{11} & R_{12} \\ R_{21} & R_{22} \end{bmatrix} + \begin{bmatrix} R_{11} & R_{12} \\ R_{21} & R_{22} \end{bmatrix}\begin{bmatrix} 0 & 1 \\ 1 & -1 \end{bmatrix} = \begin{bmatrix} 1 & 0 \\ 0 & 1 \end{bmatrix}$$

By expanding the matrix, $2R_{12} = -1$; $R_{11} - R_{12} - R_{22} = 0$; $2R_{12} - 2R_{22} = -1$

Solving gives: $R_{11} = 3/2$; $R_{12} = 1/2$; $R_{21} = 1/2$; $R_{22} = 1$

$R = \begin{bmatrix} 3/2 & 1/2 \\ 1/2 & 1 \end{bmatrix}$; $|R| \geq 0$; So R is positive definite; Equilibrium state at the origin is:

$$F = x'Rx = (x_1\, x_2)\begin{bmatrix} 3/2 & 1/2 \\ 1/2 & 1 \end{bmatrix}\begin{bmatrix} x_1 \\ x_2 \end{bmatrix} = \frac{1}{2}\left[3x_1^2 + 2x_1x_2 + 2x_2^2\right]$$

$$F = \frac{1}{2}\left[6x_1\frac{dx_1}{dt} + 2x_1\frac{dx_2}{dt} + 2x_2\frac{dx_1}{dt} + 4x_2\frac{dx_2}{dt}\right]$$

Substituting for dx_1/dt and dx_2/dt gives:

$$F = x_1^2 + x_2^2$$

41. Stability analysis of linear systems: $(SI - A) = S^n + a_1 S^{n-1} + a_2S^{n-2} + \ldots + a_{n-1} S + a_n = 0$; For discreta data system $x(K + 1) = Gx(K)$. All eigenvalues of G are to be less than unity in their magnitude. The necessary and sufficient condition for stability is that all eigenvalues of matrix A should have negative real parts.

42. Describing function: $N = \frac{Y_1}{X} \underline{|Q_1}$; N is describing function;

X = amplitude of input sinusoid; Y = amplitude of fundamental component of harmonic output; ϕ = phase shift of fundamental harmonic. Component of output. This is for non-linear behaviour. $x(t)$ (input) = $X \sin \omega t$; output, $y(t) = A_0 + \Sigma(A_n \cos nwt + B_n \sin nwt)$
$= A_0 + \Sigma Y_n \sin (nwt + \phi_n)$ where

$$A_n = \frac{1}{\pi}\int_0^{2\pi} y(t) \cos nwt \, d(wt); \; B_n = \frac{1}{\pi}\int_0^{2\pi} y(t) \sin nwt \, d(wt)$$

$y_n = \sqrt{A_n^2 + B_n^2}$; $\phi_n = \tan^{-1}(A_n / B_n)$; of non-linearity is symmetric then $A_0 = 0$; $y(t) = A_1 \cos wt + B_1 \sin wt = Y_1 \sin (wt + \phi)$

$$N = \frac{Y_1}{X} \underline{|\phi_1} = \frac{\sqrt{A_1^2 + B_1^2}}{X} \tan^{-1}\left(\frac{A_1}{B_1}\right)$$

43. Phase plane method: The geometrical representation of the system behaviour in terms of trajectories is called the phase plane representation of the system. Time coordinate is replaced by some other property of the system.

44. Microprocessors: These are equivalent to a CPU of a computer where CPU does execution of program by a set of instructions. Uses are: 1. Electronic games; 2. Composition analysers; 3. Control of sequential operations in a batch process; 4. Control of multiple variables.

45. Various pressure measuring instruments:

(a) Liquid column manometers: Less than 50 psi pressure difference. Varieties are: U-tube; Enlarged leg type; Well type; Inclined tube type; Ring type.

(b) Bourdon, Bellows, diaphragm gauges: 30 to 2000 psig.

(c) Vacuum gauges: Ionisation gauge: 3 to 0.0004 microns (gas is ionised by electrons; giving +ve ions; the +ve ions current is related to pressure of gas); Mcleoid gauge: 8 to 0.006 microns (The pressure of gas is shown by the height of mercury in the gauge); Knudsen gauge, Phillips gauge: 15 to 0.04 microns (the heat conducted away from a heated filament depends on the pressure of gas); Pirani gauge: 2 to 4000 microns (based on heat removed by gas); Thermocouple gauge; 8 to 700 microns (same as Pirani gauge); Alphatron: 3 to 40,000 microns; Bellows gauge 4000 to 10^6 microns (contraction depends on gas vacuum).

Gauge pressure = Total pressure (absolute pressure) – atmospheris pressure.

Gauge pressure = – Vacuum pressure.

46. Various flow measuring instruments:

(a) Orifice meter: Pressure drop across the office plate is related to the flow rate.

(b) Venturi: Upto 5×10^6 gal/hour and higher. Pressure drop through the venturi is related to flow rate.

(c) Pitot tube: The pressure drop is related to the velocity of fluid. This gives point velocity.

(d) Rotameters: Called as vairable area meters. The area change to produce constant pressure difference is proportional to the flow rate of fluid.

(e) Rectangular notch, V–notch, Trapezoidal notch, Parshall flume; 1 to 10 million gals/day for liquids and higher. Flow rate is proportional to the level above weir opening (weir). For V-notch, flow rate is releated to the angle and level in the notch.

(f) Velocity meters, propeller meters, cup anemometers, deflecting vane meters, hot wire anemometer for liquids and gasses. Propeller meter works on the principle that the speed of the propeller (impeller) is related to flow rate of fluid. Hot air anemometer works on the principle that the resistance of an electrically heated wire (at constant current supply) changes with flow rate of fluid.

(g) Tilting tap meter, piston meter, Rotary Vane meter, sealed drum meter, lobed impeller meter, bellows meter, tilting tap water works on the principle of centre of gravity.

(h) Conveyor weighing for dry solid materials.

47. List of liquid level measuring instruments:

(a) Float and tape method: 4 inches to 60 feet level; The float position depends on the level of liquid in tank.

(b) Float and shaft type: 1000 psi and less; the float motion is taken through a stuffing base by the shaft. The rotation of shaft makes a pointer to move on a scale, or the shaft rotation is converted to pneumatic pressure signal.

(c) Bellows: Less than 20 inches. The expansion and contraction of Bellows is used through a pointer on a scale for pressure measurement.

(d) Bubbler type: 5 inches to 400 inches. Air is bubbled into the vessel through a pipe at a constant pressure. The returned pressure from the vessel is proportional to the level of liquid.

(e) Diaphragm box: 20 inches to 250 feet. The pressure in the diaphragm box (air) is proportional to the level of liquid.

(f) Air trap: 20 inches to 250 feet. Similar to diaphragm box without diaphragm. The pressure of trapped air is proportional to the level of liquid.

(g) Differential manometers: 1 inch to 100 feet. Level is proportional to level difference in manometer.

(h) Displacement float: 14 inches to 180 inches. The float is fixed by a spring at the bottom of a tank. Based on the level of liquid the float moves up. The float position is sent through a shaft on to a scale.

(i) Pneumatic balance displacement float gauge: 2 inches to 60 feet. The force caused by float is balanced by a pneumatic servo-system.

(j) Level by weighing.

(k) Level by: Resistance method; Capactiance method; Sonic method; γ-ray method, etc.

48. Viscosity measurement: Each fluid ha some resistance to flow and is termed as viscosity. Various viscometers are: Red wood viscometer, Saybolt viscometer, U-tube viscometer, Ford-Cup viscometer, Marsh funnel viscometer, Engler viscometer.

Ostwald method: $\eta/\eta_1 = Pt/(P_1\ t_1)$; η = viscosity of unknown liquid; t = time required for liquid to fall through capillary; ρ = density of liquid; η_1, ρ_1, t_1 (correspond to known liquid say water).

Two float rotameter: One float measures flow rate and the other float varies with viscosity.

Torque method: Viscosity is a characteristic of fluid friction. So viscosity is measured by shear force or torque.

49. Density/speciic gravity measurement:
   (a) Hydrometer: A hollow glass float. The height of immersion of glass float depends on the density of liquid.
   (b) Pressure method: This method depends on the pressure difference between the liquids of two tanks (one is reference liquid and the other is unknown liquid), observed in manometer.
   (c) Weight method: Based on the principle that buoyancy force depends on density of liquid.

50. Displacement measurement:
   (a) Variable differential transform method: The output voltage depends upon the position of core (displacement) in the transformer.
   (b) Pneumetic displacement gauge: Outlet pneumetic pressure depends on the thickness of paper or cloth on a roll.

51. Speed measuring instruments: (a) Tachometer; (b) Stroboscope used when end of the shaft is in accessible. Does not require mechanical contact with the shaft. If the rate of viewing is identical with the rate of rotation of a body, movement of the body appears to be arrested. This is the principle.

52. Symbols used in instrumentation diagrams:
○ – instrument locally mounted; ⊖ – instrument at contol centre; ⊚ – instruments giving two services; ⊗ – instrument transmitting; [symbol] – pneumatic control valve; [symbol] – Electrically operated valve; [symbol] – Safety valve; [symbol] – Self operated contoller; = process concencting line; orfice flow measurements —||— ; – – – – Fluid pressure line; -------- Electric line, # # # pneumatic line; * * * capillary line.

53. Letters used in instrumentation diagrams:
D – density; *F* – Flow; *H* – Hand actuated; *L* – level; *M* – Moisture; *P* – Pressure; *T* – Temperature; *A* – alarm; *C* – Control; *E* – element; *P* – Primary; *G* – glass; *I* – Indicating; *R* – Recorder; *S* – Safety; *w* – well.

54. Recording charts: *a* – strip chart recorders (long periods of operation is possible); *b* – circular chart recorders – one day record for one chart.

55. Transmission: *a* – pneumatic transmission – for small distances, low lag time; *b* – electric transmission – for long distances, low lag time.

56. Internal modal control: The logic idea of IMC is to use a model of the process open loop, $G_M(s)$ transfer function in such a way that the selection of the specified closed loop response yields a physically realizable feedback controller.

57. Lumped parameter systems: Dynamic systems described by sets of ordinary differential equations are called as lumped parameter systems. These systems are assumed to be perfectly mixed and time only the independent variable involved. Steady state equations describing such systems are algebraic. Phase lag has a limit.

58. Distributed parameter systems: Systems described bypartial differential equations are termed as distributed parameter system. Here the process variables vary in one or more space directions within the system boundaries.
At least two independent variables are required for these systems to describe their dynamic behaviour. They are time and distance. Phase lag has no limit.

59. *Z*-transform: These are used for sampled data systems like Laplace transforms for continuous data systems. The *Z* transform is the Laplace transform of impulse modulated function.

$$f(z) = z\,f(t) \sum_{n}^{\infty} f(nT) z^{-n}$$

60. Adaptive control: In this the controller parameters like settings are adjusted automatically to account for variations in the process that it controls. Adaptive control is for some objective function like integral of the error square, gain or phase margin.

61. Inferential control: This is used when the controlled output can be measured and the disturbances that create control problems can not be measured. The conventional control actions, feedback and feed forward require measured controller output. Temperature is the common secondary measurement for un-measured composition.

62. Modified Z – transform: Used for: (a) For finding inter sample information; (b) For pulse transfer function of a system having transport lag (e^{-LS}) for which L is not equal to an integral number of sampling periods ($L \neq nT$).

QUESTIONS

SECTION—A

1. Laplace transform of unit step change:

(a) $1/S$ (b) 1
(c) $1/S^2$ (d) None

2. Laplace transform of a Ramp input of slope K:

(a) $1/S^2$ (b) K
(c) K/S^2 (d) None

3. Laplace transform of sinusoidal input of magnitude A:

(a) $AW/(S^2 + W^2)$ (b) $A/(S^2 + W^2)$
(c) $1/A$ (d) $W/(S^2 + W^2)$

4. Laplace transform of impulse input of magnitude A:

(a) 1 (b) A
(c) $1/A$ (d) None

5. Laplace transform of df/dt:

(a) $Sf(S)$ (b) $f(S)$
(c) $Sf(S) - f(0)$ (d) None

6. Laplace transform of d^2f/dt^2:

(a) $S^2f(S)$ (b) $Sf(S)$
(c) $S^2f(S) - Sf(O)$ (d) $S^2f(S) - Sf'(O) - f(O)$

7. Laplace transform of $1/(S + a)$:

(a) e^{-at} (b) e^{at}
(c) t (d) None

8. General transfer function for first order system/process:

(a) $1/TS$ (b) $1/(TS + 1)$
(c) $R/(TS + 1)$ (d) $1/(T^2S^2 + 1)$

9. General transfer function for second order system:

(a) $1/(TS + 1)$ (b) $R/(T^2S^2 + 2\xi TS + 1)$
(c) $1/S^2$

10. Laplace transform of time lag of L time units:

(a) e^{Lt} (b) e^{L}

(c) e^{-LT} (d) e^{-LS}

11. Controlled variable in a water heater:

(a) Flow rate of water (b) Temperature of outlet water

(c) Heat input

12. Manipulated variable in water heater:

(a) Flow rate of water (b) Temperature of outlet water

(c) Heat input

13. Load variable in water heater:

(a) Flow rate of water (b) Temperature of outlet water

(c) Heat input (d) Inlet temperature of water

14. Controlled variable in exothermic CSTR:

(a) Temperature of feed (b) Temperature in reactor

(c) Cooling fluid rate

15. Manipulated variable in an exothermic chemical reactor:

(a) Temperature of feed (b) Cooling fluid flow rate

(c) Temperature in reactor

16. Process degrees of freedom means:

(a) Number of variables (b) Maximum number of controllers

(c) Controllers to be used

17. In a single tank, the term related to voltage:

(a) Level (b) Flow rate

(c) Volume of liquid in tank

18. In thermal system, the quantity similar to electric current is:

(a) Temperature (b) Heat flow rate

(c) Heat content in system

19. Equation for P–D control in Laplacian form for manipulated variable:

(a) $K_C(1 + 1/T_dS)\ e(S)$ (b) $K_C(1 + T_dS)\ e(S)$

(c) $(K_C + T_d)\ e(S)$

20. Offset means:

(a) Steady state error (b) Error at any time

(c) Stabilizing time

21. Which has maximum stabilizing time in the below ones:

(a) *P* (b) *PI*

(c) *PID* (d) *PD*

22. Cascade control means:

(a) One feedback and one feed forward (b) Two feed forwad

(c) None (d) Two feedbacks or more

23. Addition of Integral control to *P* reduces:

(a) Offset (b) Stabilizing time
(c) None

24. Addition of derivative control to *P* reduces:

(a) Offset (b) Stabilizing time
(c) None

25. Bode stability method uses:

(a) Open loop transfer function (b) Closed loop transfer function
(c) Characteristic equation

26. Phase margin in degrees is:

(a) $180 + \phi$ (b) $180 - \phi$
(c) 135

27. CSTR can be considered as:

(a) Distributed parameter system (b) Lumped parameter system
(c) None

28. Distributed parameter approach gives:

(a) Partial differential equation (b) Ordinary differential equation
(c) None

29. Type of controller normally used for level control in industry:

(a) *PI* (b) *P–D*
(c) *PID*

30. An example of distributed parameter system is:

(a) CSTR (b) Tubular reactor
(c) On-off

31. For low loads the type of controller used:

(a) *P* (b) *PI*
(c) *PD* (d) On-off

32. For high loads the type of controller used is:

(a) On-off (b) Positioning controllers
(c) None

33. The type of controller used normally for flow rate change in industry:

(a) *PI* (b) *P–D*
(c) *P–I–D* (d) Mixer

34. The type of controller normally used for temperature changes in industry:

(a) *PI* (b) *PID*
(c) On-off (d) *P*

35. An example of non linear behaviour:

(a) *P* control (b) *P–I–D*
(c) *P–D* (d) On-off controller

36. Feed forward controller's transfer function is:
(a) $-G_L(S)/G_P(S)$ (b) $-G_P(S)/G_L(S)$
(c) $G_L(S)$ (d) None

37. State equation means:
(a) $x = Ax + Bm$ (b) $\dot{x} = Ax + Bm$
(c) $Y = Cx$

38. Bare mercury thermometer comes under:
(a) First order (b) Second order
(c) Zero order

39. Range of hydrogen gas in constant volume gas thermometer is:
(a) – 10 to 50°C (b) 0 to 100°C
(c) 100 to 200°C

40. Range of mercury in glass thermometer:
(a) – 20 to 100°C (b) – 100 to 300°C
(c) – 20 to 300°C

41. Range of vapour pressure thermometers:
(a) – 20 to 300° (b) – 100 to 300°C
(c) 100 to 300°C

42. Range of resistance thermometers:
(a) – 60 to 600°C (b) 400 to 800°C
(c) None

43. Range of radiation pyrometers:
(a) 800 to 2000°C (b) 400 to 800°C
(c) – 100 to 200°C

44. Range of optical pyrometers:
(a) 400 to 800°C (b) 800 to 2000°C
(c) – 100 to 300°C

45. Range of photoelectric pyrometer:
(a) 800 to 2500°C (b) 0 to 800°C
(c) 800 to 1600°C (d) None of these

46. Principle in Hg glass thermometer:
(a) Linear expansion (b) Volumetric expansion
(c) Increase of pressure with temperature

47. Principle in bimetallic thermometer:
(a) Volumetric expansion (b) Linear expansion of two different metals
(c) None

48. Principle in vapour pressure thermometers:
(a) Volumetric expansion (b) Linear expansion
(c) Increase of vapour pressure with temperature

49. Output (emf generated) in thermocouples:
(a) Volts (b) Millivolts
(c) Microvolts

50. Thermocouples use:
(a) Dissimilar metal bars (b) Similar metal wires
(c) Disimilar metal wires

51. For 400°C, the emf of iron constant thermocouple:
(a) 11 MV (b) 1 volt
(c) 22 MV (d) 33 MV

52. For 800°C, the emf of chromel-alumel thermocouple is:
(a) 31 MV (b) 45 MV
(c) 0.5 volt (d) 23 MV

53. For 600°C, the emf of $Pt - Pt$ + Rhodium termocouple is:
(a) 4 MV (b) 9 MV
(c) 5.5 MV (d) 0.2 volt

54. In thermistor resistance:
(a) Increases with temperature increase (b) Decreases with temperature increase
(c) Remains constant

55. Mass spectrometer is used for composition analysis of:
(a) Solids (b) Isotopes

56. Polorograph is used for analysis of:
(a) Solids (b) Liquids
(c) Gases

57. Combustion gas from boiler is analysed continuously by:
(a) Haldane apparatus (b) Thermal conductivity cell
(c) Mass spectrometer

58. Amount of water in air is determined by:
(a) Sling psychrometer (b) Orsat
(c) Mass spectrometer

59. Polarimeter is used for analysis of:
(a) Any liquid (b) Any solid
(c) Sugar solution

60. Good quality distilled water has specific conductance in micromhos/cm as:
(a) 300 (b) 100
(c) 10 (d) 1

61. Good quality raw water has specific conductance of, micromho/cm:
(a) 1 (b) 10
(c) 100 (d) 300

62. pH is:

(a) $\log_{10}(1/H^*)$ (b) $\log_{10}(1/H^-)$

(c) $\ln(1/H^*)$

63. Bourdon gauges are used for measurement of pressures:

(a) above 30 psia (b) below 30 psia

64. Mcleoid gauge is usd to measure:

(a) Pressure (b) Vacuum

(c) Flow rate

65. Ionisation gauge measures pressure:

(a) Above 3 microns (b) Below 3 microns

66. Alphatron measures pressure:

(a) Above 3 microns (b) Below 3 microns

67. Flow rate of sludge:

(a) Orifice (b) Venturi

(c) Open weir or Kennison nozzle

68. Composition of natural gas:

(a) Thermal conductivity cell (b) Chromotograph

(c) Orsat

69. Flow rate of a river:

(a) Vane meter (b) Pitot tube

(c) Weir

70. ...... line represents for:

(a) Electric line (b) Pipeline

(c) Pneumatic line

71. In Bode plot *AR* vs. *w* is plotted on:

(a) Log-log plot (b) Ordinary

(c) Log-log

72. In Bode plot ϕ vs *w* is plotted on:

(a) Semilog (b) Ordinary

(c) Log-log

73. Underdamped system means damping coefficient is:

(a) Greater than 1 (b) Less than 1

(c) Equal to 1

74. In fertilizer industry carbon dioxide composition is measured by:

(a) CO_2 analyser (b) Continuous orsat

(c) None

75. Critically damped means, damping coefficient is:

(a) Greater than 1 (b) Less than 1

(c) Equal to 1

76. Describing function uses:
(a) Taylor's series (b) Fourier series
(c) None

77. Phase plane method was introduced by:
(a) Bode (b) Ziegler-Nichols
(c) Poincare

78. Open loop gain of $(21 + S)/(S^2 + 4S + 5)$:
(a) 21 (b) 21/5
(c) 1/5 (d) None

79. How many steady state values a non-linear system will have:
(a) One (b) Two
(c) More than one (d) Zero

80. Line represents for; $X - X - X - X$:
(a) Electric line (b) Pneumatic line
(c) Pipeline

81. Flow rate of a small channel is measured by:
(a) Vane meter (b) Pitot tube
(c) Weir

82. Shell temperature measurement of heat exchanger:
(a) Resistance thermometer (b) Thermocouple
(c) Radiation pyrometer

83. Feed forward controller is used to account for:
(a) Load changes (b) Set point changes
(c) None

84. Signal flow graph method is developed by:
(a) Bode (b) Mason
(c) Nyquist

85. U-tube manometer comes under:
(a) Second order (b) First order
(c) Third order

86. Automatic gate valve is used for fluids containing:
(a) No solid matter (b) Solid matter
(c) Chemicals

87. In chemical plant the following method is used for transmission:
(a) Electric (b) Hydraulic
(c) Pneumatic (d) Mechanical

88. Gain margin at 180 degrees phase lag is:
(a) A.R. (b) 1/A.R.
(c) A.R. – A.R. at 180° lag

89. Mikhalov stability method uses the following:
(a) Closed loop (b) Characteristic equation
(c) Open loop

90. Nyquist method uses the following:
(a) Open loop (b) Closed loop
(c) Characteristic equation

91. Amplitude ratio of time lag:
(a) 1 (b) 0
(c) *W*

92. Working phase lag in Bode method is in degrees:
(a) – 90 (b) – 135
(c) – 125 (d) None

93. Ruth stability method uses:
(a) Open loop transfer function (b) Closed loop transfer function
(c) Characteristic equation

94. Phase lag of first order system:
(a) $\tan^{-1} (WT)$ (b) $-\tan^{-1} (WT)$
(c) WT (d) $-WT$

95. Amplitude ratio of first order system:
(a) $R/\sqrt{1+W^2T^2}$ (b) $1/TW$
(c) R (d) None

96. Servomechanism means:
(a) Feedback (b) Feed forward
(c) None

97. Solenoid valve works like:
(a) *P* control (b) *PD* control
(c) On-off

98. Which has maximum off-set in *P*, *PID*, *PD* and *PI* actions?
(a) *P* (b) *PID*
(c) *PD* (d) *PI*

99. Time constant of a mixer:
(a) V/F (b) F/V
(c) Time (d) None

100. Time constant in first order system:
(a) $R*C$ (b) $1/(R*C)$
(c) None

101. Transfer function of outlet flow rate to inlet flow rate in a single tank system:
(a) $R/TS + 1$ (b) R/TS
(c) $1/TS + 1$

102. Transfer function of level to inlet flow rate in a single tank system:

(a) $R/(TS + 1)$ (b) $1/(TS + 1)$

(c) $1/TS$

103. Laplace transform of $e^{-at} \cos kt$:

(a) $S + a/(S + a)^2 + k^2$ (b) $S - a/(S + a)^2 + k^2$

(c) $S + a/(S - a)^2 + k^2$ (d) 300

104. Output equation for unit impulse response of 1st order:

(a) $\frac{e^{-t/T}}{T}$ (b) $e^{-t/T}$ (c) $\frac{1}{T^2} e^{-t/T}$

105. Capacitance means:

(a) Driving force/flow (b) Storage/driving force

(c) None

106. Resistance means:

(a) Driving force/flow (b) Storage/driving force

(c) None

107. Overshoot comes in 2nd order system having damping coefficient:

(a) Greater than one (b) Equal to one

(c) Less than one

108. Period of oscillation in second order system comes having damping coefficient:

(a) Greater than one (b) Less than one

(c) Equal to one

109. In proportional controller $P(S)/e(S)$ is equal to:

(a) K_C (b) $1/T_i S$

(c) $K_C(1 + T_i S)$

110. Reset rate is another term used for:

(a) Derivative time (b) Integral time

(c) Dead time

111. For positive feedback closed loop transfer function of y/x is given as :

(a) $\pi_f/1 - \pi_t$ (b) $\pi_f/1 + \pi_t$

(c) π_f/π_t

112. For negative feedback, the closed loop transfer function y/x is given as:

(a) $\pi_f/1 - \pi_t$ (b) $\pi_f/1 + \pi_t$

(c) π_f/π_t

113. Servo problem means:

(a) Load changing (b) Set point changing

(c) Both load and set point changing

114. Regulator problem means:

(a) Load changing (b) Set point changing

(c) Both load and set point changing

115. Characteristic equation is the denominator of:
(a) Closed loop transfer function (b) Open loop transfer function
(c) None

116. Corner frequency is at W equal to:
(a) T (b) $1/T$
(c) 1

117. Decibles means:
(a) $20 \log_{10}$(A.R.) (b) 20 ln (A.R.)
(c) $20/\log_{10}$ (A.R.)

118. Phase angle of transportation lag is:
(a) $-WI$ radians (b) $\tan^{-1}(WT)$ degrees
(c) $-\tan^{-1}(WT)$ degrees

119. At cross over frequency phase angle in degrees is:
(a) – 135 (b) – 90
(c) – 180

120. An example of non-linear chemical system:
(a) Mixer. (b) Non-isothermal CSTR
(c) Isothermal CSTR

121. Offset means:
(a) Dynamic error (b) Steady state error
(c) None

122. Flapper-nozzle is used in:
(a) Electronic controller (b) Pneumatic controller
(c) Hydraulic controller

123. In describing function technique the following series appear:
(a) Fourier series (b) Taylor series
(c) Laurent series

124. Pyrometric cones are used in the following industry:
(a) Ceramic (b) Cement
(c) Coal gasification

125. Cost of instrumentation in modern chemical plant:
(a) 5 to 10% (b) 10 to 20%
(c) 20 to 30% (d) 30 to 40%

126. For measuring t temperature of a furnace which is the most suitable instrument:
(a) Thermocouple (b) Optical pyrometer
(c) Bimetallic thermometer (d) Resistance thermometer

127. Which of the following flow-measuring instruments is an area meter?

(a) Venturi meter (b) Pitot tube

(c) Rotameter (d) Hot wire anemometer

128. Pick out the symbol for locally mounted instrument in instrumentation diagram:

(a) Θ (b) 0

(c) ϕ (d) 8

129. Dilatometer is used to measure:

(a) Stress (b) Strain

(c) Deflection

(d) Contraction or expansion due to changes in temperature

130. Which of the following instruments is suitable for measuring the temperature of a red hot moving object (Example: steel ingot):

(a) Radiation pyrometer (b) Thermocouple

(c) Thermistor (d) None

SECTION—B

1. Output equation for a first order system with sinusoidal input of magnitude, *A* is......

2. Taylor's expansion for one variable to linearise non-linear system is......

3. Characteristic equation is......

4. Stable system means......

5. In Bode stability method, the condition for stability is......95 G.

6. A.R. overall is......

7. ϕ overall is......

8. Pneumatic controller works by the principles of......

9. Modern control theory means using......

10. Primary element of carbon dioxide anayser is......

11. Purpose of value positioner is......

12. Resistance thermometer works on the principle of......

13. Thermistors are made of......

14. Hair hygrometer is used for......

15. Refractometer works on the principle of......

16. Polarimeter works on the principle of......

17. Principle of pH meter is......

18. Bellows is made of......

19. The sinusoidal response is popular because......

20. New type of automatic controllers are......

21. ......A.R. is correct......

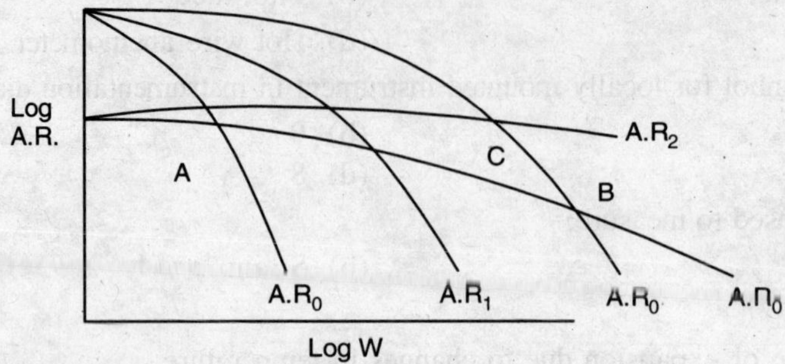

22. The figure is......input

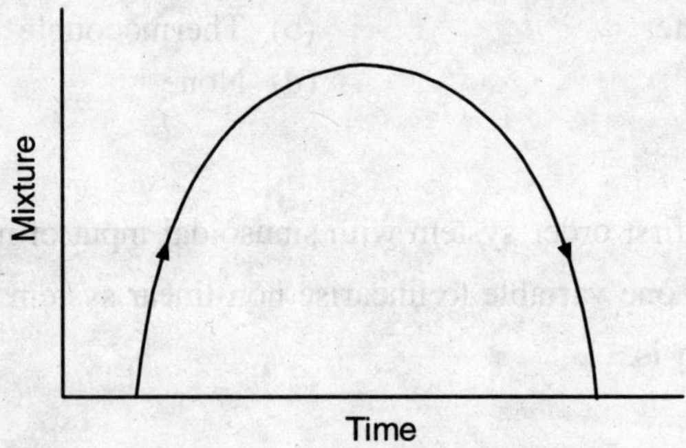

23. Characteristic equation is......

24. Equation of *PID* control is......

25. This

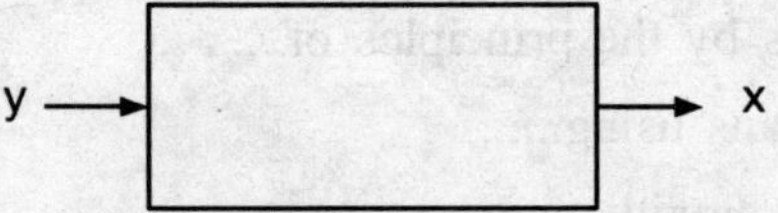

26. Output equation for first order system with unit step input is......

27. The $C(S)/R(S)$ for the below figure is......

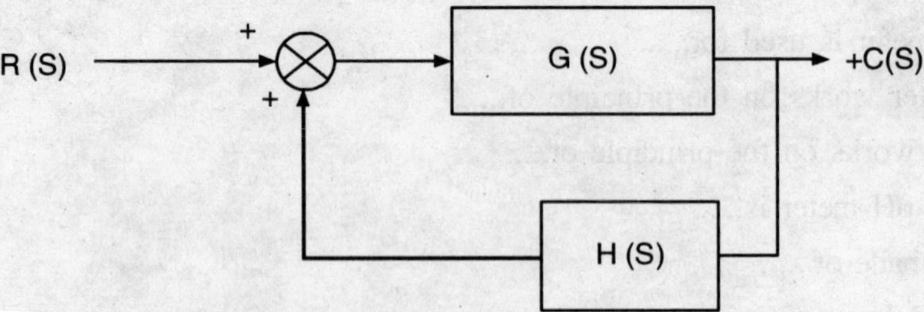

28. The above system is stable/unstable (below).

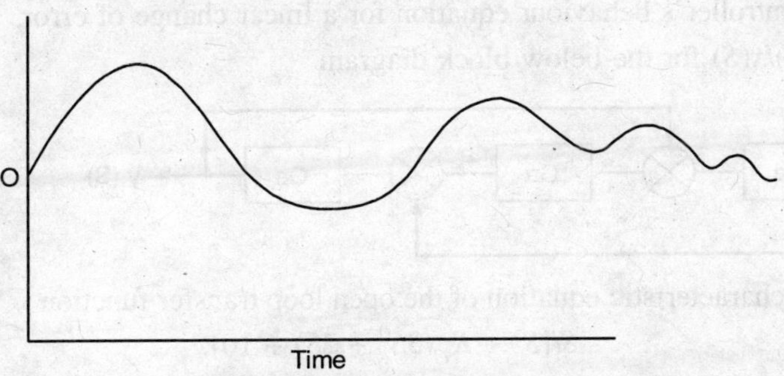

29. symbol stands for......

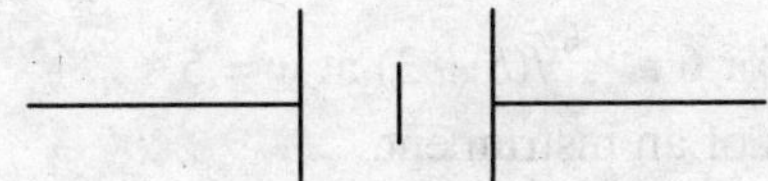

30.

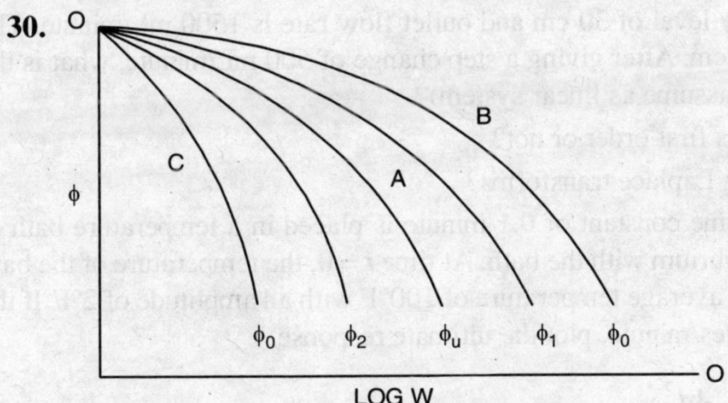

.... ϕ_0 is correct.

31. For stability while applying Ruth-method all the coefficients......

32. Phase lag of a second order system is......

33. Temperature instruments are standardized mostly with......and......of pure substances.

34. The 3 types of mechanical pressure sensing elements are: (1)...... (2)...... (3)......

35. Instrument used for moisture content of solid is......

36. ......valve is used to supply oil to burner.

37. ......valve is used for slurries.

38. pH meter has......cell and......cell.

39. Carbon dioxide analyser works on......conductivity principle.

40. Stroboscope is used for measurement of......

SECTION—C

1. Obtain *PID* controller's behaviour equation for a linear change of error.
2. Determine $y(S)/x(S)$ for the below block diagram.

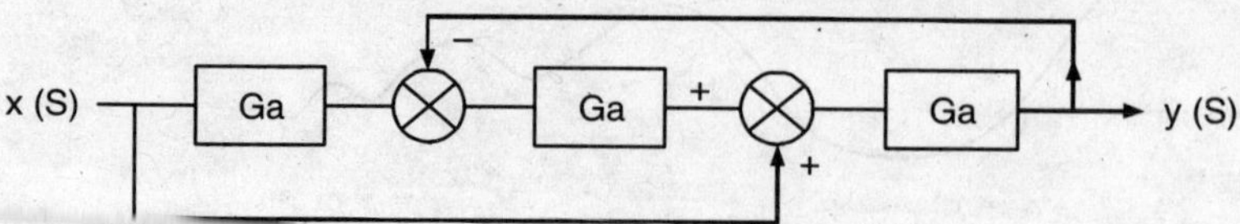

3. Determine the characteristic equation of the open loop transfer function

$$S/[S^3 + K_C(3S^2 + 2S) + 10].$$

4. Determine the cross over frequency for $\dfrac{K_C e^{-S}}{(2S+1)}$.
5. Determine the A.R. overall for $6\,e^{-0.5s}/(S + 3)$ at $w = 5$.
6. List the static characteristics of an instrument.
7. List the parts of an instrument.
8. List the dynamic characteristics of an instrument.
9. A single tank was initially at steady level of 30 cm and outlet flow rate is 1500 mL/minute. The tank cross-sectional area is 300 sq. cm. After giving a step change of 500 mL/minute, what is the calculated level after five minutes (assume as linear system)?
10. How to test whether an instrument is first order or not?
11. Solve $dx/dt + 3x = 0$; $x(0) = 2$ using Laplace transforms?
12. A mercury thermometer having a time constant of 0.1 minute is placed in a temperature bath at 100°F and allowed to come to equilibrium with the bath. At time $t = 0$, the temperature of the bath begins to vary sinusoidally about its average temperature of 100°F with an amplitude of 2°F. If the frequency of oscillation is $10/\pi$ cycles/minute, plot the ultimate response.
13. Find the laplace transform of $\dfrac{d^3y}{dt^3} + \dfrac{4d^2y}{dt^2} + 5dy/dt + 2y = 2,$

$$y(0) = dy(0)/dt = d^2y(0)/dt^2 = 0$$

14. A proportional controller is used to control temperature within the range of 60 to 100°F. The controller is adjusted so that the outlet pressure goes from 3 psi to 15 psi as the measured temperature goes from 71 to 75°F with the set point held constant. Find the gain and proportional band.
15. What is meant by stable and unstable limit cycles?
16. Obtain the response equation for a first order system for unit step input.
17. Obtain amplitude ratio and phase angle for $(S + 1)$.
18. What are the seven substances selected for international temperature scale?
19. What is the principle of thermocouple?
20. What is meant by instrumentation diagram?
21. For a single tank derive the transfer function of level to inlet flow rate. Assume linear behaviour.

22. For isothermal CSTR with first order chemical reaction derive the transfer function of outlet concentration to inlet concentration.

23. For a controlled system having open loop transfer function $2K/(5S + 6)$ determine whether the system is stable or not using Mikhalov method.

24. List the parts of an instrument.

25. What is the principle of bimetallic thermometer?

26. Write the state equation for

$\dot{x}_1 = -5x_1 - x_2 + 2M$ where M is input

$\dot{x}_2 = 3x_1 - x_2 + 5M$

$0^{(\text{output})} = x_1 + 2x_2 + 5M.$

27. What is the principle of flapper nozzle?

28. List two example of first order and second order instruments.

29. What is the principle of working of refractometer?

30. What is the principle of working of conductivity meter?

SECTION—D

1. Bode diagrams are generated from output response of the system subjected to which of the following input?

(a) Impulse (b) Step
(c) Ramp (d) Sinusoidal

2. In root locus method, a pole of transfer function, $G(S)$ is the value of S for which $G(S)$ approaches:

(a) – 1 (b) 0
(c) 1 (d) ∞

3. The reponse of two tanks of same size and resistance in series is:

(a) Under damped (b) Critically damped
(c) Overdamped (d) None of the above

4. Identify an unbounded input from four inputs whose transfer functions are given below:

(a) 1 (b) $1/S$
(c) $1/S^2$ (d) $1/(S^2 + 1)$

5. When a thermometer at 30°C is placed in a water bath at 90°C, the initial rise in thermometer temperature is found to be 2°C/sec. What is the time constant of the thermometer assuming it is a first order device with unity steady gain? What will thermometer read after one minute.

6. A first order system with transfer function $G(S) = Y(S)/X(S) = 1/(S + 1)$ is subjected to input $X(t) = t$. Derive the expression for change in output $Y(t)$ as a function of time. What is the maximum and minimum difference between input and output? At what time does these differences occur?

7. Amplitude ratio corresponding to cross-over frequency of 10 radians/sec for a process is found to be 0.10 :

(a) What is ultimate gain?

(b) What is the period of substained oscillations corresponding to ultimate gain?

(c) If a proportional controller is to be used, what is the value of proportional gain, you would recommend?

8. (a) The transfer function of a process, measuring device, controller and control valve respectively is given by:

$$G_P = \frac{K_P}{(S+2)(S+4)};\ G_m = 1,\ G_C = K_C$$

$$G_V = 1$$

The root locus diagram of the system is given in below figure:

(a) Identify location of points A and B

(b) At $K_C = 1$, the roots of the system are located at point C. Identify the location of point C.

(c) Determine the value of K_P.

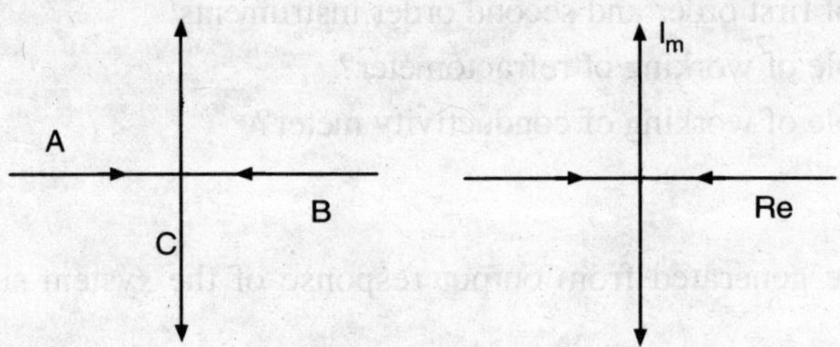

(b) The Laplace transform of a stair case function $S(b, t)$ formed by successive addition of unit step functions at 0, b, $2b$, $3b$ etc.

(a) $1/S$ (b) $1/bS$

(c) $S(1 - e^{-bS})$ (d) $\dfrac{1}{S(1-e^{-bS})}$

9. The transfer function of a process, a measuring element and a control value is given respectively as

$$G_P = \frac{2}{2S+1};\ G_m = \frac{1}{(5S+1)};\ G_V = \frac{1.5}{(3S+1)}$$

A proportional controller with a gain of $K_C = 1$ is used.

(a) Write the closed loop transfer function relating the output (Y) to the set point (Yr).

(b) What is steady state error in the output for a unit step change in the set point?

10. Using Ruth method, find K_C values to be used for the system to be stable having open loop transfer function.

$$G(S) = 5/[S^4 + 11S^3 + K_C(7S^2 + 5S) + 5]$$

11. Derive the transfer function for a single tank with non-linear behaviour (outlet level/inlet flow rate).

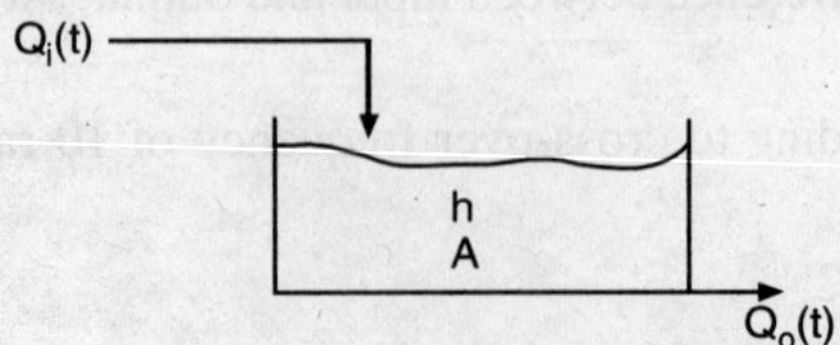

12. A thermo-couple junction may be approximated as a sphere of diameter 2 mm with thermal conductivity 30 W/(m.°C), density 8600 kg/m^3 and specific heat 0.4 kJ/(kg.°C). The heat transfer coefficient between gas stream and junction is 280 W/(m^2.°C). How long will it take for the thermocouple to record 98% of the applied temperature difference?

13. The transfer function of a first order process with time delay is:

(a) $\dfrac{e^{T_d S}}{TS+1}$ (b) $\dfrac{e^{-T_d S}}{TS+1}$

(c) $\dfrac{1}{(TS+1)(T_d S+1)}$ (d) $\dfrac{T_d S}{TS+1}$

14. The open loop transfer function of a process is $\dfrac{K(S+1)(S+4)}{(S+2)(S+3)}$. In root locus diagram, the poles will be at:

(a) – 1, – 4 (b) 1, 4

(c) – 2, – 3 (d) 2, 3

15. A proportional controller with a gain of K_C is used to control a first order process. The offset will increase if:

(a) K_C is reduced (b) K_C is increased

(c) Integral control action is introduced (d) Derivative control action is introduced

16. The transfer function of a process is $\dfrac{1}{16S^2+8S+4}$. If a step change is introduced into the system then the response will be:

(a) Underdamped (b) Critically damped

(c) Over damped (d) None of the above

17. A control system shown below:

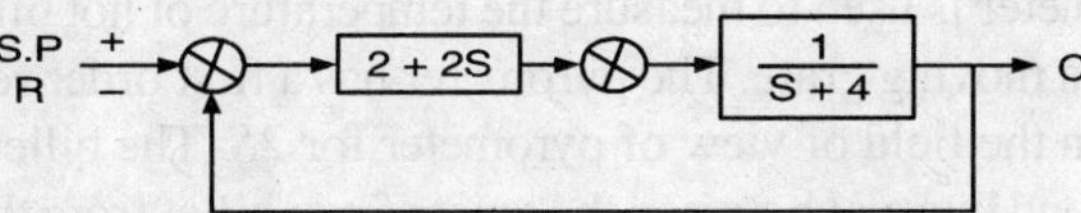

(a) Determine the variation of C with time for a unit step change in set point.

(b) What is offset?

18. The open loop transfer function of a process is $\dfrac{1}{4(3S+1)^4}$ where the time constant is in minutes. Determine (a) cross over frequency (b) ultimate gain.

19. A control system shown below:

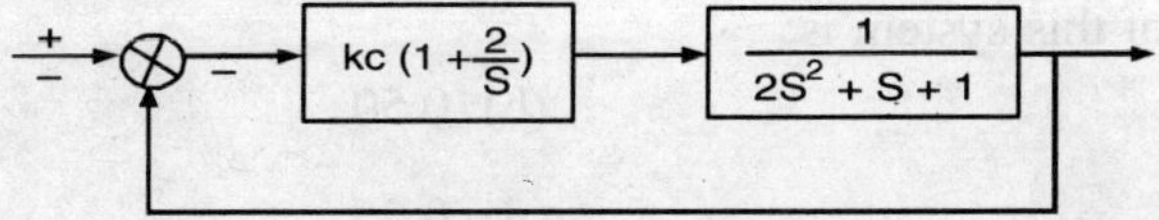

Using the Ruth test determine the value of K at which, system just becomes unstable.

20. The Root locus plot of the roots of the characteristic equation of a closed loop system having the open loop transfer function, $\frac{K(S+1)}{S(2S+1)(3S+1)}$ will have a definite number of loci for variation of K from 0 to ∞. The number of loci is:

(a) 1 (b) 3

(c) 2 (d) 2

21. A first order system with a time constant of 1 minute is subjected to frequency response analysis. At an input frequency of 1 Rad/min. The phase shift is:

(a) 45° (b) – 90°

(c) – 180° (d) – 45°

22. A proportional plus integral controller with proportional sensitivity K_C = 1 mA/mA and integral time T_i = 1 minute is initially at steady state with error signal e = 0 and controller output 1 at 10 mA. AT time t = 0 a step change of 1 mA is given in the error signal. The controller output will show one of the following responses. Identify the correct one.

(a) I jumps instantaneously to 11 mA and remains at that value thereafter.

(b) I increases at a linear rate of 1 mA/minute.

(c) I jumps to 10 mA and then falls exponentially at a rate $e^{-t/T}$.

(d) I jumps instantaneously to 11 mA and then increases linearly at the rate of 1 mA/min.

23. In a closed loop system the process to be controlled is an integrating process with transfer function $\frac{1}{2S}$. The controller proposed to be used is an integral controller with transfer function $\frac{1}{T_1 S}$. When a step change in set point is applied to such a closed system, the controlled variable will exhibit.

(a) Overdamped response (b) Underdamped response

(c) Undamped response (d) Unstable response

24. A photoelectric pyrometer is used to measure the temperature of hot billets having a temperature of 900°C and placed on a moving grate. The pyrometer has a first order response with a time constant of 1S. Each billet is in the field of view of pyrometer for 2S. The billets are so placed on the grate that there is a one second interval between the exist of one billet from the field of view of pyrometer and entry for the next billet in line. The ambient temperature is 50°C. Assume that the system is in operation for a few hours. Write equations for the rising and falling portions of the pyrometer response.

25. The closed loop poles of a stable second order system could be:

(a) Both real and positive (b) Complex conjugate with +ve real parts

(c) Both real and negative (d) One real positive and one real –ve.

26. A first order system with unity gain and time constant T is subjected to a sinusoidal input of frequency, $w = 1/T$. The A.R. of this system is:

(a) 1 (b) 0.50

(c) $1/\sqrt{2}$ (d) 0.25

27. The frequency response of a first order system has a phase shift with lower and upper bounds given by:

(a) $(-\infty, \pi/2)$ (b) $(-\pi/2, \pi/2)$

(c) $(-\pi/2, 0)$ (d) $(0, \pi/2)$

28. A mercury thermometer can be used to measure the body temperature by placing it either in the mouth or in the arm pit of the patient. The true body temperature can be taken to be the temperature in the mouth which is usually higher than the temperature in the arm pit by 0.5 K. Assume that the true body temperature of a patient is 312 K, and thermometer is intially at 300 K. Also assume that the temperature behaves like a first order system with time constant of 40 seconds.

(a) obtain a relation for the thermometer reading $T(t)$ as a function of time in terms of initial temperature $T(0)$ and body temperature (T_B).

(b) How long the thermometer be placed in the patient's mouth in order to ensure that the error in measurement is not greater than 0.05.

(c) Since the body temperature in the arm pit is less, the measurement made here using thermometer is corrected by adding 0.50 K. How long should the thermometer be placed in the arm pit in order to ensure that the error in the corrected measurement is not greater than 0.05 of the true body temperature.

29. Consider the system of two tanks in series as shown below.

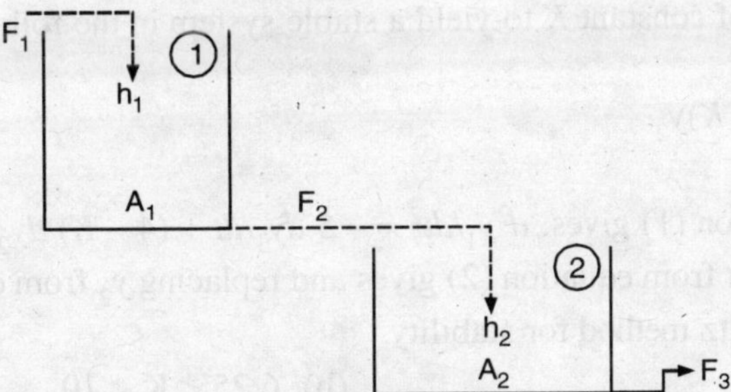

(a) Determine the transfer function of the process

(b) Compute the time constants of the process. Is the process open loop process overdamped, underdamped or critically damped?

(c) If proportional control is used with $K_C > 0$, determine K_C for which the closed loop response becomes critically damped.

30. Match the measured process variables with the list of measuring devices given below.

Measured process variables	Measuring devices
P. Temperature	1. Bourdon
Q. Pressure	2. Orifice plate
R. Flow	3. Infrared analyser
S. Liquid level	4. Displacer devices
T, Composition	5. Pyrometer

(a) $P-5, Q-1, R-2, S-4, T-3$ (b) $P-3, Q-1, R-4, S-2, T-5$

(c) $P-1, Q-3, R-4, S-2, T-5$ (d) $P-3, Q-1, R-2, S-4, T-5$

31. Suppose that given, time constant and dead time of a process with the following transfer function, $G_C(S) = 10 \exp(-0.15)/(0.5S + 1)$ are known with a possible error of ± 20% of their values. The largest permissible gain K_C of a porportional controller needs to be calculated by taking the values of process gain, time constant and dead time as:

(a) 8, 0.60, 0.12 (b) 12, 0.6, 0.12
(c) 8, 0.40, 0.12 (d) 12, 0.40, 0.08

32. Water is flowing through a series of four tanks and getting heated as shown in figure, it is desired to design a cascade control scheme for controlling the temperature of water leaving tank 4 as there is a disturbance in temperature of second stream entering the tank 2. Select the best place to take the secondary measurement for the secondary loop.

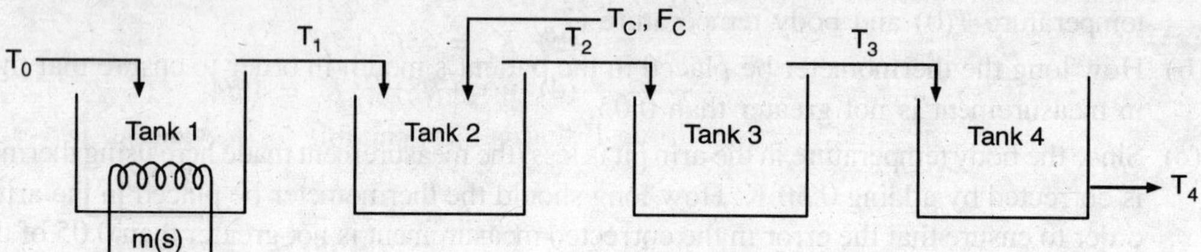

(a) Tank 1 (b) Tank 2
(c) Tank 3 (d) Tank 4

33. The range of vlaues of constant K to yield a stable system in the following set of time dependent differential equation:

$$dy_1/dt = -5\,y_1 + (4 - K)y_2 \qquad \text{...(1)}$$

$$dy_2/dt = y_1 - y_2 \qquad \text{...(2)}$$

Differentiating equation (1) gives: $d^2y_1/dt^2 = -5\,dy_1/dt + (4 - K)\,dy_2/dt$

Substituting for dy_2/dt from equation (2) gives and replacing y_2 from equation (1) results:

Applying Ruth-Hurwitz method for stability.

(a) $0 < K < 7$ (b) $6.25 < K < 10$
(c) $-6 < K < 6.25$ (d) $0 \leq K \leq 7$

34. Find the Laplace transform of the following input function shown in the figure.

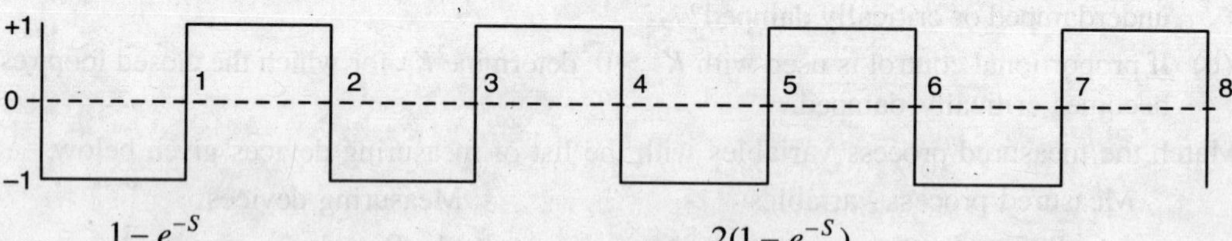

(a) $\dfrac{1 - e^{-S}}{S(1 - e^{-2S})}$ (b) $\dfrac{2(1 - e^{-S})}{S}$

(c) $\dfrac{1 - e^{-S}}{S(1 + e^{-S})}$ (d) $\dfrac{(1 - e^{-S})\,e^{-S}}{S(1 - e^{-2S})}$

35. Water is entering a storage tank at a temperature T_0 and flow rate and leaving at a flow rate Q and temperature T. There are neglible heat losses in the tank. The area of cross section of the tank is A_C. The model that describes the dynamic variation of water temperature in the tank with time is given as:

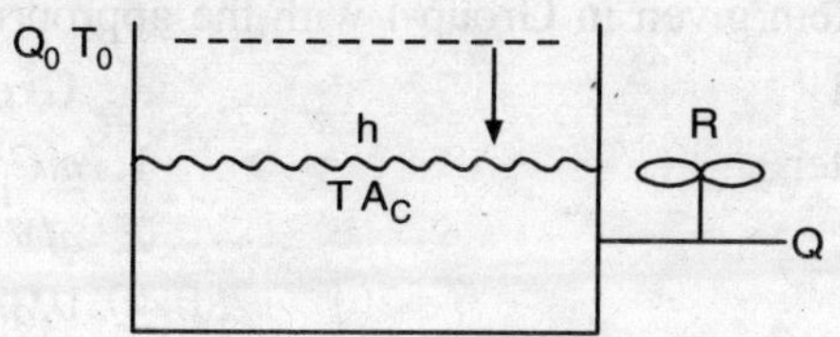

(a) $Q_0(T_0 - T) = A_C h\, dT/dt$ (b) $(Q_0 T_0 - QT) = A_C\, hdT/dt$
(c) $Q(T_0 - T) = A_C h\, dT/dt$ (d) $Q(T_0 - T) = A_C\, d(hT)/dt$

36. Find the ultimate gain and frequency for a proportional controller in the case of process having the following transfer function $G_P(S) = 1/[(4S + 1)(2S + 1)(S + 1)]$
(a) $w = 1/(14)^{1/2}$ (b) $K_C = 45/7(14)^{1/2}$
(c) $w = 1$; $K_C = 13$ (d) $w = (7/8)^{1/2}$; $K_C = 45/4$

37. Match the type of controller given in Group-II that is most suitable for each application given in Group-I.

Group-I	Group-II
P. Distillation column bottoms level to be controled with bottoms flow.	1. *P* control
Q. Distillation column pressure to be controlled by manipulating vapour flow from top plate	2. *PI* control
R. Flow control of liquied from pump by positioning the valve in the line	3. *PID* control
S. Control of temperature of a CSTR with coolant flow in the jacket.	
T. Composition	

(a) *P* – 1, *Q* – 1, *R* – 2, *S* – 3 (b) *P* – 2, *Q* – 2, *R* – 3, *S* – 3
(c) *P* – 2, *Q* – 2, *R* – 1, *S* – 1 (d) *P* – 2, *Q* – 3, *R* – 2, *S* – 3

38. In the case of feed forward control scheme, which of the following is not true?
(1) It is insesitive to modeling errors
(2) Can not cope with unmeasured disturbances
(3) It waits until the effect of the disturbance has been felt by the system before control action is taken
(4) Requires good knowledge of the process model
(5) Requires identification of all possible disturbances in their direct measurement.

39. Temperature control of an exothermic chemical reactor taking place in a CSTR is done with the help of cooling water flowing in jacket around the reactor. The types of valve and control action recommended are:
(a) Air to open valve with controller direct acting
(b) Air to close valve with the controller indirect acting
(c) Air to open valve with the controller indirect acting
(d) Air to close valve with the controller direct acting
(e) Air to open valve with the controller indirect acting
(f) Air to close valve with the controller direct acting.

40. Match the first order system given in Group-I with the appropriate time constant in Group-II.

Group-I	Group-II
P. thermometer	1. $mC_p(h A)$
Q. mixing	2. q/V
	3. V/q
	4. $h A/(mC_p)$

(a) $P - 4, Q - 2$ (b) $P - 4, Q - 3$

(c) $P - 1, Q - 2$ (d) $P - 1, Q - 3$

41. The experimental response of the controlled variable $y(t)$ for step change of magnitude P in the manipulated variable $x(t)$ is shown below. The appropriate transfer function of the process is:

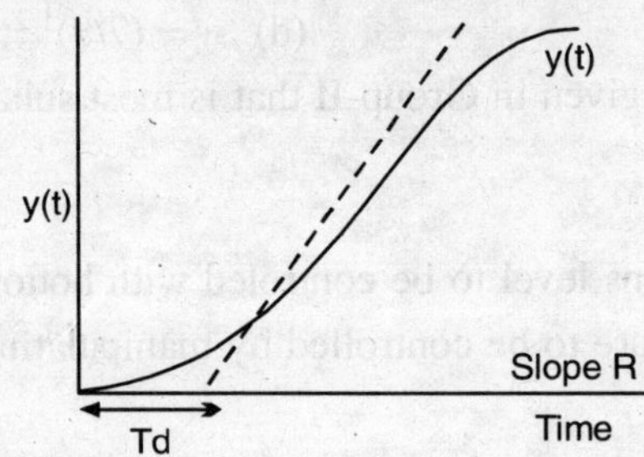

(a) $\dfrac{(Q/P)\exp(-Q/R)S}{T_d S+1}$ (b) $\dfrac{(Q/R)\exp(-T_d S)}{(Q/P)S+1}$

(c) $\dfrac{(Q/P)\exp(-T_d S)}{(Q/R)S+1}$ (d) $\dfrac{(Q/R)\exp(-P/Q)S}{(T_d S+1)}$

42. Consider the system with the open loop transfer function, $G(S) = 1/[(S + 1)(2S + 1)(5S + 1)]$. Match the range of w(frequency) in Group-I with the slope of asymptote of the log AR (amplitude ratio) versus log w plot in Group-II.

Group-I	Group-II
$P\ O < w < 0.2$	1. 5
$Q\ w > 1$	2. 3
	3. 2
	4. 1
	5. zero

(a) $P - 5, Q - 2$ (b) $P - 4, Q - 2$

(c) $P - 5, Q - 3$ (d) $P - 4, Q - 1$

43. The process and disturbance transfer functions for a system are given by:

$G_p(S) = y(S)/m(S) = 2/[(2S + 1)(5S + 1)]$;

$G_C(S) = y(S)/d(S) = 1/[(2S + 1)(5S + 1)]$. The feed forward controller transfer function that will keep the process output constant for changes in disturbance is:

(a) $\frac{2}{[(2S+1)(5S+1)]}$ (b) $\frac{(2S+1)^2 (5S+1)^2}{2}$

(c) $\frac{1}{2}$ (d) $\frac{(2S=!)}{(5S+1)}$

44. For the Block diagram shown below:

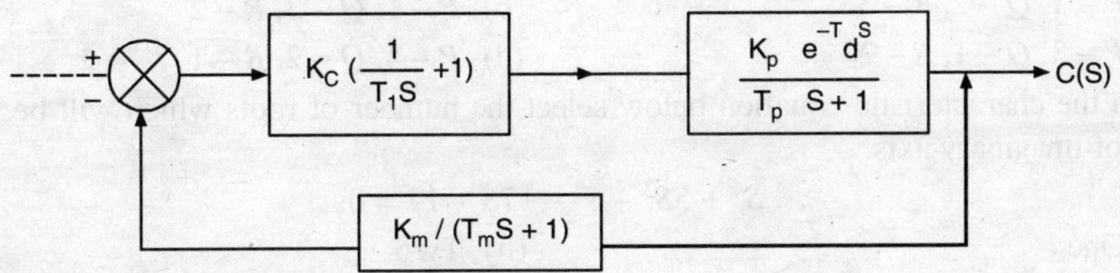

The characterstic equation is:

(a) $T_1S(T_pS+1)+K_CK_p(T_1S+1)\exp^{-T_dS=0}$

(b) $(T_mS+1)(T_pS+1)+K_mK_p \exp^{-T_dS=0}$

(c) $T_1S(T_mS+1)(T_pS+1)+K_C K_pK_m(T_1S+1)\exp^{-T_dS=0}$

(d) $(T_mS+1)(T_pS+1)+K_CK_p K_m e^{-T_dS=0}$

45. The unit step response for a first order system with time constant T and steady state gain K_p is given by:

(a) $K_p(1-e^{-t/T})$ (b) $K_p(1+e^{-t/T})$

(c) $K_p(1-e^{-2t/T})$ (d) $K_p(1+e^{-2t/T})$

46. An example for open loop second order underdamped system is:

(a) Liquid level in a tank (b) U-tube manometer

(c) Thermocouple in a thermowell

(d) Two non-interacting first order systems in series

47. The control valve characteristic is selected such that the product of process gain and valve gain:

(a) Is a linearly increasing function of manipulated variable.

(b) Is a linearly decreasing function of manipulated variable.

(c) One measurement and more than one manipulated variable.

(d) More than one measurement and more than one manipulated variable.

48. Cascade control comes under the control figuration which uses:

(a) One measurement and one manipulated variable

(b) More than one measurement and one manipulated variable

(c) One measurement and more than one manipulated variable

(d) More than one measurement and more thanone manipulated variable.

49. Match the process variables in Group-I given below with the measuring device Group-II.

Group-I	Group-II
P. High temperature	1. Orifice meter
Q. Flow	2. Chromatograph
R. Composition	3. Radiation pyrometer
	4. Bimetallic thermometer

(a) $P-1, Q-2, R-3$ (b) $P-1, Q-3, R-2$
(c) $P-3, Q-1, R-2$ (d) $P-4, Q-2, R-1$

50. Given the characteristic equation below select the number of roots which will be located to the right of imaginary axis.

$$S^4 + 5S^3 - S^2 - 17S + 12 = 0$$

(a) One (b) Two
(c) Three (d) Zero

51. Given the process transfer function, $G_p = 4/(TS + 1)^2$ and the disturbance transfer function, $G_d = 2(TS + 1)$. Select the correct transfer function for the feed forward control, for perfect disturbance rejection.

(a) $-2(TS + 1)$ (b) -1
(c) $-0.5(TS + 1)$ (d) $-(TS + 1)^2$

52. Given the process transfer function, $G_P(S) = 20/(S - 2)$ and controller transfer function, $G_C = K_C$ and assuming the transfer functions of all other elements in the control loop are unity select the range of K_C for which the closed loop response will be stable:

(a) $K_C < 1/10$ (b) $K_C < 1/100$
(c) $1/100 < K_C < 1/10$ (d) $K_C > 1/10$

53. The value of ultimate period of oscillation P_u is 3 minutes, and that of the ultimate controller gain K_{cu} is 2. Select the correct set of tuning parameters (controller gain K_c, the derivative time constant T_d in minutes, and the integral time constant T_i in minutes) for a *PID* controller, using Ziegler – Nichols controller settings.

(a) $K_c = 1.1, T_i = 2.1, T_d = 1.31$ (b) $K_c = 1.5, T_i = 1.8, T_d = 0.51$
(c) $K_c = 1.5, T_i = 1.8, T_d = 0.51$ (d) $K_c = 1.2, T_i = 1.5, T_d = 0.38$

ANSWERS

SECTION–A

1. (a)	2. (c)	3. (a)	4. (a)	5. (c)
6. (d)	7. (a)	8. (c)	9. (b)	10. (d)
11. (b)	12. (c)	13. (d)	14. (b)	15. (b)
16. (c)	17. (a)	18. (b)	19. (b)	20. (a)
21. (b)	22. (d)	23. (a)	24. (b)	25. (a)
26. (a)	27. (b)	28. (a)	29. (d)	30. (b)

31. (d)	32. (b)	33. (a)	34. (b)	35. (d)
36. (a)	37. (a)	38. (a)	39. (b)	40. (c)
41. (a)	42. (a)	43. (a)	44. (b)	45. (c)
46. (b)	47. (b)	48. (c)	49. (b)	50. (c)
51. (c)	52. (a)	53. (c)	54. (b)	55. (b)
56. (a)	57. (b)	58. (a)	59. (c)	60. (d)
61. (c)	62. (a)	63. (a)	64. (b)	65. (b)
66. (a)	67. (c)	68. (b)	69. (a)	70. (a)
71. (a)	72. (a)	73. (b)	74. (b)	75. (c)
76. (b)	77. (c)	78. (b)	79. (c)	80. (b)
81. (b)	82. (b)	83. (a)	84. (b)	85. (a)
86. (b)	87. (c)	88. (b)	89. (b)	90. (a)
91. (a)	92. (b)	93. (c)	94. (b)	95. (a)
96. (a)	97. (c)	98. (a)	99. (a)	100. (a)
101. (c)	102. (a)	103. (a)	104. (a)	105. (b)
106. (a)	107. (c)	108. (b)	109. (a)	110. (b)
111. (b)	112. (b)	113. (b)	114. (a)	115. (a)
116. (b)	117. (a)	118. (a)	119. (c)	120. (b)
121. (b)	122. (b)	123. (a)	124. (a)	125. (c)
126. (b)	127. (c)	128. (a)	129. (d)	130. (a)

SECTION–B

1. $\dfrac{A}{\sqrt{1+W^2T^2}} \sin(wt + \phi)$

2. $f(x) = f(x_s) + (x - x_s) * \dfrac{\partial f}{\partial x}\Big|_{x = x_s}$

3. Denominator of closed loop transfer function.
4. For all bounded inputs, the ouputs are also bounded.
5. For phase lag of – 180°, the amplitude ratio should be less than or equal to one.
6. $A.\ R_1 * A.\ R_2 * A\ .\ R_3......A.\ R_n.$
7. $\phi_1 + \phi_2 + \phi_3 + ... + \phi_n.$
8. Flapper-nozzle type or force type.
9. State equations and state variables.
10. Thermal conductivity cell.
11. When the control valve operates at low lift.
12. Resistance increases with temperature.
13. Oxides of salt or metal.

14. For measuring the amount of water vapour in air.
15. Refractive index/angle of refraction.
16. Polarisation of light.
17. Am e.m.f. in millivolts is generated when two solutions of different hydrogen ion concentration are separated by a thin glass wall.
18. Thin sheet of metal, plastic or leather.
19. It can be used for more than second order systems without difficulty.
20. Solid state type.
21. *C*
22. Pulse.
23. Denominator of closed loop transfer function.
24. $m(S)/e(S) = K_C(1 + T_dS + 1/T_iS)$
25. Dynamic function
26. $R(1 - e^{-t/T})$
27. $\dfrac{G(S)}{1 - G(S)\,H(S)} = \dfrac{C(S)}{R(S)}$
28. Stable.
29. Flow measuring device.
30. *C* is correct.
31. In the first column of the matrix the elements should be of same sign.
32. $\tan^{-1} 2\xi WT/[1 - (WT)^2]$
33. Boiling points and melting points.
34. (1) Diaphragm (2) Bellows (3) Bourdon tube
35. Resistance meter
36. Rotary plug valve
37. Weir
38. Measuring, reference
39. Thermal
40. Speed.

SECTION—C

1. $\dot{m} = \dfrac{K_C e}{T_i} + K_C \dot{e} + K_C t_d \ddot{e}$

or $m = \dfrac{K_C}{T_i} \int e\,dt + K_C\, e + K_C T_d\, \dot{e} + M_s$

$e = Et$ where E is constant

$m - M_s = \dfrac{K_C}{T_i} \int Et\,dt + K_C Et + K_C\, T_d\, E$

$$m - Ms = K_C E \left(\frac{t^2}{2T_i} + t + T_d \right)$$

2. $(xG_a + y)\ (G_b + x)\ G_C = y$
   Expanding, $x(G_a G_b G_c + G_c) = y(1 + G_b G_c)$
   $y(S)/x(S) = G_c + G_a G_b G_c/(1 + G_b G_c)$
3. Characteristic equation is the denominator of closed loop transfer function, $G/(1 + G)$ and equating to zero
   $$S^3 + 3S^2K_C + 2SK_C + 15 = 0$$
4. Overall = $(-\ 180\ w/3.143) - \tan^{-1}(2w)$
   By trials, $\phi = -\ 180$ degrees at $w = 1.60$ Radians/minute.
5. $A.R_{\text{overall}} = \dfrac{6 * 1}{3\sqrt{0.11\, w^2 + 1}}$

   at $w = 5$; $A\ .\ R_{\text{overall}} = 1.02$
6. 1. Range 2. span 3. accuracy
   4. reproducibility 5. sensitivity 6. dead zone
   7. drift 8. static error
7. 1. Primary element 2. secondary element
   3. manupulating 4. functioning element
8. 1. Response element 2. dynamic error 3. time lag
   4. fidality
9. $h'(t) = q_i R(1 - e^{-t/T})$
   $R = h/q = 30/1500 = 0.02$ min/cm^2
   $A = 300$ sq. cm
   $T = A * R = 300/50 = 6$ min

   $h'(t) = 500 * \dfrac{1}{50}\ (1 - e^{-5/6}) = 5.65$ cms

   $h = 30 + 5.65 = 35.65$ cms.
10. $\theta'(S)/\theta'_f(S) = 1/(TS + 1)$ for step change
    $\theta'(t) = \theta'(f)\ (1 - e^{-t/T})$
    θ = instrument reading – room temperature
    θ'_f = steady state bath temperature – room temperature
    Taking logarithms on both sides
    $\log\ (\theta'_f - \theta') = \log\ \theta'_f - t/2.303\ T$
    so by plotting $\log\ (\theta'_f - \theta')$ vs. t, if the graph is straight line then first order.
11. $Sx(S) - 2 + 3x(S) = 0$

    $x(S) = \dfrac{2}{S + 3}$; $x(t) = 2e^{-3t}$

12. $T = 0.1$, $x_s = 100°F$, $A = 2°F$, $f = 10/\pi$ cycles/min.

$W = 2\pi f = 2\pi * 10/\pi = 20$ rad/min.

$$A.R. = \frac{A}{\sqrt{T^2W^2+1}} = \frac{2}{\sqrt{4+1}} = 0.896°F$$

Phase lag $= -\tan^{-1}(2) = -63.5°F$

Response $y(t) = 0.896 \sin(20t - 63.5°)$

13. $S^3y(S) - S^2y(0) - Sy'(0) + 4[S^2y(S) - Sy(0) - y'(0)] + 5[Sy(S) - y(0)] + 2y(S) = 2/S$

Collecting common terms, $y(S) = \dfrac{2}{S(S^3+4S^2+5S+2)}$

14. Proportional band $= \dfrac{(75°F - 71°F)}{(100°F - 60°F)} * 100 = 10\%$

$$\text{Gain} = \frac{dP}{dT} = \frac{15 \text{ Psi} - 3\text{Psi}}{75°F - 71°F} = 3 \text{ Psi/°F}$$

15. Stable limit cycles are closed curves which are approached asymptotically by all nearby trajectories. Unstable limit cycles are closed curves from which all nearby trajectories diverage.

16. $y(S)/x(S) = R/(TS + 1)$, $x(S) = 1/S$

$$y(S) = \frac{R}{S}\frac{1}{(TS+1)};\ y(S) = \frac{R/T}{S(S+1/T)} = \frac{R}{S} - \frac{R}{(S+1/T)}$$

$y(t) = R(1 - e^{t/T})$

17. $G(iw) = (iw + 1)$

$$A.R. = \sqrt{(\text{Real part})^2 + (\text{Img. part})^2} = \sqrt{1+w^2}$$

Angle $= \tan^{-1}$ (Img. part/Real part) $= \tan^{-1} w$

$A.R. = \sqrt{1+w^2}$; phase angle $= \tan^{-1}(w)$.

18. 1. Oxygen (B.P.) 2. Ice (F.P.) 3. Steam (B.P.) (100°C)
4. Sulphur (B.P.) 5. Antimony (F.P.) 6. Silver (F.P.)
7. Gold

19. It consists of two dissimilar metal wires. An e.m.f. in millivolts is generated in the circuit, if the two ends are kept at different temperatures.

20. Instrumentation diagram shows the measuring instruments, automatic controllers, recorders and transmission lines present in a process. Standard symbols are used to represent instruments, controllers and transmission lines etc.

21. $q_i - q_0 = A\ dh/dt$, $R = h/q$; $q_i - q_0 = ARdq_0/dt$ in deviation form, $q_i - q_0 = A\ .\ Rdq_0/dt$. Taking Laplace transforms on both sides $q_i'(S) - q_0'(S) = SARq_0(S)$ collecting comon terms,

$$\frac{q_0'(S)}{q_i'(S)} = \frac{1}{(ARS+1)} = \frac{1}{TS+1}$$

22. $FC_i - FC_0 - KC_0V = VdC_0/dt$ in deviation form,

$FC_i' - FC_0' - KC_0'(S)\,V = SVC_0'/dt$

Taking Laplace transforms on both sides

$FC_i'(S) - FC_0'(S) - KC_0'(S)\,V = SVC_0'(S)$

Collecting common terms

$$\frac{C_0'(S)}{C_i'(S)} = \frac{F/(F+KV)}{V/(F+KV)\,S+1} = \frac{R}{TS+1}$$

23. Characteristic equation is $5S + 16 = 0$; $a_0S + a_1 = 0$

16	5	0	0
0	0	16	5
0	0	0	0
0	0	0	0

$\Delta_1 = 16 > 0$

$$\Delta_2 = \begin{vmatrix} 16 & 5 \\ 0 & 0 \end{vmatrix} = 0$$

$$\Delta_3 = \begin{vmatrix} 16 & 5 & 0 \\ 0 & 0 & 16 \\ 0 & 0 & 0 \end{vmatrix} = 0$$

24. (1) Primary element (2) Secondary element
(3) Manipulating element (4) Functioning element

25. A bimetal consists of two strips of metals welded together having different coefficients of thermal expansion. One end of bimetal is fixed and the other end is free.

When the fixed end is exposed to a high temperature, due to different coefficients of thermal expansion, the free end moves. This movement of free end is calibrated to temperature.

26. $x_1 = -5x_1 - x_2 + 2M$ where M is input

$x_2 = 3x_1 - x_2 + 5M$; 0 (output) $= x_1 + 2x_2 + 5M$

$$\begin{bmatrix} \dot{x}_1 \\ \dot{x}_2 \end{bmatrix} = \begin{bmatrix} -5 & -1 \\ 3 & -1 \end{bmatrix} \begin{bmatrix} x_1 \\ x_2 \end{bmatrix} M$$

$$O = \begin{bmatrix} 1 & 2 \end{bmatrix} \begin{bmatrix} x_1 \\ x_2 \end{bmatrix} + 5M$$

27. Flapper nozzle is a displacement type. It works by converting controlled variable and set point into corresponding displacements and converting these displacements in order to produce a pneumatic signal.

28. First order:

1. Bare mercury in glass thermometer
2. Pressure gauge having one bellows one tube only.

Second order:

1. Mercury thermometer with covering
2. Pressure gauge with one bellows, two tubes and a tank.

29. The refrectometer measures the compositions of liquid mixtures. Light rays from air passing through a liquid are deflected or refracted. The amount of refraction depends on the composition of fluid or the characteristic of the fluid. Refractive index = $\sin\theta/\sin\theta_0$ where θ is angle incidence and θ_0 is angle of refraction.

30. The electrical conductivity or resistance of a solution changes with composition of the liquid mixture.

SECTION—D

1. (d)
2. (d)

$$G(S) = \frac{1}{(S+1)\,(S+2)};\; S = -1 \text{ and } -2 \text{ are poles.}$$

3. (c)

$$\frac{H_2(S)}{Q(S)} = \frac{R_2}{T_1T_2S^2 + (T_1 + T_2 + A_1R_2)\,S + 1}$$

$$= \frac{R}{T^2S^2 + 2\xi TS + 1};\; T = T_1 = T_2;\; R_1 = R_2 = R$$

$$A_1 = A_2 = A;\; T = * R$$

$$\xi = \frac{3AR}{2RA} = 1.5 \text{ (overdamped)}$$

4. (c)

Since $L(1/S^2) = t$ (time) which goes on increasing at t increases.

5. Step response equation for a first order is:

$Y(t) = (90 - 30)\,(1 - e^{-t/T}) = 60(1 - e^{-t/T})$

$$\frac{dY(t)}{dt} = -60\, e^{-t/T} * -\frac{1}{T} = \frac{60}{T} e^{-t/T}$$

$$\left.\frac{dY(t)}{dt}\right|_{t=0} = 2 = \frac{60}{T};\; T = 30 \text{ secs.}$$

at $t = 60$ sec. (1 min); $Y(t) = 60(1 - e^{-60/30}) = 51.9$

$y(t) = 51.9 + 30 = 81.9$°C.

6. For linear input, $X(S) = K/S^2$, here $K = 1$ and $X(t) = Kt$

$$\frac{Y(S)}{X(S)} = \frac{R}{TS+1};\; Y(S) = \frac{K}{S^2}\frac{R}{TS+1}$$

$$Y(S) = \frac{K}{S^2} - \frac{KT}{S} + \frac{KT}{(S + 1/T)}$$

$Y(t) = Kt - KT + KTe^{-t/T}$; for $K = 1$

$Y(t) = t - T + Te^{-t/T} = t - T(1 - e^{-t/T})$

Minimum difference is when $1 - e^{-t/T} = 0$

i.e., $t/T = 0$; $t = 0$

Maximum difference is when $e^{-t/T} = 0$; i.e., t is ∞.

7. A.R. = 0.10 at $W_\infty = 10$ Rad/Sec.

(a) $\dfrac{1}{A.R. \text{ at } W_\infty} = \dfrac{1}{0.10} = 10$

(b) $T = 1/w = 1/10 = 0.1$ sec

(c) $K_C = 0.5\, K_v = 0.5 \times 10 = 5$

8. (a) $A = -4$(pole); $B = -2$(pole)

(b) C is breakaway point: $\dfrac{1}{S+2} + \dfrac{1}{S+4} = 0$; $S = -3$ So C is -3.

(c) By Ruth-Stability method: $G = G_C G_P G_V G_m = K_P K_C/(S + 2)$ $(S + 4)$; For $K_C = 1$, characteristic equation is: $S^2 + 6S + (8 + K_P) = 0$

$$\left|\begin{array}{cc} 1 & (8 + K_C K_P) \\ 6 & 0 \\ (8 + K_C K_P) & 0 \\ 0 & \end{array}\right| ; K_C K_P \geq 0 \text{ for stable.}$$

9. (a) K_C

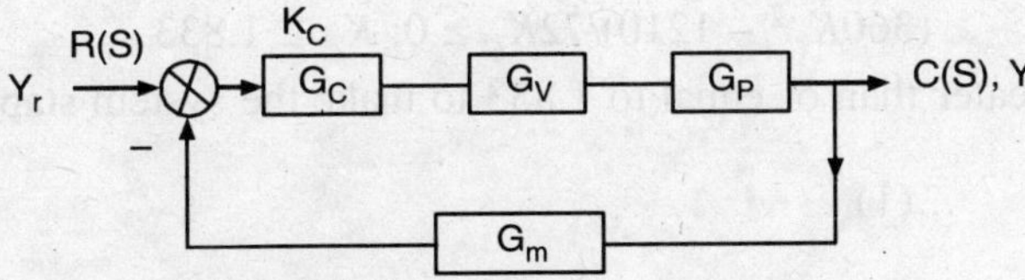

$$\frac{Y(S)}{Yr(S)} = \frac{\text{Product of elements before } Y_r \text{ and } y \text{ in Forward loop}}{1 + \text{Product of all elements in entire loop}}$$

$$= \frac{G_C G_V G_P}{1 + G_C G_V G_P G_m} = \frac{3K_C /[(2S+1)\,(3S+1)]}{1 + 3K_C /[(2S+1)\,(3S+1) * (5S+1)]}$$

$$\frac{Y(S)}{Y_r(S)} = \frac{15 + 3S}{30S^3 + 31S^2 + 10S + 4}; \; Y_r(S) = \frac{1}{S}$$

$$Y(S) = \frac{15}{4(S+1)\,(S+8i)\,(s-8i)} + \frac{3}{4S(S+1)\,(S+8i)\,(S-8i)}$$

$$Y(t) = \frac{015}{4}[......e^{-t} \sin t] + \frac{3}{4}[e^{-t} + ... \sin t + ...\cos t]$$

$t = \infty$ $Y(\infty) = \frac{3}{4}$; steady state error = $Y_r - Y(\infty)$

= 1 – 3/4 = 0.25.

10. The characteristic equation is the closed loop transfer function's denominator which is:

$S^4 + 11S^3 + K_c(7S^2 + 5S) + 10 = 0$

The standard form of characteristic equation is:

$b_n S^n + b_{n-1} S^{n-1} + ... + b_1 S + b_0 = 0$

The matrix of the given equation is:

$b_n = 1$ $b_{n-2} = 7K_C$ $b_{n-4} = 10$

$b_{n-1} = 11$ $b_{n-3} = 5K_C$ $b_{n-5} = 0$

$$\frac{b_{n-1}b_{n-2} - b_n b_{n-3}}{b_{n-1}} = \frac{11 * 7K_C - 1 * 5K_C}{1}; \frac{11 * 10 - 0 * 11}{11}$$

1	$7K_C$	10
11	$5K_C$	0
$75K_C/11$	10	0

$$\frac{(72K_C/11) * 5K_C - 11 * 10}{72K_C/11}; \frac{(72K_C/11) * 0 - 0 * 11}{72K_C/11}$$

$360K_C^2 - 1210/72K_C$	0
10	0

Ruth method says for system to be stable, all the coefficients in the first column should be of same sign. So

$$(360K_C^2 - 1210)/72K_C \geq 0; K_C \geq 1.833$$

So K_C should be greater than or equal to 1.833 to make the system stable.

11. $q_i - q_0 = A\frac{dh}{dt}$ ...(1)

$$q_i - c\sqrt{h} = A\frac{dh}{dt}$$

linearizing by Taylor's series expansion,

$$q_0 = q_{os} + \left.\frac{\partial q_0}{\partial h}\right|_{h_s}(h - h_s); \left.\frac{\partial q_0}{\partial h}\right|_{h_s} = \frac{c}{2\sqrt{h_s}}$$

$$q_{os} = c\sqrt{h}$$

So $q_0 = c\sqrt{h_s} + (h - h_s)\frac{c}{2\sqrt{h_s}}$

Eqn. (1) becomes: $q_i - c\sqrt{h_s} - (h - h_s)\dfrac{c}{2\sqrt{h_s}} = \dfrac{A\,dh}{dt}$ in deviation form:

$$(q_i - q_{is}) - \frac{C}{2\sqrt{h_s}}(h - h_s) = \frac{Ad}{dt}(h - h_s);\ Q_i - \frac{C}{2\sqrt{h_s}}H = A\frac{dH}{dt}$$

Taking Laplace Transforms on both sides and collecting common terms:

$$\frac{H(S)}{Q_i(S)} = \frac{2\sqrt{h_s}}{(2AS\sqrt{h_s} + C)}$$

Comparing with standard form of first order transfer function:

$$\frac{H(S)}{Q_i(S)} = \frac{R}{TS+1};\ T = R * A;\ R = 2h_s^{1/2}$$

12. Input-output = Rate of accumulation; $hAx - hAy = \dfrac{d}{dt}(m\,c\,y)$ where h is heat transfer coefficient, A = area of heat transfer, y = temperature indicated by thermo-couple, m = mass of thermocouple, C = specific heat in deviation form: $hAx' - hAy' = mCdy'/dt$ taking Laplace transforms and collecting comon terms,

$$\frac{y'(S)}{x'(S)} = \frac{1}{\left(\dfrac{mc}{hA}\right)S+1};\ T = \frac{mC}{hA}$$

$$T = \frac{\dfrac{4}{3}\pi \times \left(\dfrac{0.2}{2}\times 10^{-2}\right)^2 \times 8600 \times 400}{4\pi\left(\dfrac{0.2}{2}\times 10^{-2}\right)^2 \times 280} = 4.1 \text{ sec}$$

For step response: $y'(t) = |\,x'\,|\,(1 - e^{-t/T})$

for 98%; $\dfrac{y'(t)}{|\,x'\,|} = 0.98 = 1 - e^{-t/T};\ \dfrac{t}{T} = 4$

$t = 4.1 \times 4 = 16.4$ seconds

13. (b)
14. (c)
15. (a)
16. (a)

$16S^2 + 8S + 4 = 4\,(4S^2 + 2S + 1)$; standard form of second order transfer function is $T^2S^2 + 2\xi TS + 1$

Here $T^2 = 4$; $2\xi T = 2$; $T = 2$,

$\xi = 0.50$

17. $\frac{C}{R}$ (assuming load is zero) $= \frac{2(1+S)/(S+4)}{1+(2+2S)/(S+4)} = \frac{2(S+1)}{3(S+2)}$

For step change of magnitude A:

$$C(S) = \frac{2}{3}\frac{A}{S}\frac{(S+1)}{(S+2)}$$

$$C(S) = \frac{2A}{3}\left[\frac{1}{2S} + \frac{1}{2(S+2)}\right]$$

$$C(t) = \frac{A}{3}(1 - e^{-2t})$$

(b) Offset is steady state error. Here $C(\infty) = \frac{A}{3}$

18. (a) For $\frac{1}{TS+1}$; $A.R. = \frac{1}{(w^2T^2+1)^{1/2}}$; phase $= -\tan^{-1}(wT)$

For $\frac{1}{4(3S+1)^4}$; overall phase $= -4\tan^{-1}(3w)$

For cross over frequency: $-4\tan^{-1}(3w) = -180°$

(b) At $S = 0$; $A.R = \frac{1}{4}$; $w = \frac{1}{3}$ Radian

19. Open loop transfer function $[C_1(S)]$ is:

$$K_C\left(1+\frac{2}{S}\right) * \frac{1}{(2S^2+S+1)}$$

Characteristic equation is the denominator of open loop transfer function

$$= S(2S^2 + S + 1) + K_C(S+2) = 0$$

Now apply Ruth Stability method for binding K_C values.

20. (d)

21. (d)

Phase lag is: $-\tan^{-1}(wT) = -\tan^{-1}(1) = -45°$

22. (d)

For *PI* control: $m(S) = K_C\left(1+\frac{1}{T_iS}\right)$

Here $K_C = 1$; $T_1 = 1$; $e(S) = \frac{1}{S} * e(S)$

$m(S) = \frac{1}{S} + \frac{1}{S^2}$; $m(t) = 1 + t + 10 = 11 + t$

23. (d)

$$G(S) = \frac{1}{2T_1S^2} \text{ (Open loop transfer function)}$$

Closed loop TF is: $\frac{G}{1+G} = \frac{1}{1+2T_1S^2} = \frac{C(S)}{R(S)}$

For $R(S) = \frac{1}{S}$; $C(S) = \frac{1}{(2+2T_1S^2)}\frac{1}{S}$

By partial fractions:

$$C(S) = \frac{1}{2T_1}\left[\frac{A_1}{S} + \frac{A_2}{\left(S + \frac{i}{\sqrt{2T_1}}\right)} + \frac{A_3}{\left(S - \frac{i}{\sqrt{2T_1}}\right)}\right]$$

$A = 2T_1$

$C(t) = 1 + A_{22} \cos t_1 + A_{32} \sin t_1$

$C(t)$ is oscillating because of sin and cos terms.

24. For first billet, Raising $(T_1) = (900 - 50)(1 - e^{-t_1}) + 50$

For second billet, Falling $(T_2) = (900 - 50) * [1 - e^{-(t_1-1)}] + 50$

Response of pyrometer is: $850(1 - e^{-t_1}) + 50/850[1 - e^{-(t_1-1)}] + 50$

25. (c)

26. (c)

$$G(S) = \frac{1}{TS+1};\ \frac{y(S)}{x(S)} = \frac{1}{TS+1};\ x(S) = A\sin(t/T);$$

$$\text{since } w = \frac{1}{T};\ A.R. = \sqrt{\frac{1}{w^2T^2+1}};\ w = \frac{1}{T};\ A.R. = \frac{1}{\sqrt{2}}$$

27. (c)

$$\phi = -\tan^{-1}(wT) = \tan^{-1}(-wT);\ \text{tan limits are: 0 to } \frac{\pi}{2}$$

28. (a) 300 K; 312 K; $T = 40$; This is step response:

$T(t) = 300 + (312 - 300)(1 - \exp.^{-t/T})$; $= 300 + 12(1 - \exp^{-t/T})$

(b) $1 - \exp^{-t/T} = 0.9955$; By trial & error: $t/T = 7$ or 8.; Taking t/T as 8, $t = 8*40 = 320$ seconds.

(c) $T = T_a + 0.50$; $T_a(t) = 300 + [(312 - 0.5) - 300](1 - \exp^{-t/T}) = 300 + 11.5$ $(1 - \exp^{-t/T}) + 0.50 = T$; $311.844 - 300.5 = 11.5(1 - \exp^{-t/T})$; t/T is nearly 4.

So $t = 4 * 40 = 160$ seconds.

29.

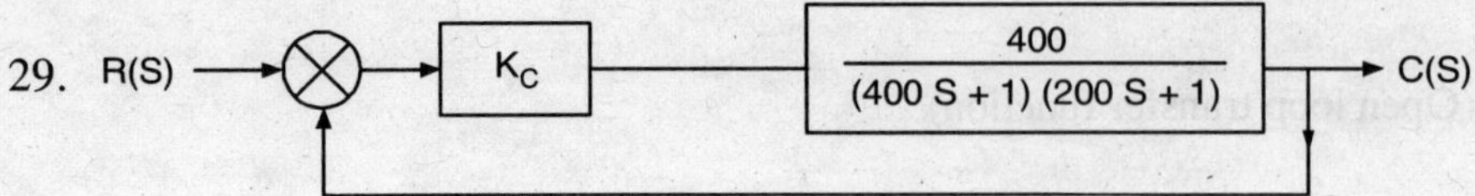

30. (a)

31. (a)

32. Feedback can be given at tank 3 only. So answer is 3. **Ans.** (c)

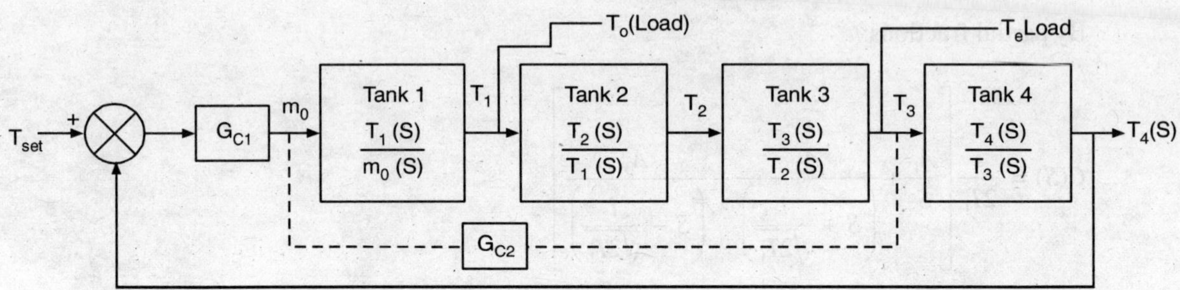

33. (c)

1	7
(6+*K*)	0
7	0
0	0

So $(6 + K) \geq 0$; K is ≥ -6 **Ans.**

34. (c)

Here for a periodic function, $Lf(t) = \dfrac{1}{1-e^{-2St}}\left[\int_0^{2T} e^{-St} f(t)\, dt\right]$

$= f(S)$; $f(t) = -1,\ T < t < 2t$; $f(t)$ is periodic with aperiod $2T$.

$= 1,\ 0 < 1 < t$

$$f(S) = \frac{1}{1-e^{-2ST}}\left(\int_0^{T} e^{-ST}(1)\, dt + \int_T^{2T} e^{-ST}(1)\, dt\right)$$

$$= \frac{(1-e^{-ST})^2}{S(1+e^{-2ST})};\ f(S) = y(S) = \frac{1(1-e^{-S})}{S(1+e^{-S})} \text{ for T = 1.}$$

35. (b)

Heat input – Heat output ± Heat loss = Rate of heat accumulation

$$Q_0 C\rho T_0 - QC\rho T \pm 0 = \frac{d}{dt}(A_C * h * C * \rho * T)$$

where ρ = density of liquid,

C = specific heat of liauid,

T = temperature,

Q = flow rate,

h = level of liquid

Assuming: ρ, C, A_C and h s constants:

$$Q_0\, T_0 - QT = A_C h\, dT/dt.$$

36. (c)

Cross over frequency: – 180 degrees = $\tan^{-1} 4w - \tan^{-1} 2w - \tan^{-1} w$.

At w is nearly one, the phase angle is: – 180 degrees.

$$AR_{overall} = \frac{1}{\sqrt{16w^2+1}}\frac{1}{\sqrt{4w^2+1}} * \frac{1}{w^2+1}\text{; At } w = 1,\ A.R_{overall} = \frac{1}{\sqrt{17}} * \frac{1}{\sqrt{5}} * \frac{1}{\sqrt{2}};$$

$$A.R_0 = \frac{1}{13.04}\text{; gain for P-controller is } = \frac{1}{A.R_0} = 13.04 \textbf{ Ans.}$$

37. (a)
38. (d)
39. (c)

Pneumatic controllers were used in the older days. The temperature in the CSTR is measured by an electric device, like thermocouple or RTD (Resistance thermometer detector). This is converted to pneumatic signal (*E*/*P* converter) and then send to a pneumatic controller,to give output. Now a days in computer controller gives an electric output. This electric output is converted to air pressure to operate a valve (changes flow rate of cooling water in jacket). Here opening the valve increases the cooling water flow rate. So answer is *C*.

40. (d)

We can check by units. Time constant units are time $\frac{mC_P}{hA}$ and $\frac{V}{q}$ give time units. Others do not give time units.

41. (c)

T_d is lag time. Q/R gives time constant units.

42. (a)
43. (c)

The feed forward controller transfer function is:

$$F_C(S) = -\frac{G(S)}{G_P(S)} = -\frac{\frac{1}{(2S+1)(5S+1)}}{\frac{2}{(2S+1)(5S+1)}} = -\frac{1}{2}.$$ Neglecting sign C is the answer.

44. (c)

45. (a)

The step response equation for a first order system with time constant, T and unit gain is: $(1 - e^{-t/T})$. So (a) is correct.

46. (b)

(U-tube manometer)

47. (a)

It is desirable that the valve opening is linear with error.

48. (b)

An example of cascade control is changing cold water flow rate in an exothermic reactor to control temperature in the reactor. This is done by measuring reactor temperature and inlet cold water temperature used for cooling (cold water temperature changes during the day).

49. (c)

High temperature is measured by Radiation pyrometer (above 800 °C). Bimetallic measures: 0 to 300 °C. Flow is measured by orfice meter only in the given list. Composition is measured by chromotograph in the list.

50. (b)

The roots are calculated by trial and error, which are $S = 1$, $S = 0.80$; $S = -3.0$, $S = -4$.

51. (c)

Feed forward controller transfer function is:

$$-\frac{G_d(S)}{G_P(S)} = -\frac{2(TS+1)^2}{(TS+1)4} = -0.5\ (TS+1).$$

52. (d)

Open loop transfer function is: $G(S) = 20\ K_C/(S-2)$;

The characteristic equation is: $1 + G(S) = 0$; $1 + \dfrac{20K_C}{(S-2)} = 0$

$S + (20K_C - 2) = 0$; Using Ruth criterion: $K_C = \dfrac{1}{10}$

53. (d)

Ziegler-Nichols settings for *P-I-D* controller from Bode stability method are:

$$K_C = 0.50\ K_u;\ T_1 = \frac{p_u}{2.0};\ T_d = \frac{P_u}{8};$$

$$\text{Here } K_C = 0.60 * 2 = 1.2;\ T_i = \frac{3}{2} = 1.50;\ T_d = \frac{3}{8} = 0.370.$$

So (d) is correct answer.

CHAPTER 11

Equipment Design and Economics

Plant design involves all engineering aspects involved in the development of a new or modified or expanded commercial process. Capital investment of a plant consists of : Fixed capital and working capital.

1. Fixed capital investments items for a chemical process are:

(a) **Direct costs:** 1. Purchased equipments; 2. Purchased equipment installation; 3. Instrumentation & controls; 4. Piping; 5. Electrical systems; 6. Buildings; 7. Yard improvements; 8. Service facilities; 9. Land.

(b) **Indirect costs:** 1. Engineering and supervision; 2. Legal expenses; 3. Con-struction expenses; 4. Contractor' s fee; 5. Contingency.

2. Types of capital cost estimates:

1. Order of magnitude estimate (ratio estimate). Probable accuracy is over 30% based on previous cost data.
2. Study estimate (factored estimate) based on cost of major items of equipment. Accuracy is ± 30%.
3. Preliminary estimate. Accuracy ± 20%. Based on sufficient data.
4. Definitive estimate: Based on complete engineering drawings data. Accuracy is ± 10%.
5. Detailed estimate (contractor's estimate): Based on complete nngineering drawing. Accuracy is ± 15%.

3. Cost index: Present cost = Original cost * index value at present/ Intial value at time of original cost was obtained.

4. Equipment cost scaling:

Cost of equipment a = cost of equipment b * X where X is times the capacity of the first.

Reactors: 0.54 to 0.56

Equipment delivery: 10%

Equipment installation: 8 to 9% of nquipment cost.

Instrumentation and controls: 26% of equipment cost,

Piping: 15 to 25%; Electrical systems: 15 to 30%; Buildings: 10 to 18%; Yard improvements: 10 to 20%; Service facilities: 8 to 20%; Health, safety and environment functions: 8.4 to 34%; Land: 4 to 8% of purchased equipment cost.

Engineering and supervision cost; 30% of equipment cost.

Legal expenses: 1 to 3 % of fixed capital investment.

Contractor's fee: 2 to 8% of direct plant cost. Contingencies: 8%.

5. Methods for estimating capital investment:

1. Detailed item estimate (+ or – 5% accurate)
2. Unit cost estimate (+ or – 10 to 20% accurate)
3. Percentage of delivered equipment cost (+ or – 20 to 30 % accuracy)
4. Lang factors for approximation of capital investment.
5. Factor* delivered equipment cost = Fixed capital investment or total capital investment for major addition to an existing plant equipment.

Type of plant	Fixed capital investment	Total capital investment
Solid	4.0	4.7
Solid-fluid	4.3	5.0
Fluid	5.0	6.0

Power factor applied to plant/capacity ratio:

$c_n = c\, f_c\, R^X$: f_C = cost index ratio at the time of cost c_n to that at the time of c ; x is 0.6 to 0.70; R = capacity of new facility/capacity of the old facility; c_n = capacity of new facility; c = capacity of constructed facility.

6. Investment cost per unit of capacity:
7. Turnover ratio:This is the ratio of gross annual sales/fixed capital investment.

6. Simple interest: $F = P(1 + i\,N)$ where F = total amount; P = principle amount; compound interest: $F = P(1+ i)^N$; i = interest rate.

The effective interest rate i_{eff} is the rate at which when compounded once per year gives the same amount of money at the end of 1 year, as does the nominal rate r compounded m times per year. $P = (1 + r/m)^m = (1 + i_{eff})^1$; $i_{eff} = (1 + r/m)^m - i$.

7. Compounding:

with A = 10,000/year; i = 0.10; N = 10. The present worth is:

$$P = \frac{A(1+i)^N - 1)}{i(1+i)^N} = 10{,}000\ (1.1^{10} - 1)/(0.1 * 1.1)^{10} = 61{,}446.$$

8. Fixed charges: These are: 1. Depreciation 2. Property taxes 3. Insurance.

9. Depreciation: The concept of depreciation is based on the fact that physical facilities deteriorate and decline in usefullness with time, thus the value of a facility decreases.

1. Physical depreciation,
2. Fundamental functional depreciation.

10. Salvage value: This is the net amount of money obtained from the sale or used property over and above any charges involved in removal and sale.

11. Recovery period:The period over which depreciation is charged is the recovery period, and this is established by tax codes.

The period over which the use of a property is economically feasible is known as the service life of the property.

12. Methods for calculating depreciation.

1. straight line method ($d = v/n$; where d = annual depreciation, v = original investment, n = length of straight line recovery period.)
2. Modified accelarated cost recovery system; The double declining balance method allows a depreciation charge in each year of the recovery period that is twice the average rate of recovery on the remaining undepreciated balance for the full recovery period.

13. Optimisation:

1. Linear programming: $a_1 x_1 + a_2 x_2 + ... = b_1$; graphical approach, computerized approach.
2. Non–linear programming: (a) Successive linear programming to approximate non-linear programming; (b) Augumented Lagrangian method (c) Successive quadratic programming.
3. Dynamic programming.
4. Optimisation solution methodologies: (a) One variable; (b) Two variables: $dc/dx = 0$; d^2c_T/dx^2 and $d^2 c_T/dy^2$ are positive gives minimum cost.

14. Materials and Fabrication selection:

Chromium is responsible for the principal corrosion resisting characteristics of the stainless steels.

Cathode protection is used for underground pipes and tanks.

Organic coatings, cladding of steel with an alloy.

15. Materials handling: 1. Pumps: $N_1/N_2 = mV_1/mV_2$ where N_1, N_2 are revolutions per minute and mV_1, mV_2 are discharge rates, kg/min. Power 1/Power 2 = $mV_1^3/mV_2^3 = Nr_1^3/Nr_2^3$.

16. Heat exchanger types available: 1. Double pipe; 2. Shell and tube; 3. Scraped surface exchangers; 4. Plate heat exchangers; 5. Spiral plate and tube exchangers; 6. Plate and fin heat exchangers; 7. Fin–Fin heat exchanger (gas-gas); 8. Condensers; 9. Evaporators.

17. Packings: 1. Raschig ring; 2. Intalox saddle (ceramic); 3. Berl Ring; 4. Pall ring; 5. Intalox saddle (metal); 8. Cascade mini ring.

QUESTIONS

SECTION—A

1. For cylindrical vessels at high pressure (above about 15 kg/cm²) it is more economical to employ:

(a) Dished head (b) Conical head
(c) Ellipsoidal head (d) Hemispherical head

2. The ultimate tensile strength of low carbon steel is usually between:

(a) 10000–15000 psi (b) 20000–30000 psi
(c) 35000–50000 psi (d) 55000–65000 psi

3. For reasons of elastic stability, the minimum shell thickness for 15 metre tanks and smaller is limited to:

(a) 3 mm (b) 5 mm (c) 7 mm (d) 10 mm

4. The thickness of the shell as given by "thin walled equation" is:

(a) $pd/2f$ (b) $pd/4fF$. (c) $pd/4f$ (d) $pd/4fEtC$

5. For plain carbon-steel, butt-welded joints may be used for shell plate thickness up to and including:
(a) 3.8 cm (b) 3.0 cm (c) 1.5 cm (d) 0.5 cm

6. The most common shape for the roof of a cylindrical tank is:
(a) Umbrella (b) Dome (c) Ellipsoid (d) Cone

7. Formed heads for cylindrical process vessels are usually used to close the vessel, when the working pressure is:
(a) Less than atmospheric (b) More than atmospheric
(c) 5–10 kg/cm^2 (d) 10–20 kg/cm^2

8. When conical heads are used as bottom heads of process equipments, the cones usually have an apex angle of:
(a) 30° (b) 45° (c) 60° (d) 90°

9. For tall thick-walled vertical vessel, ideal support is:
(a) Lug (b) Guy wire
(c) Skirt (d) None of the above

10. Horizontal cylindrical pressure vessels are commonly supported by:
(a) Guy wire (b) Saddle (c) Skirt (d) Lug

11. In case of steel pipes up to 12″ the designated size refer to :
(a) Out side meter of the pipe (b) Inside diameter of the pipe
(c) Average dimater of the pipe (d) Neither of the above.

12. Wall thickness for schedule 80 pipe is:
(a) Same as that for schedule 40 pipe (b) Less than that for schedule 40 pipe
(c) More than that for schedule 40 pipe

13. For 1″ (2.54 cm) exchanger tube with increase in BWG, the inside flow area:
(a) Remains constant (b) Increases (c) Decreases

14. For prevention of fluid leakage around moving parts, normal device used is:
(a) Stuffing box (b) Bellow
(c) Packless joint (d) Expansion loop

15. For changes in the direction of a pipeline, it is necessary to use:
(a) An union (b) A tee (c) An elbow (d) A bend

16. Plug cocks are generally used on small lines for compressed air:
(a) To control the rate of flow (b) To serve as a safety device
(c) To open or close a line completely.

17. Quick opening valves are generally of the type of:
(a) Needle valves (b) Globe valves
(c) Diaphragm valve (d) Gate valve

18. A valve that functions with lubrication and with low resistance to flow is a:
(a) Needle valve (b) Diaphragm valve
(c) Plug cock (d) Globe valve.

19. Valves are usually made of brass for pipe sizes:

(a) 5.08 cm (2″) and under (b) 2.54 cm (1′) and under
(c) Above 5.08 cm (d) Above 7.62 cm (3″)

20. For long lines of piping connected by threaded fittings, dissembling in parts for cleaning or repair is achieved by putting at frequent intervals:

(a) The tees (b) The nipples (c) The unions (d) The sockets

21. Tubings employed in flanged pipe lines are generally connected to the system by:

(a) Welding (b) Flaring
(c) Expanding (d) Threading and screwing

22. Double-pipe heat exchangers are used when the area requirement is of order of:

(a) More than 500 m^2 (b) 250–500 m^2
(c) 100–250 m^2 (d) 10–20 m^2

23. In a heat exchanger shell, with tube size and pitch remaining same:

(a) More number of tubes can be accommodated in triangular pitch arrangement.
(b) More number of tubes can be accommodated in square pitch arrangement.
(c) Equal number of rubes can be accommodated in both the above cases

24. For a heat exchanger, the effect of L/D on inside film coefficient (h_i) vanishes after a Reynold's number of:

(a) 2100 (b) 10000 (c) 100 (d) 1000

25. For heat transfer calculations, the equivalent diameter for the annulus of a double-pipe heat exchanger is given by:

(a) $D_2 - D_1$ (b) $\frac{D_2 - D_1}{4}$ (c) $\frac{D_2^2 - D_1^2}{D_1}$ (d) $\frac{D_2^2 - D_1^2}{4D_1}$

(D_2 and D_1) are I.D. of outer and O.D. of inner pipe respectively).

26. Function of baffles on the shell side of a heat exchanger is to:

(a) Decrease pressure drop (b) Reduce scale deposit
(c) Hold the tubes in position (d) Create turbulence.

27. Minimum recommended baffle spacing in a heat exchanger is equal to:

(a) I.D. of shell (b) 20% of I.D. of shell
(c) 5 cm (2″) (d) half of I.D. of shell

28. Under comparable conditions of flow and tube size, the shell side heat tansfer coefficient in case of triangular pitch is equal:

(a) To that for square pitch
(b) To nearly 25% higher than for square pitch
(c) To nearly 25% lower than for square pitch
(d) To nearly twice that of square pitch

29. Water is widely used as a coolant in heat exchange equipments, because:

(a) It has a low dirt factor
(b) It has a low viscosity

(c) It is less corrosive to materials of construction
(d) It has a comparatively high specific heat.

30. For cooling water in exchanger tubes, minimum recommended velocity per second is:
(a) 0.9 m (b) 2 m
(c) 5 m (d) None of these

31. A specific advantage of using steam as a heating medium in exchangers is its:
(a) Inexpensiveness (b) High value of latent heat
(c) High film coefficient (d) Non-corrosive condensate.

32. Minimum recommended tube pitch for heat exchanger is equal to:
(a) O.D. of rube (b) 1.25 × O.D. of tube
(c) 1.5 × O.D. of tube (d) 2 × O.D. of tube

33. Tube length to shell diameter in case of liquid exchanger is:
(a) 2 – 4 (b) 2 – 8 (c) 4 – 8 (d) 4 – 10

34. Minimum recommended ligament for square pitch arrangement in case of heat exchangers is equal to
(a) I.D. of tube (b) O.D. of tube (c) 5 cm (d) 6.5 mm

35. Sea water has a fouling factor:
(a) Equal to river water (b) Less than river water (c) More than that for river water (d) None

36. Under identical conditions, the ratio of the values of heat transfer coefficient for horizontal condenser to vertical one is nearly equal to:
(a) 1.0 (b) 2.0 (c) 3.0 (d) 4.0

37. Trombone coolers are used for the cooling of:
(a) Extremely corrosive liquids (b) Highly viscous liquids
(c) Hot slurries (d) High temperature gases

38. The chemical evaporator employing steam in the tube is a:
(a) Horizontal-tube evaporator (b) Long-tube vertical evaporator
(c) Basket-type evaporator (d) Short-tube vertical evaporator.

39. For calendria-type evaporator, the height of the tubes is usually equal to:
(a) < 1 metre (b) < 2 metre (c) 2-3 metre (d) 4-5 metre

40. In calendria-type evaporator, the tubes are held between two tube sheets:
(a) Ttop being fixed and the bottom floating
(b) Top being floating and the bottom fixed
(c) Both being floating
(d) Both being fixed

41. An accessory of evaporator installed in the vapour line to remove entrained liquid is called:
(a) Vent (b) Baffle (c) Catchall (d) Deflector

42. The evaporator for concentration of liquors with scaling tendency is of:
(a) Basket type (b) Calendria type
(c) Horizontal-tube type (d) Long-tube vertical type

43. An evaporator with an annular down take is a:
(a) Thermocompression evaporator (b) Horizontal evaporator
(c) Basket evaporator (d) Long tube-vertical evaporator

44. Long-tube vertical evaporators have excellent performance for:
(a) Viscous liquor (b) Scaling liquor
(c) Salting liquor (d) Foamy liquor

45. For calendria type evaporator, the down-take area is equal to:
(a) Cross-sectional area of the tube
(b) 50–100% of cross-sectional area of the tube
(c) 150% of cross-sectional area of the tube
(d) 200% of cross-sectional area of the tube

46. As a rough estimate, the optimum reflex ratio usurlly falls in the range of:
(a) 1.1 – 1.5 R_{Dm} (b) 1.5 – 2.0 R_{Dm}
(c) 2.0 – 2.5 R_{Dm} (d) 2.6 – 2.9 R_{Dm}
where R_{Dm} = minimum reflux ratio.

47. Maximum plate efficiency is obtained in case of:
(a) Reverse flow (b) Split flow (c) Cross flow (d) Cascade flow

48. Stepping is provided on the trays to:
(a) Decrease pressure drop (b) Increase residencetime
(c) Improve flow condition (d) Handle large flow rate of liquid

49. An acceptable total liquid gradient over a tray is:
(a) 12.5 mm (b) 25 mm (c) 35 mm (d) 50 mm

50. Diameter of bubble caps used in vacuum columns when compared to those used in identical pressure columns, are found to be:
(a) Equal (b) Smaller
(c) Bigger (d) Nothing in particular

51. Minimum recommended residence time in the down-comer is:
(a) 2 sec. (b) 3.5 sec. (c) 5 sec. (d) 10 sec.

52. The value of static submergence in case of bubble cap tray for column operating under pressure ($P \geq 1$ kg/cm^2) is:
(a) 2–5 mm (b) 4–8 mm (c) 10–20 mm (d) 12–38 mm

53. Skirt clearance for bubble caps is generally of the weir of:
(a) 10 – 20 mm (b) 12 – 38 mm (c) 15 – 45 mm (d) 20 – 80 mm.

54. It case of segmental downcomer, the length of the order is nearly equal to:
(a) 30% of D (b) 40 – 60% of D
(c) 60 – 80% of D (d) 90% of D

55. It case of a balanced tray design, the value of downcomer liquid seal is equal to:
(a) 12–38 mm (b) 15–65 mm (c) 20–80 mm (d) 40–120 mm

56. As a rough guide, bubble cap columns of diameter 120 cm and above, use caps of diameter equal to:
(a) 40 cm (b) 30 cm (c) 25 cm (d) 10 cm

57. Minimum clearance between the overflow weir and the last row of bubble caps is:
(a) 5 cm (b) 7.5 cm (c) 10 cm (d) 15 cm

58. Between the column wall and the bubble caps, it is necessary to maintain a minimum clearance of:
(a) 38 mm (b) 76 mm (c) 152 mm (d) 190 mm

59. Bubble caps are located on trays with a pitch of:
(a) (2.0 – 2.5) O.D. (b) (1.5 – 2.2) O.D.
(c) (1.3 – 2.0) O.D. (d) 1.5 O.D.
where, OD = outside dia of caps.

60. The riser in bubble-cap trays conduct the flow of:
(a) Liquid (b) Vapour (c) Both liquid and vapour
(d) None

61. The number of bubble caps to be used per tray is determined from:
(a) The vapour load (b) The liquid load
(c) Tray diameter (d) Allowable slot velocity

62. In common design practice, an approximate relation between the riser area, total slot area and passage area in the annular space under the cap is given by:
(a) Riser area = annular passage area = slot area
(b) Riser area = annular passage area = 12 slot area
(c) Riser area = 1.2 annular passage area = slot area
(d) Riser area = 1.2 annular passage area = 1.2 slot area

63. Hole diameters for the sieve trays are of the order of:
(a) 2 mm–8 mm (b) 3 mm–12.5 mm
(c) 5 mm–15 mm (d) 10–25 mm

64. For which of the following packings, is the mass transfer are maximum:
(a) Raschig ring (b) Lessing ring (c) Cross-partition ring. (d) Berl saddle

65. In packed towers, liquid distribution difficulties tend to become excessive when the ratio of tower diameter to packing diameter is:
(a) Less than 7 (b) Between 7 and 17
(c) Less than 70 (d) Between 17 and 70

66. For gas-liquid systems involving highly corrosive fluids, the ideal contact device is a:
(a) Bubble-cap plate tower (b) Packed tower
(c) Sieve plate tower. (d) Valve tray tower

67. For continuous drying of large solids of special forms and shapes, the recommended dryer is of:
(a) Drum type (b) Rotary type
(c) Vacuum shelf type (d) Tunnel type

68. The length of diameter ratio for rotary driers is of the order of:
(a) 2 – 4 (b) 3 – 6 (c) 4 – 10 (d) 5 – 15

69. In case of plate and frame filter, the thickness of plates compared to the frames is:
(a) More (b) Less (c) The same (d) None

70. The ratio of working capital to total capital investment for most chemical plants (other than non-seasonal products-based) is:
(a) 0.1 to 0.2 (b) 0.2 to 0.4 (c) 0.4 to 0.6 (d) 0.6 to 0.8

71. Installation costs for equipment is equal to:
(a) 20 – 25% of the purchased equipment cost
(b) 35 – 45% of the purchased equipment cost
(c) 50 – 60% of the purchased equipment cost
(d) 60 – 75% of the purchased equipment cost

72. For a solid-fluid processing plant the value of 'Lang' factor for fixed capital investment is:
(a) 2.2 to 2.4 (b) 2.6 to 2.8 (c) 3.6 to 3.8 (d) 4.2 to 4.4.

73. Cost of piping in percent of fixed capital investment for a fluid processing (*e.g.* distillation) is:
(a) 5 (b) 10 (c) 13 (d) 20

74. As a rough estimate, utility costs for ordinary chemical processes amount to:
(a) 5 – 10% of the total product cost (b) 10 – 20% of the total product cost
(c) 20 – 25% of the total product cost (d) 25 – 35% of the total product cost

75. While making an estimate of the total product cost the expenditures under research and development, belong to the category of:
(a) General expenses (b) Direct production costs
(c) Fixed charges (d) Plant overhead costs

76. Minimum safe distance between large storage tanks with hazardous chemicals is equal to:
(a) 1.0 – 1.5 D (b) 2.0 – 2.5 D (c) 3.0 D (d) 5.0 D
where, D = diameter of the tank.

77. For high-pressure process equipments, the connected nozzles should be:
(a) Screwed (b) Flanged (c) Welded

78. Storage tanks for volatile liquids have:
(a) Flat roof (b) Conical roof (c) Floating roof

79. A rupture disk is used as an accessory to equipment for:
(a) Creating turbulence (b) Improving mixing
(c) Relieving pressure (d) None

80. Materials like aniline, the halogens, hydrogen flouride and phosgene are toxic in concentrations as low as:
(a) 50 ppm. (b) 35ppm. (c) 15ppm. (d) 5 ppm.

SECTION—B

1. .... vessels are used for the storage of volatile liquids and gases.

2. All process vessels require nozzles for ....and ....

3. Low carbon steels are easily sheared, rolled and formed into various shapes used in fabricating vessels because they are ....and ....

4. The height-to-diameter ratio of standard steel tanks for storage at atmospheric pressure depends on ....and ....

5. The majority of tanks and vessels are cylindrical because a cylinder has ....and is ....

6. Plates with lap-welded joints are less expensive compared to butt-welded ones, because.......

7. Cylindrical storage vessels are installed preferably horizontal because for vertical vessels.......

8. ....and .... dished closures are stronger than conical or flat-plate closures for cylindrical vessels.

9. Tall vertical vessels are usually supported by .... or ....

10. Lug-supports are .... expensive than other types of supports.

11. Normal screwed fittings are used for pipes upto a diameter of.......

12. For joining pieces of larger steel pipes in process piping for high pressure service welding is preferred to screwed fittings because....... and ....

13. Tubing is differentiated from pipe by the fact that, for a given diameter, tubing wall thickness is ....

14. A reducer is used to change.......

15. Flanged fittings of steel for water, oil and gas service are obtained in different ....and ....,,,,.......

16. Safety valves are generally used to guard against .... and ....

17. Tube rolling is the method of ....

18. Removable connection of tubes to tube sheet is called a ....

19. Differential expansion problem between the tube passes and the shell of the exchanger is taken care of by using a.......or a ....

20. For removal of non-condensibiles, ...are provided in heat exchangers.

21. For very high temperature heating, the medium used are......., ...and ....

22. To prevent quick draining of the condensate from a vertical condenser-subcooler a .... is provided.

23. For condensation of vapour from a distillation column without sub-cooling generally .... condenser is preferred.

24. High pressure fluids are generally routed through the ....

25. Extremely large volumes as well as very small volumes of fluids are best routed through the .... because of the flexibility possible in ....

26. Short tubes are.......expensive per square meter of heat exchange are than the long tubes.

27. For a good performance of a heat exchanger, the LMTD correction factor should have a value of.......

28. To avoid excessive corrosion, the outlet temperature of cooling water in an exchanger should not be allowed to reach much higher than ....

29. Bleed points are provided in the evaporator to ....

30. In small columns, the liquid overflow from plate to plate is conducted through.......

31. Weep holes are provided on the trays of distillation columns to ....

32. The risers are generally connected to the plate by .... and ....

33. Tray spacing in large columns is usually determined by ....
34. Sieve trays are .... expensive than bubble-cap trays.
35. In a packed tower, the regular packings offer the advantages of....... and
36. .... towers can be designed to handle wide ranges of liquid rates without flooding.
37. The functions of flights in rotary dryer are to .... and to ....
38. The total capital investment of a chemical plant consists of the .... and the ....
39. The raw materials inventory included in working capital usually amounts to a .... supply of raw materials valued at delivered prices.
40. Purchased cost of equipments constitutes ....% of fixed capital investment.
41. Total product cost consist? .... and ....
42. Hazards in chemical plants are categorized as ......,...... and ....
43. Vents are provided for atmospheric storage tanks for ....
44. As a safety measure for moving equipments, ... are provided at all couplings, belts and chain drives.
45. .... are provided around all storage tanks containing inflammable material to prevent the spread of fire in case of hazard.

SECTION— C

1. What is brazing?
2. Why spherical-shaped pressure vessel is considered to be the most ideal?
3. What are the common head types for the cylindrical vessels?
4. What are the functions of internals for process vessels?
5. What is stress relieving?
6. What is radiographing?
7. Why shell openings greater than 5 cm for a storage tank be reinforced?
8. Which type of dished head is the strongest?
9. What is the main advantage of dished head over flat cover plates or cones as closures for cylindrical vessels?
10. What are blind flanges?
11. How are flanges connected to pipes?
12. On what factors does the optimum size of pipe for a specific situation depend?
13. What provision is made for expansion in pipelines for high temperature service?
14. What is schedule number?
15. What are nipples?
16. When are needle valves used?
17. What provisions are made to account for pipeline movement caused by thermal expansion?
18. What are the main disadvantages of hair-pin heat exchangers?
19. What are the advantages of square-pitch arrangement of heat exchanger tubes over the triangular-pitch?

20. What are the various types of baffles used in heat exchangers?

21. On what quantities does the shell side cross-flow area in a shell-and-Tube heat exchanger depend?

22. On what factors does the pressure drop on the shell-side of a heat exchanger depend?

23. What are normal lengths of heat exchanger tubes?

24. What are the two common types of floating heads that are used in heat exchangers?

25. When is the LMTD—correction factor used in heat exchanger design?

26. What characteristics of the fluid are to be considered while deciding its route in a heat'exchanger?

27. What is $(\Delta t)_{weighted}$ in case of a condenser-subcooler?

28. What are the advantages of horizontal-tube evaporators?

29. Far which service the forced circulation evaporators are normally recommended?

30. What are the different types of liquid flow encountered over bubble-cap

31. How are the bubble caps held in place over a tray?

32. What is a uniflux tray?

33. How are the holes of a sieve tray arranged?

34. What is the function of an inlet weir on a tray?

35. What is a venturi scrubber?

36. Suggest a suitable contacting device (with reasoning) for gas-liquid operations involving liquids with dispersed solids.

37. What are normal drum sizes used for rotary drum vacuum filters?

38. Give the names of a few filter media used in case of corrosive liquids.

39. What is a cost index?

40. What is six-tenths-factor-rule?

41. What is 'Lang' factor?

42. What is turn-over ratio?

43. What is 'break-even-point'?

44. Calculate the shell side equivalent diameter when 1.9 cm O.D. tubes are arranged on 2.5 cm. square pitch.

45. What will be the cross flow area for the shell side of a shell-and-tube heal exchanger for the following conditions?

I.D. of shell = 60 cm.
Tube pitch = 2.5 cm. (triangular)
O.D. of tube = 1.9 cm.
Baffle spacing = 15 cm.

SECTION—D

1. A chemical reaction vessel made of plain carbon steel is to be designed for operation at a pressure of 15 bar absolute and a temperature of 290°C. The ellipsoidal bottom will have a ratio of major axis : minor axis of 2 : 1. As safeguard it is proposed to provide a corrosion allowance of 10% of calculated thickness. Design pressure may be taken as 110% of the operating pressure. At the operating conditions the design stress (*f*) permissible will have a value of 85 N/mm . Calculate the thickness of the cylindrical portion of the reactor and the ellipsoidal bottom. The dimensions are : D = 2m; L (h) = 1m.

2. In process heat exchangers .... steam is preferred over ....steam.

3. Hot gases at moderate pressure are usually in the .... side of the shell and tube exchangers. At high pressure, however it is customary to put gas in the ....

4. Fill up blanks :

(a) The important stresses induced in the shell of a tall vertical vessels are:

(a) .... (b) .... (c) .... (d) ....

(b) The tensile stress resulting from internal pressure are (i)....... expressed as .... and (ii).....expressed as ....

5. Fill in the blanks

(a) a cylindrical shell under external pressure tends to .....as a result of external radial pressure. The external pressure at which buckling occurs depends on ..... and ...... when the length of the shell is less than the critical length.

6. Match of the following :

(i) Gasket	(a) External support
(ii) Anchor Bolts	(b) Flango Joints
(iii) Skirt	(c) Foundation
(iv) Guywire	(d) Base support

7. (a) The cost of blower in 1980 is Rs : 2000

(i) What is the cost of blower in 1988 with the same capacity? The cost index for the blower in 1980 and in 1988 respectively 250 and 300.

(ii)What is the cost of blower in 1980 with double the capacity?

(b) If the delivered costs of equipments of a fluid processing plant is Rs 4×10^6. What is the capital cost of the plant?

8. A small model reactor is to be built for scale up studies of the behaviour of a proposed large industrial stirred tank reactor having 1000 times capacity. The bigger unit of 2 m diameter will have a liquid depth of 2 m. This will be fitted with a four bladed Rushon turbine of 0.6 m diameter,

(i) Estimate the dimensions of the smaller unit.

(ii) For the optimum stirrer speed of 300 RPM observed in the smaller model, what will be the recommended speed in the industrial unit under the following conditions.

(a) Power per unit volume is kept constant.

(b) Reynolds number does not change.

(iii) What design criteria would you recommend for this type of study?

9. For a chemical plant the fixed capital investment is Rs : 4×10 and the working capital is 20% of the total capital investment. The annual total product costs are Rs : 2×10 where as the annual depreciation costs are Rs : 2×10^8. If the total annual sales are Rs : 3×10^9 and the income tax rate is 40% then determine :

(a) The percent of total capital investment returned annually as gross profit.

(b) The pay out time. Neglect start up costs.

10. A certain pressure vessel manufacturer avoids doing reinforcement calculations for openings by always providing reinforcing pad extending upto the double the diameter of the opening and of the same material and thickness as that of the shell wall. If area compensation is accepted as a code guideline, his approach leads to safe design.

(a) Only if the opening is on the spherical vessel.

(b) Only if the opening is on a vertical cylinderical vessel

(c) only if the opening is on a horizontal cylindrical vessel.

(d) Irrespective of the shape of the vessel.

11. For a cylindrical vessel of moderate height the weld joint efficiency for joints parallel to the cylinder axis is given as 1.0, while for joints along the girth (circumference) it is given as 0.80. In calculating shell wall thickness using code formula, for an internally pressurized cylindrical vessel, what value of weld join efficiency should be used?

(a) 0.80 (b) 0.90 (c) 1.0 (d) $0.8^{0.5}$

12. For a flanged point, it is proposed to use Bolts on a flange with bolt root diameter of 12.5 mm. The allowable stress for the bolt material is 150 MN/m^2. The total bolt load is 1.6 MN. Recommend a minimum number of bolts that should be provided.

SECTION—E

1. Capital cost: Annual operating cost of the pipe is given as follows:

Annual operating cost = $3 * 10^5/D^5$ Rs. /m.yr.

Estimate the most economic pipe diameter based on least annual cost approach to the nearest multiple of 10 m.m. Estimated amortzation period is 10 years.

2. Direct costs: Component of the fixed capital consists of :

(a) Contigency (b) On site and off site coste

(c) Labour costs (d) Raw material costs.

3. A series of equal payments (e.g. deposit or cost; made at equal intervals of time known as:

(a) Perpetuity (b) Capital charge factor

(c) Annuity (d) Future worth.

4. The variables required to be known in correlations used for estimating the horse power of a centrifugal pump gas compressor and hence its costs are:

P = inlet pressure, q–compressor PPM, R–delivery pressure, S–volumetric flow rate at inlet.

(a) P, Q and R (b) P & Q (c) R & S (d) P, R & S.

5. A process has a fixed capital investment of Rs. 150 lakhs. Working capital is Rs. 30 lakhs and salvage value is zero.

Annual revenues from sales are Rs. 250 lakhs, manufacturing costs 145 lakhs and other expenses are of 10% of the revenue. Assume project life span of 11 years, tax life of 5 years and interest rate to be 10%. Tax rate is 40% and sraight line depreciation ie. 20% per year is applicable.

Discount, value (to present time) of the profit before tax (for the total plant life period) in Rs. is:

(a) 228 lakhs (b) 400 lakhs (c) 520 lakhs (d) 660 lakhs

6. Discounted value of the dpreciation benefit over the tax life in Rs.

(a) 12 lakhs (b) 24 lakhs (c) 46 lakhs (d) 60 lakhs

7. Two pumps under consideration for installation at a plant have the following capital investments and salvace values, pump

(A) Rs. 40,000; C sal = Rs. 3900; Pump (B) C1 = 50,000 Rs., C sal = Rs. 20,000.

Using capitalised cost determine what could be the common life of the pumps for both to be competitive (economically equivalent). Interest rate is 10% of annum. Maintenance and operation costs are negligible.

(a) 3 years (b) 5 yrs. (c) 6 yrs. (d) 8 yrs.

8. (A) Fixed capital investment for a chemical plant is: Rs. 40 million with estimatd useful life of 6 years and a salvage value of Rs. $4 * 10^6$. The rate of interest is 15%. Tax is 25% of the annual taxable income. In the first year of operation the income from sales is Rs. 20 million and manufacturing expenses are, 5 million. The plant depreciates on a straigh line basis. The rate of return :

(a) 50% (b) 37.5% (c) 32% (d) 20%

(B) The net present value (NPV) in million Rs. at the start and at the end of first year of operation is respectively given by:

(a) Zero and – 28.9 (b) – 40 and – 28.9

(c) 40 and 12.75 (d) Zero and 12.75

ANSWERS

SECTION— A

1. c	2. d	3. b	4. c	5. a
6. d	7. b	8. c	9. a	10. b
11. d	12. c	13. b	14. a	15. c
16. c	17. d	18. b	19. a	20. c
21. a	22. d	23. a	24. b	25. c
26. d	27. c	28. b	29. d	30. a
31. b	32. b	33. c	34. d	35. b
36. c	37. a	38. a	39. b	40. d
41. c	42. a	43. c	44. d	45. b
46. a	47. c	48. d	49. a	50. b
51. c	52. d	53. b	54. c	55. a
56. d	57. b	58. a	59. c	60. b
61. d	62. a	63. b	64. c	65. a
66. b	67. d	68. c	69. b	70. a
71. b	72. d	73. c	74. b	75. a
76. a	77. b	78. c	79. c	80. d

SECTION—B

1. spherical.
2. the fluids to enter and leave, for drains, handholes and manholes
3. soft and ductile
4. processing requirements, available land area and height limitations.
5. greater structural strength and is easy to fabricate.
6. squaring of plates for lap-welded joints is not necessary.
7. the wind loads may impose the necessity of heavy foundations to prevent overturning.
8. elliptical, torispherical and hemispherical
9. skirts or lugs. **10.** less **11.** 7.6 cm(3.0″)
12. it does not weaken the pipe wall and can be made leak proof.
13. usually less than pipe wall thickness.
14. the size of pipe in a straight run.
15. size and pressure capacities.
16. excessive pressure and pressure fluctuations
17. fixing tubes to a tube sheet by expanding
18. ferrule
19. floating head rube sheet, U-bend **20.** vents
21. dowtherm, heat transfer salt, liquid metal
22. loop-seal, **23.**horizontal **24.** rubes
25. shell, baffle arrangements, **26.** more
27. 0.75 to 0.80 or greater, **28.** 50°C
29. remove non-condensible gas
30. pipes (circular-pipe downcomers)
31. drain the residual liquid on tray during shut down
32. bolts, rivets, welding
33. the need for easy access for maintenance and inspection
34. less
35. lower pressure drop and greater possible throughput of fluids.
36. packed
37. to expose the solids thoroughly to the drying action and to assist in their forward motion also.
38. fixed capital investment and the working capital.
39. 1—month,
40. 20–40%
41. manufacturing costs and general expenses
42. flammable and explosive, hygienic and mechanical
43. breathing
44. guards firewalls

SECTION—C

1. Brazing is the joining of metals with a nonferrous filler without melting the base metal.

2. Because it withstands high pressures for a given metal thickness.

3. Flat, ellipsoidal, dished, hemispherical, conical.

4. To affect the changes in the materials passing through the vessel.

5. Stress relieving is an annealing or heat treating process which relieves residual stresses.

6. Radiographing is done for all welds in order to detect defects.

7. To prevent local overstressing of the shell around the opening.

8. Hemispherical

9. It is the large reduction in the discontinuity in shape at the junction between the cylindrical vessel and the closure resulting in a reduction of discontinuity stresses at or near the junction.

10. Blind flanges are used to:

(a) Blank off pressure vessel openings like manholes and inspection ports

(b) Block off the ends of piping and valves.

11. By screwing, welding and brazing

12. Relative costs of investment, power, maintenance and stocking pipe and fittings.

13. By providing bends or loops in pipe, packed expansion joints, bellows and flexible metal hose.

14. Schedule numb $r = 1000 \dfrac{p}{s}$

where, p = gauge pressure

S = allowable working stress

15. Nipples are short pieces of pipe threaded at both ends.

16. Needle valves are generally used for high pressure throttling service in the smaller sizes of pipe and for very accurate control.

17. By use of expansion joints and changes in the direction or shape of the line.

18. The main disadvantages are:

(a) Considerable space requirement
(b) Prone to leakage
(c) High maintenance cost
(d) Comparatively less heating surface

19. (a) Easily accessible for external cleaning

(b) Low pressure drop

20. (a) Segmental baffles
(b) Disc and doughnut baffles
(c) Orifice baffles

21. Inside diameter of shell, pitch, clearance and baffle spacing.

22. Shell side mass velocity, baffle spacing, rube pitch and length, density and viscosity of the shell-side fluid.

23. 0.5, 2.5, 3.0, 4.0, 5.0 and 6.0 metres.

24. Split-ring type and pull-through type

25. For multipass shell-and-tube heat exchangers

26. Viscosity, fouling characteristic and corrosiveness.

27. $$(\Delta t)_{weighted} = \frac{Q}{\dfrac{q_c}{(\Delta t)_c} + \dfrac{q_s}{(\Delta t)_s}}$$

28. Require small head room Surface blanketing by air etc. is avoided.
29. For concentration of liquids with high viscosity and scaling characteristic.
30. Cross flow, reverse flow, split flow (double-pass flow), double-pass cascade, multi-pass flow.
31. By bolts or a hold-down bar
32. A uniflux tray consists of a series of interlocking S-shaped sections which are assembled in the form of tunnel caps slot outlets on one side only.
33. The holes are arranged in triangular pitch of 2.5 to 5 hole diameters.
34. It maintains a calming section for liquid distribution and serves as a barrier to the down coming liquid possessing large inertia.
35. In a venturi scrubber, the gas is drawn into the throat of a venturi by a stream of absorbing liquid sprayed into the convergent duct section.
36. A plate tower is recommended because the plates are more accessible for cleaning.
37. From 30 cm dia with 30 cm face to 300 cm dia with 425 cm face are common.
38. Metal cloth of monel or stainless steel, glass cloth and synthetic fibre like saran, nylon etc.
39. A cost index is merely a number for a given year showing the cost at that time relative to a certain base year.
40. According to this rule, if the cost of a given unit at one capacity is known, the cost of similar unit with X times the capacity of the first is approximately $(X)^{0.6}$ times the cost of the initial unit.
41. $$\text{Lang factor} = \frac{\text{Capital investment}}{\text{Delivered cost of major equipment}}$$
42. $$\text{Turn-over ratio} = \frac{\text{Gross annual sales}}{\text{Fixed capital investment}}$$
43. The point where the total product cost equals the total income is known as the break-even point.
44. $$D_e = \text{Equivalent diameter}$$
$$= 4 \times \frac{\text{Flow area}}{\text{Wetted perimeter}}$$
$$\text{Flow area} = P_{T^2} - \frac{\pi}{4}(O.D.)^2$$
$$= (2.5)^2 - \frac{\pi}{4}(1.9)^2 = 3.415 \text{ cm}^2$$
$$D_e = 4 \times \frac{3.415}{\pi(1.9)} = 2.288 \text{ cm.}$$
45. $$\text{Shell side cross flow area} = a_s$$
$$= \frac{I.D. \times C * B}{P_T}$$
$$= \frac{60 \times (2.5 - 1.9) \times 15}{2.5} = 216 \text{ cm}^2.$$

SECTION— D

1. **Solution** Operating Pr = $P = 1.10 \times 15 = 16.5$ bar $= 16.5 \times 10^5$ N/m² permissible stress $= f = 85$ N/(mm)² ($f = 85 \times 10^6$ N/m²)

For cylindrical portion :

$t = \dfrac{PR}{SE - 0.6P}$; P = pressure, kg/inch²; E = Joint efficiency = 1

R = I.D, inch; S = allowable stress, kg/m²;

$p = 16.5 \times 14.504 = 239.3$ pSi; $r = \dfrac{2 \times 100}{2 \times 2.54} = 39.37''$;

$f = 85 \times 10^6$ N/m² $= 850 \times (14.504$ PA) conversion factor = 123284 kg/Inch²

$$t = \frac{239.32 \times 39.37}{12328.4 - 0.6 \times 239.32} = 0.77'' = 19.64 \text{ m.m}$$

Allowance = 10%

Actual thickness for cylindriacal protion = 19.64 × 1.1 = 21.6 m.m

For Ellipsoidal heat : D = 2 m = 78.74″; D/h = 2/1 = 2

$K = \dfrac{1}{6}\left[2 + \dfrac{D}{2h}\right]$; h = 1 m = 39.37″; K = 1/2

$$t = \frac{PDK}{2SE - 0.2\,P} = \frac{239.316 \times 78.74 \times 0.5}{2 \times 12328.4 \times 1 - 0.2 \times 239.32} = 0.38'' = 9.72 \text{ m.m}$$

With 10% allowance; t = 1.1 × 9.72 = 10.7 m.m

Ellipsoidal head thickness = 10.7 m.m.

2. **Ans.** Process, exhaust
3. **Ans.** shell, tube side
4. **Ans.** (a) Hoop stress (b) Bending stress (c) Tensile stress (d) Internal stress

Ans. (*i*) Circumferential, $f_t = \dfrac{PD_i}{2t'}$; (*ii*) *axial*, $f_t = \dfrac{PD_i}{4t}$

5. **Ans.** Buckle, modulus of elasticity, thickness of shell, diameter of shell.
6. **Ans.** (i) b; (ii) c; (iii) d; (iv) a
7. **Ans.** (a) (i) $2000 \times \dfrac{300}{250} = 2400$ Rs.

(ii) $2000\left(\dfrac{2}{1}\right)^{0.60} = 3031.4$ Rs.

(Ref. plant design and economics for chemical engineers by Peters, Mcgraw Hill Co.)

(b) Delivered cost of equipments = 4 × 10⁶. If delivered cost of equipment is 100 then total capital investment is :

For fluid processmg plant = 569

For solid-fluid processmg plant = 487

For solid processing plant = 455

So capital cost = 4 * 10^6 * 569/10 = 22.76 × 10^6

8. (i) $Q_P/Q_M = L_r^{2.5} = 1000$; *P*–Prototype (Actual); *M*–model

$I_r = 16.0$; Depth = $\frac{2}{16}$ = 0.125 m = 12.5 cm; Diameter = $\frac{2}{16}$ = 0.125 m = 12.5 cm

(ii) Power ratio = $\left(\frac{P/Q}{D^5 N^3}\right)_P = \left(\frac{P/Q}{D^5 N^3}\right)_m$; for same *P/Q* : $(D^5 N^3)m = (D^5N^3)_P$

$(D_P/D_M)^5 = (N_M/N_P)$; $16^5 = (300)^3/N_P^3$; N_P = 18.75 RPM. Ans

(iii) (a) Flow rate modelling for same Reynold number

(b) Pump power modelling.

9. Ans. Fixed capital = 4 × 10^9

Working capital = 0.20 × 4 × 10^9 = 0.8 × 10^9

Annual profit before income taxes

= 3 × 10^9 – 2 × 10^8 – 2 × 10^9 = 0.8 × 10^9

(a) Annual percent return before income taxes =

$$\frac{0.8 \times 10^9}{(4 \times 10^9 + 0.8 \times 10^9)} \times 100 = 17\%$$

Annual return after income taxes = 0.8 × 10^9 × 0.6 = 0.48 × 10^9

$$\% \text{ return after taxes} = \frac{(0.48 \times 10^9)}{4.8 \times 10^9} \times 100 = 10$$

$$\text{(b) Payout time} = \frac{4 \times 10^9}{0.48 \times 10^9 + \text{Average depreciation / yr.}}$$

$$= \frac{4 \times 10^9}{0.48 \times 10^9 + 2 \times 10^8} = 6 \text{ years}$$

10. Ans. (b)

11. Ans. (c)

12. Ans. $$\frac{1.6}{150 + \frac{\pi}{4} + (12.5/1000)^2} = \frac{1.6 * 4}{150 * \pi * (0.0125)^2}$$

= 87 bolts

SECTION—E

1. **Solution:** Annual cost (A.C) = maintenance costs + operating costs + depreciation costs.

   Maintenance costs = 0.10 * 3 $D^{1.5}$ Rs./m.yr.

   operating coste 3 * $10^5/D^5$ Rs./m.yr.

   Depreciation costs = 0.30 $D^{1.5}$ Rs./m.yr.

   A.C. = 0.6 $D^{1.5}$ + 3 * $10^5/D^5$; d(A.C)/dL = 0 = 0.9 $D^{0.5}$ – 15 * $10^5/D^6$;

   $D^{6.5}$ = 16.7 * 10^5 ; D is in between 9 m.m to 10 m.m. Ans.
2. **Ans.** (b)
3. **Ans.** (c)
4. **Ans:** (a)
5. **Ans.** (b)

   Solution: $A = P\left(1+\frac{r}{100}\right)^n$ = 150 * 1.1^{11} = 400 **Ans.**
6. **Ans.** (b)
7. **Ans.** (d)

   Solution: For pump A: C1 = 40,000; C = sal = 39,00 = 40,000 $(1\text{-}10/100)^n$ = 22.2 yrs.

   For pump B: C1 = 50,000, C sal = 20,000/; 20,000 = 50000 $(1 - 10/100)^n$

   n = 8 yrs.
8. (A) **Solution:** Rate of return on investment = yearly profit*100/total intial investment. Yearly profit = income from sales – tax = 20 – 20*25/100 = 15 million. Rate of return = 15/40 * 100 = 37.5%. **Ans.** (b)

   (B) **Ans.** (d)

   Solution: Net present value at the start = 0; Let net present value of investment money at the end of first year = (40 + 5)*

   (1 + i) = 45 * 1.15 = 51.75

   51.75 = 40 + 5 – 6 + x; x = 12.75

❑❑

CHAPTER **12**

Computers

1. Describing the process step by step is Algorithm. Representing the various steps in the form of a diagram is Flow chart.
2. Adding two numbers C program:

```
/* program ADDITION
/* written by NBG
main ( )
(
int number;
float amount;
number = 100;
amount = 30.75 + 75 * 35
printf ("% d\n", number);
printf("% 5.2 f", amount);
```

3. Executing a program: The steps are :
   (a) Creating the program
   (b) Compiling the program (cc file name)
   (c) Linking the program with functions that are needed from library
   (d) Executing the program (a.out)
4. C tokens are:
   (a) Key words (float, while)
   (b) Constants (– 15.5, 100);
   (c) Strings (ABC, year)
   (d) Operators (+, –, /, *)
   (e) Identifiers (main, amount);
   (f) Special symbols
5. Constants:
   (a) Numeric constants (integer constants, real constants);
   (b) Character constants (single character constants, string constants)
6. Integer constants (decimal, octal, hexadecimal)
7. Real constants (0.08, – 0.75,435,36, + 24.70, 0.65e 4,12 e–2, –1.2 e–1)
8. Variable (average, height, total , x_1, ph – value)
9. C– program for average of numbers:

   1 ....

```
1 * .........................
1 .........................
# define N  10
main ( )
(
int count
float sum, average, number
sum = 0.0;
count = 0.0;
while (count < n)
(
Scanf ("% f", & number);
sum=sum+number;
count=count+1;
)
average=sum/N;
printf("N=%d sum=%f", N, sum);
printf("Average = % f", average);
)
1
2.3
4.67
1.42
7
3.67
4.08
2.2
4.85
8.20
Sum=39.39; average=3.94, N=10
```

10. Operators are:

(a) Arithmetic operators (+, –, /, *, %);

(b) Relational operators (<, >,=, < >)

(c) Logical operators (&& – logical AND, – OR ! – NOT)

(d) Assignment operators (a = a + 1 → a + = 1; a = a – 1 → a – = 1)

(e) Increment and decrement operators (+ + m or m + 1 → m = m + 1)

(f) Conditional operators (if a > b; x=a; else, x=b)

(g) Bit wise operators (&, bitwise AND >> shift right;)

(h) Special operators (comma operator ; value (x=10,y=5 x + y);

11. Evaluations of expressions.

```
/* EVALUATION OF EXPRESSIONS
main ( )
{
float a, b, c, x, y, z;
a=9;
b=12;
c=3;
x*(a – b)/(3 + c * 2 – 1);
y=(a – b)/(3 – c) * (2 – 1);
z=a – (b/(3 + c) * 2) – 1;
printf ("x=% f/n ", x);
print (y=% f/n " ,y);
printf (z = % f/n ", z);
}
output
x=10.000000
y=7.000000
z=4.000000
```

12. Input & output operations :

Scanff ("% 2d % 5d " & num 1, & num 2)

Data line is : 50 31426

Scanf ("% d % d,& num1, & num 2);

31426 50

Assigns 312426 as num 1 and 50 to number 2.

13. Inputting real numbers:

Scanf ("%x %f %f %f, &x, &y, &z);

Input data is: 475.89 43.21E–1 678.0

The input field is specified/separated by blank spaces.

14. Formated output:

Printf ("%7.4f", y) output is :98.7654

Printf ("% 10.2E", y) output is : bb9.88E + 01

15. Printing of strings:

% s:prints output as: NEWbDELHIb110001bbbb

16. Readability of output:

Printf("a=%d\ tb=%d, "a,b); will provide 4 blank spaces "between the two fields.

Printf("\ n outputb RESULTS\ n "); will be printing HEADINGS OR TITLES

17. Decision making:

(a) IF statement (b) Switch statement (c) Conditional operator statement

(d) GO to statement.

(a) Simple IF statement:

```
if (test expression)
(
Statement block ;
)
Statement x;
```

18. IF —ELSE statement:

```
if (test expression)
(
True block statements
)
else
(
False block statements
)
statement x.
```

19. Nested IFs:

Printing the largest of the three numbers:

```
/* SELECTING THE LARGEST OP THREE VALUES:
main ( )
(
float A.B,C;
printf ("Enter three values\ n");
scanf (% f % f % f, & A, & B, & C);
printf (" \ n Largest value is ");
if (A > B)
(
if (A > C)
printf (" % f \ n" , C );
)
else
(
     if (C > B)
printf ("% f \ n", C)
     else
```

```
printf ("%f\n", B)
)
)
      output
enter the values
2344567379  88843
Largest value is 88843 000000
```

20. Switch statement: C has a built in multiway decision statement known as switch. The switch statement tests the value of a given variable or expression against a list of case values and when a match is found, a block of statements associated with the case is executed.

21. GO TO statement. Example is:

```
main ( )
(
      double x, y;
      read;
      scan x ("% f", &x);
      if (x < 0) go to read;
      y=sqrt (x);
      printf (" % f % f \n", x, y);
      go to read;
      )
```

22. WHILE STATEMENT: This is an entry controlled loop statement. The test condition is evaluated, and if the condition is true then the body of the loop is executed.

An example is:

```
____________

____________

n=1 ;
while (n < = 10)
(
sum=sum+n*n;
n=n+1;
)
printf ("sum= % d\n", sum);
```

23. Do statement: An example is:

```
____________
i=1
sum=0;
do
(
```

```
sum=sum+1;
i=i+2
)
while (sum ≤ 40 " < 10);
printf ("%d  %d\n", I, sum);
```

---

24. The FOR statement

```
sum=0;
for(n=1; n < =10; n=n+1);
(
sum=sum+n*n;
)
printf (" sum =% d.\ n" , sum);
```

25. ARRAYS:

These are : x(1), x(2), x(3), ...x(n)

```
float height (50);
int group (10);
```

In C subscript can begin with zero, x(0),

```
E.g. /* program to calculate standard deviation
#include  math h
#define  MAXSIZB 100
main( )
(
int i, n;
float value(MAXSiZE), deviation;
sum=sums qr=n =0;
printf ("input values: input = 1 to end\n");
for (i=1; i < MAXSIZE; i++)
(
scanf("% f", & value (i));
if (value(i) = –1)
break;
      sum+=value (i);
n+ =1;
)
mean=sum/(float)n;
```

```
for (i=1; i < =n; i++)
(
deviation=value (i) – mean;
sumsqr+ =deviation*deviation;
variance = sumqr/(float)n;
std deviation=sqrt(variance);
printf("\n Number of items. %d\n" ,n);
printf("Mean: % f n," mean);
printf(Standard deviation:% f=n" , std deviation);
)
```

output

input values: input –1 to end

65 9 27 78 12 20 33 49 –1

number of items: 8

Mean:36, 62 5000

Standard deviation:23.510303

26. CHARACTER STRINGS

A string is an array of characters.

E.g. printf("/" well done ! /");

will output the string " well done!"

while the statement prints f(" WELL DONE !")

will output the string as: WELL DONE!

27. ANSI LIBRARY FUNCTIONS:

<ctype.h> character testing and conversion funstions.

<math.h> Mathematical functions.

<stdio .h> standard 1/0 library functions.

<stdlib.h)> utility functions.

<string.h> string manipulation functions.

<time.h> time manipulation functions.

QUESTIONS

1. Who was the first man to conceive the concept of a computer?
2. What was the name given to the computer by the above inventor?
3. Who has designed the first electronic digital computer?
4. What was the name of the first electronic digital computer?
5. What are called First Generation Computers?
6. What do you understand by Second Generation Computers?
7. What are known as Third Generation Computers?

8. What do you mean by Fourth Generation Computers?
9. Wha is the type of number system used in computers?
10. Who has suggested the number system?
11. What are the major types of computers?
12. What are the various units present in a computer?
13. What is called a central processing unit in a computer?
14. What is console?
15. What is a mini computer?
16. What is microcomputer?
17. How does an analog computer work?
18. What is a digital computer?
19. What are the advantages of a digital computer over an analog computer?
20. What are meant by hard-ware and soft-ware of a computer?
21. What are the various input devices used in a computer?
22. How is information stored in a computer?
23. What do you understand by internal storage and auxiliary storage in a computer?
24. What are the devices used for internal storage?
25. How are data stored in auxiliary storage?
26. What are the various output devices used in a computer?
27. What is meant by a computer program?
28. What is a computer language?
29. What is known as low level language of a computer?
30. What are known as high level languages?
31. What is a compiler?
32. What are the advantages of high level languages?
33. What are the important high level languages in use?
34. Give the expansion of the names of high level languages.
35. What is the name of digital computer manufactured by ECIL, Hyderabad?
36. What is meant by BIT?
37. How is the memory of a computer expressed?
38. What are known as source program and object program?
39. What are the various digital systems used in computers?
40. What is the most popular computer language used for scientific purposes?
41. How many characters are recognized by a digital computer?
42. What are the alphabets used in a computer language?
43. What are the numerals used in a computer language?
44. What are the special characters used for algebraic operations?
45. What are the other special characteristics recognized by a compiler?
46. What are known as 'Syntax Rules'?

47. What is called 'DEBUGGING'?

48. What is a FORTRAN constant?

49. What is a FORTRAN variable?

50. Identify the FORTRAN variables and FORTRAN constants from the following :
N, 15, XX, 20.5, M35, MMX, 2.3 NUM.

51. Name the type of FORTRAN constants and variables.

52. Define Integer Constants and Real Constants.

53. What is the other name given to Real Constants?

54. What are the various forms in which a real constant may be written?

55. Name the parts of exponential form of a real constant.

56. What are the rules followed for writting the exponential form of a real constant?

57. What is the range for the value of Exponent?

58. What is the limitation on Mantissa?

59. Identify Mantissa and Exponent in the following numbers—
– 5.400 E + 02; 0.6300 E–03, – 6.56 E+8, – .664 E + 10

60. Identify the valid and invalid numbers:

(a) 0.53 E + 3.1 (b) – .320 E (c) 3.2 E + 609
(d) 132,32 E + 5 (e) 12.6 E + 643 (f) 2.300 E – 25

61. Give the classification of FORTRAN variables.

62. How do you identify the type of FORTRAN variables?

63. What do you understand by fixed point variables and Floating point variable?

64. What are the rules to be followed in choosing a variable name?

65. Identify integer and real variables from the following :

(a) NPI (b) MUMR (c) PI (d) M35A
(e) COUNT (f) K2MAX (g) THETA

66. Identify the valid and invalid variable names. Give reasons for invalid variables :

(a) MIN + 60 (b) 70AN (c) 1ST (d) SAM, T
(e) JKLP (f) ALPHABE (g) JAYKA (h) MAS
(i) EPSILON (j) PRESS

67. What is TYPE DECLARATION statement?

68. Convert the following integer variables into real variables using TYPE DECLARATION statement
LINE, NOTE, MITE

69. How do you convert the real variables into integer variables using TYPE DECLARATION statement?

70. What are the symbols used for arithmetic operations?

71. How many types of arithmetic expressions are in use in a computer language?

72. Identify the type of arithmetic expression from the following :

(a) KA/RB (b) + XVAX (c) (9–3*k + MAS**3)
(d) (NR–NK)*3 (e) RA/RB*2AK (f) NEST*3*]AM.

73. What do you understand by heirarchy of operation by a computer?

74. Write the following expressions in FORTRAN language :

(a) $z = Sy^3 + 48^{4.6}$ (b) $y = \dfrac{x^a - z^b}{a - b}$

(c) $a = \dfrac{0.00038}{P + R}$ (d) $N = 5K - 4\ m^4$

(e) $J = 3L + 6N^4 - 4I/6M$ (f) $M = 4J^5/(3N + 6L)$

75. How are the following functions written in FORTRAN?

$\sin x, \cos x, |x|, \tan^{-1} x, e^x, \log_e x, \sqrt{x}, \tan x, \text{cotan } x.$

76. Write the following arithmetic statements in FORTRAN:

(a) $S = |B| e^{(y/\sqrt{3})} z$ (b) $H = 2 \log_e (1 + \sin \theta)$

(c) $P = \tan^{-1} e^{-5x}$ (d) $F = \cos\ (\log_e y + \tan \sqrt{\phi})$

77. What is the precaution to be followed in writing trigonometric functions like sin, cos etc.?

78. Convert the variable (T) in degrees into radians.

79. Write the above conversion in FORTRAN.

80. What are the types of statements in a FORTRAN language.

81. What is the value of J calculated in the following FORTRAN statements:

(a) J = M*2/3 + N*3 + 6 – M**3/9; (M = 3; N = 4)

(b) L = L/3*3 + 3/5 + L**2; (L = 6)

82. Write a FORTRAN Program to evaluate the following expressions —

$$s = a/b\ (b + a);\ t = sb\ ;\ r = ts/b$$

83. Write a program to determine value of A from the following equation

$A = 2.93 \times 2p\ (4p^2/q) - \log_{10}\ p(\sqrt{5} - N) + 4.3(p(\log_e 3\sqrt{3} - N) + 4.3\ q)$

$p = 2.15\ ;\ q = 2.5\ ;\ N = 0.8$

84. What is the final value of M in the following program?

```
J  = 4
N  = 5
JA = 50
JM = N*200 + IA*5
M  = JM/40 = 5*J.
```

85. Explain how the value of following integer arithmetic expression is evaluated.

```
J = 6-6/10 + 30/10 + 40/80 + (K/10)* 3 + (L/10)*5
K = L = 20.
```

86. What is the value of the following integer expression?

```
J = K/8*5 + L/16*6 + L/10**2 + K/16**3
K = L = 20
```

87. What for an INPUT statement is used?

88. What for an output statement is used?

89. What are the two parts of an input/output statements?

90. What is the purpose of each part?

91. What for READ, WRITE, PRINT and FORMAT statements are used?

92. Feed the variables F, G, H, P, Q without format to the computer.

93. Give the FORMAT free statement for obtaining the values of the above variables from the computer.

94. What are Integer and Real modes of FORMAT?

95. What are the types of real FORMAT?

96. What is called Field length?

97. What do you understand by list and record?

98. Write the input statement to feed the integer variables K, KTTE, KING, MIG, LINE, LINK, use FORMAT also.

99. Write the above input statement with blank spaces in FORMAT statement.

100. Feed the real variables R, SP, PIN, RAP, RING using FORMAT in Fractional and Exponential form.

101. What is the rule to be followed in writing the exponential form of Format?

102. Write the output statement with Format to print the variables KODE, RING, PIN, LIM, RIP.

103. Write the Format required to print the numbers 106, 0.007, 23.5×10^5, 158, 59.3, $259.6 * 10^{-8}$.

104. Write the FORTRAN statement for the numbers

$$A = 193.89 \times 10^5; \quad B = -0.593 \times 10^{-8};$$

$$C = 593.4 \times 10^9.$$

105. How can we incorporate Alphanumeric data in a Format statement for a print out.

106. Write the Fortran statement to print B (when value of B = 15.8) with caption final value of B =, in front of the value of B.

107. What is known as carriage control?

108. What are the codes used for carriage control?

109. Where are the symbols /(slash) // (two slashes) are used?

110. What is the maximum field length for input and output statements?

111. How are the following statements executed by the computer :

(a) READ 48, P, Q, R, S, T

48 FORMAT (2F 10.3/2F 10.5/F 10.5)

(b) PRINT 50 P, Q, R, S, T

50 FORMAT (2F10.3/F10.5/F10.5//F10.6)

112. How can the FORTRAN statements be classified in general? Explain with examples.

113. What are known as relational operators in FORTRAN?

114. What are the FORTRAN equivalent for the above operators?

115. Explain the types of control statement GO TO.

116. Explain the types of control statement IF.

117. What for a DO statement is used?

118. What is the general form of DO statement?

119. Show the operation of DO statement in a flow sheet.

120. What are the restrictions or the use of the DO statement?

121. What is CONTINUE statement and where is it usually used?

122. Give an example of DO statement with CONTINUE statement.

123. What is a DIMENSION statement?

124. What is the general form of DIMENSION statement?

125. Give an example of DIMENSION statement.

126. What are the restrictions to be followed for a subscripted variable?

127. Indicate the errors in the following subscripted variables :

(i) PQRSTU(15) (ii) F (0); (iii) Z(–l)

(iv) R(1.8); (v) J(R); (vi) 4L(K/2)

128. Find out the correct and incorrect DO statements from the following. Indicate the mistake in incorrect statements,

(a) DO 30 KYY = J, K, 3 (b) DO MN 5 = R, I, 2

(c) DO 2Z = J, M, M (d) DO 100 N = K, M, K.

(e) DOK = THREE T ϕ 5 (f) DO21 I = 38

(g) DO 26 J = I, I, I (h) DO 3 B = 1, N5 B, 2

129. Find out the correct and incorrect statements from the following:

(a) JAM (JUG) (b) DIMENSION X (15), F (K)

(c) X(l*2) (d) X(2*l) (e) ARR(JOHN)

(f) DIMENSION Y(10), 2 (100), P (40).

130. What is the general form of an Implied DO statement for one-dimensional array?

131. Where for an implied DO statement is used?

132. Convert the following READ statement into an equivalent. Implied DO statement

READ (5, 10) Y(l), Y(2), Y(3), Y(4), Y(5), Y(6)

133. Read and subscripted variables A(l), A(4), A(7), A(10) A(13), A(17), A(20) using implied DO statement.

134. Write the FORTRAN statement that will result in the following out-put/input:

(a) Skip 3 lines, output the values of A and B in a field of F 15.6 each.

(b) Read in 400 values X into an array. The data are punched 8 to a card, each value is 10 columns long.

135. What is a subprogram?

136. How many types of subprograms are there and what are they?

137. What is the advantage of subprogram?

138. Briefly explain the types of subprograms.

139. Write a function subprogram to obtain the absolute value of a variable X.

140. Examples of function subprograms are given below. If the statements are valid mark correct, otherwise make suitable corrections.

(a) F (X, Y, Z) = X + Y – Z
A = C – F(B(1), 2.0)

(b) XINC (A, R, N) = A* (1. = R**N)
AMT = SUM – XINC (100., 0.06, 30)

(c) ZEPO (B, E, M, 4) = B – E* M/4.0
Y = ZEPO (1.0, 2.0, 3.0, 4.0).

(d) FUNCTION FLOW (AREA, RAD, SLOPE, COEF)
Q = 1.5* AREA* (RAD" (2./3))*SQRT (SLOPE)
FLOW = Q/COEF
RETURN
END

(e) M = 6
PAM (CM, XI) = CM* (XI* (1. + XI)**M).

141. Write a function subprogram to determine the average of N numbers.

142. What is general procedure of using a subroutine?

143. Write a subrouting to add two variables A and B and store in third variable C.

144. What is the purpose of COMMON statement?

145. What is a flow sheet/diagram/chart? What is the use of it?

146. What are the various symbols used in a flow chart?

147. Construct flow charts to solve the following problems :

(a) Sum of N integers from 1 to N

(b) Average of N integers

(c) Arrangement of numbers in descending order.

148. What are a logical variable and a logical constant?

149. What are known as the logical operators? Explain their meaning.

150. How do you define logical variable.

151. What is DATA statement?

152. The following values are to be read for the variables given below. How do you make use of DATA statement B = 4.9, F = 5.2,1 = 4, M = 90.

153. What is a double precision statement? Why is it used?

154. Add the number 0.005 and 5.3 × 10 using double precision statement.

155. How do you write complex variables in FORTRAN program?

156. How do you write the following complex numbers in FORTRAN :

4.8 + i 9.8, 3.9 + i 5.4

157. Add the complex variables A = 8.9 + *i* 9.8 and B = 10.5 + *i* 5.9 using FORTRAN statement

158. What is EQUIVALENCE statement?

159. What is the general form of EQUIVALENCE statement and explain the meaning of the statement?

160. What is the FORMAT used to read Alphanumeric data a string of characters?

161. What is the general form of A FORMAT?

162. Read the following variable names using A FORMAT

(a) CART (b) CEMA

(c) SHIVCHARAN (d) ARYABHATTA

163. What is the other method used in analysing the problem instead of flow diagram?

164. Write the subroutine program for finding the roots of a quadratic equation.

165. Tabulate the function

$$f(x, y) = x^2 + y^2 + 2xy/(x^2 + y^2 + 8xy)$$

or the following set of values :

x	0	+2	4	6	8	10	12
y	−10	−8	−6	−4	−2	0	2

166. Write a program to calculate to three significant digits the roots of the equation sin x – x + 1 = 0 using Newton-Raphson technique.

167. Write a program to express a quantity in millimeters in meters, centimeters and millimeters.

168. Write a program for fitting a straight line through a set of points using least squares method.

169. Write a program to pick the largest of K numbers.

170. Write a program to sum the series of

$$y = 1 + x + x^2 + x^3 + x^n$$

using IF statement.

ANSWERS

1. Charles Babbage

2. Analytical Engine.

3. J. Ecket Jr and J.W, Manchly of U.S.A.

4. ENIAC (Electronic Numerical Integrator and Calculator)

5. There are computers using vacuum tubes for their operation.

6. These use— transistors for operation.

7. Integrated circuits (ICs) are used in these computers.

8. These computers make extensive use of large scale integrated circuits known as LSI (Large Scale Integrated Circuit) and VLSI (very large scale integrated circuits).

9. Binary number system.

10. John Von Neumann

11. Analog, digital and hybrid computers.

12. Input unit, output unit, Memory and central processing unit.

13. Control unit and Arithmetic unit comprise the central processing unit.

14. Console is the unit attached to a computer which may be used by the computer operator to communicate directly with the operating system.

15. It is with respect to size. It generally means a computer whose main frame is essentially small i.e., no more than a few cubic feet.

16. A microcomputer is one which is very small in size and its central processor called microprocessor is fabricated on a single semi-conductor device.

17. This functions by establishing similarity between an easily measurable phenomenon with another phenomenon not so easily measurable.

18. A digital computer is one which deals with discrete characteristics like digits.

19. It has better speed more precision, less error and it does not require a physical conception of the problem.

20. The various electronic circuits and the component comprising them associated with the operation of the computer is called hardware. The programs or set of instructions required to operate a computer to solve any problem is called software.

21. Punch card, punch paper tape, magnetic tape, optical character reader, console type writer and cathode ray tube.

22. Information is stored in the memory of the computer and supplied to other units whenever required.

23. Both refer to memory of a computer. Internal storage is an essential part of the memory unit and auxiliary storage supplements the main storage.

24. The devices used are magnetic core memory, thin-film memory, plated wire memory and thin rod memory.

25. The data stored in magnetic tape drive, magnetic disc drive and magnetic drums.

26. Line printer, graph plotter, visual display unit, Card punch, paper tape punch, console tape writer, magnetic tape, magnetic disc and magnetic drum.

27. A computer comprises input instructions, instructions to control the processing of input data and output instructions that specify the output information and the output device.

28. The coding of instructions in a definite language under standable by a computer is called computer language.

29. Low level languages are dependent on the design of machine and there is direct (one to one) correspondence between instruction and control circuits. These languages are also known as machine languages.

30. High level are machine independent languages and for a single instruction in the language there can be many operations involved by the computer.

31. A compiler is a program written in machine language. It translates statements written in high level language into machine language commands.

32. They relieve the programmer the task of organising storage for his program and data. They reduce the number of computer instructions by a factor of four or five. They relieve the programmer of the necessity of the following fhe detailed operation of his program by the computer.

33. FORTRAN, BASIC and COBOL.

34. FORTRAN –Formula Translation, BASIC—Beginners All purpose Symboic Instruction Code, COBOL—Common Business Oriented Language.

35. TDC—(Tronbay Digital Computer).

36. Binary Digit.

37. By tes or Kilo Bytes.

38. The initial program written in a higher level language is called source program. The corresponding machine language obtained by a compiler is called object program.

39. Binary, Octagonal, Hexagonal

40. FORTRAN

41. 47.

42. AI the capital letters of English language (A to Z, 26 in number).

43. The decimal digits 0, 1, 2, 3, 4, 5, 6, 7, 8, 9, (10 in number).

44. + (addition), – (subtraction), / also called slash and used for division), * (called STAR and used for multiplication.

45. '('[left parenthesis],')' [Right parenthesis],'/ ((coma),: (Period), = (equal), '/' (quote), bar () [Blank space].

46. These are the rules according to which only, the statements and instructions to the machine can be recognized by a machine/computer.

47. The removal of syntax errors is called DEBUGGING.

48. FORTRAN constant is the name given to a quantity whose value does not change during calculations.

49. FORTRAN variable is the one whose assigned value can go on changing as per dictates of the program.

50. FORTRAN constants – 15, 20. 5, 2.3 ; FORTRAN variables-N, XY, M 35, MAX, NUM.

51. Integer constants/variables and real constants/variables.

52. Integer constants are whole numbers without any fractional part with it. It must be written without a decimal point. Real constant may contain a whole number or a fractional part and includes a decimal point.

53. Floating point constant.

54. Fractional form and Exponential form.

55. Mantissa and Exponent.

56. The Mantissa must have at least one digital and a decimal point. The sign (+ ve) or (– ve) must precede Mantissa. The Mantissa must be followed by letter E and then the exponent. The exponent must be an integer and must have at least one digit. The exponent must be preceded by + ve or – ve sign.

57. 10^{-610} to 10^{+610}

58. The Mantissa must not contain more than seven digits.

59. Mantissa 5 A ; 0.63 ; 6.56 ; 664 ; Exponent: 02, - 03, + 8, + 10

60. Valid : (*C*), *f*: Invalid : (a), (b), (d), (e).

61. Integer variable and Real variable.

62. The first character of integer variable is one of the six letters I, J, K, L, M and N. The real variable is identified by the absence of the above six letters at the beginning of the variable name. All other alphabets A to Z are permitted.

63. The fixed point variable is the other name for integer variable while the Floating point variable is another name given for real variable.

64. The variable should start with appropriate alphabet. The variable name should not exceed five alphabets or numerals. No special characters should be present in the variable name.

65. Integer variables (a), (b), (d) and (f); Real variables : (c), (e) and (g).

66. (a) Invalid-special character+is present

(b) Invalid-Numeral is present at the beginning.

(c) Valid integer variable.

(d) Invalid special character, is present

(e) Valid-integer variable.

(f) Invalid-variable contains more than 5 characters

(g) Valid integer variable

(h) Valid real variable

(i) Invalid-contains more than five characters.

(j) Valid real variable.

67. Type declaration statement is used to convert an originally integer variable into a real variable and *vice versa.*

68. They are converted into real variables, by prefixing the word REAL, before the variable name, like REAL LINE; REAL NOTE ; REAL MITE.

69. They are converted by writing INTEGER before the variable name like INTEGER STOP, INTEGER FLOW.

70. + (Addition), – (Subtraction), *(Multiplication); /(Slash for division) and "(Exponentiation).

71. Two types : Integer Arithmetic Expression and Real Arithmetic Expression.

72. Integer-(a), (c), (d) and (f) ; Real-(b) and (e).

73. The order in which a computer executes arithmetic operations is called Heirarchy of operations.

74. (a) Z = 8.*Y**3.+48.**4.6

(b) Y = (X**A – Z**B)/(A – B)

(c) A = 0.00038/(P + R)

(d) N = 5*K – 4*M**4

(e) J = 3*L + 6*N**4 – 4*I/6*M

(f) M = 4*]**5/(3*N + 6*L)

75. Sin x – SINX, cos *x* – COS X; | X | – ABS(X)

$\tan^{-1} x$ – ATAN (X), e^x – EXP(X), Log_e x – AL ϕG(X)

$\sqrt{x}$ – SQRT(x), tan x – TAN(X), cotan x – CϕTAN(X).

76. (a) S = ABS(B)*EXP((Y/SQRT(3.))*Z)

(b) H = 2.*ALOG(1. + SIN(THETA))

(c) S = ATAN (EXP (– 5 *X))

(d) F = COS (ALOG (Y) + TAN (SQRT (PHI)).

77. The angle should always be expressed in radians.

78. $\theta\text{(in radians)} = \dfrac{\text{T (in degrees)}}{180} \times 3.143$

79. THETA = T*3.14/180.

80. (a) Arithmetic statement (b) Control statement

(c) Specification statement (d) Input/output statement.

81. (a) 19 (b) 42.

82. S = A/(B*(B+A))

T = S*B

R = T*S/B

83. REAL N

P = 2.15

Q = 2.5

N = 0.8

A = 2.93*2.*P*((4.*P**2))/Q–AL.OG 10(P* (SQRT(5.)–N)+43*(P*(ALOG (3.*SQRT (3.–N)) + 4.3*Q)

84. 51.

85. J = 6 – 0.6 + 3 + 0.5 + 2*3 + 2*5

= 6 – 0.6 + 3 + 0.5 + 6 + 10

All fractional values are made zero = 25.

86. J = (20/8)*5 + 5(20/16)* 6 + (20/10)** 2 + (20/16)** 3

= 2.25*5 + (1.25)*6 + 2**2 + (1.25)**3

All decimal numbers are truncated to integers like 2.25 is truncated to 2

= 2*5 + 1 × 6 + 4 + 1**3

= 10 + 6 + 4 + 1 = 21

87. An input statement is necessary to transfer data from one of the input units of computer to the memory of the computer.

88. It is used to transfer the results stored in the memory of the computer to one of its output units.

89. An executable part and a declarative part.

90. The executable part commands the computer to read the required data in case of input statement or to print out the results in case of output statement. The declarative part gives information to the computer on the exact form or manner in which the data will be found (say from punch cards) in case of input statement and the direction to the computer to printout the results in the form specified in the declarative part.

91. (READ) and (WRITE and PRINT) statements are execution statements for input and output respectively. FORMAT statement is the declarative part.

92. READ F, G, H, P, Q.

93. PRINT F, G, H, P, Q or WRITE F, G, H, P, Q.

94. Integer form is used for feeding integer numbers. Real form is used for feeding real numbers. These formats are used to print numbers only.

95. Fractional form and exponential form.

96. The length of space allotted to each data (variable) is called Field/Field length.

97. The group of variables specified in a READ, PRINT or WRITE statement is called LIST. The contexts of one card to be read or the group of values to be printed in one line is called a RECORD.

98. READ 25, K, KITE, KING, MIG, LINE, LINK 25 FORMAT (6 1 8).

99. READ 25, K, KITE, KING, MIG, LINE, LINK

25 FORMAT (16, 2X, 16, 3X, 17, 3X, 17, 4X, 18, 3X, 18). Note X indicates blank space. The number before X indicates the number of blank spaces.

100. READ 30, R, SP, PIN, RAP, RING

30 FORMAT (5F 16.9) (Fractional form) or

30 FORMAT (E14.5, E10.2, E14.6, E15.6, E11.4).

101. E type format specification must satisfy the condition W ≥ d + 7. It means that the total width of the field (W) must exceed the number of digits (*d*) required to the right of the decimal point by at least seven. E16.9 indicates that the total field width is 16 spaces and 9 digits can be read/printed after decimal.

102. PRINT 40, KODE, RING, PIN, LIM, RIP

40 FORMAT (16, 2F 10.4,18, F10.4).

103. FORMAT (13, F6.3, E10.3,13, F5.2, E12.4).

104. A = 193.89 E + 5; B = –0.593 E – 8;

C = 593.4 E + 09.

105. Alphanumeric data can be written in Format statement within the symbols (and).

106. PRINT 50, B

50 FORMAT (3X, 'FINAL VALUE OF B =', F10.4).

107. The first element in the list of the Format statement of a PRINT statement is the carriage control information and is an essential part of information.

108. 1X or 1H –Single space before printing.

1HO –Double space before printing.

1H1 –Skip to next page before printing.

109. They are used in FORMAT codes to deal with lists which are to be read from more than one card or which are to be printed in more than one line.

110. 80 characters for input, 120 characters for output in case of a line printer.

111. (a) The computer reads P and Q from the first card, R and S from the second card, T from the third card.

(b) P and Q are printed in the first line, R in the second line, S in the third line and T will be printed after skipping a line after third line (i.e., T will be in 5th line).

112. The FORTRAN statements may be classified into two classes in general *viz.* (a) Executable statements and (b) Non-executable statements. Ex-ecutable statements are compiled into machine instructions by the com-piler. Ex : arithmetic statements, input—output statement (READ, PRINT etc.) and control statements Non-executable statements provide information to the compiler regarding storage requirements and input and output data formats. Ex : Type declaration statement, Format statements etc.

113. The symbols >, <, ≥, ≤, = and ≠ are called relational operators.

114. > → G.T.; < → L.T.; ≥ → G.E.;

≤ → L.E.; = → EQ; ≠ → NE

115. There are three types

(a) GO TO n where n is the statement number.

(b) Computed GO TO statement written as

GO TO (n_1, n_2, n_3, ... n_n), i where

n_1, n_2 .... n_n are statement numbers.

and i is an integer variable name. The statement numbers need not be distinctive.

(c) Assigned GO TO statement written as

ASSIGN K to i

GO TO i, (p, q, r, s, t)

where k is a statement number, and i is an integer variable name, p, q, r, s, t are statement numbers.

116. There are two types IF statements

(a) Arithmetic IF (b) Logical IF

(a) Arithmetic IF statement is written as IF (a.e) m_1, m_2, m_3 where a.e stands for arithmetic expression and mj, m_2, and m_3 are statement numbers. If the arithmetic expression is negative, control is transferred to statement m_1 and if it is zero to m_2 and if it is positive to m_3.

(b) Logical IF is written as $\underset{P_2}{\text{IF}}$ (e) P_1

where e is a logical expression and P_1 and P_2 are FORTRAN statements. If e is true, control is transferred to P_1 and if false to P_2.

117. A DO statement is used when a group of instructions is to be performed again and again.

118. The DO statement has the general form DO n J = m_1, m_2, m_3 where n is a statement number. J is an integer variable, called the control variable. m_1 (initial value), m_2 (terminal value) and m_3 (increment) are unsigned integer constants or variables. If m_3 is omitted from the DO statement, it is assumed to be 1.

119.

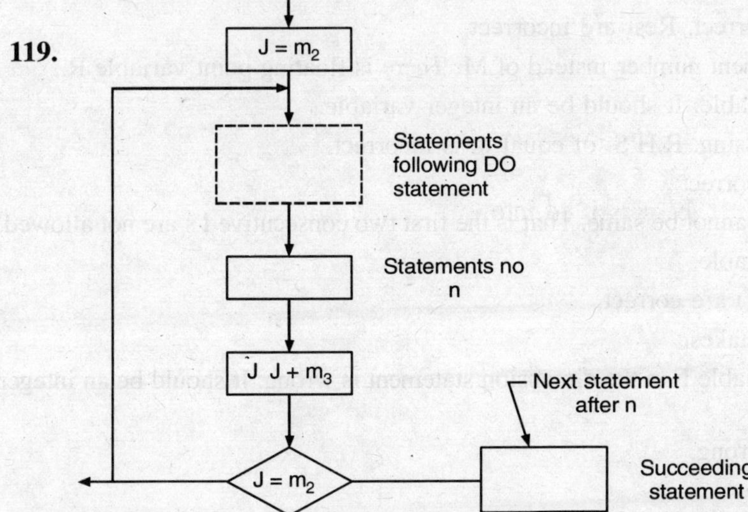

120. The restrictions are (a) we must not use any statements within the range of the DO statement that alter the value of the control variable (b). There should not be transfer of control into the DO loop from outside the loop (c). The last statement in a DO loop must not be one that causes transfer of control.

121. The CONTINUE statement is a dummy statement which causes no operation to occur. It is generally used with DO statement in order to avoid the transfer of control in the DO loop from the last statement.

122.
```
DO 5J = 5, 30, 5 ┐
.................│
.................├ Statements
.................┘
15 CONTINUE
```

123. The DIMENSION statement is a declarative statement and it must appear at the beginning of any program in which subscripted variables are used.

124. The general form of DIMENSION statement is DIMENSION name 1 (dimension 1), name 2 (dimension 2), name 3 (dimension 3).

125. DIMENSION RATE (100), CAM (150), RAG (100).

126. The subscript must always be a positive integer; 0 (zero) and negative subscripts are not allowed; Floating point subscripts are also not allowed.

127. (i) Too many characters in the name
(ii) Zero in the subscript
(iii) – ve number in the subscript
(iv) Floating point constant in the subscript
(v) Floating point variable in the subscript.
(vi) The name starts with number.

128. Statements (a) and (d) are correct. Rest are incorrect.
(b) There should be a statement number instead of M. There is floating point variable R.
(c) Z is a floating point variable. It should be an integer variable.
(e) Statement number is missing. R.H.S. of equal to is incorrect.
(f) R.H.S. of equal to is incorrect.
(g) The starting and ending cannot be same. That is the first two consecutive I s are not allowed,
(h) B is a floating point variable.

129. Statements (a), (d), (e) and (f) are correct,
(b) and (c) are incorrect mistakes.
(b) The subscript (k) for variable F in the dimension statement is wrong. It should be an integer number.
(c) The subscript (1*2) is wrong.

130. The general form of Implied DO statement is

(Name (J), J = m_1, m_2, m_3)

where Name is any array name ; J is any integer variable name: and m_1 (initial value), m_2 (terminal value), and m_3 (increment) are integer constants or variables. The last of these 'm_3' can be omitted and assumed to be 1.

131. The implied Dϕ statement is used only in the list of variables in READ or WRITE statements.

132. READ (5, 10) (A (I), I = 1, 6)

133. READ (5, 20) (A (I), 1 = 1, 20, 3)

134. (a) PRINT (6, 10) A, B

10 F ORMAT (////, 2 F 15.6)

(b) READ (5, 20) (X (I), I = 1,400)

20 FORMAT (8F10.0/8 F 10.0/8 F 10.0/8 F 10.0/8 F 10.0)

135. This is a procedure which is performed several times in a program.

136. There are four types of subprograms. They are

(a) Statement functions

(b) Intrinsic functions

(c) External function and

(d) External subroutines.

137. Once we have defined a subprogram, we can use it again and again without writing it each time.

138. A statement function is used to evaluate an arithmetic expression. This consists of a single statement. If more than one statement is necessary in defining a function, then a Function subprogram is used.

This intrinsic and external functions are FORTRAN statements. When they are used in a program, they are called in form a library of function subprograms. Ex. A B S finds the absolute value of a single floating point expression; SIN; ATAN etc. The library functions available have to be found from the manual available for a particular type of computer.

A subroutine subprogram is a more general form of a function subprogram. The difference is that while a function subprogram has the specific job of finding a value for some expression, a subroutine does not have this restriction.

139.
```
FUNCTION A B(X)
IF(X.GE.OO) GO TO 15
AB = – X
RETURN
AB = X
15 RETURN
END
```

140. (b) and (d) are correct.

(a) The second statement should be

A = C – F (1, 0, B(1), 2.0)

because there are three arguments in F

(c) ZEPO (B, E, M, P) = B–E*M/4.0

because a constant should not be in the argument of the function.

(e) M should also be present in the argument *i.e.*,

PAM (CM, XI M)

141. C–FUNCTION SUBPROGRAM FOR AVERAGE

```
FUNCTION AV (X, N)
DIMENSION X(100)
SUM = 0.0
DO 20 I = 1,N
20 SUM = SUM + X(I)
AV = SUM/N
RETURN
END
```

142. A subroutine is defined with a SUBROUTINE declaration statement. This statement has the form SUBROUTINE NAME (x_1, x_2,...... x_n) where name is any FORTRAN variable name and x_1, x_2, ... x_n are FORTRAN variables. The first statement in any subroutine must always be a SUBROUTINE statement of the above form. After this statement, we can write any instructions to tell what the subroutine is to do. These statements will be followed by RETURN and END statements.

To execute a subroutine, we use a CALL statement in a main program. This statement has the form CALL name (y_1, y_2, ..., y_n) where name is the name of a subroutine, as defined in SUBROUTINE statement, and y_1, y_2, ... y_n are any FORTRAN expressions.

This statement transfers control from the main program to the first executable statement in the subroutine name. The expressions y_1, y_2....y_n correspond to the variable names x_1, x_2, ... x_n. The subroutine will be executed with each y_1 taking the place of the corresponding x_1 in the definition of the subroutine. They must agree in number and type with the x's. The computer continues to execute the subroutine until a RETURN statement in the subroutine is executed. When this happens, control returns to the first statement after the CALL statement that called the subroutine.

143. SUBROUTINE A D D (A, B, C)

```
C = A + B
RETURN
END
```

144. The COMMON statement is a statement which allows the same storage allocation for desired variables in a subroutine, in the main program or another subroutine. This statement has the form COMMON A, B, 0, 0, 0, N.

145. A flow chart is simply a way to show a program in a diagram. It is helpful to represent what is going on in a form that is easy to read and work with. It is useful to flow chart a problem before trying to write a program to solve it.

146. The various symbols and the corresponding meanings are as below :

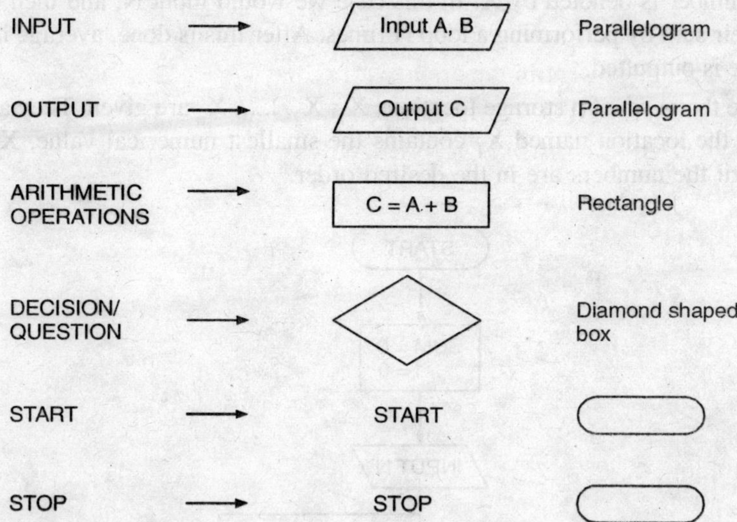

147. (a) Sum of N integers from 1 to N.

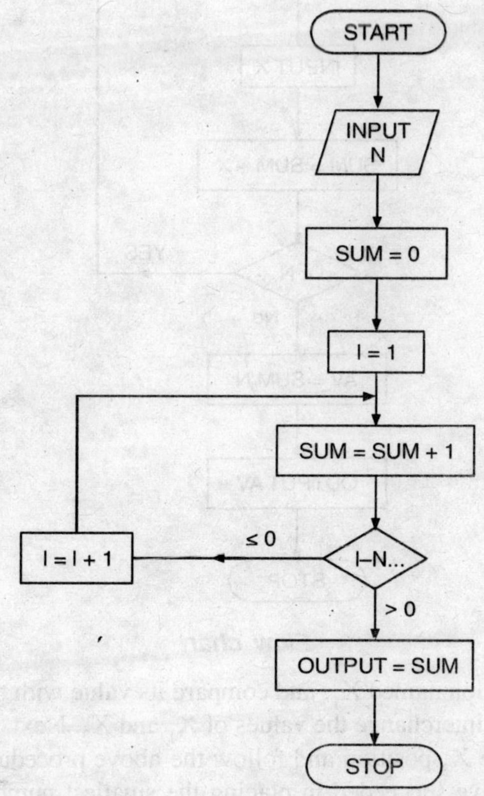

(b) Averaging of N numbers.

Each number is denoted by X. In this case we would input N, and then input the items and find their sum by performing a loop N times. After this is done, average is computed and the average is outputted.

(c) Assume that a set of n storage locations X_1, X_2, ... X_n are given. They are to be rearranged so that the location named X_1 contains the smallest numerical value, X_2 the next smallest etc. until the numbers are in the desired order.

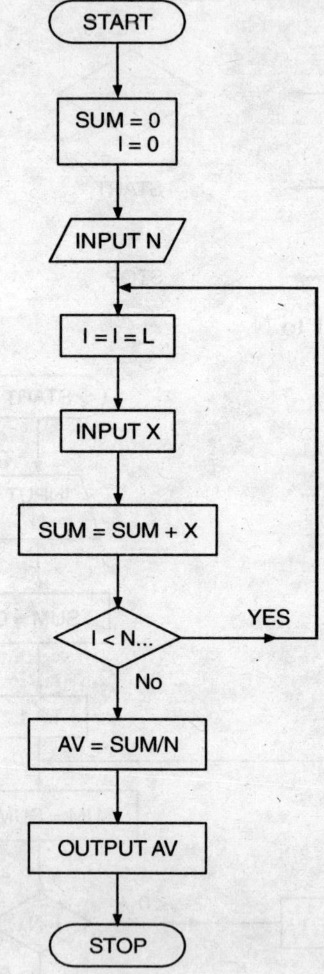

Flow chart

Start with the first location named X_1, and compare its value with that of X_2. If $X_1 \leq X_2$ make on change, but if $X_1 \geq X_2$, interchange the values of X_1 and X_2. Next, compare the number in the X_1 position with that in the X_3 position and follow the above procedure. After finishing all with all n numbers, we shall have succeeded in placing the smallest number in the first position. Next, compare the number in second position, X_2 with X_3 and either leave them alone or interchange them. Continuing as before, we eventually have the second smallest number in the X_2 position. Next, go to the X_3 position and continue as before. When we reach the n – 1 position, the procedure is complete.

A new variable KEM has been introduced to accomplish the interchange.

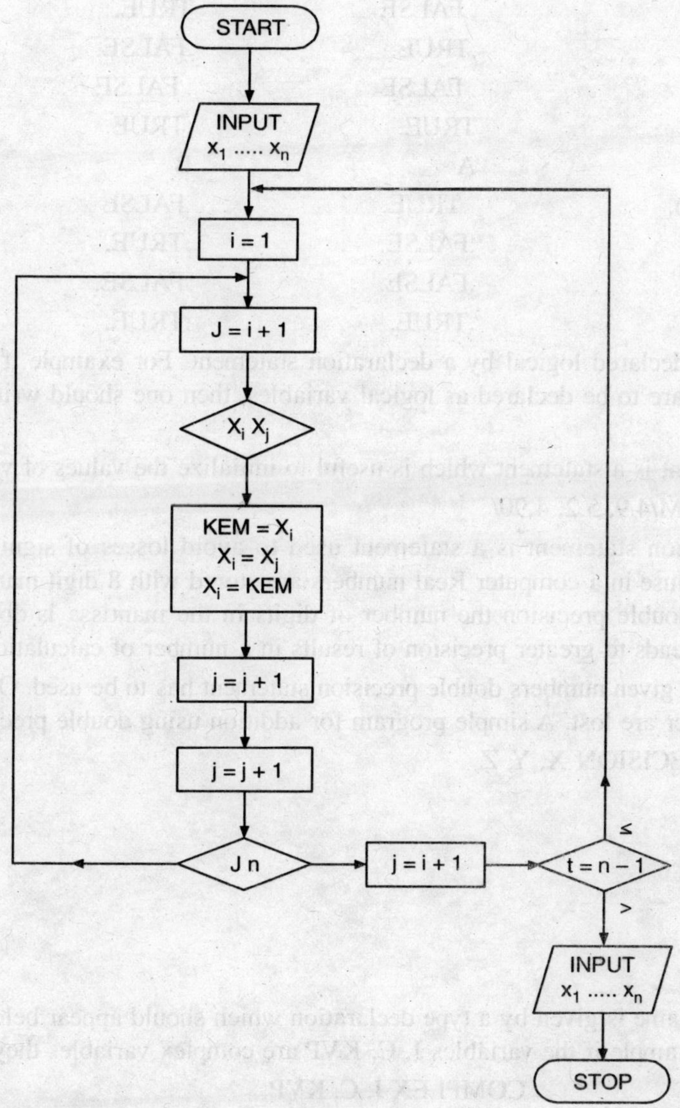

Flow chart

148. A logical variable is a two-valued variable, and it can assume one of two values TRUE or FALSE. A logical constant assumes either TRUE or FALSE.

149. The logical operators are NOT, AND, and OR. Their definition is given by examples.

.NOT	A	.NOT.A	
	.TRUE	.FALSE.	
(Compliment or Negation)	.FALSE	.TRUE	
.AND	A	B	A.AND.B
Logical	. FALSE	TRUE.	.FALSE
Intersection)	.TRUE.	.FALSE.	.FALSE.
	. FALSE	. FALSE	. FALSE
	TRUE.	.TRUE.	.TRUE.
.OR	A	B	A.OR.B
(Logical union).	.TRUE.	.FALSE	.TRUE.
	.FALSE.	.TRUE.	TRUE
	.FALSE.	.FALSE.	.FALSE
	.TRUE.	.TRUE.	.TRUE.

150. Variables are declared logical by a declaration statement. For example if the variables NAM, MAK, SEAR are to be declared as logical variables, then one should write LOGICAL, NAM, MAK, SEAR.

151. DATA statement is a statement which is useful to initialize the values of variable names.

152. DATA B, F, I, M/4.9, 5.2, 4,90/

153. Double precision statement is a statement used to avoid losses of significant digits. This is necessary because in a computer Real numbers are stored with 8 digit mantissa and a two digit exponent. In double precision the number of digits in the mantissa is doubled i.e., to sixteen digits, which leads to greater precision of results in a number of calculations.

154. For adding the given numbers double precision statement has to be used. Otherwise the digits of the first number are lost. A simple program for addition using double precision is as below.

```
DOUBLE PRECISION X, Y, Z
X = 0.005 DO
Y = 5.3 D 10
Z = X + Y
PRINT Z
STOP
END
```

155. The variable name is given by a type declaration which should appear belore the variable name is used. For example if the variables J, C, KVP are complex variables they are declared as

COMPLEX J, C, KVP

156. The number are written as

(4.8, 9.8); (3.9, 5.4)

157.
```
COMPLEX A, B, C
A = (8.9, 9.8)
B = (10.5, 5.9)
C = A + B
```

158. EQUIVALENCE statement is a statement which enables the programmer to use the same locations in memory for more than one variable at different times during the execution of a program.

159. The general form of EQUIVALENCE statement is

EQUIVALENCE (p, q, r, s),(d, e,f) where p, q, r, s, d, e,f... are all variable names. This means the variables p, q, r, s would be assigned identical memory locations so also d, e, f. So, the programmer should be careful to see that the equivalenced variables are not used simultaneously.

160. A—FORMAT

161. The general form Aw. causes w characters to be read into a memory location identified by the variable name. Normally, it is safe to allocate 4 characters per real variable name. If the number of characters in a string exceeds 4, it is convenient for use as a subscripted variable.

162. (a) READ 15 C
15 FORMAT (A 4)
Data card CART

(b) READ 20, C
20 FORMAT (A 4)
Data card CEMA

(c) READ 30, (A (I), I = 1, 3)
30 FORMAT (3 A 4)
Data card SHIVCHARAN
will store SHIV in A (1), CHAR in A (2) and ANbb in A (3)

(d) READ 40, (A (I), I = 1,3)
40 FORMAT (3 A 4)
DATA Card ARYABHATTA

163. Decision table

164.
```
SUBROUTINE ROOTS (A, B, C, XII, X12, XR1, XR2)
D = B * B–4. * A * C
IF (D) 12, 22, 32
12 XII = 0.5 * SQRT (–D)/ A
XI2 = –XI1
XR1 = –0.5 *B/A
XR2 = XR1
RETURN
22  XII  = 0
    XI2 = 0
    XR1= – B * 0.5/A.
    XR2 = XR1
    RETURN
32  XII = 0
    XI2 = 0
    XR1 = 0.5 * (– B + SQRT (D))/A
```

```
      XR2 = 0.5 * (-B-SQRT (D))/A
      RETURN
      END
```

165.
```
C     TABULATING FUNCTION
      READ (X (I), I = 1,7)
      READ (Y (I), I = 1, 7)
      N = 7
      DO 20 I = 1,N
      FXY (I) = X(I) ** 2+Y(I) ** 2 + 2 * X(I) * Y(I)/(X(I) ** 2+Y(I) ** 2+8. * X(I) * Y(I)
      X(I) * Y(1)
   20 PRINT, FXY(I), X(I), Y(I)
      STOP
      END
```

data : 0., 2., 4., 6., 8., 10., 12. – 10., – 8.,-6.,– 4., – 2.,0.,2.

166.
```
C     NEWTON-RAPHSON METHOD
      X = 0.1
    1 F = SIN(X) - X + 1.
      F DOT = COS(X) - 1.
      X1=X-(F DOT/F)
      IF (ABS(Xl-X). GT. 0.001) X=X1, GO TO 1
      PRINT, X
      STOP
```

167.
```
C     CONVERSION OF MM
      READ, AMM
      AM = AMM/1000.
      M = AM
      AM= M
      CM = (AMM-AM*1000.)/10
      ICM = CM
      CM = ICM
      MM = AMM - (AM*100.) - (CM*10.)
      PRINT, AMM, M, ICM, MM
      STOP
      END
```

168.
```
C     FITTING A STRAIGHT LINE
      REAL M, NUMR
      DIMENSION X(100), Y(100)
      READ, N
      SUM X = 0.
      SUM Y = 0.
      SUM XY = 0.
```

```
      SUM X SQ = 0.
      READ, (X(I),Y(I), I=1, N)
      DO 10 I = 1, N
      SUM X=SUM X+XM
      SUM Y=SUM Y + Y(I)
   10 SUM XY = SUM XY+X(I)*X(I)
      AN=N
      NUMR=AN*SUMXY–SUMX*SUMY
      DENR = AN*SUMXSQ–SUM X**2
      M = NUMR/DENR
      C = (SUMY – M*SUMX)/AN
      PRINT, M, C, N
      STOP
      END
```

169.
```
C     PICKING LARGEST
      READ, K
      READ, A
      BIG=A
      K = K-1
      DO 15 I=1,K
      READ, A
   15 IF (A. GT.BIGBIG=A)
      PRINT, BIG
      STOP
      END
      Data Cards
```

170.
```
C     SUMMING SERIES
      READ, X, K
      N=1
      SUM=1.
      Y=l
    2 Y = Y*X
      SUM=SUM+Y
      N=N+1
      IF(N–K)2, 2, 10
   10 PRINT, SUM
      STOP
      END
```

❑

C *denotes not to be processed by computer*
20 *denotes statement number*
15 *denotes statement number*
10 *denotes statement number*
2 *denotes statement number*
1 *denotes statement number*

CHAPTER 13

Petroleum Engineering

1. Occurrence of petroleum: Occurs in the earth's crust in all possible states in colour from light brown to dark brown or black. It is a mixure of various hydrocarbons of homologous series name paraffins, naphthalenes and aromatics. The Main elements are: C – 84 to 86%; H_2 – 11 to 14%; others 8%. Petroleum means oil obtained from earth. Middle east accounts 56.3% of crude.

2. Origin and formation:

1. Mendeleeff and Berthlet explained as inorganic reactions.
   $CaC_2 + 2H_2O \rightarrow C_2H_2 + Ca(OH)_2 : Al_4C_3 + 12H_2O \rightarrow 3CH_4 + 4Al(OH)_3$.
2. Cosmic hypothesis of V.D. Sokolov depicted that vapours of hydrocarbons were present. Precipitation of clouds, rained hydrocarbons were either absorbed or entrapped in earth's crust.
3. J.D. Haun organic theory: a – Homologous series present in petroleum are seen in organic matters; b – Nitrogen components in petroleum are also of organic nature; c – The isotopes of carbon in petroleum are also of organic nature; d – Optical activity exhibited by natural organics is exhibited by petroleum is associated with sedimentary rocks; g – The formation of petroleum is normal, continuous and does not require severe physical and chemical conditions; h – Most organisms like di–atoms are found in petroleum.
4. Organic theories: Engler (1866). That Large marine animals which dwell on the globe in the prehistoric days were the main fat contributors. After extinction, the body fat was converted to hydrocarbons. This is not proved by laboratory tests.
5. Thomas Graham (1843) was of the view that natural naphtha must be a product of action upon vegetable matter of high temperature. Chaptal (1845) of opinion that when plants become decomposed into vegetable moulds, these contained oils which escaped decomposition. Draper (1846) action of natural heat of earth on coal.
6. A bright explanation by Mikhailovosky, N. Potering and N.C. Anderson.The intial decomposition of vegetable and animal matter was due to microorganisms. Later the organic matter underwent changes due to pressure and temperature of the earth's crust.

3. Types of crudes based on time of deposit:

Tertiary crudes → $11 * 10^6 – 74 * 10^6$ years of age (Eocene Oil)

Mesozoic Era Oil → $75 * 10^6 – 200 * 10^6$ years of age

Paleozoic Era crudes → $200 * 10^6 – 500 * 10^6$ years of age.

4. Organization of the Petroleum Exporting Countries (OPEC) (13): 1 – Algeria; 2 – Iran; 3 – Iraq; 4 – Saudi Arabia; 5 – Gabon; 6 – Kuwait; 7 – Ecudor; 8 – Libya; 9 – Indonesia; 10 – Nigeria; 11 – Qatar; 12 – United Arab Emirates; 13 – Venezuela.

5. Data on oil: Total crude reserves – 2200 * 10^9 barrels; gas 250 * 10^{12} Cu. M; production of oil : 3300 Million tons;

6. Indian petroleum Industry: First Refinery is 5 bbl capacity in Pittusburgh, U.S.A., 1888. Assam oil company's refinery at Digboi, 1954 (major oil refinery). Later came Cambay oil field (1959), Ankaleswar oil field. ONGC has the major path for the country's progress on oil front presently the LPG sources are 3.15 million tons.

Reserves are: KG Offshore — 13 MT oil and 3 Billion Cu.m. gas

KG Basin — 57 MT oil

KG basin wells are giving 600 bbl/day. Gas flow rate is 52,000 Cu. M per day. Coal, oil and natural gas are the primary commercial energy resources.

7. Composition of petroleum : Petoleum occurs in nature in all the three states, solid, liquid and gas. Carbon → 84 to 86%; Hydrogen → 11 to 14%. Petroleum (crude) contains:

Paraffins ($C_n H_{2n+2}$), olefins, acetylenes, diolefins, naphthalene, aromatics, sulphur compounds.

8. Statistics/data of India:

Oil wells completions, during 1975 : 90

Gas wells completions during 1975 : 6

Average depth in meters : 2171

Utilization of natural gas (10^6 Cu.M), 1974 : 1010

Oil produced, 1975 (10^3 barrels) : 1377, 919

Crude oil reserve (million tonnes) 1980 : 366.33

Natural gas (Billion cu . meters) 1980 : 351.91

Crude oil (000 tons) 1980–81 (onshore) : 5518

Crude oil (000 tons) 1980–81 (offshore) : 10503

Natural gas (million cubic meters); 1980–81 (onshore) : 1690

9. List of Indian Refineries: 1 – Assam Oil Company, Digboi; 2 – Bharat Petroleum Corporation Ltd., Mumbai; 3 – Hindustan Petroleum Corporation, Mumbai; 4 – Hindustan Petroleum Corporation, Mumbai (Caltex Vizag); 5 – Cochin Refineries Ltd., Cochin; 6 – Madras Refineries Ltd., Chennai; 7 – Indian Oil Corporation, Guwahati; 8 – Indian Oil Corporation, Barauni; 9 – Indian Oil Corporation, Baroda; 10 – Indian Oil Corporation, Haldia; 11 – Mathura Refinery (U.P.).;

12 – Bongaigon Refinery.

10. Characteristics of crude processed in India:

1. Nahorkatia (Assam): API – 31.0; Sulphur – 0.16 wt%; Viscosity 100 CS;
2. Ankaleswar (Gujarat): API – 47.0; Sulphur – 0.05 wt%; Viscosity – 25 CS;
3. Light (Iranian): API – 34.4; Sulphur – 1.35 wt%; Viscosity – 700 CS
4. Darius crude: API – 34.1; Sulphur – 2.47 wt%; Viscosity – 560 CS.
5. Digboi: Sp. gr – 0.84; Sulphur 0.20; Four point – 29.8°C
6. Bombay high: API – 39.2; Sp gr = 0.844; Sulphur – 0.12 wt%; Four point – 27–29°C.

11. Classification of petroleum:

1. Paraffinic crude (residue containing more than 5% paraffins). 2. Naphthalene base (residue containing less than 2% paraffin). 3. Mixed base (residue containing 2 to 5% paraffin).

API (American Petroleum Institute): Degree API = $\frac{1415}{\rho} - 131.5$ where ρ is specific gravity of fraction at 15.6°C/15.6°C.

U.O.P. characterization factor $(K) = \sqrt[3]{T}/\rho = \frac{\sqrt[3]{R}}{0.827\rho}$

ρ(Sp.gr.) at 80°F; R = average boiling point in K or ρ = Specific gravity of fluid at 15.6/15.6°C.

Correlation index (C.I) = $\frac{48640}{TB} + 473.7\rho - 456.8$ where T_B = boiling point (molal) K; ρ = Sp. gravity at 15.6°C/15.6°C;

Distillation characteristics of a crude are assessed by performing a preliminary distillation called: "True Boiling Point Analysis (TBP)."

Clausius–Clapeyron equation can be used where there is data available on latent heat of vaporization.

$$\ln\left(\frac{P}{P_0}\right) = \frac{\lambda}{R}\left(\frac{1}{T_0} - \frac{1}{T}\right)$$

P_0, P vapour pressures at T_0, T absolute temperature; λ = molal latent heat of vaporization at T; R = gas constant in corresponding units.

12. Thermal properties of petroleum fractions:

(a) Specific heat: 0.30 to 0.85 with increasing density specific heat decreases.

SP. heat = $\frac{1}{\rho}(0.4024 + 0.00081\, t)$

ρ is specific gravity at 15.6/15.6°C; t = temp, 0°C.

Specific heat = [0.355 + 1280 (API) * 10^{-6} + (5.03 + 1.17 * API) (32 + 1.8 t) * 10^{-6}] (0.05 K + 0.41)

Where K = characterization factor, t = °C.

For petroleum vapors, specific heat is:

(0.045 K – 0.233) + (0.44 + 0.01777 K) (18 t + 32) * 10^{-3} – 0.153 * $(1.8\,t + 32)^2$ * 10^{-6} where K is characterisation factor; t is °C.

Heat of combustion: Heating value kJ/kg = 43.434 + 93.2 (API – 10) Wobbe No. = Heat of combustion/(Specific gravity)$^{0.50}$

For different gases like coal gas, SNG etc. Wobbe No. lies in the Range of 21.5 to 38.7. For natural gas: Low calorific value: 37.1 – 43.7; High calorific value: 43.4 – 52.4; LPG 72.0 – 83.5.

Gas modulus = $\rho^{1/2}$/Wobbe No.

Latent heat of vaporization: $L = \lambda\ L_B\ T/T_B$; where L = latent heat at any temperature (T absolute); L_B = Latent heat at temperature T_B absolute; λ = correction factor depends on reduced temperature.

$$\Delta H_V = \Delta H_{Tg}\left[\frac{T_1 - T_\Delta}{T_C - T_2}\right]^n; \quad n = 0.38; \ T_C = \text{critical temp, °C};$$

The heat of fusion is nearly half of latent heat of vaporization. Average latent heat of fusion is: 167 to 170 kJ/kg.

Spontaneous ignition temperature: This is the temperature at which a material can catch fire and burn continuously without the aid of external firing agencies.

Viscosity and viscosity index (VI):

$$VI = \left(\frac{L-U}{L-M}\right) * 100;$$ L = Viscosity of reference oil of zero. Viscosity index at 37.8°C; H = Viscosity of a reference oil of 100 viscosity index at 37.8°C; U = Viscosity of test sample at 37.8°C.

13. Important products:

The main fractions from distillation of crude are:

1. Gas/light naphtha/gasoline → Solvents/fuels/motor spirits, etc.
2. Kerosenes (light distillates) → Jet fuels/illuminating oils.
3. Gas oils or middle distillates → Diesel fuels, catalytic cracking oils, etc.
4. Heavy distillates → Lubes/waxes/fuel oils, etc.
5. Residues → Asphalt, Bitumen etc.

14. Gas: Gas from petroleum is classified as: 1. Natural gas; 2. Associated gas; 3.Dissolved gas; 4. Casing head gas; 5. Refinery off gas; 6. Liquefied petroleum gas.

1. Contains mostly methane, 85 – 98%.
2. This is from oil reservoirs and exists as a gas cap over liquid phase. Consists mainly methane, ethane and propane.
3. Gas percent upto 10% of crude petroleum.
4. Gas that has escaped from oil well. It is bye-product of oil production. Similar to natural gas.
5. Gas formed in cracking and reforming operations. Wet gas is liquefied and sold to commercial purposes like butane. Dry gas is used in refinary fuel system.
6. This is supplied in cylinders (15 kg/12 kg).

15. Important tests:

1. Vapour pressure of sample: Found by Reid vapor pressure apparatus.

   LPG composition: Propane – 24.7%; i–butane – 36.7%; butane – 38.5%;
2. Tests for gasoline: 1. ASTM distillation; 2. Reid vapour pressure; 3. Octane number; 4. Gum content; 5. Sulphur content etc.

1. If the range of boiling point is high, gasoline will not be evaporating normally. If 90% boiling is high, then crank case dilution occurs as the fuel is not volatile at the working conditions of carburettor.

2. Vapour pressure of all volatile fractions is measured by the above Reid vapour pressure apparatus (38° ± 0.1°C). LPG gasoline, naphthas, Jet fuels are tested like this. Actual vapor pressure is (1.03 to 1.45) times more of RVP. For crude upto 10.
3. Defined as percentage volume of i - octane in a mixture of i – octane and n – heptane that gives the same knocking characteristics as the fuel under consideration. Knocking is due to humidity, Engine speed, air temperature, coolant temperature spark advance and combustion chamber deposits.
4. Gum is of two kinds. One is preformed gum (existent gum and the other is potential gum (ultimate). Existing gum is estimated by A STM D 381; IP 131 test. Ultimate gum content of gasoline is evolved by oxidising the sample with oxygen under restricted conditions.
5. More TEL (tetraethyl lead) is required if sulphur components are present in gasolines. The effect of sulphur on lead can be predicted by the equation:

   $$L = \frac{(a_0 - a)}{a_0} * 100 \text{ where } a_0 = \text{TEL present in gasoline.}$$

   With sulphur compounds; a = TEL present in gasoline, free of sulpbur compounds; L = % TEL excess required to boost the (octane number. Total sulphur is estimated by Bomb method or Lamp method and mercaptan Sulphur as Silver mercaptide. Sulphur gives corrosiveness and pollution.

16. Additives added to gasoline or to petroleum fractions for different functions. They are:

1. Anti-icings and detergents: To avoid formation of ice crystals in carburettor (in cold weathers) additives like glycols, monoethers, methanol, proponol are added. For lubrication, the chemicals added are: Maleic anhydride, polyamines, poly isobutylenes, phenol, epichlorohydrin, amine products are such detergents.
2. Inhibitors (corrosive): To prevent corrosion, the following inhibitors are: Ammonium Sulphonates, organic phosphoric compounds. Magnesium and silicon compounds are used for prevention of lead corrosion.

   Oxidation inhilitors: Antioxidants like phenols, cresols, phenylene diamines, alkyl amino phenols are used to control oxidation by metallic impurities.
3. Combustion adds: Boron compounds is small amounts, organophosphorous compounds, phosphates of lead.
4. Antiknocking agents: Tetraethyl lead below 0.05%.

17. Aviation turbine, fuels, jet fuels: Modern Jet Engines use fuel similar to kerosene. The tests prescribed are:

1. Volatility; 2. Pour point; 3. Smoke point; 4. Water retention specifications are: Bromine number max – 5. Color saybolt min – 8. Recovery upto 20% – 200°C max; Fuel boiling point max – 300°C; Residue% 2. Flash point, min – 38°C; Sulphur by vol. – 0.005; Sulphur total % by wt – 0.020.

The additives are: Polysulfones; organic Boron Compounds; phosphoric compounds; Glycol, Polyhydric alcohol.

18. Naphtha : These fall in the boiling range of motor spirits used as solvents in paints, perfumery and other industries General requirements of solvent naphtha are: 1. almost water white in color;

2. Non-corrosive nature; 3. High stability; 4. High solvent power; 5. Low boiling point; 6. Low acidity. Suspended impurities are removed by filtration. Solvent powerly aniline point.

19. Kerosenes: This is the name given to the group of refined petroleum fractions employed as fuel and illuminant. Boiling range: 150–250°C.

Tests of Kerosene are:

1. Flash point and fire point: Flash point is defined as the minimum temperature at which the vapours from oil sample will give the momentary flash on application of standard flame under specific test conditions. Abel, pensky, Martins, cleveland are the test apparatus.
2. Fire point: This is the temperature when the oil vapours can burn continuously for 5 seconds when tested in Flash Point apparatus and it occurs after the flash point by 3–4°C.
3. Smoke point: This is an indication of clean lurning Quality of kerosene. This is defined as the maximum height of flame in millimeters at which the given oil will burn without giving smoke. Smoke point = 0.46P + 0.32 N + 0.20 A; where P, N, A are% of paraffins, naptha lenes and aromatics in the sample.
4. Burning quality test: This in dicates the ability of kerosene to burn steadily and cleanly over a long period. Kerosene is tested by burning in a standard wick lamp for period of 24 hours, in a room free from air currents. A good oil of 650 c.c. must burn at least for 120 hours.
5. Volatility: The nature of distillation range estimated by ASTM D-86 is of significance with regard to burning characteristics. Ten percent boiling point reveals the flash point which in turn indicates ease of ignition.
6. Sulphur content: Total sulphur is estimated by bomb method maximum permissible amount of sulphur is 0.13% in all kerosenes.
7. Aniline point: This test indicates the amounts of aromatics present in kerosene.

20. Diesel fuels: These are the fractions in the boiling range of 250–320° and fell under gas–oil fractions. The general tests for diesel fuels are:

Classification of diesel oils is done according to speed and loads on the Engine.

Low speed diesel	Below 300 RPM	For heavy loads at constant speeds.
Medium speed	300–1000 RPM	Fairly heavy loads moderately at constant speed.
High speed	above 1000 RPM	Load and speed vary.

1. Pour point: Low pour points oils are essential. At close of approach of power point (2 to 3°C), the viscosity increases very much, the result of which is high pumping cost.
2. Aniline Point: This is defined as the minimum temperature at which equal volumes of anhydrous aniline and oil mix together. Diesel index = [0.018 A.P * C + 0.32] A.P. Diesel index is a measure of ignition quality of fuel].
3. Flash point: In India flash point of diesel is kept at 50–55°C. Knowledge of this is required for safe handling and storing.
4. Calorific value: Though high calorific value is desirble, it can not be at the expense of diesel quality. Hence a fuel having calorific value of 41.83 kJ per gm is sufficient.
5. Viscosity: Ease of starting depends upon viscosity and ignition quality.

6. Diesel additives: Diesel fuels are also blended like gesoline to give satisfactory performance. Pour point depressors are: Chlorinated paraffins. Inhibitors to stop gum formation are: Phenol amines (trade name is Topanol).

21. Lube oils: The fraction that is left after lighter components, namely gasoline, kerosene, diesel oil during crude distillation. Lubes have boiling point above 350°C, obtained as products from vacuum distillation units. Lube oils are composed of paraffins, naphthalenes, aromatics, unsaturated bodies. Lube oils are classified as:

1. Machine and Engine oils—high viscosity index oils. Oils for planes, MVI oils for petrol, diesel and stationary vehicles. Oils for light equivalent.
2. Compound oils: Mixed with animal or vegetable oils. Suitable for steam engines, compressors. Emulsibility is more important than viscosity for these oils.
3. Turbine oils: Steam turbines, electric motors should have emulsi on free oils.
4. Cold test oils: Used in refrigeration and hydraulic systems. Pour point is important in this. Designed for arcitic climates.
5. Colour oils: Industrial oils used in textiles, food and paper industries.
6. Corrosive oils: Used in cutting and shaping of metals.

The tests for lube oils are: Flash point; pour point; viscocity and viscosity index; oxidation stability;carbon residue.

22. Carbon residue: The propensity of cracking is indicated by carbon residue of oil. The amount of carbon residue left during cracking provides an idea of usability of oil at high temperatures. The tests are: 1. Conradson method; 2. Ramsbottom method.

Oxidation stability: This test reveals the deterioration characteristics of oil on oxidation. All oils of high boiling point are susceptible to this natural oxidation. Inhibitors prevent this oxidation.

23. Transformer oils: Used in electrical industry for insulating, cooling proposes, oils in transformers and circuit breakers etc. belong to one class; other class of oils are for imprgenating cables and papers. Oils used in transformers are low viscous, thermally stable bit have high boiling points. Second type are highly viscous act as insulators only. The deterioration of these are: 1. Acid value; 2. Dielectric strength; 3. Flash point; 4. Pour point; 5. Sludge value; 6. Copper corrosion test.

24. Bitumen: It is the residual product obtained from crude distillation unit. It is solid at room temperature and has high viscosity.

Most of the bitumen is used in high way constructions, water proofing and coatings works, Bitumen tests are: 1. Softening point; 2. Penetration index. This should possess good ductility for surface application. Ductility is a measure of elongation and stretch.

25. Dehydration and Desalting of crudes:

Any crude that contains more than 5 kg of total salts express in terms of sodium chloride per thousand barrels may be regarded as salty crude. The following methods are used for dehydration of crudes.

1. Chemical treatment; 2. Gravity settling; 3. Centrifuge separation; 4. Electric desalter. At the oil fields, salt is removed by settling or by adding chemicals or by combination of these two. The settling techniques are not effective. Electric desalting is employed in treatment of distillate stocks.

Acid or caustic treated stocks are liable to form emulsions. These emulsions are efficiently broken by electrostatic field. Catalyst poisoning salts can also be favourably removed. In this brine in bottom stream and desalted crude from top. Desalting is maintained at 90°C and 5 to 6 kg/cm^2 pressure. Diluent is not added to avoid contamination of the products.

2. In desalting by setting method, water is added to the crude and heated to 90°C by a heat exchanger. Temperature and pressure remain constant. Desalted crude from top, Brine from bottom. An important factor for pumping crude is pour point. Additives are used to reduce pour point. Wax is removed by heating and cooling. Pipe still heaters are used for heating crude. Pipe still furnaces are used for heating crude.

26. Distillation of petroleum: Petroleum being a mixture of hydrocarbons having a boiling range of 160°C to + 1000°C or more (Pitch). Distillation crude takes place in two stages. 1. Atmospheric distillation unit 2. Vacuum distillation unit. Higher temperatures are not permitted due to degradation of crude by thermal cracking. When the crude contains a good amount of soluble gases to avoid load on ADU (atmospheric distillation unit), a prefleshing or topping column is employed.

(*a*) Atmospheric distillation unit (ADU):

Some details are: Tower diameter 5 to 8 meters; number of plates – 25 to 40; no. of plates for light naphtha to heavy naphtha (80 – 130°C) 6 to 8; Heavy naphtha to light naphtha (180°C) – 5 to 6 plates; Light distillate to middle distillate (250°C) – 4 to 6 plates; Middle distillate to gas oil (diesel) 330°C) – 3 to 4 plates; Flash zone to first draw 3 plates; Flash zone to bottoms – 3 plates; top of ADU is fitted with reflux condenser.

(*b*) Vacuum distillation unit: Pressure operating is 30 to 40 mm Hg; Trays between two draws: 3 to 4; Steam: 0.3 to 5 kg per barrel; corrosion in distillation unit is mainly due to hydrogen sulphide and chloride ion.

27. Treatment techniques: (a) Physical or mechanical impurities (dust, mud, moisture, catalyst dust etc.) to be separated; (b) Chemical impurities (mercaptans, unsaturates, sulphur compounds, nitrogen compounds and oxygen compounds). Their removals are: oxidising mercaptans to disluphides, physical extraction, destruction of sulphur bearing compounds, catalytic conversions in pressure of hydrogen.

28. Production of LPG (SNG, NG) and treatment:

(*a*) LPG production methods are:

1. Distillation at low temperatures; absorption and desorption; compression and expansion combined methods. This is separation of $C_3 - C_4$ fractions. Liquefied natural gas: The gas is compressed in 2 or 3 stages, and expanded in a turbine. The cooled gas cools the incoming gas.

Dehydration is done by passing gases through absorbers contaning Ethylene glycols.

Sweetening of feed gas is done using Ethanol amine (sulphur).

29. Gasoline treatment: Removal of sulphur compounds (sweetening by copper chloride process); Caustic methanol treatment for removal of nitrogen compounds and sulphur compounds; Dialyser process for removal of mercaptans; Lead doctoring of gasoline to remove sulphur compounds; Merox sweetering process to remove sulphur compounds; catalytic desulphurisation (natural bauxite).

30. Treatment of kerosene: (a) Removal of aromatics by liquid sulphur dioxide extraction; (b) Other solvents are: Dimethyl formamide, dimethyl sulfoxide, Ethylene glycol, Diethylene glycol (improves smoke point).

31. Treatment of Lubes: A raw lube oil having boiling point of above 350°C composed of—paraffins, naphthalenss, aromatics, wax ashphalts and resins, inorganic constitutents, sulphur compounds. The established processes for improvement of quality of lubes are: 1. Sulphuric acid treatment, clay treatment, (color removal); Solvent treatment (viscosity index improvement; Furfural, phenol, aniline); Hydro treatment.

32. Wax and purification: All heavy fractions of crude oils contain some amount of waxy material. Wax commercially known as paraffin wax is not pure wax. After stripping and bleaching, it gives white shiney look. Commercial waxes are classified on the basis of penetration index and melting point. Waste removal based on the natural principle of decreasing solubility of wax in oil at lower temperatures is commercially known as chilling and pressing. Sharpel centrifuges are in good demand for dewaxing operations. Final Removal of oil from slack wax is accomplished in sweating operations. Sweating consists of gradual rise and maintenance of temperature in large shalow sweat pans.

Solvent dewaxing is used. If natural method fails. The methods are: 1. Bansol process (ethylene dichloride and benzol solvents). 2. Ketone dewaxing (methyl ethyl ketone, methyl propyl ketone); 3. Propane dewaxing.

33. Thermal cracking: Dissociation of high molecular weight hydrocarbons into smaller fragments through heat alone is termed as thermal cracking or pyrolysis. Cracking operations are required to meet the demands. Thermal cracking reactions are:

$C_n H_{2n+2} \rightarrow C_{n/2} H_{n+2} + C_{n/2} H_n$ (around 400°C)

$C_{12} \rightarrow C_6 + C_6$; $C_6 \rightarrow C_4 + C_2/C_3 + C_3$ (one is always unsaturated). Olefins (unsaturates) crack again.

Unsaturates form dimers. Hydrogenation also occurs.

Saturates are converted to unsaturate cracking proceeds with free radical mechanism. Pressure retards cracking reactions 10 to 15 atm pr is used to minimize coke formation.

34. Catalytic cracking: This is widely used to make gesoline. Temperature range in catalytic cracking reactions is: 450–510°C, pressure does not exceed 2 atmospheres in fluidised bed reactor. Commercial cracking catalysts are: silica – aluminas and silica – mangenesias. Catalytic cracking processes; 1. Fixed bed (Houdry); 2. Moving beds; 3. Fluidised beds (Gulf Research, Kellogg Esso, Standard Oil).

35. Naphtha cracking; to produce olefins (mainly ethylene and propylene). The hydrocarbon feed stock (naphtha) is preheated and cracked in presence of steam in a tulular preheated and cracked in presence of steam in a tulular furnace (Lumus co; Kellogg co.). Temperature is 800–850°C. Cracking pressure is below 3 kg/cm^2.

36. Coking: Petroleum coke is the ultimate product of prolonged thermal cracking. This coke is used for electrodes. General methods of petroleum coke production are:

1. Hot oven method; 2. Thermal cracking; 3. Delayed coking; 4. Fluid coking; 5. Contact coking.

37. Hydrogen processes: These are: 1. Hydrocracking processes; 2. Hydrotreatment processes. Hydrocracking is a powerful tool to the Refiner in the production of Quality gasoline or mid barrel products from heavy distillates or fuel oils. Thermal and catalytic cracking methods are used.

38. Alkylation: This is the reverse of cracking, it joins an olefin with paraffin. Catalyst is hydrogen fluoride. Isomerisation processes. This converts normal paraffins into isoparaffins, to produce good quality gasoline with high octane number. The catalysts used are: platinum, aluminium chloride.

39. Polymer gasolines: Catalytic polymerisation.

40. Asphalt technology: Asphalts structure is oil → 35–50%; Rosins: 5 to 20%; Asphalteens 20 to 30; Acids etc – up to 10.

41. 1. Petroleum Refineries in India: 1. Digboi – Assam – 0.5 mil Tons/yr; 2. HPCL – Mumbai – 5.5; 3. BPCL – Mumbai 6.0; 4. HPCL – Visakhapatnam – 4.5; 5. Guwahati Assam – 0.85; 6. Barauni – Bihar – 3.2; 7. Koyali – Gujarat 9.5; 8. Cochin – Kerala 4.5; 9. MRL – Chennai 5.6; 10. Haldia – West Bengal – 2.5; 11. Bougaigon Assam – 1.4; 12. Mathura – U.P – 7.5; 13. Kamal – Haryana 6.0; 14. Mangalore – Karnataka – 6.0.

42. Sedimentary rock favours petroleum formation. Gryanic Rock, chemical rock and classic rock are sedimentary rocks. Lithufication favours sediment to tranform into rock. Wind creates cross bedding Groded bedding coarse particles below and fine particles on the top. A rock that is immature to produce petroleum is potential source work. A source rock that has expelled petroleum is effective source work. Shale is the abundant sedimentary rock in earth. Crust for geochemical exploration of oil, geothermal survey and radioactivity survey are used-oil and gas are found in shale. Reservoir rock has high permeability and high porosity. Tight packing favours lesser permeablity. Find grained rocks have high permeability and high porosity. Migration of oil occurs in pore spaces and fractures. The hydrocarbon fuel in shale oil is, Kerogen. Oganic theory states that oil and gas come from microscopic plants and small animals.

43. The oil window temperature range for hydrocarbons is 60–120°C. Petrography focuses on detailed study of rocks. Topographic map shows elevation of Earth. Base map shows location of wells. Isopach map shows thickness of petroleum reservoir. Classic rock and organic rock are the primary interest for reservoir. Seismic data at sea is obtained using hydrophones and air guns. Kerogen consists of carbon, hydrogen and oxygen with small amounts of nitrogen and sulphur paleontology deals with study of prehistoric life. Geophones are the sensors that pick up the reflected seismic waves. Porosity is the ratio of pore volume to the bulk volume. Bright spot is the hydrocarbon indicator. Sandstone rock type is the best oil reservoir. Because of heat and pressure petroleum becomes natural gas. 1. Barrel of oil is equal to 42 gallons. Saudi Arabia have largest reserves of crude. Anthracite gives of most heat.

44. Crown block and travelling block are part of drilling of crude petroleum. The topmost castings is conductor casting. Casting is steel pipe placed in oil well to prevent well from caving. The most common range of drill pipe is 27–30 feet. Spar is the Rig used to drill wells from deepest water. Tripping out is the process to pull the drill pipe out. Shear Rams are designed to cut across any pipe in the well to close the well quickly. Blowout preventers are used to close off the top of the well. Drilling jars are used to free a stuck drill pipe. Drift or inclination angle is the angle measured in degrees between the actual well path and a vertical line. Kill line is part of kill line. For drilling hard formations Diamond bit is used. Viscosity is measured using Fann V–G viscometer. Bingham fluid requires certain level of pressure gradient to produce flow. Marsh funnel viscosmeter measures also viscosity. Crown is a description for a diamnond bit.

45. Porosity is void volume/total volume. Porosity of cubic packing is 47%. Reservoir begins to fill up with oil is drainage. Rservoir begins to fill up with water is imbibition. Tortuosity is the ratio of actual path length/straight path length. If the pressure in oil is greater than the rock is water wet. If the pressure in water is greater than the rock is oil wet. The dimensionless function of saturation is called J.function. Temperature, pressure and composition effects the compressibility factor of a gas. Gas Cap reservoir drive mechanism has the highest recovery factor. Dissolved gas reservoir has a low intial gas oil ratio increasing to maximum near the end of production. Decine curve (production and time) determines the economic limit for a well. Permeability units are: Darcy. Darcy's law quantifies the rate of flow for fluids through porous media. At critical point all intensive properties are equal (Temp, Pr).

46. The process of a well ready for production is called as well completion. Christamas tree is the name for the well head and valve assembly used to control a well. The hydraulic firing head detonates at absolute pressure of 4035 PSI. When a producing well has to be shut in for Remediation operations it is called as workover. Acid fracturing does not requiring sand. For fracturing fluids. Ammonium persulphate or ammonium sulphate or diammonium peroxi disulphate are used. The control pressure for a surface safety valve is: two times valve body pressure/actuator ratio. The reservoir capability can be illustrated by: inflow performance Relationship curve. (STB/D)/PS12 is the unit of productivity index. If of the pressure drop in well bore is reduced, keeping other parameters same will result in increased production rate. In slug flow gas bubbles coalesce into large bubbles that fill the entire pipe. Intersection of IPR and IIE curves gives possible flow rate from well number of valves required in a gas lift well depends on well depth and injection pressure. New point is the point where the tubing starts to buckle.

47. Vertical separator is used for low gas oil ratio. Horizontal separator for high gas oil ratio. 2 to 4 stages are required in flash separation to separate gas and liquid. Cariolis flow meter measures mass flow rate. Bullet tanks are used for liquid petroleum gas. Sample container must not be filled in excess of 85%. The minimum number of temperatures to be taken in a tank containing more than ten feet are 3.

48. Jack up Rig is bottom supported mobile offshore drilling unit. Drill ship is floating drill Rig used in deep water. Riser connects sea bed facilities to surface facilities. Mooring is Chain/Rope system keeping floating surface facilities in place. The reasons for design of offshore structure is: wind, wave, current. World's deepest spar is in Gulf of Maxico. Rigid TTR. Riser is used in offshore drilling.

49. Formation evaluation is done for: to find Reservoir, how much hydrocarbon is present, production is economical or not. The cost of formation evolution is 20% of total well cost. Shale rocks have low resistivity. Mud log is used to detect hydrocarbons while drrilling. Density in combination with sonic and Newton logs gives lithology and shale content. Newton, density and accoustic logs are referred as porosity logs. Newton logs are porosity logs that measure the hydrogen concentration. Density log measurer electron density. Sonic log used sound waves to determine porosity. Gamma ray log measures natural radioactivity of rocks. NMR log is used to measure porosity and permeability. Natural γ-ray spectrometry is used for detailed mineral identification.

50. Formation with rock properties changing with location in the reservoir is heterogeneous formation. Linear flow equation is used for fractured wells. Pressure variation with time is a function of well geometry and reservoir properties. Well damage influences the permeability calculations. A reservoir limit test is done for: pore volume, skin factor, premeability. Drill stem test is generally done in exploratory wells. Information about permeability can be obtained from drill stem test.

51. Enhanced oil recovery (EOR) process refers to the recovery of oil through injection of fluids and energy. Water flooding is widely used post primary. Recovery depth limit for steam flooding is 3000 ft. CO_2 flooding develops miscibility to displace oil from Reservoir. Increased capillary number minimizes trapped oil saturation. Capillary number represents the Relative effect of Viscous forces and surface tension. The well pattern having 4 injectors and 1 producer is called five spot. The well pattern having 5 injectors and 1 producer is called inverted seven spot. The Recovery pecentage for steam flood is upto 65% OOIP. Capillary number is the ratio of viscous to inertial forces. Sodium hydroxide or sodium silicate or sodium carbonate are used for alkaline flooding. CO_2 flooding is used for medium and light crude oil. UTCHEM is a chemical flooding simulator. Infill drilling is best method for improved oil recovery.

QUESTIONS

SECTION—A

1. Major component of natural gas is:
(a) Methane (b) Ethane
(c) Propane (d) Dutane

2. The condensate obtained from a natural gas is called:
(a) Natural gasoline (b) Liquified natural gas
(c) Liquified natural gasoline (d) Natural condensate.

3. If the natural gas contains H_2 S gas, it is called:
(a) Foul natural gas (b) Sweet natural gas
(c) Sour natural gas (d) Sulfuretted natural gas

4. Dehydration of crude oil is primarily done to:
(a) Remove moisture (b) Remove salts
(c) Improve distillation characteristic (d) Improve calorific value.

5. Chemical additives are added to waxy crudes to:
(a) Dissolve wax (b) Precipitate and remove wax
(c) To depress pour point (d) Improve distillation properties (ASTM D92)

6. Bottoms of atmospheric distillation unit (ADU) is called:
(a) Heavy ends (b) Residual crude
(c) Reduced crude (d) Residuum.

7. Bottoms of vacuum distillation unit is called:
(a) Heavy ends (b) Reduced crude
(c) Residual crude (d) Residuum

8. Heavy residual product in a refinery is converted to coke by the process of:
(a) Cracking (b) Carbonisation
(c) Coking (d) Pyrolysis

9. Gas oil is converted to gasoline by the process of:
(a) Cracking (b) Stabilisation
(c) Isomerisation (d) Coking

10. The usual process of coking followed in Indian refineries is:
(a) Fluid coking (b) Flexi coking
(c) Contact coking (d) Delayed coking

11. A low inline point for a diesel oil indicates a high octane number:
(a) True (b) False

12. Feed stock used for obtaining cracked gasoline is:
(a) Naphtha (b) Heavy recycle oil
(c) Gas oil (d) None of these

13. A catalytic cracking operation takes place in the temperature range of:
(a) 400–450°C (b) 450–500°C
(c) 500–550°C (d) None of these

14. Which of the following hydrocarbon has the lowest octane number:
(a) *n*-paraffin (b) Iso-paraffin
(c) Naphthene (d) Aromatics

15. Cleveland apparatus is used to test .... of an oil.
(a) Fire point (b) Density
(c) Viscosity (d) Vapour pressure

16. Carbon residue of an oil is determined by:
(a) Abel apparatus (b) Engler's apparatus
(c) Dean and Stark apparatus (d) Ramsbottom apparatus

17. In the laboratory when equal volumes of aniline and the test sample were mixed at room temperature, they were found to be completely miscible. Hence the aniline point of the oil is:
(a) Same as the room temperature (b) More than the room temperature
(c) Less than the room temperature (d) Unpredictable

18. In the reforming reactor by product hydrogen gas is always recycled to:
(a) Sustain reactor pressure (b) Hydrogenate the feed stock
(c) Prevent the catalyst from poisoning (d) Control the reaction temperature

19. Feed for the reforming reactor is generally the:
(a) Reduced crude (b) Cracked gasoline
(c) Straight run gasoline (d) Gas oil

20. Gum formation in gasoline during storage is mainly due to:
(a) Presence of sulfur (b) Improper storage conditions
(c) High aromatic content (d) Oxidation and polymerisation of unsaturates

21. Smoke point of kerosene is the:
(a) Maximum height of the flame (in mm) which does not form smoke when burnt in a standard lamp
(b) Minimum height of the flame (in mm) which does not form smoke when burnt in a standard lamp

(c) Time after which smoke is observed
(d) Maximum rate of flow of kerosene which causes smoke formation

22. Petroleum coke is primarily used for:
(a) Sugar refinery (b) Manufacture of fuel gases
(c) Reduction of iron ore (d) Manufacture of carbon electrode

23. Lead susceptibility is highest in case of:
(a) Natural gasoline (b) Straight run gasoline
(c) Polymerised gasoline (d) Catalytically cracked gasoline

24. Viscosity index for lubricating oil can be greatly improved by:
(a) Clay treatment (b) Solvent extraction
(c) Acid treatment (d) Liquid SO_2 treatment

25. Removal of aromatics from kerosene is done by:
(a) Solvent extraction (b) Clay treatment
(c) Liquid SO_2 treatment (d) Acid treatment

SECTION—A A

26. Which of the following is preferred fast neutron source in neutron logging?
(a) Americum – Beryllium (b) Radium–Beryllium
(c) Deuterium – Tritium (d) Thorium – Beryllium

27. Identify the logging device that is based on the concept of longitudinal and transverse relaxation times:
(a) Thermal neutron decay (b) Induced gamma ray spectroscopy
(c) Neutron (d) Nuclear magnetic resonance (NMR)

28. The three main stages of evolution of organic matter in sediments are Catagenesis (C), Diagenesis (D) and Metagenesis (M). Their chronological order is:
(a) D – C – M (b) C – D – M
(c) D – M – C (d) C – M – D

29. For a kick off operation, a directional well has to be drilled for an arc-length of 2500 feet to achieve an inclination of 50°. The radius of curvature will be...

30. Which of the following is the most common cause for fishing job?
(a) Differential sticking (b) Use of oil based mud
(c) Lost circulation (d) Well kick

31. The figure shows the producing gas oil ratio (GOR) behaviour with time for an oil reservoir under primary production. At initial reservoir condition, P_1 is bubble point pressure of the crude oil. $P_R(t)$ represents the reservoir pressure at time, t which of the following statements is true.
(a) $P_R(t_1) > P_b$; $P_R(t_2) > P_b$ (b) $P_r(t_1) > P_b$; $P_R(t_2) < P_b$
(c) $P_R(t_1) < P_b$; $P_R(t_2) > P_b$ (d) $P_R(t_1) < P_b$; $P_R(t_2) < P_b$

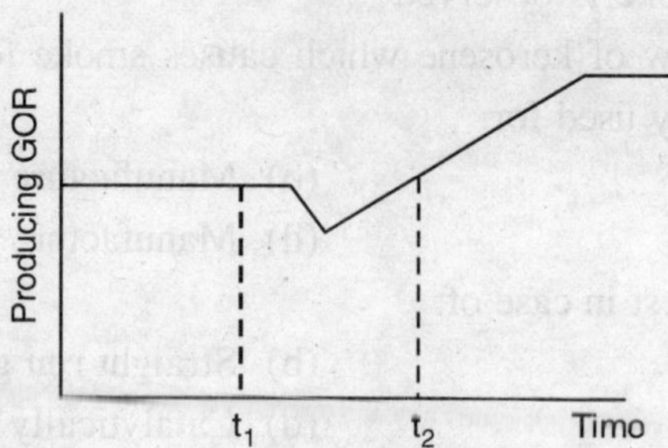

32. A core with a length of 10 cms breadth of 4 cm and width of 4 cm, weighs 282.4 g in its dry form. The core is then saturated 100% with brine of density 1.1 g/cm^3. The brine saturated core weighs 100 g. The porosity of the core is .... %.

33. A hydraulic line of a subsurface safety value has a fluid of specific gravity 1.2 to operate the valve. The valve closing pressure is 1200 psia and the recommended sefety margin is 200 psia. The maximum depth at which the valve can be positioned is ....

34. A sucker rod pump unit is designated by C – 228 D – 200 – 74. Here D represents:

(a) Double reduction gear loose (b) Diameter of sucker Rod
(c) diameter of phenger (d) Stroke length

35. The three translational motions for a floating vessel are:

(a) Roll pitch yaw (b) Heave pitch sway
(c) Surge–Sway Heave (d) Roll – Sway Heave

36. Jack up rigs are typically used for offshore drilling when the water depth is in the range:

(a) < 25 ft (b) 50 – 500 ft
(c) 100 – 20000 ft (d) > 2000 ft

37. Interference tests can be used for:

1. Determining communication between two or more cells.
2. Mature oil cells having skin damage.
3. Determining permeability in tested wells
4. Providing inputs for secondary and tertiary oil recovery methods.

(a) Only 1 and 2 (b) Only 1, 2 and 4
(c) Only 2, 3, 4 (d) 1, 2, 3 and 4

38. For an effective hydraulically fractured well, the skin factor would generally be:

(a) Negative (b) Positive
(c) Zero (d) Indeterminate

39. The maximum discharge limit of oil and grease in a marine coastal area as per environmental protection rules, 1986 in India is:

(a) 0.1 mg/L (b) 20 mg/L
(c) 500 mg/L (d) 4000 mg/L

40. Which of the following gase is not responsible for global warming:

(a) Carbon dioxide (b) Methane
(c) Water-vapor (d) Nitrogen

41. In an oil reservoir flooded with water, the volumetric sweep efficiency is 70%. The connate water saturation in the reservoir is 0.4 and the residue oil saturation for the water flood is 0.3. The overall efficiency of the reservoir is .... %.

42. Identify the pair of Correct statements for surfactant – micellar – polymer flooding:

1. It Reduces interfacial tension between crude oil and water.
2. It influences mobility ratio unfavorably.
3. It improves microscopic displacement efficiency.
4. It increases isothermal compressibility of the crude oil

(a) 1 and 2 (b) 1 and 3
(c) 3 and 4 (d) 2 and 4

43. Gas hydrate forms at:
(a) Low pressure and low temperature conditions
(b) Low pressure and high temperature conditions
(c) High pressure and low temperature conditions
(d) High pressure and high temperature conditions

44. Production of coal bed methane (CBM) is based on:
(a) Distillation (b) Underground coal gasification
(c) Desorption (d) Coal liquefaction

45. The porosities of cubic and hexagonal packings respectively are:
(a) 47.6% and 25.9% (b) 39.3% and 20.5%
(c) 47.6% and 39.5% (d) 39.5% and 25.9%

46. A section of a clean sandstone reservoir was logged and found to have a porosity of 10%. The comentation (*m*) and saturation (*n*) exponents are equal to 2. The constant A in the Archie's saturation equation is 1. The formation water resistivity is 0.036 ohm – meter and the formation resistivity is 10 ohm-meter. The water saturation in reservoir is:

47. In sonic logging, the sonic velocities in the formation and drilling mud are 50,000 ft/s and 500 ft/s respectively. The critical angle is .... Radians.

48. Match the following in Group–1 with those in Group–2.

Group–1	**Group–2**
P. Blow out preventer	1. Horizontal well problem
Q. Diamond hit	2. Reverse ballooning
R. Tubing elongation	3. Well control
S. Ecentricity	4. Crown
	5. Ballooning

(a) P – 3 ; Q – 4 ; R – 5 ; S – 2 (b) P – 5 ; Q – 4 ; R – 1 ; S – 2
(c) P – 3 ; Q – 4 ; R – 2 ; S – 1 (d) P – 4 ; Q – 3 ; R – 5 ; S – 1

49. One thousand sacks of cement are required for cementing a protection casing of setting depth of 12,000 ft (top float collor) and annular capacity of 0.40 ft^3 per linear ft. The cementing truck has a missing capacity of 20 sacks per minute. A 1015 ft^3 cycle capacity rig mud pump having an

18 inch smoke an a $6\frac{1}{2}$ inch linear operating at 60 RPM with 90% efficiency is used for the cementing job. The total cementing time is ....minutes.

50. It is desired to increase the density of 200 bbl of 10 ppg mud to 12 ppg mud using API barite of density 35 ppg. The final volume is not limited. (1 bbl = 42 gallons). The amount of API barite required is .... bbm.

51. Using the high pressure, high temperature (HPHT) filter press data given below, the estimated API filtration loss is .... cm^3.

Data given:

Time (Min)	Filtrate volume (cm^3)
1.0	6.5
7.5	14.0

52. A differential liberation experiment (DLE) end constant composition expansion (CCE)/flash liberation experiment were performed in a laboratory for a crude oil to find the formation volume factor (B_0) and the dissolved gas oil ratio (R_S). The pressure stages for both experiments were kept the same. At a pressure less than the bubble point pressure of the crude oil, which of the following statements is true?

(a) B_0 (CCE) > B_0 (DLE), R_S (CCE) > R_S (DLE)

(b) B_0 (CCE) > B_0 (DLE), R_S (CCE) < R_S (DLE)

(c) B_0 (CCE) < B_0 (DLE), R_S (CCE) > R_S (DLE)

(d) B_0 (CCE) < B_0 (DLE), R_S (CCE) < R_S (DLE)

53. The production of a gas well was found to decline exponentially. The observed production rate on Ist January 2014 was 0.60×10^{10} SCF/month and on 1st January 2015, it was 0.40×10^{10} SCF/month. The economic production limit for the well is estimated to be 0.002×10^{10} SCF/month. The remaining reserves for the cell as on Ist January, 2015 were .... $\times 10^{10}$ SCF.

54. A 30 ft thick gas reservoir has an area of 3,000 acres (1 acre = 43,560 ft^2). The porosity of the reservoir is 15% and the connate water saturation is 20%. Intial reservoir and temperature are 2,600 psig and 150°F (= 610°R) respectively. The compressibility factor (z) at initial conditions is 0.82. The gas in the reservoir can be produced till it attains the final pressure of 1000 psig (z = 0.88) under isothermal conditions. The gas recovery factor is.

55. Brine is used to measure the absolute permeability of a core plug. The rock sample is 4 cm long and its cross section is 4 cm^2. The brine has a viscosity of 2 CP and is flowing at a constant rate of 0.5 cm^3/s under a 4 atm pressure differential. The absolute permeability is .... Darcy.

56. An oil well is drilled to cover a circular drainage area of radius 700 ft. The cell is completed with a 7 inch production casing. Assume reservoir pressure of 1000 psig, permeability of 50 and, pay zone thickness of 20 ft, oil viscosity of 3 CP and oil formation volume factor of 1.25 reservoir – bbl/STB. For a flowing bottom hole pressure of 500 psig, the primary production rate is STB/day.

57. An electric submersible pump (ESP) is installed at a depth of 1000 ft from the surface. The ESP gives 20 ft water head per stage. The well head requires 100 psig pressure. Minimum number of stages of the ESP required for this well is ....

58. The schematic figure shows a two phase horizontal separator designed for an oil and water system. The oil specific gravity is 0.80. The oil pad height is h_0.

The vertical distance between the oil and water weirs (Δh) at steady state is:

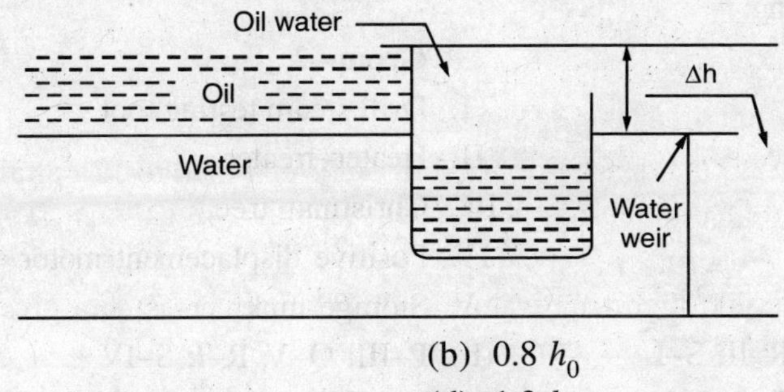

(a) 0.2 h_0 (b) 0.8 h_0
(c) 1.0 h_0 (d) 1.2 h_0

59. The vertical lift performance (VLP) and inflow performance relationship (IPR) curves are used to find the production operating conditions. If PWf is the flowing bottom-hole pressure and Q is the oil flow rate. Select the correct statement

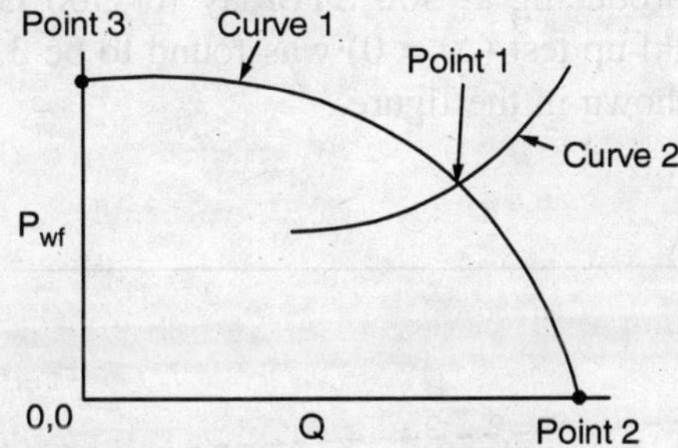

(a) Point 3 is absolute open flow curve 1 is VLP curve.
(b) Point 2 is at reservoir pressure curve is VLP curve.
(c) Point 1 is operating condition, curve 2 is IPR curve.
(d) Point 2 is absolute open flow, curve 1 is IPR curve.

60. A ground station has a pump which delivers a head of 1000 m water. It is pumping oil of specific gravity 0.80 into a horizontal pipe of diameter of 0.5 m with an average velocity of 2m/s. The efficiency of the pump is 80%. Density of water is 1000 kg/m^3 and acceleration due to gravity is 9.8 m/s^2. The power required to operate the pump is .... mega watts.

61. For a floating vessel match the correct pairs from Group–1 and Group-2 among the options given below. B = centre of buoyancy ; G = centre of gravity ; and M = metacenter.

Group–1	**Group–2**
P. M is above G	I. Stable equilibrium conditions
Q. M is below G	II. Critically stable condition
R. M is coinciding with G.	
S. B is below G	III. Unstable condition

(a) P–II; Q–III, R–I; S–II (b) P–I; Q–III, R–II; S–I
(c) P–III; Q–I, R–II; S–III (d) P–I; Q–II, R–III; S–I

62. Match the following:

Group–1	Group–2
P. Master valve	I. Drill steam testing tool
Q. Breather valve	II. Heater–treater
R. Tester Valve	III. Christmas tree,
S. Dump valve	IV. Positive displacement motor
	V. Storage tank

(a) P–III; Q–V, R–II; S–I (b) P–III; Q–V, R–I; S–IV
(c) P–II; Q–III, R–I; S–V; (d) P–I; Q–III, R–II; S–IV

63. A 50 feet thick reservoir has a porosity (ϕ) of 20% and a total isothermal compressibility (C_1) of 2.4×10^{-5} psi^{-1}. The oil in the reservoir has a viscosity of 0.75 cP and formation volume factor of 1.25 Reservoir – bbl/STB. A pressure build-up test is carried out on a well of radius 0.50 ft in the reservoir which was producing at 500 STB/day for 500 hours. The flowing bottom hole pressure at the start of build-up test ($\Delta t = 0$) was found to be 3,535 psia. The schematic of the pressure build up data is shown in the figure.

The skin factor is ....

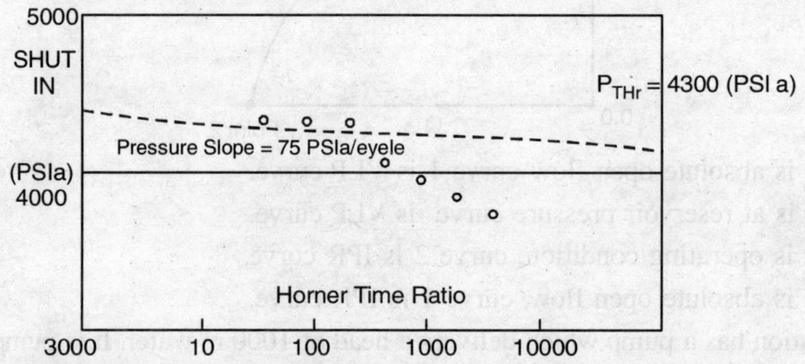

64. During a production test in an oil reservoir, the oil production rate is 2000 STB/day. The producing gas oil ratio (GOR) is 800 SCF/STB and dissolved GOR is 200 SCF/STB. The formation volume factor of gas is 001 ft/SCF and the formation volume factor of oil is 1.2 reservoir-bbl/STB. The down hole GOR is : .... ft^3/Reservoir – bbl.

65. A productivity test was conducted on a angle-phase crude oil well. The well is capable of producing 100 STB/day at a flowing bottom – hole pressure of 1000 psig. The 24 – hour shut – in static pressue is found to be 1500 psig. The maximum oil flow Rate (Q_{max}) is .... STB/day.

66. An oil well of wellbore radius 0.5 feet is shown to develop a skin due to formation damage. The damaged zone radius is 2.25 ft around the well. The formation permeability is 300 md and the permeability of the damaged zone is 100 md. The effective wellbore radius for the well is .... ft.

67. A producing well has a shut in tubing pressure of 3950 psig for errude oil of specific gravity 0.69 (1 g/cm^3 = 8.3 ppg). The kill fluid density for a work over job is at 11,600 ft (TVD) is .... ppg.

68. For a water flood operation in a one dimensional reservoir the following data are given porosity, $\phi = 0.25$, cross sectional area, A = 25,000 ft^2; Horizontal distance between the vertical production and injection well = 6000 ft. Water injection rate, s_w = 900 bbl/day. Slope of fractional flow curve at shock front water saturation = 1.97; water formation volume factor = 1.0 bbl/STB. 1 bbl = 5.615 ft^3.

The cumulative water volume injected at breakthrough .... $\times 10^5$ bbl.

69. A heavy oil reservoir is being flooded within a line drive (assume one–dimensional flooding). The functional flow of water is found to be 0.75 bbl/bbl at water saturation (Sw) of 60%. A polymer solution with twice the viscosity of water is used as displacing phase. Assume the relative permeability curves for water flooding and polymer flooding are the same. The fractional flow of polymer solution at saturation of 60% is .... bbl/ft.

70. Which of the following rocks favours petroleum formation?

(a) Sedimentary rock (b) Igneous rock
(c) Metamorphic rock (d) None of the above

71. Which type includes the principle source sediments for oil:

(a) Type–I (b) Type–II
(c) Type–III (d) Type–IV

72. Oil and gas are found in ....

(a) Shale (b) Sand
(c) Granite (d) Sulphur

73. Which of the below statement is true?

(a) Permeability increases with grain size
(b) Permeability increases as sorting becomes poorer
(c) Tight packing favours lesser permeability
(d) None of the above

74. Rock strata that have crumbled and buckled into wave like structures are called as:

(a) Faults (b) Traps
(c) Folds (d) None of the above

75. Which of the following force causes the petroleum to migrate from source rock through the carrier beds to the reservoir rock.

(a) Capillary force (b) Buoyancy force
(c) Hydrodynamic force (d) All of the above

76. Oil generation occurs when:

(a) Level of organic maturation (LOM) is 7–13
(b) Level of organic maturation (LOM) is 13–18
(c) Both (a) and (b) (d) None of the above

77. Gas generation when LOM:

(a) Is 7 – 13 (b) 13 – 18
(c) Both (a) and (b) (d) None of the above

78. The study of prehistoric life, including organisms evolution and interactions with each other and their environments is called as:
(a) Paleontology (b) Petrology
(c) Sedimentology (d) All of the above

79. Which type of kerogen has potential for mainly oil?
(a) Kerogen I (b) Kerogen II A
(c) Kerogen II B (d) Kerogen III

80. Which type of kerogen has potential for mainly gas?
(a) Kerogen I (b) Kerogen II A
(c) Kerogen II B (d) Kerogen III

81. Which Sensors pick up the reflected scismic waves and send them through cables in a recorder?
(a) Geophones (b) Reflector
(c) Refractor (d) None of the above

82. Global positioning is used for:
(a) Marine shooting (b) Land shooting
(c) Both (a) and (b) (d) None of the above

83. Which of the following basins is producing petroleum from coal rich resevoir rocks?
(a) Rajasthan basin (b) Cambay basin
(c) Cauvery basin (d) Krishna–Godavari basin

84. Which of the following rock types would most likely be the best oil reservoir?
(a) Granite (b) Shale
(c) Sandstone (d) Salt

85. One Barrel of oil is equal how many gallons?
(a) 21 (b) 42
(c) 56 (d) 100

86. The nation having the largest reserves of crude oil is:
(a) U.S.A (b) Kuwait
(c) Saudi Arabia (d) Iraq

87. Which country has maximum oil shale?
(a) USA (b) UK
(c) India (d) China

88. The topmost casing is:
(a) Surface casing (b) Intermediate casing
(c) Production casing (d) Conductor casing

89. The hosting system does not include:
(a) Draw works (b) Block and tackle
(c) Rotary table (d) None of the above

90. Steel pipe placed in oil and gas well to prevent the well bore from caving in is:
(a) Drill string (b) Annulus
(c) Casing (d) None of the above

91. What are two large blocks of steel that are designated to close the around the pipe in the well?
(a) Pipe Rams (b) Shear Rams
(c) Blind Rams (d) None of the above

92. Factors influencing bit selection include:
(a) Formation lithology (b) Rig capabilities
(c) Economic goals (d) All of the above

93. BOP consists of:
(a) Pipe Ram (b) Shear Ram
(c) Blind Ram (d) All of the above

94. Measurement while drilling (MWD) systems provide real time:
(a) Drillbit information (b) Borehole inclination
(c) Azimuth (d) All of the above

95. Which of the following lifting operation is related to fishing job:
(a) Broken well head (b) Broken Christamas Tree
(c) Broken drillbit (d) Broken storage tank

96. The depth of the well governs:
(a) The lithology that must be penetreted (b) The time required to complete the well
(c) (a) and (b) (d) None of the above

97. Which one is corect above drilling cost?
(a) It does not vary with well depth (b) It decreses exponentially with well depth
(c) It is independent of well depth (d) It increases exponentially with well depth

98. Viscous properties of oil = based mud is primarly controlled by:
(a) Viscosifier content (b) oil-water ratio
(c) Barite content (d) None of the above

99. Which of the following fluid is preferred to drill low pressure depleted zones?
(a) Water based mud (b) Oil based mud
(c) Foam (d) All

100. Which of the following is a major consideration for cutting transport in horizontal wells:
(a) Annulus size (b) Drilling rate
(c) Eccentricity (d) Rotary speed

101. One of the following is most likely a descriptor for a diamond lit.
(a) Crown (b) Journal
(c) O–Ring seal (d) Shirttail

102. The void to grain ratio of sandstone sample is 1/4. What is the porosity of this sample?
(a) 20% (b) 25%
(c) 30% (d) 40%

103. Which equipment is used for the measurement of permeability?
(a) Rusk air permeator (b) Ruska liquid permeator
(c) Hassler core holder (d) All of the above

104. Which reservoir drive mechanism has the highest recovery factor?

(a) Water drive (b) Gas cap
(c) Solution drive (d) Combination drive

105. The reservoir rocks which contain hydrocarbons that can be recovered is known as:

(a) Pay zone (b) Well
(c) Both (a) and (b) (d) None of the above

106. The gas oil ratio for ordinary black oil is:

(a) 200–700 scf/stb (b) Less than 200 scf/stb
(c) 2000 – 3200 scf/stb (d) Greater than 3000 scf/stb

107. Grid selection depends on:

(a) Minimum spacing between wells (b) Degree of reservoir hetro geneity
(c) Geometry of reservoir (d) All the above

108. What is the name for the well head and value assembly used to control well?

(a) BOP (b) Annular preventer
(c) Christmas tree (d) None of the above

109. Which of the following is a circulating device?

(a) Sliding sleeve (b) Ported collars
(c) Ported nipples (d) All of the above

110. What is the unit of productivity index?

(a) stb/psi (b) (stb/d)/psi
(c) psi/(stb/d) (d) (s + b/d)/psi

111. Productivity index can be expressed as:

(a) Production × Draw down (b) Production/Draw down
(c) Production + Drawn down (d) Production – Draw down

112. Maximum quantity of oil can be recovered in case of:

(a) Active water drive reservoirs (b) Depletion or solution gas reservoirs
(c) Gas cap driven reservoirs (d) Data insufficient

113. Very high viscous crude is lifted by:

(a) Gas lift (b) Sucker rod pump
(c) Progressive cavity pump (d) None of the above

114. Number of walls required in a gas lift well primarly depends on:

(a) Well depth and injection pressure (b) Well depth and reservoir pressure
(c) Well depth only

115. Electrical submersible pumps are:

(a) Centrifugal pumps in multistage fashion (b) Reciprocating pumps
(c) Positive displacement pumps (d) None of the above

116. Jet pump has:

(a) No moving parts (b) One moving part
(c) Some moving parts (d) None of the above

117. What are the following for the cause of excessive water production in an oil well?

(a) Channeling (b) Coning

(c) Both (a) and (b) (d) None of the above

118. What theory is used to understand the separation of oil and gas in a separator?

(a) Dalton's theory (b) Droplet settling theory

(c) Bohr's theory (d) None of the above

119. The main constituent of natural gas is:

(a) Methane (b) Carbon dioxide

(c) Carbon monoxide (d) Nitrogen dioxide

120. The recommended velocity of water in a pipeline is:

(a) 3 to 10 ft/sec (b) 50 – 75 ft/sec

(c) 75 – 100 ft/sec (d) None of the above

121. Which of the following equations is used to calculate the diameter of a pipeline?

(a) $D = 0.422\ Q^{0.4}\ (fSL/\Delta P)^{0.20}$ where D is diameter of pipe in inches; Q flow rate in g pm; f = friction factor; S = Specific gravity; L = length of pipeline; ft; ΔP = pressure drop in psi

(b) $D = 0.422\ Q^{0.20}\ (fsL/\Delta P)^{0.40}$ (c) $D = Q\ (fSL/\Delta P)$

(d) $D = 0.422\ Q^{0.20}\ (fSL/\Delta P)^{0.50}$

122. Which of the following flow meter is used for transfer of petroleum and petroleum products?

(a) Orifice flow meter (b) Coriolis flow meter

(c) Turbine meter (d) All of the above

123. Natural gas is transported mainly by:

(a) Pipeline (b) Tankers

(c) Trucks (d) None of the above

124. API gravity with .... with the densityincrease of a meterial

(a) Increases (b) Decreases

(c) Not changed (d) None of the above

125. Very high vapour pressure products are stored in .... tanks.

(a) Fixed roof (b) Spherical

(c) Floating roof (d) All of the above

126. Containers should not be filled in excess of (regardless of product):

(a) 50% (b) 85%

(c) 60% (d) 100%

127. Ballast is used for the purpose of:

(a) Keeping the cargo warm (b) Segregation of the cargo

(c) Reducing the ship's fuel consumption

(d) Maintaining the vessels stability, trim, and controlling vessel stress.

128. The temperature measuring device in a marine custody transfer is:

(a) Mercury in glass thermometer in a case (b) An in-line temperature probe

(c) An on board Saab system (d) A portable electronic thermometer

129. Free water measurements may be taken by:

(a) The image method (b) The uillage method

(c) The heimlich method (d) Both (a) and (b)

130. A pipeline carries oil at 8 ft/s through 1 ft id of pipe. At another section, the inside diameter is 2 ft. What is the velocity of oil in this section?

(a) 2 ft/s (b) 16 ft/s

(c) 4 ft/s (d) 32 ft/s; $\left(\frac{\pi D_1^2 \rho}{4} = \frac{\pi D_2^2 \rho}{4}\right)$

131. Which one is used for drilling in deep water?

(a) Drill barge (b) Jack up rig

(c) Drill ship (d) None of the above

132. A system connecting sea bed facilities to surface facilities is called:

(a) Riser (b) Well head

(c) Accumulator (d) BOP

133. A drill ship can drill in water depths upto a depth of:

(a) 12,000 ft (b) 18,000 ft

(c) 5000 ft (d) 7500 ft

134. Which of the following is the example of floating offshore structure?

(a) Jack up rig (b) Complaint tower

(c) Spar platform (d) None of the above

135. The following are used for submarine navigation:

(a) Inertial systems (b) Non–inertial systems

(c) Combine systems (d) None of the above

136. Which riser is used in offshore drilling?

(a) Rigid TTR (b) Rigid SCR

(c) Flexible riser (d) None of the above

137. Which turret system is best suited for harsh environment:

(a) External Turret systems (b) Internal turret systems

(c) Both (a) and (b) (d) None of the above

138. Which is the component of subsea production christmas tree?

(a) Tree cap (b) Jumper head

(c) Both (a) and (b) (d) None of the above

139. The cost of formation evaluation may range up to:

(a) 20% of total well cost (b) 35% of total well cost

(c) 50% of total well cost (d) 60% of total well cost

140. Which one is an example of production log?

(a) DST log (b) Neutron log

(c) Induction log (d) SP log

141. Which indicates the presence of oil in the formation?
(a) Mud and cutting analyses (b) Shale factor
(c) Gas detection (d) None of the above

142. Shale rocks have:
(a) High resistivity (b) Low resistivity
(c) Low conduction (d) None of the above

143. Which redioactive log measures natural radioactivity of rocks?
(a) gamma ray (b) Neutron log
(c) Density log (d) None of the above

144. Sonic log is used for:
(a) Open or uncased holes (b) Cased hole
(c) (a) and (b) (d) None of the above

145. Which one is of deliverability test:
(a) Flow after flow test (b) Isochronal test
(c) Modified isochronal test (d) All the above

146. Drill stem test is done in:
(a) Exploratory wells (b) Production wells
(c) Both (a) and (b) (d) None of the above

147. Which process refers to the recovery of oil through the injection of fluids and energy, not present in reservoir.
(a) EOR (b) Artificial lift techniques
(c) Work over (d) None of the above

148. Widely used primary recovery is:
(a) Water flooding (b) Polymer flooding
(c) Steam flooding (d) None of the above

149. Depth limit for steam flooding is:
(a) 3000 ft (b) 5000 ft
(c) 6000 ft (d) 10,000ft

150. Which represents the relative effect of viscous forces and surface tension acting across the interface between two immiscible liquids?
(a) Capillary number (b) Reynolds number
(c) Formation volume factor (d) None of the above

151. The well pattern having 4 injectors and one produce is called as:
(a) Five spot (b) Line drive
(c) Seven spot (d) Four spot

152. Recovery % of oil in steam flood is:
(a) Upto 65% OOIP (b) Upto 20% OOIP
(c) Upto 15% OOIP (d) Upto 25% OOIP

153. Which caustic chemical is used in alkaline flooding:
(a) Sodium hydroxide (b) Sodium silicate
(c) Sodium carbonate (d) All of the above

154. CO_2 flooding is used for:
(a) Heavy crude oil (b) Waxy crude oil
(c) Medium and light crude oil (d) None of the above

155. What is primary producing mechanism on reservoir pressure and recovery efficiency of an oil well?
(a) Liquid and rock expansion (b) Gas cap expansion
(c) Water influx (d) All of the above

156. Which is true about gravity drainage mechanism?
(a) Declines reservoir pressure rapidly and continuously
(b) Water production is absent or negligible
(c) Recovery varies from 40 to 80%
(d) All of the above

157. Short term method for improved oil recovery:
(a) Infill drilling (b) WAG
(c) FAWAG (d) CO_2 injection

158. India has huge shale potential in ....
(a) Cambay basin (b) Cauvery Basin
(c) Damoder valley (d) All of the above

159. Gas shale is:
(a) Source rock (b) Reservoir rock
(c) Seal (d) Combination of (a), (b) and (c)

160. Oil sands can be surface mined from a depth of:
(a) 75 m (b) 100 m
(c) 300 m (d) 500 m

SECTION—B

1. Distillation characteristics of a crude are assessed in the laboratory by performing.......analysis.
2. Sulfur in crude is best determined by ....method.
3. The process carried out to remove light fractions from crude is called.......
4. The characterisation factor for a paraffin base crude is ....
5. Calorific value of LPG is in the order of ....MJ/m^3.
6. Boiling point of gasoline lies between ....°C to ....°C.
7. In refineries, heating of crude is done by ....heaters.
8. The process of removal of sulfur is known as ....process.
9. A stock is said to be sweet, if it passes the .... test.
10. Diesel index number is given by the expression ....

11. Softening point of Bitumen can be found by .... test.
12. Knocking characteristic of a fuel are tested in a ....engine.
13. Upgrading of low octane gasoline can be done by the process of ....
14. Glycol is added to gasoline as a ....agent.
15. Illumination characteristics of kerosene oil is given by .... number.
16. Aniline point test indicates qualitatively the .... content of an oil.
17. Smoke point of kerosene can be improved by removing .... compounds.
18. A cracking reaction is .... in nature.
19. The most effective catalyst used for cracking is ....
20. The catalyst used to convert cracker gas into polymer gasoline is ....
21. More widely used catalyst for isomerisation reaction is ....
22. Unisol process using caustic and methanol removes .... completely from gasoline.
23. The variation of viscosity of an oil with temperature is called ....
24. .... is one of the most important properties of a jet fuel.
25. Loss of gasoline vapours during storage is called .... loss.

SECTION—C

1. What are the factors responsible for petroleum formation from organic sources?
2. What are the various types of petroleum crude?
3. What is a salty crude?
4. What are the types of hydrocarbons the petroleum crude contain?
5. What are the various types of pipe still heaters used in refineries?
6. What is smoke volatility index?
7. What is knocking?
8. Define "Smoke Point".
9. What is stabilisation?
10. What is cetane number?
11. What are the factors which affect octane number of a fuel?
12. What are the major tests for gasoline?
13. What is 'aviation gasoline' and what is its composition?
14. What are the refinery processes used to improve the quality of lubricating oils?
15. What is the effect of pressure on cracking reaction?
16. What are the various catalytic cracking processes used in the refineries?
17. Name the various dewaxing processes with use of solvents.
18. Define "Cloud Point".
19. What is isomerisation process and why it is carried out in refineries?
20. What is hydro-cracking?
21. What are the important tests carried out for transformer oils?
22. What is characterisation factor?

23. What are the major products from the vacuum distillation tower?

24. What is solvent naphtha and what are its essential characteristics?

25. What is "Deoiling" and how it is done?

ANSWERS

SECTION–A

1. a	2. a	3. c	4. b	5. c
6. c	7. d	8. c	9. a	10. d
11. b	12. c	13. b	14. a	15. a
16. d	17. c	18. a	19. c	20. d
21. a	22. d	23. c	24. b	25. c

SECTION–AA

26. (a)	27. (d)	28. (a)	29. 2685ft	30. (a)

31. (b) 32. (10) **Sol.** Core volume = 10 × 4 × 4 = 160 cm ; core weight = 282.4 g brine + core = 300 g; brine weight = 300 – 282.4 = 17.6 g. brine volume = 17.6/1.10 (density) = 16.0 cm^3;

Porosity = brine volume/total volume (core volume) = 16/160 = 0.10

Percentage = 0.10 × 100 = 10%.

33. 1850 to 2000	34. (a)	35. (c)	36. (b)	37. (d)
38. (a)	39. (b)	40. (d)	41. 35%	42. (b)
43. (c)	44. (c)	45. (c)	46. 60.0	47. 0.01
48. (c)	49. 119.0	50. 2500 lbm	51. 51.5	52. (a)
53. 12.0	54. 64	55. 0.25	56. 121	57. 61.5
58. (a)	59. (d)	60. 4.8	61. (b)	62. (b)
63. 7.0	64. 5.0	65. 300.0	66. 0.24	67. 12.25
68. 3.4	69. 0.60	70. (a)	71. (b)	72. (a)
73. (c)	74. (c)	75. (d)	76. (a)	77. (b)
78. (a)	79. (a)	80. (d)	81. (a)	82. (c)
83. (b)	84. (c)	85. (b)	86. (c)	87. (a)
88. (d)	89. (d)	90. (c)	91. (a)	92. (d)
93. (d)	94. (d)	95. (c)	96. (c)	97. (d)
98. (b)	99. (c)	100. (c)		

101. (a) 102. (a) Sol. Porosity $= \dfrac{\text{Void}}{\text{Total Volume}} = \dfrac{1}{1+4}$ = 0.20 (20%).

103. (d)	104. (b)	105. (a)	106. (a)	107. (d)
108. (c)	109. (d)	110. (b)	111. (d)	112. (a)

113. (c)	114. (a)	115. (a)	116. (a)	117. (c)
118. (b)	119. (a)	120. (a)	121. (a)	122. (d)
123. (a)	124. (b)	125. (b)	126. (b)	127. (d)
128. (d)	129. (d)	130. (a)	131. (c)	132. (a)
133. (a)	134. (c)	135. (a)	136. (a)	137. (b)
138. (c)	139. (c)	140. (a)	141. (a)	142. (b)
143. (a)	144. (a)	145. (d)	146. (a)	147. (a)
148. (a)	149. (a)	150. (a)	151. (a)	152. (a)
153. (d)	154. (d)	155. (d)	156. (d)	157. (a)
158. (d)	159. (d)	160. (a)		

SECTION–B

1. True boiling point.
2. Bomb calorimeter.
3. Stabilisation
4. 12.5.
5. 38 to 82.
6. 37 to 180.
7. Pipe still.
8. Sweetening.
9. Doctor.
10. $\dfrac{\text{Aniline point (°F)} \times \text{°API}}{100}$
11. Ball and ring.
12. Co-operative Fuel Research (CFR).
13. Reforming.
14. Anti-icing
15. Luminosity.
16. Aromatics.
17. Aromatic.
18. Endothermic.
19. Zeolite catalysts.
20. Phosphoric acid supported on inert carrier like quartz or kieselgur.
21. Aluminium chloride.
22. Mercaptans.
23. Viscosity index.
24. Freezing point.
25. Breathing.

SECTION—C

1. (a) Bacterial action. (b) Heat. (c) Pressure,
   (d) Radioactivity. (e) Catalysts
2. (a) Paraffinic base. (b) Naphthenic or asphaltic base,
   (c) Mixed base.
3. Any crude that contains more than 5 kg of total salts expressed in terms of sodium chloride per thousand barrels.
4. The hydrocarbons may be divided into two chemical classes, viz.,
   (a) Open chain or Aliphatic compounds comprising of,
      (*i*) *n*-paraffin series, $C_n H_{2n} + 2$
      (*ii*) Iso-paraffin series, C_nH_{2n+2}
      (*iii*) Olefin series and ring compounds, comprising of $C_n H_{2n}$.

(b) Naphthene series, $C_n H_{2n}$

(c) Aromatic series, $C_n H_{2n-6}$

5. (a) Box or rectangular (b) Cylindrical

(c) Radiant wall.

6. Smoke volatility index = Smoke point + 0.42 (% distilled at 204°C).
7. Knocking is the uneven pressure created in an IC engine caused due to untimely burning of fuel and results in loss of power and serious damage to engine parts.
8. Smoke point in defined as the maximum height of a flame in millimetres at which the given oil will burn without producing smoke.
9. Stabilisation is the process of removing lighter fraction from a straight run product to facilitate trouble free storing and transportation of the product.
10. It is defined as percentage by volume of *n*-Cetane (*n*-hexadecane) in a mixture of *n*-Cetane and α-methyl naphthalene, which gives the same ignition delay as the fuel under consideration, when tested in a CFR engine.
11. (i) Engine speed

(ii) Ambient weather condition

(iii) Altitude

(iv) Combustion chamber deposits

(v) Coolant temperature.

12. (a) ASTM distillation (b) Reid vapour pressure

(c) Octane number (d) Gum content and

(e) Sulphur content.

13. Aviation gasoline, with an octane number 100 or even more, is composed of about one-third alkylate, derived from isobutane and gaseous olefins, blended with catalytically cracked gasoline and suitable crude distillate, to which appropriate quantity of tetraethyl lead is added to reduce knocking tendency.
14. (i) Sulphuric acid treatment (ii) Clay treatment

(iii) Solvent refining (iv) Hydrotreatment.

15. Increased pressure has a retarding effect on cracking reaction but in practice, a positive pressure of 10–15 kgf/cm^2 is normally used to suppress coke formation. Increase of pressure tends to decrease the yield of light fractions.
16. Fixed bed process (Houdry process).

Moving bed process (Thermofor catalytic cracking process) Fluidised bed process.

(i) Fluid catalytic cracking process

(ii) Kellog (Orthoflow) process

(iii) Flexi cracking process

(iv) Ultra cracking process.

17. (i) Barisol process (ethylene dichloride and benzol)

(ii) Ketone dewaxing process (methyl ethyl ketone or methyl propyl ketone)

(iii) Propane dewaxing process.

18. Cloud point is defined as the temperature at which an oil looks hazy and cloudy, when cooled under standard conditions and specified manner. It is primarily due to the formation of microline wax crystals.
19. It is basically a process of converting *n*-paraffins to iso-paraffins in presence of a suitable catalyst. In refineries it is mostly carried out to convert butane fraction to iso-butane. The product is used to produce good quality gasoline by the alkylation process.
20. It is a process of converting heavy distillates or fuel oils of relatively refractory in nature to high quality gasoline through cracking in presence of hydrogen.
21. (i) Acid value (ii) Dielectric strength
    (iii) Flashpoint (iv) Pour point
    (v) Sludge value (vi) Copper strip corrosion test.
22. It is a constant from which the quality of a crude can be known. It can indicate necessary informations about the various fractions and their physical properties. Mathematically, it is given by the relation

$$K = \frac{3\sqrt{R}}{0.823}$$

$$R = \text{Avg. Boiling point, K.}$$

23. (i) Cracking stock
    (ii) Fuel oil mixtures (Heating Oils)
    (iii) Raw lubricating oil
    (iv) Bitumen.
24. It is produced by distilling virgin naphtha into small boiling range cuts. It is mostly used as a solvent in paints, perfumery and other industries. Its essential characteristics are,
    (i) Water white colour
    (ii) Non-corrosive nature
    (iii) High stability
    (iv) High solvent power (KB number)
    (v) Low boiling point
    (vi) Low acidity
25. A large quantity of residual oil remains in the crude wax produced in dewaxing unit. The quantity of retained oil may be as high as 50%, if chilling is done without solvent. Deoiling is the process of removal of oil from wax. It is done by any of the following processes:
    (i) Sweating process (ii) Resettling process
    (iii) Solvent extraction process.

❑

Energy Conservation and Safety

1. Occupational Safety and Health Administration (OSHA) incidence rate

$$\text{(based on injuries and illness)} = \frac{\text{Number of injuries and illness} * 200{,}000}{\text{Total hours worked by all employees during period covered}}$$

$$\text{FAR (Fatal accident rate)} = \frac{\text{Number of Fatalities} \times 10^2}{\text{Total hours worked by all employee during period covered}}$$

$$\text{Fatality rate} = \frac{\text{Number of fatalities per year}}{\text{Total number of people in applicable population}}$$

Nature of accident processes: Fires: 31% ; Explosions: 30% ; Vapour cloud explosions: 36% ; Others : 3%

2. **Significant Disasters:**
   (i) Flixborough, England–June 1974. Manufacture of caprolactum (a raw material for production of Nylon). Cause was due to ignition of cyclohexane.
   (ii) Bhopal, India – Dec. 3. 1984. The Plant produced a pesticide. Intermediate compound is methyl iso-cyanate. MIC storage tank got exposed to water and exploded.
   (iii) Sevego, Italy the product was hexachlorophene. Tetrachlorodibenzo-p-dioxin [TCDD] contaminated the soil. July 10, 1976.
   (iv) **Pasadena, Texas:** October 23, 1989. Polyethylene plant. Reason is flammable mixture of ethylene, Isobutane, hexane and hydroxide.

3. **Toxicants:** Toxicants enter our biological organisms through : *a* – Mouth to stomach ; *b* – Mouth and nose to lungs; *c* – Through cuts into skin; *d* – Through skin membrane.

   The effects of toxicants are : *a* – Respiratory systems ; *b* – Nervous systems ; *c* – Changes in skin texture, pigmentation, hair and nail appearances; *d* – blood count.

4. **Fires and Explosions :** Fires release energy slowly but explosions release energy rapidly.

   Flash point temperature is one of the major qualities used to characterise the fire and explosion hazard of liquids.

   The flash point temperature for pure liquids correlate well into the boiling point of the liquids.

   The minimum ignition energy (MIE) is the minimum energy input required to intiate combustion. All flammable materials have MIEs. MIE depends on the specific chemical or mixture, the concentration, pressure, and temperature.

The autoignition temperature of a vapour called the spontaneous ignition temperature is the temperature at which the vapour ignites spontaneously from the energy of environment.

5. **Flares:** The objective of a flare is to burn the combustible or toxic gas to produce combustion products that are neither toxic nor combustible.
6. **Fault trees:** Fault trees organised in aerospace industry and have been used estensively by the nuclear power industry to qualify and quantify the hazards and risks associated with nuclear power plants. This approach is becoming more popular in the chemical process industries as a result of successful experiences in nuclear industry.

   Advantages and disadvantages of Fault trees:

   1. Fault tree is enormous ;
   2. The developer of fault tree can never be sure that all failure modes have been considered.
   3. Fault trees developed by different individuals are different in nature.
   4. An advantage is that it begins with a top event.
7. **Relationship between fault trees and event trees :** Event trees begin with an initiating event and work toward the top event (induction). Fault trees begin with a top event and work backward toward the intiating events (deduction). The initiating events are the causes of the incident and the top events are final outcomes. The two methods are releated in that the top events for fault trees are the initiating events for the event trees. Both are used together to produce a complete picture of an incident, from its initiating causes all the way to its final outcome. Probabilities and frequencies are attached to these diagrams.
8. **QRA and LOPA:** Risk is the product of the probability of a release, the probability of exposure and the consequences of the exposure. All companies decide their levels of acceptable risk and unacceptable risk. The actual risk of a process or plant is usually determined using Quantitative risk analysis (QRA) or a layer of protection analysis (LOPA). Other methods are sometimes used, however QRA and LOPA are the methods that are most commonly used. In both methods the frequency of the release is determined using a combination of event tree, fault tree or an appropriate adaptation.

   Quantitative risk analysis is a method that identifies where operations, Engineering or management systems can be modified to reduce risk. The complexity of a QRA depends on the objectives of the study and the available information. Maximum benefits result when QRAs are used at the beginning.

QUESTIONS

SECTION—A

1. Unscientific storage of coal may result in 'wind loss' which is of the order of:

   (a) 2% (b) 5% (c) 8% (d) 10%
2. Coal, when stocked on soft ground without proper flooring, a certain thickness of the storage, is converted to 'carpet loss'. This thickness is :

   (a) 2–5 cm (b) 4–8 cm (c) 5–15 cm (d) 10–25 cm
3. The percentage of fines that can effectively burn with lumpy coal is :

   (a) 5–10 (b) 10–15 (c) 20–25 (d) 30–35

4. For a satisfactorily insulated boiler the radiation loss may be :

(a) 2–3% (b) 6–7% (c) 8–10% (d) 10–12%

5. As a means of energy conservation in a heat exchanger, the procedure adopted is :

(a) To pump more fluid.

(b) To increase tube length.

(c) To keep heat transfer surfaces clean.

6. Excess air in combustion of fuel should be economically used, because use of more excess air results in :

(a) More power requirement for blowing the air

(b) More sensible heat loss in the flue gas

(c) Uncontrolled combustion.

7. In fuel oil firing, combustion air preheating can save oil consumption up to :

(a) 5% (b) 10% (c) 15% (d) 20%

8. Cost of power expressed as a percentage of the cost per tonne in case of coal-based fertilizer is around :

(a) 10% (b) 25% (c) 30% (d) 35%

9. Consumption of power for manufacture of paper in an Indian paper industry is around :

(a) 500 kWh/tonne (b) 1000 kWh/tonne

(c) 1400 kWh/tonne (d) 2000 kWh/tonne

10. Preheating of viscous oil is a pre-requisite for perfect atomisation and improved burning efficiency. The correct preheat temperature for a fuel oil of viscosity 2000 Red wood (No. 1) seconds is :

(a) 30–40° C (b) 50–70° C (c) 80–90° C (d) 105–110° C

11. A suitably designed coal-fired furnance, if properly maintained and operated, can give overall thermal efficiency of the order of:

(a) 15 – 20% (b) 25 – 30% (c) 35 – 40% (d) 50 – 55%

12. Minimum safe distance between two liquid fuel tanks is equal to :

(a) One-sixth the height of the tank (b) One-third the height of the tank

(c) One-half the height of the tank (d) The height of the tank

13. For high capacity liquid fuel storage tanks above 450,000 litres, safe dimensions with respect to height and diameter are :

(a) $\frac{H}{D} < 2$ (b) $\frac{H}{D} < 3$ (c) $\frac{H}{D} < 1.5$ (d) $\frac{H}{D} < 1$

where, H = Height of the tank.

D = Diameter of the tank.

14. Volume of dikes used around storage vessels with hazardous material is equal to :

(a) Half the volume of the storage vessel

(b) Volume of the storage vessel

(c) Twice the volume of the storage vessel.

15. With 1350 TPD ammonia plant the specific energy consumption (in MK cal/Mt) has come down to :

(a) 12–15 (b)8–9.5 (c) 10–12

16. In order to make a naphtha-based ammonia plant energy efficient, losses in various sections can be reduced. Maximum reduction of such losses is possible in:

(a) Compressor and synthesis section
(b) Refrigeration system
(c) Primary reforming and feed preheat section
(d) HT/LT shift and methanation section

17. Oil fires can be extinguished by :

(a) Foam (b) Carbon dioxide (c) Water

18. From the point of view of safety which is preferable :

(a) A liquid phase reactor (b) A vapour phase reactor
(c) Either of the above

19. Safety or overdesign factor in heat transfer equipment expressed as percentage of the actual design is :

(a) 5–10 (b) 10–15 (c) 15–20 (d) 25–30

20. A gas is said to be non-toxic when its Threshold limit value (TLV) or maximum acceptable concentration is :

(a) 500–1000 ppm (b) 1000–5000 ppm
(c) 5000–9000 ppm (d) 10000–100000 ppm

21. From the point of view of toxicity methyl isocynate (MIC) and phosgene come under the category of:

(a) Moderatelly toxic (TLV : 100—1000 ppm)
(b) Extremely toxic (TLV : 1—10 ppm)
(c) Very extremely toxic (TLV : 1 ppm)
(d) Highly toxic (TLV : 10—100 ppm).

22. The safe distance of population centre from a plant processing a hazardous chemical of lethal concentration of $10 - 10^4$ ppm is :

(a) 5 km (b) 10 km (c) 15 km (d) 25 km

23. Polluting and hazardous industries should have an "Exclusion zone" with a green belt and public access prohibited around the plant covering a radius of:

(a) 100 m (b) 500 m (c) 1 km (d) 5 km

24. The factor which contributes most for the accidents in chemical industry is:

(a) Inadequate design/layout (b) Equipment failure
(c) Improper operating procedures (d) Unknown causes

25. Relief valve leakage for extremely hazardous material can be avoided by:

(a) Surge chamber (b) Dike (c) Rupture diaphragm

SECTION — B

1. What important factors help in achieving higher efficiency in the combustion of fuels?
2. What is 'Turn Down Ratio' for a liquid fuel burner?
3. What are the important considerations in selecting a liquid fuel burner for efficient combustion?
4. What are the advantages of preheating combustion air?
5. What is a double-decker furnace?
6. What are the important considerations for efficient insulation?
7. What is the economic thickness of insulation for a plain wall?
8. How can the liquid surfaces be insulated?
9. What is total energy system?
10. What is the size of a safe coal storage?
11. What are the main advantages of fluidized bed combustion?
12. Why elevated structures in a plant are located upwind of furnaces?
13. How are the equipments subject to possible rupture located?
14. What are dikes?
15. What safety measure is to be practised for process systems before starting cleaning and maintenance?
16. What is the general safety practice followed for most process systems prior to start up?
17. What general precautions are to be followed in hazardous chemical (a chemical harmful to human tissues) processing plant?
18. Which type of reactor is safer—a pot type or a tubular type?
19. What is a 'Sterilized zone' with respect to a hazardous chemical industry?
20. What is detonation?
21. What is a vent gas scrubber?
22. What are flame arresters and hydraulic flash arresters?
23. What is a flare tower?
24. What is'Safety Audit'?
25. What is 'Risk Assessment?

ANSWERS

SECTION A

1. a,	2. c	3. c	4. b	5. c
6. b	7. d	8. c	9. c	10. d,
11. b	12. a	13. d	14. b	15. b
16. c	17. a	18. b	19. c	20. d
21. c	22. d	23. c	24. b	25. c

SECTION — B

1. The important factors which help in achieving higher efficiency in the combustion of fuels are :
   (a) Correct fuel preparation
   (b) Supply of right quantity of combustion air
   (c) Selection of appropriate firing equipment
   (d) Efficient firing technique.
2. The range of fuel rate over which a burner operates with high burning efficiency, is called Turn Down Ratio'.
3. The important consideration in selecting a liquid fuel burner for efficient combustion are :
   (a) The temperature conditions in the combustion chamber
   (b) The shape of the combustion chamber
   (c) The required combustion intensity denoted in kilograms of fuel or litres of oil burnt per hour per unit volume of the combustion chamber
   (d) The range under which the burner will be required to operate
   (e) The 'turn down ratio'
4. The advantages of preheating combustion air are :
   (a) The flame temperature is raised, resulting in increased heat transfer
   (b) The speed of combustion is increased thus further raising the flame temperature
   (c) The amount of excess air necessary for combustion can be reduced
   (d) The heat taken away by the combustion air is returned to the furnace
   (e) The ultimate loss is reduced
5. A double decker furnace is a special type of furnace designed to avail the waste heat of the flue gas. The lower temperature gas escaping from the furnace is allowed to heat the charge in the initial stage and thus it serves the pre-heating of the charge. By doing so considerable amount of heat is saved.
6. Important considerations for efficiént insulation are :
   (a) Economic thickness of insulation
   (b) Thermal capacity of insulating material
   (c) Insulation of liquid surfaces
7. The economic thickness (t_e) of insulation for a plain wall is given as:

$$t_e = \frac{K.C_h.H(T_0 - T_a)^{0.5}}{FC_i}\left(\frac{K}{h_0}\right)$$

where, K = Thermal conductivity of insulation, $Wm^{-1} K^{-1}$
C_h = Unit cost of heat, Rs. J
H = Period of operation/year, hour
F = Factor considering the annual recovery of investment on insulation and depreciation.
C_i = Cost of insulation, RS m^3
h_0 = External heat transfer coefficient, $Wm^{-2} K^{-1}$
T_0 = Operating temperature, K
T_a = Ambient temperature, K

8. Insulation of liquid surfaces can be achieved by floating an insulating layer of hollow plastic such as poly propylene spheres on the liquid surfaces.

9. Total energy is a system in which process steam demand and electrical power requirement are achieved.

10. Height of less than 2.5 metres and a maximum capacity of 500 tonnes are the conditions of a safe coal storage.

11. Some of the main advantages of fluidized bed combustion are :

(a) It can be applied to the burning of low grade high inert-content and difficult-to-handle materials.

(b) With better gas-solid contact a higher combustion efficiency is achieved.

(c) In boiler applications, there is a bank of steam raising tubes immersed in the bed, the heat transfer to which is uniformly high in comparison to the variation of radiated heat transfer through a conventional combustion chamber.

(d) Less heat transfer surface is required for a given output.

12. Elevated structures in a plant are located upwind of furnaces to minimize :

(a) Flue gas hazard to personnel, and

(b) Dangerous concentrations of volatile combustibles.

13. Equipments subject to possible rupture should be isolated by distance, blast enclosures or pits.

14. Dikes are low walls around storage vesseles with hazardous materials.

15. Purging of process systems is an essential safety measure prior to opening up or entering for cleaning and maintenance.

16. Air must be eliminated from most process systems prior to start up. Steam is usually employed for the purpose.

17. In hazardous chemical processing plant emergency showers and eye wash stations are so be provided in all areas like storage, handling, transfer and processing.

18. A tubular reactor is safer to a pot type reactor.

19. An area with 5 km radius around a hazardous chemical industry is termed as 'Sterilized zone'. The total population in this zone should be less than 20000 and preferably in the upwind direction.

20. It is an explosion or sudden release of pressure where shock waves with supersonic velocity (> 300 metre/second) prevail.

21. Before venting, toxic materials are removed by spraying water or a suitable absorbent solution in a vent-gas scrubber.

22. A flame or flash arrester prevents flame propagation with inflammables. The hydraulic flash arrester uses a hydraulic medium for the purpose which may be water, glycol water, mineral oil or inert liquid with low vapour pressure and viscosity.

23. It is a tall tower with a pilot flame to ignite inflammables under excessive pressure.

24. 'Safety Audit' normally refers to a check of the plant hardware and operating procedures after some time in operation.

25. It is a quantification of the total risk (to life, property or production) associated with a hazardous process. Risk assessment is a general technique which can be applied to general policy decision as well as to single projects.

❑

CHAPTER 15

Optimisation

1. Optimisation, aims to find the best way to operate a system. Examples are:

(a) Minimising cost in selecting a reflux ratio in a distillation column (continuous)

(b) Selecting optimum number of effects in multiple effect evaporation.

Optimisation includes economy, making logical decisions, to find the best set of conditions.

2. Terminology:

(a) Objective function: The criterion or index is known as objective function, e.g. minimum cost of product ion, maximum profit, pay out time, rate of return, etc.

(b) Monotonic function: A function $y(x)$ is increasing or decreasing continuously with change of independent variable x.

(c) Unimodal function: Having only one maximum or minimum value.

(d) Interval of uncertainity: This is also called as the region of interest. It is the region where optimum lies.

(e) Stochastic process: Output is uncertain for a given input, e.g. ambient humidity and temperature.

(f) Deterministic process: Output depends entirely on input.

(g) Convex function: This has minimum value.

(h) Concave function: This has maximum value.

(i) Constraints are the restrictions.

(j) Feasible region: The search for optimum of a constrained optimum isation problem must be in the interior region or along the boundaries fixed by constraints. This region is feasible region.

(k) Stationary point. $dy/dx = 0$, That is first derivatives are zeroes.

(l) Slack variable: Converting inequality to equality for solving optimisation problem, when constraints are present. Comes in linear and dynamic programming.

3. In a two variable problem, checking the optimisation values.

$A = d^2 y/dx^2_1$; $B = d^2y/dx_1\, dx_2$; $C = d^2 y/dx_2{}^2$

If $B^2 - A * C < 0$, if $A < 0$, Relative maximum at x_0, y_0.

If $B^2 - A*C < 0$, if $A > 0$, Relative minimum at x_0, y_0.

If $B^2 - A*C > 0$, saddle point at x_0, y_0.

If $B^2 - A * C = 0$ nature of critical point undetermined.

At saddle point is neither maximum nor minimum.

Inflection: At this second derivative is zero, and first derivative changes in sign at this point.

Ridge: First derivatives are zeroes but may not be maximum or minimum (optimum).

4. Geometric programming: The optimum values are obtained without constratant as:

$$\sum_{j=1}^{m} C_j \, P_j \, (x) : P_j - \prod_{i=1}^{n} x_i^{a_{ij}} ; \sum a_k \, jw_j = 0$$

$$y = \pi(C_j / w_j)^{\omega_j}$$

5. Numerical search methods of one variable:

(a) Simultaneous search; (b) Sequential search; (c) Three equally spaced trials; (d) Dichotomous search; (e) Fibonacci search; (f) Golden section.

6. Numerical search methods for multi variables:

(a) Alternate variable search method; (b) Exploratory and pattern moves; (c) Powell's method; (d) Steepest ascent or descent; (e) Gradient method; (f) partan method; (g) Variable metric method; (h) Method of conjugate gradients; (i) Rosenbrock method.

Differences between direct and gradient methods:

1. Gradient (derivative) methods are more effective than direct search methods but sometimes gradient is not defined.
2. Direct search is more effective in terms of computation.
3. Gradient methods converge more rapidly.
4. Each method has some basic strength and weakness.

7. Linear programming: When the objective function is to be optimised and its constraints are linear, we speak of this optimisation process as linear programming. An example is :

A chemical plant produces A units of product 1 and B units of product 2 and C unit of product 3 having profits of 10,4,1 respectively. Due to process conditions the constraints are: $A + B \leq 5$ and $2A + B + C \leq 20$. The problem is to find the values of A, B and C so that the profit is maximum.

Dynamic programming: This is useful for multistage stage problems. Bellman introduced this and expressed as: An optimal policy has the property that whatever is the intial state and intial decisions are, the remaining decisions must constitute an optimal policy, with regard to the state resulting from the first decision.

Examples of discrete dynamic programming are: Tubular reactors, packed columns, etc.

8. Optimum values analytically:

For single variable: $y = F(x)$.

$dy/dx = 0$ gives maximum or minimum values.

d^2y/dx = negative : gives maximum value.

d^2y/dx^2 = positive gives minimum optimum value.

For two variable optimisation $y = F(x_1, x_2)$; $dy/dx_1 = 0$; $dy^2/dx^2 = 0$.

Solving both gives optimum x_1, x_2. values. Checking the optimum values for maximum or minimum is as in item 3.

9. Dichotonous search: $I_K = \frac{I_0}{2^k} + \left(1 - \frac{1}{2^k}\right)E$ (interval of uncertainity after K experiments). I_0 is intial interval of uncertainity, E is gap. $x_1 = \frac{I_0}{2} + \frac{E}{2}$; $x_2 = \frac{I_0}{2} - \frac{E}{2}$.

10. Fibonacci Search: $F_n = \frac{1}{\sqrt{5}}\left[\left(\frac{1+\sqrt{5}}{2}\right)^{n+1} - \left(\frac{1-\sqrt{5}}{2}\right)^{n+1}\right]$

$$x_3^i = \frac{F_n - (i+1)}{F_n\ (i-1)}\left[x_2^{x-1} - x_1^{i-1}\right] + x_1^{i-1}$$

$$x_4^i = \frac{F_{n-i}}{F_{n-(i-1)}}(x_2^{i-1} - x_1^{i-1}) + x_1^{i-1};\ I_{n-1} = \frac{F_1}{F_n} I_0$$

11. Golden section: $x_3^i = \frac{r-1}{r}(x_2^{i-1} - x_1^{i-1}) + x_1^{i-1}$

$$x_4^i = \frac{1}{r}(x_2^{i-1} - x_1^{i-1}) + x_1^{i-1};\ I_{n-1}/I_0 = \frac{1}{r^{n-1}}$$

12. Contour tangent equation: y = Constant or $dy = 0 = m_1\ dx_1 + m_2\ dx_2$.

13. Newton–Raphson method: $x^{i+1} = x^i - \frac{y(x_i)}{\dot{y}(x_i)}$

14. Principal of optimality: Reduces a m stage problem to m single stage problems. Recurrence relations from stage to stage are based on the principle of optimality stated by Bellman as: An optimal policy has the property whatever the intial state and intial decisions are, the remaining decisions must continue an optimal policy with regard to the state resulting from the first decision. It is convenient to calculate from last stage to first stage.

15. Bi-section method:

Step-1: Choose two points a, b such that $f'(a) < 0$ and $f'(b) > 0$. Set $x_1 = a$; $x_2 = b$; Choose a small value E.

Step-2: Calculate $z = (x_2 + x_1)/2$ and evaluate $f'(z)$.

Step-3: if $|f'(z)| \le E$ then terminate.

Else: $f'(z) < 0$; Set $x_1 = z$ and go to Step-2

If $f'(z) > 0$; set $x_2 = z$ and go to Step-2

16. Secant method: $x = x_2 - \frac{f(x_2)}{[f'(x_2) - f'(x_1)]/(x_2 - x_1)}$

QUESTIONS

SECTION—A

1. Which of the figures is concave set?

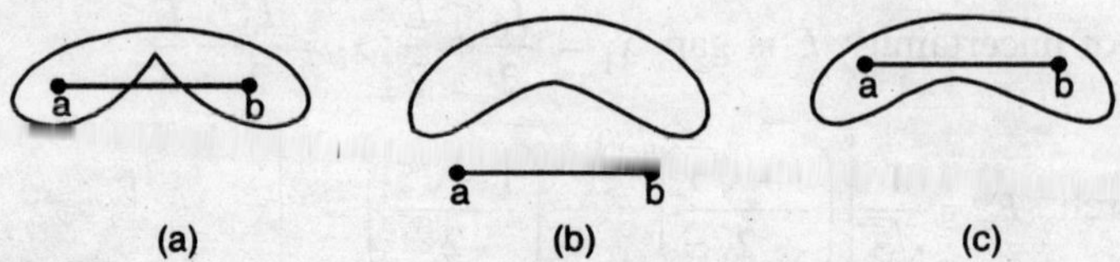

2. Which of the figures in Question (1) is convex set?

3. At stationary point:

(a) First derivative is zero (b) Second derivative is zero

(c) Third derivative is zero.

4. Saddle point means :

(a) Maximum value (b) Minimum value

(c) Neither maximum nor minimum value.

5. For $y = 1 - x + x^2$, optimum value of y is :

(a) Zero (b) 3/4 (c) –3/4 (d) None

6. For relative maximum for one variable :

(a) Second derivative should be less than zero

(b) Second derivative should be greater than zero

(c) Second derivative should be equal to zero

7. At inflection :

(a) Second derivative is greater than zero

(b) Second derivative is less than zero

(c) Second derivative is equal to zero.

8. Exploratory and pattern move is devised by :

(a) Box (b) Hookes and Jeeves (c) Lagrange

9. Rosenbrock method uses :

(a) Ist order derivatives (b) Second order derivatives

(c) No derivatives.

10. The equation $4x^2 - 12xy + 10y^2$ is :

(a) Positive definite (b) Negative definite (c) Indefinite

11. The expression $9x^2 + 2y^2 + 3z^2 - 6xy - 12xz + 2yz$ is :

(a) Positive definite (b) Negative definite (c) Indefinite

12. Alternate variable search method fails in case of :

(a) Concave function (b) Convex function and (c) Ridge

13. Powell's method of optimisation uses :

(a) First order derivatives (b) No derivatives

(c) Second order derivatives

14. In Rosenbrock method if the move was a success, in next step the distance is increased by :

(a) Twice (b) Three times (c) Four times

15. In Rosenbrock method if the move was a failure, then the distance is reduced by :

(a) Half (b) One third (c) Two third

16. Dynamic programming was introduced by :

(a) Taylor (b) Euler (c) Bellman

SECTION — B

1. Putting a number of thermocouples to search for a hot spot in a bed is called as .... search plan.
2. Interval of uncertainty after placing *n* trial points in simultaneously search plan is ....
3. The shrinkage in two equally spaced trials is ....
4. The shrinkage in three equally spaced trials is ....
5. In dichotomous search, the interval of uncertainty is ....
6. In Fibonacci search, the interval of uncertainty after *n* function evaluations is ....
7. Interval of uncertainty in Golden section after *n* function evaluations is ....
8. An example of discrete dynamic programming is ....
9. An example of continuous dynamic programming is ....
10. Montonic function .... or .... continuously with change of independent variable.
11. Unimodal function means having only one ....or ...... value.
12. Stochastic process means output is .... for a given input.
13. Deterministic process means output depends on .... entirely.
14. When responses are equal, the optimum must lie .... the two experiments.
15. In ridges are zeroes but may not be maximum or minimum.
16. Newton-Raphson relations ....
17. Stack variable is introduced to convert .... to ....
18. In linear programming the objective function and its constraints are ....
19. Full form partan is ....
20. Full form of CPM is ....
21. Full form of EVOP technique.
22. Full form of PERT is ....

SECTION — C

1. Locate the optimum for

$$y = -1/4\, x^4 - 5/3x^2 - 3/2x^3 + 9x + 3$$

2. Compute the optimum number of effects (N) for an evaporator system. Heat transfer rate is 4 million kcals/hr. Useful life is 12 years. Fixed costs are 22% per year. Days of operation per year is 200 (24 hrs. per day). Steam cost is Rs. 10.00 per 500 kg. One kg of steam will evaporate (0.8 N +1.00) kg of water. Latent heat of water is taken as 500 kcal/kg. Each effect costs 25,000 Rupees.

3. A hypothetical chemical plant consists of a tubular reactor, a product separator, compressors and a mixer. The costs depend on two variables, the pressure x_1, the recycle ratio x_2. The required production rate is 10 million pounds per year. For this rate, the annual costs in dollars are :
Compressing raw materials.......2000 x_1.
Operating reactor .... $5 \times 10^9/(x_2\, x_1^{1.5})$
Operating separator,......$2 \times 5\, x_2$
Recirculating materials $2 \times 10\, x_2$.
(a) Calculate the optimum conditions by partial differentiation.
(b) Check to determine that the value of optimum obtained is an optimum, not an inflection, ridge or saddle.

4. Solve Problem 3 by geometric programming.

5. Find optimum values by geometric programming to the function,

$$y = 1000\, x_1 + \frac{4 \times 10^9}{x_1\, x_2} + 2.5 \times 10^5\, x_2$$

6. Using Dichotomous search find the range of minimum value for $y - x^2 - 0.5x + 1.0$ with $E = 0.05$ in the interval of x, 0 to 1 using four experiments.

7. Using Fibonacci search find the interval of uncertainty after three function evaluations. The function is $y = x^2 - 0.5\, x + 1$ and beginning interval is 0 to 1.

8. Using $E = 0.05$, find the interval of uncertainty after $n = 4$ (function evaluations) for $y = x^2 - 0.5x + 1.0$ in the interval of 0 to 1 for x using Fibonacci method.

9. Calculate the number of experiments required for a Fibonacci search to reduce the interval of uncertainty to 0.10 for $E = 0.05$.

10. Using Golden section find the interval of uncertainty for 3 function evaluations for y minimum of $x^2 - 0.5x + 1$ in the interval of 0 to 1.

11. Using exploratory and pattern moves, calculate one exploration step and one pattern step for the function, $y = 100\, (x_2 - x_1^{\,3})^2 + (1 - x_1)^2$ starting search from $x_1 = -1.2$ and $x_2 = 1.0$.

12. Obtain the contour tangent equation for $y = 7x_1^{\,2} + 10x_2^{\,2}$ at $x_1 = 10$ and $x_2 = 3$.

13. Obtain minimum x for $y = 3x^2 - 2x$ with $x\,(0) = 1$.

14. Write the following methods in descending order based on function evaluations for same reduction. Golden section, simultaneous search plan, two equally spaced trials, three equally spaced trials, dichotomous search, Fibonacci search.

15. With interest at 8% calculate the present worth of (a) Rs. 540 at the end of 10 years (b) Rs. 37.3 per year for 10 years.

16. An investment of Rs. 23000 in a process produces a net income of Rs. 10,000 a year for 5 years. There is no income after this time. Using straight line depreciation plus average interest, calculate the interest rate.

17. A piece of equipment has a life of 5 years (A). Another piece of different design will perform exactly the same as the first and has a longer life. The equipment (B) costs twice as much as equipment A. Calculate the required life of B which will make it competitive with A, using a true interest rate of 10%.

(a) Assume no salvage value

(b) Assume salvage value equals 50% of first costs.

18. Draw/find critical path method.

Activity	*Description*	*Days*
1 — 2	Engineering lead time.	10
1 — 3	Time for 10% of foundation drawings	12
1 — 4	Time to complete foundation drawings	20
1 — 5	Delivery time for underground piping.	6
2 — 5	Rough grading site	3
3 — 5	Time lag from start of foundation to start of underground piping	2
3 — 4	Construct foundations	12
4 — 6	Time lag from finish of foundation to finish of underground piping	2
5 — 6	Install underground pipe and conduit	6
5 — 7	Time lag from start of undergound piping to start of paving	3
6 — 8	Time lag from finish underground to finish of paving paving	3
7 — 8	Paving	3
8 — 9	Cleanup	2

19. A distillation column is running 2,000 barrels/day to meet a market demand of 400 barrels per day (365 days/year). It costs Rs. 500 to start up the unit and. The product is valued at Rs. 5 per barrel. The company uses a rate of return of 15% for economic analysis. Storage costs are Rs. 0.25 per barrel, per year. All other costs may be neglecting. Assuming the storage costs depend on the maximum inventory, calculate the optimum number of lots per year, and the optimum number of days per operating period.

20. A pilot plant making samples for market development has a capacity of 100 gallons per day. The uniform rate of sending samples is 20 gallons per day. The optimum number of production runs per year is 10, calculate the average inventory for optimum operation.

21. Calculate the optimum thickness of insulation for the following conditions :

(a) Insulation cost factors : Capital cost: Rs. 5 per cu. ft. Depreciation rate : 10%; interest rate : 10% average; shape : flat.

(b) Heat cost factor: Constant 0.80 Rs per million BTU.

(c) Thermal factors : Conductivity of insulation = 0.05 BTU/ (ft. hr. °F)

Film coefficient on outside = 1.0 BTU/ (sq. ft. hr. °F)

22. The costs of replacing the catalyst in a catalytic reactor is Rs. 90,000. Since the catalyst will last less than a year, this should be considered as an expense, not a capital investment. Regeneration costs Rs. 1000. The time needed for regeneration of replacement is neglected in the cost analysis. During operation, the decrease in catalyst activity causes the product value to decrease at the rate of Rs. 2000 per day. Regeneration reduces this loss to 20% of the value at the beginning of the regeneration cycle. Make all regeneration cycles the same length. Find the optimum cycle of regeneration and replacement.

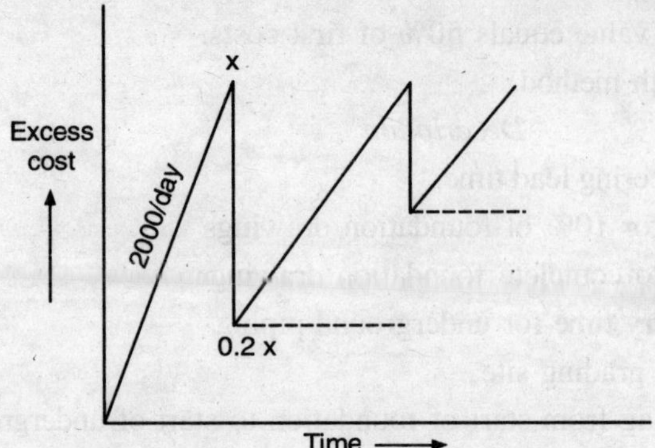

23. An evaporator costs Rs. 25,000. At the end of 10 years it will be replaced and at that time its salvage value will be Rs. 2500. With interest at 6% calculate :

(a) Annual fixed charges, exact method.

(b) Annual fixed charges by straight line depreciation plus average interest.

(c) The average interest rate.

24. A process which costs Rs. 100,000 produced an annual income of Rs. 10,000 a year for 10 years at which time its salvage value was 10,000. To what rate of interest did this correspond?

25. A reactor produces 10,000 lb/hr of product C, by reaction of $A + B$. The weight of A used per lb of C is $0.64/(1 - e^{-r})$ where r is the weight ratio of B to A. A costs Rs. 0.15 per lb and B costs Rs. 0.10 per lb. Excess reactants are not recovered. Derive an equation for optimum ratio r.

26. Fill in the squares

			X_1	X_2	S_1	S_2
Basis :	C	b	– 100	– 150	0	0
S_1	0	400	2	2	1	0
S_2	0	300	0	3	0	1
		0	100	150		

Step 2

S_1 ☐

x_2 ☐ ☐

27. Set up tableau

$P = 100\ x_1, + 50x_2$; maximise P

$x_1 + x_2 \leq 70$

ANSWERS

SECTION—A

1. a	2. c	3. a	4. c	5. b
6. a	7. c	8. c	9. c	10. a
11. c	12. c	13. b	14. b	15. a,
16. c				

SECTION—B

1. Simultaneous.
2. $2I_0/(n+1)$
3. $(2/3)^K$ where K is iterations.
4. $(1/2)^K$ where K is the number of iterations.
5. $I_K = \frac{I_0}{2^K} + \left(1 - \frac{1}{2^K}\right)E$ where K is the number of iterations and E is the gap.
6. $I_{n-1}/I_0 = I/F_n$
7. $I_n/I_0 = 1/r^{n-1}$ where $r = 1.618$
8. Multistage extractor (stagewise contact) or a number of heat exchangers in series.
9. tubular reactor or packed column.
10. increases or decreases
11. maximum or minimum
12. uncertain,
13. input
14. between
15. first derivatives
16. $x^{i+1} = x^i - \frac{y(x^i)}{\dot{y}(x^i)}$
17. inequality to equality
18. linear
19. Parallel tangents
20. Critical path method
21. Evolutionary operation
22. program evaluation review technique

SECTION—C

1. $\frac{dy}{dx} = -x^3 - 5x^2 - 3x + 9 = 0$ gives $x = 1, -3, -3$

$\frac{d^2y}{dx^2}$ at $x = 1$ is -16 and $\frac{d^2y}{dx^2} = 0$ at $x = -3$

So at $x = 1$ gives local maximum, since $\frac{d^2y}{dx^2}$ is negative

At $x = -3$ is saddle point.

2. $q = 4 \times 10^6$ kcal/hr.

$$\text{Fixed costs} = 25000 \text{ N} \times 0.22 \text{ per year}$$

$$\text{Variable cost (steam required)} = \frac{4 \times 10^6}{500(0.8\,N + 1)} \text{ kg/hr}$$

$$\text{Variable costs/year} = \frac{4 \times 10^6 \times 24 \times 200 \times 10}{500\,(0.8\,N + 1)\,500}$$

Total costs per year = variable costs per year + fixed costs/yr.

$$= 2000 * 0.22\,N + \frac{768000}{(0.8\,N + 1)}$$

$$= 5500\,N + \frac{768000}{(0.8\,N + 1)} = T.C$$

$$\frac{dT.C}{dn} = \frac{5500 - 68{,}000 \times 0.8}{(0.8N + 1)^2} = 0$$

N = 12 effects. **Ans.**

3. (a) Cost equation (C) is :

$$2000x_1 + \frac{5 \times 10^9}{x_1^{1.5}\,x_2} + 2 \times 10^5\,x_2 + 2 \times 10^5\,x_2$$

$$\frac{dC}{dx_1} = 2000 - \frac{5 \times 10^9\,x_2 x_1^0\,x_{1.5}}{x_1^3\,x^2} = 0 \quad \text{...(1)}$$

$$\frac{dC}{dx_2} = 4 \times 10^5 - \frac{5 \times 10^9\,x_1^{1.5}}{x_1^3\,x_2^2} = 0 \quad \text{...(2)}$$

Solving equation (1) and (2) gives $x_1 = 382$ and $x_2 = 1.33$

(b) Checking the value obtained is an optimum

$$A = \frac{d^2c}{dx_1^2};\ B = \frac{d^2c}{dx_1 dx_2};\ C = \frac{d^2y}{dx^2}$$

$B^2 - AC < 0$ if $A < 0$ relative maximum at (x_1, x_2)

$B^2 - AC < 0$ if $A > 0$ relative maximum at (x_1, x_2)

$B^2 - AC > 0$ saddle point at (x_1, x_2)

$B^2 - AC = 0$ nature of critical point undetermined.

A saddle point is a one which represents neither maximum nor minimum value.

Testing for saddle point :

$$\frac{d^2C}{dx_1^2} = \frac{7.5 \times 2.5 \times 10^9}{x_1^{3.5}\ x_2} = A; \frac{d^2C}{dx_2^2} = \frac{10^{10}}{x_1^{1.5}\ x_2^3} = C;$$

$$\frac{d^2C}{dx_1 dx_2} = B = \frac{7.5 \times 10^9}{x_1^{2.5}\ x_2^2}$$

at $x_1 = 382, \quad x_2 = 1.33$

$A = 12.7$, $B = 1.485 \times 10^3$, $C = 5.76 \times 10^5$

$B^2 - AC$ is less than zero. So the condition for relative minimum is satisfied.

4. $C = 2000x_1 + \dfrac{5 \times 10^9}{x_1^{1.5}\ x_2} + 4 \times 10^5\ x_2$

$$Y = \sum_{j=1}^{m} C_j\ P_j(x)$$

$$\text{expanding } Y = C_1\ P_1 + C_2\ P_2 + C_3\ P_3 + \ldots\ldots..$$

$$P_1 = \sum_{i=1}^{n} x_i\ a^{ij}$$

Here ***C***'s are positive constants.

$$C_1 = 2000, C_2 = 5 \times 10^9, C_3 = 4 \times 10^5$$

$$P_1 = x; P_2 = x_1^{-1.5}\ x_2^{-1}; P_3 = x_2$$

$$P_1 = \underset{i=1}{\overset{n}{\pi}}\ x_i^{ai1} = x_1^{a_{11}}\ x_2^{a_{21}}\ x_3^{a_{31}}$$

$$P_2 = \underset{i=x}{\overset{n}{\pi}}\ x_i^{ai2} = x_1^{a_{12}}\ x_2^{a_{22}}\ x_3^{a_{32}}$$

$$P_3 = x_1\ a^{a13}\ x_2^{a23}\ x_3^{a33}$$

at a minimum the first derivative must varnish,

$$\sum_j ak_j\ W_j = 0; K = 1, 2, \ldots\ldots\ n$$

$$\Sigma\ W_j = 1.0, W_j = \text{Weights}$$

$$\text{Expanding } \sum_j a_{1j}\ w_i = a_{11}\ W_1 + a_{12}\ W_2 + a_{13}\ W_3 = 0$$

$$\sum a_{2j} W_j = a_{21} W_1 + a_{22} W_2 + a_{23} W_3 = 0$$

$$\sum a_{3j} W_j = a_{31} W_1 + a_{32} W_2 + a_{33} W_3 = 0$$

$a_{11} = 1, a_{21} = 0, a_{31} = 0, a_{12} = -1.5, a_{22} = -1, a_{32} = 0, a_{13} = 0,$
$a_{23} = 1, a_{33} = 0;$
$W_1 - 1.5 W_2 = 0; -W_2 + W_3 = 0; W_1 + W_2 + W_3 = 1.0$
Solving thie three equations simultaneously gives,
$W_1 = 1.5/3.5; W_2 = 1/3.5; W_3 = 1/3.5$

$$\text{Minimum cost, } Y^* = \pi_j \left(\frac{C_j}{W_j}\right)^{W_j} = \frac{(2000 \times 3.5)^{1.5/3.5}}{1.5} x$$

$(5 \times 10^9 \times 3.5)^{1/3.5} \times (4 \times 10^5 \times 3.5)^{1/3.5} = 1.66 \times 10^6$
Compressor Cost $= 2000x_1 = Y^* W_1 = 1.66 \times 10^6 \times 1.5/3.5$

$$x_1 = 382 \text{ PSia} \quad \textbf{Ans.}$$

$$Y^* W_3 = 4 \times 10^5 x_2 = 1.66 \times 10^6/3.5;$$

$$x_2 = 1.33 \quad \textbf{Ans.}$$

5.

$$W_1 + W_2 + W_3 = 1.0$$

$$W_1 - W_2 = 0; \quad -W_2 + W_3 = 0$$

Solving, $W_1 = W_2 = W_3 = \frac{1}{3}$

$$Y^* = \left[\frac{(100)^{1/3}}{\frac{1}{3}}\right]\left(\frac{4 \times 10^9}{\frac{1}{3}}\right)^{1/3}\left(\frac{2.5 \times 10^5}{\frac{1}{3}}\right)^{1/3} = 3 \times 10^6$$

$$1000x_1 = 3 \times 10^6 W_1 = 10^6; x_1 = 1000. \quad \textbf{Ans.}$$

$$2.5 \times 10^5 x_2 = 3 \times 10^6 W_2 = 10^6; x_2 = 4 \quad \textbf{Ans.}$$

6.

$$x_1 = 0.5 - \frac{E}{2} = 0.5 - \frac{0.05}{2} = 0.475;$$

$$x_2 = 0.5 + \frac{E}{2} = 0.525$$

$$Y(x_1) = Y_1 = 0.9881; Y(x_2) = Y_2 = 1.0132$$

Since Y_2 is greater than Y_j, so leave the interval right of Y_2.
Remaining distance is : 0.525

$$x_3 = \frac{0.525}{2} - \frac{E}{2} = 0.2375$$

$$x_4 = \frac{0.525}{2} + \frac{E}{2} = 0.2875$$

$$Y_3 = 0.93764;\ Y_4 = 0.92704$$

Since Y_3 is greater than Y_4, so leave left of Y_3

Interval of uncertainty after four experiments

$$= 0.525 - 0.2375 = 0.2875$$

So the range of minimum now is : 0.2375 to 0.525.

The interval of uncertainty from the theoretical equations after 4 experiments is

$$(I_K) = \frac{I_0}{2^k} + \left(1 - \frac{1}{2^k}\right) E, \text{ here } k = 2$$

$$I_2 = \frac{I_0}{4} + \left(1 - \frac{1}{4}\right) E = \frac{1}{4} + \frac{3}{4} \times 0.05$$

$$= 0.2875. \quad \textbf{Ans.}$$

7. $n = 3;\ F_0 = F_1 = 1,\ F_2 = 2;\ F_3 = 3;\ F_4 = 5$

$$x_3{}^1 = \frac{F_{n-2}}{F_n}(x_2^0 - x_1{}^0) = \frac{F_1}{F_3} \times 1 + 0$$

$$= \frac{1}{3} \times 1 = 0.33$$

$$x_4{}^1 = \frac{F_{n-1}}{F_n}(x_2^0 - x_1^0) + x_1^0 = \frac{F_2}{F_3} \times 1 + 0$$

$$= \frac{2}{3} \times 1 = 0.67$$

$$Y(x_3^1) = 0.9439;\ Y(x_4^1) = 1.1116$$

Since $y(x_4^1) > Y(x_3^1)$ so leave right of 0.67

$$x_3^2 = \frac{F_{n-3}}{F_{n-1}} I_1 + x_1^0 = \frac{F_0}{F_2} \times 0.67 + 0$$

$$= \frac{1}{2} \times 0.67 = 0.335$$

$$x_4^2 = 0.67 - 0.335 = 0.335$$

1.0	0.94473	
0	0.335	0.67(1.116)

$$Y(0.335) = 0.94473.$$

Both x_3^2 and x_4^2 came at same point of 0.335, taking

$$E = 0.05$$

$$x_3^3 = 0.335 - \frac{E}{2} = 0.31.$$

$$x_4^3 = 0.335 + \frac{E}{2} = 0.36$$

$$y(0.31) = .9411; \; Y(0.36) = 0.9496$$

Since $Y(0.36)$ is greater than Y(0.31), so leave right of 0.36 theoretical interval of uncertainty is

$$= I_2 = \frac{F_{n-2}}{F_n} I_0$$

$$= \frac{F_1}{F_3} \times 1 = \frac{1}{3} \times 1 = 0.333. \quad \textbf{Ans.}$$

8. Now $n = 4$, $F_0 = F_1 = 1$; $F_2 = 2$; $F_3 = 3$; $F_4 = 5$

Ist interval of uncertaintyis I_0 and Second interval of uncertainty of $\mathbf{I}_1$ and so on.

$$I_1 = \frac{F_{n-1}}{F_n} + \frac{(-1)^n E}{F_n};$$

$$F_n = F_{n-1} + F_{n-2} \text{ for } n \geq 2$$

$$I_1 = \frac{3}{5} + (-1)^4 \frac{0.05}{5} = 0.610$$

So $x_1 = 0.61$; $x_2 = 1 - 0.610 = 0.39$

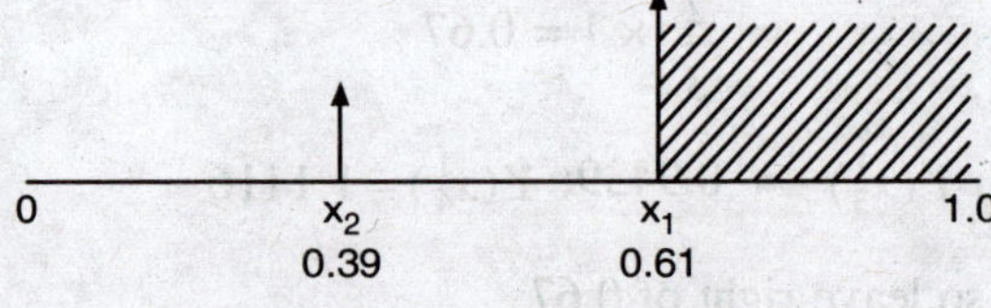

$Y(x_1)$ is greater than $Y(x_2)$ so leave interval to the right ofx_1.

next point is : $x_1 - x_2 = 0.61 - 0.39 = 0.22$

$x_2 = 0.39, \quad x_3 = 0.22$

Since $Y(x_3)$ is less than Y (x_2) so leave interval to the right of x_2.

Next point is $x_2 - x_3 = 0.39 - 0.22 = 0.17 = x_4$

Since $Y(x_4)$ is greater than $Y(x_3)$ so leave interval to the left of x_4.

Now the left interval after 4 experiments is $x_2 - x_4 = 0.39 - 0.17 = 0.22$ **Ans.**

Theoretical interval of uncertainty after 4 experiments is :

$$I_{n-1} = \frac{1}{F_n} + \frac{F_{n-2}E}{F_n} = \frac{1}{5} + \frac{2 \times 0.05}{0.5} = 0.22 \quad \textbf{Ans.}$$

9.

$$I_{n-1} = \frac{1}{F_n} + \frac{F_{n-1} E}{F_n}$$

Let $n = 6$

$$I_5 = \frac{1}{13} + \frac{5}{13} \times 0.05 = 0.096$$

for $n = 5; I_4 = \frac{1}{8} + \frac{3}{8} \times 0.05 = 0.144$

Since I_5 ($n = 6$) is near to 0.10 so n = 6 is the number of experiments.

10. $I_n = \frac{1}{r^{n-1}}$; r = 1.618, n = 3

$$I_3 = \frac{1}{1.618^2} = 0.328$$

Initial interval is 0 to 1.0 = 1.0

$$x_4^1 = x_1^0 + \frac{I_0}{r} = 0 + \frac{1}{1.618} = 0.617$$

$$x_2^1 = 1 - 0.617 = 0.383$$

$$Y(x_4^1) = 1.0744; \; Y(x_3^1) = 0.9544$$

so leave right of $x_4^1(0.617)$

next point is : 0.617 – 0.383 = 0.234

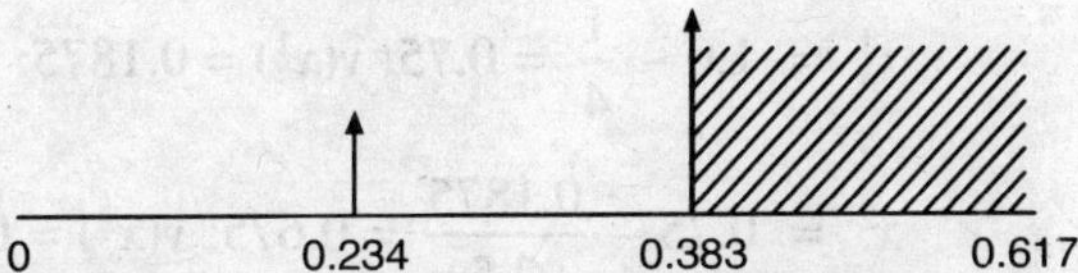

$Y(0.234) = 0.979$; $Y(0.383) = 1.0744$; $Y(0.617) = 0.9544$

$Y(0.234) = 0.0379$; $Y(0.383) = 1.0744$; $Y(0.617) = 0.9544$

Since $Y(0.383)$ is greater than $Y(0.234)$ so leave right of 0.383.

Left interval is 0 to 0.383. **Ans.**

11. Starting search form $x_1 = -1.2$ and $x_2 = 1.0$

$$Y = 100(x_2 - x_1^3)^2 + (1 - x_1)^2$$

$$B_0 = 100(1.0 + 1.2^3)^2 + (1 + 1.2)^2 = 749.04$$

let $dx_1 = 0.2$; $dx_2 = -0.2$

then point B_1 is $x_1 = -1.2 + 0.2 = -1.0$; $x_2 = 1.0 - 0.2 = 0.8$

Exploration $B_1 = 100\,(0.8 + 1)^2 + (1 + 1)^2 = 328$ (improvement)

So continue in the same way

Pattern $P_1 = (B_1 - B_0) = -1 + (-1 + 1.2) = -0.8$

$0.8 + (0.8 - 1.0) = 0.6$

So
$$p_1 = (-0.8, 0.6)$$
$$B_2 = \mathbf{P}_1 + dx = (\mathbf{x}_1 + dx_1, x_2 + dx_2) = (-0.6, 0.4)$$
$$B_2 = 40.51 \text{ (function value) – improvement to continue}$$

12.
$$\frac{dy}{dx_1} = 14x_1;\ m_1 = 14 \times 10 = 140$$
$$\frac{dy}{dx_1} = 20x_2;\ \text{m}_2 = 20 \times 3 = 60$$

contour tangent equation is : $dy = m_1\, dx_1 + m_2\, dx_2$
$$dy = 140\, dx_1 + 60\, dx_2$$
$$\text{at } x_1 = 10,\ x_2 = 3,\ y = 790$$
$$y - 790 = 140\,{*}(x_1 - 10) + 60\,(x_2 - 3)\quad \textbf{Ans.}$$

13.
$$x^{i+1} = x^i - \frac{y(x^i)}{y(x^i)}$$

for
$$i = 0$$
$$x^1 = x^0 - \frac{y(x^0)}{\dot{y}(x^0)};\ (x_0) = 1$$
$$\dot{y}(x^0) = 6x - 2 = 6 \times 1 - 2 = 4$$
$$x^1 = 1.0 - \frac{1}{4} = 0.75;\ y(x^1) = 0.1875;\ \text{y}^3(x^1) = 2.5$$
$$x^2 = 0.75 - \frac{0.1875}{2.5} = 0.675;\ y(x^2) = 0.0169$$
$$y(x^2) = 2.05$$
$$x^3 = 0.675 - \frac{0.0169}{2.05} = 0.667$$

continuing $x = 0.667$ comes.

14. (1) Simultaneous search plan; (2) Two equally spaced trials; (3) Tthree equally spaced trials; (4) Dichotomous; (5) Fibonacci search; (6) Golden section.

15. (a)
$$S = P(1 + \text{i})^{\text{n}} = P(1 + 0.08)^{10}$$
$$S = 540;\ P = 540/1.08^{10} = \text{Rs.}250\ \textbf{Ans.}$$

(b)
$$S = \frac{R[(1+i)^n - 1]}{i} = \frac{37.3(1.08^{10} - 1)}{0.08}$$
$$S = \text{Rs. } 250\quad \textbf{Ans.}$$

16. $$R = \frac{p}{n} + \frac{Pi}{2}\left(\frac{n+1}{n}\right)$$

here $R = 10{,}000; n = 5; p = 3000$

$$10{,}000 = \frac{23000}{5} + \frac{23000 \times \text{i}}{2}\frac{(5+1)}{5}$$

$$i = 0.392 \text{ or } 39.2\%$$

17. $$R = \frac{p}{n} + \frac{Pi(n+1)}{2n}$$

$$= \frac{P}{5} + \frac{P}{2}\frac{*0.1*6}{5} = 0.26p \text{ each year}$$

for B : $$0.26\,p = \frac{2P}{n} + \frac{2P}{2} * 0.1\left(\frac{n+1}{n}\right)$$

$$0.26\,P_n = 2P + (0.1n + 0.1)\,P$$

$$0.26\,n = 2 + 0.1\,n + 0.1n + 0.1$$

$$0.16\,n = 2.1;\ n = 13.1 \text{ years. } \textbf{Ans.}$$

(b) L (salvage values) $= 0.5\,P$

$$R = \frac{(P-L)}{n} + \left(\frac{P-L}{2}\right)\frac{i(n+1)}{n} + Li$$

$$R = 0.1p + 0.05p + 0.03p = 0.18p$$

equating to the other

$$0.18p = \frac{P}{n} + \frac{Pi}{2}\left(\frac{n+1}{n}\right) + Pi$$

$$0.18 + \frac{1}{n} + \frac{0.1}{2}\left(\frac{n+1}{n}\right) + i$$

$$0.18 = \frac{1}{n} + \frac{0.1}{2}\left(\frac{n+1}{n}\right) + 0.1 = \frac{1}{n} + \frac{0.1}{2} + \frac{0.05}{n} + 0.1$$

$$0.18 = \frac{1.05}{n} + 0.15;\ 0.03 = \frac{1.05}{n};\ n = \frac{1.05}{0.03}$$

$$= 35 \text{ years.}$$

18. $$2000\,tn = 400 \times 365$$

$$tn = 73\ ;\ n = 73/t$$

t = days operated/cycle

n = cycles per year

M = Maximum inventory/day

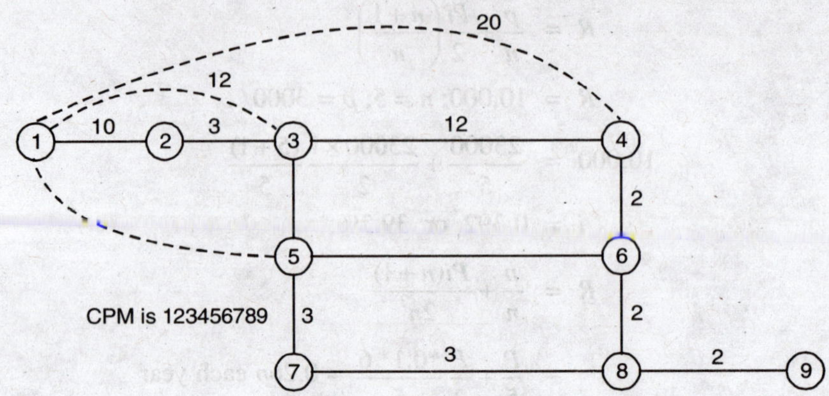

19.

$$2000\,tn = 400 \times 365$$

$$tn = 73\ ;\ n = 73/t$$

$$t = \text{days operated/cycle}$$

$$n = \text{cycles per year}$$

$$M = \text{maximum inventory/day}$$

$$A = \text{Average inventory/day}$$

$$\text{total costs} = 500\,n + 0.25\,M + 0.75A$$

$$M = 1600t\ ;\ A = M/2$$

$$TC = 500 \times \frac{73}{t} + 0.25 \times 1600t + 0.75 \times 800t$$

$$= 1000t + \frac{36500}{t}$$

$$\frac{dTC}{dt} = 800 - \frac{36500}{t^2} = 0$$

$$t = 6.042 \quad \textbf{Ans.}$$

$$n = 73/6.042 = 12.08 \quad \textbf{Ans.}$$

20.

$$t = \text{days operated per cycle}$$

$$n = \text{number of cycles} = 10$$

$$100^* t^* n = 20^* 365$$

$$t = \frac{73}{n} = \frac{73}{10} = 7.3 \text{ days/cycle}$$

$$\text{Average inventory} = \frac{80 * t}{2} = 40*7.3 = 292 \text{ gallons/cycle.}$$

21. Fixed annual costs = $5L\,(0.1 + 0.1) = L$

L = thickness of insulation in feet

Heat loss per hour per feet = UDT

$$\frac{1}{U} = \frac{1}{h} + \frac{L}{K} = \frac{1}{1} + \frac{L}{0.05} = \frac{0.05 + L}{0.05}$$

$$\text{Heat loss}/Yr = 8700^* \frac{(0.05)}{0.05 + L} dT^* \, 0.8 \times 10^{-6} \text{ Rs./year.}$$

$$\text{Total annual cost } (TC) = L + \frac{348 dT \times 10^{-6}}{(0.05 + L)}$$

$$\frac{dTC}{dL} = 1 - \frac{348 dT * 10^{-6}}{(0.05 + L)^2} = 0$$

L in inches $= (0.2244 dT^{0.5} - 0.60)$. **Ans.**

22.

$$\text{Total cost} = \int_0^t E_e dt$$

$$m = \text{days/regeneration}$$

$$n = \text{number of regenerations}$$

number of days/replacement $= m^*n$

$$n \text{ triangles are} = n\left(\frac{2000}{2}\right) * m^2 = 1000 \, m^2 n$$

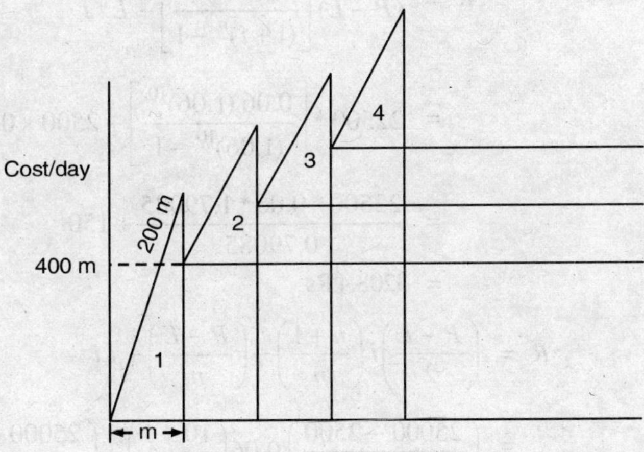

n	*rectangles*
1	0
2	1
3	$3 = n(n-1)/2$
4	6

Rectangles area = $\frac{n(n-1)}{2} m^* 400\, m = 200\, m^2 (n^2 - n)$

Total cost per replacement cycle

$$= 1000\, m^2 n + 200\, m^2 (n^2 - n) + 1000\,(n-1) + 90{,}000$$

$$\text{Total cost/day (T)} = \frac{1000\, m^2 n}{mn} + \frac{200\, m^2 (n^2 - n)}{mn} + \frac{1000(n-1)}{mn} + \frac{90{,}000}{mn}$$

$$T = 800\text{ m} + 200\, mn + \frac{100}{m} + \frac{89000}{mn}$$

$$\frac{dT}{dm} = 800 + 200\text{m} - \frac{1000}{m^2} - \frac{8900}{m^2 n} = 0$$

$$\frac{dT}{dn} = 200m - \frac{8900}{mn^2} = 0$$

Solving the above equations simultaneously gives,

$$n = 18.85; \qquad m = 1.12$$

Cycle of regeneration = 1.12 days

Days/replacement = $mn = 1.12 \times 18.85 = 21.1$ days

23. $P = 25000$ Rs. ; $L = 2500$ Rs. ; $n = 10$ yrs.; $i = 6\%$

(a)

$$R = (P-L)\left[\frac{i(1+i)^n}{(1+i)^n - 1}\right] + L * i$$

$$= 22500 * \left[\frac{0.06\,(1.06)^{10}}{(1.06)^{10} - 1}\right] + 2500 \times 0.06$$

$$= \frac{22500 * 0.06 * 1.79085}{0.79085} + 150$$

$$= 3208.4 \text{ Rs.}$$

(b)

$$R = \left(\frac{P-L}{2}\right) i \left(\frac{n+1}{n}\right) + \left(\frac{P-L}{n}\right) + Li$$

$$= \left(\frac{25000 - 2500}{2}\right) 0.06 \left(\frac{10+1}{10}\right) + \left(\frac{25000 - 2500}{10}\right) + 2500 * 0.06$$

$$= 742.5 + 2250 + 150 = 3142.5 \text{ Rs.}$$

(c)

$$i_a = \frac{i(n+1)}{2n}$$

$$i_a = \frac{0.06}{2}\left(\frac{10+1}{10}\right) = 3.3\%$$

24. $P = 100{,}000$ Rs.; $R = 10{,}000$; $n = 10$; $L = 10{,}000$ Rs

$$R = \left(\frac{P-L}{n}\right) + \left(\frac{P-L}{n}\right)\frac{1}{2}\left(\frac{n+1}{n}\right) + Li$$

Substituting the values

$$i = 16.8\%$$

25.

$$r = B/A$$

Total cost/lb. of C = cost of A/lb. of C + cost of B/lb. of C.

$$T = \frac{0.64 \times 0.15}{(1-e^{-r})} + \frac{0.64 \times 0.10r}{(1-e^{-r})}$$

$$T = \frac{0.096}{(1-e^{-r})} + \frac{0.064\,r}{(1-e^{-r})}$$

$$\frac{dT}{dr} = 0 = \frac{-0.096\,e^{-r}}{(1-e^{-r})^2} + \frac{0.064}{(1-e^{-r})}[1 - e^{-r} - re^{-r-r}] = 0$$

giving $1.5\,e^{-r} = (1-r)(1-e^{-r})$ Ans.

26. S_1 $\boxed{200}$

x_2 $\boxed{100}$ $\boxed{1}$

27.

$$-P = -100x_1 - 50x_2$$

$$x_1 + x_2 + S_1 = 70$$

			x_1	x_2	S_1
Basis	c	b	− 100	− 50	0
S_1	10	70	1	(1)	1
	− P =	0	100	50	0

❑

CHAPTER 16

Environmental Engineering

1. The wastes in Ammonia manufacture plant:

   Purge gas, hydrogen, nitrogen, ammonia, argon, catalyst FeO.

   FeO is used in iron making. Waste gases are removed by scrubbing with water (because of small amounts).

2. The wastes in nitric acid manufacture are: (1) Catalyst (platinum – rhodium), (2) Exhaust gases (NH_3), oxides of nitogen. Used up catalyst is used after melting. The gases are treated with chilled water or aqueous sodium carbonate or sodium hydroxide solution. Gases are used for making ammonium nitrate, calcium ammonium nitrate, ammonium phosphate.

3. The wastes in urea plant are: 1. Low pressure recycle gas (CO_2, NH_3, NO_2). These gases are removed by scrubbing with chilled water.

4. The wastes in Superphosphate or Triple super phosphete fertilizer plant are: 1. Silicon fluoride, 2. Hydrogen fluoride fumes, 3. Carbon dioxide. Silica is removed by sodium chloride. Other gases are removed by scrubbing with water. Pollutants are: Carbondioxide, hydrogen fluoride, silicon fluoride.

5. Composition of waste water from a fertilizer plant. ph – 7.5 to 9.5; Total dissolved solids (TDS) – 5400 mg/liter; Ammonia as nitrogen – 700 mg/L; Urea as nitrogen – 600 mg/L; Phosphate 75 mg/L; Fluoride – 15 mg/L; Arsenic – 1.5 mg/L. Acids and alkalies destroy aquatic life.

6. Treatment of Fertilizer waste water:

   Oil – Removed by gravity separator; Arsenic – Removed by chemical method; Ammonia – steam Stripping, lagooning, biological nitrification.

7. Distilleries (Breweries, distilleries) wastes are (pollutants): 1. Spent water from matt house with biochemical oxygen demand (BOD) of order of 400 to 800 mg/L; 2. fermentation residue.

   Distillery wastes are treated by an Aerobic Lagoons and Aerated Lagoons.

8. Dairy wastes are: Milk and milk products, waste water. Treatment of wastes: Aeration for a day reduces BOD and lactic acid formation. Trickling filters, activated sludge, aerated lagoon, oxidation ditch, dairy waste water after treatment in aerated lagoon can be used for Irrigation.

9. Treatment of drug industry wastes:

   1. Acidic wastes (segregation and equalisation by lime);
   2. Toxic elements like cyanide, odour producing compounds are destroyed by chlorination, heating in a furnace.

3. Cyanide salts are treated with ferrous salts to convert to non-toxic compounds.
4. Biological treatment

10. Oil refineries wastes (pollutants):

Composition of oil refinery waste water: 1. Free oil – 2000 to 3000 mg/L; 2. H_2S, RSH (Sulphides) – 10 to 220 mg/L; 3. Emulsified oil – 90–120 mg/L. 4. Phenolic compounds: 12–30 mg/L; 5 day BOD – 100 – 300; Suspended solids: 20–400 mg/L;

Treatment of wastes of refinery:

1. Physical separation of oil (gravity separators; dissolved air floatation with or without chemical floculation).
2. Chemical coagulation of emulsified oil (Coagulation), Flocculation, Sedimentation, Aeration.
3. Removal of BOD, phenol, toxic materials (biological treatment, stabilization pond, aerated Lagoon, Trickling filter, activated sludge).

11. Steel plant wastes:

The units of an integrated steel plant are: 1. Coal washery; 2. Coke ovens; 3. blast furnace; 4. Steel melting shop; 5. Rolling mills; 6. By-product units.

1. Coal washery waste composition: Total dissolved solids (TDS) – 1000 to 2500; Suspended solids – 800 to 24760 mg/L; Dissolved solids; 200–300 mg/L; Hardness – 230 mg/L of $CaCO_3$; Alkalinity 13 mg/L; pH – 4 to 7.8.
2. Waste water from Ammonia unit: pH – 7.5 to 8.0; total free ammonia – 300 to 350 mg/L; Sulphides – 10 to 20 mg/L; Total phenol – 900 to 1000 mg/l; Chlorides – 4000 to 4200 mg/L.
3. Blast furnace waste water: pH. – 7.3 to 8.2; Total suspended solids – 1000 to 10500 mg/L; TDS – 346 to 500; Total hardness – 80 to 118; Total alkalinity – 380 mg/L; Chlorides 210 to 50 mg/L.
4. Rolling mills. Composition of waste water: ph – 7.8 to 8.1; Suspended solids – 320 to 600; Hardness – 80–118; Oil and grease – 60 to 80 mg/L.

   Power plant: Waste water: pH – 10 to 11; Solids – 50–70; Phosphate 40–60; Chlorides – 70 to 140 mg/L.

   Coal washery waste treatment: Suspended solids–clarifier with or without coagulation; Coal particles are removed by froth floatation.

   Coke ovens waste treatment: Phenol by solvent recovery by (benzene, creosote oil, light oil); Ammonia liquor – oxidation with $KMnO_4$, ozone; Biological oxidation–activated sludge process/trickling filter.

   Blast furnace waste treatment: Iron oxide, silica in dust are removed by clariflocculator with coagulant like Alum.

   Scale pit waste removal: Fine scales, oil, greese, suspended solids are treated by clariflocculators.

   Pickling waste liquor: Wastes are removed by neutralization (lime) by ferrous sulphate. Acid and $FeSO_4$ are recovered by Acetone. Electrolysis of spent water to remove metals.

12. Waste treatment in sugar plants: Equalizing cum digestion pond (one day detention time); Aerobic Lagoons (detention time 6 days); Waste stabilization pond (detention time 12 days).

13. Pulp and paper pollutants (wastes): pH– 8.5 to 9.5; TDS – 4410, Suspended solids – 3300; Chemical oxygen demand (COD) – 716; BOD – 155; COD/BOD = 4.6.

Treatment: Fibres recovery by sedimentation, floatation using forced air in the tank; lime by calcination of mud color removal, BOD removal by lime treatment. Activated carbon for color removal. Suspended solids by clarifiers. BOD treatment by waste statilization ponds; Aerated lagoons and activated sludge method.

14. Textile mills: Waste water composition: pH – 9.8 to 11.8; Total alkalinity – 1735 mg/L of $CaCO_3$; BOD – 760 mg/L; COD–1418 mg/L; Total solids – 6170 mg/L; Total chromium 12.5 mg/L;

Treatment of wastes: Caustic – using dialysis for separation, dye wastes – Biological oxidation; BOD – Alum, ferrous sulphate, ferrice sulphate, ferrous chloride; pH – lime or sulphuric acid; Waste wool – treating with calcium chloride.

15. Viscose rayon waste treatment: Sodium hydroxide – by dialysis, spinning bath – Sodium sulphate recovery; Acid waste – Zinc recovery by chemical precipitation; or by ion-exchange.

16. Energy flow in ecosystems: 1. Water cycle; 2. Carbon cycle; 3. Oxygen cycle; 4. Nitrogen cycle; 5. Energy cycle.

17. Pollution is the effect of undesirable changes in own surroundings that are harmful to plants, animals and humans.

18. Air pollution: This occurs due to presence of undesirable solids or gaseous particles in air in excess of health prescribed limits.

19. Acid rain: This is due to reaction of sulphur dioxide and oxides of nitrogen with water during rain. This acid rain effects terrestrial systems, aquatic systems, damages buildings, statues, and gives asthama.

20. Various pollutions are: 1. Air pollution; 2. Water pollution; 3. Soil pollution; 4. Thermal pollution; 5. Nuclear pollution; 6. Noise pollution; 7. Marine pollution.

21. Global warming: 75% of solar energy is absorbed by Earth' surface. Rest radiates back. Some heat is trapped by greenhouse gases, mostly carbon-dioxide. This causes global warming. Without greenhouses, earth's surface should be at – 18°C. Global warming by greenhouse gases effect temperature, rainfall (floods and draughts).

22. Ozone layer depletion: Ozone is formed by the action of sunlight on oxygen. This layer is 20 to 50 km above earth. It absorbs ultraviolet rays from Sunrays not to reach earth. Ultraviolet rays react with Refrigerants giving chlorine/fluorine. Chlorine and fluorine react with ozone giving oxygn. Reduction of O_3 decreases vegetation, affecting food chain CFC, Br, hydrocarbons, nitrous oxides and oxides from fertilizers attack ozone layer.

POLLUTION CONTROL IN PETROLEUM INDUSTRIES

1. Pollution during production:

(a) Brines containing oil are discharged into seas or rivers.

(b) Oil spills during sea, Rail and road transport lead to pollution of seas, Rail yards and Road stations.

(c) Tar balls on sandy beaches is due to pollution of sees by crude.

(d) Refinery operations give maximum water pollution.

(e) Cooling water gives thermal pollution of water bodies (decrease solubility of oxygen in water).

(f) Oil bearing liquid water/waste forms a thin layer of oil over the surface of the water bodies cutting down sunlight and oxygen for flora and fauna.

(g) Phenol pollutant imparts bad odour and taste to water and is toxic to living beings.

2. Refinery processes and liquid wastes:

Refinery process	Liquid waste
	Effluent with ammonia, hydrogen
1. Distillation	Sulphide and phenols.
2. Desalting	Sodium chloride, phenols, sulphides free hydrogen sulphide.
3. Vacuum distillation	Phenol and oil.
4. Naphtha hydrotreating	Sour condensate
5. Catalytic hydrotreating	Alkalies from washings and acids from drains.
6. Solvent process	Solvents like phenol, sulphides, cooper acetate etc.
7. Treating processes	Organic and inorganic pollutants (sulphides, phenol) and emulsified oil.

3. Refineries in-treated waste characteristics: Free oil – 2000 to 3000 mg/liter,; Hydrogen sulphide and sulphides – 10 to 220 mg/L; phenols – 12 to 30 mg/L; Suspended solids – 200 to 400 mg/L; 5 day BOD at 20°C – 100 to 300 mg/L; Alkalinity – 10 to 250 mg/L.

4. Treated effluent standards:

pH – 5.5 to 9.00

Suspended solids not to excesed: 100 mg/L.

Dissolved oxygen not less than: 5 mg/L.

BOD (5 days, 20°C) not to exceed: 100 mg/L.

COD not to exceed: 250 mg/L

5. Pollution data of treated waste water from refinery:

Characteristic	**Fuel refinery**	**Lube refinery**	**Permitted**
pH	7.3 to 7.5	7.3 to 7.8	6.5 to 8.5
Oil and greese (ppm)	13 – 14	10 – 14	15
Suspended solids (ppm)	300–320	280–310	100 over influent conc.
5-day BOD at 20°C (ppm)	50–75	50–70	100
COD (ppm)	200–230	200–220	250
Disolved oxygen	3.2–3.5	3.6–3.8	4

6. General techniques followed to minimise pollution by petrochemical plant/unit.

1. Physical, biological and chemical treatment of liquid waste before disposal.
2. Treatment and recycling of process water to reduce fresh–water requirements as well as pollution.

3. Incorporation of process changes to keep the pollutant concentration at a lower level.
4. Recovery of useful chemicals.
5. Segregation of toxic liquid wastes and removal of toxic substances.
6. Equalization and proportioning of liquid wastes.

7. Refinery liquid waste treatment methods:

(a) In plant mesures:

1. Reduction of oil leakage, avoiding formation of oil water emulsion, gravity separator is used.

(b) Steam stripping: Air or steam stripping at higher pH reduces the ammonia content in water substantially. Water condensates contain ammonia, hydrogen sulphide merceptans and phenol are produced during crude distillation.

(c) Oxidation: Air is used as an oxidizing agent. Ammonia is oxidised to nitrogen and water. Sulphides are oxidised to thiosulphates, in columns. Phenols are left in the liquor, during stripping and oxidation.

(d) Removal of phenols: Adsorption of phenbol on activated charcoal is suited to refinery liquid wastes. Solvent extraction is suited for recovery of phenol. A multistage modified activated sludge process where phenol is reduced to I ppm.

(e) Biological treatment: For this, trickling filters, activated sludge oxidation ponds, aerated lagoons and anearobic digestion are used. Trickling beds are better suited for biological oxidation as compared to the activated sludge process for refinery wastes. Development is multistage sludge process.

Oxidation pond: Treatment for liquid effluents from an Indian Industry.

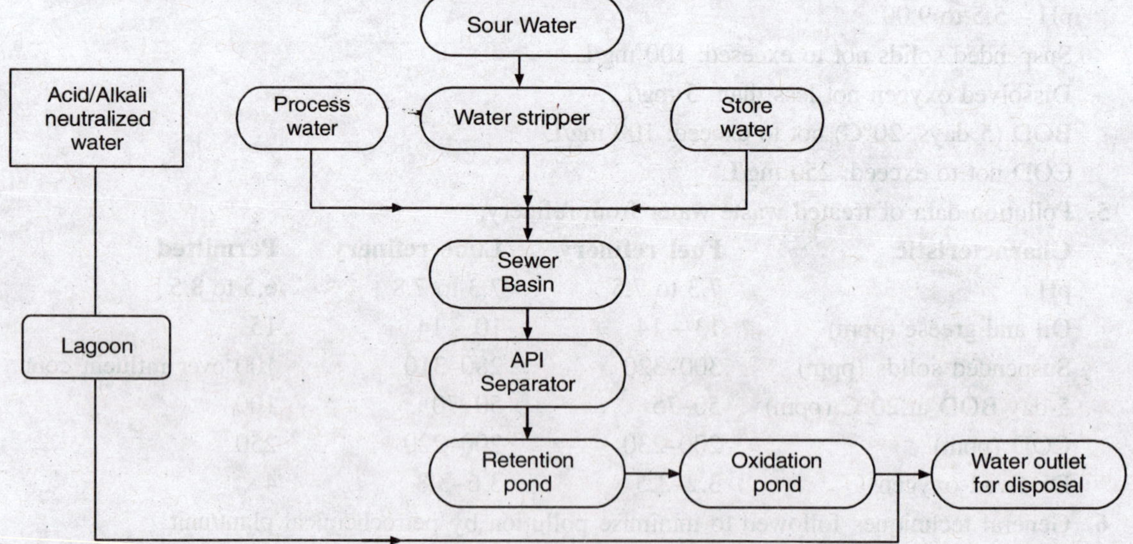

(f) Disposal of sludges: Three types: oily sludges, chemical sludges, biological sludges. Oily sludges are thickened by gravity thickeners, centrifuges or filtration. They are dried and burnt for disposed off as land fills. In India little effort is made for disposing of sludges from treated liquid effluents. Heaps of sludges in and around the factory are seen.

8. Treatment of liquid effluents from petrochemical industries: The concentration of oils, phenols, ammonia and compounds specific to the chemical plant about first be reduced or recovered by physico-chemical methods. The various processes used are:

1. Solvent extraction; 2. Inertial separation; 3. Ion-exchange; 5. Adsorption on activated carbon or polymeric resins; 5. Recovery of useful chemicals; 6. Chlorination; 7. Electrocoagulation; 8. Electro-dialysis/reverse osmosis; 9. Ozonation; 10. Biodegradation of various chemicals recovery are: $CaCO_3$ from a pickling unit polassium chloride from pharmaceutical industry. Floating oils and immiscible organics are separated by API separator.

9. Air-pollution control: Air pollutants from petroleum industries: H_2S, NH_3, SO_2, NO_2, HF, HCl, aldehydes, mercaptains, and particulate matter such as smoke, fumes, mists and dusts. Carbon monoxide and carbon dioxide are products of incomplete and complete combustion.

(a) Control of sulphur dioxide: SO_2 gas is oxidised to SO_3 and then to sulphuric acid. SO_2 is scrubbed with alkaline solution with lime generation. In the NOS–OX process, the gases are scrubbed with an alkaline solution in water with thermal regeneration. The concentration of SO_2 on the exist gases can be reduced to 100 ppm by ammonium sulphite/sulphate solution.

(b) Removal of hydrogen sulphide gas:

H_2S is absorbed (concentration high) along with CO_2 by alkaline solution at 60–65°C in a packed/plate/venturi unit. The liquor is stripped of H_2S and CO_2 and the liquid is recycled. There is a long list of chemicals used as solvents for scrubbing, starting from ethyl amines to potassium methyl aminopropionate.

Phosphate process: $K_3PO_4 + H_2S \rightleftharpoons K_3HPO_4 + KHS$

Absorption: $2Fe_2O_3 + 6H_2S \rightarrow 2Fe_2S_3 + 6H_2O$

$2Fe_2S_3 + 3O_2 \rightarrow 2Fe_2O_3 + 6S$

Sodim carbonate: $Na_2CO_3 + H_2S \rightarrow NaHCO_3 + NaHS$

Regeneration: $3NaHCO_3 + 3NaH_S + Fe_2O_3 \,.\, 3H_2O \rightarrow$

$Fe_2S_3.3H_2O + 3Na_2CO_3 + 3H_2O$

$2Fe_2\,S_3.3H_2O + 3O_2 \rightarrow 2\,Fe_2O_3.3H_2O + 6S$

(c) Removal of ammonia from gases: This is absorbed in dilute sulphuric acid. Ammonium sulphate is formed which can be separated. Ammonia can be oxidised to NO_2 which is absorbed in water to give nitric acid.

(d) Removal of particulates: Used ones are: cyclone separators, scrubber, filters and electrostatic precipitator. Flue gases are cleaned by liquid scrubbing unfortunately, in India no adequate measures are taken to reduce air pollution levels in refineries or in petrochemical industries. Implant measures include the use of floating instead of fixed roofs for oil and naphtha tanks, regular desilting of oil catchers, reduction in oil leakages and use of taller stacks at crude oil distillation plants and flare stacks.

Q. Which of the following is not responsible for global warming?

(a) Carbon dioxide (b) Methane

(c) Water vapour (d) Nitrogen

Ans. Nitrogen

QUESTIONS ON ENVIRONMENTAL ENGINEERING

MULTIPLE CHOICE QUESTIONS

1. Weight per cent of water in human body is:
(a) 59 (b) 70 (c) 30 (d) 80

2. Recent bacterial count of Ganga river at Allahabad is:
(a) 100 per mL (b) 200 per mL (c) 300 per mL (d) 400 per mL

3. About 90% of water pollution in cities is due to:
(a) Animal waste (b) Human waste (c) Rain

4. The SO_2 (sulphur dioxide) emitted out of sulphuric acid plant is:
(a) 1500 to 2500 ppm (b) 500 to 1000 ppm (c) 3000 to 4000 ppm

5. Maximum limit of sulphur dioxide in air is:
(a) 100 ppm (b) 200 ppm (c) 300 ppm (d) 500 ppm

6. Manganese dust limit in air in mg/m^3 of air is:
(a) 1 (b) 3 (c) 4 (d) 6

7. Permissible limit of mercury in blood in micrograms per 100 mL. is:
(a) 1 (b) 3 to 4 (c) 6 to 7 (d) 8 to 10

8. Noise greater than the amount of decibels which damages the earing sensation is:
(a) 50 (b) 90 (c) 100 (d) 200

9. Heat energy content of 1 ton (British) of household refuse is:
(a) 9×10^6 BTU (b) 12×10^6 BTU (c) 7×10^6 BTU

10. World Environment Day is celebrated on:
(a) 28th February (b) 1st May (c) 5th June (d) 3rd December

11. Permissible limit of mercury in drinking water is:
(a) 1 ppm (b) 0.01 ppm (c) 0.001 ppm (d) 0.0001 ppm

12. For public sewer/waste waters, maximum permissible limit of lead is:
(a) 1 ppm (b) 10 ppm (c) 100 ppm (d) 1000 ppm

13. Maximum permissible limit of sulphur dioxide in atmosphere air as per WHO specification is:
(a) 10ug/m^3 (b) 30ug/m^3 (c) 60ug/m^3 (d) 100ug/m^3

14. Threshold limit value of phosgene in air is:
(a) 0.5 ppm (b) 0.2 ppm (c) 0.1 ppm (d) .01 ppm.

15. Smoke is a mixture of air and of very fine particles formed by combustion of other chemical processes in the size range of:
(a) 100–1000 μm (b) 10–100 μm (c) 1–10 μm (d) 0.01–2 μm

16. A mixture in air of liquid droplets in the size range from 5 to 100 μm in diameter is called:
(a) Smoke (b) Fog (c) Fume (d) Mist.

17. Primary gaseous pollutants emitted from a pulp and paper plant are:
(a) Hydrocarbons (b) Nitrogenous oxides
(c) Sulphurous gases (d) Sulphurous and nitrogeneous gases

18. The gaseous pollutant, excess presence of which causes headache, mansion and dizziness is:

(a) Sulphur dioxide
(b) Carbon monoxide
(c) Carbon dioxide
(d) Nitrogen oxides

19. Maximum permissible limit of particular matter in air (as per WHO specification) is:

(a) 100 $\mu g/m^3$
(b) 90 $\mu g/m^3$
(c) 20 $\mu g/m^3$
(d) $2\mu^3 g/m^3$

20. The electrostatic precipitator is widely used for collection of particulates because:

(a) Its power requirement is low
(b) It can simultaneously collect gaseous pollutants also
(c) It has a high efficiency for all size ranges.

21. As per Indian standards the concentration of nitrogen oxides (NO_x) which constitutes a potential air pollution problem is:

(a) Above 2 ppm
(b) Above 20 ppm
(c) Above 100 ppm
(d) Above 200 ppm

22. Per capita generation of municipal solid waste in India is around:

(a) 2 kg (b) 1 kg (c) 0.4 kg (d) 0.2 kg

23. The permissible noise level in decibles to which an industrial worker can be exposed for eight hours per day is:

(a) 120 (b) 90 (c) 60 (d) 40

24. The pollutants can be classified as:

(a) Solid pollutants
(b) Liquid pollutants
(c) Air pollutants
(d) Weightless pollutants
(e) All the above

25. Solid pollutants are mainly:

(a) Sewage, garbage and dead bodies
(b) Organic and inorganic sludges, organic tars, spent catalysts and plastics
(c) Thirteen heavy metals
(d) The above
(e) None of the above

26. Liquid pollutants include:

(a) All the solid pollutants
(b) Chlorinated hydrocarbons
(c) Sea-borne spills
(d) Several chemicals
(e) All the above except

27. Air pollutants include:

(a) All the solid pollutants
(b) All the liquid pollutants
(c) Several gases
(d) All the above
(e) All the above except (a)

28. Weightless pollutants are:

(a) Radioactive rays
(b) Crowd and lighting
(c) Heat
(d) Noise
(e) All the above

29. In air pollution:

(a) Transport system contributes maximum

(b) Industrial process contributes maximum

(c) Garbage incineration contributes maximum

(d) Transport system contributes minimum

(e) None of these

30. Major techniques to assess hazards are:

(a) FTA (fault free analysis)

(b) HAZOPS (hazards and operability studies)

(c) FMEA (failure mode and effective analysis)

(d) MPDA (maximum probable danger analysis) and DA (dispersion analysis)

(e) All the above

31. To control pollution:

(a) Avoid hazardous products and select alternate safe process

(b) Select proper site and plant integrity

(c) Minimise inventories of chemical and grow green belt around factories

(d) All the above

(e) All the above except (b)

32. A chemical industry is associated with hazards of the type:

(a) Internal or occupational (b) External or pollutional

(c) All the above (d) None of the above

(e) Mostly (b)

FILL IN THE BLANKS

1. Presence of barium carbonate in water has bad effects on......,......and.......

2. Mercury and cadmium cause ....and damage......

3. Arsenic salts can create ....disease

4. Excess fluoride in water causes.......

5. Mercury causes...............and.......

6. Water containing silver causes.......disease and ....disease.

7. Aerobic means performing the experiment in the presence of.......

8. Anaerobic means performing the experiment in the absence of......

9. B.O.D. is defined as.......

10. Disinfection of water means.......

11. An example of disinfectant is ....

12. Pesticides are used to control ....

13. Polyvinyl chloride containers are harmful to use for .... materials.

14. Air pollutants in lead industry are (1)....... and (2)........

15. Air pollutants in zinc industry are (1) ....and (2).......

16. Air pollutants in copper industry are (1) ....and (2).......

17. Major sources of sulphur dioxide are (1)....... and (2).......

18. Sound level intensity is measured in units of ....

19. Effect of fluorine in atmosphere and in water gives decay of ....

20. ....treatment is used to remove fluoride and phosphate from waste waters of phosphoric acid and diammonium phosphate plants.

21. In the electrostatic precipitator the basic force which acts to separate the particles from the gas is.......

22. Particulates larger than ....settle out quickly and can be removed easily before discharge to atmosphere.

23. The photochemical smog is composed of ....

24. Waste pollution is defined as ....

25. The hazardous plant in an integrated paper plant is ....

26. Major hazards in ammonia plants are caused due to ....

27. .... is the major pollutant in heavy water plant.

28. .... is the main pollutant in urea plants using vetro coke process.

29. In chloroalkali industry using mercury cells,....... causes pollution.

30. As compared to number of deaths due to hazards and pollution, road accidents cause.......number of deaths.

31. Dust particles with size range.......microns are pollutional hazards.

32. Dust of silica causes a disease called, ...that of coal....... and of iron.......

33. ....process is the most suitable available process for reducing emission from sulphuric acid plants.

34. The pollutant of primary interest in phosphatic fertilizer plant is.......content of phosphate rock.

SHORT ANSWER QUESTIONS

1. List the methods for disinfection of water.

2. List the various sulphur dioxide abatement processes.

3. List the various types of pollution.

4. List the tolerable limits of pollutant gases.

5. List the various common processes used in waste treatment.

6. List the various dust collecting equipments.

7. List the emission standards of sulphur dioxide as per I.S. rules (Indian standard).

8. List the emission standards of nitrogen oxides as NO_2 as per I.S. rules.

9. List the various sewage standards of effluents as per I.S. rules.

10. List the types of waste water treatment.

11. List the Indian standard specification for fertilizer industries (air pollutants).

12. List the limits of solid chemical constituents in water.

13. List the contaminants of steel industry.

14. What is Air pollution index?

15. What is Ringlemann Chart?
16. What is the gaseous composition of unpolluted air?
17. List the permissible concentrations of pesticides in public water supplies.
18. List the wastes in cement plants.
19. List the wastes in chloro-alkali industry.
20. List the wastes in fertilizer industry.
21. List the wastes in glass and ceramic industry.
22. List the wastes in phosphoric acid plant.
23. List the wastes in a detergent plant.
24. List the wastes in a water treatment plant.
25. List the wastes in tanneries.
26. List the wastes in dyestuff plant.
27. List the wastes in beverage plant.
28. List the wastes in a nitric acid plant.
29. List the wastes in vegetable and fruit industries.
30. List the wastes in battery manufacturing units.
31. What is dust?
32. What is a sulphurous smog?
33. Define thermal pollution.
34. What are the thirteen metals which cause pollution?
35. List the explosive mixtures with air.
36. What are the various toxic substances?
37. Give the list of toxic levels of pollutants.
38. Give the list of pollutant levels in industrial work environments.
39. What are the pollutional impacts on workers working in industries (diseases)?
40. What are the instrumental methods used for analysing pollutants?
41. What are the participate collectors for air pollutants?
42. List the public and private benefits in using pollution control methods.

ANSWERS

MULTIPLE CHOICE QUESTIONS

1. b	2. c	3. b	4. a	5. d
6. d	7. b	8. b	9. a	10. c
11. c	12. a	13. c	14. b	15. d
16. d	17. c	18. b	19. b	20. a
21. d	22. c	23. b	24. c	25. d
26. c	27. d	28. e	29. a	30. e
31. d	32. c			

FILL IN The BLANKS

1. Veins, nerves and heart.
2. Paralysis and bones.
3. Cancer.
4. Fluorosis
5. Loss of vision, hearing and intellectual ability.
6. Liver and lung.
7. Oxygen
8. Oxygen
9. The quantity of oxygen required for oxidation of organic matter by bacterial action in presence of oxygen.
10. The destruction of water-borne pathogens.
11. Chlorine
12. Harmful organisms.
13. Food
14. Sulphur dioxide and particulate matter.
15. Sulphur dioxide and zinc dust.
16. Sulpur dioxide and particulate matter.
17. (1) Thermal power plant, (2) metallurgical smelters.
18. Decibles.
19. Bones.
20. Double lime (CaO)
21. Electrostatic attraction
22. 100 μm.
23. Parriculates and gaseous pollutants.
24. The presence on land or water of solid material, organic or inorganic which has no beneficial qualities.
25. Soda recovery boiler.
26. Ammonia spill or release of ammonia vapours.
27. Hydrogen sulphide.
28. Arsenic waste.
29. Mercury.
30. More
31. 1 to 5
32. Silicosis, pneumoconiosis, siderosis.
33. DCDA (double conversion and double absorption).
34. Fluorine.

SHORT ANSWER QUESTIONS

1. (a) By light (b) By heat
   (c) Metal ions (d) Alkalies and acids
   (e) Halogens (chlorine, bromine and iodine)
   (f) Surface active chemicals (g) Ozone
   (h) Others are coloured glasses etc.
2. (a) Double conversion and double absorption (DC/DA)
   (b) Charcoal adsorption
   (c) Alum adsorption
   (d) Sodash adsorption
   (e) Magnesium oxide absorption
   (f) Potassium silicate bisulphite
   (g) Molecular sieve
   (h) Lime scrubbing
   (i) Ammonia absortion
   (j) Tail gas conversion (SO_2 to SO_3 and then to sulphuric acid).
3. (a) Air (b) Water (c) Land (d) Noise.
4.

Pollutant	*Parts per million (PPM)*	*Pollutant*	*Parts per million (PPM)*
Acetaldehyde	100	Ozone	0.1
Ammonia	50	Benzene	10
Bromine	0.1	Formaldehyde	5
Carbon disulphide	20	Hydrogen cyanide.	10
Carbon monoxide	50	Carbon dioxide	500
Chlorine	1	Phenol	5
Hydrogen sulphide	10	Toluene	200
Nitrogen oxide	5	Xylene	100
Sulphur dioxide	5	Phosgene	0.10
		MIC	0.02

5. (a) Activated sludge process
   (b) Aerated lagoon process
   (c) Trickling filter process
   (d) Waste stabilization pond process.
6. (a) Dry inertial collectors: Settling chambers, baffle chambers, cyclone, multiple cyclone, electrostatic precipitators.
   (b) Wet scrubbers: Gravity, spray, centrifugal, impingement, packed bed, jet, venturi .
   (c) Fabric filtered : Intermittent, continuous.

7. (a) Existing plants up to 200 tons/day – 16 kg/ton of 100% sulphuric acid
   (b) Existing plants above 200 tons/day – 12 kg/ton of 100% sulphuric acid
   (c) New plants up to 200 tons per day – 12 kg/ton of 100% sulphuric acid
   (d) New plants above 200 tons per day – 4 kg/ton of 100% sulphuric acid.
8. Nitrogen oxides as NO_2:
   (a) Existing plants, 24 kg/ton of 100% nitric acid
   (b) new plants, 3 kg/ton of 100% nitric acid.
9. (a) pH — 6 to 8.5
   (b) Suspended solids — 30 mg/litre
   (c) B.O.D. in 5 days at 20°C — 20 mg/litre
   (d) C.O.D. — 250 mg/litre
   (e) Dissolved oxygen — 5 mg/litre
   (f) Oils and grease — 5 mg/litre
   (g) Phenolic compounds as phenol — 1 mg/litre
   (h) Fluorides as fluorine — 1 mg/litre
   (i) Zinc — 2 mg/litre
   (j) Copper — 5 mg/litre
   (k) Lead — 3 mg/litre
   (1) Iron — 0.1 mg/litre
   (m) Manganese — 3 mg/litre
   (n) Cadmium — 3 mg/litre
10. (a) Biological – decomposition of sewage
    (b) Physical/chemical – coagulation and solid sedimentation
    Adsorption, ion exchange, membrane processes.
11. (a) SO_2–up to 200 tons per day plants-16 kg/ton of 100% H_2SO_4 above 200 per day plants–12 kg/ton of 100% H_2SO_4
    (b) SO_3–exiting plants-5 kg/ton of 100% H_2SO_4
    (c) Fluorine–1.5 kg/ton of P_2O_5
    (d) Particulate matter–500 mg/m^3
    (e) Nitrogen oxide as NO_2 for old plants–4 kg/ton of 100% HNO_3
    Nitrogen oxide as NO_2 for new plants–3 kg/ton of 100% HNO_3
    (f) Hydrocarbons as methane–5 kg/ton of Ammonia.
12.

Constituent	*Concentration (mg/litre)*
Arsenic	0.05
Barium	1.0
Cadmium	0.01
Chromium	0.05
Cyanide (fluorine)	0.025

Fluoride	1.0
Lead	0.05
Selenium	0.01
Silver	0.05

13. (a) Particulates

    (b) Gases-sulphur compounds, carbon monoxide, cyanides, fluorides and benzene compounds.

14. Air pollution index (API) = 20 [SO_2 + 1 (CO) + 2 (Smoke shade)]

    Average value of API is 12.0. An index reading of 50.0 is considered adverse and a cause for alarm.

15. Ringlemann chart is used in air pollution evaluation for assigning an arbirary number referred to as smoke density, to smoke emanating from any source.

16. On wet basis:

Nitrogen	756,500 ppm (vol)
Oxygen	202,900
Water	31,200
Argon	9000
Carbon dioxide	305
Neon	17.4
Helium	5.0
Methane	1 to 1.2
Krypton	1.0
Nitrous oxide	0.50
Hydrogen	0.50
Xenon	0.10
Organic vapours	0.02 μg (micro-gram)

17. Permissible concentrations in μg/litre for organic pesticides in public water supplies.

Aldrin	17.0
Chlordane	3.0
DOT	42.0
Dieldrin	17.0
Endrin	1.0
Heptachor	18.0
Heptachor epoxide	18.0
Lindane	56.0
Methoxychlor	35.0
Organic phosphates and Carbamates	100.00
Toxaphene	5.0

18. Chromium, dust, fluoride, sulphur dioxide.
19. Chloro-alkali industry: Brine, calcium, chloride, mercury
20. Fertilizer: Ammonia, nitrates, oxides of nitrogen, phosphates.
21. Glass and ceramic: Arsenic, barium, manganese, selenium.
22. Phosphoric acid : Calcium sulphate, hydrofluoric acid.
23. Detergents: Boric acid, phosphates, sulphates.
24. Water treatment: Calcium salts, filtered solids.
25. Tanneries : Arsenic, chromium, fibres, hair, lime.
26. Dyestuffs: Aniline, chromium, phenol, selenium.
27. Beverages : Alkali, B.O.D. waste, carbon dioxide, cullet detergent.
28. Nitric acid : Ammonia, oxides of nitrogen.
29. Vegetable and fruit: Alkali, Bleach, B.O.D. waste, oil solvent, wax.
30. Batteries : Cadmium, lead, manganese, mercury, nickel, zinc.
31. Dust is a mixture in air of irregular-shaped mineral particles in the size range from 1 to 200 μ m formed by various operations.
32. Sulphurous smog is a mixture in air of sulphur dioxide and other sulpur compounds along with particulates including water droplets, and the reaction products which occur under these conditions.
33. Thermal pollution is the discharge into the environment of a stream of air or water which is at a different temperature from that of the environment at the point of discharge or downstream of this point.
34. Antimony, arsenic, berryllium, copper, cobalt, chromium, cadmium, lead, manganese, mercury, molybdenum, nickel and zinc.
35. Explosive mixtures with air:

	Lower limit %	*Upper limit %*
CH_4	5.3	14.0
C_2H_6	3.2	12.5
C_3H_8	2.4	9.5
C_4H_{10}	1.9	8.5 (butane)
C_6H_6	1.5	8.0
Toluene	1.5	7.0
H_2S	4.3	45.0
NH_3	12.5	74.0
CO	4.0	75.0
H_2	6.0	41.0
HCN	6.0	41.0
Propylene	2.4	10.3
Vinyl chloride	4.0	22.0

36. List of toxic substances

Arsenic, mercury, cadmium, thallium, beryllium, chromium, lead, phenolic compounds, cyanide, isocyanates, organo-halogenator, chlorinated solvents, organic solvents, biocides and phyto-pharmaceutical substances, tars, pharmaceutical compounds, peroxides, chlorates, perchlorates and oxides, ethers, asbestos, selenium and compounds, tellurium and compounds, polycyclic aromatic hydrocarbons, metal - Carbonyls and soluble copper compounds.

37. Toxic levels of some substances

Substance	*Occurrence*	*mg/litre*
Acridin	Coal tar wastes	0.7 to 1.0
Aldrin	Insecticide	0.02
Free ammonia	Sewage effluent, gas works effluent	2.0 to 3.0
Chorine	Chlorinated effluents	0.05 to 0.2
Copper sulphate	Metal processing	0.1 to 2.0
Cyanide	Gas liquor, plating wastes	0.04 to 0.10
DDT	Insecticide	less than 0.01
Synthetic detergents	Sewage effluent	15.0 to 18.0
Fluoride	Aluminium smelting	2.5 to 6.0
Gammaxene	Insecticide	0.035
Parathion	Insecticide	0.20
Hydrogen	Bottom muds	
Sulphide	Sludge, sewage, several industries	0.5 to 1.0
Methyl mercaptan	Oil refineries	1.0
Naphthalene	Coal tar wastes	10.0 to 20.0
Zinc	Galvanising, Rayon manf.	1 .0 to 2.0
Sevin	Pesticide	0.015
Temic	Pesticide	0.010

38. Pollutant level in industrial work environments :

Industry	*Pollutant*	*Pollution levels*	*Permissible level*
Sulphuric acid plants	SO_2	12.2 to 35.0 ppm	5 ppm
Steel plant	Dust containing silica,	3 to 1	5 mppcf
	refractory dust	2–17 mppcf (Milli parts per cf)	20 mppcf (Milli parts per cf)
	Carbon monoxide	20–200 ppm	50 mppcf
	Aromatic hydro carbons	20–50 ppm	35 ppm

Chemical works	Oxides of nitrogen (HNO_3 plant)	11–31 ppm	5 ppm
	Chromic acid mist (chromic acid plant)	0.5 to 2.1 mg/m^3	0.1 mg/m
	Bauxite dust	8 to 198 mppcf	50 mppcf
Blending of pesticides	Soap stone dust	33–134 mppcf	20 mppcf
	DOT	15.3 to 17.8 mg/m^3	1.0 mg/m^3
	BHC	25.2 to 29.8 mg/m^3	0.5 mg/m^3
Pharma-ceutical laboratory	Talc dust	60–90 mppcf	20 mppcf
	Ethyl alcohol	482–7812 mg/m^3	1900 mg/m^3
	iron compound dust	0.67 to 5.6 mg/m^3	1.00 mg/m^3
Heavy Engineering Factory	Alumina dust	15 to 32.4 mppcf	30 mppcf
	Silica	37.1–4157 mppcf	30 mppcf
	Coal dust	28.0 to 35.3 mppcf	25 mppcf
	Iron dust	18 to 52 mppcf	30 mppcf
	Sand bearing free silica Asbestos	23.5–101 mppcf	
		847 to 1518 fibres/CC	6 mppcf 5 fibres/CC
	Oil mist noise	4–158 mg/mm^3	5 mg/m^3
		87 to 120 d BA	85rfBA

39. Pollution impact on workers in some industries

Industry	*Polluting chemical*	*Result*
1. Dichromate manufacture	Dust and mist bearing Hexavalent chromium	Skin ulcers, nasal Perforations
2. Hand drilling in mica mining	Dust bearing free silica	Silicosis
3. Metal grinding	Dust	Silicosis
4. Refractories	Dust bearing free silica	Silicosis
5. Viscose rayon	Carbon disulphide sulphur dioxide	Psychic, nurological signs and silicosis eye problems, high cholesterol
6. Lead and zinc mining	Dust bearing free silica	Silicosis

7. Coal mine	Coal dust	Pneumoconiosis bronchial spasm, cough breathlessness on exertion
8. Ferro-manganese	Manganese	Manganese poisoning pneumonities, pigmentation bronchitis
9. Leather tanning	Tail pit liquid hydrogen sulphide gas	—All fatal
10. DDT manufacture	DDT, chlorine and benzene	Respiratory ailments
11. Nitric acid manufacture	Noise	Mild to moderate heating
12. Pesticide formulation	Parathion	Execessive sweating Headache, giddiness Blurring vision
13. D.C. meters repairing	Metallic mercury vapours	Mercury poisoning mercury intoxication
14. Foundry	Dust bearing free silica	Chest pain, lowered pulmonary function
15. Leather industry	Biological dust	Chronic branchities
16. Leather tanning	Tanning agents, solvents,	Dermatities, other skin diseases
17. Pesticides formulation	BHC	Skin and eye irritation, constipation.
18. Chloro-alkali plant	Metallic mercury	Gingivities, stomatities, loss of memory, insomnia, nervousness
19. Cotton and textile mills	Cotton dust	Byssinosis

40. On-stream instrumental analysis

Parameter	*Method*
1. Alkalinity	Automatic analyser
2. Bromide	Specific ion electrode
3. Ammonia	Specific ion electrode, automatic analyser, continuous spectrophotometer.
4. Bromide	Specific ion electrode
5. Cadmium	Specific ion electrode
6. Calcium	Sspecific ion electrode

7. Chloride	Specific ion electrode
8. Chlorine	Automatic analyser, polarographic device, colorimetric analyser
9. Colour	Spectro-photometric analyser
10. Copper	Polorographk device, specific ion electrode, automatic analyser and colorimetric.
11. Cyanide	Specific ion electrode, automatic analyser
12. Dissolved solids	Specific conductance
13. Filtrable solids	Inductance or capacitance
14. Fluoride	Specific ion electrode, colorimetric analyser.
15. Hardness	Automatic analyser, polyvalent ion electrode.
16. Iron	Automatic analyser, colorimetric analyser, spectrophotometer.
17. Mercury	Flameless ultra-violet absorption.
18. Nitrates, citric	Specific ion electrode, spectro photo; meter, automatic analyser.
19. Nitrogen (Kjeldahl)	Automatic analyser
20. Oil	Ultraviolet light transmittance, light reflectance.
21. Oxidation, reduction	Use pH equipment with substitution of platinum electrode for glass membrane electrode.
22. Oxygen demand, chemical oxygen demand (C.O.D.) total organic carbon (T.O.C) or total carbon (T.C).	Wet oxidation, combustion, wet oxidation with persulphate.
23. Total oxygen demand (T.O.D), biochemical oxygen demand (B.O.D)	Combustion, respirometer.
24. Oxygen (dissolved)	Thallium and reference electrode, polarographic system.
25. Phenols and other dissolved organics	Ultraviolet spectrophotometry, automatic analyser.
26. Phosphorous	Flame emission photometry, automatic analyser, colorimetric.
27. ph	Potentiometric (electrode device)
28. Silver	Spectrophotometric
29. Sulphide	Specific ion
30. Suspended solid	Light scattering capacity
31. Temperature	Thermistors, thermocouples
32. Turbidity	Light scattering or capacity, reflectance.

41. Particulate collectors for air pollutants

1. Electrostatic precipitators	Dry plate type Dry-pipe type Wet type
2. Mechanical collectors	Tangential inlet type axial inlet type
Cyclones	(Multicyclones)
3. Fabric filters	Fibrous bed Granular bed
4. Wet scrubbers.	

Emission control equipment for gases and odours

1. Wet absorbers (i) scrubbers (ii) absorption towers.
2. Dry adsorbers...............fixed bed, dynamic bed, dispersed bed.
3. Incinerators.

42. Public versus private benefits

(a)	(b)
1. Improved healths	1. Low absenteeism
2. Reduced safety hazards	2. Reduced risk o civil damage suit.
3. Reduced health risks to men and animals.	3. Better employee relations
4. More comfortable enjoyment of life and property	4. Better public relations
5. Reduced property damage	5. Reduced maintenance cost
6. Increased property values	6. Increased property value
7. Less vegetation damage	7. Product recovery
	8. New markets for new products
	9. Reduced product contamination or damage.

❑

CHAPTER

17 Summary of Equations

A. Instrumentation and Process Control.

1. Laplace transform of step input of magnitude A, $= A/S$.
2. Laplace transform of impulse of magnitude, $B = B$.
3. Laplace transform of sinusoidal input of magnitude,

$$A = \frac{AW}{(S^2 + W^2)}$$

4. First order transfer function $= R/(TS + 1)$.
5. Second order transfer function

$$= R/(T^2 S^2 + 2\zeta TS + 1).$$

6. $L(y') = Sy(S) - y(0)$
7. $L(y'') = S^2 y(S) - Sy(0) - y'(0)$
8. P-action : $m'(S) = Kc\ e(S)$
9. $P\text{-}D.m'(S) = Kc\ e(S)(1 + T_d S)$.
10. $P\text{-}I : m'(S) = Kc\ e(S)\ (1 + 1/T_i S)$
11. $P\text{-}I\text{-}D : m'(S) = Kc\ e(S)\ (1 + T_d S + 1/T_i S)$
12. Closed loop transfer function : Feedback positive :
    $Y/X = \pi_f/(1 - \pi_T)$ where π_f is for forward loop and π_T is total loop.
13. Closed loop transfer function for negative feedback :

$$Y/X = \pi_f/(1 + \pi_T)$$

14. Response of first order system to step input of magnitude

$$A : RA\ (1 - e^{-t/T})$$

15. Response of first order to impulse input of magnitude

$$B : (BR/T)\ e^{-t/T}$$

16. Response of first order to sinusoidal input of magnitude.

$$A : AR/\sqrt{1 + W^2T^2}\ \sin(wl + \phi) \text{ where } \phi = -\tan^{-1}(WT)$$

17. 2nd order system : Amplitude ratio :

$$\frac{1}{[(1 - W^2T^2)^2 + (2\xi WT)^2]^{1/2}}$$

Phase lag : $\tan^{-1} [-2\zeta WT/(1 - W^2T^2)]$

18. Time lag : $AR = 1$:Phase lag $= -WL$ radians.

$$= \frac{-W * 180 * L(\text{Degrees})}{\pi}$$

19. P-action : $AR = Kc$; Phase lag = 0 degrees.

20. $P\text{-}I : AR = (1 + 1/W^2\, T_i^2)^{1/2}$

Phase lag $= -\tan^{-1} (1/WT_i)$

21. $P\text{-}D : AR = K_C\,(1 + W^2\, T_d^2)^{1/2}$

Phase lag $= \tan^{-1} (WT_d)$

22. $P\text{-}I\text{-}D : AR = K_c\,[(WT_d - 1/WT_i)^2 + 1)]^{1/2}$

Phase lag : $\tan^{-1} (WT_d - 1/WT_i)$.

23. For closed loop values from open loop values :

$$M = \frac{1}{\left[\left(1+\frac{\cos\phi}{A}\right)^2 + \left(\frac{\sin\phi}{A}\right)^2\right]^{1/2}}$$

$$\theta = \tan^{-1}\frac{\sin\phi/A}{(1+\cos\phi/A)}$$

24. Gain margin = $1/AR$ at 180° phase lag.

25. Phase margin = 180° + Phase angle.

26. Ultimate period : $P_u = 2\pi/WC_0$

27. Mason's formula $= 1 - \Sigma M_{aa} + \Sigma M_b M_c - \sum_{d,e,f} M_d M_e M_f$

28. State equation : $x(t) = Ax\,(t) + Bm\,(t)$.

29. Output equation : $0(t) = Cx(t) + Dm\,(t)$.

30. $\phi(t) = e^{At} = L^{-1} [(SI - A)^{-1}]$.

31. Discrete data : $x(K + 1)\,T = \phi\,(T)\,x\,(KT) + \theta\,(T)\,m\,(KT)$

32. Discrete data output equation : $\theta\,(T) = Cx\,(KT) + Dm\,(KT)$

33. $\theta(T) = \int_0^\tau \phi(T - \tau)\,BD\,\tau.$

34. $L\,(d^n f/dt^n) = S^n f\,(S) - S^{n-1} f(0) - S^{n-2} f'(0) \ldots - Sf^{n-2}\,(0) - f^{n-1}\,(0)$.

35. Linearizing using Taylor series :

$$f(x) = f\,(x_s) + \left.\frac{df}{dx}\right|_{xs} (x - x_s) + \ldots$$

36. For second order system, step response :

(a) $\xi < 1 : y(t) = 1 - \frac{1}{\sqrt{1-\xi^2}} e^{-\xi t/T} \sin[\sqrt{1-\xi^2}\; t/T + \tan^{-1}\sqrt{1-\xi^2}/\xi]$

(b) $\xi = 1$; $y(t) = \left[1 - \left(1 + \frac{t}{T}\right) e^{-t} / T\right]$

(c) $\xi > 1$: $y(t) = 1 - e^{-\xi t/T} \left[\cosh \sqrt{\xi^2 - 1}\, t/T\right] + \frac{\xi}{\sqrt{\xi^2 - 1}} * \sin h \sqrt{\xi^2 - 1}\, t/T$

37. Second order system, impulse response :

(a) $\xi = 1$, $y(t) = \frac{t\, e^{-t/T}}{T^2}$

(b) $\xi < 1$, $y(t) = \frac{1}{T\sqrt{1 - \xi^2}}\, e^{\xi t/T} \sin \sqrt{1 - \xi^2}\, \frac{t}{T}$

(c) $\xi > 1$, $y(t) = \frac{1}{T\sqrt{\xi^2 - 1}}\, e^{-\xi t/T} \sin h \sqrt{\xi^2 - 1}\, \frac{t}{T}$.

38. Optimum controller settings (Ziegler-Nichols)

	K_c,	T_i	T_d
P	$0.\ K_u$	—	—
PI	$0.45\ K_u$	$P_u/1.2$	—
PID	$0.6\ K_u$	$P_u/2$	$P_u/8$

where $P_u = 2\pi / W_{C_e}$;

W_{c_e} = frequency at – 180° phase angle

K_u = $1/A$ where A is amlitude at 180° phase lag.

B. Mass Transfer

1. Fick's first law : $J_A = -D_A\, \partial C_A / \partial z$.

2. Fick's second law : $\partial C_A / \partial \theta = D_{AB}\ \partial^2 C_A / \partial z^2$

3. Diffusion of A through non-diffusing B (stagnant case) :

$$N_A = \frac{D_{AB} P_t}{RTZ\, P_{B_m}} (P_{A_1} - P_{A_2}) \text{ for gases}$$

$$= \frac{D_{AB}}{Zx_{B_m}} \left(\frac{\rho}{M}\right)_{av} (x_{A_1} - x_{A_2}) \text{ ... liquids.}$$

4. Equimolal counter diffusion :

$$N_A = \frac{D_{AB}}{RTZ} (P_{A_1} - P_{A_2}) \text{ ... for gases.}$$

$$= \frac{D_{AB}}{Z} \left(\frac{\rho}{M}\right)_{av} (x_{AI} - x_{A2}) \text{ ... liquids.}$$

5 Individual mass transfer coefficients :

$$N_a = K_y(Y_{AG} - Y_{Ai}) = K_x(X_{Ai} - X_{AL})$$

6. (Overall mass transfer coefficients $= 1/K_y = (1/k_y + m/k_x)$

$$1/K_x = (1/k_x + 1/mk_y)$$

7. Humidity, $Y = P_A M_A/(P_t - P_A) M_B$

$$Y = \frac{18}{29}\left(\frac{pw}{(P_t - pw)}\right) \text{ for water vapour-air.}$$

8. Enthalpy of air, $H = C_s(t - t_0) + Y\lambda_0$ where $C_s = 0.24 + 0.45\,Y$ for water and t_0 is reference temperature.

9. Wet bulb temperature : $(t_G - t_w) = \lambda_w (Y_w - Y)\, hG/ky$.

10. Height of humidication tower : $Z = H_{to\,G} * N_{to\,G}$

$$H_{to\,G} = \frac{G_s}{Ky_a};\ N_{toG} = \int_{tL_1}^{tL_2} \frac{dt_L}{(H^* - H)} = \int_{H_1}^{H_2} \frac{dH}{(H^* - H)}$$

11. Raoult's law : $P^* = Px$.

12. Henry's law : $y^* = mx$

13. Material balance equation for absorber :

$Gs\,(Y_1 - Y) = Ls\,(X_1 - X)$ where X and Y are ratios.

14. Adiabatic humidifier : $N_{to\,G} = \ln\left(\frac{Y_a S - Y_1}{Y_a S - Y_2}\right)$

$$H_{to\,G} = Gs/K_{ya}.$$

15. $H_{to\,G}$ in absorption $= \dfrac{G}{K_{G_a}\; P_t(1 - y^*)\; L_m}$

$$N_{to\,G} = \int_{Y_2}^{Y_1} \frac{dY}{(Y - Y^*)} + \frac{1}{2}\ln\left(\frac{1 + Y_2}{1 + Y_1}\right)$$

$$H_{to\,L} = \frac{L}{K_{L_a}\; C(1 - x) * Lm}$$

$$N_{to\,L} = \int_{X_2}^{X_1} \frac{dX}{(X^* - X)} + \frac{1}{2}\ln\left(\frac{1 + X_1}{1 + X_2}\right)$$

16. Relative volatility $= \dfrac{y^*/(1 - y^*)}{x/(1 - x)};\ y^* = \dfrac{\alpha x}{1 + (\alpha - 1)x}$

17. Flash vaporization : $\dfrac{W}{D} = \dfrac{x_F - y_D}{x_F - x_w} = \dfrac{(H_F + Q/F) - H_D}{(H_F + Q/F) - H_w}$

18. Differential distillation : $\ln\left(\frac{F}{W}\right) = \int_{xw}^{x_F} \frac{dx}{(y^* - x)}$

For constant relative volatility, $\log\left(\frac{Fx_F}{Wx_W}\right)$

$$= \alpha \log \frac{F(1 - x_F)}{W(1 - Xw)}$$

19. Fenske's equation (at total reflux) :

$$N_{m+1} = \frac{\log \frac{XD(1 - Xw)}{(1 - XD)\, Xw}}{\log \alpha_{av}}$$

20. Enriching section operating line (O L) :

$$Y_{n+1} = \frac{R}{R+1} x_n + \frac{x_D}{R+1}; R = L/D.$$

21. Stripping section : $y_{m+1} = \frac{L}{L-W} x_m - \frac{W}{L-W} x_w$

22. q-line : $y = \frac{q}{q-1} x - \frac{x_F}{q-1}$; $q = 1$ for saturated liquid feed; $q = 0$ for saturated vapor feed.

23. Drying : Time of drying, $\theta = -\frac{L_s}{A} \int_{X_2}^{X_1} \frac{dX}{N}$;

C (constant rate period) = $\frac{Ls}{ANc}(X_1 - X_c)$

(θ, Falling rate) = $\frac{Ls(X_c - X_2)}{A(N_c - N_2)} \ln\left(\frac{N_c}{N_2}\right)$

if $N = mX + b$.

(Falling rate) $Ls\left(\frac{X_c - X^*}{N_c A}\right) \ln\left(\frac{X_1 - X^*}{X_2 - X^*}\right)$ if entire falling rate curve is straight line.

24. Chilton-Colburn analogy :

$$J_H = \frac{N_u}{Re\,.\,Pr} P_r^{2/3} = \frac{f}{2} = J_D = \frac{S_h}{Re\,.\,Sc.} S_c^{2/3}$$

25. Liquid-extraction : Material balance equation is :

$$F + S_1 = M_1 = E_1 + R_1 \; ; \; E_1 y_1 + R_1 x_1 = M_1 * x_{M1} \; ;$$

For insoluble liquids : $\dfrac{-A}{B_n} = \dfrac{y'_s - y'_n}{x'_{n-1} - x'_n}$

$$N_{\text{to } R} = \int_{X_2}^{X_1} \frac{dx}{(x - x^*)} + \frac{1}{2} \ln\left(\frac{1 - x_2}{1 - x_1}\right) + \frac{1}{2} \ln \frac{x_2(r-1)+1}{x_1(r-1)+1}$$

r is ratio of molecular weights of non-solute to solute,

$$N_{\text{to } E} = \int_{y_2}^{y_1} \frac{dy}{y^* - y} + \frac{1}{2} \ln\left(\frac{1 - y_1}{1 - y_2}\right) + \frac{1}{2} \ln \frac{y_2(r-1)+1}{y_2(r-1)+1}$$

Ist 2 terms in mole fractions and all 3 terms for weight fractions.

$$H_{\text{to } R} = \frac{R}{KRa\,(1-x)^*\, lm} ; H_{to\, E} = \frac{E}{KEa(1-y)^*\, lm}$$

26. Adsorption : Freundlich equation : $C^* = K\,[V\,(C_0 - C^*)]^n$

C. Chemical Reaction Engineering

1. Arrhenius equation : $K = Ae^{-E/RT}$

2. Collision theory : $K = AT^{1/2}\, e^{-E/R}$

3. Transition state theory : $K = ATe^{-E/RT}$

4. Analysis of total pressure data in a constant volume system :

$$p_R = C_R\, RT = PR_0 + \frac{r}{\Delta_n}(\pi - \pi_0)$$

$$p_A = C_A\, RT = p_{A_0} - \frac{a}{\Delta_n}(\pi - \pi_0)$$

For the reaction $aA + bB + ... = rR + sS + ...$

$\Delta n = (r + s + ...) - (a + b + ...)$

π = Total pressure of the system at any time

π_0 = Initial pressure of the system.

5. Rate equation in terms of pressure :

$$r = \frac{1}{RT}\frac{dp}{dT}$$

6. Differential and integral rate equations (Single reactions) :

(i) Constant volume systems :

Type	Order	Differential equation	Integral
1. $A \rightarrow$ product	0	$-\frac{d\,C_A}{dt} = K$	$(CA_0 - C_A) = K * t$
2. $A \rightarrow$ products	1	$-\frac{d\,C_A}{dt} = KC_A$	$-\ln\frac{C_A}{CA_0} = Kt$
3. $2A \rightarrow$ products	2	$-\frac{d\,C_A}{dt} = +KC_A^2$	$\frac{1}{C_A} - \frac{1}{C_{A_0}} = Kt$
4. $A + B \rightarrow$ product	2	$-\frac{dC_A}{dt} = \frac{-dC_B}{dt}$	$-\frac{\ln C_B C_{A0}}{C_{B0} C_A} = (C_{B0} - C_{A0})Kt$ For $C_{B0} \neq C_{A0}$
	2	$-\frac{dC_A}{dt} = KC_A\,C_B$	
5. aA + $bB \rightarrow$ products	$a + b$;	$-\frac{d\,C_A}{dt} = KC_A^a C_B^b$	$\ln\frac{C_A}{C_{A0}} = \frac{bC_{A0} - a\,C_{B0}}{a}\,Kt + \ln\frac{C_{A0}}{C_{B0}}$
6. $n_A \rightarrow$ products		$-\frac{dC_A}{dt} = KC_A{}^n$	$\frac{1}{C_A{}^{n-1}} - \frac{1}{C_{A0}{}^{n-1}}$ $= (n - 1)\,Kt;\ n \neq 1$
7. $A \underset{K_2}{\xrightarrow{K_1}} R$		$-\frac{dCA}{dt} = K_1 C_A - K_2 C_A$	$-\ln\left(1 - \frac{X_A}{X_{A0}}\right) = (K_1 + K_2)t$
8. $A + B \underset{K_2}{\xrightarrow{K_1}} R + S$		$-\frac{dC_A}{dt}$ $= K_1 C_A\,C_B - K_2\,C_R\,C_S$	$\ln\frac{X_{Ae} - (2X_{Ae} - 1)X_A}{X_{Ae} - X_A}$ $= 2K_1\left(\frac{1}{X_{Ae}} - 1\right)C_{A0}t$
$2A \underset{K_2}{\overset{K_1}{\rightleftarrows}} R + S$			
$2A \underset{K_2}{\overset{K_1}{\rightleftarrows}} 2R$			
$A + B \underset{K_2}{\overset{K_1}{\rightleftarrows}} 2R$			

$C_{A_0} = C_{B_0}; C_{R_0}$ $= C_{S_0} = 0$ $X_{Ae} = \frac{K}{1+K}$ $K = \frac{K_1}{K_2}$ $1 - X_A = CA/C_{A_0}$		

(ii) Variable volume systems :

$$V = V_0 (1 + \varepsilon_A X_A);\ \text{where } \in_A = \frac{V_{XA=1} - V_{XA=0}}{V_{XA=0}}$$

$$\frac{C_A}{C_{A_0}} = \frac{1 - X_A}{1 + \in_A X_A}; -r_A = \frac{-1}{V}\frac{dN_A}{dt}$$

$$= \frac{C_{A_0}}{1 + \in_A X_A}\frac{dX_A}{dt}$$

Zero-order Reactions : $A \to$ products

$$-r_A = -\frac{1}{V}\frac{dN_A}{dt} = \frac{C_{A_0}}{1 + \in_A X_A}; -\frac{dX_A}{dt} = K$$

On integration : $\frac{C_{A_0}}{C_A} \ln \frac{V}{V_0} = Kt$

First Order : $A \to$ products

$$-r_A = \frac{1}{V}\frac{dN_A}{dt} = \frac{C_{A0}}{1 + \in_A X_A}; \frac{dX_A}{dt} = KC_A$$

On integration : $-\ln(1 - X_A) = -\ln\left(1 - \frac{V}{\in_A V_0}\right) = Kt$

Second Order : 2A $\to$ products or $A + B \to$ products $C_{A0} = C_{B0}$

$$-r_A = KC_A^{\,2} = KC_{A0}^2\left(\frac{1 - X_A}{1 + \in_A X_A}\right)^2$$

On integration : $\frac{(1 + \in_A) X_A}{1 - X_A} + \in_A \ln(1 - X_A) = KC_{A0}t$

7. Multiple reaction systems :

(a) Parallel reactions : $A \xrightarrow{K_1} R;\ A \xrightarrow{K_2} S$

$$\frac{-dC_A}{dt} = (K_1 + K_2)\, C_A;\ \ln \frac{C_A}{C_{A0}} = (K_1 + K_2)\, t;$$

$$\frac{C_R - C_{R0}}{C_S - C_{S0}} = \frac{K_1}{K_2}$$

(b) Homogeneous catalyzed reaction :

$$R \xrightarrow{K_1} S\ ;\ R + C \xrightarrow{K_2} S + C;$$

$$-\frac{dC_R}{dt} = (K_1 + K_2)\, C_R$$

$$-\ln \frac{C_R}{C_{R0}} = -\ln (1 - X_R) = (K_1 + K_2)\, t = K't$$

(c) Autocatalytic reactions :

$$A + R \rightarrow R + R;\ \frac{dC_A}{dt} = KC_A\, C_R$$

$$\ln \frac{C_{A0}\, C_R}{C_A\, C_{R0}} = (C_{CA0} + C_{R0})\, Kt$$

(d) Series reactions :

$$P \xrightarrow{K_1} Q \xrightarrow{K_2} S.$$

$$r_P = \frac{dC_P}{dt} = -K_1 C_P;\ r_Q = \frac{dC_Q}{dt} = K_1\, C_P - K_2\, C_Q$$

$$r_s = \frac{dC_s}{dt} = +K_2 C_Q;\ \frac{C_P}{C_{P0}} = \exp.(-K_1 t)$$

$$\frac{C_Q}{C_{P0}} = \frac{K_1}{K_2 - K_1}(e^{-K_1 t} - e^{-K_2 t});$$

$$\frac{C_s}{C_{P0}} = 1 + \frac{K_2}{K_1 - K_2} e^{-K_1 t} + \frac{K_1}{K_2 - K_1} e^{-K_2 t}$$

$$\frac{C_{Q\max}}{C_{P0}} = \left(\frac{K_1}{K_2}\right)^{K_2/(K_2 - K_1)}$$

8. Half-life for an nth order reaction

$$t_{1/2} = \frac{2^{n-1} - 1}{K(n-1)}\, C_{A0}^{\ 1-n} \text{ for } n \neq 1$$

9. Ideal reactors: Design equations; $V = V_0 (1 + \epsilon_A X_A)$

$\epsilon_A = O$ for constant V

Reactor type	*Constant volume*	*Variable volume*
Batch	$t = C_{A0} \int \frac{dX_A}{(-r_A)}$	$t = C_{A0} \int \frac{dX_A}{(1 + \epsilon_A X_A)(-r_A)}$
Plug-flow	$\tau = C_{A0} \int \frac{dX_A}{-r_A}$	$t = C_{A0} \int \frac{dX_A}{(-r_A)(1 + \epsilon_A X_A)}$
Catalytic reactor (tubular reactor)	$\frac{W}{F} = \int \frac{dX_A}{-r_A}$	
Back mix	$\tau = \frac{C_{A0} X_A}{-r_A}$	$\bar{t} = \frac{C_{A0} X_A}{(-r_A)(1 + \epsilon_A X_A)}$

10. Space time $= \tau = \frac{\text{Volume of the reactor}}{\text{Volumetric flow rate}} = \frac{V}{v_0}$

Space velocity $= S = \frac{1}{\tau} = \frac{v_0}{V}$ (time)$^{-1}$

11. Non-ideal reactors :

(i) (a) F-curve : Rsponse to a step input (C/Co) *vs* time)

(b) C-curve : Response to a pulse input (C/Co) *vs* time)

(c) $C = E = \frac{dF}{dt}$

(d) Mean for a C *vs* t curve : $\frac{\int_0^\infty t\, Cdt}{\int_0^\infty Cdt}$

Variance (σ) *vs* t curve $= \sigma^2 = \frac{\int_0^\infty t^2\, Cdt}{\int_0^\infty Cdt} - t^{-2}$

(ii) **Residence time distribution.**

Plug flow reactor : $F(\theta) = u\,(t - \bar{t})$ where

$$U\,(t - \bar{t}) = 0, t < \bar{i}$$
$$= i, t > \bar{i}$$

CSTR : $F(\theta) = 1 - e^{-\theta}$, $C(\theta) = e^{-\theta}$

(*iii*) Dispersion model

(a) Basic differential equation :

$$\frac{\partial C}{\partial \theta} = \left(\frac{D}{UL}\right)\frac{\partial^2 C}{\partial Z^2} - \frac{\partial C}{\partial Z}$$

Dispersion number : D/UL : D/UL = O for plug flow

$D/UL = \infty$ for CSTR.

(b) R.T.D. in a tube with longitudinal dispersion

$$F(\theta) = \frac{1}{2}\left[1 - erf\left[\frac{1}{2}\sqrt{\frac{UL}{D}}\frac{1-\theta}{\sqrt{\theta}}\right]\right]$$

(c) For small extent of dispersion :

$$C(\theta) = \frac{1}{2\sqrt{(D/UL)}}\exp.\left[-\frac{(1-\theta)^2}{4(D/UL)}\right]$$

(d) Relationship between dispersion number and variance of the C-curve for different set-ups.

Systems	*Mean (θc)*	*Variance (σ^2)*
Closed vessels	1	$\frac{D}{UL} - 2\left(\frac{D}{Ul}\right)^2\left(1 - e\frac{U}{LD}\right)$
Open vessels	$1 + 2\frac{D}{UL}$	$2\frac{D}{UL} + 8\left(\frac{D}{UL}\right)^3$
Open-closed vessels	$1 + \frac{D}{UL}$	$2 + \frac{D}{UL} + 3\left(\frac{D}{UL}\right)^2$
Small $-\frac{D}{UL}$	1	$1 - \frac{D}{Ul}$

(iv) Tanks-in series model:

(a) $F(\theta) = 1 - e^{-N\theta}\left[1 + N\theta + \frac{1}{2l}(N\theta)^2 + ... + \frac{(N\theta)^{N-1}}{(N-1)!}\right]$

(b) $E(\theta) = \frac{N^N - \theta^{N-1}\, e^{-N\theta}}{(N-1)!}$

(c) For large N : $\frac{N_{\text{tanks}}}{V_{\text{plug}}} = 1 + \frac{K\tau}{N}$

and $\dfrac{C_A N\,.\,\text{Tanks}}{C_{A,}\ \text{plug}} = 1 + \dfrac{(KT)^2}{N}$

12. Heterogeneous reaction systems.

(a) Fluid-Particle reactions : (non-catalytic)

Unreacted core model.

(i) Diffusion through gas-film controls (For a reaction : $A + bB \rightarrow$ products)

$$\frac{t}{\tau} = 1 - \left(\frac{r_C}{R}\right)^x = X_B$$

where t is time : τ = Time for complete conversion

R = initial radius of the particle

r_e = radius of the unreacted core; X_0 – Conversion

(ii) Diffusion through ash layer control

$$\frac{t}{T} = 1 - 3\,(1 - X_B)^{2/3} + 2(LX_B)$$

(iii) Chemical reactions controls.

$$\frac{t}{\tau} = 1 - \frac{r_c}{R} = 1 - (1 - X_B)^{1/3}$$

(iv) Combination of resistance :

$$-\frac{1}{Se^x}\frac{d\overline{NA}}{dt} = \overline{Ks}C_A = \frac{C_A}{\dfrac{1}{K_R} + \dfrac{R}{2D_e} + \dfrac{1}{K_S}}$$

Shrinking spherical particle model :

(Reaction : aA (g) + bB(s) $\rightarrow rR$ (g)

Gas film and diffusion control :

$$\frac{K_g d_p y}{D} = 2 + 0.6\,(Sc)^{1/3}\,(Re)^{1/2}$$

$y \rightarrow$ mole fraction of a component which is being transferred.

$$K_g = \frac{1}{d_p} \text{ for small } d_p \text{ and } u$$

$$K_g = \frac{u^{1/2}}{d_p^{1/2}} \text{ for large } d_p \text{ and } u.$$

Stokes Regime : (small particles)

$$\frac{t}{\tau} = 1-\left(\frac{R}{R_O}\right)^2 = 1-(l-X_B)^{2/3}$$

R_0 = original size of the particle
R = size of the particle at time t

Combination of resistances:

$$-\frac{1}{S_{ex}}.\frac{dNA}{dt} = \frac{1}{\frac{1}{K_g}+\frac{1}{K_1}}C_A$$

(b) Solid-catalyzed reactions.

Surface area : Void volume and pore volume determination.

Surface area determination;

Langmuir Isotherm : $\theta = \frac{V}{V_m} = \frac{K_p}{1+K_p}$

Frendlich Isotherm : $\theta = C(p)^{1/n} : n > 1$

Brunauer–Emmett–Teller equation :

$$\frac{p}{V(p_0-p)} = \frac{1}{V_m}+\frac{(C-1)p}{CV_m p_0}$$

$$S_g = \left(\frac{V_m N_0}{V}\right)\alpha : \alpha = 1.09\left[\frac{M}{N_0\rho}\right]^{2/3}$$

For N_2 at – 195.8° C; $\alpha = 1.62 \times 10^{-7}$ cm^2

Void volume ($\in$) = $V_g\ \rho_s/(V_g\ \rho_s + 1) = \rho_p\ V_g$

Average pore radius : $\bar{a} = 2V_g / S_g$

where V_m = volume of one monomolecular layer of gas at STP
V = volume of gas absorbed at pressure p_1 per gram of catalyst.
S_g = total surface area of catalyst (per gram)
α = projected area on the catalyst surface of one molecule of adsorbed gas.
ρ_g = density of liquified gas at temperature of adsorption.
ρ_s = density of catalyst particle.
N_0 = Avogadro's number, 6.023×10^{23} molecules/mole.
ρ_p = density of pellet.
V_g = Volume of catalyst per gram.

Effective Diffusivity (D_e)

$$\frac{1}{D_e} = \frac{1}{D_k} + \frac{1}{D_M}$$

D_M – Molecular diffusivity

D_K – Knudsen diffusion $= 9.7 \times 10^3 \sqrt{\frac{T}{M}}$

r – pore radius

M – molecular weight

D_M – from Chapman-Enskog equation

$$D_M = 0.0018583 \frac{T^{3/2}[1/M_A + 1/M_B]^{1/2}}{P_t \sigma_{AB^2}\, \Omega_{AB}}$$

T – Temperature (K)

M_A, M_B – Molecular weights of gases A and B.

P_t – total pressure of the gas mixture ; atm

σ_{AB}, $\in_{AB}$ – Constants in the Lennard-Jones potential energy function for the molecular pair AB σ_{AB} in Å.

Ω_{AB} – Collision integral, which would be unity if the molecules are rigid spheres and is a function of $K_B\, T/\in_{AB}$ for real gases (K_B = Boltzmann's constant).

Approximate relations :

$$\sigma = 1.18\ V_b^{1/3}$$

$$\frac{K_B T}{\in} = 1.30\ \frac{T}{T_c}$$

where Tc = critical temperature

V_b = volume per mole at normal boiling point : (cm^3/g mole)

$$\sigma_{AC} = \frac{1}{2}(\sigma_A + \sigma_c); \in_{AC} = (\in_A \in_C)^{1/2}$$

Effectiveness factor ($\in$) :

$$\frac{\text{Actual reaction rate within the pore}}{\text{Rate if not slowed by diffusion}}$$

For first order reactions :

$$\in = \frac{\tan h_m L}{m_L}$$

where $m = \sqrt{\frac{K}{D}}$, L = Length of the pore

K = Reaction rate constant, D = Diffusivity

Non-Isothermal systems

(i) Batch reactor:

$$\int_{T_1, X_1}^{T_2, X_2} C_p dT + \frac{C_0}{\rho_0} (\Delta H_r)\, dX_j = \int_1^2 \frac{q}{m}\, dt \quad \begin{pmatrix}\text{constant}\\ \text{pressure}\end{pmatrix}$$

$$\int_{T_1, X_1}^{T_2, X_2} C_v dT + \frac{C_0}{\rho_0} (\Delta u_r)\, dX_j = \int_1^2 \frac{q}{m}\, dt \quad \begin{pmatrix}\text{constant}\\ \text{volume}\end{pmatrix}$$

(ii) Plug-flow reactor :

$$G_m \left[C_p\, dT + \frac{C_0}{\rho_0} (\Delta H_r)\, dX_j \right] = \frac{US}{R_h} (T_C - T)\, dz$$

(iii) Backmix reactor :

$$C_p (T_2 - T_1) + \frac{C_0}{\rho_0} (\Delta H_r)(X_2 - X_1) = \frac{UA}{G_m} (T_C - T_2)$$

where q = heat added to the system; U = heat transfer coefficient, G_m = molar flow rate; S = surface area; ΔH_r = heat of reaction ; m = mass of the mixture.

D. Thermodynamics

1. Basic equation

$H = U + PV$ $dU = TdS - pdV$

$G = H - TS$ $dH = TdS + VdP$

$A = U - TS$ $dG = -SdT + VdP$

$dU = dQ - dW$ $dG = -SdT + VdV$

$dS = (dQ/T)p,\ t,\ v.$

2. Fundamental property relation :

$d(nU) = Td\,(nS) - Pd\,(nU) + \Sigma\mu_i\, dn_i$

3. Maxwell relations :

$$\left(\frac{\partial T}{\partial V}\right)_{s\, ni} = -\left(\frac{\partial p}{\partial S}\right)_{V\, ni} ; \left(\frac{\partial T}{\partial p}\right)_{tis,\, ni} = \left(\frac{\partial V}{\partial S}\right)_{p,\, ni}$$

$$\left(\frac{\partial S}{\partial V}\right)_{T,\, ni} = \left(\frac{\partial p}{\partial T}\right)_{V,\, ni} ; \left(\frac{\partial S}{\partial P}\right)_{T,\, ni} = \left(\frac{\partial V}{\partial T}\right)_{p,\, ni}$$

4. Thermal capacity relations :

$$C_p = \left(\frac{dQ}{dT}\right) = \left(\frac{\partial H}{dT}\right)_p = T\left(\frac{dS}{dT}\right)_p$$

$$C_v = \left(\frac{dQ}{dT}\right)_v = \left(\frac{du}{dT}\right)_v = T\left(\frac{dS}{dT}\right)_v$$

$$c_p = C_p/\Sigma n_i ;$$
$$c_v = C_v/\Sigma n_i \text{ (molar heat capacities)}$$

$$\alpha \text{ (compresibility factor)} = \frac{1}{V}\left(\frac{\partial V}{\partial T}\right)_p$$

$$\text{k (Bulk modulus)} = -\frac{1}{V}\left(\frac{\partial V}{\partial P}\right)_T$$

$$-\left(\frac{\partial K}{\partial T}\right)_p = \left(\frac{\partial \alpha}{\partial P}\right)_T$$

$$(C_p - C_v) = -T\left(\frac{\partial p}{\partial T}\right)_v\left(\frac{\partial V}{\partial p}\right)_T$$

$$= -T\left(\frac{\partial V}{\partial T}\right)_p^2\left(\frac{\partial p}{\partial V}\right)_T = TV\frac{\alpha^2}{K}$$

$$\frac{C_p}{C_v} = -\left(\frac{\partial V}{\partial T}\right)_p\left(\frac{\partial T}{\partial p}\right)_v\left(\frac{\partial p}{\partial V}\right)_s = \gamma$$

$$\left(\frac{\partial C_p}{\partial P}\right)_T = -T\left(\frac{\partial^2 V}{\partial T^2}\right)_p ; \left(\frac{\partial c_v}{\partial V}\right)_T = T\left(-\frac{\partial^2 p}{\partial T^2}\right)_v$$

5. Gibbs-Helmholtz relations :

$$\left[\frac{\partial(\Delta A/T)}{T}\right]_{v,\,ni} = -\frac{\Delta U}{T^2}; \left[\frac{\partial(\Delta G/T)}{\partial T}\right]_{p,\,ni} = -\frac{\Delta H}{T^2}$$

6. $\Delta s = \int_{T_1}^{T_2} C_p \, d\ln T - R\ln\frac{p^2}{p^1}; U = \int_0^T pdT + RT + U_0$

$$\int_{T_1}^{T_2} C_p \, d\ln T - R\ln\frac{p^2}{p'}; U = \int_0^T C_p dT + RT + U_0$$

$$\left[\frac{dQ}{T}\right] < \Delta S; H = \int_0^T C_p dT + RT + H_0$$

Properties of fluids

Ideal behaviour : Gases

Boyle's law : $\left[P \propto \frac{1}{V}\right]_T$; Charle's law : $[V \propto T]_P$

Gas law $PV = n\,RT$

Real gases : Van der Waals equation

$$\left(p + \frac{a}{V^2}\right)(V - b) = RT$$

$$a = \frac{9}{8}\frac{RT_c}{P_c}; b = \frac{RT_c}{8pc}$$

$$(C_p - C_v) = R\left(1 + z\frac{ap}{R^2T^2}\right)$$

Redlich-Kwong equation

$$p = \frac{RT}{(V-b)} - \frac{a}{T^{1/2}\,V(V+b)}$$

Virial equation of state :

$$\frac{pV}{RT} = 1 + B/V + C/V^2 + D/V^3$$

B, C and D are virial coefficients.

Compressibility factor : $Z = pV/nRT$.

Enthalpy, entropy and heat capacity

$$\left(\frac{H^* - H}{T_C}\right)_T = \left[RT_r{}^2\int_0^{pr}\left(\frac{\partial T}{\partial T_r}\right)P_r\frac{dp_r}{p_r}\right]_T$$

$$(S_p^* - S_p)_T = \left[\frac{H^* - H}{T_C T_r} + R\ln\left(\frac{f}{p}\right)\right]_T$$

$$= R\int_0^{pr}\left[Z - 1 + T_r\left(\frac{\partial Z}{\partial T_r}\right)P_r\right]\frac{dp_r}{p_r}$$

$$\frac{U^* - U}{T_C} = \frac{H^* - H}{T_c} - (1 - Z)RT_r$$

$$C_p^* - C_p = \frac{\partial\left(\frac{H - H^*}{T_e}\right)}{\partial T_r}$$

*—indicates ideal behaviour

Properties of mixtures of ideal gases are

$$V_m = \Sigma y_i V_i; U_m = \Sigma y_i U_i$$

$$H_m = \Sigma y_i H_i;$$

$$S_m = \Sigma y_i \, S_i + R \, \Sigma y_i \ln \frac{1}{y_i}$$

Joule-Thomson coefficient :

$$\beta = \left(\frac{\partial T}{\partial P}\right)_H = \frac{1}{C_p}\left(\frac{\partial H}{\partial p}\right)_T \quad \text{(ideal gas)}$$

$$= \frac{1}{C_p}\left[\frac{2a}{RT} - b - \frac{3abp}{R^2T^2}\right] \quad \text{(Van der Waals gas)}$$

Vapour-pressure functions :

Clausius equation : $\ln p = A - B/T$

Antonie equation : $\ln p = A + B/(T + C)$

Heat effects :

$$C_p^* = \alpha + \beta T + \gamma T^2 + ...$$

$$\int_{T_1}^{T} C_p^* \, dT = \alpha\,(T_2 - T_1) + \beta/2\,(T_2^2 - T_1^2) + \gamma/3\,(T_2^3 - T_1^3) + ...$$

Heat of reaction, $\Delta H = H_{\text{products}} - H_{\text{reactants}}$

$$\Delta H_2 - \Delta H_1 = \Delta C_p \, (T_2 - T_1)$$

$$\Delta C_p = \Delta\alpha + \Delta\beta T + \Delta\gamma T^2$$

$$\Delta\alpha = \Sigma\,\alpha_{\text{products}} - \Sigma\,\alpha_{\text{reactions}}$$

$$\Delta H_T^{\text{o}} = \Sigma[(H_T^{0} - H_0^{0}) - (\Delta H_f^{\,o})_0]_p - \Sigma[(H_T^{\,o} - H_0^{\,o}) - \Delta H_f^{\,0}]_r$$

$$\left(\frac{\Delta G^0}{T}\right)_T = \Sigma\left[\frac{G_{T^0} - H_{0^o}}{T} + \frac{(\Delta\, H_{f^o})}{T}\right] - \Sigma\left[G_{T^o} - \frac{H_{0^0}}{T} + \frac{(\Delta\, H_{f^0})_0}{T}\right]_T$$

Expansion and compression of gases : (Ideal)

Isothermal non-flow processes :

$$W_{\text{(work done)}} = P_1V_1 \ln\frac{V_2}{V_1} = P_1V_1 \ln\frac{P_1}{P_2}$$

$$= RT_1 \ln\frac{V_2}{V_1}$$

Isothermal flow processes :

$$W = P_1V_1 \ln\frac{V_2}{V_1} = P_1V_1 \ln\frac{P_1}{P_2} = RT_1 \ln\frac{V_2}{V_1}$$

Adiabatic non-flow process : PV^{γ} = constant

$$W = \frac{R}{\gamma - 1}(T_1 - T_2) = \frac{P_1 V_1 - P_2 V_2}{\gamma - 1}$$

$$= \frac{RT_1}{\gamma - 1}\left(1 - \frac{V_1}{V_2}\right)^{\gamma - 1}$$

$$= \frac{RT_1}{\gamma - 1}\left[1 - \left(\frac{P_2}{P_1}\right)\right]^{\gamma - 1/\gamma}$$

Adiabatic flow processes :

$$W = \frac{RT_1\gamma}{\gamma - 1}\left[1 - \left(\frac{P_2}{P_1}\right)^{\gamma - 1/\gamma}\right]$$

$$= \frac{RT_1}{\gamma - 1}\left[1 - \left(\frac{V_1}{V_2}\right)\right]^{\gamma - 1}$$

Thermal efficiency of cyclic processes

η = efficiency : ρ = expansion ratio :

P = compression ratio)

Carnot process $\eta = 1 - (T_2/T_1)$

Otto process : $\eta = 1 - (1/P^{\gamma - 1})$ (constant volume)

Diesel pressure

$$\text{(constant pressue)} ; \eta = 1 - \frac{1}{P^{\gamma - 1}} \frac{P^{\gamma - 1}}{\gamma^{\rho - 1}}$$

For a 2-stage compressor with an interstage pressure p and volume, V, the total work is

$$-W = -\frac{\gamma RT}{\gamma - 1}\left[\left(\frac{P}{P_1}\right)^{(\gamma - 1)/\gamma} - 1\right]$$

$$+ \frac{rRT}{\gamma - 1}\left[\left(\frac{P_2}{P}\right)^{(\gamma - 1)/\gamma} - 1\right]; p = \sqrt{p_1 p_2}$$

Refrigeration :

Coefficient of performance $= T_2/(T_1 - T_2)$

One ton of refrigeration capacity = Capable of absorbing in one day the heat evolved in freezing 1 ton of water at 0°C.

Physical equilibrium

Chemical potential : $\mu = \bar{G} = \bar{H} - \bar{T} \times \bar{S}$

$\bar{G}, \bar{H}, \bar{T}$ and $\bar{S}$ are molar quantities

$$\mu_i = \left(\frac{\partial U}{\partial n_i}\right)_{s,v,nj} = \left(\frac{\partial G}{\partial n_i}\right)_{T,p,nj}$$

$$= \left(\frac{\partial H}{\partial n_i}\right)_{s,p,nj}$$

$$= \left(\frac{\partial A}{\partial n_i}\right)_{T,v,nj}$$

Fugacity :

$$[dm_i = dG_i = RT\, d \ln f_i]_T$$

$$\left(\frac{\partial \ln f}{\partial P}\right)_T = \frac{V}{RT}; \left[\frac{\partial \ln f}{\partial T}\right]_p = 0$$

$$= \frac{H^* - H}{RT^2}; \left[\frac{\partial \ln f^*}{\partial T}\right]_P = 0$$

*—indicates ideal behaviour.

Fugacity of pure component;

$$\ln \frac{f}{P} \int_0^P \frac{z-1}{P} = \int_0^{Pr} \frac{z-1}{Pr}\, d\,Pr$$

Activity (a) : $\dfrac{(f)}{(f^o)_T}; \bar{G} - \bar{G}^0 = RT \ln a.$

—indicates partial molal property

0—indicates standard state.

Activity coefficient : (γ)

$$\gamma_i = a_i/x_i;\ x_i - \text{mole fraction}$$

Gibbs-Duhem equation :

$$\sum_{i=1}^{n} x_i\, (\partial G_i / \partial x_i)_{P,T} = 0;$$

$$\Sigma x_i (\partial \ln f_i / \partial x_i)_{P,T} = 0$$

$$\Sigma x_i^* \,(\partial \ln p_i / \partial x_i)_{P,T} = 0 \text{ (at low pressure)}$$

Vapour-liquid equilibrium

Gibbs-Phase Rule : $F = C + 2 - P$

P – number of phases, C – number of components

F = degrees of freedom

Lewis-Randall rule ; $f_i = \bar{f}_i\, x_i$

$\bar{f}_i$ refers to fugacity of pure component.

Raoult's law : $p_i = P^* x_i;\ f_i = \bar{f}_i x_i$

$$\gamma_i = a_i/x_i = f_i(f_i\, x_i)$$

Clausius-Clapeyron equation : $dp/dT = L/T\,\Delta V$

L refers to latent heat of vaporization

$$\ln \frac{p_2}{p_1} = -\frac{L}{R}\left[\frac{1}{T_2} - \frac{1}{T_1}\right];\ \frac{d \ln p}{dT} = \frac{L}{RT^2}$$

Activity coefficients in binary solutions :

Margules equations :

$$\log \gamma_1 = X_2^2[A + 2(B - A)\,X_1] = (2B - A)\,X_2^2 + 2(A - B)\,X_2^3$$

$$\log \gamma_2 = X_1^2[B + 2(A - B)\,X_2] = (2A - B)\,X_1^2 + 2(B - A)\,X_1^3$$

Van Laar equations :

$$\log \gamma_1 = \frac{A\,x_2^2}{[(A/B)\,x_1 + x_2]^2};\ \log \gamma_2 = \frac{Bx_1^2}{[x_1 + (B/A)x_2]^2}$$

Consistency test :

$$RT\int_0^1 \ln(\gamma_1/\gamma_2)\,dx_1 = G^E_{x=1} - G^E_{x=0} = 0 \text{ (Isobaric conditions)}$$

where $G^E = RT\,\Sigma n_i \ln \gamma_i$ = the excess free energy.

Chemical equilibrium :

$$\Delta G = \Sigma v_i\,\mu_i = 0 \text{ criterion for equilibrium}$$

$$v = \text{stoichiometric coefficients}$$

Perfect mixture : $\mu_i = \text{m}_i^{\text{o}} + RT \ln p_i;\ K_p = \pi.\ p_i\ v_i$

Imperfect mixture : $m_i = m_i^{\text{o}} + RT \ln f_i$

$d\ln k_p/dT = \Delta H/RT^2$ (Vant Hoff's equation)

$$\Delta G^{\text{o}} = -\Delta G_T + \Delta H_0 - \Sigma v_i\left(\alpha T \ln T + \frac{\beta}{2}T^2 + \frac{\gamma}{6}T^3\right)$$

C and ΔH_0 are integration constants.

$\Delta G^{\text{o}} < 0$; Reaction is promising

$0 < \Delta G^o < +10000$: Reaction is doubtful, needs further study.

$G^o > +10000$; unfavourable

$K_c = \pi_i \; C_i \Sigma v_i = K_p RT^{\Sigma v_i}$

E. Mechanical Operations

1. Range of operations :

Grizzlies—300 to 40 mm

Trommels—200 to 2.5 m.m.

Screens—100 to 0.05 mm

Hydraulic classifiers—2 to 0.15

Pneumatic classifiers—0.8 to 0001

Rake classifiers—0.8 to 0.1 mm

Hydraulic cyclones—0.4 to 0.004 mm

Hydroseparators—0.4 to 0.006 mm

Centrifuges—0.15 to 0.001 mm

2. Screen effectiveness :

$$Sc = \frac{(x_f - x_O)(x_d - x_f)\, x_d\, (1 - x_b)}{(x_d - x_b)^2\, (1 - x_f)\, x_f}$$

x_f, x_d, x_b refer to mass fraction of material in feed, over-flow and under-flow respectively.

3. Critical speed of trommel : $N = 42.3/D^{1/2}$ where N is in RPM and D is m.

4. ISI specifications for test sieves :

d = wire diameter, mm ; W = aperture size, mm.

d	1.80	1.60	1.40	1.25	1.12	1.00	0.90	0.80	0.71	0.63
W	5.56	4.74	4.00	3.35	2.80	2.36	2.00	1.70	1.40	1.18
d	0.56	0.53	0.45	0.425	0.355	0.30	0.25	0.212	0.18	0.15
W	1.00	0.85	0.71	0.60	0.50	0.425	0.355	0.300	0.250	0.212
d	0.125	0.106	0.09	0.075	0.06	0.053	0.045	0.038	0.032	—
W	0.18	0.15	0.125	0.106	0.09	0.075	0.053	0.038	0.045	—

5. Centrifuge critical speed = $W = \sqrt{i/M}$; i = ratio of restoring force/deflection, M = mass of basket and contents.

6. Shape factors : $V_p = ad^3$, $S_P = 6bd^2$, $\lambda = dS_p/6V_P$ Spheres, cubes, short cyclinders — 1.0 (shape factor)

Round sand	– 1.2	Cost dust	– 1.4
Crushed glass	– 1.5;	Flue dust — 1.8; Mica flakes	– 3.6

7. Equivalent diameter, $d_e = d/\lambda = 6V_p/S_P$

8. Law of comminution : $dE/dL = -CL^P$

Rittinger's law : $E = K(1/L_2 - 1/L_1)$; $P = -2$

Kick's law : $E = K_2 \log_e (L_1/L_2)$; $P = -1$

Bond's law : $E = K_3\sqrt{100/L_2}$

$$\left[1-\left(\frac{L_2}{L_1}\right)^{1/2}\right]; P = -3/2$$

L = Charactristic dimension of particle, K_1, K_2, K_3 are constants.

9. Capacity of crushing rolls : $Q = pNdbZ$

10. Critical speed of Ball mill : $N = 42.3/(D - d)^{1/2}$ where

D = diameter of mill in m, d = diameter of balls, m.

or Kick's law : $P/T = K_k \log (D_{FAVG}/D_{PAVG})$

Rittinger's law : $P/T = K_R\left(\frac{1}{D_{PAVG}} - \frac{1}{D_{FAVG}}\right)$

Bond's law : $P/T = K_B\left(\frac{1}{D_{PAVG}^{1/2}} - \frac{1}{D_{FAVG}^{1/2}}\right)$

F. Heat Transfer

1. Conduction equation (Fourier) : $q = -KA\ dT/dx$

2. Conduction through multilayer wall :

$$\frac{q}{A} = \frac{(dT)\text{ overall}}{\left(\frac{dx}{K}\right)_a + \left(\frac{dx}{K}\right)_b + \left(\frac{dx}{K}\right)_c + ...}$$

3. Conduction through cylinder :

$$q = \frac{2\pi KL(dT)}{\ln (r_2/r_1)}$$

4. Conduction through hollow sphere :

$$q = \frac{4\pi K(dT) r_1 r_2}{(r_2 - r_1)}$$

5. Range of individual coefficients :

(a) Steam, dropwise condensation : 25,000 – 100,000 W/m^2 K

(b) Steam, filmwise condensation : 5,00 – 15,000 W/m^2 K

(c) Water, boiling : 1,500 – 4,500 W/m^2 K
(d) Liquid metals, heating or cooling : 1,500 – 5,000 W/m^2 K
(e) Organic vapour, condensation : 1,000 – 2,000 W/m^2 K
(f) Water, cooling or heating : 250 – 15,000 W/m^2 K
(g) Oils, heating or cooling : 50 – 1,500 W/m^2 K
(h) Air, cooling or heating : 1 – 50 W/m^2 K

6. Overall heat transfer coefficient :

$$\frac{1}{U_o} = \frac{A_o}{h_i A_i} + \frac{A_o}{h_{di} A_i} + \frac{x_m A_o}{K_m A_{lm}} + \frac{1}{h_O} + \frac{1}{hd_O}$$

7. Free or natural convection : $N_u = c\,(G_r \,.\, P_r)^n$.

8. Flow through pipe : $N_u = 1.86\,(R_e)^{1/3}\,(P_r)^{1/3}$; (Laminar flow)

9. Turbulent flow : $N_u = 0.023\, R_e^{\,0.8}\, P_r^{\,0.33}$

10. Flow through helical tube ; h (helical)

$$= h \text{ (st. tube)} \left(1 + 3.54\frac{d}{D_s}\right)$$

11. Flow past a sphere : $N_u = 2.0 + 0.6\,(R_e)^{1/2}\,(P_r)^{1/3}$

Flow past a cylinder : $N_u = CR_e^{\,n}$

13. For heating or cooling through jackets

$$\frac{hD_t}{K} = 0.40\left(\frac{D_a^2\, n\rho}{\mu}\right)^{2/3} (P_r)^{1/3}\,(\mu/\mu w)^{0.14}$$

D_t = diameter of tank, D_a = diameter of agitator.

14. Agitated vessels : For heating or cooling through coils.

$$\frac{hD_t}{K} = 1.01\left(\frac{D_a^2\, n\rho}{\mu}\right)^{0.62} (P_r)^{1/3}\,(\mu/\mu w)^{0.14}$$

15. Radiation : Emissive power of a black body, $w = \sigma \in A\, T^4$

G. Fluid Flow and Flow Measurement

1. Mechanical energy balance : (Euler equation for real fluld)

$$g\,\Delta Z + \frac{1}{2}\Delta u^2 + \frac{\Delta P}{\rho} - F + W = 0$$

2. Bernoulli's equation (mechanical energy balance with no shaft work and friction loss)

$$g\Delta Z + \frac{1}{2}\,\Delta u^2 + \frac{\Delta P}{\rho} = 0$$

3. Velocity distribution :

(a) Laminar flow in circular tubes

$$u = 2V\,[1 - (r/r_w)^2]$$

$$u_{max} = 2V;\ V = \text{average velocity}$$

(b) Laminar flow in annulus :

$$u = \frac{2V\left[r_2^2 - r^2 - 2r_{max}^{\ 2}\ln\dfrac{r_2}{r_1}\right]}{r_2^2 + r_1^2 - 2r_{max}^{\ 2}}$$

$$r_{max} = \frac{r_2^2 - r_1^2}{2\ln\dfrac{r_2}{r_1}}$$

(c) Turbulent flow in smooth circular tubes :

Laminar layer : $u^+ = Y^+$, $Y^+ < 5$

Buffer layer : $u^+ = -\,3.05 + 3.0\ \ln Y^+$, $5 < Y^+ < 30$

Turbulent core : $u^+ = 5.5 + 2.5\ \ln Y^+$, $Y^+ > 30$

$$u^+ = u/u^*\ .\ y^+ = \frac{u^*Y_\rho}{\mu}$$

$$u^* = \sqrt{\frac{\tau_w g_c}{\rho}} = V\sqrt{f/2}$$

Y = distance measured from the wall.

(d) Turbulent flow for rough circular tubes

$$u^+ = 8.5 + 2.5\ \ln\frac{Y}{e}$$

e = height of wall roughness.

4. Flow through pipes and fittings :

(a) Hagen-Poiseuille equation (for laminar flow).

$$\Delta p = \frac{32LV\mu}{d^2}$$

(b) Head loss due to friction :

$$Hfs = \frac{\Delta P}{\rho} = \frac{2fLV^2}{d}$$

(c) Fanning friction factor :

Laminar flow (smooth/rough pipe).$f = 16/R_e$

Turbulent flow (smooth pipe) :

$$f = 0.0014 + 0.125\ Re^{-0.32}$$

$$1/\sqrt{f} = 4.0 \log (R_e \sqrt{f}) - 0.40$$

$$f = 0.046\, R_e^{-0.2} \text{ (used in heat transfer)}$$

Turbulent flow (rough pipe)

$$1/\sqrt{f} = 4.0 \log (d/Z_e) + 3.48$$

(d) Enlargement losses .

$$H_{fe} = \frac{V^2}{2}\left[1 - \frac{A_1}{A_2}\right]^2$$

(e) Contraction losses :

$$H_{fc} = V^2/2\,(1/C_c - 1)^2$$

For an usual value of $C_c = 2/3$, $H_{fc} = V^2/8$.

(f) Value of hydraulic radius (R_H) for various cross-section :

Shape of cross-section,	R_H
Pipe and ducts running full : circle, dia = D	$D/2$
Annulus, inside dia = d, outside dia = D	$\frac{D-d}{4}$
Square, side = D	$D/4$
Rectangle, sides a, b	$ab/2\,(a+b)$

(g) Friction of valves and fittings :

Values and fittings.	*Equivalent pipe diameter, (add to length of straight pipe to get equivalent length)*	*Resistance (in terms of velocity head lost).*
45° elbow	15	0.3
90° elbow	30–40	0.6–0.8
90° square elbow	60	1.2
Entry from leg of %-piece	60	1.2
Entry into leg of T-piece	90	1.8
Unions and couplings	Negligible	Negligible
Globe valve (fully open)	60–300	1.2–6.0
Gate valve (fully open)	7	0.15
75% open	40	1
50% open	200	4
25% open	800	16

5. Metering of fluids :

(a) Manometer equation (simple) :

$$\Delta P = R_m g\ (\rho_m - \rho)$$

Manometer equation (inclined)

$$\Delta P = R \sin \alpha . g\ (\rho_m - \rho)$$

R is reading along the inclined leg.

(b) Pilot tube—

$$v = C\sqrt{\frac{2(P_s - P_0)}{\rho}}$$

(c) Flow meters—

Velocity through throat of venturi-meter or through orifice

$$u_0 = CY\sqrt{\frac{2\,\Delta P}{\rho(1-\beta^4)}}$$

$$Y = \left\{\gamma^{2/K}(K/K-1)\left[1 - \frac{\gamma^{(K-1)K}}{1-\gamma}\left[\frac{1-\beta^4}{1-\beta^4\gamma^{2/K}}\right]\right]\right\}$$

$$Y = \text{expansion factor, unity for liquids}$$

$$\beta = \frac{\text{Diameter of orifice/throat}}{\text{Diameter of pipe}}$$

$$\gamma = P_2/P_1,\ K = C_p/C_v$$

(d) Flow nozzle :

Critical flow conditions (for $\beta \le 2$)

$$W_{max} = \text{CA}_2\left\{P_2(p_1)K(2/K+1)\frac{K+1}{K-1}\right\}^{1/2}$$

6. Boundary layer flow

(a) Laminar boundary layer on a flat plate

Boundary layer thickness ; $\delta = 4.64\ x\sqrt{Re_x}$

Displacement thickness : $\delta^* = 1.74\ x\sqrt{Re_x}$

Local drag coefficient : $fx = 0.648/x\sqrt{Re_x}$

Total drag coefficient : $f_L = 1.296/\sqrt{R_{eL}}$

$$Re_x = xV\rho/\mu,\ Re_L = LV\ \rho/\mu$$

(b) Turbulent boundary layer on flat plates :

Boundary layer thickness,

$$\delta = 0.376\ x/(Re_x)^{0.2};\ 5 \times 10^5 < Re_x < 10^7$$

$$\delta = \frac{0.129x}{(Re_x)^{0.14}};\ Re_x > 10^7$$

Local drag coefficient.

$$f_x = 0.0585 / Re_x^{0.2};\ 5 \times 10^5 < Re_x < 10^7$$

$$f_x = 0.0262 / Re_x^{0.14};\ Re_x > 10^7$$

Total drag coefficient,

$$f_L = \frac{0.074}{Re^{0.2}};\ 5 \times 10^5 < Re_L < 10^7$$

$$f_L = \frac{0.0306}{Re^{0.14}};\ Re_L > 10^7$$

7. Flow in packed and fluidized beds :

(a) Packed beds,

$$f_b = \frac{150}{Re_p} + 1.75$$

$$f_y = \frac{(\Delta H)\in^3}{L u_0^2 (1-\in)};\ Re_p = \frac{d_p u_0 \rho}{\mu}$$

$$d_p = 6 / S_v \left(S_v = \frac{\text{Area of particle surface}}{\text{Volume of the particle}} \right)$$

(b) Fluidized beds,

$$Re_{mf} = d_p U_{mf\rho} = [(33.6)^2 + 0.048\ G_a - 33.7]^{1/2}$$

$$U_{mf} = \text{minimum fluidization velocity}$$

$$G_a = \frac{d_p^3 \rho(\rho_s - \rho) g}{\mu^2}$$

(c) Terminal setting velocities

Drag force, $F_D = (C_D\ \pi \rho\ u_o^2\ d_p^2)/8$

Stoke's law region $Re_p < 1$,

$$u_t = (\rho_s - \rho)\frac{g d_p^2}{18\mu},\ Re_p = \frac{d_p m_0 \rho}{\mu}$$

$$C_D = 24/Re_p$$

Intermediate region ($1 < Re_p < 10^3$)

$$u_t = \frac{0.153\ g^{0.71}\ d_p^{1.14}\ (\rho_s - \rho)^{0.71}}{\rho^{0.29}\ \mu^{0.42}}$$

$$C_D = \frac{18.5}{Re_p^{0.6}}$$

Newton's law region ($Re_p > 500$)

$$u_t = 1.74 \left[g \frac{d_p(\rho_s - \rho)}{\rho} \right]^{1/2},\ C_D = 0.44$$

❑

CHAPTER 18 Mathematics

DETERMINANTS

1. The minor of an element in a determinant is the determinant obtained by deleting row and column which intersect in that element. For example in:

$$\Delta = \begin{vmatrix} a_1 & b_1 & c_1 \\ a_2 & b_2 & c_2 \\ a_3 & b_3 & c_3 \end{vmatrix}; \text{ minor of } b_3 = \begin{vmatrix} a_1 & c_1 \\ a_2 & c_2 \end{vmatrix}$$

The cofactor of any element in the determinant is its minor with proper sign. The sign of an element in the *i*th row and *j*th. column is $(-1)^{i+j}$.

2. For expansion of determinant (Laplace's expansion) multiply each element of the row (or column) by its cofactors and then add up all the terms. Expanding by Ist row, $\Delta = a_1A_1 + b_1B_1 + c_1C_1$.

3. Multiplication of two determinants :

$$\begin{vmatrix} a_1 & b_1 & c_1 \\ a_2 & b_2 & c_2 \\ a_3 & b_3 & c_3 \end{vmatrix} \times \begin{vmatrix} l_1 & m_1 & n_1 \\ l_2 & m_2 & n_2 \\ l_3 & m_3 & n_3 \end{vmatrix} = \begin{vmatrix} a_1l_1 + b_1m_1 + c_1n_1; a_1l_2 + b_1m_2 + c_1n_2; a_1l_3 + b_1m_3 + c_1n_3 \\ a_2l_1 + b_2m_1 + c_2n_1; a_2l_2 + b_2m_2 + c_2n_2; a_2l_3 + b_2m_3 + c_2m_3 \\ a_3l_1 + b_3m_1 + c_3n_1; a_3l_2 + b_3m_2 + c_3n_2; a_3l_3 + b_3m_3 + c_3n_3 \end{vmatrix}$$

4. Solve the below equations by Cramer's rule :

$$\begin{aligned} 3x + y + 2z &= 3 \\ 2x - 3y - z &= -3 \\ x + 2y - z &= 4 \end{aligned}$$

$$\text{here } \Delta = \begin{vmatrix} 3 & 1 & 2 \\ 2 & -3 & -1 \\ 1 & 2 & 1 \end{vmatrix} = 3(-3 * 1 - 2 * -1) - 2(1 * 1 - 2 * 2) + 1(1 * -1 + 3 * 2) = 8$$

$$x = \frac{1}{\Delta}\begin{vmatrix} 3 & 1 & 2 \\ -3 & -3 & -1 \\ 4 & 2 & 1 \end{vmatrix} = 1; \; y = \frac{1}{\Delta}\begin{vmatrix} 3 & 3 & 2 \\ 2 & -3 & -1 \\ 1 & 4 & 1 \end{vmatrix} = -1$$

$$z = \frac{1}{\Delta}\begin{vmatrix} 3 & 1 & 3 \\ 2 & -3 & -3 \\ 1 & 2 & 4 \end{vmatrix} = -1$$

5. Differentiation: Differentiate each row (or column) of the given determinant separetely giving n new determinants. Add each of these n determinants.

For example, $\Delta = \begin{vmatrix} e^x & \sin x \\ x & x^2 \end{vmatrix}$ with respect to x.

Differentiating by rows, $\begin{vmatrix} e^x & \cos x \\ x & x^2 \end{vmatrix} + \begin{vmatrix} e^x & \sin x \\ 1 & 2x \end{vmatrix}$

$$\begin{vmatrix} 2e^x & \sin x + \cos x \\ x+1 & x^2 + 2x \end{vmatrix} = 2e^x (x^2 + 2x) - (x + 1)(\sin x + \cos x)$$

MATRICES

1. Find the inverse of $\begin{bmatrix} 1 & 1 & 3 \\ 1 & 3 & -3 \\ -2 & -4 & -4 \end{bmatrix}$

A^{-1} = adjoint A/| A |

$$|A| = \begin{vmatrix} 1 & 1 & 3 \\ 1 & 3 & -3 \\ -2 & -4 & -4 \end{vmatrix} = \begin{vmatrix} a_1 & b_1 & c_1 \\ a_2 & b_2 & c_2 \\ a_3 & b_3 & c_3 \end{vmatrix}$$; Let A_1, A_2 ...... be cofactors of a_1, a_2 ... in | A | then,

$$A_1 = \begin{vmatrix} 3 & -3 \\ -4 & -4 \end{vmatrix} = -24 ; A_2 = -\begin{vmatrix} 1 & 3 \\ -4 & -4 \end{vmatrix} = -[-4+12] = -8$$

$$A_3 = \begin{vmatrix} 1 & 3 \\ 3 & -3 \end{vmatrix} = -3 - 9 = -12$$

Similarly $B_1 = 10$; $B_2 = 2$; $B_3 = 6$; $C_1 = 2$, $C_2 = 2$, $C_3 = 2$

$|A| = a_1 A_1 + a_2 A_2 + a_3 A_3 = -8$

$$\text{adj } A = \begin{vmatrix} A_1 & A_2 & A_3 \\ B_1 & B_2 & B_3 \\ C_1 & C_2 & C_3 \end{vmatrix} = \begin{bmatrix} -24 & -8 & -12 \\ 10 & 2 & 6 \\ 2 & 2 & 2 \end{bmatrix}$$

$$A^{-1} = \text{adj } A/\Delta = -\frac{1}{8}\begin{bmatrix} -24 & -8 & -12 \\ 10 & 2 & 6 \\ 2 & 2 & 2 \end{bmatrix} = \begin{bmatrix} 3 & 1 & \frac{3}{2} \\ -\frac{5}{4} & -\frac{1}{4} & -\frac{3}{4} \\ -\frac{1}{4} & -\frac{1}{4} & -\frac{3}{4} \end{bmatrix}$$

2. A matrix A is orthogonal if
$AA^{-1} = I$ (unit matrix)

3. Eigenvalues and Eigenvectors of matrix :

$$A = \begin{vmatrix} 8 & -6 & 2 \\ -6 & 7 & -4 \\ 2 & -4 & 3 \end{vmatrix}$$

The characteristic equation is : $|A - \lambda I| = 0$

$$\begin{vmatrix} 8-\lambda & -6 & 2 \\ -6 & 7-\lambda & -4 \\ 2 & -4 & 3-\lambda \end{vmatrix} = \lambda(\lambda - 3)(\lambda - 15) = 0; \lambda = 0, 3, 15$$

The Eignevalues of A are 0, 3, 15.

$$[A - lI]\, X = \begin{vmatrix} 8-\lambda & -6 & 2 \\ -6 & 7-\lambda & -4 \\ 2 & -4 & 3-\lambda \end{vmatrix} = 0 \text{ Putting } \lambda = 0 \text{ gives}$$

$8x - 6y + 2z = 0$; $-6x + 7y - 4z = 0$; $2x - 4y + 3z = 0$
Solving gives $x = 1$, $y = 2$, $z = 2$
For $\lambda = 3$ gives, $x = 2$, $y = 1$, $z = -2$
For $\lambda = 15$ gives, $x = 2$, $y = -2$, $z = 1$
The Eigenvectors are : (1, 2, 2) ; (2, 1, – 2) ; (2, – 2, 1)

SERIES

Checking the series $\sum_{n=1}^{\infty} \frac{n^2}{3^n}$ convergent or divergent.

$$u_n = \frac{n^2}{3n}; u_{n+1} = (n+1)^2 / 3^{n+1}$$

$$\text{Ratio test : } \frac{u_n + 1}{u_n} = \frac{(n+1)^2 3^n}{3^{n+1} n^2} = \frac{1}{3}\left(1 + \frac{1}{n}\right)^2$$

$= \frac{1}{3}$ as $n \to \infty$, So convergent.

Raabe's test : In the series Σu_n, it $\underset{n \to \infty}{\text{Lt.}}\ n\left(\frac{u_n}{u_n+1} - 1\right) = K$. The series Converge for $K > 1$, diverges for $K < 1$ but fails for $K = 1$.

Logarithmic test : $\underset{n \to \infty}{\text{Lt.}}\ n\left(n \log \frac{u_n}{u_n+1}\right) = K$. The series converges for $K > 1$ and diverges for $K < 1$ but fails for $K = 1$.

Cauchy's Root test : A positive term series Σu_m converges if

$\underset{n \to \infty}{Lt}\ (v_n)^{1/n} = \lambda < 1$ and diverges $\underset{n \to \infty}{Lt}\ (v_n)^{1/n} = \lambda > 1$.

COORDINATE GEOMETRY

1. Distance between two points : (x_1, y_1, z_1), (x_2, y_2, z_2) is $[(x_2 - x_1)^2 + (y_2 - y_1)^2 + (z_2 - z_1)^2]^{1/2}$
2. Perpendicular distance of the point $(x_1, y_1\ z_1)$ from plane $ax + by + cz + d = 0$ is
$$\frac{ax_1 + by_1 + cz_1 + d}{(a^2 + b^2 + c^2)^{1/2}}$$
3. Equation of tangent plane at any point (x_1, y_1, z_1) of the sphere $x^2 + y^2 + z^2 = a^2$ is $xx_1 + yy_1 + zz_1 = a^2$
4. Ellipsoid : $\frac{x^2}{a^2} + \frac{y^2}{b^2} + \frac{z^2}{c^2} = 1$
5. The equation of tangent plane and the normal line to the surface, $z^2 = 4\ (1 + x^2 + y^2)$ at (2, 2, 6).
Solution : $F(x, y, z) = 4x^2 + 4y^2 - z^2 + 4$ $\quad \partial F/\partial x = 8x$; $\partial F/\partial y = 8y$;

$\partial f/\partial z = -2z$ *at* (2, 2, 6), $\frac{\partial F}{\partial x} = 16$; $\frac{\partial F}{\partial y} = 16$; $\frac{\partial F}{\partial Z} = -12$. The equation of the tangent plane at (2, 2, 6) is $16\ (X - 2) + 16\ (Y - 2) - 12\ (Z - 6) = 0$ in : $4X + 4Y - 3Z + 2 = 0$.

The equation of the normal line at (2, 2, 6) being perpendicular to $4X + 4Y - 3Z + 2 = 0$ is

$$\frac{X-2}{4} = \frac{Y-2}{4} = \frac{Z-6}{-3}$$

PARTIAL – DIFFERENTIATION

1. Transform $\frac{\partial^2 u}{\partial x^2}$ from cartesian to cylindrical coordinates.

$x = r \cos \theta$; $y = r \sin \theta$; $z = z$; $r = \sqrt{x^2 + y^2}$;

$$\theta = \tan^{-1}(y/x); \frac{\partial r}{\partial x} = \frac{x}{(x^2+y^2)} = \cos\theta;$$

$$\frac{\partial \theta}{\partial x} = \frac{-y}{x^2+y^2} = \frac{-\sin\theta}{r}$$

$$\frac{\partial u}{\partial x} = \frac{\partial u}{\partial r}.\frac{\partial r}{\partial x} + \frac{\partial u}{\partial \theta}.\frac{\partial \theta}{\partial x} = \cos\theta\frac{\partial u}{\partial r} - \frac{\sin\theta}{r}\frac{\partial u}{\partial \theta}$$

$$\text{i.e.: } \frac{\partial}{\partial x} = \left(\cos\theta\frac{\partial}{\partial r} - \frac{\sin\theta}{r}\frac{\partial}{\partial\theta}\right)$$

$$\frac{\partial^2 u}{dx^2} = \frac{\partial}{\partial x}\left(\frac{\partial u}{\partial x}\right) = \left(\cos\theta\frac{\partial}{\partial r} - \frac{\sin\theta}{r}\frac{\partial}{\partial\theta}\right)\left(\cos\theta\frac{\partial v}{\partial r} - \frac{\sin\theta}{r}\frac{\partial u}{\partial\theta}\right)$$

$$\frac{\partial^2 u}{\partial x^2} = \cos^2\theta\frac{\partial^2 u}{\partial r^2} - \frac{2\sin\theta\cos\theta}{r}\frac{\partial^2 u}{\partial r\,\partial\theta} + \frac{\sin^2\theta}{r^2}\frac{\partial^2 u}{\partial r^2} + \frac{\sin^2\theta}{r}\frac{\partial u}{\partial r} + \frac{2\sin\theta\cos\theta}{r^2}\frac{\partial u}{\partial r}$$

$$+\frac{2\sin\theta\cos\theta}{r^2}\frac{\partial u}{\partial\theta}$$

2. A rectangular box open at the top is to have volume of 48. Find the dimensions of box requiring least material for its construction.

Let a, b, c be edges of box and S is surface.

$S = ab + 2bc + 2ca$; $abc = 108$; $c = 48/(ab)$

$$S = ab + 2b\times\frac{108}{ab} + \frac{2\times108}{ab} = ab + \frac{216}{a} + \frac{216}{b}$$

$$\frac{\partial S}{\partial a} = 0 = b\frac{-216}{a^2}; \frac{\partial S}{\partial b} = a - \frac{216}{b^2}; b = 216/a^2$$

$a = 216/b^2$; Solving, $b = 6 = a$.

COMPLEX NUMBERS

1. Find modulus and argument of $(1 + 2i)/(l - 3i)$

$$\frac{1+2i}{1-3i} = \frac{(1+2i)}{(1-3i)(1+3i)} = \frac{-5+5i}{10} = (-1+i)/2$$

$$\text{Modulus} = (\text{R.P}^2 + \text{Img. part}^2)^{1/2} = \left[\frac{(-1)^2+(1)^2}{4}\right]^{1/2} = \sqrt{1/2}$$

Amplitude (argument) = $\tan^{-1}$ (Img. part/Real part)

$$= \tan^{-1}\left(\frac{1/2}{-1/2}\right) = -45°$$

2. Express $1 + \sin\alpha + i\cos\alpha$ in modulus-amplitude form :

$1 + \sin\alpha = r\cos 0$; $\cos\alpha = r\sin\theta$

$r^2\cos^2\theta + r^2\sin^2\theta = r^2 = (1 + \sin\alpha)^2 + \cos^2\alpha = 2 + 2\sin\alpha$

$r = \sqrt{2}\,(1 + \sin a)$; $\tan\theta = (1 + \sin\alpha)/\cos\alpha$

Thus $1 + \sin\alpha + i\cos\alpha = r(\cos\theta + i\sin\theta)$

3. De Moivre's theorem : $(\cos\theta + i\sin\theta)^n = \cos n\theta + i\sin n\theta$

4. Using De Movire's theorem solve $x^2 + 1 - 0$

$x^5 = -1 = \cos\pi + i\sin\pi = \cos[(2n+1)\pi] + i\sin[(2n+1)\pi]$

$= C\text{ is }[(2n+1)\pi]$; $x = C\text{ is }[(2n+1)\pi]^{1/5} = C\text{ is }(2n+1)\dfrac{\pi}{5}$

The values are C is $\pi/5$, C is $3\pi/5$ C is $\dfrac{7\pi}{5}$, C is $9\pi/5$.

or $\cos\dfrac{\pi}{5} + i\sin\dfrac{\pi}{5}$; $\cos\dfrac{3\pi}{5} + i\sin\dfrac{3\pi}{5}$;

$-1, \cos\dfrac{3\pi}{5} - i\sin\dfrac{3\pi}{5}, \cos\dfrac{\pi}{5} - i\sin\dfrac{\pi}{5}$

Rejecting – value, the required, roots are:

$$\left(\cos\frac{\pi}{5} + i\sin\frac{\pi}{5}\right), \left(\cos\frac{3\pi}{5} \pm i\sin\frac{3\pi}{5}\right)$$

5. Write $\sin^7\theta$ in a series of sines of multiples

$x = \cos\theta + i\sin\theta$; $x - \dfrac{1}{x} = 2i\sin\theta$; $x^P - \dfrac{1}{x^P} = 2i\sin P\theta$

$(2i\sin\theta)^7 = \left(x - \dfrac{1}{x}\right)^7$; By binomial expansion,

$$\left(x - \frac{1}{x}\right)^7 = x^7 - 7C_1x^6\frac{1}{x} + 7C_2x^5\frac{1}{x^2} - 7C_3x^4\frac{1}{x^3} + 7C_4x^3\frac{1}{x^4} - 7C_5x^2\frac{1}{x^5} + 7C_6x\frac{1}{x^6} - \frac{1}{x^7}$$

$$= \left(x^7 - \frac{1}{x^7}\right) - 7C_1\left(x^5 - \frac{1}{x^5}\right) + 7C_2\left(x^3 - \frac{1}{x^3}\right) - 7C_3\left(x - \frac{1}{x}\right)$$

$$= 2i\sin 7\theta - 7(2i\sin 5\theta) + 21(2i\sin 3\theta) - 35(2i\sin\theta)$$

$$\sin^7\theta = -\frac{1}{64}[\sin 7\theta - 7\sin 5\theta + 21\sin 3\theta - 35\sin\theta]$$

6. Prove that $i^i = e^{-(4n+1)\pi/2}$

$$i^i = e^{i\log i} = e^{i(2in\pi + \log i)} = e^{-2n\pi + i\log e^{i\pi/2}}$$

since $i = \cos\frac{\pi}{2} + i\sin\frac{\pi}{2} = i\frac{\pi}{2} = e^{i\pi/2}$

$$= e^{-2n\pi} + i(i\pi/2) = e^{-(2n+1/2)\pi}$$

7. Sum the series : $1 + \cos\theta + \cos 2\theta + \ldots.\ \cos(2n-1)\theta$

Let $e = 1 + \cos\theta + \cos 2\theta + \ldots + \cos(n-1)\theta$

Let $S = \sin\theta + \sin 2\theta + \ldots + \sin(n-1)\theta$

$c + iS = 1 + cis\,\theta + cis\,2\theta + \ldots. + cis\,(n-1)\theta$

$$= 1 + e^{i\theta} + 2^{2i\theta} + \ldots.. + e^{i(n-1)\theta} = \frac{1-e^{n\,i\theta}}{1-e^{i\theta}}$$

$$= \frac{(1-e^{-i\theta})}{1-e^{i\theta}-e^{-i\theta}+1} = \frac{1 - cis\,n\theta - (\cos\theta - i\sin\theta) + cis(n-1)\theta}{2-2\cos\theta}$$

Equating real and imaginary parts,

$c = 1 - \cos n\theta - \cos\theta + \cos(n-1)/[2(1-\cos\theta)]$

8. Cauchy-Riemann equations : $\frac{\partial u}{\partial x} = \frac{\partial v}{\partial y}$ and $\frac{\partial u}{\partial y} = -\frac{\partial v}{\partial x}$

9. Taylor series : if $f(Z)$ is analytic inside a circle C with centre a, then for Z inside C,

$$f(z) = f(a) + f'(a)(z-a) + \frac{f''(a)}{2}(a)(z-a)^2 + \ldots.$$

VECTORS

A scalar has magnitude only. A vector has magnitude and direction.

1. Point of division : The point R dividing the join of points A and B in the ratio of $m : n$ is given by,

$$R = (mB + nA)/(m+n)$$

2. Scalar or dot product: $A\,.\,B = ab\cos\theta$ where θ is angle between vectors A and B.

$$I.\,I = 1,\ I\,.\,J = 0$$

$$(A+B)\,.\,C = A\,.\,C + B\,.\,C;\ A\,.\,B = a_1 b_1 + a_2 b_2 + a_3 b_3.$$

3. Vector or cross product: $A \times B = ab\sin\theta$

$$A \times B = -B \times A;\ I \times I = 0;\ I \times I = K;\ J \times I = -K$$

$$A \times B = \begin{vmatrix} i & j & k \\ a_1 & a_2 & a_3 \\ b_1 & b_2 & b_3 \end{vmatrix}$$

4. Work done (scalar product) : $F\,.\,AB$ where F is Force vector, AB is displacement vector.

5. Normal flux: $V.N$ where V is velocity, N is normal vector.

6. Moment of a force : about A is $AP \times$ F where F is vector of force F acting at point P about point A.

7. $ABC = BCA = CAB = V = -BAC$ (Scalar product of three vectors)

$$[ABC] = \begin{vmatrix} a_1 & a_2 & a_3 \\ b_1 & b_2 & b_3 \\ c_1 & c_2 & c_3 \end{vmatrix}$$

8. The volume of tetrahedron is $\frac{1}{6}$ $[AB, AC, AD]$ if $A = (1, 1, 3)$; $B = (4, 3, 2)$; $C = (5, 2, 7)$; $D = (6, 4, 8)$ the tetrahedron volume is : $AB = OB - OA = (4i + 3j + 2k) \quad (i + j + 3k) = 3i + 2j - k$
$AC = (4i + j + 4k)$; $AD = OD - OA = 5i + 3i + 5K$

$$\text{Volume} = \frac{1}{6}[AB, AC, AD] = \frac{1}{6}\begin{vmatrix} 3 & 2 & -1 \\ 4 & 1 & 4 \\ 5 & 3 & 5 \end{vmatrix} = \frac{14}{3}$$

9. Vector product of 3 vectors :

$$(A \times B) \times C = (A \cdot C)\, B - (C \cdot B)\, A.$$

10. Equation of a line through point A and parallel to the vector B is : $R = A + tB$
Equation of a line through the point A and parallel to the vector B is : $R = A + tB$.

11. $\Delta F = i\frac{\partial f}{\partial x} + j\frac{\partial f}{\partial y} + k\frac{\partial f}{\partial z}$

A unit vector normal to the surface, $x\,y^3\,z^2 = 4$ at the point $(-1, -1, 2)$ is :

$\nabla f = \nabla(xy^3\,z^2) = y^3\,z^2\,i + 3xy^2\,z^2\,j + 2xy^3\,zk$ at $(-1; -1, 2)$,

$\nabla f = -4i - 12j + 4k.$

The unit vector normal to surface is :

$$\frac{-4i - 12j + 4k}{\sqrt{(-4)^2 + (-12)^2 + (4)^2}} = \frac{1}{\sqrt{11}}(i + 3j - k)$$

12. $\text{div } F = \nabla . F = i . \frac{\partial F}{\partial x} + j . \frac{\partial F}{\partial y} + k . \frac{\partial F}{\partial z}$ (scalar function)

13. $\text{Curl } F = \nabla \times F = i \times \frac{\partial F}{\partial z} + \lambda \times \frac{\partial F}{\partial y} + k \times \frac{\partial F}{\partial z}$

Curl $V = 0$ (irrotational)

Ω (angular velocity) $= \frac{1}{2}$ Curl V where V is velocity of particle.

14. Stokes theorem: $\int_c F . dR = \int_S \text{curl } F . NdS$

15. Green's theorem : $\int_c (\phi\, dx + \psi\, dy) = \iint_E \left(\frac{\partial \psi}{dx} - \frac{\partial \phi}{\partial y}\right) dx\, dy^2$

16. Solenoid field : $\nabla . F = 0; F = \nabla \times V = \text{Curl } V$

ORDINARY/PARTIAL DIFFERENTIAL/NUMERICAL EQUATIONS

1. Fourier Series : $f(x) = \frac{a_o}{2} + \Sigma a_n \cos nx + \Sigma b_n \sin n x$

$$a_0 = \frac{1}{\pi}\int_0^{2\pi} f(x); \; a_n = \frac{1}{\pi}\int_0^{2\pi} f(x) \cos n x \, dx$$

$$b_n = \frac{1}{\pi}\int_0^{2\pi} f(x) \sin n x \, dx$$

2. Half Range : $f(x) = \Sigma b_n \sin \frac{n \pi x}{l}; \; b_n = \frac{2}{l}\int_0^l f(x) * \sin \frac{n \pi x}{l} dx$

$$f(x) = \frac{a_0}{2} \Sigma a_n \cos \frac{n \pi x}{l}; \; a_0 = \frac{2}{l}\int_0^l f(x) dx; \; a_n = \frac{2}{l}\int_0^l f(x) \cos \frac{n \pi x}{l} dx.$$

3. Methods of solving linear differential equations :

(a) Variable separable

(b) $y = Vx$ (substitution)

(c) Leibnitz equation : $\frac{dy}{dx} + py = Q$; Solution is $y^{e\int pdx} = \int Qe^{\int Pdx} dx + C$

(d) Bernoulli equation : $\frac{dy}{dx} + Py = Qy^n$; divide by y^n then Leibnitz form.

(e) Exact form : $M\,dx + N\,dy = 0$; $\frac{\partial M}{\partial y} = \frac{\partial N}{\partial x}$ for exact differential.

Solution is : $\int M\,dx$ (y constant) $+ \int$(terms of N not having x) $dy = c$

(f) Complimentary functions:

(1) $(D - m_1)(D - m_2)(D - m_3) = ... = 0$; $y = C_1 e^{m_1 x} + C_2 e^{m_2 x} + ...$

(2) $(D - m_1)(D - m_1)(D - m_2) = ... = 0$; $y = (C_1 + C_2 x)\, e^{m_1 x} + C_3 e^{m_2 x} + ...$

(3) $(D - m_1)(D - m_1)(D - m_1) ... = 0$; $y = (C_1 + C_2 x + C_3 x^2)\, e^{m_1 x} + ...$

(4) If one pair of roots are imaginary : $m_1 = \alpha + i\beta$; $m_2 = \alpha - i\beta$; $y = e^{ax}(C_1 \cos \beta x + C_2 \sin \beta x) + ...$

(5) If two pairs of imaginary roots be equal : $m_1 = m_2 = \alpha + i\beta$; $m_3 = m_4 = \alpha - i\beta$
$y = e^{ax}[(C_1 x + C_2) \cos \beta x + (C_3 x + C_4) \sin \beta x] + ...$

(g) Particular solutions :

1. $\frac{1}{D} \times = \int \times dx$

2. $\frac{1}{D-a} \times e^{ax} = \int \times e^{-ax} dx$

3. $\frac{1}{f(D)} e^{ax} = \frac{1}{f(u)} e^{ax}$

4. $\frac{1}{f(D)^2} \sin(ax+b) = \frac{1}{f(-a^2)} \sin(ax+b)$

5. $\frac{1}{f(D)^2} \cos(ax+b) = \frac{1}{f(-a^2)} \cos(ax+b)$

6. $\frac{1}{f(D)^2} x^m = [f(D)]^{-1} x^m$

7. $\frac{1}{f(D)} e^{ax} V = \frac{1}{f(D+a)} V e^{ax}$

4. Numerical solutions :

(a) Picard's method : $\frac{dy}{dx} = f(x, y)$

$$y_1 = y_0 + \int_{x_0}^{x} f(x_1, y_0)\, dx; \; y_2 = y_0 + \int_{x_0}^{x} f(x, y_1)\, dx$$

(b) Euler's method : $y_n = y_{n-1} + hf(x_{n-1}, y_{n-1})$

(c) Modified Euler Method

$$y_1^{(1)} = y_0 + \frac{h}{2}\left[(f(x_0, y_0) + f(x_1, y_1^{(1)})\right]$$

$$y_1^{(1)} = y_0 + h\, f(x_0, y_0)$$

(d) Runge-Kutta method : $k_1 = f(x_0, y_0)\, h$

$$k_2 = hf\left(x_0 + \frac{h}{2}\right), y_0 + k_1/2; \; k_3 = hf\left(x_0 + \frac{h}{2}, y_0 + \frac{k_2}{2}\right)$$

$$K_4 = hf(x_0 + h, y_0 + k_3); \; dy = \frac{1}{6}(k_1 + 2k_2 + 2k_3 + k_4)$$

5. $\Gamma(n+1) = \Gamma(n)\, n = \int_0^{\infty} e^{-x}\, dx\, x^n$

6. $\beta(m, n) = \Gamma(m)\Gamma(n)/F(m+n)$

7. Series solutions : $y = a_0 + a_1 x + a_2 x^2 + \ldots + a_n x^n$

$y = a_0 x^m + a_1 x^{m+1} + a_2 x^{m+1} + \ldots$

8. Wave equation : $\frac{\partial^2 y}{\partial t^2} = \frac{C^2 \partial^2 y}{\partial x^2}$;

Solution is : $y = (C_1 \cos Px + C_2 \sin Px)(C_3 \cos Cpt + C_4 \sin Cpt)$

9. Heat flow equation : $\frac{\partial u}{dt} = \frac{c^2 \partial^2 u}{\partial x^2}$

Solution is : $u(x, t) = (C_1 \cos Px + C_2 \sin Px)\ e^{-c^2 p^2 t}$

10. $\frac{\partial^2 u}{\partial x^2} + \frac{\partial^2 u}{\partial y^2} = 0$; $u(x, y) = (C_1 \cos Px + C_2 \sin Px)(C_3 e^{Py} + C_4 e^{-Py})$

11. Newton's Forward interpolation formula :

$$f(a + nh) = \left(f(a) + n\,\Delta f(a) + \frac{n(n-1)}{2}\Delta^2 f(a) + \ldots \right)$$

12. Newton's Backward interpolation formulae :

$$y_n = y_0 + n\,\nabla\, y_0 + \frac{n}{2}(n+1)\,\nabla^2 y_0 + \frac{n(n+1)(n+2)}{3}\nabla^3 y_0$$

13. Lagrange Formula : $f(x) = \frac{(x - a_1)(x - a_2)(x - a_3)\ldots}{(a_0 - a_1)(a_0 - a_2)(a_1 - a_3)\ldots}$

$$f(a_0) + \frac{(x - a_0)(x - a_2)(x - a_3)\ldots.}{(a_1 - a_0)(a_1 - a_2)(a_1 - a_3)\ldots.} f(a_1) + \frac{(x - a_0)(x - a_1)(x - a_3)\ldots}{(a_2 - a_0)(a_2 - a_1)(a_1 - a_3)\ldots} f(a_2)$$

14. Trapezoidal rule : $\int_a^b f(x)\,dx = \Delta x\left(\frac{y_1}{2} + y_2 + \ldots + \frac{y_n + 1}{2}\right)$

15. Simpson's Rule : $\int_a^b f(x)\,dx = \frac{h}{3}(f_1 + 4f_2 + f_3) + \frac{h}{3} + (f_3 + 4f_4 + f_5) + \frac{h}{3}(\ldots) + \ldots.$

LAPLACE TRANSFORMS

$$Lf(t) = f(S) = \int_0^\infty e\,f(t)^{-St\,dt}$$

$L(t) = 1/S^2$

$L(1) = 1/S$

$L(e^{at}) = \frac{1}{(S - a)}$

$L(\sin at) = a/(S^2 + a^2)$

$L(\cos at) = S/(S^2 + a^2)$

$L(\sin h\, at) = a/(S^2 - a^2)$

$L(\cos h\, at) = S/(S^2 - a^2)$

$L(e^{at} t^n) = n\,!/(S - a)^{n+1}$

$$L(e^{at} \sin bt) = \frac{b}{(S-a)^2 + b^2}$$

$$L(e^{at} \cos bt) = \frac{S-a}{(S-a)^2 + b^2}$$

$$L(e^{at} \sin h\, bt) = \frac{b}{(S-a)^2 - b^2}$$

$$L(e^{at} \cos h\, bt) = \frac{S-a}{(S-a)^2 - b^2}$$

$L[f'(t)] = S\, f(S) - f(0)$

$L[f^n(t)] = S^n f(S) - S^{n-1} f(0) - S^{n-2} f'(0) \ldots$

$$L[t^n f(t)] = (-1)^n \frac{d^n}{dS^n} [f(s)]$$

$$L\left[\frac{1}{t} f(t)\right] = \int_S^{\infty} f(S)\, dS$$

$L[f''(t)^S] = S^2 f(S) - Sf(0) - f'(0)$

STATISTICS

Mean = $\Sigma x/n$

Mean = $\Sigma fx/N$ where f is frequency

Histogram is drawn by erecting rectangles over the class intervals.

Mean deviation : $\frac{1}{n} \Sigma f \mid x - A \mid$, A is mean

Standard deviation : $\sigma = \sqrt{\frac{\Sigma f(x-\bar{x})^2}{n}}$

Least squares method :

Fitting a straight line : $y = a + bx$;

$$\Sigma y = na + b\,\Sigma x \qquad \ldots(1)$$

$$\Sigma xy = a\Sigma x + b\Sigma x^2 \qquad \ldots(2)$$

Solving equations (1) and (2) gives a and b.

Fitting a parabola : $y = a + bx + ex^2$

$$\Sigma y = na + b\,\Sigma x + C\,\Sigma x^2$$
$$\Sigma x\,y = a\,\Sigma x + b\,\Sigma x^2 + C\,\Sigma x^3$$
$$\Sigma x^2\,y = a\,\Sigma x^2 + b\,\Sigma x^3 + C\,\Sigma x^4$$

Solving the three above equations gives *a, b, c* values

Find the probability of throwing 9 with two dice :

There are 6 × 6 = 36 possible ways of throwing two dice. The number of ways getting 9 would be (4, 5) (5, 4), (6, 3), (3, 6) is : 4. The required probability is = 4/36 = 1/9.

G = geometric mean = $\left[x_1^{f_1}\, x_2^{f_2}\, \ldots\ldots\, x_p^{fp}\right]^{1/n}$

H = Harmonic mean = $\dfrac{n}{\Sigma f.\dfrac{1}{x}}$

Mean – mode = 3 (mean – median)

$P = m/(m + n),\ q = n/(m + n)$

Arithmetic mean of binomial distribution = nP

Standard deviation of binomial distribution = $\sqrt{npq}$

Probables error = 0.6745 $\sqrt{npq}$

DIFFERENTIAL FORMULAE

1. $d(u/v) = (v\,du - u\,dv)/v^2$;
2. $d(u^n) = nu^{n-1}$;
3. $d(u^v) = v^{v-1}\,u\,du + u^v\,(\log u)\,dv$;
4. $d\,(a^u) = a^u \log a\,du$;
5. $d(\log u) = du/u$
6. $d\,(\log a^u) = (\log a^e)\,du/u$;
7. $d(\sin u) = \cos u\,du$;
8. $d(\cos u) = -\sin u\,du$;
9. $d\,(\tan u) = \sec^2 u\,du$;
10. $d(\cot u) = -\operatorname{cosec}^2 u\,du$;
11. $d(\sec u) = \tan u \sec u\,du$
12. $d\,(\operatorname{cosec} u) = \cot u \operatorname{cosec} u\,du$;
13. $d(\tan^{-1} u) = du/(1 + u^2)$;
14. $d(\sin^{-1} u) = du/(1 - u^2)^{1/2}$
15. $d\,(\sin h\,u) = \cos h\,u\,du$;
16. $d(\cos hu) = \sin hu\,du$
17. $d(\tan h\,u) = \sec^2 hu\,du$;
18. $d(\sin \mathrm{h}^{-1} u) = du/(u^2 + 1)^{1/2}$
19. $d(\cos h^{-1} u) = du/(u^2 - 1)^{1/2}$; $d(\tan^{-1} u) = du/(1 - u^2)$

INTEGRATION FORMULAE

1. $\int dx/x = \ln x$;
2. $\int e^x\,dx = e^x$;
3. $\int a^x\,dx = e^x/\ln a$;
4. $\int dx/(a + bx) = -\dfrac{1}{b}\ln(a + bx)$;
5. $\dfrac{dx}{a + bx^2} = -\dfrac{1}{b}\,\dfrac{1}{(a + bx)}$
6. $\int \dfrac{x\,dx}{(a + bx)} = \dfrac{x}{b} - \dfrac{a}{b^2}\ln(a + bx)$;

7. $\int (a+bx)^{1/2}\, dx = \frac{2}{3b}(a+bx)^{3/2};$

8. $\int \frac{x}{(a+bx)}\, dx = 2(3bx-2a)\,(a+bx)^{3/2}/15b^3$

9. $\int \frac{dx}{(a+bx)^{1/2}} = 2(a+bx)^{1/2}\, b;$

10. $\frac{\int dx}{(a+bx^2)^2} = \frac{x}{2a(a+bx^2)} + \frac{1}{2a}\int \frac{dx}{(a+bx^2)};$

11. $\frac{\int dx}{a+bx^2} = \frac{-1}{(ab)^{1/2}} \tan^{-1} x\sqrt{b/a}$

12. $\frac{\int dx}{a+bx+cx^2} 2 = \frac{2}{\sqrt{-q}} \tan^{-1} \frac{2cx+b}{(1-q)} * \frac{1}{2}(b^2 < 4ac) = \frac{1}{\sqrt{-q}} \log\left[\frac{2cx+b-\sqrt{-q}}{2cx+\sqrt{-q}+4ac}\right]$

13. $\int (x^2 \pm a^2)^{1/2} dx = \frac{1}{2}\left[x(x^2 \pm a^2)^{1/2} - a^2 \log x + (x^2+a^2)^{1/2}\right]$

14. $\int \frac{dx}{(x^2+a^2)^{1/2}} = \log\left[x + \sqrt{x^2 \pm a^2}\right]$

15. $\int (a^2 \pm x^2) dx = \frac{1}{2}\left[x(x^2-a^2) + a^2 \sin^{-1} x/a\right]$

16. $\int \frac{dx}{x^2+a^2} = \sin^{-1}(x/a)$

OTHER EQUATIONS

1. Area of rectangle = $a * b (a, b$ are sides)
2. Area of a square = $a * a$ (a is side)
3. Area of a parallelogram = $ab \sin\theta$ (a, b are sides)
4. Area of a rhomlus = 1/2 * product of digonals;
5. Area of a triangle = $1/2 * b * h$;
6. Area of a trapezium = $1/2 * (a * b) * h$;
7. Area of a polygon = $\frac{1}{4} * n\, a^3 \cot(180/n)$ (n equal sides);
8. area of a circle = πr^2;
9. Area of a circular sector = $(\pi/360)\, \theta r^2$;
10. Area of ellipse = 3.14 * a * b
11. Area of a sphere = $4\pi r^2$;
12. Area of cylinder = $2\pi rh$;
13. Area of a prism = $p * h$ (cone/pyramid) = $\frac{1}{3} p$ * Slant height
14. Area of a spherical segment = 2 * 3.142 * $r * h$
15. Volume of a rectangle/parallepiped = $a * b * c$;
16. Volume of a sphere; $\frac{4}{3} * 3.142 * r^3$;
17. Volume of cone pyramid = $3.142\, r^2\, h/2$.

TRIGNOMETRIC RELATIONS

sin (90 – x) = sin x; sin (90 + x) = cos x

cos (90 – x) = sin x; cos (90 + x) = – sin x

sec (90 – x) = cosec x; cosec (90 – x) = sec x

tan (90 – x) = cot x; cot (90 – x) = tan x

sin (180 + x) = sin x; cos (180 – x) = – cos x

tan (180 – x) = tan x; cot (190 + x) = cot x

sin (– x) = – sin x; cos (– x) = cos x

tan (– x) = – tan x; sin h (– x) = – sinh x; tan h (– x) = tan hx

sin ($x \pm y$) = sin x cos y ± cos x sin y

cos ($x \pm y$) = cos x cos y ∓ sin x sin y

sin 2x = 2 sin x cos x; cos 2x = 2 $\cos^2 x$ – 1;

SERIES

$$e^{-x} = 1 - x + \frac{x^2}{2!} - \frac{x^3}{3!} + \frac{x^4}{4!}$$

$$\sin x = x - \frac{x^3}{3!} + \frac{x^5}{5!} - \frac{x^7}{7!} + \ldots$$

$$\cos x = 1 - \frac{x^2}{2!} + \frac{x^4}{4!} - \frac{x^6}{6!} + \ldots$$

$$\tan x = x + \frac{x^3}{3!} + \frac{2x^5}{25} + \frac{17x^7}{315} + \ldots$$

EMPIRICAL EQUATIONS

1. $y = a + bx$; plot y versus x. b is slope, a is intercept.

2. $y = ax^n$; plot log y versus log x. n is slope, log a is intercept.

3. $y = c + ax^n$; plot ln ($y - e$) versus ln x. n is slope and ln a is intercept.

4. $y = ab^x$; plot ln y versus x. b is slope, ln a is intercept.

5. $y = a\, e^{bx}$; plot ln y versus x. ln a is intercept, and b is slope.

6. $y = a + \frac{b}{x}$; plot y versus $\frac{1}{x}$. a is intercept and b is slope.

7. $y = \frac{x}{a + bx}$; plot x/y versus x OR $\frac{1}{y}$ versus $\frac{1}{x}$. a is slope and b is intercept.

8. $y = a + bx + cx^2$; plot $\frac{y - y_n}{x - x_n}$ versus x . x_m, y_n are the coordinates on smooth curve.

9. $y = \dfrac{x}{a + bx} + c;\ y - c = \dfrac{x}{a + bx}$; plot $\dfrac{1}{(y - c)}$ versus $\dfrac{1}{x}$. a is slope, b is intercept.

Rayleigh's method : $X = K \times {}_1^a \times {}_2^b \times {}_3^c \ldots$

Variation of parameters for two variables :

$y = c_1 y_1 + c_2 y_2;\ y_p = u y_1 + v y_2$

$$u' = -y_2 \frac{Q(x)}{(y_1 y_2 - y_2 y_1)};\ V' = \frac{y_1 Q(x)}{(y_1 y_2' - y_2 y_1')}$$

Continuity equation : $\rho A V_x$ = Constant OR

$$\frac{d\rho}{\rho} + \frac{dA}{A} + \frac{dV_x}{V_x} = 0$$

Unsteady state mass balance equation:

Rate of mass in – Rate of mass out ± mass change due to reaction = Rate of mass accumulation

Series solution : For second and higher order :

$y = a_0 x^m + a_1 x^{m+1} + a_2 x^{m+2} + \ldots + a_n x^{m+n}$

For $m = 0$; $y = a_0 + a_1 x + a_2 x^2 + \ldots + a_n x^n$

Equation of continuity in 3 dimensions : $\dfrac{D\rho}{Dt} = -\rho(\nabla \cdot V)$

where $\dfrac{D}{Dt} = \left(\dfrac{d}{dt} + V_x \dfrac{d}{dx} + V_y \dfrac{d}{dy} + V_z \dfrac{d}{dz}\right)$

Equation of heat conduction equation in 3 dimensions :

$$\frac{dT}{dt} = C^2 \Delta^2 T;\ \Delta^2 T = d^2T/dx^2 + d^2T/dy^2 + d^2T/dZ^2.$$

In cylindrical coordinates : $\dfrac{d^2T}{dx^2} + \dfrac{d^2T}{dy^2} + \dfrac{d^2T}{dz^2} = \dfrac{d^2T}{dr^2} + \dfrac{1}{r^2}\dfrac{d^2T}{d\theta^2} + \dfrac{1}{r}\dfrac{dT}{dr} + \dfrac{d^2T}{dz^2}$.

In spherical coordinates, $\nabla^2 T = \dfrac{1}{r^2}\dfrac{d}{dr}\left(r^2 \dfrac{dT}{dr}\right) + \dfrac{1}{r^2 \sin^2}\dfrac{d}{d\theta}\left(\sin\phi \dfrac{dT}{dp}\right) + \dfrac{1}{r^2}\sin^2\phi \dfrac{d^2T}{d\theta^2}$;

Fourier Series: $f(x) = \dfrac{a_0}{2} + \Sigma\, a_n \cos nx + \Sigma b_n \sin nx$

$$a_0 = \frac{1}{\pi}\int_0^{2\pi} f(x)\,dx;\ a_n \frac{1}{\pi}\int_0^{2\pi} f(x)\cos nx\,dx;\ b_n = \frac{1}{\pi}\int_0^{2\pi} f(x)\sin nx\,dx$$

Half Range Fourier Series: $f(x) = \dfrac{a_0}{2} + \Sigma a_n \cos\dfrac{n\pi x}{l}$

OR $f(x) = \Sigma b_n \sin\dfrac{n\pi x}{l};\ a_0 = \dfrac{2}{l}\int_0^l f(x)\,dx;\ a_n = \dfrac{2}{l}\int_0^l f(x)$ *

$\cos\left(\frac{n\pi x}{l}\right)dx;\ b_n = \frac{2}{l}\int_0^l f(x)\sin\frac{n\pi x}{l}dx.$

Separation variable equation for heat conduction :

$u(x, t) = (C_1 \cos px + C_2 \sin px)\ e^{-c^2p^2t/l^2};\ u(x_1\ t) = u_t(x, t) + u_s(x)$

SECTION – A

1. An apple costs Rs. 10. An onion costs Rs. 8. (GATE)

Select the more suitable sentence with respect to grammer and usage.

(a) The price of an apple is greater than an onion.

(b) The price of an apple is more than onion.

(c) The price of an apple is greater than that of an onion.

(d) Apples are more costlier than onions. **Ans.** (c)

2. The Buddha said, "Holding on to anger is like grasping a hot coal with the intent of throwing it at someone else. You are the one who gets burnt. Select the word below which is close in meaning to the statement given.

(a) Burning (b) Igniting

(c) Clutching (d) Flinging **Ans.** (d)

3. *M* has a son *Q* and a daughter *R*. He has no other children. *E* is the mother of *P* and daughter in-law of *M*. How is *P* related to *M*?

(a) *P* is the son-in law of *M* (b) *P* is the grand child of *M*

(c) *P* is the daughter in law of *M* (d) *P* is the grand father of *M* **Ans.** (b)

4. The smaller the least fits this set (324, 441, 97 and 64) is:

(a) 324 (b) 441

(c) 97 (d) 64 **Ans.** (d)

5. It takes 10 seconds and 15 seconds respectively for two trains travelling at different constant speeds to completely pass a telegraphic post. The length of first train is 120 m and that of second train is 150 m. The magnitude of the difference in the speed of the two trains is (in m/s) is:

(a) 2.0 (b) 10.0

(c) 12.0 (d) 22.0 **Ans.** (a)

6. The velocity *V* of a vehicle along a straight line is measured in m/s and plotted with respect to time in seconds. At the end 7 seconds, how much will the odometer reading increase by in m?

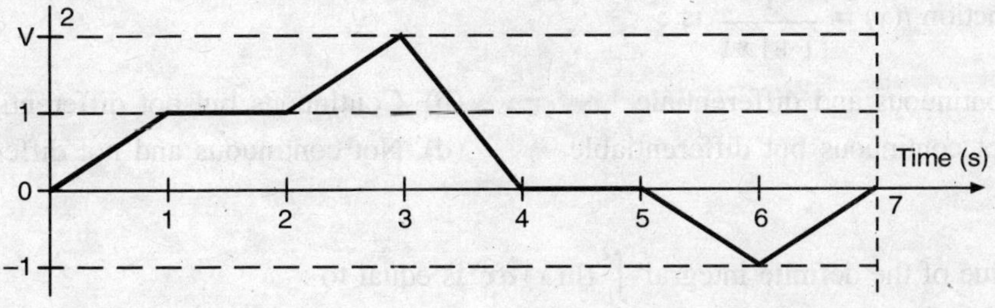

(a) 0 (b) 3

(c) 4 (d) 5 **Ans.** (d)

Solution. Calculating areas: $\frac{1}{2} \times 1 \times 1 + 1 \times 1 + \frac{1}{2} \times 1 \times 1 + 1 + \frac{1}{2} \times 2 \times 1 + \frac{1}{2} \times 1 \times 1 + \frac{1}{2}$

$\times 1 \times 1 = \frac{1}{2} + 1 + \frac{1}{2} + 1 + 1 + 1 = 5$ **Ans.**

7. The over whelming number of people infected with rabies in India has been flaggedly the world health organisation as a source of concern. It is estimated that inoculating 70% of pets and stray dogs against rabies can load to a significant reduction in the number of people infected with rabies.

Which of the following can be logicallyinferred from the above sentences?

(a) The number of people in India infected with Rabies is high

(b) The number of people in other parts of the world who are infected with Rabies is low.

(c) Rabies can be eradicated in India by Vaccinating 70% of stray dogs.

(d) Stray dogs are the main source of Rabies world wide. **Ans.** (a)

8. A flat is shared by four first year undergreduate students. They agreed to allow the oldest of them to enjoy some extra space in the flat. Manu is 2 months older than Sravan, who is three months younger than Trideep. Pavan is one month older then Sravan. Who should occupy the extra space in the flat?

(a) Manu (b) Sravan

(c) Trideep (d) Pavan **Ans.** (c)

9. Find the area bounded by the lines : $3x + 2y = 14$; $2x - 3y = 5$ in first quadrant.

(a) 14.95 (b) 15.25

(c) 15.70 (d) 20.35 **Ans.** (b)

10. A straight line is fit to a data set ($\ln x$, y). This line intercepts the abscissa at $\ln x = 0.10$ and has a slope of – 0.02. What is the value of y at $x = 5$ from the fit?

(a) – 0.030 (b) – 0.014

(c) 0.014 (d) 0.030 **Ans.** (a)

Solution. $y = m \ln x + c$; at $y = 0$, $0 = - 0.02 * 0.10 + c$

$C = 0.002$; $y = - 0.02 \ln x + 0.002$ at $x = 5$; $y = - 0.02 \ln 5 + 0.002 = - 0.030$

11. The value of Lt $x \to \left(\frac{e^x - 1}{\sin x}\right)$ is equal to 1.0.

12. The function $f(x) = \frac{1}{1 + |x|}$ is :

(a) Continuous and differentiale (b) Continuous but not differentiable

(c) Not continuous but differentiable (d) Not continuous and not differentiable

Ans. (b)

13. The value of the definite integral $\int_1^e (\ln x)\, dx$ is equal to : **Ans.** 1.0

14. For a complete number $z = \left(\frac{1}{2} + \frac{\sqrt{3}}{2}i\right)$, the value of z^6 is :

(a) $-\left[\frac{1}{2} + \frac{\sqrt{3}}{2}i\right]$ (b) $-i$

(c) $\left(\frac{1}{2} - \frac{\sqrt{3}}{i}i\right)$ (d) 1 **Ans.** (d)

15. The Laplace transform of the function e^{-2t} is :

(a) $1/2S$ (b) $2/S$

(c) $\frac{1}{S+2}$ (d) e^{-2S} **Ans.** (c)

16. The divergence of the velocity field $\bar{V} = (x^2 + y)i + (z - 2xy)\,j + (xy)\,k$ at (1, 1, 1) is 0.0.

17.

x	0.0	0.2	0.4	0.60	0.80	1.0
$f(x)$	1.00	1.24	1.56	1.96	2.44	3.00

The value of the integral $\int_0^1 f(x)\,dx$ according to trapezoidal rule is 1.8 to 1.9.

18. A box has a total of ten identical sized balls. Seven of these balls are black in colour and the rest three are red. Three balls are picked from the lose one after another without replacement. The probability that two of the balls are black and one red is equal to 0.50 to 0.55. **Ans.**

19. Consider the matrix, $M = \begin{bmatrix} 5 & 3 \\ 3 & 5 \end{bmatrix}$. The normalized given vector corresponding to the smaller eigenvalue of the matrix M is:

(a) $\begin{pmatrix} \sqrt{3}/2 \\ 1/2 \end{pmatrix}$ (b) $\begin{pmatrix} \sqrt{3}/2 \\ -1/2 \end{pmatrix}$

(c) $\begin{pmatrix} \frac{1}{\sqrt{2}} \\ \frac{-1}{\sqrt{2}} \end{pmatrix}$ (d) $\begin{pmatrix} \frac{1}{\sqrt{2}} \\ \frac{1}{\sqrt{2}} \end{pmatrix}$ **Ans.** (c)

20. For the differential equation : $x^2 \frac{d^2y}{dx^2} - 2x\frac{dy}{dx} + 2y = 0$ the general solution is :

(a) $y = C_1\,x + C_2\,e^x$ (b) $y = C_1 \sin x + C_2 \cos x$

(c) $y = C_1\,e^x + C_2\,e^{-x}$; (d) $y = C_1\,x^2 + c_2\,x.$ **Ans.** (d)

GATE EXAM QUESTONS AND ANSWERS

1. The rank of matrix $\begin{vmatrix} 1 & 0 & 0 \\ 0 & 2 & 0 \\ 3 & 0 & 0 \end{vmatrix}$

(a) 0 (b) 1 (c) 2 (d) 3. **Ans.** (c)

2. The angle between two vectors $2i - j + k$ and $j + i + 2k$ is:

(a) 0° (b) 30° (c) 45° (d) 60° **Ans.** (d)

3. A function $f(x) = 12x - x^3$ has a maximum value at x =......

(a) –2 (b) 0 (c) 2 (d) $\sqrt{12}$ **Ans.** (c)

4. $\underset{x \to 0}{\alpha t} \dfrac{\tan h\, x}{h}$ is :

(a) ∞ (b) 1 (c) 0 (d) – 1 **Ans.** (c)

5. The second order Taylor series expansion for a function, $f(x) = x^2$ at $x = 1$ is

(a) x^2 (b) $1 + x^2$ (c) $1 + x + x^2$ (d) $1 - x + x^2$

Ans. (a)

6. The average value of a function $f(x) = x^3$ in the interval $0 < x < 2$ is :

(a) 1 (b) 2 (c) 4 (d) 8 **Ans.** (b)

7. Match :

(a) $y = x^2$ (A) Linear O.D.E.

(b) $\dfrac{dy}{dx} = 2x$ (B) Non-linear O.D.E.

(C) Linear algebraic equation

(D) Non-linear algebraic equation **Ans.** a – D; b – B

8. Match :

(a) $\dfrac{dy}{dx} + 5y = 0$; $y(0) = y_0$ (A) $y = y_0 + 5x$

(b) $\dfrac{dy}{dx} + 5 = 0$; $y(0) = y_0$ (B) $y = y_0 - 5x$

(C) $y = y_0 e^{-5x}$

(D) $y_0 e^{5x}$ **Ans.** a-C; b-B

9. Match the items :

(*i*) cosh (*at*) (A) $a/(S^2 + a^2)$

(*ii*) sin (*at*) (B) $a/(S^2 - a^2)$

(C) $S/(S^2 - a^2)$

(D) $S/(S^2 + a^2)$

Ans. (*i*)-C; (*ii*)-A.

10. The rate of increase of number of bacteria under certain conditions is proportional to the number of bacteria. If the number doubles in 4 hours, how much increase in number of bacteria can be expected after 12 hours?

11. Match the following:

(i) $\dfrac{dy}{dx} = x^2 + y^2$ (a) Linear first order O.D.E. with constant coefficient

(ii) $\dfrac{dy}{dx} = x^2 + y$ (b) Linear O.D.E. with variable coefficient

(c) First order non-linear O.D.E

(d) Linear second order O.D.E

Ans. (i) (c) (ii) (b)

12. Find Eigenvalues and Eigenvectors of matrix $\begin{vmatrix} 4 & -5 \\ 1 & -2 \end{vmatrix}$

13. The differential equation $\dfrac{d^2y}{dx^2} + \dfrac{dy}{dx} + \sin y = 0$ is :

(a) Linear (b) Non-linear (c) Homogeneous (d) Of degree two

Ans. (d)

14. The inverse of matrix $\begin{vmatrix} a & 0 \\ 0 & b \end{vmatrix}$ is :

(a) $\begin{vmatrix} ab & 0 \\ 0 & 1 \end{vmatrix}$ (b) $\begin{vmatrix} b & 0 \\ a & b \end{vmatrix}$ (c) $\begin{vmatrix} 1/a & 0 \\ 0 & 1/b \end{vmatrix}$ (d) $\begin{vmatrix} a & 0 \\ 0 & 1/b \end{vmatrix}$

Ans. (c)

15. The Eigenvector of the matrix : $\begin{pmatrix} 0 & 0 & \alpha \\ 0 & 0 & 0 \\ 0 & 0 & 0 \end{pmatrix}$ a ≠ 0 are :

(a) (0, 0, 0) (b) (α, 0, 0) (c) (0, 0, 1) (d) (0, α, 0)

Ans. (a)

16. The limit of $f(x) = x/\sin x$ as x tends to zero is:

(a) 0 (b) 1 (c) 2 (d) ∞ **Ans.** (b)

17. Simpson's rule for integration gives when $f(x)$ is a polynomial of degree:

(a) 1 (b) 2 (c) 3 (d) 4 **Ans.** (b), (c)

18. Integrating factor for the differential equation is $\dfrac{dy}{dx} + P(x)\,y = Q(x)$.

(a) $\exp[\int P dx]$ (b) $\exp[-\int P dx]$ (c) $\int P dx$ (d) dP/dx

Ans. (a)

19. If i, j, k are the unit vectors in rectangular coordinates then the curl of the vector $iy + jy + kz$ is :

(a) k (b) $-k$ (c) $j + k$ (d) $i + k$.

Ans. (b)

20. The solution of the differential equation $\frac{d^2y}{dt^2} + 5\frac{dy}{dt} + 6y = 0$ is :

(a) $C_1e^{-2t} + C_2e^{3t}$ (b) $C_1 \sin 2t + C_2 \cos 2t$

(c) $C_1e^{2t} + C_2e^{-3t}$ (d) $C_1e^{-2t} + C_2e^{-3t}$

Ans. (d)

21. Which of the following is (are) valid Fortran – 77 statement (S)?

(a) DO13I = 1 (b) A = DIM ***7

(c) READ = 15.0 (d) Go To 3 = 10

Ans. (c)

22. The Taylor's series expansion of $f(x)$ around $x = a$ is ....

23. For a differential function $f(x)$ to have a maximum value $\frac{df}{dx}$ should be .... and $\frac{d^2f}{dx^2}$ should be ....

Ans. zero, negative

24. For a differential function $f(x)$ to have minimum $\frac{df}{dx}$ should be .... and $\frac{d^2f}{dx^2}$ should be..........

Ans. zero, positive

25. $Mdx + N\,dy$ is an exact differential when ......

26. The integral of $x \sin x$ is ....

27. If a is a scalar and b is a vector, then $\nabla \times a \cdot b =$ ...... **Ans.** Vector

28. The differential equation $\frac{d^2y}{dx^2} + y = 0$ with conditions $y(0) = 0$ and $y(1) = 1$ is called a .... value problem.

Ans. Boundary.

29. The series $1 + x + x^2 + x^3 + ...$ for $x < 1$ is divergent. True or False and reason.

Ans. False

30. Fourier series of the periodic function (period 2π) defined by : $f(x) = x, 0 < x < \pi$ is $\frac{\pi}{2} - \frac{4}{\pi}\left[\cos x + \frac{1}{3^2}\cos 3x + ...\right] + 2\left[\sin x - \frac{1}{2}\sin 2x + 1\right]$ the sum of the series, $1 + \frac{1}{3^2} + \frac{1}{5^2} + \frac{1}{7^2} + ...$ is :

(a) $\pi^2/4$ (b) $\pi^2/6$ (c) $\pi^2/8$ (d) $\pi^2/12$ **Ans:** (c)

31. Which of the following improper integrals in (are) convergent?

(a) $\int_0^1 \frac{\sin x}{(1-\cos x)} dx$ (b) $\int_0^\infty \frac{\cos x}{1+x} dx$

(c) $\int_0^\infty \frac{x}{1+x^2} dx$ (d) $\int_0^1 \frac{(1-\cos x)}{x^{\frac{5}{2}}} dx$ **Ans.** (b)

32. The function $f(x, y) = x^2y - 3xy + 2y + x$ has :

(a) No local extremum

(b) One local minimum but no local maximum

(c) One local maximum but no local minimum

(d) One local minimum and one local maximum

Ans : (d)

33. $\text{Limit}_{x \to 0} \frac{x(e^x - 1) + 2(\cos x - 1)}{x(1-\cos x)}$ is 1

34. The radius of convergence of the power series :

$\sum_0^\infty \frac{(3m)!}{(m!)^3} x^m$ is .... **Ans.** Infinity

35. If the linear velocity $\bar{V}$ is given by $\bar{V} = x^2\, yi + xyz\, j - yz^2k$, the angular velocity $\bar{w}$ at the point (1, 1, – 1) is ......

36. Given the differential equation y′ = x – y with the initial condition y (0) = 0. The values of y (0.1), calculated numerically upto third place of decimal by the second order Runge-Kutta method with step size h = 0.1 is....**Ans.** 0.00405

37. For $X = 4.0$, the value of I in the Fortran – 77 statement :

$I = -2.**2 + 5.0 * 3. + 3./4.$ is ..... **Ans.** – 11.

38. The value of the double integral $\int_0^1 \int_x^{1/x} \frac{x}{1+y^2} dx\, dy$ is ....

39. If $A = \begin{bmatrix} 1 & 0 & 0 & 1 \\ 0 & -1 & 0 & -1 \\ 0 & 0 & i & i \\ 0 & 0 & 0 & -i \end{bmatrix}$ the matrix A^4, calculated by the use of Cayley-Hamilton theorem or otherwise is ......

40. Given $\bar{V} = x\cos^2 y\, i + x^2 e^2 j + z\sin^2 yk$ and S the surface of a unit cube with one corner at the origin and edges parallel to the coordinate axes, the value of the integral $\iint_S \bar{V}.n\ dS$ is ....

Ans. 1

41. The differential equation $y'' + y = 0$ is subjected to the boundary conditions, $y(0) = 0$; $y(\lambda) = 0$. In order that the equation has non-trivial solution (S), the general value of λ is .....

Ans. $n\pi - x$

42. The Laplace transform of the periodic function $f(t)$ described by the curve below, *i.e.* : $f((t) = \sin t$ if $(2n - 1)\pi \le t \le 2n\pi$ (n = 1, 2, 3) 0 otherwise is ....

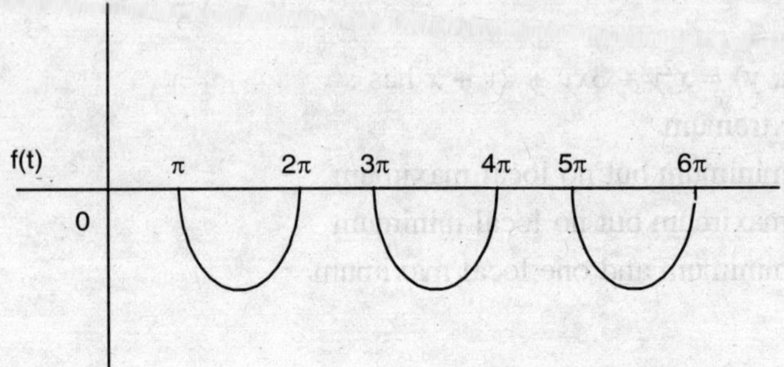

43. The energy balance equation over a tabular reactor under transient conditions is :

(a) Ordinary differential equation

(b) An algebraic equation

(c) A linear partial differential equation

(d) A non-linear partial differential equation

Ans. (c)

44. Solve $d^2y/dx^2 - m^2y = 0$ subject to $y = 1$ at $x = 0$ and $dy/dx = 0$ at $x = 1$;

45. Given $\overline{\overline{A}} = \begin{bmatrix} -1 & -2 \\ 3 & 4 \end{bmatrix}$

(a) Write down the characteristic equation

(b) Compute $(\overline{\overline{A}})^4$ without direct multiplications

46. Solve $\dfrac{dy}{dx} + 0.6\,y = 6e^{-0.5x}$ using I. F. method, giving $y = 1$ at $x = 0$

47. The sum of infinite series :

$$3 + 1 + \frac{1}{3} + \left(\frac{1}{3}\right)^2 + \ldots + \left(\frac{1}{3}\right)^n \text{ is}$$

(a) 9 (b) $\dfrac{9}{2}$ (c) $\dfrac{15}{2}$ (d) infinity

Ans. (b)

48. Limit $x \to \infty$ $\dfrac{x^2 + 1}{2x^2 - 80x + 1}$

(a) 0 (b) 1/2 (c) 1 (d) infinity

Ans. (a)

49. Given $f(x, y) = x^2 + y^2$; $\Delta^2 f$ is :

(a) 4 (b) 2 (c) 0 (d) $4(x^z + y^2)$

Ans. (a)

50. The cubic equation $x^3 - x + 10$ has a root in the interval

(a) – 1, 0 (b) 0, 1 (c) –3, –1 (d) 3, 4 **Ans.** (c)

51. The Fourier series of the function:

$$f(x) = 1;\ 0 \leq x \leq \pi\ ;\ = -1;\ -\pi \leq x \leq 0$$

$$f(x + 2\pi) = f(x) \text{ is extended periodically}$$

(a) A sine series (b) A cosine series

(c) A mixed series (d) A power series

52. Find the values of λ such that the equation given below has a non-zero solution.

$$\frac{d^2y}{dx^2} + \lambda^2 y = 0;\ y(0) = y(l) = 0$$

53. A unit normal to the plane $2x + 2y + 2z = 6$ can be expressed in the vector form as :

(a) $3i + 2j + 2k$ (b) $i\frac{2}{3} + j\frac{1}{3} + k\frac{2}{3}$

(c) $i\frac{1}{3} + j\frac{1}{2} + k\frac{1}{2}$ (d) $-i\frac{2}{3} + j\frac{1}{3} - k\frac{2}{3}$ **Ans:** (b)

54. $\underset{x \to 0}{Lt} \dfrac{x - \sin 2x}{x + \sin 3x}$ has the value :

(a) 1 (b) –1/4 (c) 0 (d) ∞ **Ans:** (b)

55. The function $z = (x - 1)^2 - 2y^2$ has

(a) A stationary point which is a minimum.

(b) A stationary point which is a maximum

(c) A stationary point at which no extremum exists. **Ans:**(c)

(d) No stationary point.

56. The integral $\int_0^1 \frac{dx}{x^P}$ is convergent for :

(a) No value of P (b) $P > 1$

(c) $P < 1$ (d) all values of P

57. The mass balance equations of a blender are : $0.2F_1 + 0.8F_2 = 0.5F$

$$F_1 + F_2 = F$$

Express the equation set given above in a matrix form $AX = B$ where

$$X = \begin{bmatrix} \frac{F_1}{F} \\ \frac{F_2}{F} \end{bmatrix}$$

Find A^{-1}. Use A^{-1} to calculate F_1/F and F_2/F.

58. Find the dimensions of a hollow cylinder with both ends closed which can hold 27 $\frac{\pi}{4}$ m^3 of water and has minimum outer surface area.

OTHER QUESTIONS

59. Using variation of parameters method, find the particular integral of :

$$d^2y/dx^2 + 4y = \tan x.$$

60. When a gas flows isentropically through a closed channel, the pressure P, density ρ and velocity v are related by the Bernoulli's equation, $\frac{dP}{dv} = -\rho v$ along a stream line. In addition it is known that $P = k\rho^r$ where k and r are constants. If the speed of sound in a gas given by $a^2 = rP/\rho$ prove that,

$$\frac{a^2}{r-1} + \frac{v^2}{2} = \text{constant}$$

61. A compound dissolves in water at a rate proportional to the amount undissolved * difference in concentration of saturated solution and instantaneous concentration. 20 kg of solute is dumped into a tank containing 120 kg of solvent, and at the end of 12 minutes, the concentration is observed to be 1 /30. Find the amount of solute in solution at any time, t if saturation concentration is 1/30.

62. Derive an expression to relate x moles of c formed with time t for the following reaction between A and B, $mA + nB = PC$. Assume the rate of formation of (C is proportional to the product of concentrations A and B. Initially a moles of A and b moles of B are taken in the reactor. Given $m/n = a/b$.

(b) In a reaction, $A + 2B = C$, initially 40 moles of A and 80 moles of B are charged to the reactor 12 moles of C are formed in 20 minutes, determine the time required for formation of 15 moles of C (assume the rate of formation of c is proportional to the product of concentration of A and B.

63. A substance evaporates at a rate proportional to the exposed surface. If spherical naphthalene ball of radius 0.5 cm has a radius of 0.4 cm after 6 months, how long will it take, (a) for radius to be 0.25 cm (b) volume of naphthalene ball to be half of what it was originally.

64. In the experimental study of the saponification of methyl acetete by sodium hydroxide, it is found that 25% of ester is converted to alcohol in 12 minutes when initial concentration of both ester and caustic are 0.01 m. What conversion of ester would be obtained in one hour if the initial concentration of ester is 0.025 m and initial caustic concentration is 0.015 m.

65. A tank contains 100 m^3 fresh water. $2m^3$ of brine having a concentration of 1 kg/m^3 of salt is run into the tank per minute and the mixture is kept uniform by mixing runs out at the rate of 1 m^3/ minute. What will be the exit brine concentration when the tank contains 150 m of brine?

66. In a nunclear reaction, the following equations are formulated :

$\frac{dx}{dt} + \lambda x = 0, \frac{dz}{dt} = \mu y : x + y + z = r$ where λ, r and μ are constants. $Z = \frac{dZ}{dt} = 0$ when $t = 0$, show that $Z = r + \frac{r}{(\lambda - \mu)}(\mu\, e^{-\lambda t} - \lambda\, e^{-\mu t})$

67. For the figure given below, calculate the amount of salt in tank B after 1 hour.

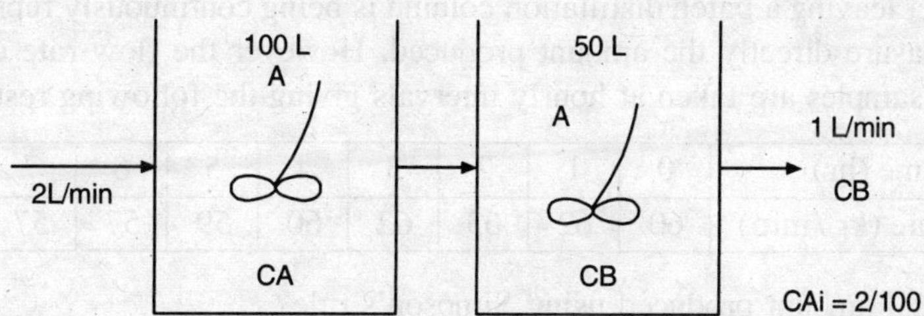

68. A small distillation unit is separating propane and butane at 135°C and initially contains 100 kg moles of a mixture whose composition is $x = 0.30$ (mole fraction of butane). Additional mixture is fed ($x_F = 0.30$) at the rate of 5 kg.moles/hr. If the total volume, of liquid in the still is constant and the concentration of vapour from the still (xp) is related to X_S (mole fraction of butane) remaing in the still as follows. $x_D = x_S/(1 + x_S)$. How long will it take for the value of x_S to change from 0.30 to 0.40. What is the steady state value of x_s in still?

69. Prove that the limiting velocity of spherical particles in a medium is given by $V_\infty = g/3K$; $dr/dt = kr$.

70. It is desired to produce a substance B from raw raaterial A in a continuous stirred tank reactor of effective volume, V ft^3. If ft^3/minute of a solution of A of concentration C_0 is fed to the empty reactor and the chemical reaction of A is represented by, $A \underset{K_2}{\overset{K_1}{\longrightarrow}} B \overset{K_3}{\longrightarrow} C$ in which all the reactions are of first order, show that the number of moles of B in the initial discharge from the reactor is given by the solution of the differential equation, $\dfrac{d^2NB}{dt^2} + \dfrac{PdN_B}{dt} + RN_B = S$.

where $S = QC_0\, K$, $R = K_1\, K_3$ and $P = K_1 + K_2 + K_3$.

71. Obtain series solution of : $dy/dt = ty$; $y(0) = 1$

72. Solve $\dfrac{dy}{dx} = x + 0.1\, y^2$, $x_0 = 1.8$, $y_0 = 0$ for $x_0 = 1.8$ to 3.0 with $\Delta x = 0.30$ using Euler method.

73. Solve the following differential equation using numerical method subject to the conditions that when $x = 0$, $y = 2$ and $dy/dx = -1$. Calculate the values of y for $x = 0$ to 0.60 with an increment of 0.20 for x and tabulate y versus x.

74. Using Lagranges interpolation formula, fit a polynomial with the following data.

x	0	1	3	4
y	−12	0	5	12

75. Solve $y' = 2x^3 - 2xy$, $y(0) = 0$ using Runge-Kutta method.

76. Solve using Taylor's expansion series for $dy/dx = 2x^3 - 2xy$, $y(0) = 0$

77. Using Trapezoidal rule find the area

x	0	1	2	3	4	5	6	7
y	2	6	18	38	66	102	146	198

78. The top product leaving a batch distillation column is being continuously reprocessed so that it is difficult to measure directly the amount produced. However the flow rate can be continuously measured and samples are taken at hourly intervals giving the following results.

Time (hr)	0	1	2	3	4	5	6	7	8
Flow rate (kg/min)	60	62	63	62	60	59	57	57	57

What is the total amount produced using Simpson's rule?

79. Using least squares method, fit a straight line to the data below.

W	50	70	100	120
P	12	15	21	25

80. Two ends A and B of a copper rod of 40 cm long have their temperatures kept at 0°C and 80°C respectively, until steady state conditions prevail. The temperature of end B is suddenly reduced to 40°C and kept so, while that of end A is kept at 0°C. Find the temperature $u(x, t)$ as a function of length (l) and time (t), the time elapsed after sudden change of temperature at end B.

81. A thin rod of length, l has its ends at A and B maintained at 0°C and 100°C respectively until steady state conditions are reached. If B is suddenly reduced to 0°C and maintained at 0°C, find the temperature T at a distance x from A at anytime t.

82. Find Fouries series of : $f(x) = \pi x;\ 0 < x < 1$
in the interval 0 to 2. $= \pi(2 - x);\ 1 < x < 2$

83. Find inverse Laplace transforms of :

(a) $\log\left(\dfrac{1+S}{S}\right)$ (b) $\dfrac{1}{S^2(S^2+a^2)}$ (c) $\dfrac{S}{(S+a)^2}$ (d) $\dfrac{1}{S^2(S+2)'}$

84. Find inverse Laplace transform of : $\dfrac{2S+18}{S^2+9} + \dfrac{24-30\sqrt{S}}{S^4}$

85. Solve the following differential equation by Laplace transform method.

$$y^{ll} + 2y^{l} + 5y = e^{t} \sin t\ ;\ y(0) = 0,\ y^{l}(0) = 1$$

86. Derive Laplace transform for the step function.

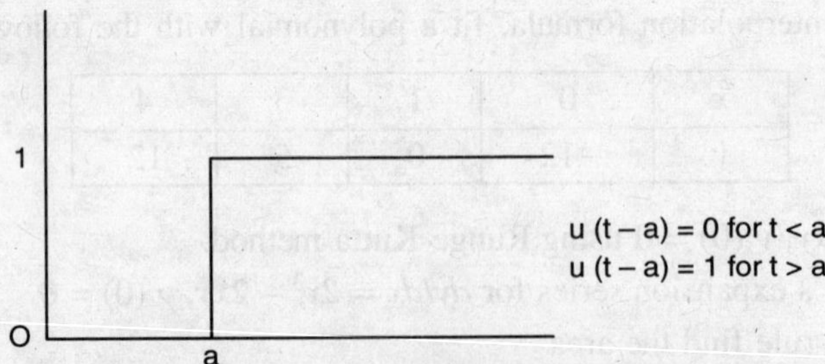

87. A pipe 20 cm. diameter contains steam at 150°C and is provided with covering of 5 cm thick for which $k = 0.0025$. Temperature of the outer surface of covering is 40°C. Find the temp. half way through covering at steady state.

DESCRIPTIVE ANSWERS (GATE EXAM QUESTIONS)

1. A matrix is said to be rank r when (i) it has at least one non-zero minor of order (ii) every minor of order higher than r vanishes. The rank of matrix is the largest order of any non-vanishing minor of the matrix. Here the third order matrix vanishes (determinant is zero). The second order minors formed also vanish except $\begin{vmatrix} 1 & 0 \\ 0 & 2 \end{vmatrix}$ and $\begin{vmatrix} 0 & 2 \\ 3 & 0 \end{vmatrix}$. So the order (Rank) is 2. **Ans.c**

2. $(2i - j + k) \,.\, (i + j + 2k) = |A| \, |B| \cos \theta$

$$3 = (2^2 + 1^2 + 1^2)^{1/2} (1^2 + 1^2 + 2^2)^{1/2} \cos \theta$$

$$\cos \theta = 3/6; \theta = 60°$$ **Ans.d**

3. $dy/dx = 12 - 3x^2 = 0; x = \pm 2$

For maximum value, $d^2y/dx^2 = -6x$ should be negative. So, $x = 2$ **Ans.c**

4. $\dfrac{\tan hx}{h} = \dfrac{x \tan h}{h}; Lt \dfrac{\tan h}{h} = 1$

So $\underset{x \to 0}{\text{Limit}} \dfrac{\tan hx}{h} = \underset{x \to 0}{Lt} = 0$ **Ans.c**

5. $f(x) = f(x_0) + f'(x_0)(x - x_0) + f''(x_0)(x - x_0)^2/2$ here $x_0 = 1$

$f(x) = 1 + 2(x - 1) + 2(x - 1)^2/2 = x^2$ **Ans.a**

6. $\int_0^2 x^3 \, dx / \int_0^2 dx = \left. \dfrac{x^4/4}{x} \right|_0^2 = 2$ **Ans.b**

7. (a) $y = x^2$ non-linear algebraic equation.

(b) $\dfrac{dy}{dx} = 2x$. This is of form $\dfrac{dy}{dx} + P(y) = Q(x)$ which is linear Ist order ordinary differential equation. **Ans.a–D, b–B**

8. (a) $dy/dx + 5y = 0; dy/y = -5dx$; integrating : $\ln y = -5x + C$; at $x = 0$; $y = y_0$; $y = y_0 e^{-5x}$.

(b) $\dfrac{dy}{dx} + 5 = 0; dy = -5\, dx; y = -5x + c$; at $x = 0$, $y = y_0$; $y = y_0 - 5x$ **Ans.a–C, b–B**

9. Laplace transform of $\sin at = \dfrac{a}{S^2 + a^2}$

Laplace transform of $\cos h\, at = \dfrac{S}{S^2 - a^2}$ **Ans. i–C, ii–A**

10. $\dfrac{dN}{dt} = KN; \dfrac{dN}{N} = K\, dt$; integrating, $\ln N = Kt + C$ at $t = 0$, $N = N_0$

$\ln N = Kt + \ln N_0$; $\ln (N/N_0) = Kt$

$N/N_0 = e^{kt}$; at $t = 4$ hours, $N/N_0 = 2$ giving $K = 0.347$ at $t = 12$ hr;

$N/N_0 = e^{0.347 \times 12} = 64.02$ **Ans.**

11. (i) $\frac{dy}{dx} - y^2 = x^2$. This is first non-linear ordinary differential equation.

(ii) $\frac{dy}{dx} - y = x^2$; This is of Leibnitz form, $\frac{dy}{dx} + P(y) = Q(x)$. So first order O.D.E. with variable coefficient. **Ans.i–C, ii–B**

12. $\begin{bmatrix} (4-\lambda) & -5 \\ 1 & (-2-\lambda) \end{bmatrix} = (4-\lambda)(-2, -\lambda) - 1 * 5 = \lambda^2 - 2\lambda - 3 = 0$

Putting $\lambda - 3$, we have Eigenvectors $\begin{bmatrix} (4-3) & -5 \\ 1 & (-2-3) \end{bmatrix} = \begin{bmatrix} 1 & -5 \\ 1 & -5 \end{bmatrix} = x - 5y = 0$; or $x = 5y$

Putting $\lambda = -1$; $\begin{bmatrix} (4-1) & -5 \\ 1 & (-2+3) \end{bmatrix} = \begin{bmatrix} 5 & -5 \\ 1 & -1 \end{bmatrix} = 5x - 5y = 0 - x = y$. Eigenvectors are : (1, 5) and (1, 1) **Ans. d**

13. of degree two.

14. $A = \begin{vmatrix} a & 0 \\ 0 & b \end{vmatrix}$; Inverse = adjoint of A.determinant $A_{adj} A = \begin{bmatrix} b & 0 \\ 0 & a \end{bmatrix}$;

$$\text{Inverse} = \frac{1}{ab}\begin{bmatrix} b & 0 \\ 0 & a \end{bmatrix} = \begin{bmatrix} 1/a & 0 \\ 0 & 1/b \end{bmatrix}$$ **Ans. c**

15. $|A - \lambda I| = \begin{vmatrix} 0-\lambda & 0 & \alpha \\ 0 & 0-\lambda & 0 \\ 0 & 0 & 0-\lambda \end{vmatrix} = -\lambda(\lambda^2 - 0) - 0 \times 0 + a \times 0$

$$= -\lambda^3 = 0; 1 = 0, 0, 0$$ **Ans. a**

16. $f(x) = x/\sin x; \frac{1}{f(x)} = \frac{\sin x}{x} = F(x)$

$\underset{x \to 0}{\text{Lt}} \frac{\sin x}{x} = 1$; So $F(x) = 1$; $f(x) = 1$ **Ans. b**

17. $y = ax^2 + bx + c$

$y = ax^3 + bx^2 + cx + d$ **Ans. b,c**

18. $e^{\int Pdx} \frac{dy}{dx} + ye^{\int Pdx} = e^{\int Pdx} Q$

$\frac{d}{dx}(e^{\int Pdx\, y}) = Qe^{\int Pdx\, y}$

Integrating on both sides, $y\, e^{\int P\, dx} = \int Q e^{\int P\, dx\, y} + \text{constant}$ **Ans. a**

19. $\left(i\frac{\partial}{dx}+j\frac{\partial}{\partial y}+k\frac{\partial}{\partial z}\right)\times(ix+jy+kz)$

Using $i \times i = 0; j \times j = 0; k \times k = 0$

$= j \times i + j \times j + k \times k = - k$ **Ans. b**

20. $\frac{d^2y}{dt^2}+\frac{5\,dy}{dt}+6y=0;\ D^2+5D+6=0;\ D=-2,-3,\ y=C_1\,e^{-2t}+C_2\,e^{-3t}$ **Ans. d**

21. READ is a real variable. 15.0 is also a real value, so same mode. So READ = 15.0 is correct.

Ans. c

22. $f(x) = f(x_0) + f'(x_0)(x - x_0) + f''(x_0)(x - x_0)^2/2$ at $x_0 = a$

$f(x) = f(a) + f'(a)(x - a) + f''(a)(x - a)^2/2$

23. zero, negative

24. zero, positive

25. $\frac{\partial M}{\partial y}=\frac{\partial N}{\partial x}$

26. $\int x\sin x\,dx;\ \int u\,dV = uv - \int V\,du;\ u = x;\ dV = \sin x\,dx$

V = $-\cos x$

$\int x\sin x\,dx = -x\cos x + \int \cos x\,dx = -x\cos x + \sin x$

27. $\left(i\frac{\partial}{\partial x}+j\frac{\partial}{\partial y}+k\frac{\partial}{\partial z}\right)\times a(b_1 i+b_2 j+b_3 k)$

Using $i \times i = 0; j \times j = 0; k \times k = 0$

$=\left(a\,i\times j\frac{\partial b_2}{\partial x}+a\,i\times k\frac{\partial b_3}{\partial x}\right)+\left(aj\times\frac{i\,\partial b_1}{\partial y}+aj\times\frac{k\,\partial b_3}{\partial y}\right)+\left(ak\times\frac{i\,\partial b_1}{\partial z}+ak\times\frac{j\,\partial b_2}{\partial z}\right)$

$=\frac{ak\,\partial b_2}{\partial x}-\frac{aj\,\partial b_3}{\partial x}-\frac{ak\,\partial b_1}{\partial y}+\frac{ai\,\partial b_2}{\partial y}+\frac{aj\,\partial b_1}{\partial z}-\frac{ai\,\partial b_2}{\partial z}$ is a vector. **Ans. Vector**

28. Boundary

29. This is geometric series having ratio x.

Sum = $\frac{1-x^n}{1-x}$; for $n > 1; 1 - x^n = 1$

So sum = $\frac{1}{1-x}$ **Ans. False**

30. $f(x)\frac{-\pi}{2}-\frac{4}{\pi}\left[\cos x+\frac{1}{32}\cos 3x\right]+2\left[\sin x-\frac{1}{2}\sin 2x+...\right]=x$ at x = 0

$0=\frac{\pi}{2}-\frac{4}{\pi}\left[1+\frac{1}{3^2}+...\right]$ giving $1+\frac{1}{3^2}+\frac{1}{5^2}+...=\pi^2/8.$ **Ans. c**

31. (a)*** $\int \frac{\sin x\, dx}{2\sin^2 x/2} = \int_0^1 \cot x/2\, dx = 2 \ln \sin \frac{x}{2}\Big|_0^1$

= 2 [ln sin (1/2) – ln 0]. This is not convergent.

(b) $\frac{\cos x}{1+x}$ = 1 at $x = 0$, = 0 at $x = \infty$. So convergent

(c) $\int \frac{x\, dx}{1+x^2}$; Let $x = \tan y$ giving integral as

$\int \tan y\, dy = \ln \cos y\Big|_0^\infty$ = ln cos ∞ which is not convergent

(d) $\frac{1-\cos x}{x^{5/2}}$ = 1 – cos 1 at x = 1 is indeterminate at x = 0. **Ans. b**

32. $x^2 y - 3xy + 2y + x = f(x, y)$

$\frac{\partial f}{\partial x} = 2xy - 3y + 1; \frac{\partial f}{\partial y} = x^2 - 3x + 2$

Equating to zero and solving ; $x^2 - 3x + 2 = 0$

$(x - 1)(x - 2) = 0$; x = 1 or 2

For x = 1, from $2xy - 3y + 1 = 0$ gives y = 1

For x = 2, from $2xy - 3y + 1 = 0$ gives y = – 1

$f(x)$ = 1 for x = 1, y = 1; f(x) = 2 for x = 2, y = – 1

$\frac{\partial^2 f}{\partial x^2}$ = 2y from $2xy - 3y + 1$ is 2 at y = 1 and is – 2 at y = – 1; $\frac{\partial^2 f}{\partial x^2}$ = 0 from $x^2 - 3x + 2$. So one local maximum and minimum.**Ans. d**

33. Lt $\frac{x(e^x - 1) + (2\cos x - 1)}{x(1-\cos x)}$; Putting x = 0 gives $\frac{0}{0}$ so taking derivatives of numerator and denominator, $e^x - 1 + x e^x - 2 \sin x \rightarrow / (1 - \cos x + x \sin x)$

Lt $x \rightarrow 0$ is : (1 – 1)/(1 – 1) = $\frac{0}{0}$, so differentiating again gives,

$(2e^x + e^x + xe^x + 2 \sin x)/(2 \cos x - x \sin x + \cos x)$

Lt $x \rightarrow 0$ gives, (2 + 1 + 0 + 0)/(2 – 0 + 1) = 3/3 = 1. **Ans.**

34. $\frac{C_m + 1}{C_m} = \frac{(3m+1)!}{[(m+1)!]^3} \frac{(m!)^3}{(3m!)} = \frac{3m+1}{(m+1)^3}$

$\frac{3m+1}{(m+1)^3}$ can be written as $\frac{3(m+1)}{(m+1)^3} - \frac{2}{(m+1)^3}$

$$= \frac{3}{(m+1)^2} - \frac{2}{(m+1)^3} = 0$$ as m → ∞ which means Radius of convergence is ∞. (Infinity)

35. $\bar{w} = \frac{1}{2} \nabla \times \bar{V} = \frac{1}{2} \text{Curl } \bar{V} = \frac{1}{2} \begin{vmatrix} i & j & k \\ \frac{\partial}{\partial k} & \frac{\partial}{\partial y} & \frac{\partial}{\partial} z \\ x^2 y & xyz & -yz^2 \end{vmatrix}$

$$= \frac{1}{2} i(-z^2 - xy) \frac{-j}{2}(0-0) + \frac{k}{2}(yz - x^2)$$

$$= \frac{-i}{2}(z^2 + xy) + \frac{k}{2}(yz - x^2)$$ **Ans.**

36. $A_n = hf(x_n, y_n) = 0.1\, f(0, 0) = 0$

$$B_n = hf\left(x_n + \frac{h}{2}, y_n + \frac{A_n}{2}\right) = 0.1\, f(0.05, 0) = 0.005$$

$$C_n = hf\left(x_n \frac{h}{2}, y_n + \frac{B_n}{2}\right) = 0.1\, f(0.05, 0.0025) = 0.00225$$

$$D_n = h\, f(x_{n+1}, y_n + C_n) = 0.1\, \text{f}(0.1, 0.00225) = 0.009775.$$

$$y(0.10) = y_n + \frac{1}{6}(A_n + 2\, B_n + 2C_n + D_n) = 0.004046$$

37. I = – 4.0 + 20.0/4.0 + 3.0 + 3/4.0

= – 4.0 + 5.0 * 3.0 + 0.75 = – 4.0 + 15.0 + 0.75

= 11.75

I = 11 **Ans.**

38. $\int_0^1 \int_x^{1/x} \frac{x}{1+y^2}\, dx\, dy = \int_x^{1/x} \text{x} \tan^{-1} y\, dx \Big|_0^1$

$$= \int x\, dx\left(\frac{\pi}{4} - 0\right) = \int_x^{1/x} \frac{\pi}{4} x\, dx = \frac{\pi}{4} \frac{x^2}{2}\Big|_0^{1/x}$$

$$= \frac{\pi}{8}\left(\frac{1}{x^2} - x^2\right) = \frac{\pi}{8x^2}(1 - x^4)$$ **Ans.**

39. $|A - \lambda I| = 0 = \begin{vmatrix} 1-\lambda & 0 & 0 & 1 \\ 0 & -(1+\lambda) & 0 & -1 \\ 0 & 0 & i-\lambda & i \\ 0 & 0 & 0 & -(i+\lambda) \end{vmatrix} = 0$

The determinant being upper triangular,

$|A - \lambda I| = (1 - \lambda)^* - 1\ (1 + \lambda)\ (i - \lambda)\ ^* - 1\ (i + \lambda) = 0$

$(1 - \lambda^2)\ (i - \lambda)^2 = 0;\ \lambda^4 - (i + 1)\ \lambda^2 + i = 0$ **Ans.** λ^4

40. By Green's theorem :

$$\int_0 F\,.\,NdS = \int_0 diV\,.\,Fdv$$

$$= \left(i\frac{\partial}{\partial x} + j\frac{\partial}{\partial y} + k\frac{\partial}{\partial z}\right).(x\cos^2 y\, i + x^2 e^z j + z\sin^2 yk)$$

$$= \frac{\partial}{\partial x}(x\cos^2 y) + \frac{\partial}{\partial y} + (x^2 e^z) + \frac{\partial}{\partial z}(z\sin^2 y)$$

$$= \cos^2 y + \sin^2 y = 1$$ **Ans. 1**

41. $D^2 + 1 = 0;\ D = \pm\, i;\ y = C_1 \cos x + C_2 \sin x$

$$y = C_1 \cos x - C_2 \sin x = 0;\ C_1 \cos x = +\, C_2 \sin x$$

$$\tan x = C_1/C_2 \quad ...(1)$$

$$y(\lambda) = C_1 \cos \lambda + C_2 \sin \lambda = 0;\ C_1 \cos \lambda = -\, C_2 \sin \lambda$$

$$\tan \lambda = -C_1/C_2 \quad ...(2)$$

From (1) and (2) : $\tan \lambda + \tan x = 0$

$\tan \lambda + \tan x = 0;\ \tan(-\lambda + x) = 0$

$$x + \lambda = 0, \pi, 2\pi, 3\pi, ... = np \text{ for } n = 0, 1, 2...$$

$$\lambda = n\pi - x$$ **Ans**

42.

$$L(f) = \frac{1}{1 - e^{-2\pi S}} \int_{\pi}^{2\pi} e^{-ST} \sin t\, dt$$

$$= \frac{1}{1 - e^{-2\pi S}} \int_{\pi}^{2\pi} e^{(-S + j)t}\, dt = -\frac{1}{(j - S)} \frac{e^{(-S + j)t}}{1 - e^{-2\pi S}}\Bigg|_{\pi}^{2\pi}$$

$$= \frac{1}{S - j}[e^{(-S + j)\pi} - 1]\frac{e^{(-S + j)\pi}}{1 - e^{-2\pi S}} = \frac{1}{(S^2 + 1)} \frac{e^{-\pi S}}{(1 - e^{-\pi S})}$$ **Ans**

(taking the imaginary part).

43. The ratio $\int_{x_1}^{x_2} \frac{dx/y}{(x_2 - x_1)/y \text{ at } x_1 = x_2}$ $(x_2 > x_1)$ where $1/y$ is a monotonically increasing function of x is :

(a) Less than unity

(b) Equal to unity

(c) Greater than unity

(d) Less than zero. **Ans.** (c)

Since $\int_{x_1}^{x_2} dx / y = (x_2 - x_1)/y \Big|_{x_1}^{x_2}$; The ratio is $\dfrac{x_2 - x_1}{y\big|_{x_1}^{x_2}(x_2 - x_1)}$;

Since $| x_2 \; is > y \,|_{x_1}^{x_2}$; so the ratio is greater than 1.

44. Solve $\dfrac{d^2y}{dx^2} - m^2 y = 0$, subject to $y = 1$ at $x = 0$ and $\dfrac{dy}{dx} = 0$ at $x = 1$

$D^2 - m^2 = 0$; $D = \pm m$; $y = c_1 e^{mx} + c_2 e^{-} mx$

at
$$x = 0, y = 1 = C_1 + C_2 \quad ...(1)$$
$$\frac{dy}{dx} = C_1 \, m \, e^{mx} - C_2 \, me^{-\,mx} = 0;\; C_1 = C_2 \, e^{-\,2m} \quad ...(2)$$
$$C_1 = 1 - C_2 = C_2 e^{-2m};\; C_2 = \frac{1}{1+e^{2m}};\; C_1 = \frac{e^{-2m}}{(1+e^{-2m})}$$
$$y = \frac{1}{1+e^{-2m}}\,[e^{-mx} + e^{-m(2-x)}] \; \textbf{Ans.}$$

45. Given matrix $A = \begin{bmatrix} -1 & -2 \\ 3 & 4 \end{bmatrix}$;

(a) Write down characteristic equation

(b) Compute $(A)^{-4}$ without direct multiplication

$\begin{bmatrix} (-1-\lambda) & -2 \\ 3 & (4-\lambda) \end{bmatrix} = 0$; expanding, $(-\lambda - 1)(4 - \lambda) - (-2) * 3 = 0$

$\lambda^2 - 3\lambda + 2 = 0$; $(1 - 1)(\lambda - 2) = 0$; $\lambda = 1, 2$ characteristic equation is : $\lambda^2 - 3\lambda + 2 = 0$. **Ans.**

46. Solve $\dfrac{dy}{dx} + 0.6\,y = 6e^{-0.5x}$, given $y = 1$ at $x = 0$

$$P = 0.6;\; \int P dx = 0.6x$$
$$\int_e P \, dx_y = \int Q \, e^{\int P \, dx} + C$$
$$e^{0.6x} y = \int 6e^{-\,0.5x} \, e^{0.6\,x} + C = \int 6e^{0.1x} \, dx + C$$
$$y = 60e^{-0.5x} + C\,e^{-\,0.6\,x}\text{; using } y = 1 \text{ at } x = 0 \text{ gives}$$
$$c = -59;\; y = 60\,e^{-\,0.5x} - 59\,e^{0.6x} \quad \textbf{Ans.}$$

47. Ans. $3 + 1 + \frac{1}{3} + ...$ is geometric series. The sum is : $\frac{1}{1-q}$ for $q < 1$. Sum is $3 + \frac{1}{1-q}; q = \frac{1}{3}$.

Sum = $\frac{9}{2}$ **Ans. b**

48. Ans. (a)

49. $\nabla^2 f = \frac{\partial^2 f}{\partial x^2} + \frac{\partial^2 f}{\partial y^2} = 2 + 2 = 4$ **Ans. a**

50. Ans. (c)

51. Ans. (b)

52. Ans. $y = C_1 \cos \lambda x + C_2 \sin \lambda x$; using $y(0) = 0 = C_1$

$y(1) = C_2 \sin \lambda = 0; \lambda = \pi. 2\pi. ... n\pi$ **Ans.**

53. Ans. Here $a = 2\ kj + 2k$; $C = 6$;

$| a | = (2^2 + 1^2 + 2^2)^{1/2} = 3$; unit vector = $\frac{a}{|a|} = n = \frac{2}{3}i + \frac{1}{3}j + \frac{2}{3}k$ **Ans. b**

54. Ans. (b) Taking derivatives of numerator and denominator

$$\underset{x \to 0}{\text{Lt}} \frac{1 - 2\cos 2x}{1 + 3\cos 3x} = \frac{1-2}{1+3} = -\frac{1}{4}$$ **Ans. b**

55. Solution $\quad A = \frac{\partial^2 z}{\partial z^2} = 2; C = \frac{\partial^2 z}{\partial y^2} = -4$

$$B = \frac{\partial^2 z}{\partial z\, \partial y} = 0; B^2 - AC = 0 + 4 \times 2 = 8$$

Since $B^2 - AC > 0$, so saddle point.

56. (b)

57. Ans. $\quad AX = B$

$$\begin{bmatrix} 0.2 & 0.8 \\ 1 & 1 \end{bmatrix} \begin{bmatrix} F_1/F \\ F_2/F \end{bmatrix} = \begin{bmatrix} 0.5 \\ 1 \end{bmatrix}; A = \begin{bmatrix} 0.2 & 0.8 \\ 1 & 1 \end{bmatrix}$$

$$A^{-1} = \frac{-1}{0.6} \begin{bmatrix} 1 & -0.8 \\ -1 & 0.2 \end{bmatrix} = \begin{bmatrix} -1.67 & 1.33 \\ 1.67 & -0.33 \end{bmatrix}$$

$$\begin{bmatrix} F_1/F \\ F_2/F \end{bmatrix} = \begin{bmatrix} -1.67 & 1.33 \\ 1.67 & -0.33 \end{bmatrix} \begin{bmatrix} 0.5 \\ 1 \end{bmatrix}; F_1/F = 0.50$$

$F_2/F = 0.50$ **Ans.**

58. Ans. $\quad V = \pi r^2 h = (27/4)\pi; h = 27/(4r^2)$

$$S = 2\pi rh + 2\pi r^2 = 2\pi\left(\frac{27r}{4r^2} + r^2\right)$$

$$\partial S/\partial r = 0 \text{ gives } r = 4\sqrt{27/(8\pi)}$$

$$h = \frac{27}{4r^2}; h = \frac{\sqrt{54\pi}}{2} \quad \textbf{Ans.}$$

59. Ans. Complimentary solution $(y_c) = c_1 \cos 2x + c_2 \sin 2x$
$y_1 = \cos 2x$; $y_2 = \sin 2x$; $y_1{}' = -2 \sin 2x$, $y_2{}' = 2 \cos 2x$;

$$u' = \frac{-y_2\,Q(x)}{y_1 y_2{}' - y_2 y_1{}'}; v = \frac{y_1\,Q(x)}{y_1\,y_2{}' - y_2 y_1{}'};$$

$$u' = -\sin 2x \tan x/2;\ u = -\frac{x}{2} + \frac{\sin 2x}{4}$$

$$v = \cos 2x \tan x/2;\ v = \frac{\cos 2x}{4} + \frac{1}{2} L_n \cos x$$

$$y_p = uy_1 + vy_2$$

$$y = y_c + y_p = c_1 \cos 2x + c_2 \sin 2x + \ln \frac{\cos x \sin 2x}{2} - x \cos 2x/2$$

60. Ans. $p = k\,\rho^r; \rho = p^{1/r}\,k^{-1/r}; \dfrac{dP}{dv} = -P^{1/r}\,k^{-1/r}\,v$

Integrating, $\int P^{-1/r}\,d_P = -k^{-1/r} \int v\,dv$

$$\frac{rP^{r-1/r}}{r-1} = k^{-1/r}\frac{v^2}{2} + \text{constant.}$$

$$a^2 = k^{-1/r}$$

$$\frac{r}{r-1}p^{r-1}/r + k^{-1/r}\frac{v^2}{2} = \text{constant.}$$

$$\frac{a^2}{(r-1)\,k^{1/r}} + k^{-1/r}\frac{v^2}{2} = \text{constant}$$

$$\frac{a^2}{r-1} + \frac{v^2}{2} = \text{constant.}$$

61. Ans. $\dfrac{d}{dt}(w_i - w) \propto w\left(\dfrac{1}{3} - \dfrac{w}{s}\right)$

$w_i - w$ = amount dissolved in water; w = amount
Present at any time, t (i.e. undissolved amount)
s = śolvent, kg.; w_i = constant

$$\frac{-dw}{dt} = kw\left(\frac{1}{3} - \frac{w}{120}\right) = \frac{kw}{120}(40 - w)$$

Integrating, $\frac{dw}{dt} = \frac{kw}{120}(w-40)$

$$\frac{dw}{40\,(w-40)} - \frac{dw}{40w} = \frac{kdt}{20};\ 40\ln\left(\frac{w-40}{w}\right) = \frac{kt}{120} + c$$

at $t = 0$, $w = 20$; giving $c = 0$, using $t = 12$ Mins, Conc = 1/30 (w/120)

gives $w = \frac{40}{[1 + e^{0.5493t/3}]}$

62. Ans. $mA + nB = Pc = \left(a - \frac{mx}{P}\right)\left(b - \frac{nx}{P}\right)$

$$\left(a - \frac{mx}{P}\right)\left(b - \frac{nx}{P}\right)$$

$$\frac{dx_c}{dt} = k\left(a - \frac{mx}{p}\right)\left(b - \frac{nx}{P}\right)$$

$$\frac{dx_c}{ab - x\left(\frac{an}{P} + \frac{bm}{P}\right) + \frac{mn}{P}x^2} = Kt + \text{const} = \frac{dx_c}{A + Bx + C^2x}$$

Satisfies, $B^2 = 4AC$ Relation.

Integrating, Kt + const = $\frac{2P^2}{(2anP - 2an^2\,x/b)}$

Using at $t = 0$, $x = 0$ gives

$$x = \frac{aP}{m} - \frac{-P^2a}{m(kt\,a\,n + P)}$$

(b) $m = 1$, $n = 2$, $P = 1$, $a = 40$, $b = 80$, $m/n = a/b$

from part a, $t = \frac{P^2}{k(amp - mnx)}\frac{-P}{kan}$

$t = 20$, $c = 12$ moles, giving $k = 0.43/1600$

at $c = 15$, $t = \frac{1}{kan}\left[\frac{80}{(80-30)} - 1\right]$ = 28 Minutes.

63. $dw/dt = K4\pi r^2$; w = weight of substance, gm.
t = time, months; ρ = density of ball, g/c.c

$$w = \frac{4}{3}\pi r^3\rho;\ \frac{dw}{dt} = 4\pi r^2\ \rho\frac{dr}{dt} = k4\pi r^2$$

$\rho \frac{dr}{dt} = k$; $dr = k\, dt/\rho$; integrating and using at $t = 0$, $r = 0.5$ gives

$r = 0.5 - 0.0167\, t$

(a) for $r = 0.25$ cm, $t = 15$ months.

(b) Volume = $4/3\, \pi r^3$; for volume change to be 0.5,

$$r_f/r_i = 0.7937,\ t = 6.18 \text{ months.}$$

64. Ans. $\underset{(a-x)}{\text{Me AC}} + \underset{b-x}{\text{Na OH}} = \underset{x}{\text{Na AC}} + \underset{x}{\text{Me OH}}$

$\frac{dx}{dt} = K(a - x)\,(b - x)$; if $a = b = 0.01$

$\frac{dx}{(0.01 - x)^2} = K\,dt$; integrating, $\frac{1}{0.01 - x} = Kt + const.$

Using $t = 0$, $x = 0$ gives const = 100

Using $t = 12/60$ Hrs and x = 0.25 × 0.01 = 0.0025 m

gives $k = 166.7$

In second case : $\frac{dx}{dt} = 166.7\,(0.025 - x)\,(0.015 - x)$

Integrating gives, $\ln(0.015 - x) - \ln(0.025 - x) = -1.66\, t + \text{const}$

Using $x = 0$ at $t = 0$ gives

$$\ln\left(\frac{0.015 - x}{0.025 - x}\right) = \frac{0.015}{0.025} e^{-1.66t}$$

at $t = 1$ hr, $x = 0.01373$, % Conversion = $\frac{0.01373}{0.025} = 55\%$

65. Ans.

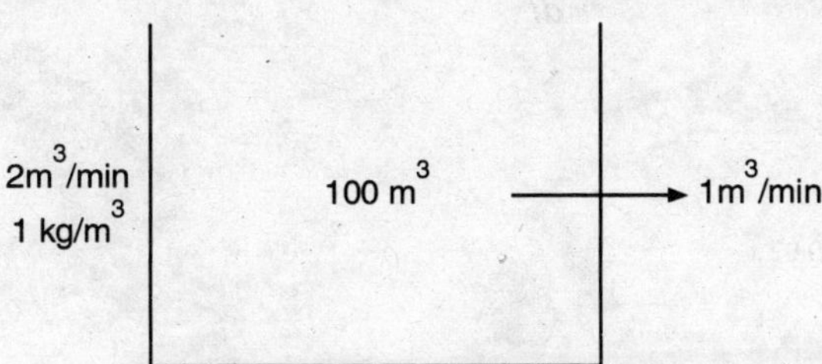

By overall balance : $2 - 1 = dV/dt$; $dV/dt = 1$; $V = 100 + t$

Component balance : $2 \times 1 - 1 * C = \frac{d}{dt}\,[(100 + t)\, C]$

$$2 - 2c = \frac{dC}{dt}(100 + t);\ \frac{dC}{(2 - 2C)} = \frac{dt}{100 + t}$$

Integrating $\ln\left(\frac{2}{2-2c}\right) = 2\ln\left(\frac{100+t}{100}\right)$

at $V = 150\text{ m}^3,\ t = 50\text{ min}$

$C = 0.555$ **Ans.**

66. Ans. Solving $\frac{dx}{dt} + \lambda x = 0$ gives $x = C_1\, e^{-\lambda t}$ where C_1 is a constant. $x + y + z = r$

$$\frac{dz}{dt} = \mu\,(r - x - z) = r\mu - \mu\, C_1\, e^{-\lambda t} - \mu z$$

Complimentary solution of $z = C_2\, e^{-\mu t}$ where C_2 is a constant.

Particular integral is : $\frac{r\mu}{D+\mu}$ is r

$$z = C_2\, e^{-\mu t} - \frac{\mu C_1 e^{-\lambda t}}{(\mu - \lambda)} + r;\ \text{using } z = 0 \text{ at } t = 0$$

$$C_2 = \frac{C_1\mu}{(\mu-\lambda)} - r \to 1;$$

$$\frac{dz}{dt} = -\mu C_2 + \frac{C_1\,\mu\lambda}{(\mu-\lambda)} = 0 \to 2$$

Solving equations 1 and 2 gives $C_1 = r,\ C_2 = \frac{r\lambda}{(\mu-\lambda)}$

$$z = \frac{r}{(\rho-\mu)}(\mu\, e^{-\lambda t} - \lambda\, e^{-\mu t}) + r$$

67. Ans. For tank A component balance : $2 * 0 - 2 * C_A = \frac{d}{dt}(V_A\, C_A)$

$$\frac{dC_A}{dt} = \frac{-2}{100} C_A$$

Integrating and using $C_A = 0.02$ at $t = 0$

$$C_A(\text{t}) = 0.02\, e^{-0.02\,\text{t}} \qquad ...(1)$$

For tank-B component balance is :

$$2C_A - C_B = \frac{d}{dt}\left[(50+t)\, C_B\right]$$

Re arranging, $\frac{dC_B}{dt} + \frac{2C_B}{(50+t)} = \frac{0.04e^{-0.02\,t}}{(50+t)}$

This is of Leibnitz form. Solution for

$$\frac{dy}{dx} + Py = Q \text{ is } y\,e^{\int Pdx} = \int Qe^{\int Pdx}\,dx + const.$$

Here $$P = \frac{2}{(50+t)};\, Q = \frac{0.04\,e^{-0.02\,t}}{(50+t)}$$

$$C_B\,(50+t)^2 = \int \frac{0.04\,e^{-0.02t}}{(50+t)}\,(50+t)^2\,dt + \text{const}$$

Using at $t = 0$, $C_B = 0$; giving constant = 200

$$C_B\,(50-t)^2 = \left[\frac{e^{-0.02t}}{0.0004} + \frac{(50+t)\,e^{-0.02t}}{0.02}\right] 0.04 + 200$$

C_B at t = 60 min is 0.00828

Amount of salt in tank B after one hour = (60 + 50) × 0.00828 = 0.9103 kg. salt.

68. Ans.

Rate of input – Rate of output = Rate of change of Accumulation

$5 - D = 0$; So $D = 5$

Component balance :

$5 \times 0.30 - 5y_D = 100\,dx\,w/dt$

$$dt = 100\,dx_w/(1.5 - 5y_D)$$

F
F = 0.30
5 kg Moles
Pr
0.30
100 kg
Moles

Replacing x_D by x_S; $dt = \dfrac{100\,dx_S}{(1.5 - 3.5\,x_S)} + \dfrac{100\,x_S\,dx_S}{(1.5 - 3.5\,x_S)}$

Integrating and using at $t = 0$, $x_S = 0.30$

$$t = 4.09 \ln \frac{4.5}{(1.5 - 3.5\,x_S)} - 28.6\,x_S + 8.58$$

at $x_S = 0.40$, $t = 5.87$ hour.

(b) $1.5 - 5_{yD} = x_D = x_S/(1 + x_S)$; $x_S = 0.430$

69. Ans. By force balance :

Gravitational force – Drag force or buoyant force = Rate of change of momentum.

$$\frac{4}{3}\pi r^3 \rho g - 6\pi\,\mu\,Vr = \frac{d}{dt}\left(\frac{4}{3}\pi r^2 \rho V\right)$$

$$g - \frac{9}{2}\frac{\mu V}{\rho r^2} = \frac{3r^2 V}{r^3}\frac{dr}{dt} + \frac{dV}{dt};\ \frac{dr}{dt} = kr\,;\, r = C_0\,e^{kt}$$

$$\frac{dV}{dt} + 3KV + \frac{9}{2}\frac{\mu V}{\rho C_1\,e^{2kt}} = g \text{ (Leibnitz form)}$$

Solution is : I.F × $V = \int g\,e^{3kt} + const. = \dfrac{g\,e^{3kt}}{3k}$ + const.

Since e^{-2kt} becomes zero at $t = \infty$, so left one is

$$V = \frac{g}{3K} e^{3kt} e\left(-3kt + \frac{2.25}{\rho C_1 K} e^{-2kT} \right) + Const.\, e^{(-3kt + ...)}$$

This at $t = \infty$ is $V = \dfrac{g}{3K}$

70. Ans :

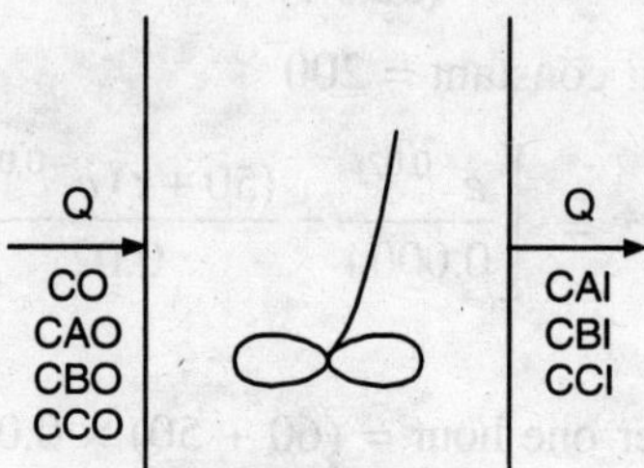

The material balance equations for the components are

$$\frac{VdcA_1}{dt} = qC_{AO} - [qCA_1 + VK_1 CA_1 - VK_2 CB_1] \qquad ...(1)$$

$$\frac{VdcB_1}{dt} = qC_{BO} - (qCB_1 + VK_2\, CB_1 + VK_3\, CB_1 - VK_1\, CB_1) \qquad ...(2)$$

$$\frac{VdCC_1}{dt} = qCC_0 - [qC_{C1} - VK_3\, C_{B1}] \qquad ...(3)$$

differentiating equation (2) w.r. to time, t and substituting for CA_1 from equation (2), gives

$$\frac{Vd^2C_{B1}}{dt^2} = -\frac{VdCB_1}{dt}\left[\frac{2q}{V} + K_1 + K_2 + K_3\right] - VC_{B1}$$

$$\left[\frac{q^2}{v^2} + \frac{q}{v}K_2 + \frac{q}{V}K_3 + \frac{q}{V}K_1 + K_1K_3\right] + K_1 qC_0$$

Terms containing q/V do not come until steady state is reached or reactor is filled completely.

So $\dfrac{d^2NB}{dt^2} + \dfrac{dN_B}{dt} \times (K_1 + K_2 = K_3) + N_B\, R = S$

71. Ans. $\int dy = \int ty\, dt$, using $y\,(0) = 1$

$$\int dy = \int_0^t t\, dt + y(0) = y_1 = 1 + \frac{t^2}{2}$$

$$y_2 = 1 + \int_0^t ty_1\, dt = 1 + \int_0^t t\left(1 + \frac{t^2}{2}\right)$$

$$y_2 = 1+\frac{t^2}{2}+\frac{t^4}{8};\ y_3 = 1+\int_0^t t\, y_2\, dt$$

$$y_3 = 1+\frac{t^2}{2}+\frac{t^4}{2.4}+\frac{t^5}{2.4.6}$$

So, $y = 1+\dfrac{t^2}{2}+\dfrac{t^4}{2.4}+\dfrac{t^6}{2.4.6}+......+\dfrac{t^{2n}}{2.4.6.2n}$

72. Ans. $y_1 = y_0 + \left(\dfrac{dy}{dx}\right)_{x_0,\, y_0} (x_1 - x_0)$

$$(dy/dx)_{x_0,\, y_0} = 1.8 + 0.1 * 0 = 1.8$$

$$y_1 = 0 + 1.8 \times 0.3 = 0.54$$

$$(dy/dx)_{x_1/y_1} = 2.1 + 0.1 \times 0.54^2 = 2.1292$$

$$y_2 = y_1 + \left(\frac{dy}{dx}\right)_{x_1,\, y_1} (x_2 - x_1);$$

$= 0.54 + 2.1292 \times 0.30 = 1.179$, Like this continuing.

73. Ans. $d^2y/dx^2 + 2\ dy/dx + 2y = \sin x$; x is in radians.

$$d^2y/dx^2 = -2dy/dx - 2y + \sin x$$

$$n = 0, x = 0;\ y_0'' = 0 - 2 \times 2 - 2 \times -1 = -2$$

$y_1 = y_0 + h\, y_0' + \dfrac{h^2}{2} y_0'' = 2 + 0.2 \times -1 + \dfrac{0.2 \times 0.2}{2} \times -2 = 1.76$ ($h = 0.20$)

$x_1 = 0.20$, $y_1 = 1.76$; $y' = y_0' + h\, y_0'' = -1 + 0.2 \times -2 = -1.4$

$y_1'' = \sin 0.2 - 2 \times 1.76 - 2 \times -1.4 = -0.5213$

$y_2 = y_1 + h\, y_1' + \dfrac{h^2}{2} y_1'' = 1.76 + 0.2 \times -1.4 + \dfrac{0.2 \times 0.2}{2} * -0.5213$

$= 1.4696$

$y_2' = 1.504$; $y_2'' = 0.4588$; $y_3 = 1.1779$

x	0	0.2	0.4	0.6
y	2.0	1.76	1.4696	1.1779

74. Ans. $y = a_1(x - x_2)(x - x_3)(x - x_4) + a_2(x - x_1)(x - x_3)$
$(x - x_4) + a_3(x - x_1)(x - x_2)(x - x_4) + a_4(x - x_1)(x - x_2)(x - x_3)$.

$$a_1 = \frac{y_1}{(x_1 - x_2)(x_1 - x_3)(x_1 - x_4)} = -1$$

$$a_2 = \frac{y_2}{(x_1 - x_2)(x_2 - x_3)(x_2 - x_4)} = 0$$

$$a_3 = \frac{y_3}{(x_3 - x_1)(x_3 - x_2)(x_3 - x_4)} = -1$$

$$a_4 = \frac{y_4}{(x_4 - x_1)(x_4 - x_2)(x_4 - x_3)} = 1$$

$$y = 1(x-1)(x-3)(x-4) + 0 - 1(x-0)(x-1)(x-4)$$
$$+ 1(x-0)(x-1)(x-3) = x^3 - 7x^2 + 18x - 12$$

75. Ans.

$$F = 2x^3 - 2xy, \text{y(0)} = 0$$
$$K_0 = 0.1\ F(x_0, y_0) = 0.1\ F(0, 0) = 0$$
$$K_1 = 0.1\,F\left(x_0 + \frac{h}{2}, y_0 + \frac{K_0}{2}\right)$$
$$K_1 = 0.1\ F[0 + 0.05, 0 + 0] = 0.00025$$
$$K_2 = hF\ (0 + 0.05, 0.00025/2) = 0.000024875$$
$$K_3 = 0.1(0.10, 0.00024875) = 0.0009995$$
$$y_1 = y_0 \frac{1}{6}(K_0 + 2k_1 + k_2 + k_3) = 0.0000054$$

76. Ans. $y^{I} = 2x^3 - 2xy$; $y^{II} = 6x^2 - 2xy^{I} - 2y$

$y^{III} = 12x - 2xy^{II} - 4y^{I}$; $y^{IV = 12 - 2xy}III - 6yII$

Taylor's expansion is : $y(x) = y_0 + y^1(0)\,x + y^{II}(0)\,\frac{x^2}{2} + ...$

at x = 0, y(0) = 0, $y^{I}(0) = 0$; $y^{II}(0) = 0$, $y^{III}(0) = 0$

$$y^{IV}(0) = 12$$
$$y(x) = 12x^4/24 = 0.50x^4$$

77. Ans. $I = \int_{x=0}^{7} y dx = \Delta x\left(\frac{y_0}{2} + y_1 + y_2 + y_3 + y_4 + y_5 + y_6 + \frac{y_7}{2}\right)$

$$= 1\left(\frac{2}{2} + 6 + 18 + 38 + 66 + 102 + 146 = \frac{198}{2}\right) = 476$$

78. $\frac{h}{6}(y_1 + 4y_2 + y_3) + \frac{h}{6}(y_3 + 4y_4 + y_s) + \frac{h}{6}(y_5 + 4y_6 + y_7)$

$$= \frac{1}{3}(60 + 4 \times 62 + 63) + \frac{1}{3}(63 + 4 \times 62 + 62) + \frac{1}{3}(62 + 4 \times 60 + 59)$$
$$+ \frac{1}{3}(59 + 57 \times 4 + 57) = 483$$

Total product in hours = 483 × 60 = 28,980 kg.

79. Ans. $y = mx + C; P = bw + a$

$$nC + (\Sigma x)\, m = \Sigma y$$
$$(\Sigma x)\, C + (\Sigma x^2)\, m = \Sigma (xy)$$

n	$w = x$	y	x^2	xy
1	50	12	2500	600
2	70	15	4900	1050
3	100	21	10,000	2100
4	120	25	14,000	3000

$$\Sigma x = 340, \Sigma y = 73, \Sigma x^2 = 31{,}800, \Sigma xy\ 6750$$
$$4\,C + 340\,m = 73\ ;\ 340\,C + 31800\text{ m} = 6750$$
$$\text{giving } m = 0.1724;\ C = 3.596$$
$$P = 0.1724\, W + 3.596$$

80. Ans. $u(x, t) = u_t(x, t) + u_s(x)$; given $u_s(x) = 40\, x/l$

$$u_s(0) = 0;\ u_s(t);\ \text{given } u(0, t) = 0;\ u(l, t) = 40$$
$$u_t(x, t) = u(x, t) - u_s(x);\ u_t(0, t) = u(0, t) - u_s(0) = 0 - 0 = 0$$
$$u_l(l, t) = u(l, t) - u_s(l) = 40 - 40 = 0$$
$$u_t(x, 0) = u(x, 0) - u_s(x) = \frac{80x}{l} - \frac{40x}{l} = \frac{40x}{l}$$
$$u_t(x, t) = (C_1 \cos Px + C_2 \sin px)\ e^{-C^2 p^2 t};\ u_t(0, t) = 0 \text{ gives } C_1 = 0$$
$$u_t(0, t) = 0 = C_2 \sin Pl\ e^{-C^2 p^2 t};\ \sin pl = n\,p\ ;\ P = n\pi/l$$
$$u_t(x, t) = \Sigma\, C_n \sin\left(\frac{n\pi x}{l}\right) e^{-C^2\, n^2\, \pi^2\, t\,/\,l^2}$$
$$u_t(x, 0) = \frac{40x}{l} = \Sigma\, C_n \sin\left(\frac{n\,\pi\, x}{l}\right)$$
$$C_n = \frac{2}{l}\int_0^l \left(\frac{40x}{l}\right) \sin\left(\frac{n\,\pi x}{l}\right) dx = \frac{-80}{n\pi} \cos n\,\pi$$
$$C_1 = \frac{-80}{\pi}(-1);\ C_2 = \frac{-80}{\pi} \times \frac{1}{2}$$
$$u_t(x, t) = \frac{-80}{n\pi} \cos n\pi \frac{\sin n\pi x}{l}\ e^{-C^2 n^2 \pi^2 t/l^2}$$
$$u(x, t) = u_s(x) + u_t(x, t) = x + \frac{80}{\pi}\sum_{n=1}^{\infty} \frac{(-1)^{n+1}}{n} \sin\left(\frac{n\pi x}{40}\right) e^{\frac{-C^2\pi^2 t n^2}{1600}}$$

81. Ans. Heat conduction equation is : $\frac{\partial u}{\partial t} = \frac{C^2 \partial^2 u}{\partial x^2}$ ...(1)

at steady state, $\frac{\partial u}{\partial t} = 0$; So $u = 100\, x/l = u(x, 0)$ (I.C.)

B.C.S are : $u(0, t) = 0$; $u(l, t) = 0$ for all t values.

Solution of equation (1) is : $u(x, t) = (C_1 \cos px + C_2 \sin px)\, e^{-C^2 p^2 t}$

Using $u(0, t) = 0$ gives $C_1 = 0$

Using $u(l, t) = 0$ gives; $\sin pl = 0$; $pl = n\pi$; $P = n\pi/l$

$$u(x, t) = b_n \sin \frac{n\pi x}{l}\, e^{-C^2 n^2 \pi^2 t/l^2} \quad \text{where } b_n = C_2$$

$$u(x, t) = \sum_{n=1}^{\infty} b_n \sin\left(\frac{n\pi x}{l}\right) e^{-C^2 n^2 \pi^2 t/l^2}$$

$$u(x, 0) = \sum_{n=1}^{\infty} b_n \sin \frac{n\pi x}{l} 100x/l$$

$$b_n = \frac{200}{n\pi}(-1)^{n+1}$$

$$u(x, t) = \frac{200}{\pi} \sum_{n=1}^{\infty} \frac{(-1)^{n+1}}{n} \frac{\sin n\pi x}{l}\, e^{-c^2 n^2 \pi^2 t} \Big/ l^2$$

where $C^2 = K/C_p \cdot \rho$

82. Ans. $f(x) = \frac{a_0}{2} + \sum_{n=1}^{\infty} a_n \cos nx + \sum_{n=1}^{\infty} b_n \sin nx$

$$a_0 = \int_0^1 \pi\, x\, dx + \int_0^1 \pi\,(2 - x)\, dx = \pi$$

$$a_n = \int_0^1 \pi\, x \cos n\,\pi\, x\, dx + \int_0^1 \pi\,(2 - x) \cos n\,\pi\, x\, dx$$

$$a_n = \frac{1}{\pi n^2}(2 \cos n\pi - 1 - \cos 2n\pi)$$

$$b_n = \int_0^1 \pi x\, dx \sin n\pi\, x + \int_0^1 \pi(2 - x) \sin n\pi\, x\, dx$$

$$= 0$$

$$f(x) = \frac{\pi}{2} - \frac{4}{\pi}\left[\frac{\cos \pi x}{1^2} + \frac{\cos 3\pi\, x}{3^2} + \frac{\cos 5\pi\, x}{5^2} + \ldots\right]$$

83. Ans. (a) $\log\left(\frac{1+S}{S}\right) = \log[(1+S) - \log S] = \int_S^{\infty} \frac{dS}{(S+1)} - \int_S^{\infty} \frac{1}{S}\, dS$

$$L\left[f(t)\frac{1}{t}\right] = \int_S^{\infty} f(S)\, dS; \text{So } f(S) = -\frac{1}{S} \text{ hence f(t)} = -1$$

$$L^{-1}\left[-\int_S^{\infty} \frac{1}{S} dS\right] = L^{-1}\left[\left(-\frac{1}{S}\right)1\right] = -1$$

$$L^{-1}\left[\int_S^{\infty} \frac{dS}{(S+1)}\right] = L^{-1} \frac{1}{(S+1)} = e^{-t}$$

$$\int_S^{\infty} \frac{dS}{S+1} - \int_S^{\infty} \frac{dS}{S} = \frac{e^t - 1}{t}$$

(b) $\frac{1}{S^2(S^2+a^2)} = \left[\frac{1}{S^2} - \frac{1}{S^2+a^2}\right]\frac{1}{a^2}$

$$\frac{1}{a^2}\left[L^{-1}\left(\frac{1}{S^2}\right) - L^{-1}\left(\frac{1}{S^2+a^2}\right) = \frac{1}{a^2}\left[t - \frac{\sin at}{a}\right]\right]$$

84. Ans. $\frac{2S}{S^2+9} + \frac{18}{S^2+9} + \frac{24}{S^4} - \frac{30\sqrt{S}}{S^4}$

$$L^{-1}\left(\frac{2S}{S^2+9}\right) = 2\cos 3t;\ L^{-1}\left(\frac{18}{S^2+9}\right) = 6\sin 3t$$

$$L^{-1}\left(\frac{24}{S4}\right) = 4t^3;\ L^{-1}\left(\frac{30\sqrt{S}}{S4}\right) = L^{-1}\frac{30}{S^{3+\frac{1}{2}}} = \frac{8t^{5/2}+30}{15\sqrt{\pi}}$$

$$= 2\cos 3t + 6\sin 3t + 4t^3 - \frac{16t^{5/2}}{\sqrt{\pi}}$$

85. Ans. $L(Y^{II}) = S^2\, y(S) - S\, y(0) - y^{I}(0) = S^2\, y(S) - 1$

$$L(2y^{I}) = 2\, S\, y(S) - 2y(0) = 2Sy(S)$$

$$L(y) = y(S)$$

$$L(e^t \sin t) = \frac{1}{(S+1)^2+1}$$

$$S^2 y(S) - 1 + 2S\, y(S) + 5y(S) = \frac{1}{(S+1)^2+1}$$

$$y(S) = \frac{1}{3}\left[\frac{1}{(S+1)^2+1^2}\right] + \frac{2}{3}\left[\frac{1}{(S+1)^2+2^2}\right]$$

$$y(t) = \frac{1}{3}e^{-t}\sin t + \frac{1}{3}e^{-t}\sin 2t$$

86. Ans.

$$L\,[u(t)] = \int_0^{\infty} e^{-St}\,u(t)\,dt$$

$$= \int_0^{a} e^{-st}\,0\,dt + \int_a^{\infty} e^{-St}\,1\,dt = +\frac{e^{-aS}}{S}$$

87. [Figure: circle with centre O, radius arrow labelled 5 cms; 150°C]

Q = – KA dt/dx = – K (2πx*1) dt/dx

$$Q = \frac{-kA\,dT}{dx} = -k(2\pi x * 1)\frac{dT}{dx}$$

$$dT = \frac{Q}{2\pi}\frac{dx}{k} \text{ Integrating} : T = \frac{Q}{2\pi x}x + c$$

at x = 10, T = 150°C; T = 40°C at x = 15

$$150 = -\frac{q}{2\pi k}\ln 10 + C \qquad \text{...(1)}$$

$$40 = -\frac{q}{2\pi k}\ln 15 + C \qquad \text{...(2)}$$

$$\text{at } x = 12.5,\ t = -\frac{Q}{2\pi k}\ln 12.5 + C$$

Solving the above equations, T = 89.5°C.

SECTION – B

1. A box contains 6 red balls and 4 green balls, one ball is randomly picked and then a second ball is piked without replacement of the first ball. The probability that both the balls are green is:

(a) 1/15 (b) 2/25 (c) 2/15 (d) 4/25 **Ans.** (c)

Sol. 4 *C1*/(10 *C1*) * 3 *C1*/(9*C1*) = 4/10 * 3/9 = 2/15 **Ans.**

2. The directional derivative of $f(x, y, z) = x^2 + y^2 + z^2$ at the point (1, 1, 1) in the direction $i - k$ is:

(a) 0 (b) 1 (c) $\sqrt{2}$ (d) $2\sqrt{2}$ **Ans.** (a)

Sol. $F(x, y, z) = x^2 + y^2 + z^2$

$Af - (x, y, z) = 2xi + 2yj + 2z\,k = A$ (say); $B = i - k$; $\overline{B} = 1\ (i - k)/\sqrt{2}$

$A \cdot B = 2\ (x - z)/\sqrt{2}$; $A \cdot\cdot\ \overline{B} = 2(1-1)/\sqrt{2} = 0$ **Ans.**

3. The Taylors series expansion of the function : $F(x) = x/1 + x$ around x is:

(a) $x + x^2 + x^3 + ...$ (b) $1 + x + x^2 + x^3$

(c) $2x + 4\,x^2 + 8x^3 + 16\,x^4\,x$ (d) $x - x^2 + x^3 - x^4 + ...$

Sol. $f(x) = f(0) + f'(0)\ (x - 0) + f''\ (0)\ (x - 0)^2/2 + ...$

$= 0 + x - x^2 + x^3 - ...$ **Ans.** (d)

4. The range of values for a constant k to yield a stable system in the following set of time dependent differential equation as:

$dy_1/dt = -5y_1 + (4 - k)\,y_2$; $dy_2/dt = y_1 - 2y_2$

(a) $0 < k < 7$ (b) 6.25

(c) $-6 < k < 6.25$ (d) $-0 < k < 7$

Sol. $dy1/dt = -5\,y_1 + (4 - k)\,y_2$

$dy2/dt = y1 - 2\,y_2 \rightarrow 2$ ...(3)

Putting $d/dt = D$, we have from equation (1),

$Dy1 + 5y2 - (4 - k)\,y2 = 0$

$(D + 5)\,y1 - (4 - k)\,y2 = 0$ ...(3)

From equation (2), $D\,y_1 = y1 - 2y2$

$(D + 2)\,y2 - y1 = 0$; $y1 = (D + 2)\,y2$; Putting in equation 3, we have:

$(D^2 + 7D + (6 + K))\,y2 = 0$; $K \leq 6.25$ **Ans.** (c)

5. The value of y as $t \rightarrow \infty$ for the following differential eqution for an intial value of $y(1) = 0$ is.

$(4\,t^2 + 1)\,dy/dt + 8\,y\,t - t = 0$

(a) 1 (b) 1/2 (c) 1/4 (d) 1/8 **Ans.** (d)

Sol. Rearranging the given equation,

$dy/dt + (8\,t/4t^2 + 1)\ y = t/4\,t^2 + 1$ ...(1)

This is of Leibnitz form; I.F. $= \int 8t/(4t^2+1)$

$= 4\,t^2 + 1$; Solution of equation is: $y(4t^2 + 1) = \int \frac{t}{4t^2 - 1}(4t^2+1)dt + c = t^2/2 + c$;

$$y = \frac{t^2}{2(4t^2+1)} + \frac{c}{4t^2+1} \quad ...(2)$$

Now $y(1) = 0$; giving $c = -1/2$

Putting in equation (2), we have; $y = \frac{t^2}{2(4t^2+1)} - \frac{1}{2(4t^2+1)}$

y at $t = \infty = \lim_{t\to\infty} \frac{-1}{2(4t^2+1)} + \frac{1}{2(4+1/t^2)} = 1/2\ (4 + 0) = 1/8$ **Ans.**

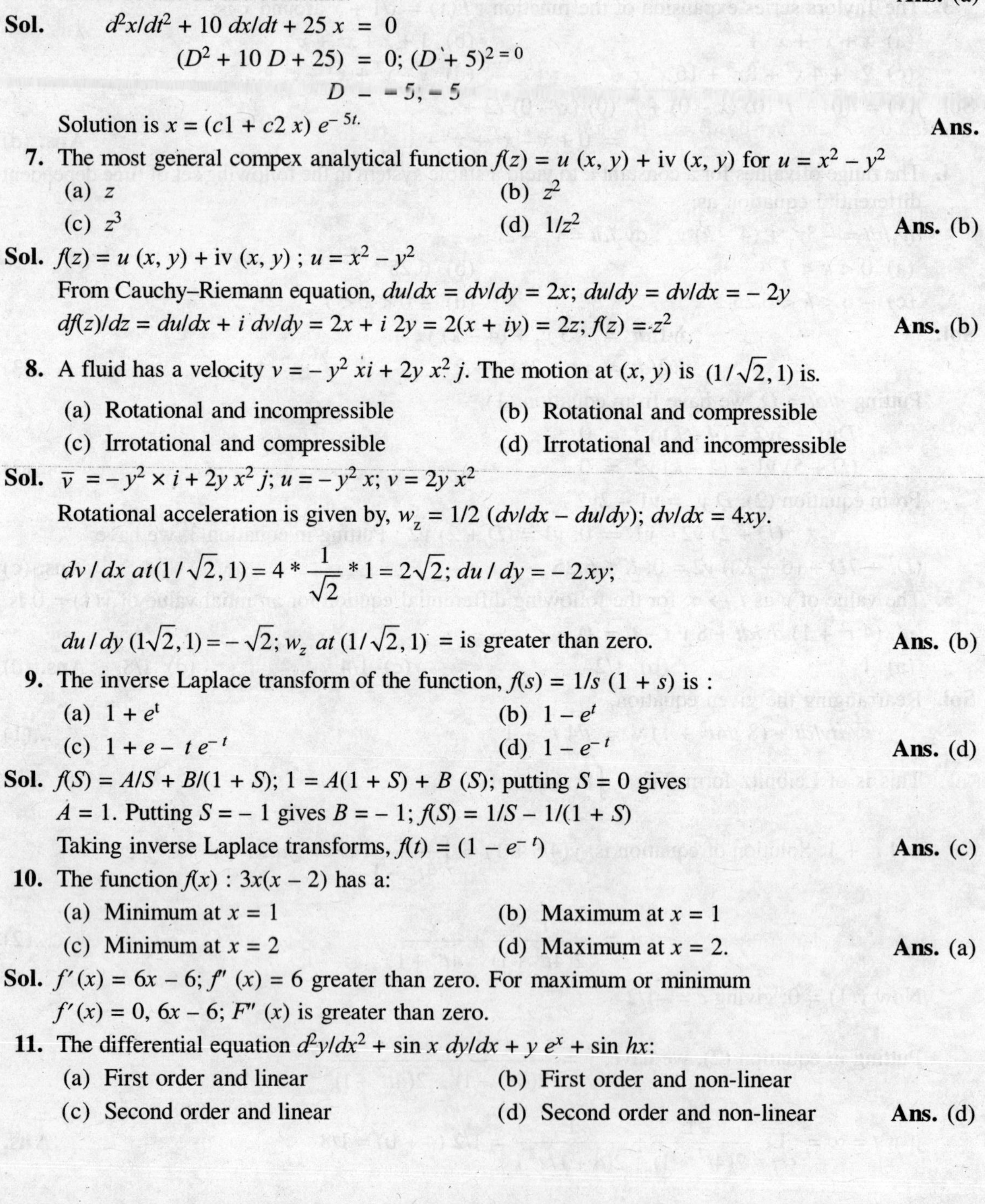

6. The differential equation: $d^2x/dt^2 + 10\ dx/dt + 25\ x = 0$ will have a solution of the form.

(a) $(c1 + c2\ t)\ e^{-5t}$ (b) $c1\ e^{-2t}$

(c) $c1 + c2\ e^{5t}$ (d) $c1\ e^{-5t} + x2\ e^{2t}$.

where $c1$ and $c2$ are constants. **Ans.** (a)

Sol. $d^2x/dt^2 + 10\ dx/dt + 25\ x = 0$

$(D^2 + 10\ D + 25) = 0;\ (D + 5)^2 = 0$

$D = -5, -5$

Solution is $x = (c1 + c2\ x)\ e^{-5t.}$ **Ans.**

7. The most general compex analytical function $f(z) = u\ (x, y) + \text{iv}\ (x, y)$ for $u = x^2 - y^2$

(a) z (b) z^2

(c) z^3 (d) $1/z^2$ **Ans.** (b)

Sol. $f(z) = u\ (x, y) + \text{iv}\ (x, y)\ ;\ u = x^2 - y^2$

From Cauchy–Riemann equation, $du/dx = dv/dy = 2x;\ du/dy = dv/dx = -2y$

$df(z)/dz = du/dx + i\ dv/dy = 2x + i\ 2y = 2(x + iy) = 2z;\ f(z) = z^2$ **Ans.** (b)

8. A fluid has a velocity $v = -y^2\ xi + 2y\ x^2\ j$. The motion at (x, y) is $(1/\sqrt{2}, 1)$ is.

(a) Rotational and incompressible (b) Rotational and compressible

(c) Irrotational and compressible (d) Irrotational and incompressible

Sol. $\bar{v} = -y^2 \times i + 2y\ x^2\ j;\ u = -y^2\ x;\ v = 2y\ x^2$

Rotational acceleration is given by, $w_z = 1/2\ (dv/dx - du/dy);\ dv/dx = 4xy.$

$$dv / dx\ at (1/\sqrt{2}, 1) = 4 * \frac{1}{\sqrt{2}} * 1 = 2\sqrt{2};\ du / dy = -2xy;$$

$du / dy\ (1\sqrt{2}, 1) = -\sqrt{2};\ w_z\ at\ (1/\sqrt{2}, 1)$ = is greater than zero. **Ans.** (b)

9. The inverse Laplace transform of the function, $f(s) = 1/s\ (1 + s)$ is :

(a) $1 + e^t$ (b) $1 - e^t$

(c) $1 + e - t\ e^{-t}$ (d) $1 - e^{-t}$ **Ans.** (d)

Sol. $f(S) = A/S + B/(1 + S);\ 1 = A(1 + S) + B\ (S)$; putting $S = 0$ gives

$A = 1$. Putting $S = -1$ gives $B = -1;\ f(S) = 1/S - 1/(1 + S)$

Taking inverse Laplace transforms, $f(t) = (1 - e^{-t})$ **Ans.** (c)

10. The function $f(x) : 3x(x - 2)$ has a:

(a) Minimum at $x = 1$ (b) Maximum at $x = 1$

(c) Minimum at $x = 2$ (d) Maximum at $x = 2$. **Ans.** (a)

Sol. $f'(x) = 6x - 6;\ f''(x) = 6$ greater than zero. For maximum or minimum

$f'(x) = 0,\ 6x - 6;\ F''(x)$ is greater than zero.

11. The differential equation $d^2y/dx^2 + \sin x\ dy/dx + y\ e^x + \sin hx$:

(a) First order and linear (b) First order and non-linear

(c) Second order and linear (d) Second order and non-linear **Ans.** (d)

12. The complex number $2(1 + i)$ can be represented in polar form as:

(a) $2\sqrt{2}\, e^{\pi/4}$ (b) $-\frac{\pi}{4} e^{2i\sqrt{2}}$

(c) $\sqrt{2}\, e^{i\pi/4}$ (d) $\frac{\pi}{4} e^{i\sqrt{2}}$ **Ans.** (a)

Sol. $2(1 + i) = 2 + 2i$; $r^2 \sin\theta = 2$; $r^2 \cos\theta = 2$; $r = 2\sqrt{2}$

$\tan\theta = r^2 \sin\theta / r^2 \cos\theta = 1$; $\theta = \pi/4$

Hence the polar form is: $(2 + 2i) = 2\sqrt{2}\, e^{i\pi/4}$

13. The sum of the Eigenvalues of the matrix:

$\begin{pmatrix} 3 & 4 \\ x & 1 \end{pmatrix}$ for real and negative values of x is: **Ans.** (a)

(a) Greater than zero (b) Less than zero

(c) Zero (d) Dependent on values of x.

Sol. $\begin{vmatrix} 3-\lambda & 4 \\ x & 1-\lambda \end{vmatrix} = 0$; Since x is real and negative, put $x = -k$, where k is a positive constant.

$(3 - l)(1 - \lambda) + 4k = 0$; $\lambda^2 - 4\lambda + 3 + 4k = 0$

If $\lambda 1$ and $\lambda 2$ be the solutions of the above equations, then $\lambda 1$ and $\lambda 2$ are eigen values. Sum of eigen values = sum of roots of the above equation.

$\lambda 1 + \lambda 2 = -(-4)/1 = 4$ (greater than zero)

14. The system of equations :

$$4x + 6y = 8 \quad \text{.... (1)}$$
$$7x + 8y = 9 \quad \text{....(2)}$$
$$3x + 2y = 1 \quad \text{....(3)}$$

(a) No solution (b) Only one solution

(c) Two solutions (d) Infinite number of solutions. **Ans.** (b)

Sol. The determinedant value of the given system of equations is:

$$A = \begin{vmatrix} 4 & 6 & 8 \\ 7 & 8 & 9 \\ 3 & 2 & 1 \end{vmatrix} = 4(8 - 18) - 6(7 - 27) + 8(14 - 24) = 0$$

Since the determinant value is zero, so only one solution or alternatively, adding equations (1) and (3);

$$4x + 6y = 8 \quad \text{.....(1)}$$
$$3x + 2y = 1 \quad \text{.....(3)}$$
$$\overline{7x + 8y = 9} \quad \text{Which is Equation (2)}$$

Thus there are two equations and two unknowns. So only one solution.

15. A box containing three blue balls and another identical box contains two blue balls and five red balls. One ball is picked at random from one of the two boxes and it is red. The probability that it came fromthe first box is:

(a) 2/3 (b) 4/9

(c) 4/7 (d) 2/7 **Ans.** (d)

Sol. Number of boxes are 2. Probability of picking ball from one box = 1/2 total number of balls = blue balls + red balls = 3 + 4 = 7 probability of red ball = 4/7; probability of getting red ball from one box 1/2 * 4/7 = 2/7 **Ans.**

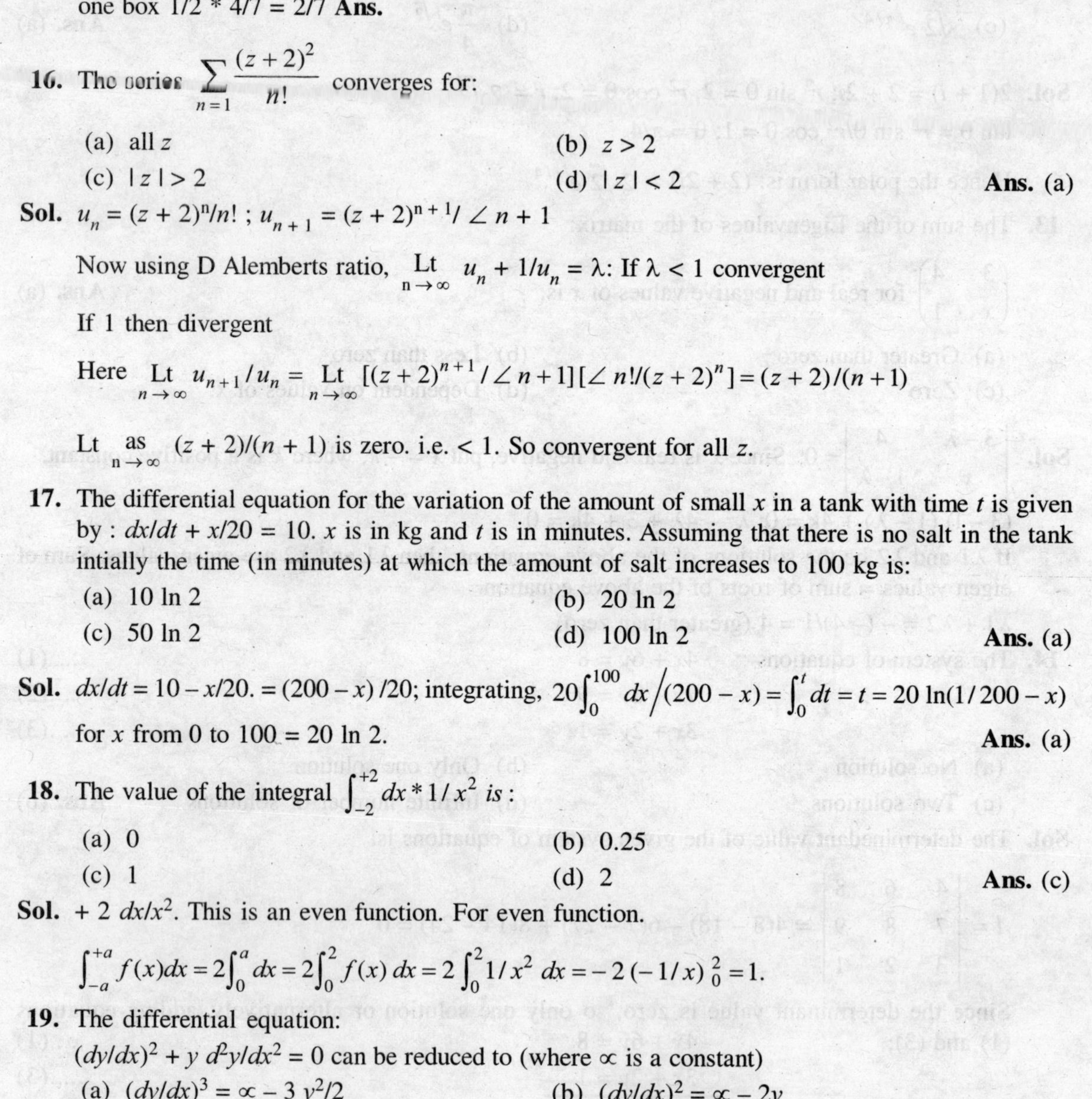

16. The series $\sum_{n=1} \frac{(z+2)^2}{n!}$ converges for:

(a) all z (b) $z > 2$

(c) $|z| > 2$ (d) $|z| < 2$ **Ans.** (a)

Sol. $u_n = (z+2)^n/n!$; $u_{n+1} = (z+2)^{n+1}/\angle n+1$

Now using D Alemberts ratio, $\underset{n\to\infty}{\text{Lt}}\ u_n + 1/u_n = \lambda$: If $\lambda < 1$ convergent

If 1 then divergent

Here $\underset{n\to\infty}{\text{Lt}}\ u_{n+1}/u_n = \underset{n\to\infty}{\text{Lt}}\ [(z+2)^{n+1}/\angle n+1][\angle n!/(z+2)^n] = (z+2)/(n+1)$

$\text{Lt}\ \text{as}_{n\to\infty}\ (z+2)/(n+1)$ is zero. i.e. < 1. So convergent for all z.

17. The differential equation for the variation of the amount of small x in a tank with time t is given by : $dx/dt + x/20 = 10$. x is in kg and t is in minutes. Assuming that there is no salt in the tank intially the time (in minutes) at which the amount of salt increases to 100 kg is:

(a) 10 ln 2 (b) 20 ln 2

(c) 50 ln 2 (d) 100 ln 2 **Ans.** (a)

Sol. $dx/dt = 10 - x/20. = (200 - x)/20$; integrating, $20\int_0^{100} dx\Big/(200-x) = \int_0^t dt = t = 20\ln(1/200 - x)$ for x from 0 to 100 = 20 ln 2. **Ans.** (a)

18. The value of the integral $\int_{-2}^{+2} dx * 1/x^2\ is:$

(a) 0 (b) 0.25

(c) 1 (d) 2 **Ans.** (c)

Sol. + 2 dx/x^2. This is an even function. For even function.

$$\int_{-a}^{+a} f(x)dx = 2\int_0^a dx = 2\int_0^2 f(x)\,dx = 2\int_0^2 1/x^2\,dx = -2\,(-1/x)\,_0^2 = 1.$$

19. The differential equation:

$(dy/dx)^2 + y\,d^2y/dx^2 = 0$ can be reduced to (where $\propto$ is a constant)

(a) $(dy/dx)^3 = \propto - 3y^2/2$ (b) $(dy/dx)^2 = \propto - 2y$

(c) $dy/dx = \propto/y^2$ (d) $dy/dx = \propto/y$

Sol. Let $dy/dx = p$; Then $d^2y/dx^2 = dp/dx = dp/dy * dy/dx = p\,dp/dy$ inegrating $p^2 + py\,dp/dy = 0$; $p = -y\,dp/dy$; Inegrating, $\log p = -\log y + \log c$ where c is a constant. $p = c/y = \propto/y$; but $p = dy/dx$; so $dy/dx = \propto/y$ **Ans.** (d)

20. The value of $\underset{x \to 9}{\text{Lt.}} (\sqrt{x} - 3)/(x^2 - 81)$ is:

(a) 0 (b) 1/27

(c) 1/108 (d) 0 **Ans.** (c)

Sol. $\underset{x \to 9}{\text{Lt.}} (\sqrt{x} - 3)/(x-9)(x+9) = \underset{x \to 9}{\text{Lt.}} \dfrac{(\sqrt{x} - 3)}{(\sqrt{x} - 3)(\sqrt{x} + - 3)(x+9)}$

$\underset{x \to 3}{\text{Lt.}} \dfrac{1}{(\sqrt{x} + 3)(x+9)} = \text{Lt.} = 1/(3+3)(9+9) = 1/108$ **Ans.** (c)

21. The ratio $\int_{x_1}^{x_2} dx / y = (x2 - x1)/(y \; at \; x2)$ where y is monotonically increasing function of.

(a) Less than one (b) Equal to 1

(c) Greater than one (d) Less than zero **Ans.** (c)

Sol. $_{x1}^{x2}\int dx / y = x2 - x1 / y \; at \; x2.$ The ratio is $(x2 - x1)/y$ at $x2 = y$ at $x2/(x2 - x1)$

Since y at $x2$ is greater than $y\big|_{x1}^{x2}$; so that ratio is greater than one.

22. The dimensions of a cone are: radius 4 cm, height is 6 cm. What is the error in its volume if the scale is used in taking the measurement is short by 0.01 cm per cm.

Sol. V(Volume of a cone), $V = (1/3)\pi r^2 h$; $dv = 1/3\pi \; dh + \pi h/3 \; (2r \; dr)$

$dV/V = dh/h + 2dr/r$;

$dv/V = dh/h + 2dr/r$; $dV/v = 0.01 + 2 * 0.01 = 0.03$ or 3%. **Ans.**

❑

CHAPTER 19

Material Science and Engineering

SUMMARY

Avogadro's number = $N = 6.023 * 10^{23}$ per mole.
Boltzmann constant = $k = 1.380*10^{-23}$ J/K
Gas constant = $R = 8.34$ J/mole.K
Planck constant $h = 6.626 * 10^{-14}$ J.S.
Electronic charge = $e = 1.602 * 10^{-19}$ C
Electron mass = $9.109 * 10^{-31}$ kg.
Velocity of light = $C = 2.998 * 10^{8}$ m/sec.
Permeability of free space = $4\pi * 10^{-7}$ H/m
Faraday constant = $F = 96.49$ K C/mole.
Atomic mass unit (a m u) = $1.660 * 10^{-27}$ kg
Accelaration due to gravity = $g = 9.81$ m/sq.sec.

1. A space lattice is an infinite array of points, all with identical surroundings.
2. Space lattices are limited to fourteen, but known crystal structures run into thousands.
3. Crystal directions and crystal planes are denoted by Miller indices.
4. The Bragg law gives the condition for diffraction by a crystal.
5. The powder method is the commonly used X-ray diffraction technique.
6. A crystal structure is obtained by combining a space lattice with a basis.
7. The atom consists of a nucleus and surrounding electrons. Each electron occupies a quantum state with a unique set of quantum numbers.
8. The order of occupation of quantum states by electrons is determined by Pauli exclusion principle, Hund's rule and minimum energy criterion.
9. Ionisation potential is the energy required to remove an electron from the outer orbital of an atom.
10. Bond energy is the magnitude of energy released when two atoms come together from large distance of separation to equilibrium distance.
11. Primary bond energies are in the range of 100 to 1000 KJ/mole. Secondary bond energies are 1-50 KJ/mole.

12. Sharing of electrons between neighbouring atoms is covalent bond.

13. A metal is an array of positive ions which are held tigether in a cloud of free electrons.

14. Mellting and boiling points of materials- increase with increasing bond length. Strong and directional bonds result in hard and brittle solids. Free electrons are responsible for the high thermal and electrical conductivities of metals.

15. More than half of the elements are metals with simple crystal structure such as FCC, HCP, BCC.

16. Polymers are classified as thermoplasts and thermosets. The thermoplast, comprise plastics, fibres and elastomers.

17. Crystal imperfections are : point imperfections,edge and screw dislocations.

18. The phase rule gives the degrees of freedom that are permissible for a system of a given number of components and phases in equilibrium.

19. Brasses are more economical and have better mechanical properties than copper.

20. Fick's law is : $N_A = -D\nabla C_A$

21. Fick's 2nd law describes the unsteady state flow.

22. Rubber like elasticity arises from entropy changes during streaching when the coiled long chain molecules uncoil.

23. The majority of failures of structural and machine components are : fatigue.

24. The formation of an adherent oxide films is necessary to improve the oxidation resistence of a metal.

25. The common alloying elements added to steel for improving its oxidation resistance are : chromium, aluminium, and nickel.

26. The standard electrode potential measures the tendency of a metal to go into solution.

27. Passivation refers to the formation of a thin high resistance oxide layer which reduces the corrosion current by two or three orders of magnitude.

28. Corrosion prevention methods : (a) Avoiding contact between dissimilar metals; (b) Inhibitors; (c) Metallic and non-metallic coatings; (d) Cathode protection; (e) Passivation.

29. The Fermi level is defined as that level which has 50% probability of occupation by an electron.

30. Solids are classified as : Metals, semiconductors, insulators, according to their band structure. Insulators and semiconductors have forbidden gap.

31. A p–n junction acts as a rectifier, passing current in one direction only n-p junction is a transistor, amplifies electric signals.

32. The charge stored in a capacitor increases if a dielectric material is inserted between the capacitor plates. This is due to polarisation in the dielectric.

33. Polarization processes are : electronic, ionic, orientation, space charge polarization.

34. The breakdown of a dielectric material : Impurities, surface flaws (cracks and pores) absorption of moisture and gases on surface.

35. Ferroelectric crystals have large dielectric constants.

QUESTION AND ANSWERS

1. A space lattice with two lattice parameters belong to the crystal system.
(a) Tetragonal (b) Rhombhohedral
(c) Hexagonal (d) Triclinic. **Ans.** (a), (b), (c)

2. The Miller indices of the fifth reflection in an FCC crystal is :
(a) 331 (b) 222 (c) 311 (d) 400 **Ans.** (b)

3. The electron affinity of He in KJ/mole is :
(a) 369 (b) 704 (c) 86 (d) 0 **Ans.** (d)

4. Primary bonds have energy range of KJ/mole :
(a) 1000 – 5000 (b) 100 – 1000 (c) 10 – 100 (d) 1 – 10 **Ans.** (b)

5. Thermal expansion of material arise from :
(a) Strong bonds (b) Weak bonds
(c) Thermal vibrations (d) Aasymmetry of potential energy curve
Ans. (a)

6. Hydrogen bonds are stronger than :
(a) Van der Waal's bonds (b) Metallic bonds
(c) Ionic bonds (d) Covalent bonds **Ans.** (a)

7. Tick the close packing arrangements in the following :
(a) AB AB AB AB (b) ABC ABC ABC A
(c) AB ABC BC ABC (d) ACCC BCA BCA BC **Ans.** (a), (b), (c)

8. The factors that promote non-crystallinity in polymers are :
(a) Large random side groups (b) Branching
(c) Addition of plasticisers (d) Copolymerisation **Ans.** (a), (b), (c), (d)

9. Zone refining is more efficient, if the rate of impurity in the solid to that in liquid is :
(a) 0.001 (b) 0–0.1 (c) 0–0.4 (d) 0–1.0
Ans. (a)

10. For a spherical particle of Radius r, the volume to suface area is :
(a) $3/r$ (b) $r/3$ (c) $3r$ (d) $r/3$ **Ans.** (b)

11. Fine grain sizes are obtained by :
(a) Slow cooling (b) Increasing nucleation rate
(c) Decreasing growth rate (d) Fast cooling **Ans.** (b), (c), (d)

12. During overaging hardness :
(a) Decreases (b) Increases
(c) Is constant (d) Increases abruptly **Ans.** (a)

13. High elastic modulus in materials arises from :
(a) High strength of bonds
(b) Weak bonds
(c) Sharp curvature at the minimum potential energy
(d) Shallow potential well. **Ans.** (a), (c)

14. Covalent and Ionic solids are not suitable as structural components, because :
(a) They have weak bonds (b) They are ductile
(c) They are brittle (d) They have high elastic modulus **Ans.** (c)

15. On heating an elastomer under tensile load, it shrinks :
(a) To maximise the enthalpy (b) To maximise the entropy
(c) To maximise the free energy (d) To avoid breaking **Ans.** (b), (c)

16. On heating a rubber under tensileforce it :
(a) Shrinks (b) Expands
(c) Expands rapidly (d) Shows no change **Ans.** (a)

17. In BCC iron, residual carbon occupies the mid points of :
(a) 100 (b) 110 (c) 111 (d) none **Ans.** (a)

18. In tensile test, the Engineering stress corresponding to the maximum load is called :
(a) Yield strength (b) Tensile strength
(c) UTS (d) Upper yield stress **Ans.** (b), (c)

19. Crystals like diamond and silicon are brittle, beccause :
(a) They contain no dislocations (b) They are non crystalline
(c) The stress to move a dislocation is high in them
(d) They contain very few dislocations **Ans.** (c)

20. In Griffith equation, the fraction stress is proportional to :
(a) $\sqrt{c}$ (b) $1/\sqrt{c}$ (c) 1/c (d) 1/2c **Ans.** (b)

21. Tick the elements added to iron to improve its oxidation resistance:
(a) Zinc (b) Chromium
(c) Magnesium (d) Aluminium **Ans.** (b), (d)

22. During melting the relative dielectric constant :
(a) Always increases (b) Always decreases
(c) May increase or decrease **Ans.** (c)

23. Pure silicon at 0 K is :
(a) Intrinsic semiconductor (b) Extrinsic semiconductor
(c) Metal (d) Insulator **Ans.** (d)

24. The two common doping elements in silicon are :
(a) Phosphorous (P) (b) Boran (B)
(c) Sb (d) Bi (bismuth) **Ans.** (a), (b)

25. To increase mechanical strength of aluminium conductor we use ;
(a) Solute strengthing (b) Cold working
(c) Doping (d) Steel reinforcement **Ans.** (d)

26. The unit of electrical conductivity is :
(a) ohm.m (b) $mho^{-1}\ m^{-1}$
(c) $m^{-3}\ C^{-2}$ (d) $ohm^{-1}\ m^{-1}$ **Ans.** (d)

27. The magnetisation of super conductor is :
(a) 0 (b) – B
(c) – C (d) – H **Ans.** (d)

❑

CHAPTER 20

Corrosion Engineering

SUMMARY/BACKGROUND

1. **Corrosion Engineering:** is the application of science and art to prevent or control corrosion damage economically and safely.
2. **Definition of corosion:** Corrosion is defined in several ways.
   (a) Destruction or deterioration of a material because of reaction with environment.
   (b) Destruction of materials by means of other than straight mechanical.
   (c) Extractive metallurgy in reverse.
3. **Corrosion classification:** (a) Wet corrosion; (b) Dry corrosion. Example of wet corrosion is corrosion of steel by water. Example of dry corrosion is attack on steel by furnace gases.
4. **Weight loss of metal specimen during corosion test is:**
   $mpy = \frac{534w}{DAT}$; w = weight loss, mg; D = Density of specimen gms/cm^3; A = Area of specimen, sq. inches; T = exposive time, hr.
5. **Electrochemical aspects:**
   $Zn + 2HCl \rightarrow Zn\,Cl_2 + H_2$ ....(1)
   $Zn + 2H^+ \rightarrow Zn^2 + H_2 \rightarrow 2;$ ....(2)
   Eqn. (2) can be written as:
   $Zn \rightarrow Zn^{2+} + 2e$ (oxidation, anodic reaction) ....(3)
   $2H^+ + 2e \rightarrow H_2$ (reduction, cathodic reaction) ....(4)
   During metallic corrosion, the rate of oxidation equals the rate of reduction. Any reaction which can be divided as oxidation reaction and reduction reaction is electrochemical reaction. Polarization is: 1. activation polarization; 2. Concentration polarization (Refers to electrochemical reactions which are controlled by diffusion in the electrolyte). Active polarization is controlled by reaction sequence at the metal electrolyte interface.
   Passivity refers to the loss of chemical reactivity expressed by cetain metals and alloys under particular environmental conditions.
6. **Environmental aspects:**
   Temperature increases the rate of chemical reactions. Oxygen and oxidisers increase corrosion rate. Velocity also increases corrosion rate. In some materials like lead, corrosion rate increases with corrosive concentration (sulphuric acid). For some materials, the corrosion rate increases

to a maximum with increasing concentration of corrosive and then falls or decreases. The effect of galvanic coupling is identical to that of adding an oxidiser to a corrosive solution.

7. **Metallurgical aspects:**

Effect of purity of aluminium on corrosion by hydrochloric acid.

% Aluminium	Relative corrosion rate
99.998	1
99.97	1000
99.2	30,000

Ring worm corrosion: During hot forming or upsetting operation steel is subjected to large temperature gradients. This thermal gradient causes variations in metallurgical structure along the pipe in the section exposed to intermediate temperatures, the iron carbides coalesce forming spheroids. During service exposure, this narrow section corrodes preferentially. This is Ringworm corrosion.

8. **Galvanic corrosion:** Because of electric currents and dissimilar metals involved this form of corrosion is called as galvanic or two metal corosion.

Prevention: (a) Select the combination of metals as close as posible in the galvanic series; (b) Avoid large cathode and small anode; (c) Insulate dissimilr metals; (d) Add inhilitors; (e) Apply coatings with caution; (f) Avoid threaded joints; cathodic protection: The protection of a metal structure in making it the cathode of the galvanic cell. Galvanized (zinc coated) steel is an example of cathodic protection of steel.

9. **Crevice corosion (deposit or gasket corrosion):**

Corrosion in shielded areas, within crevices on metal surfaces exposed to corrosives:

Remedies: 1. Use welded butt points; 2. Close crevices; 3. Complete drawings facilities; 4. Remove deposits frequently; 5. Remove solids in suspension; 6. Remove wet packing metarials; 7. Provide uniform environment; 8. Use non-absorbent gaskets like teflon; 9. Weld in sheets instead of tubes.

10. **Fitting corrosion:** This is a localized attack that results in holes in the metal. These holes are small.

11. **Intergranular corrosion:** This is caused by impurities at the grain boundaries. Depletion of chromium in the grain boundary regions results in intergranular corrosion of stainess steels.

12. **Erosion corrosion:** This is the acceleration or increase in rate of deterioration or attack on a metal because of relative movement between a corrosive fluid and the metal surface.

13. **Stress corosion:** Refers to cracking caused by the simultaneous presence of tensile stress and a specific corrosive medium.

Remedies: 1. Lowering the stress below the threshold valve; 2. Eliminating the critical environmental species; 3. Changing the alloy; 4. Applying cathodic potection; 5. Adding inhibitor to the alloys.

14. **Corrosion prevention methods:**

Materials selection: 1. Using alloys. 2. Using purified metals. 3. Using carbon and graphite; wood; linings or coverings with rubber; plastics, ceramics, carbon and graphite and wood. 4. Alteration of environment: (a) Changing mediums; (b) Lowering temperature; (c) Decreasing velocity; (d) Removing oxygen or oxidizer; (e) Changing concentration of material which gives corrosion; (f) Inhibitors: An inhibitor is a substance which when added in small concentrations to an environment, decreases corrosion rate. Adsorption type inhibitors: These are organic compounds which adsorb on metal surface and supress metal dissolution and Reduction Reaction.

(g) Hydrogen evolution poisons: Substances such as arsenic and antimony ions specifically retard the hydrogen evolution reaction. (h) Scavengers: These substances act by removing corrosive reagents from solution. Examples of this type of inhibitor are sodium sulphite and hydrazine which remove dissolved oxygen from aqueous solutions. (i) oxidizers: Substances as chromate, nitrate and ferric salts also act as inhibitors in many systems.

15. Designs to reduce/avoid corrosion:

(a) Wall thickness: Since corrosion is a penetrating action, it is necessary to make allowances for this reduction in thickness in designing pipes, tanks and other components.

(b) Design rules: Weld rather than rivet tanks and other containers; Design tanks and other container for easy draining and easy cleaning; design for easy replacement of components that are expected to fail rapidly in service. Avoid excessive mechanical stresses and concentrations in components exposed to corrosive medium; Avoid electrical contact between dissimilar metals to avoid galvanic corrosion; Avoid sharp bends etc for fluid direction changes to avoid corrosion; Avoid hot spots to avoid stress corrosion; Design to exclude air (oxygen reduction is one of the most common calhodic reactions); Design to avoid heterogenity in design attempt to make all conditions as uniform as possible throughout the entire system).

16. Cathodic protection:

$M \rightarrow M^+ + ne$; $2H^+ + e \rightarrow H_2$

Cathodic protection is achieved by supplying electrons to the metal structure to be protected. The two methods adapted in cathodic protection are: 1. By an external power supply. 2. By appropriate galvanic coupling.

Examples are: 1. Cathodic protection of an underground tank using impressed currents; 2. Cathodic protection of a domestic hot water tank using a sacrificial anode; 3. Protection of an underground pipeline with a magnesium anode.

17. Anodic protection: Anodic protection is based on the formation of a protective film on metals by externally applied anodic currents.

18. Coatings:

1. Thin coatings of metallic and inorganic materials can provide a satisfactory barrier between metal and its environment. The chief function of such coating is to provide as effective barrier. Examples of metal coated articles automobile bumpers and trim, household appliances.
2. Electrodeposition: This process called as electroplating consists of immersing a part to be coated in a solution of the metal to be placed and passing a direct current between the part and another elelctrode.
3. Flame spraying: This process also called metallizing consists of feeding a metal wire or powder through melting frame to the metal, in fairly divided liquid particles, is blown on to the surface to be produced. Cladding: This involves a surface layer of sheet metal usually

put on by rolling two sheets of metal together. For example a nickel and steel sheet are hot rolled together to produce a composite sheet.

4. Cladding: This involves a surface layer of sheet metal usually put on by rolling two sheets of metal together.
5. Hot dipping: Hot dip coatings are applied to metal by immersing them in a molten metal bath of low melting point metals, chiefly zinc,lead, tin and aluminium.
6. Vapor deposition: This is accomplished in a high vacuum chamber. The coating metal is vaporized by heating electrically and the vapor deposits on the parts to be coated.
7. Diffusion: Diffusion coatings involve heat treatment to cause alloy formation by dilution of one metal into the other. This is also termed as surface alloying.
8. Chemical conversion: Coatings from chemical conversion are produced by corroding the metal surface to form and adhere and protective corrosion producet.
9. Organic coatings: These involve retalively thin barrier between substrate and the environment. Paints, varnishes lacquers and similar coatings protect metal on a tonnage basis.

19. Corrosion rate measurements: (a) Total extrapolation method; (b) Linear polarization technique.

20. Thermodynamics in corrosion: In any electro-chemical reaction the most negative or active half cell tends to be oxidised and the most positive or noble half cell tends to be reduced. There is a definite relation between the free energy change and the cell potential of an electrochemical reaction. In most instances, the actual magnitude of the free energy change is relatively un important in corrosion application. The sign of the free energy change for a given reaction indicates whether or not the reaction is spontaneous.

21. Passivity: This is an unusual phenomena observed during the corrosion of certain metals and alloyus. It can be defined as a loss of chemical reactivity under certain environmental conditions.

22. Activation polarization: This refers to electrochemical reactions which are controlled by a slow step in the reaction sequence.

23. Concentration polarization: At low reduction rates the distribution of hydrogen ions in the solution adjacent to electrode surface is relatively uniform. At very high reduction rates, the region adjacent to the electrode surface will become depleted of hydrogen ions. If the reduction rate is increased further, a limiting rate will be reached which is determined by the diffusion rate of hydrogen ions to the electrode surface.

both activation polarization and concentration polarization usually occur at an electrode. At low reaction rates, activation polarization controls, while at high reaction rates concentration polarization becomes controlling. The total polarisation of an electrode is the sum of the contributions of activation polarization and concentration polarization.

❑

CHAPTER 21

Petroleum Engineering

(Multiple Choice Question's)

1. Petroleum geology refers to the specific set of geological disciplines that are applied to the search for hydrocarbons.
2. Oil and gas fields are geological features that result from the coincident occurrence of four types of geologic features; (1) oil and gas source rocks, (2) migration, (3) reservoir rocks, (4) seals and traps.
3. Source rocks refer to rocks from which hydrocarbons have been generated or are capable of being generated.
4. Source rocks form one of the necessary elements of a working petroleum system. They are organic rich sediments that may have been deposited in a variety of environments including deep water marine, lacustrine and deltaic.
5. Oil shale can be regarded as an organic rich but immature source rock from which little or no oil has been generated and expelled.
6. When temperatures of the organic rich sedimentary rocks exceed 120°C (250°F) the organic remains within the rocks begin to be "cooked" and oil and natural gas are formed from the organic remains and expelled from the source rock. It takes millions of years for these source rocks to be buried deeply enough to attain these maturation temperatures and additional millions of years to cook (or generate) sufficient volumes of oil and natural gas to form commercial accumulations as the oil and gas are expelled from the source rock into adjacent reservoir rocks.
7. If the organic materials within the source rock are mostly wood fragments, then the primary hydrocarbons generated upon maturation are natural gas.
8. If the organic materials are mostly algae or the soft parts of land plants, then both oil and natural gas are formed.
9. Gas can be generated in two ways in the natural systems; directly from woody organic matter in the source rocks or it can be derived by thermal breakdown of previously generated oils at high temperatures.
10. Oil window: Oil maturation begins at 120°F (50°C) peaks at 190°F (90°C) and ends at 350°F (175°C).
11. Above and below oil window, natural gas is generated.
12. At higher temperatures above 500°F (260°C), the organic material is carbonized and destroyed as a source material. So, if source beds become too deeply buried, no hydrocarbons will be produced.
13. Migration is the process of the oil and gas moving away from the source rock. This is a slow process.
14. Migration is caused by burial, compaction, and increase in volume and separation of the source rock constituents.
15. Primary migration is the process of movement from source rock.
16. Secondary migration is movement to or within the reservoir entrapment. Once the water, oil

and gas migrate into the trap, it separates according to density. Gas being the lightest, goes to the top of the trap to form the free gas cap. Oil goes to the middle and water that is always present, on the bottom.

17. Reservoir rock is a rock that contains connected pore spaces used to reserve the fluid inside it. Most reservoir rocks are initially saturated with saline groundwater.
18. Seal (Cap rock) is a relatively impermeable rock that forms a barrier, cap or seal above and around reservoir rock so that fluids cannot migrate beyond the reservoir. The permeability of a cap rock must equal zero. Some examples are shales, evaporites such as anhydrite and salt, & zero-porosity carbonates.
19. Traps are configuration of rocks suitable for containing hydrocarbons and sealed by a relatively impermeable formation through which hydrocarbons will not migrate.
20. Structural Traps: It is formed where the space of petroleum is limited by a structural feature.
21. Anticline traps are the trap whose closure is controlled by the presence of an anticline.
22. Fault traps are the traps in which closure is controlled by the presence of at least one fault surface.
23. Stratigraphic traps are the traps created by the limits of reservoir rock itself, without any structural control. They are formed by changes in rock type or pinch-outs, unconformities, or sedimentary features such as reefs.
24. Pinch-out trap: The termination by thinning or tapering out ("pinching out") of a reservoir against a nonporous sealing rock creates a favorable geometry to trap hydrocarbons.
25. Unconformity trap: Trap whose closure is controlled by the presence of an unconformity.
26. Reef trap: Sedimentary rock, most commonly produced by organisms that secrete shells such as corals. Because the rocks that surround reefs can differ in composition and permeability, porous reefs can form stratigraphic traps for hydrocarbons.
27. Combination traps: It is a combination trap is where two (or more) trapping mechanisms come together to create the trap.
28. Body waves: A wave that propagates through a medium rather than along an interface. It is faster than surface waves.
29. P-wave: An elastic body wave in which particles motions are parallel to the direction the wave propagates. It`s velocity is faster than S-wave. P-waves incident on an interface at other than normal incidence can produce reflected and transmitted S-waves, in that case known as converted waves.
30. S-wave: An elastic body wave in which particles motions are perpendicular to the direction the wave propagates. S-waves are generated by most land seismic sources, but not by air guns.
31. Surface waves: A wave that propagates at the interface between two media.
32. Rayleigh wave: It is a surface wave in which particles move in an elliptical path. Because Rayleigh waves are dispersive, with different wavelengths traveling at different velocities, they are useful in evaluation of velocity variation with depth.
33. Love wave: It is a surface wave in which particles oscillate horizontally and perpendicularly to the direction of wave propagation.
34. Stoneley wave: It is a surface wave generated by a sonic tool in a borehole. It can propagate along a solid-fluid interface, such as along the walls of a fluid-filled borehole. It can allow estimation of

the locations of fractures and permeability of the formation. It is a major source of noise in vertical seismic profiling data.

35. Tube waves: It occurs in cased wellbores when Rayleigh waves encounters a wellbore and perturbs the fluid in wellbore. It suffers little energy loss and very high amplitude which interferes with reflected arrivals occurring later in time on vertical seismic profile (VSP) data.

36. Seismic anisotropy is the variation of seismic velocity with the direction or with wave polarization.

37. Transverse isotropy is the most common and important type of anisotropy in seismic studies.

38. Mediums affect on seismic waves: 1-Geometrical spreading: The energy intensity decreases when wave front gets farther from the source; 2-Absorption: Transformation of elastic energy to heat as seismic wave passes through a medium, causes amplitude to decrease; 3-Dispersion: is dependence of seismic velocity on the frequency (is negligible for body waves but very important for surface waves); 4-Interface related effects: When a wave finds an abrupt change in elastic properties, some of energy reflected and some of energy refracted.

39. Seismic noises: It is anything other than desired signal. Noise includes disturbances in seismic data caused by any unwanted seismic energy.

40. Random noise: Random on all traces and includes wind, rain, human and machines (Environment noise).

41. Coherent noise: Include surface waves, refractions, diffractions and multiples.

42. Seismic trace: It is the seismic data recorded for one channel. A seismic trace represents the response of the elastic wave field to velocity and density contrasts across interfaces of layers of rock or sediments as energy travels from a source through the subsurface to a receiver or receiver array.

43. Seismograph: It is the instrument that measure motions of the ground, including those of seismic waves generated by earthquakes, nuclear explosions, and other seismic sources.

44. Seismogram: It is a graph output by a seismograph. It is a record of the ground motion at a measuring station as a function of time.

45. Primary events (Reflections) : It is generated by waves that have been reflected from interface. It carries useful information about velocity structure. Anything else but reflections considered as unwanted noise.

46. Non-primary events (Direct waves) (T-X curve is straight line with intercept = 0) : It is P-wave that travel directly from source to receiver along earth`ssurface. It arrives before the reflected wave of first layer. It is usually attenuated by muting and stacking.

47. Seismic sources:
   - Land sources :
      - o Impulsive sources: Which are divided to Explosive sources such as Dynamite (common in Petroleum exploration), and non-explosive such as weight drop and hammers (common in shallow seismic investigation);
      - o Non- impulsive sources: The main common is Vibroseis which is a designed vehicle lift its weight on large plate in contact with ground surface in sweeps.
      - o Up sweep: Frequency begins low and increase with time.
      - o Down sweep: Frequency begins high and decrease with time.

– Marine sources (Air gun): The common in offshore survey (first produced in 1960). This gun releases highly compressed air into water. It uses a compressed air at 2000- 5000 PSI to produce an explosive blast of air. Several air guns with different sizes are fired to enhance their initial pulses & reduce their bubble effects.

48. Seismic detectors:

– Land detectors (Geophone): It is a device is used to detect the sound waves. It consists of coil of wire suspended from spring and surrounded by (W) shaped magnet. Upward energy from seismic source is recorded as electrical current generated by movement of coil.

– Marine detectors (Hydrophone): It is a device used to detect the pressure waves. Upward energy is recorded as electrical current generated by piezoelectric device (which generates a voltage if acted with pressure).

49. Vertical Seismic Profiling (VSP): A class of borehole seismic measurements used for correlation with surface seismic data, for obtaining images of higher resolution than surface seismic images and for looking ahead of the drill bit; also called a VSP. Purely defined, VSP refers to measurements made in a vertical wellbore using geophones inside the wellbore and a source at the surface near the well. Most VSPs use a surface seismic source, which is commonly a vibrator on land and an air gun in offshore or marine environments.

50. Logging While Drilling (LWD): Instead of sensors being lowered into the well at the end of wireline cable, the sensors are integrated into the drill string and the measurements are made while the well is being drilled. While wireline well logging occurs after the drill string is removed from the well, LWD measures geological parameters while the well is being drilled. However, because there are no wires to the surface, data are recorded downhole and retrieved when the drill string is removed from the hole.

51. Measuring While Drilling (MWD): It is the evaluation of physical properties, usually including pressure, temperature and wellbore trajectory in three-dimensional space, while extending a wellbore. MWD is now standard practice in offshore directional wells. The measurements are made downhole, stored in solid-state memory for some time and later transmitted to the surface.

52. Logging Tools: These are electronic devices that records data over depth. The tool is attached to the end of wireline cable and lowered to the borehole.

53. Cartridge: The section of a wireline logging tool that contains the telemetry, the electronics and power supplies for the measurement, as distinct from the sonde that contains the measurement sensors.

54. Caliper Tool: The Caliper Tool is a three-armed armed device that measures the internal diameter (I.D.) of casing or open borehole completions.

55. The sonic log is used to evaluate porosity in liquid filled pores. The tool is only capable of measuring travel time.

56. Density Tool: These tools have a source emitting gamma rays, a short-spaced and a long-spaced detector. Gamma ray enters the formation, then scatters and loses some of its energy then get absorbed by a formation. Then, the detectors detect γ ray which emitted from excited atoms which related to the formation.

57. The neutron log is sensitive mainly to the amount of hydrogen atoms in a formation. Its main use is in the determination of the porosity of a formation. The tool operates by bombarding the

formation with high energy neutrons. These neutrons undergo scattering in the formation, losing energy and producing high energy gamma rays. The scattering reactions occur most efficiently with hydrogen atoms. The resulting low energy neutrons or gamma rays can be detected, and their count rate is related to the amount of hydrogen atoms in the formation.

58. The gamma ray log measures the total natural gamma radiation emanating from a formation. This gamma radiation originates from potassium-40 and the isotopes of the Uranium-Radium and Thorium series

59. Resistivity logging is a method of well logging that works by characterizing the rock or sediment in a borehole by measuring its electrical resistivity.

60. The spontaneous potential log (SP) measures the natural or spontaneous potential difference (sometimes called self-potential) that exists between the borehole and the surface in the absence of any artificially applied current.

61. Porosity (φ) is defined as a percentage or fraction of void to the bulk volume of a material.

62. Porosity of commercial reservoirs may range from about 5% to about 30% of bulk volume.

63. Primary porosity is defined as a porosity in a rock due to sedimentation process.

64. Secondary porosity is defined as a porosity in a rock which happen after sedimentation process, for example fracturing and re-crystallization.

65. The permeability of a rock is a measure of the ease with which fluids can flow through a rock. This depends on how well the pore spaces within that rock are interconnected.

66. The unit for permeability (k) is darcy named after a French scientist, Henry Philibert Gaspard Darcy who investigated flow of water through filter beds in 1856.

67. One darcy is defined as the permeability that will permit a fluid of 1 centipoise viscosity to flow at a rate of 1 cubic centimeter per second through a cross sectional area of 1 square centimeter when the pressure gradient is 1 atmosphere per centimeter.

68. Capillary pressure is the pressure difference existing across the interface separating two immiscible fluids.

69. Oil reservoir contains mainly oil with or without free gas (gas cap).

70. Undersaturated Oil Reservoir ($P_{res} > P_b$) - no free gas exists until the reservoir pressure falls below the bubble point pressure.

71. Saturated Oil Reservoir ($P_{res} < P_b$) - free gas (gas cap) exists in the reservoir.

72. Primary Recovery: The recovery of hydrocarbons from the reservoir using the natural energy of the reservoir as a drive.

73. Secondary Recovery: The recovery aided or driven by the injection of water or gas from the surface.

74. Drive Mechanism: The natural energy of the reservoir used to transport hydrocarbons towards and out of the production wells.

75. There are five important drive mechanisms (or combinations).; Solution Gas Drive, Gas Cap Drive, Water Drive, Gravity Drainage., Combination or Mixed Drive.

76. Displacement efficiency is that fraction of original oil saturation that has been displaced from the pores by water.

77. Contact factor is that fraction of the swept volume that has been reached, or contacted by the injection fluid.

78. Lowering the viscosity of the reservoir oil will lower the mobility ratio and improve sweep efficiency.
79. Enhanced recovery methods which are useful in improving sweep efficiency are polymer flooding and thermal methods. Polymers increase injected water viscosity. Thermal methods apply heat to the reservoir and lower viscosities of heavy crude oils.
80. Any operation which increases the productivity of a well is called well stimulation.
81. In acidizing, acid is used to remove damage near the well bore in all types of well.
82. Acid washing is an operation designed to remove acid soluble scales present in the well bore or to open perforation. It my involve nothing more than spotting a small quantity of acid at the desired position in the well bore and allowing it to react, without any external agitation, with scale deposits on the formation.
83. If the injection rates are below fracture pressure it is termed as Matrix acidizing.
84. Matrix acidizing is applied primarily to remove skin damage caused by drilling, completion, work over and well killing fluids
85. In sandstone acidizing, a mixture of HCl and HF are used. It is called Mud acid. Other additives in mud acid are : (a) Corrosion Inhibitors; (b) Intensifier; (c) Stabilizers; and (d) Sequestering agents.
86. In carbonate acidizing only HCl is used. HF cannot be used in the stimulation of carbonate reservoirs because formation of insoluble reaction product which results in severe pore plugging
87. If the injection rates are above fracture pressure it is termed as acid fracturing. It is not applicable to sandstone wells
88. Acid fracturing is the most widely used acidizing technique for stimulating limestone or dolomite formations.
89. In this method, high viscosity fluid (fracturing fluid) is pumped into the well at a sufficient pressure to cause tensile failure of the formation. The resulting fracture grows in width and length as the pumping continued. The frac. fluid required to initiate the fracture is called the pad. When the frac. fluid is sufficiently wide, sand is blended into the frac fluid. The resulting sand/ fluid mixture is called the slurry. At the end of treatment the fracture is substantially filled with sand. When the pumping is stopped and the pressure is relieved. The sand props the fracture open resulting in a highly permeable sheet extending a large distance out into the reservoir.
90. In nitro-shooting, the explosives are lowered against the perforated interval and then exploded. The explosive particularly nitroglycerine (a liquid) is converted into a gaseous state at such a speed that tremendous , energy is created and communicated to the surrounding rocks in vibrating waves of great intensity and at a very high velocity as the result of which the formation is fractured.
91. Solid propellant fracturing produces multiple fractures.
92. The solid-propellant fracturing tool generates high-pressure gases at a rate that creates fractures dramatically different from either high explosives or hydraulic fracturing.
93. In dilatant technology, the rock-volume is permanently changed through the use of explosives. Porosity can be increased 200% or more as a result of controlled microfracturing/cracking.
94. In this method a cylindrical shaped powder charge is lowered to the appropriate depth in the wellbore via wireline or tubing. Upon receiving an electrical signal, the generator ignites almost instantaneously along its length, creating high pressure and temperature. This is a deflagration process and not an explosion. The powder charge disintegrates completely and the well is ready to be returned to production within hours.

95. The most common methods in removing paraffins from wells are: (1) Mechanical; (2) Solvent; (3) Heat; and (4) Dispersants.
96. Surfactants can bring about the following changes in reservoir fluids and reservoir rocks: (a) Raise or lower surface and interfacial tension (b) Make, break, weaken or strengthen an emulsion (c) Change the wettability of reservoir rocks and casing tubing or flow lines and (d) Disperse or flocculate clay or other fines.
97. Artificial lift system involves the use of artificial means to raise oil to the surface through a well after reservoir pressure has declined to the point at which the well no longer produces by means of natural energy.
98. It is oldest and the most commonly used method. It comprises o a bottom hole pump with a plunger and cylinder. The plunger is moved up and down by a sucker rod string which is attached to the surface unit. The surface unit provides reciprocating motion to the plunger through the sucker rod string which pumps the oil up.
99. Electrical Submersible Pumping Unit comprises a subsurface unit comprising of a multi-stage centrifugal pump and an electrical motor.
100. Hydraulic pumping has a down hole pumping unit which has a reciprocating pump and engine. The down hole engine is driven by high pressure power fluid which is pumped from surface through the tubing. The power fluid drives the down hole engine which in turn drives the pump which displaces the oil upwards normally through the annulus. A surface pump is used to pump the power fluid which is prepared in an elaborate surface system.
101. In continuous gas lift continuous volume of high pressure gas is introduced into the tubing to aerate or lighten the fluid column until reduction of the bottom hole pressure allows a sufficient differential across the sand face, causing the well to produce at the desired rate of flow.
102. The intermittent lift, gas is injected at regular intervals by the intermitter (a motor valve operated by a connecting timing device that permits selective cycling with controlled gas injection into the casing annulus). The cycling is regulated to coincide with fluid fill- in rate from the producing formation into the well bore.
103. Klinkenberg postulated that liquids had a zero velocity at the sand grain surface while gases exhibited some finite velocity at the sand grain surface. And this slippage at the sand grain surface has resulted in higher flow rate for the gas at a given pressure differential.
104. Fluid saturation is defined as the fraction of pore volume occupied by a particular fluid.
105. Wettability is defined as the tendency of one fluid to adhere or spread on a solid surface in presence of other immiscible fluids.
106. Absolute permeability relates to permeability with one fluid present at 100 percent saturation. It is also called as specific permeability or base permeability.
107. Effective permeability is the permeability to a given phase when more than one phase saturates the porous medium. The effective permeability is a function of saturation.
108. Relative permeability to a given phase is defined as the ratio of effective permeability to the absolute or, in some cases, a base permeability. Relative permeability is also a function of saturation.
109. For an oil-water reservoir, the base permeability, k is taken as effective permeability to oil at irreducible water saturation. For a gas reservoir, the base permeability will be effective permeability to gas in the presence of irreducible water.

110. Perforating is used to provide the basic fluid communication between the reservoir and the well bore.
111. Ported and scalloped guns are hollow carrier guns and are relatively heavy-walled, sealed tubular conduits that are used to convey the explosive train inside them, and to protect the explosives from the surrounding well bore environment.
112. The primary purpose of under balance perforating is to minimize perforation damage.
113. Retrievable gun: Consisting of a cylindrical, hollow steel charge carrier.
114. Semi-expendable gun: the charges are conveyed into the well on a retrievable metal strip or wire carrier (this is used in through-tubing operations where gun size is an important factor) .
115. Fully expendable gun: the cases and carrier linkage disintegrate and only the wireline is retrieved.
116. Novel drilling system where the annulus pressure at the seafloor is reduced to near seawater HSP is called dual gradient drilling.
117. Tripping operation is running in or pulling out of tubing or both in or out of the well. Pulling out of all tubing along with BHA from the well and running back in the well with old serviced or new BHA completely is termed as a "Making a complete trip".
118. Reverse circulation means to pump kill fluid through annulus and gets returns from tubing.
119. Bull heading method consists of pumping kill fluid to the well and forcing the well fluids back into the formation without pumping at a rate which will fracture the formation.
120. Tripping operation is running in or pulling out of tubing or both in or out of the well. Pulling out of all tubing along with BHA from the well and running back in the well with old serviced or new BHA completely is termed as a "Making a complete trip".
121. While pulling out a pipe in well, swab force is dominant. Swabbing is caused by the inability of the circulating fluid to fall past the work string as fast as the string can be pulled. Swabbing create a suction effect near well bore and thus induce flow of formation fluid into well bore even if well is full of correct fluid weight. Swab pressure is negative and reduces the effective hydrostatic pressure throughout the well below the tubing shoe. If this pressure reduction lowers the effective hydrostatic pressure below the formation pressure, a potential kick is developed.
122. Fishing" refers to the application of tools, equipment and techniques for the removal of junk, debris or fish from a wellbore.
123. Coiled tubing units are second generation of workover rig with hydraulic system for well servicing under pressure.
124. Perforating is the process of establishing a flow path between the reservoir and the wellbore. Perforations must penetrate the casing, cement and formation.
125. Acid diversion is a major problem in matrix acidizing as the acid tends flow where the permeability is the highest, leaving the low permeability regions of rock untreated.
126. The primary function of a packer is to create a seal between the production tubing and the casing.
127. A sliding side door or SSD is designed as a means of providing communication between the tubing string and the annulus.
128. Blast Joint is used mainly in multiple string completions. It is an extra thick joint of tubing designed to withstand the external erosion forces or blast from adjacent perforations.
129. Flow Couplings are placed above and below safety valves or restrictions within the tubing. They

are extra thickshort joints of tubing to withstand internal erosion created by turbulent flow from restrictions.

130. Fluid flow in porous media is governed by the diffusivity equation.

131. The diffusivity equation can be derived by combining the law of conservation of mass, Darcy's law and an equation of state.

132. When a well is opened at surface, the first flow at the wellhead is due to the expansion of wellbore fluid alone. This expansion continues after the reservoir fluid starts to contribute to the production, until the sandfaceflowrate equals the surface flowrate (when expressed at the same conditions) This effect is called wellbore storage.

133. Nodal analysis is the application of systems analysis to the complete well system from the outer boundary of the reservoir to the sand face, across the perforations and completion section to thetubing intake, up the tubing string including any restrictions and downhole safety valves, the surface choke, the flow line and separator.

134. It uses a combination of well inflow performance, downhole multipurpose flow conduit performance (vertical or directional conduit performance) and surface performance (including choke, horizontal or inclined flow performance and separator).

135. Spherical flow occurs when the flow streamlines converge to a point. This flow regime occurs in partially completed wells and partially penetrated formations. For the case of partial completion or partial penetration near the upper or lowerbed boundary, the nearest impermeable bed imposes a hemispherical flow regime.

136. Steady state implies that pressure in the well drainage volume does not vary in time at anypoint and that the pressure gradient between any two points in the reservoir is constant.

137. Whereas a build-up test is performed by closing a valve (shut-in) on a producing well, a drawdown test is performed by putting a well into production.

138. In an isochronal test, the well is flowed at four or more differentrates for periods of equal duration. Between flow periods, the well is shut-in until static conditions are reached. The last flowing period is extended until stabilized flowing conditions are reached.

139. In practice, the true isochronal test is usually replaced by a modified test sequence with flow and shut-in periods of equal duration. The modified isochronal test is faster because it is not necessary to wait for stabilization.

140. All fluids used in the wellbore during drilling operations are called drilling fluids (muds).

141. Water based mud consist of water, clay, weight material and chemicals which influence and control the behavior of clays (colloidal materials).

142. Oil emulsion muds are oil-in-water type emulsion, in which small droplets of oil are dispersed in the continuous phase water.

143. In water-in-oil emulsion muds, oil is the continuous phase and water is in the form of small droplets is the dispersed phase. This is also called inverted emulsion mud.

144. Lost circulation materials are used in the mud when losses of mud occur in the cracks of formation exposed in the well bore.

145. Most of the drilling fluids exhibit thixotropic behavior. It develops a gel like structure during quiescence which reverts to a liquid suspension upon agitation.

146. Fan VG viscometer is used for the measurement of viscosity. The apparent viscosity of drilling fluid at the standard of 600 rpm is one half of the scale reading. Plastic viscosity is the difference of the scale reading at 600 and 300 rpm.
147. Low water loss muds are desirable so that thick filter cakes will not be formed in the hole on exposed faces of permeable wells, which might stick the drill pipe.
148. The main equipment used in the rotary drilling operations includes swivel, Kelly, rotary table, drill pipe, drill collar and bit.
149. The swivel suspends the Kelly and drill string and it permits their free rotations. It also provides a rotary hole and a passage way for the flow of drilling fluid into the top of the Kelly and drill string. Kelly is a hollow section whose upper end is suspended from the swivel.
150. The flow of drilling fluid passage downwards through the Kelly and into the top of the drill pipe.
151. Rotary table transmits power and impart rotary motion in the Kelly and drill sting.
152. The lower section of rotary drill string is composed of drill collar.
153. Bit is attached to the lower end of the drill fluid. Drag bits have no independently moving parts and term is applied specifically to the bits of the blade type.
154. Rotary cutter bits are classified as cone type bits and cross roller bits. Diamond bits are similar to drag bits and have no independently moving parts. They drill by direct abrasion and scrapping against the bottom of the well.
155. Primary function of hoisting equipment is to get necessary equipment in and out of the hole as rapidly as is economically possible. The principal components of hoisting operations are block and tackle system, derrick, draw works and miscellaneous operating hoisting equipment such as hooks, elevators and weight indicators.
156. Oil well cementing involves placing pumpable slurry of Portland cement, additives, and water into a wellbore. The cementing take place soon after the lowering of casing is called the primary cementations.
157. Most commonly techniques used for primary cementing are single stage cementing, multistage cementing and liner cementing.
158. Any cementing operation that occurs after the primary cementing operations is referred to as secondary cementing. The cement squeezing and cement plugging are the main two types of secondary cementing process.
159. The different cementing equipment and accessories are cementing plug, well scratchers, centralizers, first collar, casing cementing head and cone jet mixer.
160. To prevent the cement from setting too quickly retarder must be added to the neat cement slurries, especially in deep wells. The chemical used as retarders are lignin retarders, calcium lignosulphonate, carboximethyl hydroxyethyl cellulose, borax and saturated salt etc.
161. Low temperature formations require accelerators to shorten thickening time and to increase early strength, particularly at a temperature below 300°F. Different accelerators used are calcium chloride, sodium chloride, gypsum, sodium silicates etc.
162. The process of directing the wellbore along some trajectory to a predetermined target is caked directional drilling.

163. Drift or inclination angle is the angle measured in degree between the actual well path at some depth and a vertical line directly below the rig site. The minimum desired drift angle is 12 to 15°. The upper limit of drift angle is 35°.
164. Kickoff point is the depth at which wellbore path will be intentionally diverted from the vertical position.
165. Build rate is the rate at which angle changes during deviation.
166. The different types of well trajectories used in directional wells are continuous build, build and hold, S type (build-hold-drop) and modified S type (build-hold-drop-hold).
167. Reamer is a tool to smooth the wall of the wellbore, enlarge the hole, stabilize the bit and strengthen the wellbore where dog legs are encountered.
168. Stabilizer is a tool placed in the drilling assembly to change or maintain the inclination angle in a wellbore by controlling the location of the contact point between the hole and the drill collar; centre the drill collar near the bit to improve drilling performance and prevent wear and differential sticking of drill collar.
169. Spud bit a special bit to change the direction and inclination of the wellbore during drilling.
170. Side tracking is an operation to redirect the wellbore by starting a new hole at a position above the bottom of the original hole.
171. Offshore rigs are referred to as mobile offshore drilling units or MODU's.
172. The jackup rig is the most common offshore drilling rig. A jackup can have 3 or 4 legs. There are two types of jackup rigs, the independent-leg type, usually three legs with lattice construction and the mat type, in which the legs are attached to a very large mat that rests on the ocean bottom. Mat types may be more limited in depth (>250 ft).It is used rom approximately 20 to 30 ft minimum depth to 350 to 550 ft maximum depth. Most textbooks cite 350 ft maximum depth.
173. If drilling from a bottom supported unit is not possible, the semisubmersible is the most popular floating unit. Generations 4 and 5 are the deeper water rigs, with generation 4 being able to drill in 3500 to 4,000 ft and generation 5 being able to drill in 5000 ft plus depths. Generation 1, 2 and 3 use mooring systems and operate in waters less than 3500 ft. It is used in aproximately 150 to 8000 ft using dynamically mooring system.
174. Mooring systems and dynamic positioning systems have greatly improved the drillship's capability of drilling in harsh environments. Drill ships are used in approximately 100 to at least 8000 ft.
175. FPSOs are large ships equipped with processing facilities and moored to a location for a long period. The main types of floating production systems are: FPSO (floating production, storage, and offloading system), FSO (floating storage and offloading system), and FSU (floating storage unit).
176. Fixed platforms are used upto 1500 ft. Compliant towers are used from 1500 to 3000 ft. Sea stars are used from 500 to 3500 ft. Tension leg platforms (TLPs) are used in water depths from 1500 to 7000 ft. Spar platforms are used in water depth of 2000 to 10000 ft.
177. The main function of blow out preventer is to close the annular space between drill pipe and casing.

Multiple Choice Questions (MCQ)

1. ________ is a type of bit.
   a) Roller cone b) Two cone
   c) Fixed cutter d) All of the above
2. What is the name for the longest string of casing ran in a well?
   a) Surface casing b) Intermediate casing
   c) Production casing d) Conductor casing
3. ________are designed to cut across any pipe in the well to close the well quickly.
   a) Shear rams b) Pipe rams
   c) Blind rams d) None of the above
4. The connection between joints of the drill pipe is called:
   a) Tool joint b) Pipe joint
   c) Blast joint d) None of the above
5. "Journal Angle" is related to:
   a) Rig design b) Casing design
   c) Drill string design d) Bit design
6. Measurement while drilling (MWD) systems provide real-time:
   a) Drillbit information b) Borehole inclination
   c) Azimuth d) All of the above
7. Which of the following shows the correct order of solid removal equipment used when unweighted mud is reconditioned?
   a) Centrifuge > shale shaker > desander > desilter
   b) Desilter > centrifuge > shale shaker >desander
   c) Shale shaker > desander > desilter >centrifuge
   d) Desander > desilter > centrifuge > shale shaker
8. The tendency of differential pressure sticking is high when:
   a) Overbalance pressure is low
   b) Mud cake thickness is low
   c) Both overbalance pressure and mud cake thickness are high
   d) Annular diameter ratio is low
9. Which of the following bits are recommended for drilling hard formations in the deep section of the well?
   a) Roller bit with tungsten carbide insert b) Roller bit with mill tooth
   c) PDC bit d) Diamond bit
10. Which of the following is a major consideration for cutting transport in horizontal wells:
    a) Annular size b) Drilling rate
    c) Eccentricity d) Rotary speed

11. Kick volume is normally determined from:
    a) Shut in drill pipe pressure
    b) Density of the kick
    c) Pit gain
    d) Shut in casing Pressure
12. Geo-steering is a tool to control:
    a) Formation pressure
    b) Rheology of drilling fluid
    c) Drilling trajectory
    d) None of the above
13. If the maximum weight on bit is anticipated to be 20 tonne, the average drill collar length is 8.2 m, the drill collar weight is 0.320 tonne/m and buoyancy factor of the mud is 0.847, The number of drill collars required will be:
    a) 9
    b) 10
    c) 11
    d) 12
14. A well has a proposed kick-off point of 12,000 ft MD (12,000 ft TVD) with 0 degrees of deviation and with a final proposed total depth of 13,500 ft MD (12,100 ft TVD) at 90 degrees of deviation. The 90 degrees of deviation was initially achieved at 12,157 ft MD. The directional drilling technique that most nearly addresses the desired wellbore trajectory is:
    a) Short radius
    b) Medium radius
    c) Long radius
    d) Extended reach
15. Given data: bbl/stk, of a 5-1/2-in. by 14-in. duplex pump @ 100% efficiency. Rod diameter = 2.0 in. Determine the pump output in bbl/stk?
    a) 0.30892
    b) 0.10892
    c) 0.40892
    d) 0.60892
16. Determine the hydrostatic pressure decrease when pulling dry pipe out of the hole for the following given data:

    Number of stands pulled = 5

    Average length per stand = 92ft

    Pipe displacement = 0.0075 bbl/ft

    Casing capacity = 0.0773 bbl/ft

    Mud weight = 11.5ppg

    a) 18.5 psi
    b) 21.5 psi
    c) 29.5 psi
    d) 35.5 psi
17. Determine the hydrostatic pressure decrease when pulling wet pipe out of the hole for the following data:

    Number of stands pulled = 5

    Average length per stand = 92ft

    Pipe displacement = 0.0075 bbl/ft

    Pipe capacity = 0.01776 bbl/ft

    Casing capacity = 0.0773 bbl/ft

    Mud weight = 11.5ppg

a) 133.5 psi b) 152.5 psi
c) 165.5 psi d) 173.5 psi

18. Determine the maximum anticipated surface pressure (MASP), psi:
Given: Proposed casing setting depth = 4000ft
Estimated fracture gradient = 14.2ppg
Gas gradient = 0.12 psi/ft
Safety factor = 1.0ppg
Assume 100% of mud is blown out of the hole.
a) 1682psi b) 2682psi
c) 4682psi d) 3682psi

19. If a kick has been observed at 3000 m depth and the shut-in drill pipe pressure is 40 kg/cm^3 with 1.2 g/cc in the wellbore, what mud weight is required to balance the formation pressure?
a) 1.33 b) 1.22
c) 1.44 d) None of the above

20. Absolute permeability of a rock saturated with oil and water is 100 mD. If the relative permeability to water is 0.40, the effective permeability to oil is:
a) 60 mD b) 40 mD
c) 50mD d) None of the above

21. Decreasing the saturation of the wetting phase is called:
a) Drainage b) Imbibition
c) Wettability d) None of the above

22. The viscosity of crude oil measured at a pressure above bubble point and system temperature condition is known as:
a) Under-saturated oil viscosity b) Saturated oil viscosity
c) Kinematic viscosity d) Effective viscosity

23. The values of flash liberation and differential liberation will not be equal for:
a) $P \geq P_b$ b) $P < P_b$
c) $P = P_b$ d) $P=0$

24. Capillary pressure is a measure of the tendency of the rock to:
a) Suck in the wetting fluid phase and to repel the non-wetting phase
b) Suck in the non-wetting fluid phase and to repel the wetting phase
c) Suck in wetting fluid phase and non-wetting phase
d) Repel wetting fluid phase and non-wetting phase

25. Which of the following statement is true about the wet gas reservoir?
a) It contains colorless gas
b) It primarily contains methane and ethane
c) Its initial GOR is 20,000 - 100,000 scf/stb
d) All of the above

26. What is the original oil in place in stb of a reservoir with the following parameters?
    Area = 80 acres, net pay thickness = 50 ft
    Average porosity = 20%,
    Initial water saturation, Swc = 10%,
    Initial gas saturation, Sg = 10%,
    Initial Oil Formation Volume Factor, Boi = 1.42 rb/stb,
    Initial Gas Formation Volume Factor, Bgi = 0.8 rb/mscf.
    a) 3,496,563 b) 3,933,633
    c) 6,206,400 d) 300,000,000
27. If by applying an artificial lift, the bottom hole pressure is reduced to 3000 psi, at what rate oil will produce?
    a) 6707 bbl/day b) 7707 bbl/dy
    c) 8707 bbl/day d) 9707 bbl/day
28. The following informations are given for a oil reservoir: 160 acres, 20 ft net pay, 10% porosity, 25% water saturation, FVF =1.0 and 35% recovery factor. The amount of oil reserve would be:
    a) 651672 bbls b) 605172 bbls
    c) 321762 bbls d) None of the above
29. A reservoir produces gas at an initial gas/liquid ratio (GLR) of 20,000 Scf/STB. After some period of production at constant GLR, the producing GLR begins to increase with time. Essentially all the produced liquids are gas condensate.
    The reservoir fluid is most likely which one of the following types?
    a) Black oil b) Dry gas
    c) Gas condensate (retrograde gas) d) Wet gas
30. A volumetric (no water influx) dry gas reservoir had an initial pressure of 4,750 psia with a gas deviation factor of 0.975. At an abandonment reservoir pressure of 1,500 psia with a gas deviation factor of 0.892, the gas recovery factor (%) would be most nearly:
    a) 34.5 b) 65.5
    c) 68.4 d) 71.1
31. A gas well is producing with a sand face pressure above the dew point pressure. The surface producing gas/liquid ratio is 15,000 scf of surface gas (γg = 0.67) per STB of condensate (γo = 0.759, Mo = 124). The specific gravity of the produced reservoir gas (fraction, air = 1.0) is most nearly:
    a) 0.675 b) 0.705
    c) 0.855 d) 1.015
32. The recovery factor is:
    a) A ratio of the OOIP to the amount of oil in the reservoir at the economic limit
    b) A ratio of the OOIP to the amount of gas in the reservoir at the economic limit
    c) A ratio of the OOIP to the amount of total reserves at the economic limit.
    d) None of the above

33. The maximum temperature above which liquid cannot be formed regardless of pressure is called:
    a) Cricondentherm b) Critical temperature
    c) Bubble point temperature d) None of the above
34. Which of the following is true about black oil simulation?
    a) Assumes constant composition for each phase (oil, gas and water) and assumes isothermal conditions
    b) Fluid and rock properties function of pressure only
    c) Considers gravitational and capillary forces
    d) All of the above
35. Which of the following is true about packer?
    a) It provides a seal between the tubing and the tubing casing annulus
    b) It protect casing as well as prevent excessive tubing movement
    c) It isolates the producing intervals
    d) All of the above
36. In which type of multilateral well completion, both the main well and the secondary wells are open hole?
    a) Level 1 b) Level 2
    c) Level 3 d) None of the above
37. Which of the following is flow control equipment?
    a) Surface and subsurface chokes b) No-go nipples
    c) Landing nipples d) All of the above
38. The liner geometry of deep penetrating shape charge is:
    a) Conical b) Parabolic
    c) Hemispherical d) None of the above
39. What type of down hole tool is needed to complete a high pressure squeeze cementing job?
    a) Packer b) Liner
    c) Casing d) Tubing
40. Which of the following acid is used in the main flush of matrix acidizing for carbonates?
    a) HCl b) 3 wt % HF and 12 wt % HCl
    c) H_2SO_4 d) None of the above
41. Which of the following operation does not require sand?
    a) Hydrajet perforation b) Acid Fracturing
    c) Hydraulic Fracturing d) None of the above
42. PKN fracture model predicts:
    a) Short and wider fracture b) Longer and thinner fracture
    c) Both (a) and (b) d) None of the above
43. Surface chokes are usually sized for ___ flow
    a) Critical b) Sub-critical
    c) Both (a) and (b) d) None of the above

44. Subsurface chokes are usually sized for_____ flow
    a) Sub-critical b) Critical
    c) Both of these d) None of the above
45. Which of the following is a circulating device:
    a) Sliding sleeve b) Ported collars
    c) Ported nipples d) All of the above
46. Which of the following is true about Tension-set packers?
    a) The change in pressure from below pushes on tubing and sets packer harder
    b) Used on shallow applications.
    c) Set by rotate and pull tubing
    d) All of the above
47. Tubing elongation most likely occurs with:
    a) Ballooning b) Buckling
    c) Reverse ballooning d) Temperature reduction
48. In ________ flow the larger gas bubbles become unstable and collapse, resulting in a highly turbulent flow pattern with both phases dispersed.
    a) Slug flow b) Churn flow
    c) Bubble flow d) Annular flow
49. The better suited Inflow Performance Relationship for wells flowing below bubble point is:
    a) Straight line relationship b) Vogels IPR
    c) None of the above d) Orkiszewski's IPR
50. Intersection point of IPR and TIC curves indicates:
    a) Possible flow rate from the well b) Quantity of formation damage in the well
    c) None of the above d) Reservoir pressure
51. Dynagraph is a record of :
    a) Rod load and rod position b) Rod torque and power
    c) Rod torque and rod position d) None of the above
52. The basic characteristics of intermittent gas lift wells are:
    a) Low BHP and low PI or high PI b) Low producing rate
    c) Deep with low fluid level d) All of the above
53. ESP belongs to the following class of mechanical pumps:
    a) Positive displacement type b) Centrifugal pump
    c) Both of the above d) None of the above
54. A well is operating on intermittent gas lift. The injection gas pressure is 700 psi and valve depth is 4500 ft. The cycle frequency for intermittent lift operation would be:
    a) 152 cycle per day b) 107 cycle per day
    c) 204 cycle per day d) 45 cycle per day
55. A well is producing from a saturated reservoir with an average reservoir pressure of 2500 psig. Stabilized production test data indicated that the stabilized rate and wellbore pressure are 350 STB/day and 2000 psig, respectively.

The oil production rate at flowing bottom hole pressure of 1850 psi assuming constant PI would be :

a) 1067.1 STB/day b) 967.1 STB/day
c) 1167 STB/day d) 1267 STB/day

56. For a well, a pump with a $1\frac{3}{4}$ inch plunger is set on 4560 ft of $\frac{3}{4}$ inch rods. The fluid level is known to be low, and the tubing is anchored. With pumping at 20.5 spm and a 64-inch stroke length, production is 355 bbl per day of fluid of specific gravity 0.87. The plunger stroke length is 57.7 inch. The volumetric efficiency (E) for the pump is:

a) 0.60 b) 0.72
c) 0.82 d) 0.50

57. Which of the following method is used to control sand production problem?

a) Mechanical sand control b) Downhole sand consolidation
c) Downhole emulsification d) All of the above

58. Uniformity coefficient is determined by:

a) Comparing 40 percentile size with 90 percentile size
b) Comparing 40 percentile size with 80 percentile size
c) Comparing 40 percentile size with 70 percentile size
d) None of the above

59. Which of the following is the main cause of excessive water production in an oil well?

a) Channelling b) Coning
c) Both (a) and (b) d) None of the above

60. The point where the tubing starts to buckle is known as:

a) Neutral Point b) Critical point
c) Middle Point d) None of the above

61. For low gas oil ratio, the most suitable separator is

a) Vertical separator b) Horizontal separator
c) Spherical separator d) None of the above

62. The orifice plate in a gas meter should be installed with the sharp edge:

a) Upstream b) Downstream
c) Parallel to the flow stream d) It does not matter

63. Which of the following flow meter is based on Bernoulli's Theorem?

a) Orifice meter b) Coriolis flow meter
c) Sonic flow meter d) None of the above

64. BS and W stands for:

a) Basic sediment and water b) Basic sand and water
c) Basic scale and water d) None of the above

65. _______ are used to store the products having very high vapour pressure.

a) Spherical tanks b) Fixed roof tanks

c) Floating roof tanks d) None of the above

66. The LNG storage tanks have ability to store LNG at:

a) The very low temperature of –162°C b) the temperature of 30°C

c) The temperature of –30°C d) None of the above

67. What safety device is used in the storage tanks to regulate fast pressure changes?

a) Breather valve b) Check valve

c) Stuffing box d) None of the above

68. A top sample is taken from____

a) The surface of the liquid b) 6" below the surface

c) The mid-point of the upper third of the liquid

d) Anywhere in the top portion of the tank

69. A sample that is to be tested for Reid vapor pressure [RVP] should be taken with___

a) A zone sampler

b) A glass bottle with suitable cage or weight

c) A bomb sampler

d) A Tulsa Thief

70. A shell and tube heat exchanger is constructed using single pass horizontal tubes to condense the steam at 54°C on the outside of the tube. The cooling water enters at 20°C and leaves at 36°C. What will be the value of log mean temperature difference?

a) 15.16°C b) 35.16°C

c) 45.16°C d) 25.16°C

71. Which of the following rig is used for deep water?

a) Drill Barge b) Jack-Up Rig

c) Drill Ship d) None of the above

72. Which of the following platform is anchored to the seabed and hold most of the equipments above the surface?

a) Jack up rig b) Complaint towers

c) Both (a) and (b) d) Drillship

73. TlPs are used in water depths up to about the depth of:

a) 6,000 feet b) 8000 feet

c) 9500 feet d) 10500 feet

74. ______________ is used to transmit the control fluid and/or electrical current necessary to control the functions of the subsea production and safety equipment.

a) Subsea manifold and connection systems b) Jumpers

c) Risers d) Umbilical systems

75. Which of the following is used to specifically monitor pressures and inject fluids from the host facility to critical areas within the subsea production equipment?

a) Umbilical systems b) Jumpers
c) Risers d) None of the above

76. _______________ is the portion of the flowline that resides between the host facility and the seabed adjacent to a host facility.
a) Risers b) Umbilical systems
c) Jumpers d) None of the above

77. Which of the following configuration of the risers are used in offshore oil and gas drilling and production system?
a) Top tension risers b) Compliant risers
c) Steel catenary risers d) All of the above

78. Which of the following installation method is used for pipelines/risers?
a) S-Lay method b) Z-lay method
c) C-laying method d) None of the above

79. Heave causes major stability problems for _______
a) Jack up rig b) Compliant towers
c) Semi submersibles d) None of the above

80. What is the major limitation of Steel Catenary Riser?
a) Sensitive to waves and current loads b) Sensitive to vessel motion
c) Both (a) and (b) d) None of the above

81. Which of the following statement is true about Single Buoy Mooring?
a) It is used for ship shaped vessels
b) It comprises of a round buoy fixed to the seabed by chains or anchors
c) The buoy body provides the necessary buoyancy and stability to the entire unit
d) All of the above

82. Which of the following is the example of radioactive log?
a) Induction log b) SP log
c) Density log d) None of the above

83. In the SP curve the deflection is left if:
a) Formation water salinity is greater than mud filtrate salinity
b) Formation water salinity is less than mud filtrate salinity
c) Formation water salinity is equal to mud filtrate salinity
d) None of the above

84. SP log is measure in:
a) Kilo volts b) Mega volts
c) Milli volts d) None of the above

85. The electrical logs are used to determine:
a) Fluid saturations b) Viscosity
c) Density d) None of the above

86. Resistivity logs are used to determine:
a) Hydrocarbon bearing zone vs water
b) Bearing zone to indicate permeable zones
c) Porosity
d) All of the above

87. NMR logging is used for the effective measurement of:
a) Porosity b) Permeability
c) Both (a) and (b) d) None of the above

88. Which log is extremely useful for seismic imaging?
a) Density log b) Gamma ray log
c) SP log d) Sonic log

89. A density log in a shaly sand records a formation bulk density, ρ_b = 2.20 g/cm^3. The correction factor for mudcake and hole rugosity is $\Delta\rho$ = 0.051g/cm^3. From measurements on cores and produced fluids it is known that the matrix and fluid densities are, ρ_{ma} = 2.68 g/cm^3 and ρ_f = 1.2 g/cm^3. What is the density log porosity, %?
a) 27 b) 29
c) 32 d) 36

90. The pressure transient analysis assumes that:
a) Reservoir is homogeneous b) Reservoir is heterogeneous
c) Both (a) and (b) d) None of the above

91. Which of the following statement is true about semi-steady state flow:
a) Pressure does not change with time
b) Pressure is constant with time
c) Pressure variation with time is a function of well geometry and reservoir properties
d) None of the above

92. Characteristics of constant wellbore storage effect include:
a) Unit slope straight line on the log log plot
b) P is a linear function of log (time)
c) Both (a) and (b) d) None of the above

93. Which of the example of deliverability test:
a) Flow after flow test b) Isochronal test
c) Modified Isochronal test d) All of the above

94. Isochronal test requires a:
a) Static reservoir pressure, flow period of fixed duration followed by shut-in until pressure stabilizes again
b) Static reservoir pressure, varying flow period followed by shut-in until pressure stabilizes again
c) Static reservoir pressure, then flow and shut in periods of equal duration
d) None of the above

95. Modified Isochronal test requires a:
   a) Static reservoir pressure, flow period of fixed duration followed by shut-in until pressure stabilizes again
   b) Static reservoir pressure, varying flow period followed by shut-in until pressure stabilizes again
   c) Static reservoir pressure, then flow and shut in periods of equal duration
   d) None of the above

96. Multi rate deliverability test:
   a) Indicates the effect of pressure loss due to turbulence
   b) Doesn't indicate the effect of pressure loss due to turbulence
   c) Doesn't indicate the effect of pressure loss due to production of fluid
   d) None of the above

97. Drill stem test is generally done in
   a) Exploratory wells b) Production wells
   c) Both (a) and (b) d) None of the above

98. A build up test was carried out on an oil well with an estimated drainage radius of 2,460 ft. Before shut-in, the well had produced at a stabilized rate of 4900 STB/day for hours. Known reservoir data are given below:

Depth = 10,476 ft	$r_w = 0.354$ ft	$c_t = 22.6 \times 10^{-6}$ psi^{-1}
$B_o = 1.55$ bbl/stb	$\mu_o = 0.20$ cp	$Q_o = 4900$ STB/D
$K_o = 60$ md	H = 482 ft	$p_{wf}(\Delta t{=}0) = 4000$ psi
Ø = 0.09	$t_p = 310$ hours	casing ID = 0.523 ft
m = 40 psi/cycle	$p_{1hr} = 3266$ psi	

   The average permeability of the reservoir is:
   a) 12.8 md b) 55.5 md
   c) 70.0 md d) 90.0 md

99. A well is producing oil at a rate of 800 bbl/day with a bottom hole pressure of 850 psi from a reservoir with very strong water drive (assume steady state). What is the improved permeability if a pressure drawdown analysis indicates that ΔP_{skin} at this rate is 150 psi? Other data are: Well spacing = 40 acres, $r_w = 6.5$ inch, µo = 2.0 cP, $B_o = 1.255$ rb/stb; h = 45 ft; and $p_e = 1350$ psi. The effective radius (re) is:
   a) 745 ft b) 845 ft
   c) 1045 ft d) 545 ft

100. Reservoir and well data are given below:

$P_s = 2800$ psi	$p_e = 2800$ psi
C_t = 10-6 psia-1	$r_e = 2000$ ft
$r_w = 0.5$ ft	$\mu_o = 0.65$ cP
$p_w = 3.050$ psi	$q_o = 900$ stb/day
$B_o = 1.150$ rb/stb	$\mu = 3.75$ cP

   $k = 45$ mD, $c_{ti} = 5.5 \times 10^{-5}$ psi^{-1}, ø = 12.5%, A = 40 acres

What will be the time to reach pseudo-steady state for an oil well drilled at 40 acre spacing?

a) 15.6 days b) 18.5 days

c) 20.5 days d) 22.5 days

101. Following data are given for an oil well:

P_i = 1550 psi, μ = 3.75 cP, k = 45 mD, c_{ti} = 55×10^{-5} psi^{-1}, ø = 12.5%, A = 40 acres

The time required to reach pseudo-steady sate for an oil well drilled at 160 acre spacing is:

a) 2983 days b) 1983 days

c) 5983 days d) 2283 days

102. EPA stands for:

a) Environmental Protection Agency b) Environmental Petroleum Agency

c) Environmental Pollution Act d) None of the above

103. Which of the following is true about the flaring?

a) It is controlled burning of natural gas that cannot be processed for sale or use because of technical or economical reasons

b) It is done during emergencies like power failures, equipment failures etc. to prevent any accident

c) Both (a) and (b)

d) None of the above

104. The primary treatment of the effluent water is usually done by:

a) Sand filtration b) Biofiltration

c) Clarifier d) All of the above

105. The pH requirement of effluent water for disposal (as per MINAS) is

a) 6.0 – 8.5 b) 9.0 – 11 .5

c) 3.0 – 7.5 d) 6.5 – 8.0

106. The permissible exposure level of hydrogen sulfide is?

a) 10 ppm b) 15 ppm

c) 25 ppm d) 50 ppm

107. Which of the following are the solid removal equipments for spent drilling fluids:

a) Shale shaker b) Desander

c) Desilter & centrifuge d) All of the above

108. Naturally Occurring Radioactive Materials (NORM) can be found in many components of operating oil and gas facilities including:

a) Piping runs, including down hole piping,

b) Well heads, production manifolds,

c) Gas/oil separator flow lines, dehydrator and desalinator, valves, storage tanks

d) All of the above

109. Reservoir volumes of oil mobilized by Enhanced Oli Recovery (EOR) agent divided by reservoir volumes contacted by EOR agent is called:

a) Displacement efficiency b) Volumetric sweep efficiency

c) Vertical sweep efficiency d) None of the above

110. The surfactant/polymer flooding doesn't improve:
a) Displacement efficiency
b) Volumetric sweep efficiency
c) Both (a) and (b)
d) None of the above

111. The well pattern having 6 injectors and 1 producer is called:
a) Five spot
b) Line drive
c) Seven spot
d) Four spot

112. The recovery efficiency of a tertiary micellar polymer flood is the product of:
a) Volumetric sweep efficiency, displacement efficiency
b) Areal efficiency, displacement efficiency
c) Volumetric sweep efficiency, displacement efficiency, and mobility buffer efficiency
d) None of the above

113. Capillary number is the ratio of:
a) Viscous to inertial forces
b) Viscous to gravity forces
c) Inertial to viscous forces
d) None of the above

114. CO_2 flooding method is used for:
a) Heavy crude oil
b) Waxy crude oil
c) Medium and light crude oil
d) None of the above

115. Which of the following is true about simultaneous water and gas injection (SWAG)?
a) In SWAG process, water and gas injection take place simultaneously
b) In SWAG process, water, oil & gas injection take place simultaneously
c) Both (a) ans (b)
d) None of the above

116. Which of the following is best short term method for improved oil recovery?
a) Infill drilling
b) WAG
c) FAWAG
d) CO_2 Injection

117. The coal desorption characteristics and isotherm are the function of:
a) Coal rank
b) Cleat concentration
c) Ash and moisture content
d) All of these

118. Langmuir pressure is the pressure at which:
a) Gas storage capacity of coal is equal to ½ of the Langmuir volume
b) Gas storage capacity of coal is equal to of the Langmuir volume
c) Gas storage capacity of coal is double of the Langmuir volume
d) None of the above

119. Which of the following is thermodynamic hydrate inhibitor?
a) Methanol
b) Ethylene glycol
c) Both (a) and (b)
d) None of the above

120. Multistage hydraulic fracturing can be accomplished by:
a) Plug and Perf method
b) Sliding sleeve method
c) Both (a) and (b)
d) None of the above

121. Gas shale is:
   a) Source rock
   b) Reservoir rock
   c) Seal
   d) Combination of source rock, reservoir rock and seal
122. Which of the following is true about oil shales?
   a) Oil shales are sedimentary rocks in a type of organic matter called kerogen
   b) Oil shale is heated to about 660° and then kerogen is transformed into shale oil
   c) Both (a) and (b)
   d) None of the above
123. Which of the following artificial lift technique is used in SAGD operations?
   a) Gas lift technique b) Electrical submersible pumping system
   c) Both (a) and (b) d) None of the above
124. The oil having API gravity 21-31 are classified as:
   a) Heavy crude oil b) Extra heavy crude oil
   c) Medium heavy crude oil d) Bitumen
125. Which of the following EOR method is used for heavy oil?
   a) Cyclic steam stimulation method b) In-situ combustion method
   c) Both (a) and (b) d) None of the above

Answer

1.	d	16.	c	31.	c
2.	c	17.	a	32.	a
3.	a	18.	b	33.	a
4.	a	19.	a	34.	d
5.	b	20.	a	35.	d
6.	d	21.	a	36.	a
7.	c	22.	a	37.	d
8.	c	23.	b	38.	a
9.	d	24.	a	39.	a
10.	c	25.	d	40.	a
11.	c	26.	a	41.	b
12.	c	27.	a	42.	b
13.	a	28.	a	43.	a
14.	a	29.	c	44.	a
15.	b	30.	b	45.	d

46.	d	73.	a	100.	a
47.	c	74.	d	101.	b
48.	b	75.	a	102.	a
49.	b	76.	a	103.	c
50.	a	77.	d	104.	a
51.	a	78.	a	105.	a
52.	d	79.	c	106.	a
53.	b	80.	c	107.	d
54.	b	81.	d	108.	d
55.	a	82.	c	109.	a
56.	c	83.	a	110.	d
57.	d	84.	c	111.	c
58.	a	85.	a	112.	c
59.	c	86.	d	113.	a
60.	a	87.	c	114.	c
61.	a	88.	d	115.	a
62.	a	89.	c	116.	a
63.	a	90.	a	117.	d
64.	a	91.	b	118.	a
65.	a	92.	c	119.	c
66.	a	93.	d	120.	c
67.	a	94.	a	121.	d
68.	b	95.	c	122.	c
69.	b	96.	a	123.	c
70.	d	97.	a	124.	c
71.	c	98.	a	125.	c
72.	c	99.	a		

❑❑❑

APPENDIX 1

DIMENSIONIESS GROUPS/NUMBERS

1. Reynold'Number = $DV\rho/\mu$ = *Re. No.*
D = diameter of pipe, V = velocity through pipe,
P = density, viscosity.
Significance: Inertial forces/viscous forces.
This is used for dynamic similarity.

2. Dean Number = $(dV\rho/\mu)\Big/\left(\dfrac{d}{D}\right)^{0.5}$
D = diameter of tube, D = diameter of helix.
Significance: (Centrifugal force/inertial force)* *Re. No.*
This is used for flow through helical pipes.

3. Power number = $P_{NO} = P/L^5\,\rho n^2$
P = power, n = rotation rate, L = length of paddle.
Significance: Drag forces on paddle/inertial forces.
This is used for power consumption in agitated vessels.

4. Weber number = $We = \sigma/D^2\rho g$
σ = surface tension of liquid, D = diameter of tube, ρ = density,
g = acceleration due to gravity.
Significance: Surface tension forces/Inertial forces.
This is for bubble formation, break up of liquids, jets.

5. Mach. No. = **Ma** = u/u_c
u = velocity of fluid, u_c = velocity of sound
Significance = velocity/sonic velocity.
This is for supersonic flow.

6. Froude No = $F_r = V^2/gL$
Significance : Internal forces/Gravitational forces.
This is useful for fluid-particle interaction.
V = Velocity of fluid; g = Acceleration due to gravity; L = Length

7. Galileo Number

$= Ga = L^3(\rho_s - \rho)\rho_g/u^2 = \frac{3}{4}\ Cd\ Re.\ No^2$

ρ_s = solid density, ρ_g = gas density
This is useful for fluid-particle interaction.

8. Fanning's friction factor $= \frac{dP}{\rho V^2}\frac{d}{L}\frac{g}{2}$

d = diameter of pipe, L = length of the pipe

Significant $= \frac{d}{L}\frac{\text{pressure force}}{\text{inertial forces}}$

This is useful for fluid friction in conduits.

9. Euler No $= Eu = dP/\rho\ V^2$
Significance = pressure forces/inertial forces.
This is used for fluid friction in conduits.

10. Drag coefficient $= C_d = 2\ F_d/\rho\ V^2\ A$
F_d = drag force, $A = \pi d^2/4$.
Significance: Drag forces/inertial forces.
This is useful for free settling velocities.

11. Prandtl Number $= P_r = Cp\mu/K$
Cp = specific heat at constant temperature
K = Thermal conductivity, μ = viscosity.
Significance : Effect of viscosity on heat transfer, viscous forces/thermal forces.

12. Schmidt number $= \mu/(\rho D) = Sc$
μ = viscosity, ρ = density, D = diffusion coefficient.
This comes in mass transfer, similar to Prandtl number in heat transfer.

13. Lewis Relation $= h_G/K_yC_s$
h_G = heat transfer coefficient, K_y = mass transfer coefficient, C_s = humid heat (specific heat).
This is used in psychrometry to relate adiabatic saturation lines and wet bulb temperature lines.

14. Sherwood number $= \frac{K_G P_{BM} RTL}{P_t D_{AB}} = Sh$

This comes in mass transfer :

K_G = mass transfer coefficient,
D_{AB} = diffusion coefficient,
R = gas constant,
T = absolute temperature,
L = length of pipe/diameter.
P_t = total pressure
P_{BM} = log mean pressure of B.

15. Nusselt number = hD/K or $hL/K = Nu$

h = heat transfer coefficient, D = diameter of pipe
K = thermal conductivity.

This is used in forced convection heat transfer.

16. Stanton number = $h/C_pG = Nu/(Re \,.\, Pr)$

h = heat transfer coefficient, C_p = specific heat, G = mass velocity.
This is used in forced convection heat transfer.

17. Peclet number = $Pe = Re \,.\, Pr = \dfrac{(DV\rho)}{\mu} \times \left(\dfrac{C_p\mu}{K}\right)$

18. Brinkman number = $Br = \dfrac{\mu V^2}{K(T_b - T_a)}$

This is a measure of the extent to which viscous heating is important relative to the heat flow resulting from the impressed temperature difference $(T_b - T_a)$
K = thermal conductivity of liquid, μ = viscosity, V = velocity.

19. Archimedes number = $\dfrac{D^3\rho^2 g}{\mu^2}\left(\dfrac{K_1 - K_g}{K_1}\right)$

K_1 = thermal conductivity of liquid,
K_g = thermal conductivity of gas,
D = diameter, ρ = density,
g = acceleration due to gravity, μ = viscosity.

20. Grashof number = $G_r = \dfrac{gl^3\, d\rho}{\rho}\left(\dfrac{\rho}{\mu}\right)^2_{avg}$

ρ = density, μ = viscosity,
$d\rho$ = difference of density,
g = acceleration due to gravity,
l = length of tube.

This is used for heat transfer in natural convection.

21. Condensation number = $\dfrac{D^3\rho^2 g}{K_1\mu\,\Delta t}$

D = diameter of tube,
K_1 = thermal conductivity of liquid,
Δt = Temperature difference.

22. Kutateladze number = $\dfrac{\lambda}{C_P\,\Delta T}$

λ = latent heat of condensation,
C_P = specific heat of liquid,
ΔT = temperature difference.

23. Dispersion number = $D/V * L$ where
D = dispersion coefficient,
v = volumetric flow rate,
L = length of reactor.

24. Archimedes No. = $gL^3/(\rho_P - \rho)\ \rho/\mu^2$
Inertial force ∗ buoyancy forces/(viscous forces)^2

25. Blake No. = $VP/[\mu(1 - \in)^3]$ (inertia force/viscous force)

26. Bond No. = $(\rho_L - \rho_C)\ L^2\ g/\sigma$ (gravitational force/surface tension)

27. Cauchy No. = $\rho V^2/\beta$ (Inertia force/Compressibility force)

28. Knudsen No: l/d (Pore Dia.)

29. Capillary No: $\mu V/\sigma$ (Viscous force/surface tension rorce)

30. Deborah No: = μ_w (Fluid relaxation time/Flow characteristic time)

Appendix II

CONVERSION OF UNITS AND CONSTANTS (SI)

Length	**meter**	**Mass**	**kg.**
1 foot	0.3048	1 lb	0.4535
1 inch	0.0254	1 grain	6.48×10^{-5}
1 mile	1609	**Velocity**	**m/sec**
1 Å	10^{-10}	1 ft/hour	8.467×10^{-5}
1 micron	10^{-6}	1 ft/sec.	0.3048

Pressure	**bar**
1 N/M^2	10^{-5}
1 atm.	1.0133 = 101, 325 pascals
1 kg/cm^2	0.9807
1 mm Hg	1.333×10^{-3}
1 m H_2O	0.0981
1 Psi	0.069
1 inch Hg	0.0339

1 bar = 10^5 pascals.

Force	**N**
1 dyne	10^{-5}
1 kg	9.8067
1 lb	4.4482

Energy	**kJ**
1 NM	10^{-3}
1 kcal	4.1868
1 kWh	3600

1 Kgf. m	9.81×10^{-2}
1 BTU	1.0551
1 ft. lb	1.356×10^{-3}
1 erg.	10^{-10}
1 therm	1.055×10^{5}
1 hP—Hr.	2.648×10^{3}
Area	**m^2**
1 ft^2	0.0929
1 $inch^2$	6.452×10^{-4}
Volume	**m^2**
1 ft^3	2.8317×10^{-2}
1 $inch^3$	1.6387×10^{-5}
Volume flow rae	**m^3/sec.**
1 ft^3/hr.	7.8658×10^{-6}
1 U.S. gal/hr.	1.055×10^{-6}
1 barrel/day	1.84×10^{-6}
Density	**kg/m^3**
1 lb/ft^3	16.018
1 g/c.c.	1000
Viscosity	**NS/m^3**
1 C.P.	10^{-3} pascals. sec.(Pas)
1 kg/m hr.	0.2778
1 lb/ft.hr	4.134×10^{-7}
1 lb/ft. sec.	1.4882
1 stoke (1 cm^2/sec.)	10^{-4} m^2/sec.
1 m^2/hr.	2.778×10^{-4} m^2/sec.
1 ft^2/hr.	2.581×10^{-5} m^2/sec.
Surface tension	**N/m**
1 dyne/cm.	10^{-3}
1 kg f/m	9.814
1 lb/ft.	1.46
Enthalpy	**kJ/kg.**
1 kcal/kg.	4.1868
1 BTU/lb	2.326

Specific heat	**KJ/(Kg. K)**
1 kcal/(kg °C)	4.1868
1 BTU/(lb. °F)	4.1868

Power	**kJ/sec.**
1 kW	1
1 h.P.	0.7355
1 BTU/h.	0.293×10^{-3}
1 kcal/h.	1.163×10^{-3}
1 erg./sec.	10^{-10}

Thermal conductivity	**W/(m.K)**
1 kcal/(hr. m. °C)	1.163
1 BTU/(h. ft. °F)	1.7308

Coefficient of expansion	**kg/(m³.K)**
1 lb/(ft³ . °C)	16.018
1 lb/ft³ . °F)	28.833

Heat transfer coefficient	**W/(m².K)**
1 kcal/(h. m² . °C)	1.163
1 BTU/(h. ft² . °F)	5.6784

Heat flux	**kW/m²**
1 kcal/hr. m²	1.163×10^{-3}
1 BTU/h ft²	3.153×10^{-3}

Mass flux	**kg/(m² . sec)**
1 lb/hr. ft²	1.356×10^{-3}

Mass transfer coefficient

1 lb/hr. ft²	=	4.8823 kg/h. m²
1 kg/hr. m² . atm.	=	0.98687 kg/hr. m² . bar
1 kg/S.N.	=	3.6×10^{8} kg/hr. m² . bar
1 lb/hr. ft² . atm.	=	4.8182 kg/hr. m² . bar

PHYSICAL CONSTANTS

Gas constant. R : = 8.314 kJ/kg. mol. K

= 1.987 kcal/kg. mol. K

= 0.08206 m^3 atm./(kg mole, K)

= 1.545 ft-lb/(lb. mole °R)

Avogadro's number 3 = 6.023 × 10^{23} atoms/g. mole

Standard gas conditions : at NTP (273 K, 1.033 bar = 1 atm

= 22.42 m^3/kg. mole = 359.01 ft^3/lb. mole

Stefan–Boltzmann constant = 1.381 × 10^{-23} J/K = 5.65 × 10^{-24} ft. Ib f/R.

Planck's constant = 6.625 × 10^{-34} JS

Gravitational constant = 9.81 kg/kgf. s^2

= 32.17ft. Ib/lb. sec^2

Viscosity of water = 0.001 kg/(m.s)

Viscosity of air = 1.8 × 10^{-5} kg/(m.s)

Accelaration due to gravity (g) = 9.81 $\frac{m}{s^2}$

Atmospheric composition

Vol. %—Nitrogen = 78.05, Oxygen = 21.0, Argon = 0.93,
Carbon dioxide = 0.02 – 0.03

Wt.%—Nitrogen = 75.5, Oxygen = 23.2, Argon = 1.25,
Carbon dioxide = 0.50 – 0.40

Appendix III

MATHEMATICAL FORMULAE GRAPHICAL REPRESENTATION OF EXPERIMENTAL DATA FOR STRAIGHT LINE

1. $y = a + bx$	: Plot y versus x
2. $y = ax^n$	: Plot log y versus log x
3. $y = ax^n + e$	: Obtain c as intercept on plot of y versus x, then plot log $(y - c)$ versus log x
4. $y = ae^{nx}$	: Plot log y versus x
5. $y = ae^{nx} + c$	: Plot log $(y - c)$ versus x
6. $y = ab^x$	: Plot log y versus x
7. $y = x/(a + bx)$	: Plot x/y versus x or $1/y$ versus $1/x$
8. $y = x/(a + bx) + c$	: Plot $(x - x_i)/(y - y_i)$ versus x where x_i, y_i are the coordinates of any point on a smoth curve through the experimental points.
9. $y = a + bx + cx^2$	: Plot log $(y - y_i)/(x - x_j)$ versus x
10. $\log y = a + bx + cx^2$	: Plot $(\log y - \log y_i)/(x - x_i)$ versus x

Length : Chord of circle $= 2\,r \sin \theta/2$

Arc of circle $= \pi\, r\, \theta/180$

Perimeter of a circle $= 2\,\pi\, r$

Triangle : $a^2 = b^2 + c^2 - 2bc \cos A$

Perimeter of ellipse $= 2\pi\sqrt{0.5\,(a^2 + b^2)}$

Rectangle (sides a, b) $= a \times b$ (area)

Area of parallelogram $= ab \sin \theta$

Area of rhombus $= \frac{1}{2} \times$ Product of diagonals

Area of triangle $= \frac{1}{2} \times h \times b$

Area of trapezium $= \frac{1}{2}(a \times b)\, h$

where a, b are parallel sides.

Area of Polygon of n equal sides of length a :

$\frac{1}{4}\, na^2 \cot(180/n)$

Area of circle $= \pi r^2$

Circular sector $= (\pi/360)\, \theta\, r^2$

Ellipse $= \pi a \times b$

Surfaces : Sphere $= 4\pi r^2$, Cylinder $= 2\pi\, r\, h$

Prism $= Ph$, Cone or Pyramid

$= 1/3$ P × slant height

Spherical segment $= 2\pi rh$

Volume : Rectangular Parallelopiped $= a \times b \times c$

Sphere $= 4/3\, \pi r^3$

Cone or Pyramid : $\pi r^2\, h/2$

Simpson's rule : $1/3h\, [(y_0 + y_n) + 4\,(y_1 + y_2 + y_3 + ...) + 2\,(y_2 + y_4 + y_6) + ...]$

Trapezoidal Rule : $h\left[\frac{1}{2}(y_0 + y_n) + y_2 + y_3 + ... = y_{n-1}\right]$

Trigonometric relations : $\sin(90 + x) = \cos x$;

$\cos(90 - x) = \sin x$; $\text{cosec}\,(90 - x) = \sec x$

$\sec(90 - x) = \text{cosec}\, x$; $\tan(90 - x) = \cot x$

$\cot(90 - x) = \tan x$; $\sin(180 + x) = \sin x$, $\cos(180 - x) = -\cos x$

$\tan(180 - x) = -\tan x$, $\cot(180 - x) = -\cot x$

$\sin(90 + x) = \cos x$; $\cos(90 + x) = -\sin x$

$\tan(90 + x) = -\cot x$; $\cot(90 + x) = -\tan x$

$\sin(180 + x) = -\sin x$; $\cos(180 + x) = -\cos x$

$\tan(180 + x) = \tan x$; $\cot(180 + x) = \cot x$, $\sin(-x) = -\sin x$

cc s $(-x) = \cos x$, $\tan(-x) = -\tan x$

$\sin(x \pm y) = \sin x \cos y \pm \cos x \sin y$

$\cos(x \pm y) = \cos x \cos y \mp \sin x \sin y$

$\sin 2x = 2 \sin x \cos x$, $\cos 2x = 2\cos^2 x - 1$

sin h $(-x) = -\sin hx$, $\cos h\,(-x) = \cos hx$; $\tan h\,(-x) = -\tan hx$

Series : $e^{-x} = 1 - (x/1) + (x^2/2) - (x^3/3)!) + ...$

$\sin x = x - (x^3/3\,!) + (x^5/5\,!) - (x^7/7\,!) + (x^9/9\,!)$

$\cos x = 1 - (x^2/2\,!) + (x^5/4\,!) - (x^6/6\,!) + ...$

$\tan x = x + (x^3/3) + (2x^5/25) + (17x^7/315) + ...$

$\sin x = x + (x^3/3\,!) + (x^5/5\,!) + (x^7/7\,!) + ...$

$\cos hx = 1 + (x^2/2\,!) + (x^4/4\,!) + (x^6/6\,!) + ...$

Roots of $ax^2 + bx + c = 0$, $x = \dfrac{-b \pm \sqrt{b^2 - 4ac}}{2a}$

Scalar product of 2 vectors : $V_1 \,.\, V_2 = |\,V_1\,|\,|\,V_2\,| \cos\theta$

Divergence of $V = \nabla \,.\, V = \text{div.}\, V = (\partial Vx/\partial x) + (\partial Vy/\partial y) + (\partial Vz/\partial z)$

Coordinates transformation:

Cartesian	**Cylindrical**	**Spherical**
$x = x$	$x = \text{r}\cos\theta$	$x = r\cos\theta\sin\phi$
$y = y$	$y = r\sin\theta$	$y = r\sin\theta\sin\phi$
$z = z$	$z = z$	$\text{z} = r\cos\phi$

Differentials

$$d(u/v) = (vdu - udv)/v^2$$
$$d(u^n) = nu^{n-1}\, du$$
$$d\,(u^V) = Vu^{V-1}\, du + u^V (\log u)\, dV$$
$$d\,(a^u) = a^u (\log a)\, du$$
$$d\,(\log u) = (du/u)$$
$$d\,(\log a^u) = (\log a^e)\, du/u$$
$$d\,(\sin u) = \cos u\, du$$
$$d\,(\cos u) = -\sin u\, du$$
$$d\,(\tan u) = \sec^2 u\, du$$
$$d\,(\tan u) = \sec^2 u\, du$$
$$d\,(\cot u) = -\text{cosec}^2\, \mu\, \text{du}$$
$$d\,(\sec u) = \tan u \sec u\, du$$
$$d\,(\text{cosec}\, u) = -\cot u\, \text{cosec}\, u\, du$$
$$d\,(\sin^{-1} u) = du / \sqrt{1 - u^2}$$
$$d\,(\tan^{-1} u) = du/(1 + u^2)$$
$$d\,(\sin h\, u) = \cos h\, u\, du$$
$$d\,(\cos h\, u) = \sin h\, u\, du$$
$$d\,(\tan h\, u) = \sec^2 u\, du$$
$$d\,(\sin h^{-1} u) = du/(u^2 + 1)^{1/2}$$
$$d\,(\cos \text{h}^{-1} u) = du/(u^2 - 1)^{1/2}$$
$$d\,(tan^{-1} u) = \text{du}/(1 - u^2)$$

Integrals

$$\int dx / x = \log_e x$$

$$\int e^x \, dx = e^x ; \int a^x \, dx = a^x / \log_e a$$

$$\int \frac{dx}{(a+bx)} = \frac{1}{b} \log (a+bx)$$

$$\int \frac{dx}{a+bx^2} = \frac{-1}{b} \frac{1}{(a+bx)}$$

$$\int \frac{x \, dx}{(a+bx)} = \frac{x}{b} - \frac{a}{b^2} \log (a+bx)$$

$$\int (a+bx)^{1/2} \, dx = \frac{2}{3b} (a+bx)^{3/2}$$

$$\int x (a+bx)^{1/2} \, dx = \frac{2\,(3bx-2a)\,(a+bx)^{3/2}}{15\,b^3}$$

$$\int \frac{dx}{\sqrt{a+bx}} = 2\sqrt{(a+bx)} / b$$

$$\int \frac{dx}{(a+bx^2)^2} = \frac{x}{2a(a+bx^2)} + \frac{1}{2a} \int \frac{dx}{(a+bx^2)}$$

$$\int \frac{dx}{a+bx^2} = \frac{1}{\sqrt{ab}} \tan^{-1} (x\sqrt{b/a}, \; a \text{ and } b \text{ are + ve}$$

$$= \frac{1}{2}\sqrt{-ab} - \log \frac{\sqrt{a} + x\sqrt{-b}}{\sqrt{a} - x\sqrt{-b}}, \; a + ve, \; b - \text{ve}$$

$$\int \frac{dx}{(a+bx+cx^2)} = \frac{2}{\sqrt{-q}} = \tan^{-1} \frac{2cx+b}{\sqrt{-q}} ; b^2 < 4ac$$

$$= \frac{1}{\sqrt{-q}} \log \frac{2cx+b-\sqrt{-q}}{2cx+b\sqrt{-q}} ; b^2 > 4ac$$

$$= \frac{-2}{(2cx+b)}, b^2 = 4ac$$

where $q = 4ac - b^2$

$$\int \sqrt{x^2 \pm a^2} = \frac{1}{2}\left[x\sqrt{x^2 \pm a^2} - a^2 \log x + \sqrt{x^2 \pm a^2} \right]$$

$$\int \frac{dx}{(x^2 \pm a^2)^{1/2}} = \log(x + x^2 \pm a^2)$$

$$\int (a^2 - x^2)\, dx = \frac{1}{2}\left(x\sqrt{a^2 - x^2} + a^2 \sin^{-1}\frac{x}{a} \right)$$

$$\int \frac{dx}{a^2 - x^2} = \sin^{-1} x/a$$

$$\int \sqrt{\frac{1+x}{1-x}}\, dx = \sin^{-1} x - \sqrt{1-x}$$

$$\int \tan x\, dx = -\log \cos x$$

$$\int \cot x\, dx = \log \sin x$$

$$\int \sec x\, dx = \log(\sec x + \tan x)$$

$$\int \operatorname{cosec} dx = \log(\operatorname{cosec} x - \cot x)$$

$$\int \frac{dx}{1 + \sin x} = \mp(\tan \pi/4 \mp x/2)$$

$$\int \frac{dx}{1 + \cos x} = \tan x/2$$

$$\int \sqrt{1 - \cos x}\, dx = -2\sqrt{2} \cos x/2$$

$$\int \sqrt{1 + \cos x}\, dx = 2\sqrt{2} \sin x/2$$

$$\int \log x\, dx = x \log x - x$$

$$\int x \log x\, dx = \frac{x^2}{2} \log x - \frac{x^2}{4}$$

Laplace transforms :

$f(t)$	$f(s)$
1	$1/S$
e^{at}	$1/(S - a)$
$\frac{t^n - 1}{(n-1)!}$	$1/S^n$
$1/\sqrt{\pi t}$	$1/\sqrt{S}$

$\frac{2t^{1/2}}{\sqrt{\pi}}$	$1/S^{3/2}$
$\frac{1}{a}\sin at$	$\frac{1}{S^2+a^2}$
cos at	$S/(S^2 + a^2)$
$1/a$ (*sin h* at)	$1/(S^2 - a^2)$
cos *h* at	$S/(S^2 - a^2)$
$\frac{1}{b}e^{at}\sin bt$	$\frac{1}{(S-a)^2+b^2}$
e^{at} cos bt	$(S - a)/[(S - a)^2 + b^2]$
$\frac{1}{2\sqrt{\pi t^3}}(e^{bt}-e^{at})$	$(S - a)^{1/2} - (S - b)^{1/2}$
$\frac{1}{(a-b)}(e^{at}-e^{-bt})$	$\frac{1}{(S-a)(S-b)}, a \neq b$
$\frac{1}{(a-b)}(ae^{at}-be^{bt})$	$\frac{S}{(S-a)(S-b)}, a \neq b$

Atomic Weights

Element	*Symbol*	*Atomic Weight*	*Element*	*Symbol*	*Atomic Weight*
Aluminium	Al	26.98	Mercury	Hg	200.61
Antimony	Sb	121.76	Molybdenum	Mo	95.95
Argon	A	39.944	Neon	Ne	20.183
Arsenic	As	74.91	Nickel	Ni	58.69
Barium	Ba	137.36	Nitrogen	N	14.008
Beryllium	Be	9.013	Oxygen	O	16.00
Bismuth	Bi	209.00	Phosphorous	P	30.975
Boron	B	10.82	Platinum	Pt	195.23
Bromine	Br	79.916	Plutonium	Pu	242.0
Cadmium	Cd	112.41	Potassium	K	39.10
Calcium	Ca	40.08	Radium	Ra	226.05
Carbon	C	12.01	Rhodium	Rh	102.95
Cerium	Ce	140.13	Selenium	Se	78.96
Chlorine	Cl	34.457	Silicon	Si	28.09
Chromium	Cr	52.01	Silver	Ag	107.88
Cobalt	Co	58.94	Sodium	Na	22.997

Contd...

Element	*Symbol*	*Atomic Weight*	*Element*	*Symbol*	*Atomic Weight*
Copper	CU	63.54	Stromium	Sr	87.63
Fluorine	F	19.00	Sulphur	S	32.006
Gallium	Ga	69.72	Thorium	Th	232.12
Germanium	Ge	72.60	Tin	Sn	118.70
Gold	Au	197.2	Titanium	Ti	47.90
Helium	He	4.003	Tungsten	W	183.92
Hydrogen	H	1.008	Uranium	U	238.07
Iodine	I	126.91	Vanadium	V	50.95
Iridium	Ir	193.1	Xenon	Xe	131.3
Iron	Fe	55.85	Zinc	Zn	65.38
Krypton	Kr	83.80	Zirconium	Zr	91.22
Lead	Pb	207.21			
Lithium	Li	6.940			
Magnesium	Mg	24.32			
Manganese	Mn	54.93			

Appendix IV

DIFFUSIVITY OF GASES AT ATMOSPHERIC PRESSURE

System	Temp °C	Diffusivity, sq. cm/sec
Air-toluene	25.9	0.086
	59.0	0.092
Air-ethanol	0	0.102
Air-*n* butanol	25.9	0.087
	59.0	0.104
Air ethyl acetate	25.9	0.074
	59.0	0.090
Air-aniline	25.9	0.074
	59.0	0.090
Air-chlorobenzene	25.9	0.074
	59.0	0.090
$H_2 - CH_4$	0	0.626
$O_2 - N_2$	0	0.181
$CO - O_2$	0	0.185
$CO_2 - O_2$	0	0.139
Air-ammonia	0	0.198
Air- water	25.9	0.258

DIFFUSIVITIES OF LIQUIDS

Solute	Solvent	Temp. °C	Solute Conc, g moles/liter	Diffusivity sq. cm/sec* 10^5
Methanol	Water	15	0	1.28
Acetic acid	Water	12.5	1.0	0.82
		18.0	1.0	0.96
		12.5	0.01	0.91
Ethanol	Water	10.0	3.75	0.50

Contd...

Solute	*Solvent*	*Temp. °C*	*Solute Conc, g moles/liter*	*Diffusivity sq. cm/sec* 10^5*
			0.05	0.83
			2.0	0.90
n-butanol	Water	15	2.0	0.77
Chloroform	Ethanol	20	2.0	1.25
CO_2	Ethanol	17	0	3.2
Cl_2	Water	16	0.12	1.26
HCI	Water	0	9	2.7
		10	9	3.3
		10	2.5	2.5
		16	0.5	2.44
NH_3	Water	5	3.5	1.24
		15	1.0	1.77
$C0_2$	Water	10	0	1.46
		20	0	1.77
NaCI	Water	18	0.05	1.26
		18	5.4	1.54

TERNARY EQUILIBRIUM DATA

	Water layer		*Isopropyl ether layer*		
Wt. % acetic acid 100 x	*Water*	*Isopropyl ether*	*Acetic acid 100 y*	*Water*	*Isporopyl*
0.69	98.1	1.2	0.18	0.5	99.3
1.41	97.1	1.5	0.37	0.7	98.9
2.89	95.5	1.6	0.79	0.8	98.4
6.42	91.7	1.9	1.93	1.0	97.1
13.30	84.4	2.3	4.82	1.9	93.3
25.5	71.1	3.4	11.40	3.9	84.7
36.7	58.9	4.4	21.60	6.9	71.5
44.3	45.1	10.6	31.10	10.8	58.1
46.4	37.1	16.5	36.2	15.1	48.7

VISCOSITIES OF GASES

Gas	*Temp. °C*	*Viscosity C.P.*
Air	0	0.0163
	10	0.0177
Hydrogen	0	0.00805
	10	0.00862
Oxygen	0	0.01816
	10	0.01976

Contd...

Gas	*Temp. °C*	*Viscosity C.P.*
Nitrogen	0	0.01579
	10	0.01711
Carbon monoxide	0	0.01561
	10	0.01698
CO_2	0	0.01283
	10	0.01416
Water vapour	45	0.01192

THERMAL CONDUCTIVITY OF METALS AND ALLOYS

Metal	*Temp. °C*	*Thermal conductivity. BTU/hr.ft°C*	*Melting point, °C*
Aluminium	0	117	630
	200	124	
Brass	0	56	
	200	63	940
Cast iron	0	32	
	200	28	1275
Copper	0	224	
	200	215	1083
Lead	0	20	327.5
	200	18	
Nickel	0	36	1452
	200	33	
Steel (mild)	100	26	1375
Steel (stainless)	100	9.4	—
Tin	0	36	231.85
	200	33	
Zinc	0	65	419.4
	200	62	

THERMAL CONDUCTIVITY OF LIQUIDS

Liquid	*Temp. °C*	*Thermal conductivity BTU/(ft. hr. °F)*
Acetic acid, 100%	20	0.399
Acetic acid, 50%	20	0.999
Acetone	60	0.0846
	45	0.106
Benzene	18	0.0853
	50	0.0792

Contd...

Liquid	*Temp. °C*	*Thermal conductivity BTU/(ft. hr. °F)*
Carbon tetrachloride	20	0.0665
	50	0.0564
Chlorobenzene	40	0.0812
	80	0.0866
Ethyl ether	80	0.0994
	16	0.0752
Gasoline	30	0.078
Kerosene	16	0.086
	75	0.081
Mercury	60	5.59
Oils	30	0.079
Toluene	16	0.0779
Xylene (c)	16	0.090
Turpentine	15	0.074
Ethylene glycol	0	0.146
Heptane (n)	30	0.081
NaOH 20% aqueous	20	0.369
KOH 20% aqueous	20	0.346
H_3PO_4 20% aqueous	20	0.322
NaCl 20% aqueous	20	0.334
Na_2SO_4 20% aqueous	20	0.348
Glycerine 20% aqueous	20	0.301
Methanol 20% aqueous	20	0.280
KCI 20% aqueous	20	0.323
$CuSO_4$ 20% aqueous	20	0.334
Ammonia 20% aqueous	20	0.279

PROPERTIES OF WATER AND SATURATED STEAM

Temp °C	*Absolute pressure lbs/sq**	*Latent heat of evaporation, BTU/lb*	*Sp. Vol. of steam cuft/lb*	*Viscosity of liquids water C.P.*	*Thermal conductivity of liquid water BTU/(hr ft.°F)*	*Prandl no. of liquid water*
0	0.0885	1075.8	3.306	1.786	0.320	13.60
24	0.4298	1051.0	740.0	0.913	0.349	6.32
46	1.471 ‘	1028.7	231.9	0.585	0.368	3.83
78	6.715	993.3	55.78	0.357	0.387	2.24
100	14.696	970.3	26.80	0.281	0.393	1.74
121	29.82	945.5	13.821	0.229	0.396	1.42
150	67.01	910.1	6.466	0.185	0.396	1.16

VISCOSITIES OF LIQUIDS AT 25°C AND DENSITIES AT 20°C

Liquid	*Viscosity Centipoises*	*Density g/cc*	*Liquid*	*Viscosity Centipoises*	*Density g/cc*
Acetaldehyde	0.21	0.783	Ethyl acetate	0.45	0.901
Acetic acid	1.20	1.049	Ethyl alcohol	1.10	0.789
Acetic anhydride	0.90	1.832	Ethyl benzene	0.70	0.867
Acetone	0.32	0.948	Ethyl ether	0.25	0.708
Allyl alcohol	1.30	0.854	Ethylene glycol	19.0	1.113
Ammonia 100%	0.105	0.817 (– 79°C)	Formic acid	1.60	1.220
Amyl acetate	0.80	0.879	Glycerol, 100%	100 (60°C)	1.261
Aniline	4.0	1.022	Heptane	0.40	2.684
Amyl alcohol	0.80	0.817	Hexane	0.31	0.659
Benzene	0.60	0.879	Kerosene	2.30	0.78 to 0.82
Bromne	1.0	3.119	Mercury	1.80	13.55
Butyl acetate	0.60	0.882	Methanol 100%	0.55	0.792
Butyl alcohol	2.60	0.810	Nitrobenzene	1.90	1.205
Butyric acid	1.5	0.964	Phenol	7.0	1.071
Carbon dioxide	0.05	1.101 (–37°C)	Prophyl alcohol	2.20	0.804
Carbon disulphide	0.35	1.263	Sulphuric acid 98%	22.0	1.836
Carbon tetrachloride	0.95	1.595	Toluene	0.55	0.866
Chlorobenzene	0.80	1.107	Vinyl acetate	0.44	0.932
Chloroform	0.55	1.489	Water	0.95	0.998
Cresol, meta	18.0	1.034	Xylene ortho	0.75	0.831
Cyclohexanol	50.0	0.962			

SPECIFIC HEATS OF GASES AT 1 ATM. PRESSURE

Gas	*Temp. °C*	*Sp. ht. BTU/lb. °F*
Air	0	0.239
	150°C	0.243
Hydrogen	0	0.364
	150	0.461
Oxygen	0	0.218
	150	0.226
Nitrogen	0	0.248
	150	0.250
CO	0	0.248
	150	0.251

Contd...

Gas	Temp. °C	Sp. ht. BTU/lb. °F
CO_2	0	0.190
	150	0.228
Water vapour	0	0.443
	150	0.457
Acetylene	0	0.41
	150	0.49
Ammonia	0	0.51
	150	0.56
Methane	0	0.60
	150	0.68
Sulphur dioxide	0	0.15
	150	0.17
Hydrogen chloride	0	0.19
	150	0.20

THERMAL CONDUCTIVITIES OF GASES

Substance	Temp. °C	Thermal conductivity BTU/(ft. hr. °F)
Acetone	0	0.0057
	46	0.0074
	100	0.0099
Acetylene	0	0.0108
	50	0.0140
	100	0.0172
Air	– 100	0.0095
	0	0.014
	100	0.0183
	200	0.0226
	300	0.0265
Ammonia	0	0.0128
	50	0.0157
	100	0.0185
Benzene	0	0.0052
	46	0.0073
	100	0.0103
	212	0.0176
Carbon dioxide	– 50	0.0068
	0	0.0085
	100	0.0133
	200	0.0181
	300	0.0228

Contd...

Substance	*Temp. °C*	*Thermal conductivity BTU/(ft. hr. °F)*
Carbon monoxide	– 191	0.0041
	0	0.0133
Carbon tetrachloride	46	0.0041
	100	0.0052
	184	0.0065
Chlorine	0	0.0043
Chloroform	0	0.0038
	46	0.0046
	100	0.0058
	184	0.0077
Dichlorodifluoromethane	0	0.0048
	50	0.0064
	100	0.0080
	150	0.0097
Ethane	– 70	0.0066
	– 34	0.0086
	0	0.0106
Ethyl alcohol	0	0.0075
	20	0.0085
	100	0.0124
Ethyl ether	0	0.0077
	46	0.0099
	100	0.0131
	212	0.0152
Hydrogen	– 100	0.065
	– 50	0.083
	100	0.129
	300	0.178
Mercury	200	0.0197
Methyl alcohol	0	0.0083
	100	0.0128
Nitrogen	– 100	0.0095
	0	0.0140
	50	0.0160
	100	0.0180
Oxygen	– 100	0.0095
	– 50	0.0119
	0	0.0142
	100	0.0185
Propane	0	0.0087
	100	0.0151

Contd...

Substance	*Temp. °C*	*Thermal conductivity BTU/(ft. hr. °F)*
Sulphur dioxide	0	0.0050
	100	0.0069
Water vapour	46	0.0120
	100	0.0137
	200	0.0187
	300	0.0248

SPECIFIC HEATS OF LIQUIDS AT 25°C

Liquid	*Sp. ht. BTU/ib. °F*	*Liquid*	*Spht. BUT/lb. °F*
Acetic acid	0.49	Freon-ll	0.22
		Glycerol	0.58
Acetone	0.53	Heptane	0.51
Ammonia	1.0 (– 100°C)	Hexane	0.54
Amyl alcohol	0.57	Isopropyl alcohol	0.66
Amyl acetate	0.48	Isopropy lether	0.51
Aniline	0.49	Methyl alcohol	0.60
Benzene	0.41		
Butyl alcohol	0.58	Nitrobenzene	0.36
Carbon disulphide	0.24	Propyl alcohol	0.60
Carbon tetrachloride	0.21	Sulphuric acid 98%	0.35
Chlorobenzene	0.32		
Chloroform	0.23	Toluene	0.41
Dowtherm A	0.38	Water	1.0
Ethyl acetate	0.46	Xylene, ortho	0.40
Ethyl alcohol	0.58	Sulphur dioxide	0.34
		Pyridine	0.41
Ethyl ether	0.56	Ethyl benzene	0.43
Ethylene glycol	0.58	Brine 25% NaCI	0.815

APPROXIMATE OVERALL DESIGN COEFFICIENTS

Hot fluid	*Cold fluid*	*Overall U, BTU (hr.ft² °F)*
Water	Water	250–500
Methanol	Water	250–500
Ammonia	Water	250–500
Aqueous solutions	Water	250–500
Light organics	Water	75–150
Medium organics	Water	50–125
Heavy organics	Water	5–75
Gases	Water	2–50
Water	Brine	100–200
Light organics	Brine	40–100

HEATERS

Hot fluid	*Cold fluid*	*Overall U*
Steam	Water	200—700
Steam	Methanol	200—700
Steam	Ammonia	200—700
Steam	Aqueous solutions	100—500 (> 2 C.P.)
		200—700 (< 2 C.P)
Steam	Light organics	100—200
Steam	Medium organics	50—100
Steam	Heavy organics	0—60
Steam	Gases	5—50

EXCHANGERS

Hot fluid	*Cold fluid*	*Overall U*
Water	Water	250—500
Aqueous solutions	Aqueous solutions	250—500
Light organics	Light organics	40—75
Medium organics	Medium organics	20—60
Heavy organics	Heavy organics	10—40
Heavy organics	Light organics	30—60
Light organics	Heavy organics	10—40

PHYSICAL PROPERTIES OF PURE SUBSTANCES

Substance	*Formula*	*Formula wt.*	*Sp. gr.*	*M.P°C*	*B.P.°C*
Ammonia	NH_3	17.03	0.817 (–79°C)	– 77.7	– 33.4
Ammonium chloride	NH_4CI	53.50	1.53	350	520 (subl. 1)
Barium chloride	$BaCI_2$	208.27	3.856	925	1560
Calcium chloride	$CaCl_2$	110.99	2.152	772	7160
Calcium carbonate	$CaCO_3$	100.99	2.711	1339	—
Calcium sulphate	$CaSO_4$	136.14	2.96	1450	—
CO	CO	28.01	0.814 (– 195°C) liq.	– 207	– 192
CO_2	CO_2	44.01	1.101 (– 37°C liq.)	– 56.6	– 78.5
Copper sulphate	$CuSO_4$	159.63	3.606	> 600	cuoat 650°
Hydrogen	H_2	2.016	0.0709 (– 252.7 liq.)	– 259.1	– 252.7
Magnesium sulphate	$MgSO_4$	120.38	2.66	1185	—
Nitrogen	N_2	28.02	0.808 (– 95.8°)	– 209.86	– 195.8
Oxygen	O_2	32.0	1.14 (– 183°)	– 218.4	– 183
Potassium acetate	$KC_2H_3O_2$	98.14	1.8	292	—
Potassium sulphate	K_2SO_4	174.25	2.662	588	—

Contd...

Substance	*Formula*	*Formula wt.*	*Sp. gr.*	*M.P°C*	*B.P.°C*
Silver	Ag	107.88	10.5	960.5	1950
Sodium nitrate	$NaNO_3$	85.01	2.257	308	380
Sulphuric acid	H_2SO4	98.08	1.834	10.49	340
Acetic acid	CH_3COOH	60.05	1.049	16.7	118.1
Acetone	CH_3COCH_3	58.08	0.792	– 94.6	56.6
Aniline	$C_2H_5NHC_5H_{11}$	192.25	0.992	—	261
Anthracene	C_6H_4	178.22	1.25	218	342
Benzene	C_6H_6	78.11	0.879	5.5	80.1
Benzoic acid	C_6H_5COOH	112.12	1.266	121.7	249.2
Carbon disulphide	CS_2	76.13	1.263	108.6	46.3
Carbon monoxide	CO	28.01	0.81 (– 195°)	– 207	– 192
Carbon tetrachloride	CCl_4	153.84	1.595	– 22.6	76.8
Chloroform	$CHCl_3$	119.39	1.489	– 63.5	61.2
Citric acid	C_3H_4OH $(CO_2H)_3$	192.12	1.542	153	—
Ether	$(CH_3CH_2)_2O$	74.12	0.708	– 116.3	34.6
Ethyl acetate	$CH_3CO_2C_2H_5$	88.1	0.901	– 82.4	77.1
Ethy benzene	$C_6H_6C_2H_5$	106.16	0.867	– 94.4	136.2
Ethylene oxide	$(CH_2)_2O$	44.05	0.887	– 111.3	13.5
Formic acid	HCO_3H	46.03	1.220	8.6	100.8
Furfural	$C_4H_3O.CHO$	96.08	1.159	– 38.7	161.7
Glycard	CH_2OHCHO HCH_2OH	92.09	1.260	17.9	290
Glycol	$CH_2OHCHOH$	62.07	1.113	– 15.6	197.4
Hexane (m)	$CH(CH_2)_4CH_3$	86.17	0.659	– 94	69
Heptane (m)	$CH_3\ (CH_2)_5\ CH_3$	100.20	0.684	– 90.6	98.4
Maleic acid	$HO_2CCH,$ $CHCO_2H$	116.07	1.609	130.5	135
Methyl acetate	CH_3COOCH_3	74.08	0.924	– 98.7	57.1
Methyl alcohol	CH_3OH	32.04	0.792	– 97.8	64.7
Nitrobenzene	$C_6H_5NO_2$	123.11	1.205	5.7	210.9
Oleic acid	$C_3H_{17}CH;CH$ $(CH_2)_7CO_2H$	282.45	0.854	14	288
Phenol	C_6H_5OH	94.11	1.071	43	181.4
Pyridine	$CH(CHCH)_2N$	79.10	0.982	– 42	116
Starch	$(C_6H_{10}O_5)$	162.14	1.50		—
Sucrose	$C_{12}H_{22}O_{11}$	342.3	1.588	170.86 Dec.	—
Succinic acid	$HO_2C(CH_2)_2$ CO_2H	118.09	1.572	189.90	235
Toluene	$C_6H_5CH_3$	92.13	0.866	– 95	110.8

Contd...

Substance	Formula	Formula wt.	Sp. gr.	M.P°C	B.P.°C
Urea	NH_2CONH_2	60.06	1.335	132.7	
Vinyl acetate	CH_3CO_2CH : CH_2	86.09	0.932	<– 60	73
Xylene (o)	$C_6H_4(CH_3)_2$	106.16	0.881	– 25	144

LATENT HEATS OF VAPORIZATION

Subtance	*Temp. °C*	*Cal/g*
Acetic acid	118.3	96.75
Acetaldehyde	21	136.17
Acetone	20	131.87
Methane	– 161.6	121.87
Ethane	– 88.9	116.87
Benzene	25	103.57
Toluene	25	98.55
Ethyl benzene	25	95.11
Cyclohexane	25	93.81
Air	—	51.0
Aniline	183	103.68
Carbon disulphide	0	89.35
Carbon tetrachloride	0	52.06
Chloroform	0	64.74
Ethyl acetate	0	102.01
Ethyl alcohol	78.3	204.26
Ethyl chloride	4.7	92.93
Ethylene glyool	197	191.12
Ethly ether	34.6	83.85
Formic acid	101	119.93
Furfural	160.5	107.51
Methyl alcohol	0	284.29
Nitrobenzene	210	79.08
Pyridine	114.1	107.36
Tetrachloroethane	145	55.07
Ethyl benzene	25	95.11
Propane	25	81.76
Chlorobenzene	130.6	77.59
Cresol (m)	202	100.58
Dichlorodifluoromethane	– 29.8	40.4
Ethylene chloride	0	85.29
Methyl acetate	0	113.96

VAPOUR PRESSURES

Substance	*Temp, range C*	*A*	*B*	*C*
Acetic acid	0 to36	7.80307	1651.4	225.0
	36 to170	7.18807	1416.7	211.0
Acetone	—	7.02447	1161.0	224.0
Ammonia	– 83 to 60	7.55466	1002.711	247.885
Aniline	—	7.24179	1675.3	200.0
Benzene	—	6.90565	1211.033	220.790
Bromine	—	6.83298	1133.0	228.0
Cyclohexane	0 to 97	7.34139	6178.81	235.0
Ethyl benzene (o)	20 to 97	7.30701	1831.9	216.62
Toluene (P)	10 to 85	7.22838	1743.67	218.0
	85 to 280	7.00762	1612.35	206.36
Butyric acid (n)	0 to 82	7.85941	1800.7	200.0
	82 to 210	7.38423	1542.6	179.0
Carbon disulphide	– 10 to 160 °C	6.85145	1122.50	236.46
Carbon tetrachloride	—	6.93390	1242.43	230.0
Chtorobenzene	0 to 42	7.10690	1500.0	224.0
	42 to 230	6.94504	1413.2	216.0
Cresol (o)	—	6.97943	1479.4	170.0
Cyclohexane	– 50 to 200	6.84498	1203.526	
Diethylether	—	6.78574	994.195	222.863
Dimethyl ether	—	6.73669	791.184	220.0
Ethyl acetate	– 20 to 150	7.09808	1238.71	230.0
Ethyl alcohol	—	8.04494	1554.3	217.0
Ethyl benzene	—	6.95719	1424.255	222.65
Ethylene $C_2H_6O_2$	25 to 112	8.2621	2197.0	212.0
Glycol	112 to 340	7.8808	1957.0	193.8
n-heplane	—	6.90240	1268.115	216.900
n-hexane	—	6.87776	1171.530	224.366
Methyle acetate	—	7.20211	1232.83	228.0
Methyl alcohol	– 20 to 140	7.87863	1472.11	130.0
Phenol	—	7.13617	1518.1	175.0
Proponic acid	0 to 60	7.71558	1690	210.0
	60 to185	7.71558	1497.775	194.12
Propyl alcohol (n)	—	7.99733	1569.70	209.5
Tolune	—	6.95464	1344.800	219.482
Water	0 to 60	8.10765	1080.286	235.0
	60 to150	7.96681	1668.21	228.0

Equation : $\log_{10} P = A - B/(C + t)$; P is vapor pressure in mm Hg, t is temp, in °C.

Appendix V

Major Chemical and Allied Industries In Our Country

1. Bharat Aluminium Co, Ltd. Factories at Korba (M.P), Kayna (Maharashtra). Product : Aluminium metal.
2. Cement Corporation of India Ltd, Factories at Mandher (M.P) Kurgunta (Karnataka).
   Product : Cement
3. Delhi Cloth and General Mills Co. Ltd. Kota and Delhi.
   Product : Sulphuric acid, Chlorine, Sodium hydroxide, Urea, PVC resins.
4. Fertilizer Corporation of India Ltd. Factories at Sindri (Bihar), Nangal (Punjab), Trombay (Maharashtra), Gorakpur (U.P), Namrup (Assam), Durgapur (W.B).
   Products : Sindri (ammonium Sulphate, ammonium sulphite nitrate and urea), Nangal (Calcium ammonium nitrate), Trombay (Urea and Ammonium nitrate phosphate), Gorakpur (Urea), Namrup (urea and ammonium sulphate).
5. Hindustan Copper Limited, Khetri (Rajasthan)
   Products : Copper metal, Sulphuric acid and fertilizers.
6. Hindustan Insecticides Ltd, Factories in New Delhi, Alway (Kerala).
   Products : DDT, BHC
7. Indian Drugs and Pharmaceuticals Ltd. Rishikesh and Hyderabad (closed).
   Products : Antibiotics, drugs.
8. J.K Rayon, Kanpur
   Products : Carbon disulphide, Viscose rayon.
9. Mohan Meakins Brewaries Ltd, Factories of Gaziabad (U.P), Lucknow (U.P), Simla (H.P)
   Products : Beverages, Industrial alcohal, Carbon dioxide.
10. Shriram Rayons, Kota (Rajasthan)
    Products : Viscose rayon, Sodium sulphate, Sodium sulphide, Carbon disulphide
11. Indial Oil (IOC) (Mathura)
12. Allied Chemicals (Kanpur, U.P). Na_2S, Manganese Sulphate, $CaCl_2$.
13. Bharat Chemical Industries, Mathura, (U.P.) KNO_3, KCl, Nacl, Chemical manures.
14. Asian Chemicals, Kanpur (U.P.), Acetates, Organic chemicals.
15. Agro and Industrial Chemicals, Nanital.

EAST ZONE

1. Alkali and Chemical Corporation of India Ltd; Hoogly, W.B.
   Products: Polyethylene, rubber chemicals, sodium hydroxide, resins and coatings.
2. Bengal Chemicals and Pharmaceutical Works Ltd, factories in Kolkata, Mumbai and Kanpur.
   Products : Pharmaceuticals, heavy and fine chemicals
3. Dunlop India Ltd, Factories at Hoogly and Chennai. Products : Rubber tyres, tubes, rubber products.
4. Goodyear India Ltd, Kolkata.
   Products: Automobile Tyres, Tubes.
5. Gwalior Rayon Silk Manufacturing Co. Ltd. Kolkata.
   Products : Rayon fibre, Sodium Sulphate, Sulphuric acid, Carbon disulphide.
6. Hindustan Steel Limited (SAIL) factories at Rourkela (Odisha), Bokaro (Bihar), Durgapur (W.B), Bhilai (M.P), Burnpur (W.B).
7. Imperial Chemical Industries (India) Pvt. Ltd, Kolkata.
8. Indian Aluminium Company Ltd; Muri (Bihar).
   Product: Aluminium metal.
9. Jayshree Chemicals and Fertilizers (W.B); Odisha.
   Products : Superphosphate, fertilizer mixtures, sulphuric acid.
10. Orient Paper Mills Ltd, Brajrajnagar (Odisha).
    Product: Paper, boards, Sodium hydroxide.
11. Oster Chemical and Pharmaceutical Works Pvt. Ltd., Kolkata.
    Products : Sulphuric acid, Nitric acid, Hydrochloric acid, Acetic acid, Chemicals.
12. Sahu Chemicals and Fertilizers, Varanasi (U.P.)
    Products : Soda ash, anhydrons ammonia.
13. Tata Iron & Steel Co. Ltd., Tatanagar (Bihar)
    Products: Sulphuric acid, Benzol, Benzeno, Steel, Ammonium Sulphate.
14. Titaghur Paper Mills Co. Ltd., Factories at Titaghur (W.B.) and Choudwar (Odisha).
    Products : Sodium hydroxide, Paper, Paper boards.
15. Indian Oil (Guwahati, Barauni, Haldia).
16. Bongaigon (Assam) Refinary & Petrochemicals Ltd.
17. Jayashree Chemicals LTD., Odisha (NaOH, Chlorine).
18. Bharat Chemical Works, Kolkata (HNO_3, Na_2SO_4, HCl, H_3PO_4)
19. Bihar State Superphosphate Factory, Sindri, Bihar (H_2SO_4) Superphosphate)
20. Aluminium Corporation of India LTD. Kolkata (Carbon anodes, Alumina and Aluminium metal).
21. Bengal Tarproducts, Dhan BAD, Bhilai (Coaltar, Naphthalene Balls, BHC, DDT).
22. Calcutta Chemical Co. LTD; Kolkata (Antiseptics, Drugs).
23. Acetochemicals Pvt. Ltd. Kolkata (Ethyl, Amyl, Butyl acetates, Betanaphthol, Sodium acetate).
24. Allied Resin and Chemicals LTD; Kolkata (Formaldehyde, Synthetic Resins and Adhesives).

SOUTH ZONE

1. Andhra Fertilizers, Tadepalle (A.P).
   Products : Superphosphates, Sulphuric acid.
2. Andhra Pradesh Paper Mills Ltd, Rajahmundry, A.P.
   Products : Paper, Paper boards.
3. Andhra Sugars Ltd, Tanuku (A.P).
   Products : Sodium hydroxide, Acetic acid, Sulphuric acid, Hydrochloric acid. Sugar, Ethyl alcohol.
4. Caltex Oil Refining (India) Ltd, Visakhapatnam, A.P. (HPCL).
   Products : Petroleum Products
5. Coramandal Fertilizers Ltd, Visakhaparnam (A.P).
   Products : Diammonium phosphate, urea.
6. EID Parry Ltd, Factories at Ranipet and Ennore (Tamil Nadu), Tadepalle (A.P).
   Products : Diammonium phosphate, super phosphate, fertilzer mixtures and acids.
7. Fertilizers and Chemicals Travancore Limited, (FACT), Alwaye, Kerala. Products : Sulphuric acid, ammonia, ammonium sulphate, ammonium phosphate, ammonium chloride, superphosphate, sulphur dioxide.
8. Hindustan Photo-films Manufacturing Co. Ltd, Ooty (Tamil Nadu). Product : Films.
9. Indian Detonators Ltd, Factories at Hyderabad (A.P), Rourkela (Odisha) Products : Detonators, explosives.
10. Madras Fertilizers Ltd, Manal (T.N).
    Product : Urea, Complex fertilizers.
11. Mettur Chemical and Industrial Corp. Salem (T.N).
    Products are : Sodium hydroxide, Bleaching powder, HCI, Common salt.
12. Mysore Paper Mills Ltd, Bhadravathi (Karnataka).
    Products : Sodium hydroxide, chlorine, pulp and paper.
13. Neyveli Lignite Corporation Ltd. Neyveli (T.N).
    Products : Lignite, clay, urea.
14. Premier Fertilizers Ltd, Chennai
    Products : Sulphuric acid, Superphosphate, NPK fertilizers.
15. Shaw Wallace and Company Ltd, Tamil Nadu
    Products : Potash, sulphur, insecticides, mixed fertilizers.
16. Sirpur Paper Mills Ltd., Sirpur (A.P).
    Products : Sodium hydroxide, Chlorine, Paper
17. South India Viscose Ltd, Coimbatore (T.N).
    Products : Sulphuric acid, Carbon disulphide, Viscose rayon.
18. Travancore Rayons Ltd. Kerala.
19. Travancore Cochin Chemicals Ltd, Alwaye (Kerala).
    Products : Hydrochloric acid, Sodium hydroxide, Chlorine, Sodium sulphide.

20. Alkali Metals (P) Ltd. Hyderabad Sodium Metal

21. Amrutanjan Chemicals (Balms), Chennai.

WEST ZONE

1. Amar Dye Chem. Ltd, Kalyan, Mumbai.
Products : Naphthols, dyes, pigments.

2. Asian Paints (India) Ltd, Mumbai.
Products : Resins, paints, varnishes, enamels.

3. Association Cement Companies Ltd, Mumbai.

4. Baroda Reyon Corporation Ltd, Surat.
Products : Rayon Yarn, Sulphuric acid, Carbon disulphide, Sodium sulphate.

5. Burmah Shell Refineries Ltd, Mumbai (BPCL)

6. Ceat Tyres of India Ltd, Mumbai. Products : Automobile tyres and tubes.

7. Century Rayon, Kalyan (Mumbai)
Products : Carbon disulphide, Viscose rayon

8. Firestone Tyre and Rubber Company of India Ltd, Mumbai.
Products : Automobile tyres, tubes, rubber products.

9. Godrej Soaps Pvt. Ltd, Vikhroli, Mumbai
Products : Glycerine, soaps

10. Gujarat State Fertilizers Company Ltd, Baroda.
Products : Ammonia, Urea, Sulphuric acid, Phosphoric acid.

11. Heridilla Chemicals Ltd, Mumbai
Products : Phenol, acetone, alcohol, phthalic anhydride

12. Hindustan Antibiotics Ltd, Poona (Pimpri)
Products : Antibiotics like pencillin.

13. Hindustan Lever Limited. Factories at Mumbai, Gaziabad (U.P), Etah (U.P), Trichy (Kerela).
Products : Washing and toilet soaps, detergents, edible fats.

14. Hindustan Organic Chemicals Ltd, Kolaba (Maharashtra)
Products : Sodium hydroxide, Chlorine, Ammonia, Nitric acid and aniline.

15. J.K. Chemicals Ltd, Mumbai.
Products : Sulphuric acid, Nitric acid, Hydrochloric acid, Sodium hydroxide.

16. Nepa Chemicals, Nagpur
Products : Sodium hydroxide, Chlorine.

17. Nirlon Synthetic Fibres and Chemicals Pvt. Ltd, Goregon (Mumbai)
Products : Nylon, Polyester.

18. Sarabhai Chemicals, Ahmedabad (Gujarat) Products : Fine Chemicals.

19. Swastik Oil Mills Ltd., Mumbai
Products : Hydrogen, oxygen, soaps, detergents, oils.

20. Tata Chemicals Limited, Mumbai.
Products : Soda ash, Sodium hydroxide, Chlorides, Bromine.

21. Union Carbide India Ltd., Mumbai.
Products : Polyethylene, acetic acid, ethyl acetate, butyl alcohol.

22. IOC (Baroda)

23. Acids and chemicals, Jabalpur, M.P. (H_2SO_4) HCl, HNO_3 NH_3 Liquor)

24. Alambic Chemical Works Co. Ltd. Baroda (Pharmaceuticals, Fine Chemicals, Fertilizers).

25. Amol Chemical Industries (P) Ltd. Thane, Maharashtra (Fe, Cu, Mercury, Nickel and Tin Salts).

26. Anil Starch Products Ltd. Ahmedabad (Starch, Sugars, Fertilizers, Superphosphates).

27. Apex Chemicals, Mumbai (HCl, HNO_3, H_2SO_4) Cadmium, Bismuth, Cobalt, Mercury, Ni, Zn Salts).

28. Bayer (India) Ltd., Thane, Mumbai (Rubber Chemicals, Pesticides, Pharmaceuticals).

29. Bharat Chemicals Co; Mumbai (NaOH, Soda ash, Bleaching powder, Solvents, Fertilizers).

VAPOUR PRESSURES MM. HG.

Substance	1	2	5 *Temp. °C*	10	20	30	40	50	60
Acetic acid	118.1	143.5	180.3	214.0	252.0	276.5	297.0	312.5	—
Acetone	56.5	78.6	113.0	144.5	181.0	205.0	214.5	—	—
Aniline	184.4	212.8	254.8	292.7	342.0	375.5	400.0	422.4	—
Benzene	80.1	103.8	142.5	178.8	221.5	249.5	272.3	290.3	—
Butyric acid	163.5	188.3	225.0	257.0	295.0	319.0	338.0	352.0	—
Carbon disulphide	46.5	69.1	104.8	136.3	175.5	201.5	222.8	240.0	256.0
Carbon tetrachloride	76.7	102.0	141.7	178.0	222.0	251.2	276.0	—	—
Chlorobenzene	132.2	160.2	205.0	245.3	292.8	324.4	349.8	—	—
Diethyl ether	34.6	56.0	90.0	122.0	159.0	183.3	—	—	—
Dimethyl ether	– 23.7	– 6.4	20.8	45.5	75.7	96.0	112.1	125.2	—
Ethyl acetate	77.1	100.6	236.6	169.7	209.5	235.0	—	—	—
Ethyl benzene	136.2	163.5	207.5	246.3	294.5	326.5	—	—	—
Methyl acetate	57.8	79.5	113.1	144.2	181.0	205.0	225.5	—	—
Phenol	181.9	208.0	248.2	283.8	328.7	358.0	382.1	400.0	418.7
Propionic acid	141.1	160.0	186.0	203.0	220.0	228.0	233.0	238.0	—
Toluene	110.6	136.5	178.0	215.8	262.5	292.8	319.0	—	—

Appendix VI

CONSTANT PRESSURE (1 ATM) VAPOR-LIQUID EQUILIBRIHM DATA OF BINARY

Component		Temperature	Mole Fraction of	
A	B	°C	Liquid (x)	Vapor (y)
Acetone	Methanol	64.65	0.0	0.0
		59.60	0.190	0.012
		56.96	0.401	0.505
		55.75	0.519	9.631
		55.29	0.756	0.760
		55.54	0.895	0.880
		56.21	1.000	1.000
Acetone	Water	74.80	0.050	0.6381
		65.26	0.150	0.7716
		61.87	0.900	0.3124
		59.95	0.500	0.8387
		58.29	0.700	0.8712
		56.30	0.950	0.9627
Ethanol	Benzene	76.1	0.80	0.137
		70.8	0.100	0.307
		68.4	0.245	0.390
		67.9	0.450	0.447
		68.7	0.680	0.528
		69.5	0.766	0.566
		76.9	0.984	0.937
Ethanol	Water	95.5	0.019	0.170
		86.7	0.0966	0.4375
		84.1	0.1561	0.5089
		82.3	0.2608	0.5580
		80.7	0.3965	0.6122
		79.7	0.5198	0.6599
		78.41	0.7472	0.7815
		78.15	0.8943	0.8943

			x	y
Water	Acetic acid	116.3	0.00	0.00
		107.8	0.3084	0.4467
		104.3	0.5195	0.6580
		102.8	0.6750	0.7797
		101.5	0.7951	0.3671
		100.5	0.9134	0.9409
		100.0	1.00	1.00
Ethyl acetate	Ethanol	78.3	0.00	0.00
		75.5	0.100	0.187
		72.8	0.300	0.389
		71.8	0.500	0.516
		71.9	0.600	0.576
		73.0	0.800	0.726
		76.0	0.950	0.914
		77.1	1.000	1.000
Methanol	Benzene	70.67	0.026	0.67
		62.87	0.088	0.457
		58.64	0.333	0.559
		58.10	0.699	0.633
		59.90	0.898	0.760
		62.71	0.973	0.9017
Methanol	Water	100.0	0.00	0.00
		93.5	0.040	0.230
		89.3	0.030	0.365
		84.4	0.150	0.517
		78.0	0.300	0.665
		73.1	0.500	0.770
		69.5	0.700	0.870
		66.0	0.900	0.958
		64.5	1.00	1.00

REFERENCES

STOICHIOMETRY

1. Vora and Bart, *'Stoichiometry'*, Tata McGraw Hill Co.
2. Hougen, Watson and Ragatz, *'Chemical Process Principles'*, Vol. 1, Asia Publishing House.

MECHANICAL OPERATIONS

1. J.M.Coulson and J.F. Richardson; *'Chemical Engineering'* Vol.2, Pergamon Press Inc.
2. G.G. Brown and Associates, *'Unit Operations'*, Asia Publishing House, New Delhi (1959).
3. W.L. Mc Cabe and T.C. Smith, *'Unit Operations of Chemical Engineering'*, McGraw Hill Co.
4. Perry's *'Chemical Engineer's Handbook'*, McGraw Hill Co.

FUELS AND COMBUSTION

1. Samir Sircar, *'Fuels and combustion'*, Orient Longmans.
2. Perry, *'Chemical Engineer's Handbook'*, McGraw Hill Co.

CHEMICAL TECHNOLOGY

1. C.E. Dryden, *'Outlines of Chemical Technology'* Affiliated East-West
2. N.Shreeve, *'Chemical Industries'* McGraw Hill Co.
3. Groggins, *'Unit Processes in Organic Synthesis'*, McGraw Hill Co.

FLUID MECHANICS

1. Mc Cabe and Smith, *'Unit Operations of Chemical Engineering'*, McGraw Hill Co.
2. Perry, *'Chemical Engineer's Handbook'*, McGraw Hill Co.
3. Kuni and Levenspiel *'Fluidization Engineering'*, Wiley Eastern.

HEAT TRANSFER

1. Mc Cabe and Smith, *'Unit Operations of Chemical Engineering'*, McGraw Hill Co.
2. Knudsen and Katz, *'Fluid Dynamics and Heat Transfer'*, McGraw Hill Co.
3. Bennet and Myers, *'Momentum, Heat and Mass Transfer', McGraw Hill Co.*

MASS TRANSFER

1. R.E. Treybal. *'Mas Transfer Operations'* McGraw Hill Co.
2. Perry, *'Chemical Engineer's Handbook'* McGraw Hill Co.
3. Mass Transfer Operations; A. Suryanarayan, New Age Int. Pub. 2017. (2nd. Ed.)

THERMODYNAMICS

1. *'Introduction to Chemical Engineering Thermodynamics'* J.M. Smith and H.C.Vanness, McGraw Hill, Co, 3rd Edition.
2. *'Thermodynamics for Chemical Engineers',* Harold C. Weber and Herman P. Meissher, John Wiley and Sons, 2nd Edition.
3. *'Engineering Thermodynamics',* P.K. Nag, Tata McGraw Hill Co.
4. The Principles of Chemical Equilibrium, D. Denbigh, Cambridge University Press.
5. *'Calculations of Chemical Equilibria'.* A.S. Kazanskaya & V.A. Skoblo, Mir Publishers, Moscow.
6. An Introduction to Chemical Thermodynamics, R.P. Rastogi and R.R: Misra, Vikas Publishing House Pvt, Ltd. 1978.

INSTRUMENTATION AND PROCESS CONTROL

1. D.R. Coaghanour and L.B. Koppel, *'Process Systems Analysis and Control'* McGraw Hill Co.
2. D.P. Eckman, *'Industrial Instrumentation',* John Wiley & Sons Co.
3. A. Suryanarayana, *'Outlines of Chemical Instrumentation and Process Control';* Khanna Publishers.

PETROLEUM ENGINEERING

1. Samir Sarkar, "Fuels and Combustion", Orient Longmans.
2. W.L. Nelson, "Petroleum Refinery Engineering" McGraw-Hili International Book Company.
3. B.K. Bhaskara Rao, "Modern Petroleum Refining Process", Oxford and IBH Publishing Company.

ENERGY CONSERVATION AND SAFETY

1. Chemical Engineering Handbook, 4th Edition, McGraw Hill Co.
2. Energy Conservation National Productivity Council Publication, First Edition – 1979.
3. Chemical Age of 31, 6, (1980) p. 697.
4. Chemical Engg. World, XIX, 7 (1984)
5. Chemical Age of India, 36, 7, (1985)
6. *Ibid*, 36,8, (1985)
7. *Ibid*, 36,9, (1985)
8. *Ibid*, 36,10, (1985)
9. *Ibid*, 37,1, (1986)
10. *Ibid,* 37,2, (1986)

KINETICS

1. O Levenspiel; *'Chemical Reaction Engineering',* John Wiley & Sons.
2. J.M. Smith, *'Chemicil Engineering Kinetics',* McGraw Hill Co.
3. Y. B.G. Varma, K. Subba Raju and Ch. Durga Prasada Rao, *'Mathematics, Physical—Chemical Data and Essential Formulae for Chemical Engineers',* Chemical Engineering Education Development Centre, Indian Institute of Technology, Madras.
4. Stanley M. Walas. *'Reaction Kinetics for Chemical Engineers',* McGraw Hill—Kogakusha Ltd., 1959.

❑❑❑